POWER SYSTEMS-I

AUTHOR

Dr. Suresh Mikkili

Associate Professor & Head
Electrical and Electronics Engineering,
National Institute of Technology Goa.

REVIEWER

Prof. M. K. Verma

Professor,
Electrical Engineering,
Indian Institute of Technology (BHU), Varanasi

KHANNA BOOK
PUBLISHING CO. (P) Ltd.
4C/4344, Ansari Road, Darya Ganj,
New Delhi, India

All India Council for
Technical Education
Nelson Mandela Marg, Vasant Kunj,
New Delhi, India

Dear Readers,

To prevent the piracy, this book is secured with HIGH SECURITY HOLOGRAM on the front title cover. In case you don't find the hologram on the front cover title, please write us to at contact@khannabooks.com or whatsapp us at +91-99109 09320 and avail special gift voucher for yourself.

Specimen of Hologram on front Cover title:

Moreover, there is a SPECIAL DISCOUNT COUPON for you with EVERY HOLOGRAM.

How to avail this SPECIAL DISCOUNT:

Step 1: Scratch the hologram

Step 2: Under the scratch area, your "coupon code" is available

Step 3: Logon to www.khannabooks.com

Step 4: Use your "coupon code" in the shopping cart and get your copy at a special discount

Step 5: Enjoy your reading!

ISBN: 978-93-5538-494-2

Book Code: UG481EN

Power Systems - I

by Dr. Suresh Mikkili

First Edition: 2026

Originally Published by:
All India Council for Technical Education, Ministry of Education, GOI

Exclusively Published & Distributed by:
Khanna Book Publishing Co. (P) Ltd.
4C/4344, Ansari Road, Daryaganj,
New Delhi (INDIA)

Visit us at: **www.khannabooks.com**
Write us at: **contact@khannabooks.com**

To view complete list of books,
Please scan the QR Code:

Printed in Bharat.

प्रो. टी. जी. सीताराम
अध्यक्ष
Prof. T. G. Sitharam
Chairman

अखिल भारतीय तकनीकी शिक्षा परिषद्
(भारत सरकार का एक सांविधिक निकाय)
(शिक्षा मंत्रालय, भारत सरकार)
नेल्सन मंडेला मार्ग, वसंत कुंज, नई दिल्ली–110070
दूरभाष : 011–26131498
ई–मेल : chairman@aicte-india.org

ALL INDIA COUNCIL FOR TECHNICAL EDUCATION
(A STATUTORY BODY OF THE GOVT. OF INDIA)
(Ministry of Education, Govt. of India)
Nelson Mandela Marg, Vasant Kunj, New Delhi-110070
Phone : 011-26131498
E-mail : chairman@aicte-india.org

FOREWORD

Engineers are the backbone of any modern society. They are the ones responsible for the marvels as well as the improved quality of life across the world. Engineers have driven humanity towards greater heights in a more evolved and unprecedented manner.

The All India Council for Technical Education (AICTE), have spared no efforts towards the strengthening of the technical education in the country. AICTE is always committed towards promoting quality Technical Education to make India a modern developed nation emphasizing on the overall welfare of mankind.

An array of initiatives has been taken by AICTE in last decade which have been accelerated now by the National Education Policy (NEP) 2020. The implementation of NEP under the visionary leadership of Hon'ble Prime Minister of India envisages the provision for education in regional languages to all, thereby ensuring that every graduate becomes competent enough and is in a position to contribute towards the national growth and development through innovation & entrepreneurship.

One of the spheres where AICTE had been relentlessly working since past couple of years is providing high quality original technical contents at Under Graduate & Diploma level prepared and translated by eminent educators in various Indian languages to its aspirants. For students pursuing 2nd year of their Engineering education, AICTE has identified 87 books, which shall be translated into 12 Indian languages - Hindi, Tamil, Gujarati, Odia, Bengali, Kannada, Urdu, Punjabi, Telugu, Marathi, Assamese & Malayalam. In addition to the English medium, books in different Indian Languages are going to support the students to understand the concepts in their respective mother tongue.

On behalf of AICTE, I express sincere gratitude to all distinguished authors, reviewers and translators from the renowned institutions of high repute for their admirable contribution in a record span of time.

AICTE is confident that these outcomes based original contents shall help aspirants to master the subject with comprehension and greater ease.

(Prof. T. G. Sitharam)

ACKNOWLEDGEMENT

The author is grateful to the AICTE authorities, particularly Prof. T. G. Sitharam, Chairman; Dr. Abhay Jere, Vice-Chairman; Prof. Rajiv Kumar, Member-Secretary; and Dr. Sunil Luthra, Director, Training and Learning Bureau, for their planning to publish this book on Power Systems-I.

Prof. M. K. Verma, Professor of Electrical Engineering at the Indian Institute of Technology (BHU), Varanasi, deserves my heartfelt gratitude for his contributions as a book reviewer. His efforts have made this book more accessible to learners and effectively enhanced it, resulting in a better overall shape.

Prof. O. R. Jaiswal, Director of the National Institute of Technology Goa, deserves my sincere gratitude for a firm support and encouragement.

I would also like to acknowledge the reference books that have been instrumental in the preparation of this book: "Power System Analysis" by J. Grainger and W. D. Stevenson; "Modern Power System Analysis" by D. P. Kothari and I. J. Nagrath; "Electric Energy Systems Theory" by O. I. Elgerd; "Electrical Power Systems" by C. L. Wadhwa; "Principles of Power Systems" by V. K. Mehta and Rohit Mehta "A Course in Power Systems" by J. B. Gupta. Additionally, I extend my gratitude to my students, over the years, who have made significant contributions that have enriched my experience and expertise in this subject.

This book is an outcome of various suggestions of AICTE members, experts and authors who shared their opinion and thought to further develop the engineering education in our country. Acknowledgements are due to the contributors and different researchers in this field whose published books, review/research articles, magazines, footnotes and other valuable information enriched me at the time of writing this book.

I am extremely thankful to my Ph.D. scholars, Mr. Abhilash, Mr. Gaurav, Mr. Abhinav, and Dr. Praveen Kumar, for their invaluable assistance in drawing the diagrams.

I am truly grateful to my parents Mr. Kantharao and Mrs. Suvarna Pushpa; my wife Mrs. Alekhya, and my children Sathvik and Jaswanth Mikkili, for their unwavering patience throughout the completion of this project; without their support, this work would not have materialized.

Above all, I would like to thank The Almighty God for the wisdom and perseverance He has bestowed upon me during this endeavour, and indeed throughout my life as well.

……Dr. Suresh Mikkili

PREFACE

Welcome to the world of power systems. This book focuses on the topics recommended by AICTE in a very systematic and orderly manner, and it serves as a thorough reference for comprehending the principles, concepts, and applications of power systems. Whether you are a student starting out on a learning journey or a professional looking to update your skills, this book will give you with a firm foundation in this stimulating subject.

This book will explore the complexities of the power system structure, encompassing generation, transmission, and protection. We will examine the modelling of overhead transmission lines, the construction of underground cables, three-phase transformer connections, synchronous machines, the causes of over-voltages, circuit breakers, neutral grounding methods, and both symmetrical and asymmetrical faults in power systems. We will also study various protective strategies for alternators, transformers, busbars, and transmission lines. We are going to explore HVDC transmission, photovoltaic systems, wind energy systems, a list of LCC and VSC HVDC projects globally, and a list of solar and wind farms in India.

During the preparation period, I carefully reviewed numerous standard textbooks, journal articles, magazines, and prepared components including critical questions, solved problems, and exercises. This book covers a wide range of medium and high-level challenges, organized in a logical and systematic manner. This book includes significant illustrations and examples, as well as solved examples in each unit to enhance a thorough comprehension of the topics. Each chapter also includes summary, short answer questions, and a "Know More" section which involves QR codes. QR codes contain the most recent research from journal articles, websites, videos, MATLAB/Simulink, experiments, and more.

I would like to express my gratitude to all the researchers, educators, and engineers who have contributed to the field of power systems over time. Their combined efforts have prepared the way for the progress we see today. I hope this book will honour their efforts and inspire you to continue exploring the amazing potential of power systems.

......Dr. Suresh Mikkili

OUTCOME BASED EDUCATION

For the implementation of an outcome-based education the first requirement is to develop an outcome-based curriculum and incorporate an outcome-based assessment in the education system. By going through outcome-based assessments evaluators will be able to evaluate whether the students have achieved the outlined standard, specific and measurable outcomes. With the proper incorporation of outcome-based education there will be a definite commitment to achieve a minimum standard for all learners without giving up at any level. At the end of the programme running with the aid of outcome-based education, a student will be able to arrive at the following outcomes:

PO1. Engineering knowledge: Apply the knowledge of mathematics, science, engineering fundamentals, and an engineering specialization to the solution of complex engineering problems.

PO2. Problem analysis: Identify, formulate, review research literature, and analyze complex engineering problems reaching substantiated conclusions using first principles of mathematics, natural sciences, and engineering sciences.

PO3. Design/development of solutions: Design solutions for complex engineering problems and design system components or processes that meet the specified needs with appropriate consideration for the public health and safety, and the cultural, societal, and environmental considerations.

PO4. Conduct investigations of complex problems: Use research-based knowledge and research methods including design of experiments, analysis and interpretation of data, and synthesis of the information to provide valid conclusions.

PO5. Modern tool usage: Create, select, and apply appropriate techniques, resources, and modern engineering and IT tools including prediction and modelling to complex engineering activities with an understanding of the limitations.

PO6. The engineer and society: Apply reasoning informed by the contextual knowledge to assess societal, health, safety, legal and cultural issues and the consequent responsibilities relevant to the professional engineering practice.

PO7. Environment and sustainability: Understand the impact of the professional engineering solutions in societal and environmental contexts, and demonstrate the knowledge of, and need for sustainable development.

PO8. Ethics: Apply ethical principles and commit to professional ethics and responsibilities and norms of the engineering practice.

PO9. Individual and team work: Function effectively as an individual, and as a member or leader in diverse teams, and in multidisciplinary settings.

PO10. Communication: Communicate effectively on complex engineering activities with the engineering community and with society at large, such as, being able to comprehend and write effective reports and design documentation, make effective presentations, and give and receive clear instructions.

PO11. Project management and finance: Demonstrate knowledge and understanding of the engineering and management principles and apply these to one's own work, as a member and leader in a team, to manage projects and in multidisciplinary environments.

PO12. Life-long learning: Recognize the need for, and have the preparation and ability to engage in independent and life-long learning in the broadest context of technological change.

Course Outcomes

After completion of the course the students will be able to:

CO-1: Comprehend the concepts of power systems.

CO-2: Analyse various power system components.

CO-3: Understand the causes of overvoltage and insulation coordination.

CO-4: Evaluate fault currents for different types of faults.

CO-5: Understand basic protection schemes, HVDC power transmission and renewable energy generation.

CO-6: Analyse the power system numerically

Course Outcomes	Expected Mapping with Programme Outcomes *(1- Weak Correlation; 2- Medium correlation; 3- Strong Correlation)*											
	PO-1	PO-2	PO-3	PO-4	PO-5	PO-6	PO-7	PO-8	PO-9	PO-10	PO-11	PO-12
CO-1	3	2	2	2	1	2	1	1	1	-	1	1
CO-2	3	2	2	1	1	1	1	-	-	-	-	1
CO-3	3	3	2	1	-	1	-	1	-	1	-	1
CO-4	3	3	3	2	2	1	1	-	-	1	-	1
CO-5	3	3	2	1	2	1	1	1	-	-	-	1
CO-6	3	3	3	2	2	2	-	-	-	-	1	3

GUIDELINES FOR TEACHERS

To implement Outcome Based Education (OBE) knowledge level and skill set of the students should be enhanced. Teachers should take a major responsibility for the proper implementation of OBE. Some of the responsibilities (not limited to) for the teachers in OBE system may be as follows:

- Within reasonable constraint, they should manoeuvre time to the best advantage of all students.
- They should assess the students only upon certain defined criterion without considering any other potential ineligibility to discriminate them.
- They should try to grow the learning abilities of the students to a certain level before they leave the institute.
- They should try to ensure that all the students are equipped with the quality knowledge as well as competence after they finish their education.
- They should always encourage the students to develop their ultimate performance capabilities.
- They should facilitate and encourage group work and team work to consolidate newer approach.
- They should follow Blooms taxonomy in every part of the assessment.

Bloom's Taxonomy

Level	Teacher should Check	Student should be able to	Possible Mode of Assessment
Create	Students ability to create	Design or create	Mini project
Evaluate	Students ability to justify	Argue or defend	Assignment
Analyse	Students ability to distinguish	Differentiate or distinguish	Project/Lab Methodology
Apply	Students ability to use information	Operate or demonstrate	Technical Presentation/ Demonstration
Understand	Students ability to explain the ideas	Explain or classify	Presentation/Seminar
Remember	Students ability to recall (or remember)	Define or Recall	Quiz

GUIDELINES FOR STUDENTS

Students should take equal responsibility for implementing the OBE. Some of the responsibilities (not limited to) for the students in OBE system are as follows:

- Students should be well aware of each UO before the start of a unit in each and every course.
- Students should be well aware of each CO before the start of the course.
- Students should be well aware of each PO before the start of the programme.
- Students should think critically and reasonably with proper reflection and action.
- Learning of the students should be connected and integrated with practical and real-life consequences.
- Students should be well aware of their competency at every level of OBE.

Abbreviations and Symbols

List of Symbols

Symbol	Meaning
s	Accelerating torque
ρ	Air density
δ	Air density factor
ω_n	Angular frequency of oscillation
S_{base}	Apparent power
a	Area of cross-section
Z_{base}	Base impedance
$kW_{1\emptyset}$	Base power in kW
$MW_{1\emptyset}$	Base power in MW
K	Boltzmann's constant
C	Capacitance
X_C	Capacitive reactance
q	Charge
L_{11}	Coil primary self-inductance
L_{22}	Coil secondary self-inductance
$\oint$	Contour integral
I	Current in amperes
X_{dw}	Damper winding reactance
K	Degree of compensation

Symbol	Meaning
η	Efficiency
f	Electrical frequency in Hz
T_e	Electromagnetic torque
X_f	Field winding reactance
$d\phi$	Flux
ψ	Flux linkages in weber-turns
ϕ_f	Flux/pole
F	Force
f	Frequency
f_d	Frequency of damped oscillation
a	Ideality factor of diode
L	Inductance
E	Induced EMF of alternator
V_1	Input voltage
ρ	Insulation resistance
X_1, X_2	Leakage reactance
l	Length
P_{loss}	Line losses
X_L	Line reactance
V_L	Line-to-line voltage
δ	Load angle or power angle
X_L	Loop reactance

Symbol	Meaning
H	Magnetic field intensity
$\emptyset_{1l}$	Magnetic flux
I_m	Magnetizing current
M	Mass flow rate of the air
f_m	Mechanical frequency
ΔP_{MMPL}	Mismatch power loss
J	Moment of inertia
$\emptyset_{21}$	Mutual flux component
L_{21}	Mutual inductance between the coils
f_n	Natural frequency of oscillation
H	Net head height in meters
I_n	Neutral current
N_P	Number of parallel connected modules
P	Number of poles
N_s	Number of series connected modules
N_1	Number of turns of coil 1
N_2	Number of turns of coil 2
T	Operating temperature of PV cell
$V_{PV\ cell}$	Output voltage of PV cell in [V]
X_a	Parallel with armature reactance
$I_{p.u}$	Per unit current
$Z_{p.u}$	Per unit impedance

Symbol	Meaning
ε_0	Permittivity of free space
V_{Ph}	Phase voltage
$I_{Ph\ cell}$	Photon current due to irradiance
PSM	Plug-setting multiplier
V	Potential difference between the conductors in volts
δ_2	Power Angle at receiving end
δ_1	Power Angle at sending end
C_p	Power Coefficient
ϴ	Power factor angle
$I_{Pv\ array}$	PV array output current
$V_{Pv\ array}$	PV array output voltage
RRRV	Rate of Rise of Restriking Voltage
$\cos\varphi_R$	Receiving end power factor
V_R	Receiving end voltage
R	Resistance
ρ	Resistivity
I_r	Reverse saturation current of diode [A]
N	Rotor speed
P_S	Sending end power
$\cos\varphi_S$	Sending end power factor
V_S	Sending end voltage

Symbol	Meaning
R_s , R_{sh}	Series and shunt resistances of PV cell
Z	Series impedance
Y	Shunt admittance
X_d	Steady state reactance of short circuit
X_d''	Sub-transient reactance of short circuit
Z_C	Surge impedance
P_R	Surge impedance loading
α_0	Temperature coefficient at 0° C
$I_{PV\ cell}$	Terminal current of PV cell in [A]
$V_{T\ cell}$	Terminal voltage of PV cell
t	Time
τ	Torque
$\frac{N_s}{N_p}$	transformer's turns ratio
X_d'	Transient reactance of short circuit
λ	Wavelength
I_{a0}, I_{b0}, and I_{c0}	Zero sequence components of currents
X_{a0}, X_{b0}, and X_{c0}	Zero sequence components of reactance

List of Abbreviations

Abbreviation	Full form
AAR	Adaptive Array Reconfiguration
AMI	Advanced Metering Infrastructure
AT&C	Aggregate Technical and Commercial
ABCB	Air Blast Circuit Breaker
ACSR	Aluminium Conductor Steel Reinforced
BIL	Basic Impulse Insulation Level
BL	Bridge Linked PV Configuration
CEA	Central Electricity Authority
CB	Circuit Breaker
CAES	Compressed Air Energy Storage
CSC	Current Source Converter
CT	Current Transformer
DETC	De-Energized Tap Changer
DR	Demand Response
DER	Distributed Energy Resources
DG	Distributed Generation
DS	Dominant Square PV Reconfiguration
LLG	Double Line to Ground
DFIG	Doubly Fed Induction Generators
EV	Electric Vehicle
EAR	Electrical Array Reconfiguration
EPS	Electrical Power System
ERP	Enterprise Resource Planning
EHV	Extra High Voltage
FAME	Faster Adoption and Manufacturing of Hybrid and Electric Vehicles

Abbreviation	Full form
FF	Fill Factor
GMPP	Global Maximum Power Point
GPS	Global Positioning System
HV	High Voltage
HVAC	High Voltage Alternating Current
HVDC	High Voltage Direct Current
HC	Honeycomb PV Configuration
HAWT	Horizontal Axis Wind Turbine
IPDS	Integrated Power Development Scheme
IoT	Internet of Things
IE	Irradiation Equalization
KCL	Kirchhoff's Current Law
KVL	Kirchhoff's Voltage Law
LA	Lightning Arrester
LCC	Line Commutated Converter
LG	Line to Ground
LL	Line to Line
LBB	Local Breaker Backup
LMPP	Local Maximum Power Point
MS	Magic Square PV Reconfiguration
MPPT	Maximum Power Point Tracking
MNRE	Ministry of New and Renewable Energy
MMPL	Mismatch Power Loss
NTPC	National Thermal Power Corporation
NGT	Neutral Grounding Transformer
NLTC	No Load Tap Changers

Abbreviation	Full form
OCTC	Off Circuit Tap Changer
OCB	Oil Circuit Breaker
OLTC	On Load Tap Changers
OC	Open Circuit
OH	Overhead Tx. line
P	Parallel PV Configuration
PSC	Partial Shading Condition
PMSG	Permanent Magnet Synchronous Generator
PDP	Phasor Data Processor
PMU	Phasor Management Unit
PV	Photovoltaic
PSM	Plug Setting Multiplier
PCC	Point of Common Coupling
PT	Potential Transformer
PGCIL	Power Grid Corporation of India Ltd.
PWM	Pulse Width Modulation
RF	Radio Frequency
RRRV	Rate of Rise of Restriking Voltage
RTU	Remote Terminal Unit
R-APDRP	Restructured-Accelerated Power Development and Reforms Program
SEIG	Separately Excited Induction Generator
S	Series PV Configuration
SP	Series Parallel PV Configuration
SC	Short Circuit
SCP	Short Circuit Power
SCR	Short Circuit Ratio

Abbreviation	Full form
SECI	Solar Energy Corporation of India
SPV	Solar Photovoltaic
SCIG	Squirrel Cage Induction Generator
STC	Standard Test Conditions
STATCOM	Static Synchronous Compensator
SVC	Static VAR Compensator
SCADA	Supervisory Control and Data Acquisition
SIL	Switching Impulse Insulation Level
SSR	Sub Synchronous Resonance
TCR	Thyristor Controlled Reactor
TSR	Tip Speed Ratio
TCT	Total Cross Tied PV Configuration
LLL	Triple Line (3 phase)
UHV	Ultra-High Voltage
V2G	Vehicle to Grid
VAWT	Vertical Axis Wind Turbine
VSC	Voltage Source Converter
WRSG	Wound Rotor Synchronous Generator

List Of Figures

Unit 2

Unit 3

Unit 4

Unit 5

Unit 6

Unit 7

Unit 8

Unit 9

Unit 10

Unit 11

Unit 12

Unit 13

List Of Tables

TABLES OF CONTENTS

01 BASIC CONCEPTS

Unit specifics: In this unit, the following topics have been discussed for a basic understating of basic concepts:

- Evolution of power systems and its past, present and future scenario, single-line diagram
- Structure of bulk power grid and microgrid and aims of IPDS (erstwhile R-APDRP)
- Smart grid Vs Microgrid, Generation from conventional and renewable energy sources
- Transmission and distribution (ring and meshed)
- Calculation of node voltages in radial and meshed feeder
- Power flow in a three-phase system with a balanced and unbalanced load
- Calculation of shunt capacitor to improve the power factor of the load

Rationale: In this unit, students will be introduced to the evolution of power system, its past, present and future scenarios, single line diagrams, bulk grids, smart grids, conventional, and distributed energy sources, energy storage, transmission, and distribution systems using radial and meshed distribution, supply fed from single end, supply fed through both the ends, synchronous and asynchronous grid connection, HVDC links in India, with the help of necessary diagrams, derivations and examples.

Pre-Requisites: Basic knowledge of electrical engineering.

Unit Outcomes: The list of outcomes of this unit is as follows:

U1-O1: To understand the evolution of past, present, and future power systems.
U1-O2: To understand the structure of the power system bulk and microgrid.
U1-O3: Analyse the conventional and distributed energy sources and energy storage.
U1-O4: To understand the transmission and distribution system, radial and meshed distribution system.
U1-O5: To analyse the synchronous and asynchronous grid interconnection.
U1-O6: To analyse the simple three-phase circuit, power flow and role of reactive power.

Unit-1 outcomes	**Expected mapping with course outcomes** (1: Weak, 2: medium, and 3: strong correlation)					
	CO-1	**CO-2**	**CO-3**	**CO-4**	**CO-5**	**CO-6**
U1-O1	2	1	-	-	-	-
U1-O2	2	-	-	-	-	-
U1-O3	2	3	1	-	-	-
U1-O4	2	-	1	-	1	3
U1-O5	2	3	-	-	2	-

1.1 Introduction:

The evolution of power systems has been a remarkable journey marked by technological advancements, changing energy sources, and the development of sophisticated infrastructure. In today's modern era, understanding the structure and dynamics of power systems is crucial for ensuring efficient generation, transmission, and distribution of electricity. This unit provides an overview of key aspects of power systems including bulk power grids, micro-grids, generation from both conventional and renewable sources, distributed energy resources (DERs), energy storage, transmission and distribution systems, synchronous and asynchronous interconnections, three-phase systems, and power transfer in AC circuits. Power systems are the backbone of modern society, enabling the reliable delivery of electricity to homes, businesses, industries, and essential services. Over time, power grids have evolved from simple localized networks to huge interconnected systems that span continents. The integration of renewable energy sources, advancements in energy storage technologies, and the adoption of smart grid solutions have transformed the way electricity is generated, transmitted, distributed, and consumed.

1.2 Evolution of Power Systems:

The power system is a complex network that contains different components like generators, transmission lines, protection devices, feeders, substations etc. The generation of power is far from the load locations. Power is transmitted through the transmission lines at higher voltages. The advantages of higher voltage transmission are to reduce the losses and increase the transmission efficiency.

Electricity's commercial utilisation commenced during the late 1870s, initially serving for lighthouse illumination and street lighting through arc lamps.

First complete electric power system – 1882:

Thomas Alva Edison pioneered the first complete electric power system in history, establishing the iconic Pearl Street Station in New York City. Operational from September 1882, this system was based on DC technology and included a generator, cable, fuse, meter, and loads. The station's DC generator, powered by a steam engine, supplied electricity to 59 customers within a radius of about 1.5 km at 110 volts through an underground cable. Edison's innovation quickly spread, leading to similar electric power systems in major cities worldwide. The addition of electric motors by Frank Sprague in 1884 marked a significant milestone, expanding the utility of these systems and laying the foundation for the global electric power industry.

Introduction of AC systems – 1886:

Despite the widespread use of DC systems initially, they were gradually replaced by AC systems. The limitations of DC systems became increasingly evident by 1886, as they could only transmit power over short distances from the generators. Maintaining acceptable levels of power loss and voltage drop in DC systems required high voltage levels for long-distance transmission, which was not feasible for power generation and consumption. The need for voltage transformation led to the development of transformers and AC transmission by L. Gaulard and J.D. Gibbs in Paris, France. In the United States, George Westinghouse gained rights to these advancements. In 1886, William Stanley, a Westinghouse partner, created and tested a commercially feasible transformer and AC distribution system capable of lighting 150 lamps in Great Barrington, Massachusetts. Nikola Tesla's creation of poly-phase systems strengthened the appeal of AC systems. By 1888, Tesla held numerous patents related to AC motors, generators, transformers, and transmission systems. Westinghouse acquired these patents, forming the foundation of modern AC power systems.

AC vs DC [Tesla vs Edison]:

During the 1890s, a significant debate emerged regarding the standardization of the electric utility industry on either DC or AC. This period witnessed heated arguments between Edison, a proponent of DC, and Westinghouse, who favoured AC technology. The AC versus DC controversy ended with victory for the AC system.

By the turn of the century, the AC system had won out over the DC system for the following reasons:

- AC systems offer easy adaptability of voltage levels, facilitating the utilization of varying voltages for power generation, transmission, and consumption.
- AC generators exhibit simplicity compared to DC generators due to the absence of brushes, leading to reduced maintenance requirements.

Frequency – 50 Hz vs 60 Hz:

During the initial phase of AC power transmission, there was no standardization of frequency. Various frequencies like 25 Hz, 50 Hz, 60 Hz, 125 Hz, and 133 Hz were utilised, leading to interconnection challenges. However, at higher frequencies, they realized that the transmission losses increased because of increased inductive reactance. And if the frequency is too low like 25 Hz or 30 Hz, then the lamp is not glowing continuously. So at 50 Hz or 60 Hz, there is reduction of transmission losses and lamp flickering. Eventually, North America settled on a 60 Hz standard, while many other nations adopted 50 Hz. The growing demand for transmitting greater power across extended distances incentivized the adoption of higher voltage levels.

Early AC Systems:

The early AC systems used 12, 44, and 60 kV (RMS line-to-line).

- From 1922 to 1965, the voltage levels increased gradually: 165 kV in 1922, 220 kV in 1923, 287 kV in 1935, 330 kV in 1953, and finally 500 kV in 1965.
- Hydro Quebec successfully activated the initial 735 kV power line in 1966, whereas the United States implemented the 765 kV power line in 1969.
- The industry has standardized voltage levels to prevent the proliferation of a limitless number of voltages.
- The standards for the high voltage (HV) class are 115, 138, 161, and 230 kV, while the extra-high voltage (EHV) class standards are 345, 500, and 765 kV.

HVDC Transmission Systems – 1950s:

The advent of mercury arc valves in the early 1950s marked a significant milestone, making high-voltage DC HVDC transmission systems economically viable for specific applications.

- High-voltage direct current (HVDC) transmission is ideal for delivering large amounts of power across long distances. DC transmission may become a viable option over AC transmission after a distance of approximately 500 kilometres for overhead lines and 50 kilometres for subterranean or submarine cables.
- HVDC transmission allows for asynchronous connections between systems where AC connectivity is not feasible due to system stability constraint or different nominal frequencies.

The advent of HVDC transmission in modern commercial settings traces back to 1954, marked by a 96 km submarine cable linking the Swedish mainland to Gotland Island. The integration of thyristor valve converters has notably augmented the allure of HVDC transmission. In 1972, the first high-voltage direct current (HVDC) system featuring thyristor valves was implemented at Eel River, establishing an asynchronous connection between Quebec and New Brunswick's power systems via a back-to-back configuration.

The adoption of HVDC transmission has witnessed a consistent rise, driven by declining costs and compactness of conversion equipment alongside enhanced reliability. Interconnecting nearby utilities has notably bolstered system security and operational efficiency. The utilities' ability to offer mutual emergency assistance further enhances security, while economies of scale are achieved through reduced production reserve capacity requirements for each system.

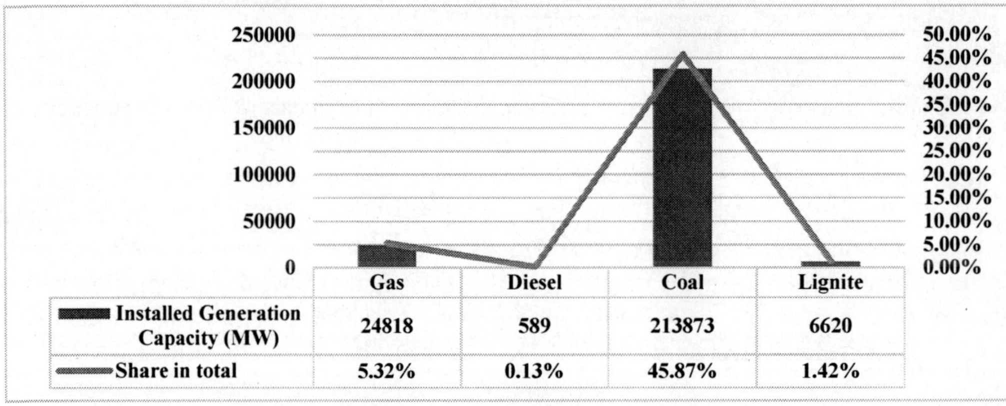

Fig. 1.1 Installed capacity of fossil fuels as on 31-01-2025 [1].

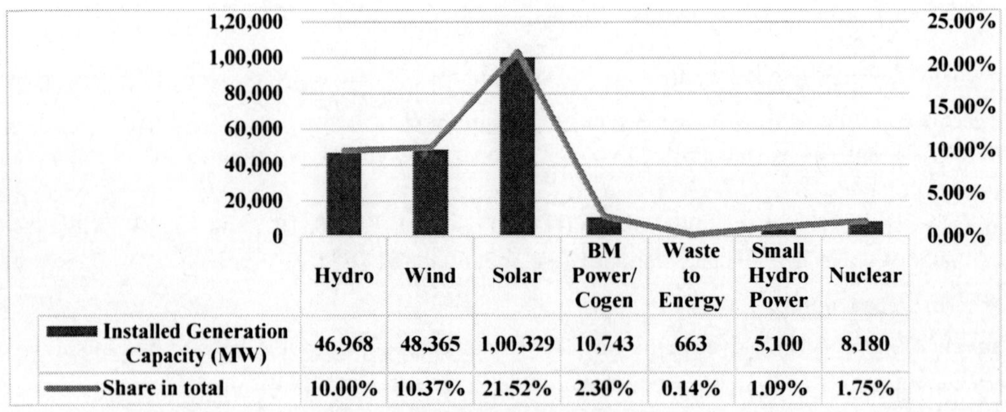

Fig. 1.2 Installed capacity of non-fossil fuels as on 31-01-2025 [1].

Furthermore, the interconnection facilitates the utilities in conducting economic transfers, so enabling them to leverage the most cost-effective sources of power. The aforementioned advantages have been acknowledged since the beginning, and the interrelationships persist in expanding. In the United States and Canada, the majority of utilities have been integrated into a unified and interconnected system. The outcome is an exceedingly vast system of immense intricacy. The design and secure operation of such a system pose significant challenges. Fig. 1.1 shows the sharing of fossil fuel-based generation installed capacity in India. It shows that coal sharing is higher than others. Gas-operated power plants are sharing 5.32% of total generation. Fig. 1.2 shows the sharing of non-fossil fuel-based generation—solar generation shares 21.52% of total renewable energy sources.

1.3 Power systems past, present-day and future scenario:

1.3.1 Past Electric Grid:

The origins of power systems trace back to the late 19th century, marked by the establishment of the initial electric power grids predominantly utilizing DC technology. Yet, the constraints of DC systems, including limited transmission distances, prompted the embrace of AC systems championed by Nikola Tesla and George Westinghouse.

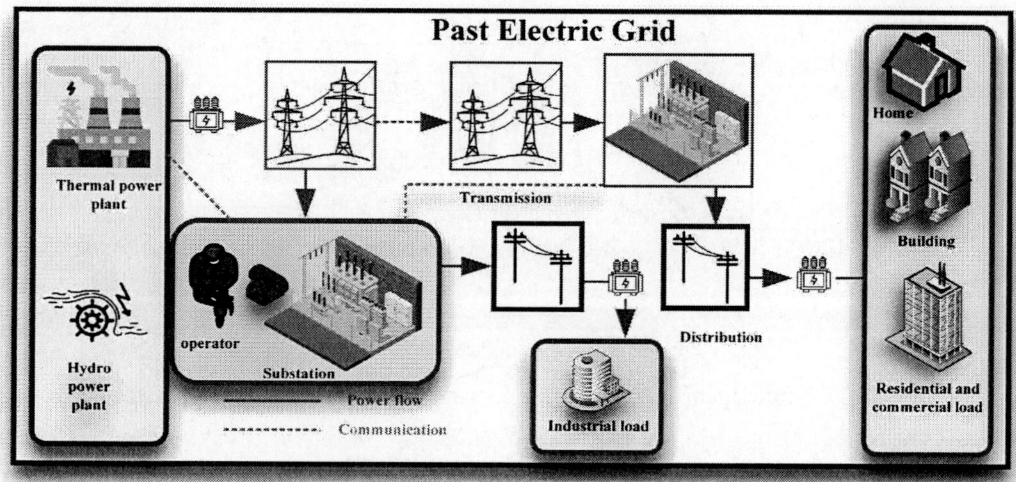

Fig. 1.3. Architecture of Past Electric Grid.

Fig. 1.3 depicts the historical landscape of the power system, characterized by electricity generation via thermal power plants operating between 11kV to 25 kV. To transmit this power to the load centres, the supply voltage is stepped up to higher voltages using transformers at the generation site. Power is transmitted by the transmission lines at 110 kV, 220 kV, 400 kV and 765 kV. The generating site has step-up transformers to boost the voltage before transmission. Further, the higher voltage is reduced for distribution to 132 kV, 66 kV, 33 kV, 11 kV and 400 volts. The communication between the substations was telephonic. An operator is used to communicate to other substations to pass the useful data like voltage and current.

1.3.2 Present Electric Grid:

The modern active distribution network tends to be information technology enabled. It represents a departure from traditional unidirectional power transport in distribution grids. With the integration of Distributed Generation (DG), today's distribution networks have evolved into active systems, facilitating bidirectional power flows. This evolution not only involves managing demand but also incorporates DG integration. These components collectively contribute to the development of robust and flexible active distribution networks capable of handling bidirectional power flows and integrating various distributed energy resources effectively shown in Fig. 1.4. Today, power systems have evolved into vast and interconnected networks that span continents, supplying electricity to homes, industries, and infrastructure.

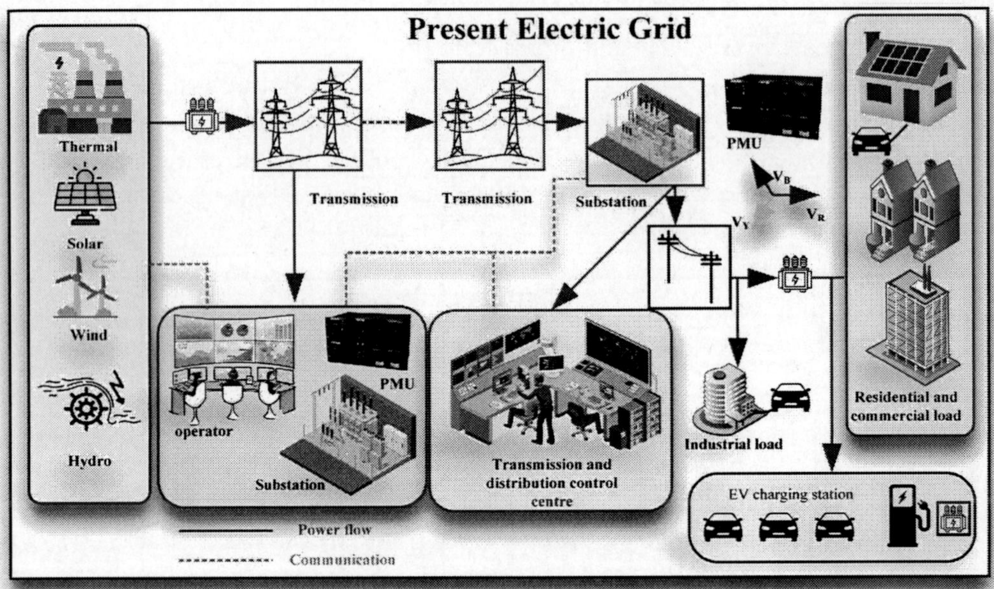

Fig. 1.4 Architecture of Present Electric Grid.

To ensure efficient and intelligent operation and control of active distribution networks, several key elements are essential:

- Implementation of adaptive protection and control mechanisms.
- Integration of wide-area active control strategies.
- Deployment of advanced sensors and measurement technologies.
- Utilization of network management tools and systems.
- Real-time network simulation capabilities for improved planning and operation.
- Development of a distributed communication network to facilitate data exchange.
- Utilization of intelligent methods for knowledge extraction and data analysis.
- Adoption of new and modern design approaches for transmission and distribution infrastructure.

The integration of renewable energy sources, such as solar and wind, has gained momentum, contributing to a more diverse and sustainable energy mix. Smart grid technologies, including advanced metering, real-time monitoring, and demand-response systems, have enhanced grid efficiency, reliability, and resilience. At present in the substations, people are using SCADA, phasor measurement unit (PMU), sensors and other advanced devices to monitor and control the power system. India has been operating on a single frequency since 2013. It has one national grid. On the distribution side, consumers have the penetration of electric vehicles also. In a growing country like India, people are using EVs to improve transportation and reduce CO2 emissions. The Indian government is also supporting EV purchases. The government has initiatives like FAME India schemes for electric and hybrid vehicles. At present, we are collecting the data from the grid within 15 minutes. That means in one hour 96 samples of data, from the grid.

1.3.3 Future Smart Electric Grid:

The future of power grids is noble for significant transformation driven by technological advancements, environmental imperatives, and evolving consumer needs. One of the most prominent trends is the increasing integration of renewable energy sources like solar, wind, and hydroelectric power, facilitated by declining costs and improved technologies. This shift towards renewables will necessitate smart grid solutions capable of managing the variability of these sources and optimizing grid operations. Additionally, there will be a notable move towards decentralization, with microgrids and DERs playing a vital role in enhancing energy resilience and accessibility shown in Fig. 1.5.

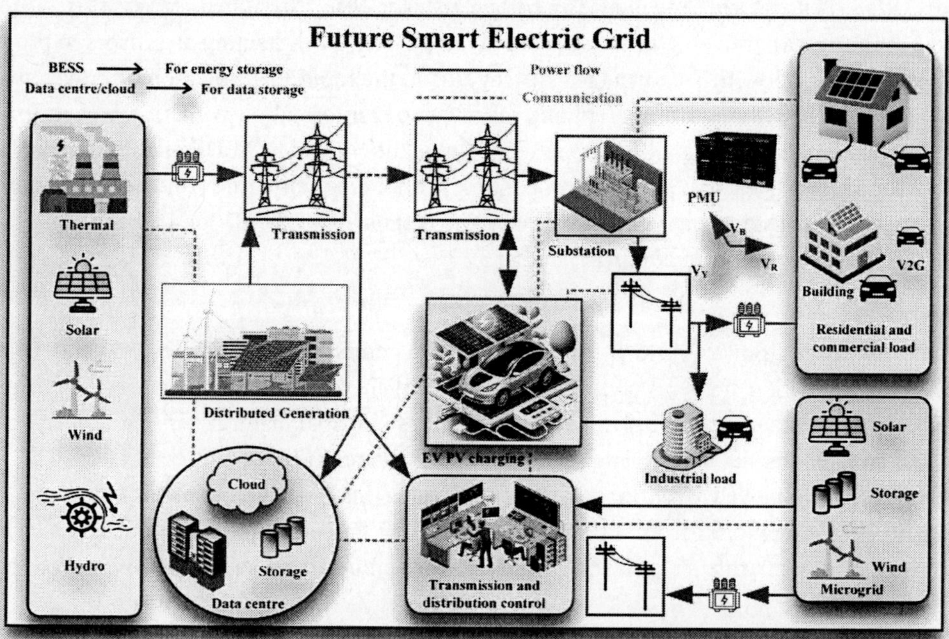

Fig. 1.5 Architecture of Future Smart Electric Grid.

Energy storage solutions will also become indispensable for balancing supply and demand, integrating intermittent renewables, and ensuring grid stability. Moreover, the electrification of transportation, particularly with the rise of EVs, will impact grid dynamics and necessitate smart charging solutions. Cybersecurity, digitalization, and climate resilience will be key focus areas, with grid operators investing in robust measures to safeguard infrastructure, optimize operations, and adapt to changing climatic conditions. Overall, the future power grid will evolve into a more decentralized, flexible, and sustainable ecosystem, catering to the demands of a cleaner and smarter energy future. EVs will be charged or discharged to the grid. EVs can share their power to the grid during peak demand or it can be used as emergency backup. This is called a V2G connection. The EV charging stations can be installed in the transmission network and distribution network. A microgrid is installed near the load centre. The residential buildings and houses are equipped with rooftop solar panels.

PMU plays a crucial role in electrical systems by precisely measuring current and voltage in terms of both amplitude and phase at specific points within the transmission network. These measurements, known as synchro phasors, are synchronized with high accuracy using GPS time synchronization, enabling comparisons of data from distant substations. This capability allows for real-time assessment of the system's condition and dynamic events like power swings.

All devices in the SIPROTEC 5 series can function as PMUs. When configured as a Phasor Measurement Unit, these devices capture current and voltage phasors, timestamp them accurately, and transmit them, along with other relevant data such as frequency and its rate of change, using the IEEE C37.118 communication protocol. This data is typically forwarded to the central control centre for analysis. The SIGUARD PDP (Phasor Data Processor) is a comprehensive system designed for Wide Area Monitoring and Grid Monitoring. It utilizes synchro phasors obtained from PMUs acting as sensors to provide real-time insight into the network's status. This system aids in the rapid identification of power swings and transient phenomena, offering transparent and instantaneous information without the need for network topology replication. By storing all measurement results, the SIGUARD PDP enables quick analysis of power system disturbances, assisting control centre operators in making informed decisions and taking appropriate actions during critical grid situations. The details about the PV and its applications and the EV standards are given in the upcoming chapters.

1.4 Structure of a power system:

The architecture of a power system encompasses the comprehensive framework and constituents involved in the generation, transmission, and distribution of electricity. It commences with the generation phase, where power plants convert diverse energy sources like coal, natural gas, nuclear, hydroelectric, wind, solar, and geothermal energy into electrical energy using turbines and generators. The produced electricity is then conveyed over long distances through high-voltage transmission lines, typically operating at voltages ranging from 110 kV to 800 kV or higher. Within the transmission network, substations play a vital role in adjusting voltages as required and serve as connection points for transmission lines.

Moving into the distribution phase, distribution substations further step-down voltages to levels suitable for delivery to homes, businesses, and industries. Distribution lines, operating at voltages below 400 V for residential areas and up to a few kV for commercial and industrial areas, transport electricity from substations to end-users. Consumers utilize electricity for various purposes such as lighting, heating, cooling, appliances, machinery, and other electrical devices. In some cases, localized areas may have microgrids that can operate independently or in parallel with the main grid, incorporating renewable energy sources, energy storage systems, and smart grid technologies.

At the operational level, control centres managed by power system operators oversee and manage the grid's operations, including generation, transmission, and distribution, optimizing performance and responding to contingencies. Protection systems are integral, utilizing devices like relays, circuit breakers, surge arresters, and grounding systems to detect faults, isolate faulty equipment, and prevent damage. Additionally, smart grid technologies enhance grid management with advanced metering infrastructure (AMI) enabling real-time monitoring and management of electricity usage, grid automation through SCADA systems and intelligent devices, and integration of renewable energy sources for a more sustainable and resilient grid structure.

1.4.1 Line diagrams:

A single-line diagram is a graphical representation of an electrical power system that simplifies the complex network into a single line with symbols. It depicts the connections and components of the power system using standardized symbols to represent generators, transformers, transmission lines, distribution lines, loads, switches, and protective devices. The diagram typically shows the flow of electrical power from the power source, such as a generator or utility grid, through various components like transformers and circuit breakers, to the loads such as motors, lights, and appliances.

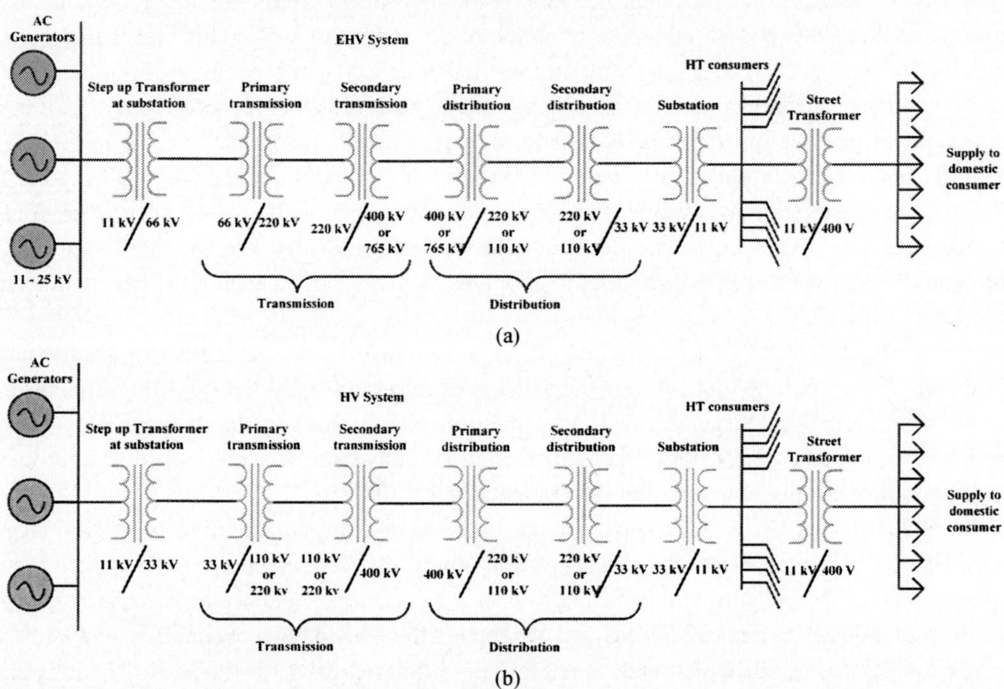

Fig. 1.6 (a) Single line diagram of EHV transmission system. (b) Single line diagram of HV transmission system.

Single-line diagrams are essential tools for engineers, technicians, and operators to understand the overall structure of the electrical system, identify key components, analyse power flow, and troubleshoot faults or abnormalities. They are used in the design, planning, operation, and maintenance of electrical power systems across industries including power generation, transmission, distribution, and industrial facilities. Fig. 1.6 (a) shows the single line diagram of an EHV transmission line, which operates at 765 kV. Fig. 1.6 (b) shows the single line diagram of a high voltage line, which operates on 400 kV.

The generating voltage of the AC alternators is 11 kV to 25 kV, which is being stepped up to 66 kV or 33 kV. Further 33/66 kV voltage is stepped up using a transformer up to 400/765 kV for high voltage transmission. Such high voltage is reduced at primary distribution and secondary distribution. High voltage like 33 kV or 11 kV is supplied to the high voltage industrial consumer. The voltage is reduced up to 400 V for the commercial and domestic consumer.

1.5 Bulk Power Grids and Micro-grids:

1.5.1 Background of power grid:

During the 20th century, local power networks gradually extended and eventually networked with each other to improve economic sustainability and dependability. By the 1960s, the electric grids in developed nations had transformed into huge, mature, and intricately interconnected networks. These networks comprised multiple 'central' generation power plants that provided electricity to large load centres through high-capacity power lines. Subsequently, these lines were divided and partitioned to accommodate the needs of smaller industrial and home consumers within the supply region. The grid architecture of the 1960s was based on the principle of economies of scale. This means that big power stations, fuelled by coal, gas, or oil and ranging from 1 GW to 3 GW in capacity, were found to be cost-effective due to the efficiency improvements that could be achieved at larger sizes.

The strategic placement of power stations near fossil fuel reserves, hydroelectric dams in mountainous regions, and nuclear power plants near water sources significantly influenced the grid's layout. Initial fossil fuel-fired power stations, despite their pollution, were positioned at a distance from populated areas once electricity distribution networks permitted such placement. By the late 1960s, the electricity infrastructure had grown to serve the bulk of the population in developed countries, leaving only distant regional areas 'off-grid.'

Effective metering of electricity consumption on a per-user basis became essential for accurate billing. However, technological limitations during the grid's growth period led to fixed-tariff arrangements and dual-tariff structures. These structures facilitated the utilization of low-cost night-time electrical power for applications like 'heat banks,' which helped smooth out daily demand variations, thus improving the utilization and profitability of generation and transmission facilities.

From the 1970s to the 1990s, escalating demand necessitated the construction of additional power stations. However, inadequate supply during peak times resulted in compromised power quality, including blackouts, power cuts, and brownouts. This heightened dependence on electricity for various purposes, from industry to domestic needs, led consumers to demand higher reliability standards.

In the late 20th century, predictable trends in energy use, primarily driven by household heating and air-conditioning requirements, resulted in daily surges in demand. These peaks were matched by 'peaking power generators' that functioned intermittently for brief durations each day. Nevertheless, the limited usage of these generators, along with the grid's need for backup systems, resulted in elevated expenses for power providers, which were then transferred to customers through higher bills.

1.5.2 Modernization of power grid:

Opportunities for modernization in the electrical grid have emerged since the early 21st century, obligations to advancements in electronic communication technology. These improvements offer solutions to the limitations and costs associated with the traditional grid. With technological advancements, the need to average out peak power prices and distribute them equally to all consumers is no longer necessary. Concurrently, growing concerns about environmental harm from fossil fuel power plants have prompted a transition toward increased reliance on renewable energy sources.

Nevertheless, prominent renewables like wind and solar power display considerable variability, underscoring the need for advanced control systems to enable their seamless integration into the grid. The rise of power generation from photovoltaic cells and, to a lesser degree, wind turbines has disrupted the conventional paradigm of large-scale, centralized power stations. Additionally, the declining costs associated with these technologies indicate a significant shift from a centralized grid structure to a highly distributed one, where power is generated and consumed at the grid's limits. Fig. 1.7 shows the future grid with information technology, PV integration, cloud storage and energy storage.

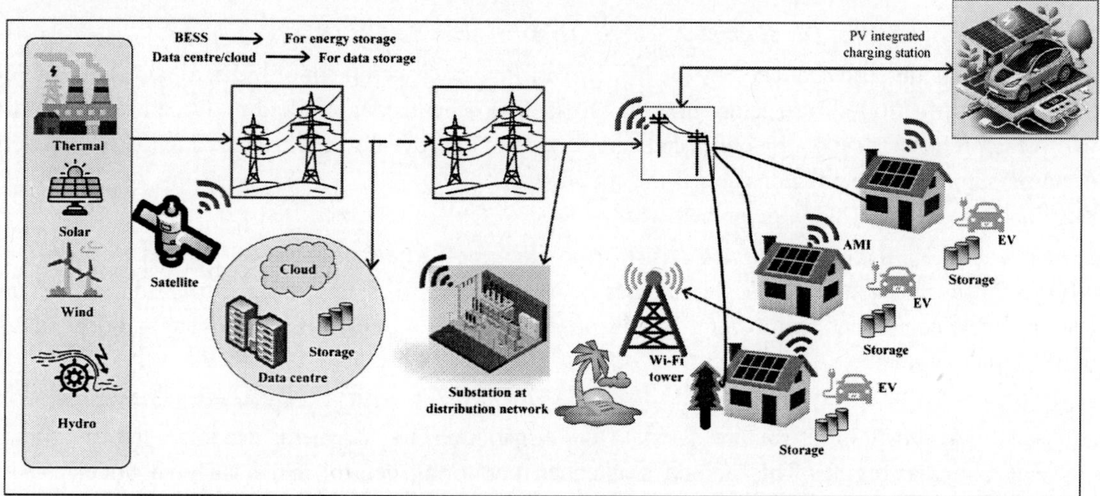

Fig. 1.7 Future grid with IT, cloud storage, rooftop solar, energy storage and EVs penetration [2], [3].

Both smart grids and microgrids are examples of electrical systems in operation today. Both methods can be used in a variety of contexts in today's society. There are many benefits and drawbacks to using them. Both grids have the necessary components for bidirectional power exchange. Learning about the smart grid and microgrid is crucial. In the conventional grid, transmission and distribution losses are 22.5% in India. The generation loss is about 4 to 5%.

The Restructured Accelerated Power Development and Reforms Program (R-APDRP) is a Government of India initiative focused on modernizing and enhancing the nation's power distribution infrastructure. Commencing in 2008, the program aims to diminish aggregate technical and commercial (AT&C) losses, elevate power distribution efficiency, and bolster the quality and reliability of electricity supply specifically in urban regions.

1.5.2.1 Aim of R-APDRP:

The R-APDRP aims to achieve several key objectives in the power sector:

- Reduce Aggregate Technical and Commercial (AT&C) losses
- Automated sustainable systems for the collection of baseline data
- Adoption of information technology in energy accounting
- Consumer care - improving quality and reliability of power supply
- Strengthening of distribution network

Main Points about the microgrid:

- Grids are decentralized energy networks that can function on their own or in tandem with the larger power grid.
- By continuing to operate independently during grid disruptions or emergencies, microgrids boost resilience and energy security.
- Microgrids allow for decentralized power generation and storage, while smart grids use real-time data to improve energy usage and delivery.

1.5.2.2 Integrated Power Development Scheme (IPDS) (erstwhile R-APDRP) scheme:

The IPDS is an initiative launched by the Ministry of Power, Government of India, aimed at enhancing the power distribution infrastructure in urban areas. The scheme was notified on December 3rd, 2014, with a comprehensive set of objectives and components to improve the efficiency, reliability, and quality of power supply in urban regions across the country.

Key components of IPDS are given below:

Strengthening Sub-Transmission and Distribution Networks: The main goal of IPDS is to strengthen urban sub-transmission and distribution networks. This involves upgrading current infrastructure, installing new equipment, and integrating advanced technologies to establish a resilient and dependable power supply system.

Metering of Distribution Transformers, Feeders, and Consumers: IPDS emphasizes the installation of Advanced Metering Infrastructure (AMI) for distribution transformers, feeders, and individual consumers in urban regions. This facilitates enhanced monitoring, control, and management of electricity usage, resulting in increased operational efficiency and decreased losses.

IT Enablement of Distribution Sector: The scheme emphasizes the integration of IT solutions in the distribution sector. This involves implementing enterprise resource planning (ERP) systems, data analytics, real-time monitoring tools, and other digital platforms to streamline operations, enhance decision-making, and improve customer service.

Underground Cabling and Smart Metering: IPDS supports initiatives such as underground cabling to address additional demand and reduce transmission losses. It also promotes the deployment of smart metering solutions, enabling automated meter reading, demand-side management, and integration of renewable energy sources at the consumer level.

Solar power integration: As part of its renewable energy focus, IPDS encourages the installation of solar panels on government buildings with net metering arrangements. This promotes clean energy generation, reduces carbon emissions, and contributes to the overall sustainability of the power sector.

Financial support and funding: The scheme provides financial assistance to power distribution companies (Discoms) for implementing the identified projects under IPDS. The funding includes grants from the Government of India, additional grants based on performance milestones, and support for procurement of equipment and services.

Quality control and standardization: IPDS emphasizes the use of standardized equipment, adherence to quality control measures, and compliance with energy efficiency standards. This ensures that the infrastructure deployed under the scheme is reliable, durable, and environmentally sustainable.

1.5.3 Motivation for the smart grid:

These motivating factors provide a comprehensive overview of the challenges and concerns that drive the transition towards a smarter grid infrastructure:

Global warming/greenhouse gas emissions: The urgent need to combat climate change and reduce greenhouse gas emissions is a primary motivator for adopting smart grid technologies.

Low reserve margins: Traditional grids often struggle with low reserve margins, which can lead to power shortages during peak demand periods or unexpected events. Smart grids address this challenge by implementing demand response programs, energy storage solutions, and grid automation to ensure a reliable and stable power supply.

High AT&C losses: Aggregate Technical and Commercial (AT&C) losses, which include transmission and distribution losses, pose challenges to power system efficiency. Smart grid technologies like Advanced Metering Infrastructure (AMI), distribution automation, and grid analytics aid utilities in minimizing losses, boosting revenue collection, and optimizing operational efficiency.

Poorly planned distribution network: Many existing distribution networks suffer from inefficiencies due to inadequate planning and outdated infrastructure. Smart grids leverage digital technologies, real-time monitoring, and predictive analytics to optimize the distribution network, improve grid performance, and enhance overall system reliability.

Low metering and bill collection efficiency: Manual meter reading, billing errors, and payment collection challenges can impact revenue streams for utilities. Smart grid solutions like smart meters, automated billing systems, and mobile payment platforms streamline metering processes, reduce billing inaccuracies, and enhance customer service.

Power theft and pilferage: Unauthorized power consumption, power theft, and pilferage are significant issues that affect energy efficiency and revenue for utilities. Smart grids employ advanced metering, tamper detection features, and data analytics to detect anomalies, identify potential theft, and improve energy accountability.

Ever-increasing load and peak demand management: The increasing electricity demand, especially during peak times, necessitates efficient load management tactics. Smart grids facilitate demand response initiatives, time-of-use pricing, and load forecasting methods to handle peak demand, alleviate grid strain, and optimize energy consumption.

Deteriorating grid discipline: Grid stability, reliability, and operational discipline are critical for maintaining a robust power system. Smart grid technologies enhance grid monitoring, control capabilities, and grid resiliency measures to address grid discipline challenges and ensure a reliable supply of high-quality power.

1.5.4 Microgrid vs Smart Grid:

The smart grid is a massive electricity distribution system, in contrast to the much smaller scale of a microgrid. Table 1.1 contains the difference between conventional and smart grids. The smart grid is based on the technologies used to power entire communities. A microgrid, on the other hand, is a more compact power grid. The microgrid is meant to serve localized communities. However, they are equally distributed networks. The smart grid is an advanced power system with multiple functions and energy-saving features. Smart grids are primarily used to distribute electricity and provide two-way digital communication. Smart grid operations are digital. Communications can be analysed, controlled, and monitored with the aid of the smart grid. The smart grid is integrated into the supply chain, where it helps to maximize productivity.

Table 1.1 Difference between conventional and smart grid.

Aspect	Conventional Grid	Smart Grid
Technology integration	Relies on traditional power generation sources like coal, gas, and nuclear plants with limited integration of renewable energy sources. Limited use of digital technology for monitoring and control.	Integrates renewable energy sources such as solar and wind power with advanced digital technologies like smart meters, sensors, automation, and communication systems for real-time monitoring, control, and optimization.
Grid flexibility	Operates on a centralized, one-way power flow model with limited flexibility to adapt to changing demand patterns or integrate distributed energy resources (DERs).	Enables bidirectional power flow, allowing for the integration of DERs like rooftop solar panels, energy storage systems, and electric vehicles. Offers greater flexibility to manage demand-response programs and optimize grid operations.
Data and analytics	Relies on manual data collection and basic analytics for grid management and planning, leading to limited visibility into real-time grid conditions.	Leverages cutting-edge data analytics, machine learning, and artificial intelligence algorithms to analyze extensive data from smart meters, sensors, and devices. Empowers predictive maintenance, outage management, and grid optimization through data-driven insights
Energy efficiency	Limited ability to promote energy efficiency due to lack of real-time consumption data, inefficient grid operations, and reliance on conventional generation methods.	Promotes energy efficiency through demand-side management programs, time-of-use pricing, and energy consumption feedback to consumers. Enables dynamic pricing, load balancing, and peak shaving to reduce energy waste and costs.
Reliability and resilience	Vulnerable to single-point failures, limited visibility into grid disturbances, and slower response times to outages or grid disturbances.	Enhances grid reliability and resilience through self-healing capabilities, predictive maintenance, grid automation, and real-time monitoring. Reduces outage durations, improves fault detection, and enhances grid stability during extreme weather events or emergencies.
Cybersecurity and privacy	Basic cybersecurity measures with limited protection against cyber threats and data breaches. Limited control over data privacy and consumer information.	Implements robust cybersecurity protocols, encryption techniques, and secure communication networks to protect against cyber-attacks, data manipulation, and unauthorized access. Prioritizes consumer privacy and data confidentiality.
Grid management and optimization	Relies on manual grid management, reactive maintenance, and traditional grid planning methods with limited situational awareness and control capabilities.	Enables proactive grid management, real-time monitoring, predictive analytics, and optimized grid operations. Supports grid modernization, renewable energy integration, and efficient resource allocation.

The smart grid has the potential to drastically cut costs and increase visibility throughout the supply chain. The smart grid was developed to address the shortcomings of traditional power systems. The smart grid employs smart-net meters. Microgrids are small, independent power systems. The microgrid is compatible with the larger grid. In island mode, the microgrid can operate independently. The island mode is what sets the microgrid apart from others. Power may be switched between the island and linked modes on the microgrid. By switching between several modes, it makes the supply more reliable. Using a microgrid in an off-grid setting is an example of its autonomous application.

1.5.5 Understanding of Smart Grid:

Definition: The smart grid represents the fusion of an electrical power system with a communication network, sophisticated sensing capabilities, advanced metering and measurement infrastructure, comprehensive decision support systems, as well as human interfaces in both software and hardware forms. Its purpose is to effectively monitor, control, and manage all aspects of energy generation, distribution, storage, and consumption.

Features of the smart grid are given below:

- Fiber optic routers are utilized in the advanced metering infrastructure of the smart grid. The circuit breakers and distribution routes are also high-tech.
- The smart grid is efficiently communicated by the smart home controls.
- The smart grid relies heavily on electronic power conditioning and production control.
- Smart grid technology exemplifies the electrical grid's underlying technical architecture. The security of smart meters is an important problem.
- Digital communication technology has enabled the flow of data in the smart grid. The smart grid can detect, respond, and react to issues, giving it self-healing capabilities.
- It is capable of addressing several problems simultaneously. One business that invests much in IT is switching to the smart grid.
- The smart grid has many advantages, including the reduction of energy costs, the identification of fraud, improved customer service, and lower overall costs.

1.5.6 Understanding of Microgrid:

One viable solution for rural communities is to install a microgrid. A microgrid can improve the functionality of a tiny geographical island. The layout of microgrid system is shown in Fig. 1.8. The following points are given about the microgrids:

- A microgrid integrates distributed generation and renewable sources. However, its primary challenge lies in regulating and securing the system, which is its major drawback.
- Microgrid services face difficulties at low-short circuit levels. Maintaining security under low-voltage short-circuit is difficult.
- It can be used in two modes of operation. It can be used as a grid-connected mode and an islanding mode of operation.
- The system's multiple energy requirements are met by the microgrid. It not only generates power but also heats and cools the home. Microgrids have very efficient energy use.
- The microgrid also includes a bidirectional AC-DC converter, DC-DC converter, filters and controller.

- Thousands of users rely on their community's microgrid for their daily power needs. The electricity for the community microgrid comes from a few strategic locations. AC and DC converters are present in the centralized microgrid.
- Microgrid can be one of five distinct varieties. Microgrids can be found in a variety of settings, including college campuses, neighbourhoods, military bases, and businesses.
- Microgrid can be either AC or DC or a mix of the two. Local energy generation, consumption, energy storage, and a point of common coupling are the foundations of microgrids.

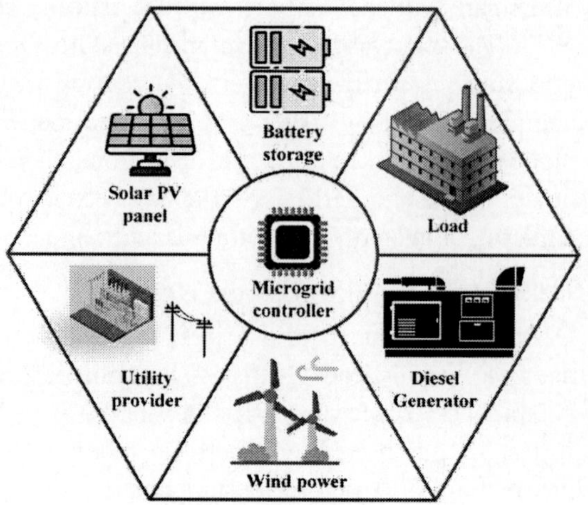

Fig. 1.8 Layout of the microgrid system.

1.5.6.1 Difference between smart grid and microgrid:

Table 1.2 contains the differences between microgrids and smart grids in different aspects. The features of the smart grid are given below:

i. Real-time Simulation: This refers to the process of simulating dynamic systems, such as power systems, in real-time or at a speed that matches the actual system operation. This technique involves using advanced computer models and simulation software to replicate the behaviour of the physical system and predict its performance under different conditions.

ii. Wide-area reliability: It refers to the ability of an interconnected power system to maintain a stable and reliable electricity supply across a large geographic area, typically spanning multiple states or regions. This concept is crucial in modern power systems where extensive networks connect various generation sources, transmission lines, substations, and distribution systems.

iii. Network optimization: It entails harnessing advanced technologies and tactics to boost the efficiency, reliability, and sustainability of power distribution systems. Smart grids incorporate digital communication and control features into conventional power grids, allowing for real-time monitoring, data analytics, and intelligent decision-making.

iv. Customer participation: In the context of smart grids and modern power systems refers to the active involvement of energy consumers in various aspects of electricity management, decision-making, and resource utilization. Here are key aspects of customer participation in the energy sector:

a. Demand response programs: Consumers engage in demand response (DR) initiatives by voluntarily modifying their electricity consumption based on price signals, grid conditions, or utility directives. This aids utilities in handling peak demand, alleviating grid pressure, and circumventing expensive infrastructure enhancements.

Table 1.2 Differences between microgrid and smart grid in different aspects.

Aspect	Microgrid	Smart grid
Definition	A localized group of interconnected loads and DERs that operates autonomously or can connect to the main grid [4].	The utility grid with two-way communications and distributed intelligent devices. These two-way flows of electricity and information enhance the efficiency and reliability of the utility grid [5].
Scale	Typically smaller in scale, serving localized areas such as campuses, communities, or industrial sites.	Larger in scale, covering wide geographic regions and serving entire cities, regions, or countries.
Independence	Can operate independently or connected to the main grid ("Islanded" or "Grid-tied" modes).	An integral part of the main utility grid, ensuring integration with centralized power generation and distribution.
Control	Localized control with a focus on optimizing local energy generation, consumption, and storage.	Centralized control with monitoring and management of multiple energy sources, demand-response systems, and grid infrastructure.
Resilience	Provides increased resilience and dependability, particularly during grid failures or emergencies, thanks to its autonomous operational capability.	Provides improved resilience through real-time monitoring, predictive maintenance, and rapid fault detection and response.
Renewable integration	Enables smooth incorporation of renewable energy sources such as solar, wind, and storage within the microgrid confines.	Supports the integration of renewables at both local and grid-wide levels, managing fluctuations and optimizing renewable energy utilization.
Efficiency	Optimizes local energy generation and consumption patterns, reducing energy losses and improving overall efficiency.	Enhances grid efficiency by implementing demand-side management, smart meters, and grid optimization technologies.
Use Cases	Ideal for remote or off-grid areas, critical infrastructure (hospitals, military bases), and communities seeking energy independence.	Suitable for urban areas, large-scale power systems, and regions requiring advanced grid monitoring, control, and management capabilities.
Cost	Initial deployment costs can be high, especially for standalone microgrids with extensive DERs and storage systems.	Requires significant investment in infrastructure, digital technologies, and grid modernization, but offers long-term benefits in grid optimization and sustainability.

b. Time-of-use pricing: Consumers have the option to choose time-of-use pricing schemes, where electricity prices fluctuate depending on the time of day or grid conditions. By moving their high-energy tasks to non-peak periods, customers can reduce their electricity costs and support grid stability.

c. Energy efficiency programs: Utilities engage customers in energy efficiency initiatives by offering incentives for adopting energy-efficient appliances, home upgrades, and behaviour changes. Customers can save energy, reduce costs, and lower their environmental impact through energy conservation measures.

d. Smart home technologies: Customers can install smart home devices like thermostats, lighting controls, and appliances connected to the Internet of Things (IoT). These devices enable automated energy management, remote monitoring, and optimization of energy usage based on preferences and real-time data.

e. Renewable energy adoption: Customers have the option to install rooftop solar panels, and wind turbines, or participate in community solar programs to generate their renewable energy. Through net metering or feed-in tariffs, customers can sell excess energy back to the grid and contribute to renewable energy integration.

f. Electric vehicles: With the rise of electric vehicles, customers can participate in V2G programs where EV batteries can store energy from the grid and discharge it when needed. This two-way flow of electricity supports grid stability, enhances renewable energy utilization, and provides backup power during outages.

g. Energy data access: Customers have access to their energy consumption data through smart meters and online portals. This data transparency allows customers to track their usage patterns, identify energy-saving opportunities, and make decisions about their electricity consumption.

h. Microgrids and community energy projects: Customers can collaborate in microgrid initiatives or community energy projects where local energy generation, storage, and distribution are managed collectively. This community-centric approach promotes energy resilience, self-sufficiency, and peer-to-peer energy trading.

i. Feedback and engagement: Utilities engage customers through educational programs, energy audits, and feedback mechanisms to raise awareness about energy conservation, efficiency best practices, and sustainable behaviours.

j. Grid services participation: Advanced customers with DERs like solar panels, battery storage, or flexible loads can participate in grid services such as frequency regulation, voltage support, or demand flexibility. This active participation contributes to grid stability, reliability, and overall system efficiency.

1.5.7 Concept of Nano Grid:

A Nano grid is a localized energy system that operates independently or in coordination with a larger power grid. Unlike microgrids, which typically serve a single building or campus, Nano grids are even smaller in scale, often catering to a single device or a small cluster of devices. They integrate renewable energy sources, energy storage systems, and advanced control technologies to optimize energy usage and reliability at the micro-level.

One of the key features of Nano grids is their ability to operate autonomously or in concert with the main power grid, depending on factors like energy demand, availability of renewable resources, and grid stability. They can function as standalone units during grid outages or peak demand periods, ensuring continuous power supply to critical loads. Additionally, Nano grids contribute to energy efficiency by locally generating and consuming power, reducing transmission losses and improving system resilience.

1.6 Generation: Conventional and Renewable Energy Sources:

1.6.1 Thermal power plant:

The thermal power generation of India increased by 8.21% in the 2022-2023 financial year [6]. The total thermal generation including coal, lignite, gas and diesel is 245.90 GW as on 31-01-2025. The calorific values of the major coal used in thermal power plants are given in Table 1.3. Bituminous coal is mostly used for electricity generation in India.

Table 1.3 Types of coal with their calorific values.

Type of coal	kJ/kg	kCal/kg
Peat	8000	1912
Lignite	20000	4780
Bituminous	27000	6453
Anthracite	30000	7170

1.6.1.1 Working principle of thermal power plant:

In a thermal power plant, the primary working fluid is water and steam, forming what's known as the feed water and steam cycle. This cycle is based on the Rankine Cycle, a thermodynamic process used to convert thermal energy into mechanical work, which in turn drives electrical generators. The process begins with the combustion of fuel in the furnace, producing hot gases that heat water in the boiler. The boiler's function is to generate dry, superheated steam at the required temperature. This steam is then directed to a steam turbine, where it expands and drives a synchronous generator to produce electrical energy.

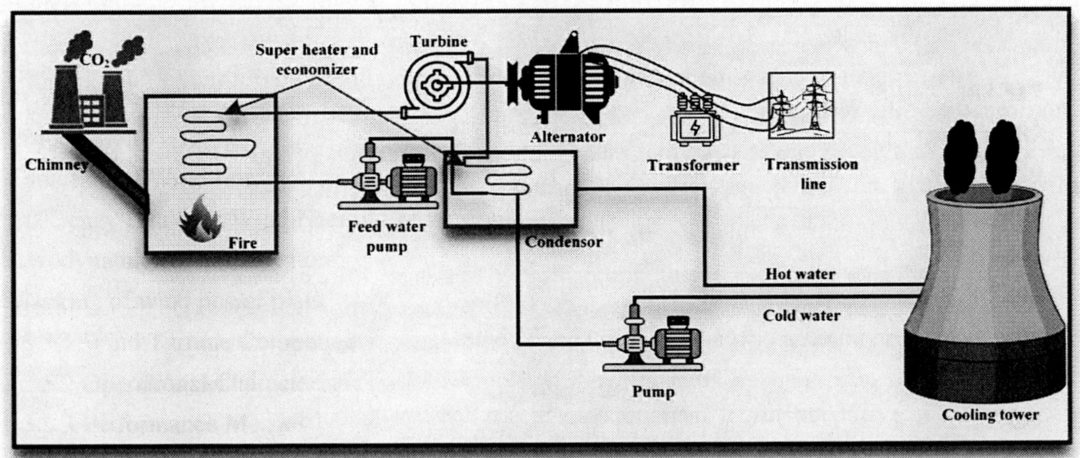

Fig. 1.9 Layout of the thermal power plant.

After passing through the turbine, the exhaust steam enters the condenser, where it is condensed back into water using cooling water. This condensation creates suction and allows for steam expansion in the turbine at low pressure, increasing overall efficiency. The condensed water, along with fresh makeup feed water, is then pumped back into the boiler by a boiler feed pump to maintain the water-steam cycle. A cooling water circuit cools and condenses the steam in the condenser, with the cooling water being recycled through a cooling tower. The air and flue gas circuit involves the intake of ambient air for combustion in the furnace, and the exhaust of flue gas into the atmosphere through stacks. This circuit is maintained by Forced Draught and Induced Draught fans. Fig. 1.9 shows the layout of the thermal power plant.

Within the boiler, different heat exchangers like the economizer, evaporator, superheater, and reheater are utilized to enhance heat exchange and efficiency. A thermal power plant harnesses energy released from fuel combustion to generate steam, powering turbines and generators for electricity production. Table 1.4 outlines the pros and cons of thermal power plants.

Table 1.4 Advantages and disadvantages of thermal power plant.

Advantages of Thermal Power Plants	Disadvantages of Thermal Power Plants
✓ Reliability	× Environmental Impact
✓ Quick Start-Up	× Water Usage
✓ Fuel Availability	× Land Requirements
✓ Energy Efficiency	× Resource Depletion
✓ Large-Scale Generation	× High Capital Costs

1.6.1.2 Components of Thermal Power Plant:

Boiler: The boiler is a crucial component where heat from burning fuel is transferred to water tubes surrounding the flames. Water is kept circulating through the tubes with the help of a pump. The temperature of the fire tube boiler varies from 315 degrees C to 575 degrees C and the pressure is limited from 17 kg/cm^2 to 125 kg/cm^2.

Drum: This part contains water and steam at high pressure, producing steam for the turbine. It receives water from the boiler-feed pump.

High-Pressure Turbine: Converts thermal energy into mechanical energy as steam expands through turbine blades. The steam may pass through a reheater to improve thermal efficiency.

Medium-Pressure Turbine: Similar to the high-pressure turbine but larger to allow further steam expansion.

Low-Pressure Turbine: Removes remaining energy from the steam, often comprising two identical sections. It can be an impulse or reaction turbine or a combination.

Condenser: Causes steam condensation by passing it over cooling pipes. Cold water flowing through these pipes absorbs heat, creating a vacuum pressure that drives a condensate pump.

Reheater: A heat exchanger that raises the temperature of feed water using steam bled from the high-pressure turbine, improving overall efficiency.

Burner: Supplies and controls the fuel (gas, oil, or coal) injected into the boiler. Coal is typically pulverized before injection, while heavy oil is preheated and atomized for efficient combustion.

Forced-Draft Fan: Supplies the large amount of air needed for combustion.

Induced-Draft Fan: **Carries combustion gases and by-products towards the cleansing apparatus and out through the stack.**

i. Generator: **Converts mechanical energy from turbines into electrical energy. The synchronous generator is used to generate electricity. The speed of the generator will be 3000 RPM for two two-pole salient-type generators.**
ii. Cooling Towers: **Utilizes evaporation to cool the condenser. Water is broken into droplets, and air is passed through to facilitate cooling.**

1.7. Distributed Energy Resources (DER):

DERs encompass a diverse array of small-scale power generation technologies positioned near the point of consumption. These resources comprise renewable sources like solar PV panels, wind turbines, biomass generators, micro-hydro turbines, and geothermal systems, alongside energy storage systems such as batteries and flywheels. DERs also encompass combined heat and power systems, fuel cells, and demand response technologies. Fig. 1.10 illustrates the advantages of distributed energy resources. The integration of DERs into the energy grid offers several benefits:

Improved energy resilience: **DERs enhance the reliability and resilience of the energy grid by decentralizing power generation. They can continue to supply electricity during grid outages or disruptions, ensuring a more robust and stable energy supply.**

Reduced greenhouse gas emissions: **Numerous DERs, particularly renewable sources like solar and wind, generate electricity without emitting greenhouse gases. Harnessing these resources aids in decreasing the overall carbon footprint of the energy system, thereby supporting climate change is necessary.**

Promotion of energy independence: **By generating electricity locally, DERs contribute to greater energy independence and reduced reliance on centralized power plants. This decentralization can enhance energy security and diversify the energy supply mix.**

Energy cost savings: **DERs can help reduce energy costs for consumers by generating electricity onsite and offsetting the need to purchase power from the grid at higher rates. This is particularly beneficial during periods of peak demand when electricity prices are typically higher.**

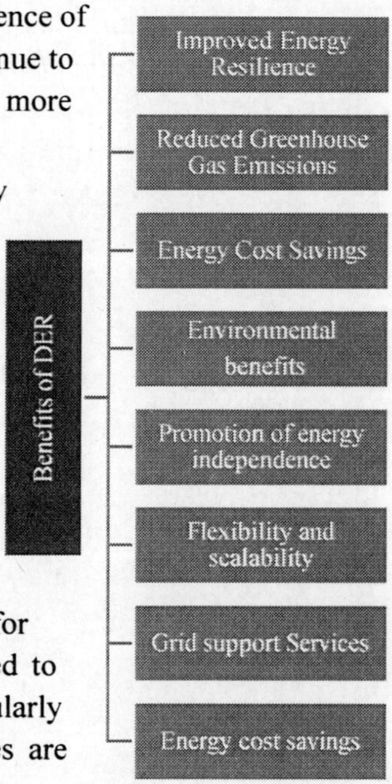

Fig. 1.10 Benefits of DER.

Grid support Services: **Some DER technologies, such as energy storage systems, can provide grid support services like frequency regulation, voltage control, and peak shaving. These services help optimize grid operations and improve overall system efficiency.**

Flexibility and scalability: **DERs offer flexibility and scalability in power generation and distribution. They can be easily installed in various locations, including residential, commercial, and industrial settings, allowing for tailored solutions based on specific energy needs and requirements.**

Environmental benefits: In addition to reducing greenhouse gas emissions, DERs often have lower environmental impacts compared to traditional fossil fuel-based power generation. This includes reduced air and water pollution, preservation of natural resources, and support for sustainable energy practices.

Overall, the adoption and integration of DERs offer numerous advantages, ranging from environmental sustainability and energy cost savings to enhanced grid reliability and resilience. These benefits make DERs a valuable component of modern energy systems determined for efficiency, sustainability, and flexibility.

1.7.1 Impact of DER on the utility grid:

DERs have a profound impact on the utility grid, introducing both opportunities and challenges. Fig. 1.11 contains the impact of DER on the grid. Here are some key impacts of DERs on the utility grid:

Grid reliability: DERs can enhance grid reliability by providing localized power generation and reducing dependency on centralized power plants. They offer backup power during grid outages, especially when combined with energy storage systems like batteries. However, integrating DERs requires careful planning to ensure grid stability and resilience.

Grid flexibility: DERs, particularly renewable energy sources like solar and wind, introduce variability and intermittency to the grid. This variability can challenge grid operators in maintaining a balance between electricity supply and demand. Advanced control and management systems are needed to optimize DERs and improve grid flexibility.

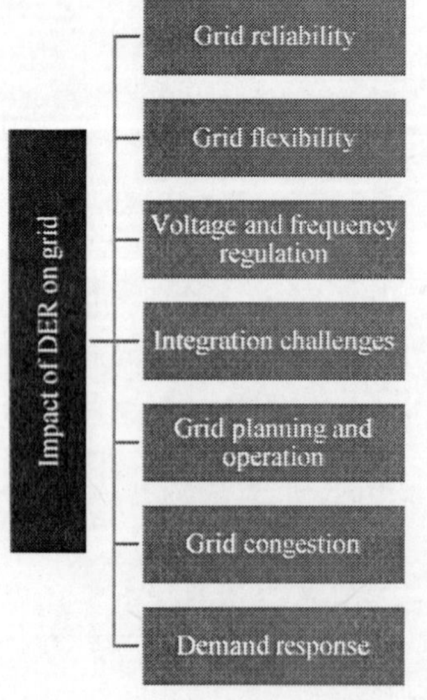

Fig. 1.11 Impact of DER on grid.

Integration challenges: Integrating DERs into existing grid infrastructure poses technical, regulatory, and economic challenges. Issues such as interconnection standards, tariff structures, grid codes, and market design need to be addressed for seamless DER integration.

Grid planning and operation: Utilities need to adapt their grid planning and operation strategies to accommodate DERs effectively. This includes modelling DER impacts, forecasting renewable generation, optimizing DER dispatch, and implementing grid-friendly DER standards and protocols.

Voltage and frequency regulation: DERs can impact voltage and frequency levels on the grid, especially during periods of high penetration. Inverter-based DERs, such as solar PV systems, may influence voltage profiles, requiring voltage regulation mechanisms. Frequency regulation becomes crucial as more intermittent renewables are integrated into the grid.

Demand response: DERs enable demand response programs where consumers can adjust their electricity usage based on price signals or grid conditions. This demand-side flexibility can help in load management, peak shaving, and overall grid optimization.

Grid congestion: In areas with high DER penetration, grid congestion can occur, leading to voltage violations and distribution system overloads. Grid upgrades and smart grid technologies are essential to mitigate congestion challenges and accommodate growing DER installations.

1.8 Renewable energy:

As of the latest available data, renewable energy technologies have experienced significant growth and adoption worldwide. One of the notable developments is in wind power, with global capacity surpassing 700 GW by the end of 2020, marking a substantial increase from the 95 GW reported in 2007. This expansion reflects the growing emphasis on clean and sustainable energy sources. Additionally, the overall contribution of renewables to the world's total primary energy has notably increased, reaching around 26% of global electricity generation in 2020. Solar photovoltaic PV technology has also witnessed rapid growth, with installed capacity exceeding 800 GW by the end of 2020. Hydropower remains a major player in renewable energy generation, boasting a global capacity exceeding 1,300 GW in 2020. This shift towards renewable energy sources has been driven by various factors, including environmental concerns, supportive policies, technological advancements, and declining costs of renewable technologies. As a result, renewables, including hydro, solar, wind, geothermal, and biomass, contributed approximately 35.4% to global electricity generation as on 31-01-2025, highlighting the significant progress and evolution of the renewable energy landscape.

1.8.1 Hydropower plant:

Hydro-energy is often regarded as a traditional renewable energy source, hinging on the natural flow of water from higher to lower land surfaces, which constitutes its potential. To harness this potential into usable electric energy, the water flow is directed to drive a hydraulic turbine. This process converts hydro energy into mechanical energy, which subsequently powers a connected generator to transform the mechanical energy into electric energy. Unlike coal power generation, where the exploitation of fuel and the generation of electricity are separate processes, hydropower combines these steps simultaneously, offering advantages over thermal power generation. In India, the share of hydroelectric power is 11.6 % of total installed capacity. The installed generation capacity of hydroelectric is 46,968 MW till 31-01-2025. India has the 5th position in hydropower generation in the world behind China, Brazil, the USA and Canada. Table 1.5 contains the details of major hydropower projects located in India. Fig 1.12 outlines the layout of the hydropower plant.

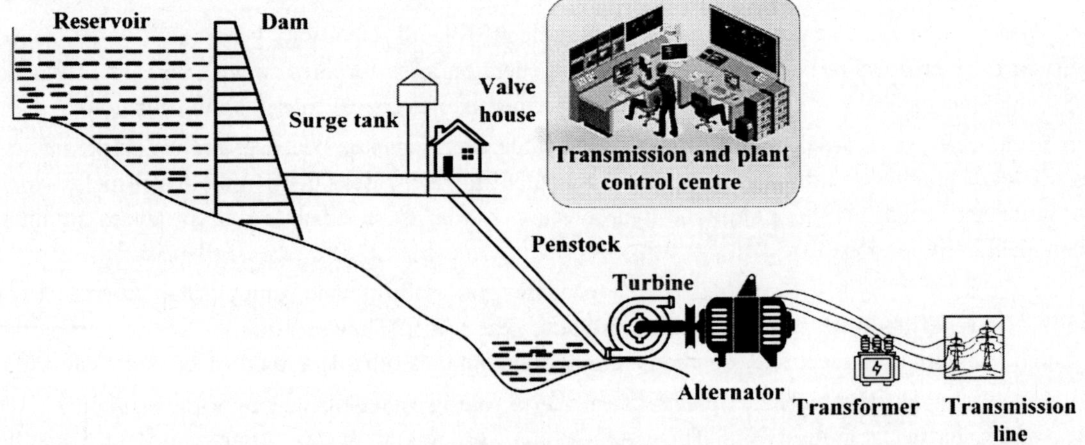

Fig. 1.12 Layout of the hydropower plant.

The electric power generation using the head and flow of water can be calculated as equation 1.1.

$$kW = 9.81 \times Q \times H \times \eta \qquad \text{……… (1.1)}$$

Where Q = water flows from the turbine in meter cubic per second

H = net head height in meters,

η = efficiency of the power plant

Table 1.5 Detail of major hydropower projects located in India.

Hydropower Plant	State	Capacity (MW)	River
Tehri Dam and hydropower complex	Uttarakhand	2400	Bhagirathi
Koyna hydroelectric project	Maharashtra	1,960	Koyna
Nathpa Jhakri dam	Himachal Pradesh	1,530	Satluj
Sardar Sarovar dam	Gujarat	1450	Narmada
Bhakra Nangal dam	Punjab/Himachal Pradesh	1325	Sutlej
Indira Sagar Dam	Madhya Pradesh	1000	Narmada
Baglihar Hydroelectric Power plant	Jammu and Kashmir	900	Chenab
Nagarjuna Sagar dam	Andhra Pradesh/Telangana	800	Krishna
Idukki hydroelectric project	Kerala	780	Periyar
Salal Hydroelectric power station	Jammu and Kashmir	690	Chenab
Dulhasti Hydroelectric plant	Jammu and Kashmir	390	Chenab

At the heart of this complex is the Tehri Dam, a massive rock and earth-fill embankment structure towering 260.5 meters high. Its primary function is water storage, creating the expansive Tehri reservoir with a capacity exceeding 2.6 billion cubic meters. This reservoir is a lifeline, supporting irrigation, and municipal water supply, and acting as a buffer against floods. It is the 13th tallest dam in the world.

Power generation estimation from hydropower plant

The main purposes of a hydroelectric plant are given below:

Electricity Generation: The primary goal of a hydroelectric plant is electricity generation. Water from a reservoir or river flows through turbines linked to generators. The turbines convert the water's kinetic energy into mechanical energy, subsequently transformed into electrical energy by the generators.

Renewable Energy: Hydroelectric power is a renewable energy source because it relies on the natural water cycle, including rainfall and river flows, to replenish water reservoirs. Unlike fossil fuels, which are finite and contribute to pollution, hydroelectric power is sustainable and produces minimal greenhouse gas emissions.

Grid Stabilization: Hydroelectric plants can provide grid stability and support the integration of intermittent renewable energy sources like wind and solar power. They can quickly adjust their output to match fluctuations in electricity demand, helping to maintain a reliable and balanced electrical grid.

Water Management: Hydroelectric plants often play a role in water management by regulating river flows and water levels in reservoirs. This can help control flooding, ensure water supply for irrigation and drinking purposes, and manage seasonal variations in water availability.

1.8.1.1 Components of the hydropower plant:

Table 1.6 shows the details of turbine type, head height and suitable head. The main components of the hydel plant are detailed below:

Dam: A dam is a structure built across a river or waterway to create a reservoir by impounding water. It helps regulate the flow of water and ensures a consistent supply for power generation.

Reservoir: The reservoir is a large body of water created by the dam. It stores water and provides the necessary head (height difference) for driving turbines.

Intake Structure: This is the opening in the dam or reservoir where water is captured and directed into the power plant.

Penstock: A penstock is a large pipe or conduit that carries water from the reservoir or intake to the turbines. The pressure of the water in the penstock helps drives the turbines.

Surge tank: A surge tank acts as a pressure stabilizer and safeguards against water hammer effects. Positioned strategically along the penstock route, usually near the powerhouse or at bends in the pipeline, surge tanks are designed to manage sudden fluctuations in water flow rates. This is particularly important during turbine start-ups, shutdowns, or changes in load demand, as these events can cause pressure spikes that may damage pipes and equipment. The surge tank's design allows it to absorb excess water during high-flow periods, preventing pressure from rising too rapidly.

Turbines: Turbines are machines that convert the energy of flowing water into mechanical energy. Different types of turbines, such as Francis, Pelton, and Kaplan turbines, are used based on factors like water flow rate and head.

Table 1.6 Detail of turbine type, head height and suitable head.

Turbine type	Suitable head	Height (meter)
Propeller of Kaplan	Low	Less than 30
Kaplan or Francis	Medium	30 - 300
Francis or Pelton	High	More than 300

Generator: The mechanical energy produced by the turbines is transferred to a generator. The synchronous generator is used in the electricity generation. The generated voltage is 11 kV to 25 kV. The used generator is operated at a low RPM compared to the generator used in the thermal power plant. The number of poles in the rotor can be calculated using the below formula given in equation 1.2. P is the number of poles, f is the supply frequency.

$$P = \frac{120f}{RPM\ of\ the\ turbine} \quad \text{……… (1.2)}$$

Transformer: The electricity generated by the generator is initially produced at a relatively low voltage. Transformers are used to step up the voltage to a higher level for efficient transmission over long distances.

Switchyard: The switchyard is a facility where the electricity from the generator is connected to the electrical grid. It contains switches, circuit breakers, and other equipment for controlling and directing the flow of electricity.

Control room: The control room is where operators monitor and control the operation of the hydropower plant. They manage water flow, turbine speed, and electrical output to optimize efficiency and ensure safety.

1.8.2 Photovoltaic (PV) energy:

The installed capacity of solar energy is 100.32 GW as of 31-01-2025 in India. It shares 21.52% of the total installed capacity. India holds the 4th position in solar energy generation in the world. PV energy operates on the principle of converting sunlight directly into electricity using solar panels. Fig. 1.13 shows the map of major solar parks available in India.

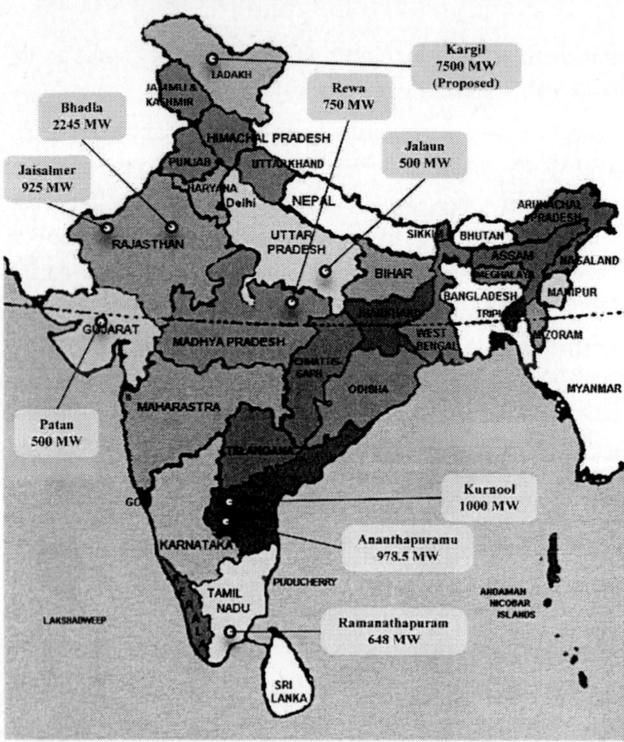

Fig. 1.13 Major solar parks installed in India as on 31-01-2025

At the core of this technology are photovoltaic cells made from semiconductor materials such as silicon. When sunlight strikes these cells, it transfers its energy to electrons within the material, causing them to become energized and move, creating an electrical current. This current flows within the solar panel and is captured by conductive metal plates, generating DC electricity. The power generated by the PV depends on the temperature and irradiance. When the DC load is connected to the panel, it does not transfer the maximum power because of impedance mismatching. To transmit the maximum power from PV to load it is required maximum power point tracking algorithm.

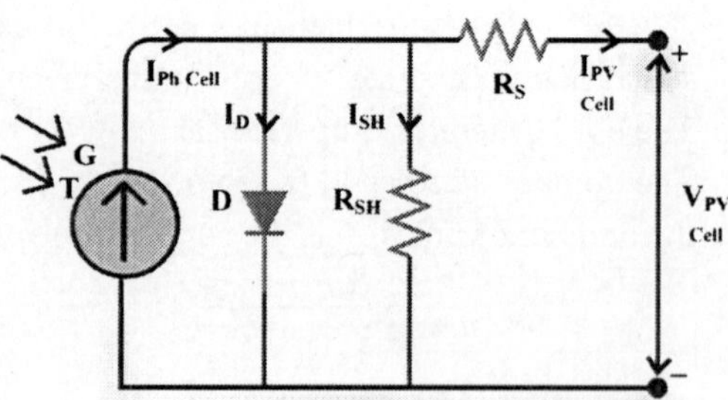

Fig. 1.14 Equivalent circuit diagram of solar photovoltaic cell.

The equivalent circuit diagram of a solar PV cell is shown in Fig. 1.14. When light falls on the cell then photocurrent is generated in the p-n junction. It exists on a diode and shunt resistance in parallel with the cell. A series resistance also exists in the circuit. The details about solar PV energy are explained in the upcoming chapter. Table 1.7 shows the solar projects, location, capacity, year of operation, and operator of the plant. Fig. 1.15 shows the various applications of photovoltaics.

The Bhadla Solar Park in Rajasthan, India, is a testament to the nation's aspiring renewable energy goals and its dedication to curbing greenhouse gas emissions. Spanning an impressive 56 square kilometres with a massive capacity of 2,245 MW, it holds the title of the world's largest solar park as of 2023. This ground-breaking project, initiated in 2015, received substantial funding amounting to $2.175 billion from diverse sources, including the Climate Investment Fund and other entities.

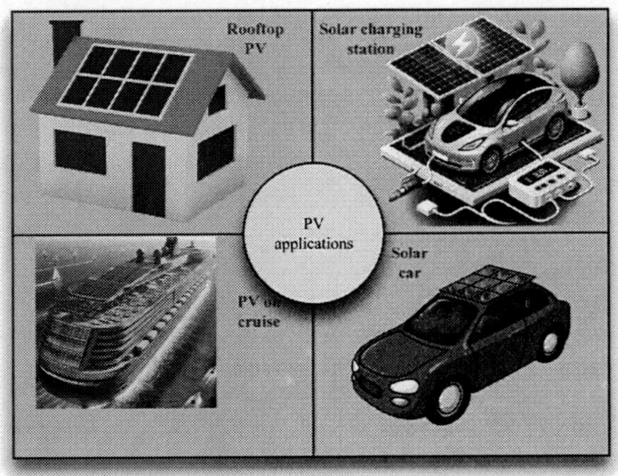

Fig. 1.15 Photo-Voltaic Applications.

Rajasthan Renewable Energy Corporation Limited (RRECL), a collaboration between the Government of Rajasthan and the Ministry of New and Renewable Energy (MNRE), spearheaded this initiative. The park's development unfolded in four phases, characterized by successful auctions and partnerships with leading developers and organizations. Distinguished milestones include NTPC Limited's Phase I. Involvement, auctioning 420 MW, and Solar Energy Corporation of India (SECI) overseeing Phase II auctions. By December 2018, the park had achieved an impressive total capacity of 2,055 MW.

Beyond its numerical achievements, it significantly contributes to India's National Solar Mission, aiming for 100 GW of solar power by 2022, and aligns with international commitments like the Paris Agreement. Moreover, it encourages electricity supply reliability and quality while substantially reducing fossil fuel dependence, averting approximately 4 million tons of CO2 emissions annually.

Table 1.7 Detail of major solar power projects located in India as on 31-01-2025.

Solar power plant	Location	Capacity (MW)	Years of operation	Operator
Bhadla solar park	Jodhpur, Rajasthan	2,245	2017	Solar Energy Corporation of India
Pavagada solar park	Tumkur, Karnataka	2,050	2017	Solar Power Developers India Limited
Kurnool ultra-mega solar park	Kurnool, Andhra Pradesh	1,000	2017	NTPC
NP Kunta Ultra Mega Solar Park	Ananthapuramu, Andhra Pradesh	978.5	2018	Andhra Pradesh Solar Power Corporation Private Limited
Nokh solar park	Jaisalmer, Rajasthan	925	2022	NTPC
Rewa Ultra Mega Solar Park	Rewa, Madhya Pradesh	750	2018	Rewa Ultra Mega Solar Limited
Kamuthi solar power project	Ramanathapuram, Tamil Nadu	648	2016	Adani Green Energy Limited
Charanka solar park	Patan, Gujarat	500	2012	Gujarat Solar Power Corporation Limited
Dhursar solar park	Dhursar, Rajasthan	125	2014	NTPC vidyut vyapar nigam

1.8.3 Wind energy:

The installed capacity of wind energy is 48.36 GW as on 31-01-2025. It shares 10.37% of total installed capacity. It holds the 4th position in the world. Wind energy harnesses the kinetic energy of moving air to generate electricity through wind turbines. Fig. 1.16 shows the layout of wind energy system. The working principle of wind energy is relatively simple: Wind turbines have large blades mounted on a tall tower designed to capture the kinetic energy of the wind. As the wind blows, it causes the turbine blades to rotate. The angle and design of the blades are optimized to capture as much wind energy as possible. These rotating blades are connected to a rotor, which is connected to a shaft inside the turbine. Inside the turbine, the shaft is linked to a generator.

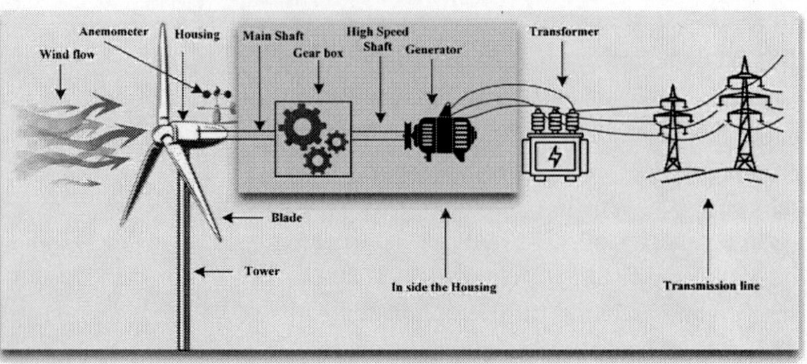

Fig. 1.16 Layout of wind energy system.

As the shaft turns due to the wind's force on the blades, it spins the generator, converting mechanical energy into electrical energy. The generator produces electricity in the form of AC, which is then transmitted through cables inside the turbine and sent to a transformer. The transformer converts the electricity to a higher voltage suitable for transmission over long distances and feeds it into the electrical grid for distribution. Control systems in modern wind turbines monitor wind speed and direction, adjusting the angle of the blades or shutting down the turbine in high winds to prevent damage, ensuring efficient and safe operation. Major wind energy systems installed in India as on 31-01-2025 is shown in Fig. 1.17. The details of the wind power generation are explained in the further chapter.

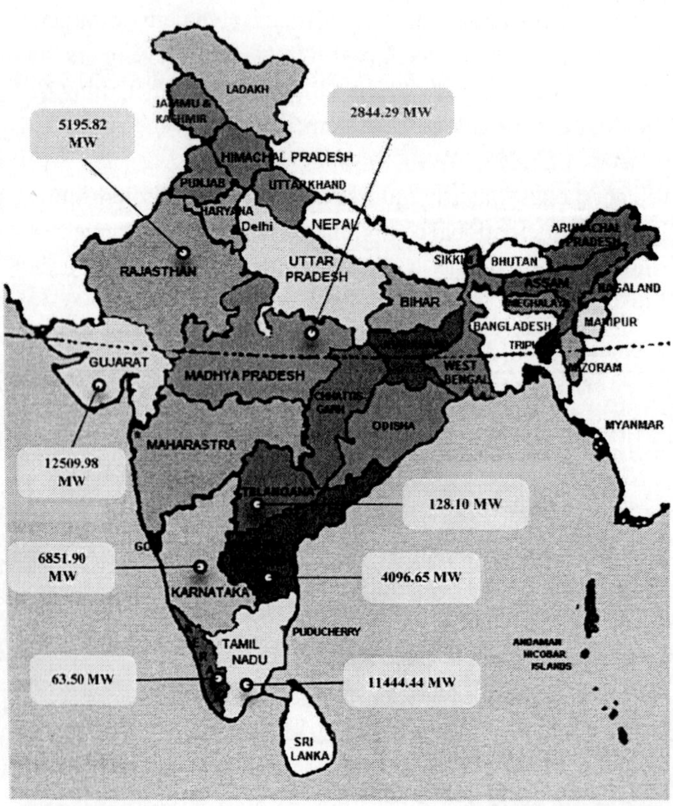

Fig. 1.17 Major wind energy systems installed in India as on 31-01-2025 [7].

1.9 Energy Storage Technology:

Grid-scale energy storage refers to large-scale systems that store electrical energy on the grid for later use. These storage solutions are designed to address challenges related to the intermittent nature of renewable energy sources, demand variability, and the need for grid stability. There are various technologies to store energy like Li-ion batteries, pumped storage plants, flywheels, and compressed air. Grid-scale energy storage systems offer several advantages that contribute to the stability, reliability, and efficiency of the electrical grid. Some key advantages include:

Load balancing: Energy storage allows for the balancing of supply and demand fluctuations in real time. It can store excess energy during periods of low demand and release it during peak demand hours, helping to maintain grid stability.

Integration of Renewable Energy: Energy storage facilitates the integration of renewable energy sources like solar and wind power into the grid. It stores surplus renewable energy generated during optimal conditions and delivers it when renewable generation is low, ensuring a continuous and reliable power supply.

Peak shaving: By discharging stored energy during peak demand periods, grid-scale storage systems reduce the need for expensive and less efficient peak plants. This helps to lower overall electricity costs and improve grid efficiency.

Backup Power and Resilience: Energy storage systems provide backup power during grid outages or emergencies, enhancing grid resilience and reliability. They can also support critical infrastructure and essential services during disruptions.

Frequency Regulation: Grid-scale storage helps in regulating grid frequency by injecting or absorbing power as needed. This improves the stability of the grid and reduces the risk of frequency fluctuations.

Reduced Transmission and Distribution Costs: By reducing the need for additional transmission and distribution infrastructure upgrades, energy storage can lower overall grid infrastructure costs and improve system efficiency.

Demand Response: Energy storage enables demand response programs by allowing utilities to manage demand fluctuations more effectively. This encourages consumers to shift their electricity usage to off-peak hours, reducing peak demand pressures.

Environmental Benefits: Utilizing energy storage with renewable energy sources reduces greenhouse gas emissions and promotes a cleaner and more sustainable energy system.

1.9.1 Pumped storage plant:

A pumped storage plant is a type of hydroelectric power facility that plays a crucial role in balancing and stabilizing the electrical grid. Table 1.8 shows the performance measure of the pumped storage plant. Here's an overview of how it works:

Basic principle: A pumped storage plant operates by using water to store and generate electricity. It consists of two reservoirs positioned at different elevations - an upper reservoir and a lower reservoir. When there is excess electricity in the grid or during times of low demand, the plant uses this surplus energy to pump water from the lower reservoir to the upper reservoir, effectively storing energy in the form of potential energy.

Table 1.8 Performance measure of pumped storage plant.

Performance measures	Pumped hydro storage
Efficiency (%)	70-85
Cycle Lifetime (cycles)	N/A
Expected Lifetime (years)	30-60
Specific Energy (Wh/kg)	0.5-1.5
Specific Power (W/kg)	-
Power Capacity Cost ($/kW)	600-2,000
Energy Capacity Cost ($/kWh)	0-23

Generating electricity: During periods of high electricity demand or when additional power is needed in the grid, water from the upper reservoir is released and allowed to flow back down to the lower reservoir through turbines. As the water descends, it spins the turbines, which are connected to generators, converting the kinetic energy of the moving water into electrical energy. This electricity is then fed into the grid to meet demand.

Energy storage: One of the key advantages of pumped storage plants is their ability to store large amounts of energy for later use. By pumping water uphill during off-peak hours when electricity is abundant and cheaper, these plants essentially store surplus energy that can be tapped into during peak hours or periods of high demand when electricity prices are typically higher.

Grid stabilization: Pumped storage plants also play a vital role in grid stability and reliability. They can respond quickly to fluctuations in demand or unexpected changes in renewable energy output (such as solar or wind), providing rapid injections of electricity or absorbing excess power as needed to maintain a balanced grid.

Efficiency and environmental impact: Pumped storage plants are known for their high efficiency in energy conversion, with typical efficiencies ranging from 70% to 85%.

Advantages and disadvantages of hydropower plants:

Advantages:

Clean and renewable: Hydropower is a clean and renewable energy source that doesn't pollute the air like fossil fuel power plants. It relies on the water cycle, making it a sustainable option.

Domestic energy source: Hydropower is produced domestically in countries with suitable water resources, reducing dependence on imported fuels.

On-demand power generation: Engineers can control the flow of water through turbines, allowing for electricity production as needed, making hydropower a reliable energy source.

Additional benefits: Hydropower plants create reservoirs that offer recreational opportunities like fishing, swimming, and boating. They also provide water supply and flood control benefits.

Disadvantages:

Impact on fish population: Impoundment dams can obstruct fish migration, impacting fish populations. Measures like fish ladders and screens are used to mitigate these impacts.

Water quality and flow: Hydropower plants can affect water quality and dissolved oxygen levels, harming riverbank habitats. Techniques like aeration and maintaining minimum flows help address these issues.

Vulnerability to drought: Hydropower production relies on water availability, making plants susceptible to drought conditions.

Environmental impact: New hydropower facilities can disrupt local ecosystems and compete with other land uses, potentially leading to habitat loss and cultural impacts.

1.9.2 Lithium-Ion batteries:

Lithium-ion batteries are rechargeable energy storage devices that have become abundant in various applications, from portable electronics to electric vehicles and grid-scale energy storage systems. Lithium-ion batteries consist of positive and negative electrodes (cathode and anode) separated by an electrolyte. The cathode is typically made of lithium metal oxides like lithium cobalt oxide ($LiCoO_2$), lithium iron phosphate ($LiFePO_4$), or lithium manganese oxide ($LiMn_2O_4$). The anode is usually graphite, and the electrolyte is a lithium salt dissolved in an organic solvent. These batteries are used with a proper control circuit called a Battery Management Circuit (BMS). The performance measures are given in Table 1.9. The advantages and disadvantages of Li-ion batteries are given in Table 1.10.

Table 1.9 Performance measure of Li-ion battery.

Performance measure	Range
Efficiency (%)	85-98
Cycle lifetime (cycles)	1,000-10,000
Expected lifetime (years)	5-15
Specific energy (Wh/kg)	75-200
Specific power (W/kg)	150-315
Power capacity cost ($/kW)	175-4,000
Energy capacity cost ($/kWh)	500-2,500
Balance of Plant (BOP) Cost ($/kWh)	120-600
PCS Cost ($/kW)	0
Operation & Maintenance (O&M) Cost ($/kW-yr)	12-30

Table 1.10 Advantages and disadvantages of Li-ion battery.

Aspect	Advantages	Disadvantages
High Energy Density	Provides more energy storage capacity for size and weight	Limited lifespan
Low Self-Discharge	Holds charge for longer periods when not in use	High cost of manufacturing
No Memory Effect	Does not require full discharge before recharging	Temperature sensitivity
Fast Charging	Quick recharge times	Safety concerns
Wide Range of Applications	Versatile use in various devices and systems	Complex recycling process
Lightweight	Suitable for portable devices and electric vehicles	Voltage fades over time

1.9.3 Compressed Air Energy Storage (CAES):

CAES is a technology that stores energy in the form of compressed air.

Operating Principle: CAES systems typically involve compressing air using electricity during off-peak hours when energy demand is low. The compressed air is stored in underground caverns or tanks. When electricity demand increases, the compressed air is released and heated using natural gas or other fuels. The heated air expands and drives a turbine connected to a generator, producing electricity. Table 1.11 contains the performance measure of compressed air energy storage.

Table 1.11 Performance measure of compressed air energy storage.

Performance Measures	Compressed Air Storage
Efficiency (%)	57-85
Cycle Lifetime (cycles)	N/A
Expected Lifetime (years)	20-40
Specific Energy (Wh/kg)	30-60
Power Capacity Cost ($/kW)	400-800
Energy Capacity Cost ($/kWh)	2-140
BOP ($/kWh)	270-580
PCS ($/kW)	46-190
O&M ($/kW)	1.6-29
Maturity	Commercial

1.10 Transmission of power:

The transmission of power is done using the transmission lines at high voltages. There should be minimum transmission losses and high transmission efficiency. The main objectives of the transmission of power are given below:

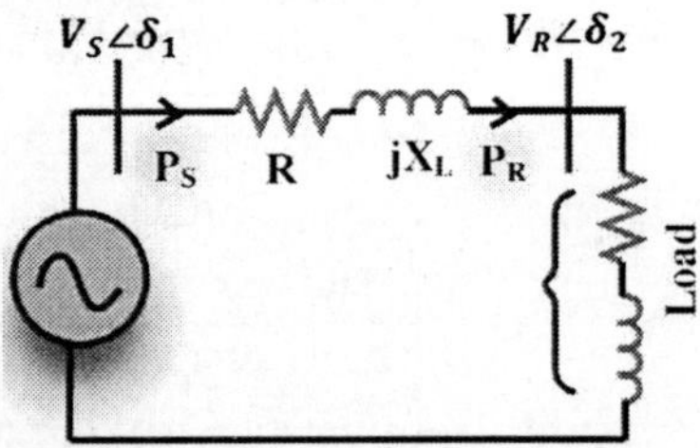

Fig. 1.18 Equivalent circuit diagram of short transmission line.

i. Transmission efficiency should be maximum: This is achieved by using high-voltage transmission lines, which reduce the amount of current flowing through the lines and consequently lower the energy lost as heat.

Assume the sending end voltage is V_S, the receiving end voltage is V_R, the line resistance is R and the line reactance is X_L as shown in Fig. 1.18.

Transmission efficiency, η we may get

$$\eta = \frac{Reciving\ end\ active\ power\ (P_R)}{Sending\ end\ active\ power\ (P_S) + \ line\ losses\ (P_{loss})} \times 100 \quad \text{... ... ... (1.3)}$$

To improve η, P_R should be maximum and the line losses should be minimum.

ii. Maximize the receiving end power: Neglecting line resistance, the receiving end power is given as equation 1.4

$$P_R = \left|\frac{V_R V_S}{X_L}\right| \sin(\delta_1 - \delta_2) \qquad \text{......... (1.4)}$$

To maintain the voltage regulation is zero, $|V_R| = |V_s|$ put in equation 1.4

$$P_R = \left|\frac{V_S{}^2}{X_L}\right| \sin(\delta_1 - \delta_2) = P_{max} \sin(\delta_1 - \delta_2) \qquad \text{......... (1.5)}$$

To maximize the P_{max}, V_S should be higher. It is directly proportional to the square of the sending end voltage and inversely proportional to the line reactance. So, this is the reason to choose the higher voltages for transmitting the power. The maximum power is given as equation 1.6.

$$P_{max} = \frac{V_S{}^2}{X_L} \qquad \text{......... (1.6)}$$

Advantages of higher V_S:

✓ The receiving end maximum active power is increased by the square of the sending voltage.

✓ Transmission line losses can be reduced.

$$P_R = V_S I_R cos\varphi \qquad \text{......... (1.7)}$$

$$I_R = \frac{P_R}{V_S \times cos\varphi} \qquad \text{......... (1.8)}$$

Ohmic loss of transmission line $P_{loss} = I_R{}^2 R$ .... (1.9)

From equation (1.8), $P_{loss} = (\frac{P_R}{V_S \times cos\varphi})^2 R$ .... (1.10)

Hence, from equation 1.10, we can conclude that P_{loss} is inversely proportional to the square of the receiving end voltage.

✓ The area of the cross-section and the volume of the conductor reduces

From equation 1.10, we can say

$$\frac{1}{R} = \frac{P_R{}^2}{P_{loss}(V_S \times cos\varphi)^2} \qquad \text{......... (1.11)}$$

Area of the conductor, $A = \frac{\rho l}{R} = \frac{\rho l\,(P_R)^2}{P_{loss}(V_R \times cos\varphi)^2}$ .... (1.12)

The area of the conductor is inversely proportional to the square of the receiving end voltage.

The volume of the conductor, $V = A \times$ length of the conductor (l)

$$V = \frac{\rho l^2\, P_R{}^2}{P_{loss}(V_R \times cos\varphi)^2} \qquad \text{......... (1.13)}$$

From equation 1.13, we can say that the volume of the conductor reduces as the receiving end voltage increases.

However, there is the limitation of higher voltages because as the voltage increases the insulation requirement of the equipment also increases.

The methods to minimize the line reactance and maximize P_R are given below:

✓ By using a parallel transmission line

In parallel combination, the reactance of the line will be $\frac{X_L}{2}$. Hence maximum power will be twice the previous case.

$$Maximum\ active\ power\ of\ parallel\ transmisssion\ line = \frac{2V_S{}^2}{X_L} = 2P_{max} \qquad \text{......... (1.14)}$$

- ✓ By using a bundle conductor reactance can be reduced

  Inductance of the conductor, $L = 2 \times 10^{-7} \ln(\frac{d}{0.7788r})$ Henry/meter

  As the radius of the conductor increases inductance reduces and reactance also reduces, thus P_{max} increases.
- ✓ By inserting the capacitor in series with line inductance, the overall reactance of the transmission line reduces.

iii. Voltage regulation of the transmission line should be minimum: It refers to the ability of the transmission line to maintain a consistent voltage level despite changes in load or other factors. Zero voltage regulation means that the voltage remains stable throughout the transmission process, ensuring a flat voltage profile.

iv. There should be less breakdown of transmission line: The transmission system should be robust and resilient, with minimal breakdowns or interruptions. This is achieved through proper maintenance, quality protective equipment, and effective monitoring and control systems.

1.10.1 Types of transmission line:

The generation of electricity is far from the load centres. The power is transported using the transmission lines. There are three types of transmission lines. These are categorized based on the length of the line.

i.	Short transmission line	: length < 80 km	: voltage < 33 kV
ii.	Medium transmission line	: 80 < length < 200 km	: 33 kV < voltage <132 kV
iii.	Long transmission line	: length > 200 km	: voltage > 132 kV

The details about the transmission line are given in the upcoming chapter.

1.11 Distribution systems:

The generation of power is at remote locations. To transport the power from generating locations to the load centres, transmission lines are used. The distribution of power is done using the feeders. There are two types of feeders, these are given below:

i. Radial feeder: Power flows in one direction, typically from a single source (such as a substation) to multiple endpoints (such as homes, businesses, or other loads). The source can be connected at the one end of the feeder or both ends of the feeder. The voltage received at the far-end consumer will be minimal due to line loss.

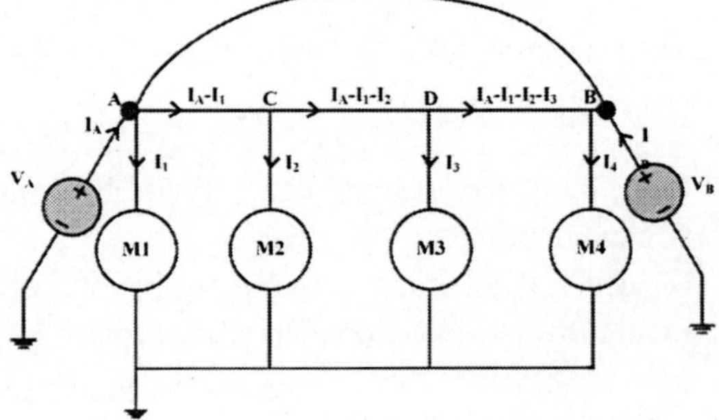

Fig. 1.19 Meshed feeder with both the ends supply fed.

ii. Meshed or ring feeder: A meshed or ring feeder is a system where power can flow in multiple paths, creating a redundant network. Unlike radial feeders, which have a single point of supply, a meshed or ring feeder provides multiple paths for power to reach its destination. Comparison between radial and meshed distribution systems are given in Table 1.12.

In the meshed feeder, both ends are connected, which means the voltage between connected points is zero as shown in Fig. 1.19.

The voltage difference between $A\ and\ B,$

$$V_A - V_B = 0$$

The procedure of finding I_A is the same as both sides of supply-fed radial feeders. The remaining steps to find the minimum potential node and potential at each node are also the same.

1.11.1 Supply fed from one end:

The voltage source is connected at point A of the feeder as shown in Fig. 1.20. There are four nodes A, B, C and D. Assume that,

r = resistance of each conductor Ω/m.

2r = resistance of both conductors Ω/m.

Resistance of section AB = $R_{AB} = 2rl_1\,\Omega$

Resistance of section BC = $R_{BC} = 2rl_2\,\Omega$

Resistance of section CD = $R_{CD} = 2rl_3\,\Omega$

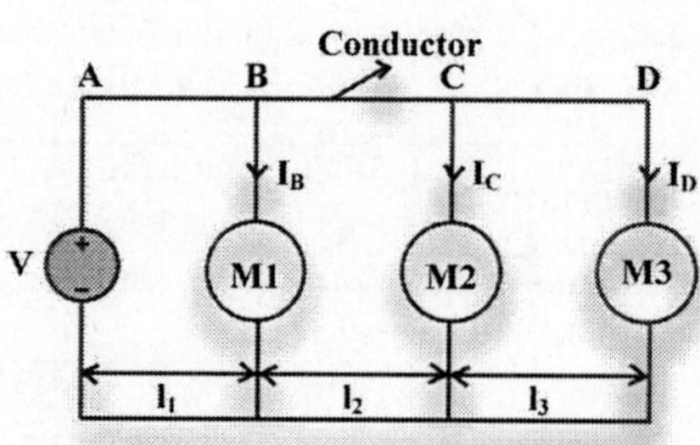

Fig. 1.20 Radial feeder having supply at one end.

The voltage drop between A and D is, applied K.V.L.

$$V_{AD} = R_{AB}\,(I_B + I_c + I_D) + R_{BC}(I_c + I_D) + R_{CD}I_D \qquad \dots\dots\dots(1.15)$$

The voltage at node D $\qquad V_D = V_A - V_{AD}$

Similarly, the voltages of other nodes can be calculated.

1.11.2 Supply fed at both ends:

The radial feeder with both ends supply fed is shown in Fig. 1.21. Let us assume that I_A current is supplied by the source connected at node A.

For both ends supply-fed radial feeder; to find out the minimum potential node follow the following steps:

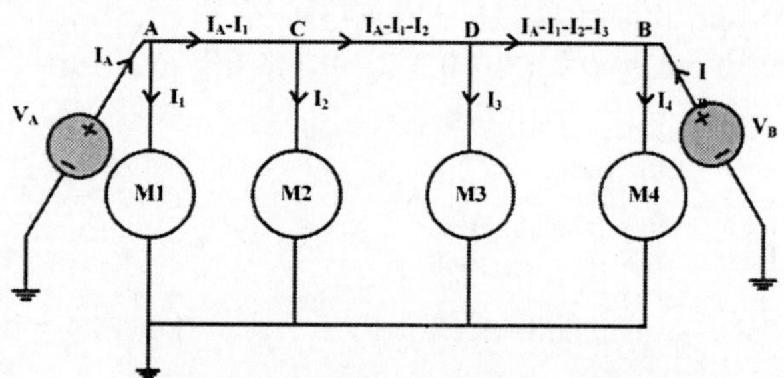

Fig. 1.21 Radial feeder having supply at both the ends.

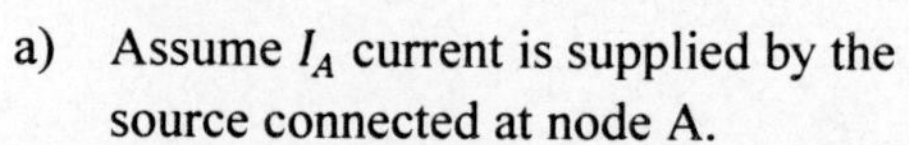

a) Assume I_A current is supplied by the source connected at node A.

b) Calculate I_A using the voltage drop between both ends. By applying KVL, we get

$$V_A - V_B = (I_A - I_1)R_{AC} + (I_A - I_1 - I_2)R_{CD} + (I_A - I_1 - I_2 - I_3)R_{DB} \qquad \dots\dots\dots(1.16)$$

c) Substitute I_A in $(I_A - I_1), (I_A - I_1 - I_2)\ and\ (I_A - I_1 - I_2 - I_3)$ and check the node of sign change.

d) Node of sign change (current) = minimum potential node.

e) Calculate minimum potential using KVL.

Table 1.12 Comparison between radial and meshed distribution systems.

Aspect	Radial distribution system	Meshed distribution system
Power flow	Unidirectional: Power flows from source to loads	Bidirectional: Power can flow in either direction along the ring
Redundancy	Limited redundancy; if a fault occurs, affected areas may experience a power outage until the fault is cleared	Enhanced redundancy; if a fault occurs, power can be rerouted, minimizing disruptions
Fault tolerance	Lower fault tolerance; a single fault can affect multiple downstream loads	Higher fault tolerance; faults are localized, and power can be restored quickly to unaffected areas
Reliability	Moderate reliability; reliability depends on fault management and restoration times	High reliability; the system can self-heal by rerouting power in case of a fault
Common Applications	Common in rural and residential areas with fewer loads	Common in urban and industrial areas with high load density
Loop Configuration	Not applicable; power flows from source to loads in a linear fashion	Closed-loop configuration; power can circulate the ring, enhancing system stability
Cost	Lower initial cost due to simpler design	Higher initial cost due to added infrastructure for ring configuration
Flexibility	Less flexible in terms of rerouting power or adding new connections	More flexible for adding new connections or modifying the network topology

Example 1.1. For the given network, find the voltage at each node. The distributed resistances are given in the network.

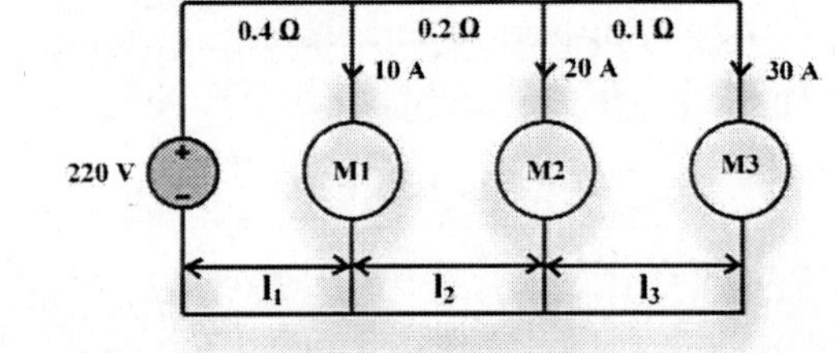

Ans. $V_{AB} = R_{AB}\,(I_B + I_c + I_D)$

$V_{AB} = 0.4 * (10 + 20 + 30) = 24$ V

$V_{BC} = R_{BC}(I_c + I_D)$

$V_{BC} = 0.2 * (20 + 30) = 10$ V

$V_{CD} = R_{CD} I_D$

$V_{CD} = 0.1 * (30) = 3$ V

The voltage at node B, $V_B = V_A - V_{AB} = 220 - 24 = 196\ V$

The voltage at node C, $V_C = V_B - V_{BC} = 196 - 10 = 186$ V

The voltage at node D, $V_D = V_C - V_{CD} = 186 - 3 = 183$ V

Example 1.2. A 2-wire distribution network is given. If the permissible voltage drop in the given feeder is 20 volts, with resistivity $1.55 \times 10^{-8}\ \Omega$ m. Find the cross-sectional area of the distributor.

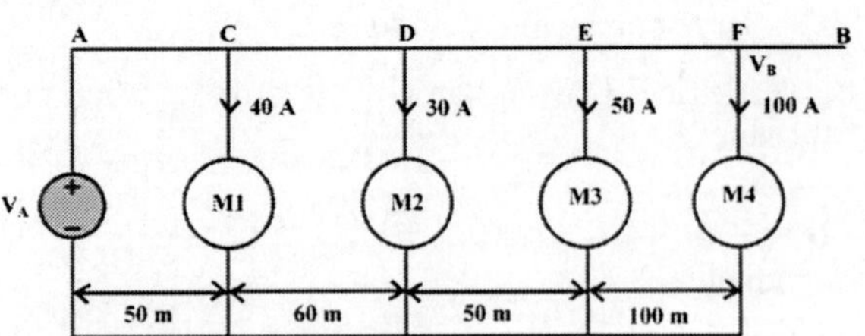

Ans. Let us assume the total resistance is $r/100$ Ω/ m.

$Resistance\ for\ AC,\ R_{AC} = r\frac{50}{100}\Omega = 0.5\, r\, \Omega$

$Resistance\ for\ CD, R_{CD} = r\frac{60}{100}\Omega = 0.6\, r\, \Omega$

$Resistance\ for\ DE, R_{DE} = r\frac{50}{100}\Omega = 0.5\, r\, \Omega$

$Resistance\ for\ EF,\ R_{EF} = r\frac{100}{100} = r\, \Omega$

Voltage drop between AB

$$V_{AB} = (I_C + I_D + I_E + I_F)R_{AC} + (I_D + I_E + I_F)R_{CD} + (I_E + I_F)R_{DE} + I_F R_{EF}$$

$$V_{AB} = (40 + 30 + 50 + 100) \times 0.5\,r + (30 + 50 + 100) \times 0.6\,r + (50 + 100) \times 0.5\,r + 100r$$

$$V_{AB} = 110r + 108r + 75r + 100r = 393r$$

The voltage drop between AB is given 20 volts

$20 = 393$ r

r = 0.050 Ω per *100* meter

The resistance of single wire $\left(\frac{r}{2}\right) = 0.0254\ \Omega$ per 100 m

The resistance of the wire = $\frac{\rho l}{a} \Rightarrow \frac{r}{2} = 0.0254 = \frac{1.55 \times 10^{-8} * 260}{a}$

The cross – sectional area of the conductor $= a = 1.58\ cm^2$ per conductor

Example 1.3. Find the minimum potential node for the given network. If the potential of node A is 220 volts and node B is 215 volts.

Ans. The voltage drop between A and B

$$V_A - V_B = (I_A - I_1)R_{AC} + (I_A - I_1 - I_2)R_{CD} + (I_A - I_1 - I_2 - I_3)R_{DB}$$

$$220 - 215 = (I_A - 10)0.1 + ((I_A - 30)0.2 + (I_A - 60)0.3$$

$$5 = 0.6\,I_A - 25$$

$$I_A = 50\ A$$

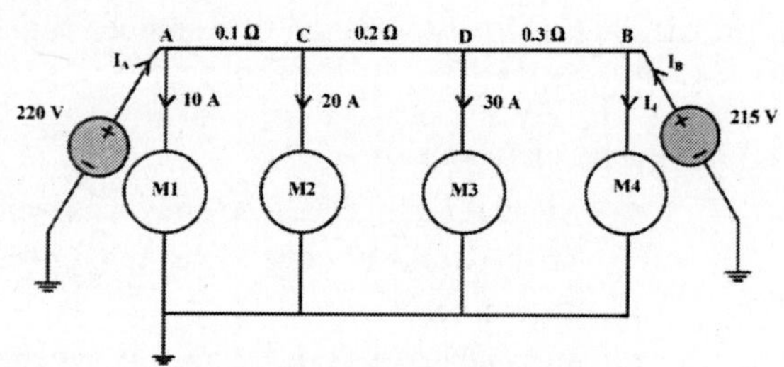

$(I_A - 10) = 40\ A$ at node A

$(I_A - 30) = 20\ A$ at node C

$(I_A - 60) = -10\ A$

Sign changes of current at node D

So, D is the minimum potential node.

Example 1.4. The resistance of the go and return conductor is *(0.3+0.2j)* Ω/km. Calculate the voltage drop between points A and C of the given network.

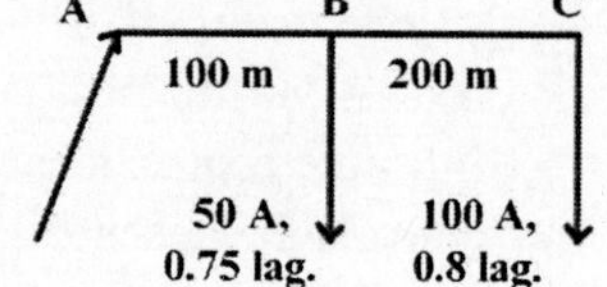

Ans. Resistance of AB = $R_{AB} = \frac{100(0.3+0.2j)}{1000} = 0.03 + 0.02j\Omega$

Resistance of BC = $R_{BC} = \frac{200(0.3+0.2j)}{1000} = 0.06 + 0.04j\ \Omega$

Current in the section AB = $50\angle - \cos^{-1}(0.75) + 100\angle - \cos^{-1}(0.8)$

$= 149.89\ 4\angle{-38.37^\circ}$ A

The voltage drop across points A and C

$$V_{AC} = R_{AB} \times (I_B + I_C) + R_{BC} \times I_C$$

$$V_{AC} = (0.03 + 0.02j) \times (149.89\ 4\angle{-38.37^\circ}) + (0.06 + 0.04j) \times (100\angle{-36.86^\circ})$$

$$V_{AC} = 12.614\angle{-3.81^\circ}\ V$$

1.12 Synchronous Grids and Asynchronous (DC) interconnections:

The electricity grid in India is a complex network of power generation, transmission, and distribution infrastructure that supplies electricity to various regions across the country. India has five regional grids and one national grid. The nation is operating at a single frequency. For the 50 Hz system, the frequency must lie between 48.5 Hz to 51.5 Hz to maintain the stability of the grid.

i. Regional grid: A regional grid refers to an interconnected network of power generation, transmission, and distribution systems that serves a specific geographic region. India has five regional grids. Each regional grid is responsible for managing the generation, transmission, and distribution of electricity within its respective region. These are given below:
   - Northern grid (NR)
   - Eastern grid (ER)
   - Western grid (WR)
   - Southern grid (SR)
   - North-eastern grid (NER)

ii. National grid: It is formed by interconnecting the different regional grids. These regional grids, also known as the Northern, Western, Southern, Eastern, and North-Eastern grids, cover specific geographical areas within the country. By interconnecting these regional grids through high-voltage transmission lines, the national grid enables the exchange of electricity between regions, thereby ensuring optimal utilization of resources and enhancing grid reliability and stability on a national scale.

1.12.1 Synchronous Grids:

- Synchronous grids, also known as synchronous interconnections or synchronous AC grids, are networks of power lines that operate at the same frequency and are synchronized with each other.
- In synchronous grids, power is generated, transmitted, and distributed in the form of AC.
- These grids are interconnected through high-voltage transmission lines and substations, allowing for the transfer of electricity over long distances.

1.12.2 Asynchronous (DC) Interconnections:

- Asynchronous interconnections, also known as high-voltage direct current (HVDC) interconnections, involve the transmission of electricity in the form of DC rather than AC.
- HVDC technology is used to connect grids that operate at different frequencies or have incompatible AC systems.
- In HVDC systems, power is converted from AC to DC at the sending end, transmitted over long distances with minimal losses, and then converted back to AC at the receiving end.
- HVDC interconnections are particularly useful for transmitting power across large distances, underwater, or through regions with different AC standards.

Table 1.13 has the details of HVDC link projects in India.

Table 1.13 Detail of HVDC link projects in India.

HVDC Link (Operator company)	From	To	Voltage rating	Capacity (MW)	Commissioning Year	Line distance (km)
Chandrapur-Padghe (ABB and BHEL)	Chandrapur	Padghe (near Mumbai)	500 kV	1500	1999	752
Talcher-Kolar (Siemens)	Talcher (Odisha)	Kolar (Karnatak a)	500 kV	2000	2003	1450
Bhadrawati-Chandrapur (GEC-Alstom)	Bhadrawati	Chandrap ur	205 kV	1000	1997	20
Raigarh-Pugalur (Hitachi and BHEL)	Raigarh (Chhattisgarh)	Pugalur (Tamil Nadu)	800 kV	6000	2014	1830
Champa-Kurukshetra (GE)	Champa (Chhattisgarh)	Kurukshet ra (Haryana)	800 kV	3000	2017	1365
Pugalur-Thrissur (ABB and BHEL)	Pugalur (Tamil Nadu)	Thrissur (Kerala)	320 kV	2000	2019	250
Raipur-Pugalur (ABB and BHEL)	Raipur (Chhattisgarh)	Pugalur (Tamil Nadu)	800 kV	6000	2020	1760

1.12.3 Back-to-back HVDC link in India:

Back-to-back HVDC links are a type of HVDC transmission system that connects two AC grids without any intermediate DC transmission line. In India, several back-to-back HVDC links play a crucial role in enhancing grid stability and power transfer capabilities. Some key back-to-back HVDC links in India are given in Table 1.14.

Table 1.14 Detail of major HVDC back-to-back links in India.

HVDC Link	Location	Purpose	Capacity (MW)	Operator	Operational Year
Vizag-Chennai	Vizag (Andhra Pradesh) - Chennai (Tamil Nadu)	Enhances power transfer capability between Southern and Eastern grids for stability.	1000	Power Grid Corporation of India Ltd (PGCIL)	2013
Tiruvalam-Pugalur	Tiruvalam (Tamil Nadu) - Pugalur (Kerala)	Connects the southern grid with the western grid, aiding power transfer and balance.	2,000	PGCIL	2019

Chandrapur-Padghe	Chandrapur (Maharashtra) - Padghe (Maharashtra)	Improves power flow control and stability in the Maharashtra region.	1,500	Maharashtra State Electricity Transmission Co. Ltd (MSETCL)	2016
Bhiwadi-Ajmer	Bhiwadi (Rajasthan) - Ajmer (Rajasthan)	Supports power transfer and stability in Northern India, linking with N & W grids.	500	PGCIL	2014
Vindhyachal	Vindhyachal (Madhya Pradesh) - Vindhyachal (Madhya Pradesh)	Interconnects Northern and Western grids, enhancing power transfer capabilities.	500	Power Grid Corporation of India Ltd (PGCIL)	1989

1.13 Review of three-phase systems:

A three-phase system is a fundamental aspect of electrical engineering, playing a crucial role in various applications across industries. It comprises three ACs with equal magnitudes but 120 degrees out of phase with each other. This arrangement results in a balanced system that offers several advantages over single-phase systems. One of the key benefits is smoother power delivery, as the phases work together to provide continuous and efficient electrical power. Three-phase systems are extensively used in power generation, especially in large-scale power plants and industrial generators, due to their ability to handle high power loads effectively. It produces the rotating magnetic field in the air gap of the three-phase induction motor. There is no frequency doubling effect (instantaneous power, single-phase) in a three-phase system.

Some key points of the three-phase system are given below.

a) Line current = phase current : For star connection
b) Line voltage = $\sqrt{3}$ phase voltage : For star connection
c) Line current = $\sqrt{3}$ phase current : For delta connection
d) Line voltage = phase voltage : For delta connection

Example 1.5. A three-phase 400 V, 50 Hz balanced power supply is connected to a load of 5 kW, 10 kW, and 15 kW, distributed across phases R, Y and B respectively. Find the neutral current and draw the phasors of voltages and currents.

Ans. phase voltages $V_{RN} = \frac{400}{\sqrt{3}} = 231\angle 0^{\circ}\ V$, $V_{YN} = \frac{400}{\sqrt{3}} = 231\angle{-120^{\circ}}\ V, V_{BN} = \frac{400}{\sqrt{3}} = 231\angle{-240^{\circ}}\ V$

$$P_R = 5000\ \text{W},\ P_y = 10000\ \text{W, and}\ P_B = 15000\ \text{W},$$

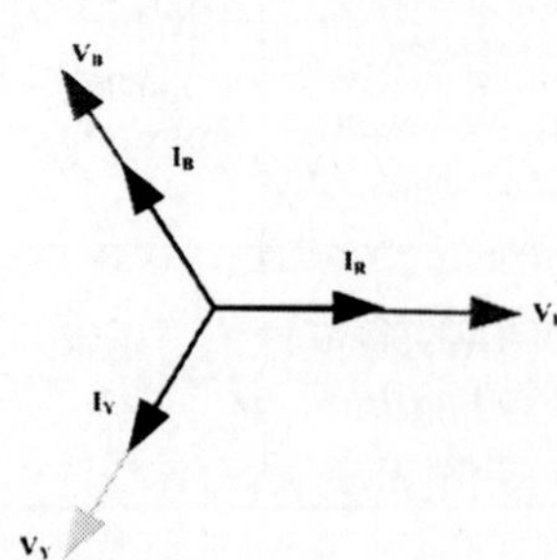

p. f. of R, Y, B phases is $\cos\varphi = 1 \Rightarrow \varphi = \cos^{-1} 1 = 0^{\circ}$

Active power $|P_R| = |V_{RN}\, I_R \cos\varphi| = 231 \times I_R \times 1$

Phase current $I_R = \frac{|P_R|}{|V_{RN}\ \cos\varphi|} = \frac{5000}{|231*1|} = 21.645\angle 0^{\circ}$ A

Similarly, $|P_Y| = |V_{YN}\, I_Y \cos\varphi| = 231 \times I_Y \times 1$

$$I_Y = \frac{|P_Y|}{|V_{YN}\ \cos\varphi|} = \frac{10000}{|231*1|} = 43.290\angle{-120^{\circ}}\ \text{A}$$

$$|P_B| = |V_{BN}\, I_B \cos\varphi| = 231\times I_B \times 1$$

$$I_B = \frac{|P_B|}{|V_{BN}\cos\varphi|} = \frac{15000}{|231*1|} = 64.93\angle{-240^\circ}\text{ A}$$

Neutral current $I_N = I_R + I_Y + I_B = 21.645\angle 0^\circ + 43.290\angle{-120^\circ} + 64.93\angle{-240^\circ}$

$$I_N = 37.485\angle 150^\circ\text{ A}$$

Example 1.6. A three-phase 400 V, 50 Hz balanced power supply is connected to a load of 5 kW at unity p.f., 10 kW at 0.86 p.f. lagging, and 15 kW at 0.86 p.f. leading, distributed across phases R, Y and B respectively. Find the neutral current.

Ans. Phase voltages $V_{RN} = \frac{400}{\sqrt{3}} = 231\angle 0^\circ\ V$,

$$V_{YN} = \frac{400}{\sqrt{3}} = 231\angle{-120^\circ}\ V \text{ and}$$

$$V_{BN} = \frac{400}{\sqrt{3}} = 231\angle{-240^\circ}\ V$$

p. f. of R phase is $\cos\varphi = 1 \Rightarrow \varphi = \cos^{-1}1 = 0^\circ$

p. f. of Y phase is $= \cos\varphi = 0.86\ lagging \Rightarrow \varphi = \cos^{-1}0.86 = -30.68^\circ$

p. f. of B phase is $= \cos\varphi = 0.86\ leading \Rightarrow \varphi = \cos^{-1}0.86 = 30.68^\circ$

Phase current $I_R = \frac{|P_R|}{|V_{RN}\cos\varphi|} = \frac{5000}{|231*1|} = 21.645\angle 0^\circ\text{ A}$

Phase current $I_Y = \frac{|P_Y|}{|V_{YN}\cos\varphi|} = \frac{10000}{|231*0.86|} = 50.33\angle(-30.68 - 120)^\circ = 50.33\angle{-150.68^\circ}\text{ A}$

Phase current $I_B == \frac{|P_B|}{|V_{BN}\cos\varphi|} = \frac{15000}{|231*0.86|} = 75.5\angle(30.68 - 240)^\circ = 75.5\angle{-209.32^\circ}\ A$

Note: Since I_Y and I_B are calculated with respect to V_Y and V_B.

Neutral current $I_N = I_R + I_Y + I_B = 88.92\angle 172.03^\circ\ A$

1.13.1 Importance of reactive power:

Reactive power helps in the energy conversion in AC machines like induction motors, generators, and transformers as shown in Fig. 1.22. The energy conversion occurs in the presence of a magnetic domain provided by the inductor coil. It requires reactive power to produce the magnetic field. Inductors and capacitors consume the reactive power. The importance of reactive power is given below:

Voltage regulation: Reactive power plays a pivotal role in voltage regulation within AC circuits. Particularly in systems featuring inductive loads like motors and transformers, reactive power serves to offset the voltage drop induced by inductive reactance. This compensation mechanism helps maintain voltage levels within permissible boundaries across the entire network, ensuring stable and reliable operation.

Power factor correction: Reactive power plays a crucial role in enhancing the power factor of AC circuits. The power factor, which signifies the ratio of real power to apparent power, holds significance in efficient power transmission. Low power factors can result in inefficient energy utilization and heightened losses within the system. Power factor correction mechanisms, employing devices like capacitors or synchronous condensers, work by injecting or absorbing reactive power. This process aids in ameliorating the overall power factor, thereby fostering improved energy efficiency within the system.

Optimal utilization of transmission lines: Reactive power management is crucial for the efficient operation of transmission lines. By carefully controlling the flow of reactive power, voltage profiles along the transmission lines can be maintained at optimal levels, minimizing losses and ensuring reliable power transfer over long distances.

Prevention of voltage instabilities: Reactive power control helps in preventing voltage instability, voltage sag and voltage flicker. Voltage instabilities can lead to equipment malfunction, disruptions in power supply, and damage to sensitive electronic devices. By maintaining voltage stability through proper reactive power management, the reliability of the power system is enhanced.

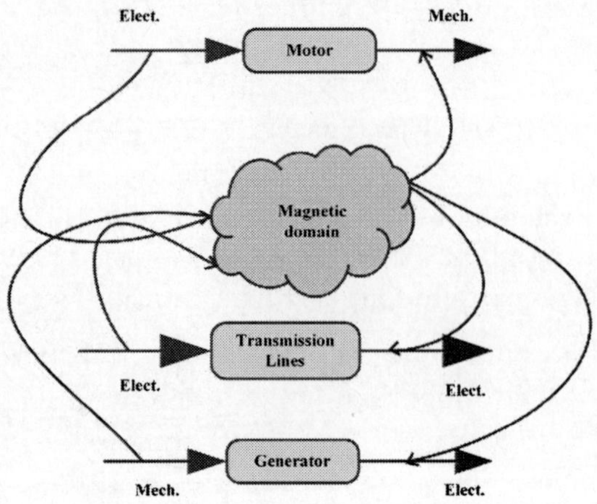

Fig. 1.22 Role of reactive power in power system.

1.13.2 Determining the Shunt Capacitor for Power Factor Improvement:

In Fig. 1.23(a) the lagging load is connected to the receiving end of the feeder. The reactive power requirement of the load is Q_1 and the supply of reactive power is Q_2 as shown in Fig 1.23 (a). To improve the power factor of the load a shunt capacitor is required to feed the reactive power to the load and the overall power factor is improved as shown in Fig. 1.23 (b).

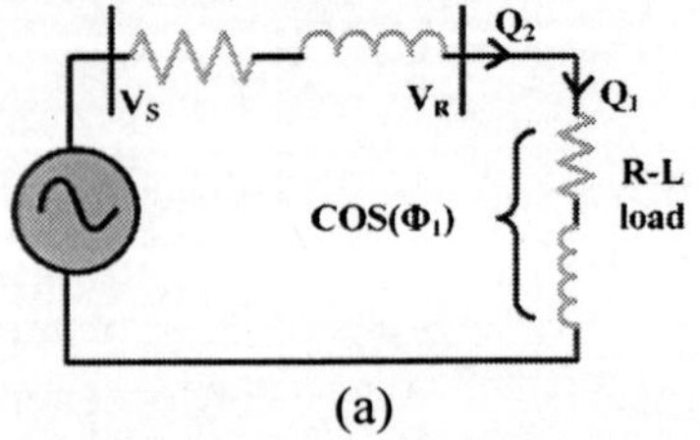

(a)

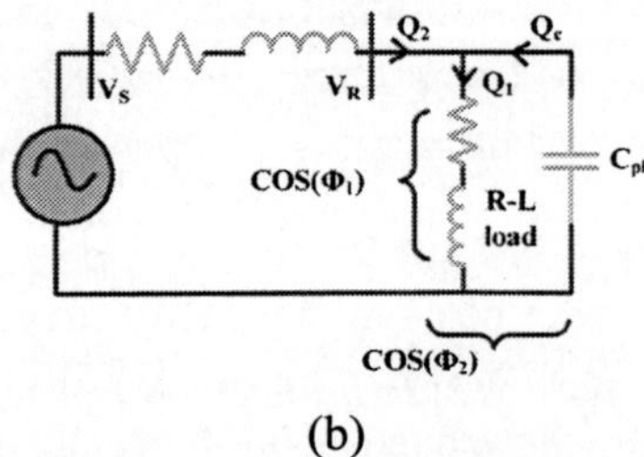

(b)

Fig. 1.23 (a) Circuit diagram without reactive power compensation (b) Circuit diagram with shunt capacitor compensation

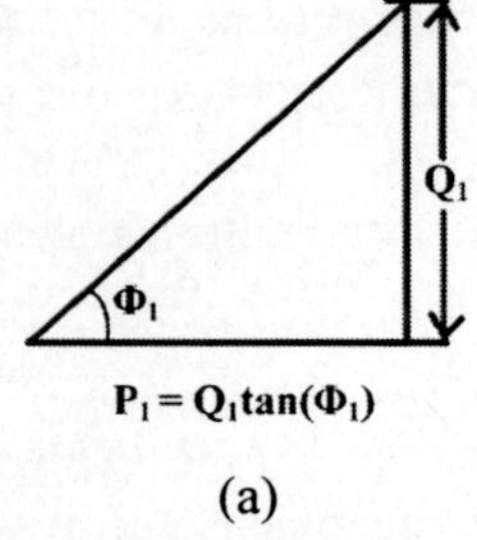

(a)

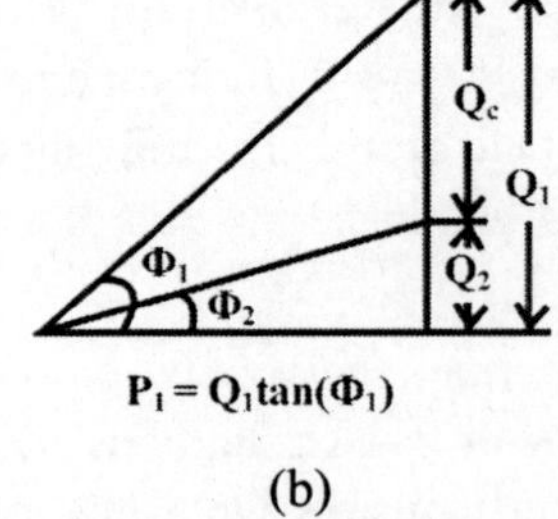

(b)

Fig 1.24 (a) Power triangle of the load without shunt capacitor (b) Power triangle of the combination of shunt capacitor and load.

Assume if $Q_1 > Q_2$ that means $|V_S| > |V_R|$, $|V_R|$ is the receiving end voltage and $|V_S|$ is the sending end voltage.

Without C_{ph}, the power factor angle of the load is φ_1 and with C_{ph} *power factor* angle is φ_2. P_1 is the active power consumed by the load. The load power tringle is shown in Fig 1.24 (a) and 1.24 (b). Reactive power balance equation at receiving end from Fig. 1.23 (b).

$$Q_1 = Q_2 + Q_c \qquad \text{......... (1.17)}$$

Where Q_c is the reactive power supplied by the capacitor

From the power triangle of the load

$$Q_c = Q_1 - Q_2 \qquad \text{......... (1.18)}$$

$$Q_c = P_1[\tan\varphi_1 - \tan\varphi_2] \qquad \text{......... (1.19)}$$

Example 1.7. The facility currently consumes a total apparent power of 500 kVA with a power factor of 0.7 lagging. To avoid penalties from the utility company, the facility manager aims to improve the power factor to 0.95 lagging. Find the reactive power supplied by the shunt capacitor.

Ans. The apparent power of the load = 500 kVA

Power factor of the load without shunt capacitor = 0.7 lagging

Reactor power supplied by the shunt capacitor

$$Q_c = \text{Load active power} \times [\tan\varphi_1 - \tan\varphi_2]$$

$$Q_c = 500 \times 0.7\,[\tan(\cos^{-1}0.7) - \tan(\cos^{-1}0.95)]$$

$$Q_c = 242.67\ kVAr$$

Example 1.8. A single-phase load with an apparent power of 5 kVA and a power factor of 0.6 lagging needs to be corrected to a power factor of 0.9 lagging. We need to calculate the capacitance of the shunt capacitor required for this correction. The system voltage is 230 volts.

Ans. Load voltage = 230 volts

Reactive power supplied by the shunt capacitor

$$Q_c = \text{Load active power} \times [\tan\varphi_1 - \tan\varphi_2]$$

$$Q_c = 5 \times 0.6\,[\tan(\cos^{-1}0.6) - \tan(\cos^{-1}0.9)]$$

$$Q_c = 2.54\ kVAr$$

$$Q_c = V^2 * 2\pi * f * C$$

$$\text{Capacitance } (C) = \frac{Q_c}{V^2 * 2 * \pi * f} = \frac{2.54 * 10^3}{230^2 * 2 * 3.14 * 50} = 152.84\ \mu\text{F}$$

Example 1.9. A load of $(3 + 4j)\ \Omega$ per phase is connected in a star across a line voltage of 415 volts shown in the network. Find the following:

a. Phase current and line current in each phase.

b. Three-phase power consumed by the load.

c. How much of the capacitance should be connected in parallel with load per phase to make the unity power factor?

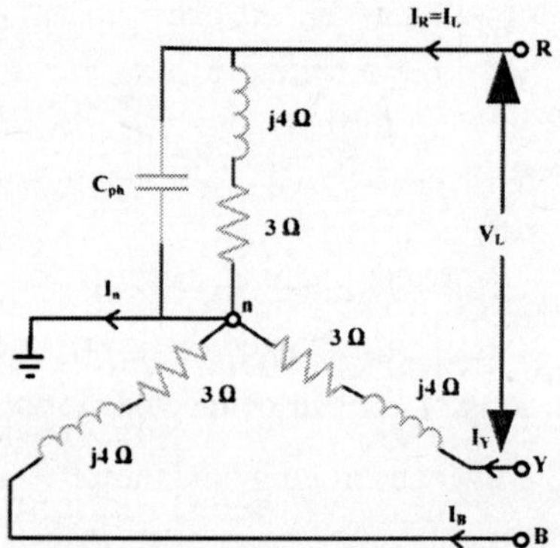

Ans. Phase current $= I_{ph} = \frac{V_{ph}}{Z}$

$$I_{ph} = \frac{V_{ph}}{Z} = \frac{\frac{415}{\sqrt{3}}}{3+4j} = \frac{239.60}{3+4j} = 47.92\ \angle{-53.13}^{\circ}\ \text{amps}$$

For the star connection phase current is equal to the line current $I_{ph} = I_L$

$$I_R = I_{ph} = I_L = 47.92\ \angle{-53.13}^{\circ}\ \text{A}$$

$$I_Y = I_{ph} = I_L = 47.92\ \angle(-53.13 - 120)^{\circ} = 47.92\ \angle{-173.13}^{\circ}\ \text{A}$$

$$I_B = I_{ph} = I_L = 47.92\ \angle(-53.13 + 120)^{\circ} = 47.92\ \angle 66.87^{\circ}\ \text{A}$$

b. The active power consumed by the load per phase

$$P_{ph} = |V_{ph}||I_{ph}| \cos\varphi$$

$$P_{ph} = \frac{415}{\sqrt{3}} * 47.92 \cos(53.13)$$

$$P_{ph} = 6.889\ \text{kW}$$

3- Phase power consumed by the load = $3P_{ph}$= 20.66 kW

Reactive power consumed by the load = Three-phase active power × tan(53.13) = 27.54 kVAr

c. $\vec{I_L} = jI_c + \vec{I_{ph}}$

$$\vec{I_L} = I_L \angle\varphi = I_L \cos\varphi + j\, I_L \sin\varphi$$

For the unity factor, $\cos\varphi = 1,\ \varphi = 0^{\circ}$ and the imaginary part of I_L is zero.

$$I_L = jI_c + 47.92\ \angle{-53.13}^{\circ}$$

$$= jI_c + 28.75 - j38.33$$

Since, the imaginary part of $I_L = 0$

$$I_c = 38.33\ A$$

$$I_c = \frac{V_{ph}}{X_c} = V_{ph}(2\pi * f * C_{ph})$$

$$38.33 = 239.6 * 2 * 3.14 * 50 * C_{ph}$$

$$C_{ph} = \frac{38.33}{239.6 * 2 * 3.14 * 50} = \frac{38.33}{75234.4} = 509.47\ \mu F$$

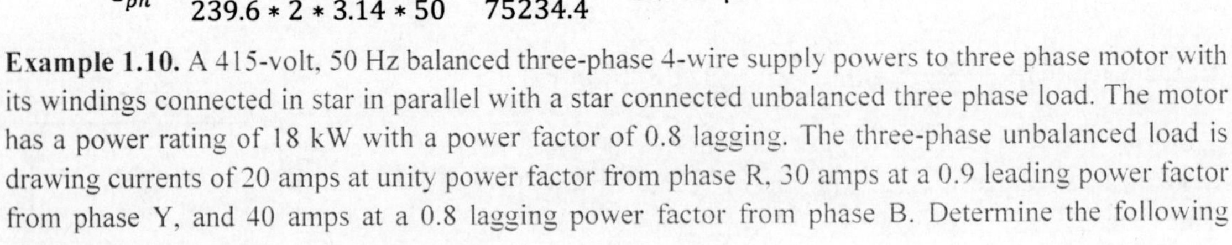

Example 1.10. A 415-volt, 50 Hz balanced three-phase 4-wire supply powers to three phase motor with its windings connected in star in parallel with a star connected unbalanced three phase load. The motor has a power rating of 18 kW with a power factor of 0.8 lagging. The three-phase unbalanced load is drawing currents of 20 amps at unity power factor from phase R, 30 amps at a 0.9 leading power factor from phase Y, and 40 amps at a 0.8 lagging power factor from phase B. Determine the following parameters:

i. Find the line currents and neutral current.
ii. The required capacitance per phase to obtain the unity power factor.

Ans. i. for star connected system line current is equal to phase current

Power absorbed by the motor = $\frac{18000}{3} = 6000\ W$

$$6000 = \frac{415}{\sqrt{3}} * I_{motor} * 0.8$$

$$6000 = 239.6 * I_{motor} * 0.8$$

$$I_{R\,motor} = \frac{6000}{239.6*0.8} = 31.30\angle(0-36.86)^{\circ} = 31.30\angle-36.86^{\circ}A$$

$$I_{Y\,motor} = 31.30\angle(-120-36.86)^{\circ} = 31.30\angle-156.86^{\circ}A$$

$$I_{B\,motor} = 31.30\angle(120-36.86)^{\circ} = 31.30\angle83.14^{\circ}A$$

Currents flowing to the unbalanced load

$$I_{R\,Load} = 20\angle(0-0)^{\circ} = 20\angle0^{\circ}\text{ A}$$

$$I_{Y\,Load} = 30\angle(-120+25.84)^{\circ} = 30\angle-94.16^{\circ}A$$

$$I_{B\,Load} = 40\angle(120-36.86)^{\circ} = 40\angle83.14^{\circ}A$$

Current flowing through the load connected to the R phase

$$I_{R\,Load} = 20\angle0^{\circ}\text{ A},\quad I_{R\,motor} = 31.30\angle-36.86^{\circ}A$$

Phase current in the R phase I_R= Motor current ($I_{R\,motor}$) + unbalanced load R phase current ($I_{R\,Load}$)

$$I_R = 31.30\angle-36.86^{\circ} + 20\angle0^{\circ} = 45.045 - 18.76j = 48.799\angle-22.62^{\circ}A$$

Phase current in the Y phase

$$I_{Y\,motor} = 31.30\angle-156.86^{\circ}A$$

$$I_{Y\,Load} = 30\angle-94.16^{\circ}A$$

I_Y= Motor current ($I_{Y\,motor}$) + unbalanced load Y phase current ($I_{Y\,Load}$)

$$I_Y = 31.30\angle-156.86^{\circ} + 30\angle-94.16^{\circ} = -30.95 - 42.22j = 52.35\angle-126.25^{\circ}A$$

Phase current in the B phase

$$I_{B\,motor} = 31.30\angle83.14^{\circ}A$$

$$I_{B\,Load} = 40\angle83.14^{\circ}A$$

I_B= Motor current ($I_{B\,motor}$) + unbalanced load B phase current ($I_{B\,Load}$)

$$I_B = 31.30\angle83.14^{\circ} + 40\angle83.14^{\circ} = 8.51 + 70.78j = 71.3\angle83.14^{\circ}A$$

Neutral current $I_N = I_R + I_Y + I_B$

$$I_N = 48.799\angle-22.62^{\circ} + 52.35\angle-126.25^{\circ} + 71.3\angle83.14^{\circ}$$

$$I_N = 24.64\angle23.44^{\circ}A$$

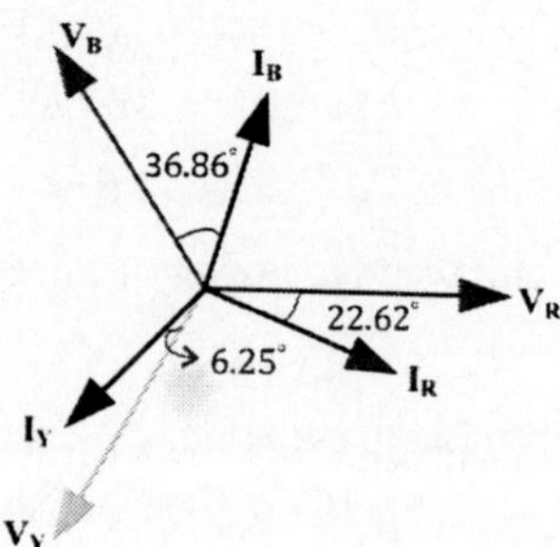

ii. Required capacitance in the R phase to make power factor unity

Line current in R phase $= jI_{CR} + I_R = jI_{CR} + (45.045 - 18.76j)$

By equating the imaginary part to zero $\quad 0 = j(I_{CR} - 18.76)$

$$I_{CR} = 18.76\text{ A}$$

$$I_{CR} = \frac{V_{ph}}{X_{CR}} = V_{ph}(2\pi * f * C_R)$$

$$C_R = \frac{18.76}{239.6 * 2 * 3.14 * 50} = 249.22\,\mu F$$

Required capacitance in the Y phase to make power factor unity

$$I_Y = -30.95 - 42.22j = 52.35\angle{-126.25^\circ}\,A$$

However, I_Y corresponds to V_Y, which is 120° lagging, 120° is added in the phase angle of the current to make it at zero degree reference because the capacitor current (jI_{CY}) is taken with respect to zero degree reference.

$$I_Y = 52.35\angle(-126.25 + 120)^\circ = 52.35\angle{-6.25^\circ} = 52.03 - 5.72j\ A$$

$$\text{Line current in Y phase} = jI_{CY} + I_Y = jI_{CY} + (52.03 - 5.72j)$$

By equating the imaginary part to zero $0 = j(I_{CY} - 5.72)$

$$I_{CY} = 5.72\text{ A}$$

$$I_{CY} = \frac{V_{ph}}{X_{CY}} = V_{ph}(2\pi * f * C_Y)$$

$$C_Y = \frac{5.72}{239.6 * 2 * 3.14 * 50} = 76.02\ \mu F$$

Similarly for the B phase

$$I_B = 71.3\angle83.14^\circ = 71.3\angle(83.14 - 120)^\circ = 71.3\angle{-36.86^\circ}$$

$$\text{Line current in B phase} = jI_{CB} + I_B = jI_{CB} + (57.04 - 42.77j)$$

By equating the imaginary part to zero $0 = j(I_{CB} - 42.77)$

$$I_{CB} = 42.77\text{ A}$$

$$I_{CB} = \frac{V_{ph}}{X_{CB}} = V_{ph}(2\pi * f * C_B)$$

$$C_B = \frac{42.77}{239.6 * 2 * 3.14 * 50} = 568\ \mu F$$

Alternate Method to find the Capacitance required in each phase to achieve unity power factor:

Total reactive power requirement of the load connected to the R-phase

$$Q_R = V_{RN} I_R \sin\varphi_R = 239.6 \times 48.79 \times \sin(22.62) = 4496.21\ VAr$$

Formula to calculate the capacitance

$$Q_R = V_{RN}^{\ 2}\omega_{CR} = V_{RN}^{\ 2} \times 2 \times \pi \times f \times C_R$$

$$4496.21 = 239.6^2 \times 2 \times 3.14 \times 50 \times C_R = 18035305.37 \times C_R$$

$$C_R = 249.30\ \mu F$$

Total reactive power requirement of the load connected to the Y-phase

$$Q_Y = V_{YN} I_Y \sin\varphi_Y = 239.6 \times 52.35 \times \sin(6.25) = 1365.52\ VAr$$

Formula to calculate the capacitance

$$Q_Y = V_{YN}^{\ 2}\omega_{CY} = V_{YN}^{\ 2} \times 2 \times \pi \times f \times C_Y$$

$$1365.52 = 239.6^2 \times 2 \times 3.14 \times 50 \times C_Y = 18035305.37 \times C_Y$$

$$C_Y = 75.71\ \mu F$$

Total reactive power requirement of the load connected to the B-phase

$$Q_B = V_{BN} I_Y \sin\varphi_B = 239.6 \times 71.3 \times \sin(36.86) = 10247.72\ VAr$$

Formula to calculate the capacitance

$$Q_B = {V_{BN}}^2 \omega_{CB} = {V_{YN}}^2 \times 2 \times \pi \times f \times C_B$$

$$10247.72 = 239.6^2 \times 2 \times 3.14 \times 50 \times C_B = 18035305.37 \times C_B$$

$$C_B = 568.20\ \mu F$$

Example 1.11. A load of (3+4j) Ω/phase is connected in star to the bus bar of 33 kV. Find the following:

i. Reactive power required to improve the p.f. to 0.9 lagging.
ii. Capacitance/phase required for star-connected capacitor bank.
iii. Capacitance of each cell if the bank contains 10 cells/phase.

Ans. i. Total reactive power required to improve the p.f.

$$Q_c = P_1(\tan\varphi_1 - \tan\varphi_2)$$

For star-connected load, phase current $= \frac{V_{ph}}{Z_{ph}} = \frac{\frac{33000}{\sqrt{3}}}{3+4j} = \frac{19052.55}{5\angle 53.13} = 3810.51\angle{-53.13}^\circ\ A$

$$P_1 = \sqrt{3}\, V_L I_L \cos\varphi_1$$

$$P_1 = \sqrt{3} \times 33000 \times 3810.51 \times \cos\left(tan^{-1}\left(\frac{4}{3}\right)\right) = 130.68\ MW$$

$$Q_c = 130.68[\tan(53.13) - \tan(\cos^{-1} 0.9)] = 110.953\ MVAr$$

Per phase, Reactive power required $Q_{cph} = Q_c/3 = 36.984\ MVAr$

ii. Capacitance per phase $C_{ph} = \frac{Q_{cph}}{2\times\pi\times f\times {V_{ph}}^2} = \frac{36.984\times10^6}{2\times3.14\times50\times(\frac{33000}{\sqrt{3}})^2} = 324.3\ \mu F$

iii. Capacitance of each cell = number of cells per phase × C_{ph}

$$= 10 \times 324.3 = 3243\ \mu F$$

Example 1.12. The shunt reactor absorbs reactive power of 25 MVAr at the rated voltage. If the voltage rating is raised by 10% and the frequency of the power supply is decreased by 5%, what will be the updated magnitude of the reactive power consumed by the shunt reactor?

Ans. New reactive power absorbed by the reactor $= Q_{L1}[\left(\frac{V_2}{V_1}\right)^2 \left(\frac{f_1}{f_2}\right)]$

$$= 25\left(\frac{1.1V_1}{V_1}\right)^2\left(\frac{f_1}{0.95 f_1}\right) = 31.842\ MVAr$$

1.14 Unit Summary:

- Power system started in the 1870s, and initially provided power to street arc lamps.
- First power system completed in 1882, supplied electricity to 59 customers at 110 V DC in Pearl Street Station in New York City.
- In 1886 it had been started AC distribution because of several advantages of AC.

- AC vs DC debate started in the 1890s, AC won in this debate because of voltage variation using a transformer and less distribution loss.
- The generation voltage is 11 kV to 25 kV.
- Transmission voltages are 110 kV, 132 kV, 220 kV, 400 kV, 765 kV.
- HVDC transmission is used for bulk power transfer for long distances. HVDC started in 1954.
- As on 31-01-2025, the installed capacity of fossil fuel is 245.9 GW and it shares 52.74% of the total installed capacity in India.
- As on 31-01-2025, the installed capacity of non-fossil fuel is 220.35 GW and it shares 47.2% of the total installed capacity in India.
- In past scenarios power systems had only conventional generation with thermal and hydro power plants, with telephonic data transfer using an operator. There was a lack of sensors and proper monitoring of the power system.
- At present scenario power system has thermal, solar, wind and hydro generation with power line communication. PMUs, RTUs and SCADA systems are used to monitor the power system. It is operating at one frequency throughout India. It has decentralized generation using PV and wind.
- The future power systems will have centralized and decentralized generation, two-way communication, sensors for automation, SCADA for monitoring, storage at transmission and distribution, and integrated EV charging stations available at generation and distribution.
- Bulk grid is a combination of generation, transmission and distribution.
- Aims of IPDS are strengthening substations, solar power integration, underground cabling, advanced metering and updation in the distribution network.
- Smart grid consists of the communication network, sensors, AMI, cybersecurity etc.
- Nano grid is smaller in scale compared to the microgrid. It is used for a single building or campus. It can be used as a single or cluster.
- Microgrid is a decentralized power generation consisting of storage, PV, wind and diesel generators.
- Thermal power plant works on Rankine cycle. As on 31-01-2025, the installed capacity of thermal power generation is 245.90 GW. It shares 52.74% of the total installed capacity.
- As on 31-01-2025, hydro power generation is 46.96 GW in India. It shares 11.6% of the total installed capacity.
- As on 31-01-2025, solar PV energy installed capacity is 100.32 GW in India. It shares 21.52% of the total installed capacity.
- As on 31-01-2025, Bhadla Solar Park is the world's largest solar park with 2245 MW of installed capacity, located in Rajasthan.
- As on 31-01-2025, the installed capacity of the wind energy is 48.3 GW in India. It shares 10.37% of the total installed capacity. Gujarat has an installed capacity of 11.72 GW. It is generating the highest wind energy.
- Energy storage technologies have advantages like load balancing, peak shaving, backup power, frequency regulation, demand response etc.
- Pumped storage plant has 70-85% efficiency. It helps to support the grid dynamics.

- Lithium-ion batteries are rechargeable energy storage which have abundant use from portable electronic items to electric vehicles. It has 85-95% efficiency.
- There are short, medium and long transmission lines to transport the power. The receiving end reactive power depends on the reactive power requirement of the load and transmission line. The active power flow depends on the power angle of the voltages.
- For higher sending voltages, line losses are reduced.
- Bundle conductors have low leakage reactance compared to normal conductors.
- Voltage regulation of the transmission line should be low.
- Short transmission lines are less than 80 km. Medium transmission line length is 80 to 200 km. The long transmission line length is more than 200 km.
- The radial feeder has a source at one end. It has unidirectional power flow.
- Meshed feeder is a bidirectional power flow feeder.
- There are five regional grids and one national grid in India.
- Reactive power is required for the transmission lines, transformers, motors, and generators.
- The power factor of the load can be improved by connecting the capacitor in parallel with the load.
- To maintain the flat voltage profile of the transmission line capacitor and inductor banks are required at the substation.
- Capacitor absorbs the leading VAr and delivers the lagging VAr.
- The inductor absorbs the lagging VAr and delivers the leading VAr.
- Synchronous grids are networks of power lines that operate at the same frequency and are synchronized with each other.
- Asynchronous grids operate on different frequencies. These are connected with two different frequency grids. Asynchronous grids are connected with an HVDC link.
- Back-to-back HVDC links connect the two AC grids which are operating at different frequencies.
- For 3 phase balanced supply
    - Line current = phase current : For star connection
    - Line voltage = $\sqrt{3}$ *phase voltage : For star connection
    - Line current = $\sqrt{3}$ *phase current : For delta connection
    - Line voltage = phase voltage : For delta connection

Short and Long Answer Questions

1. Draw the architecture of past, present and future power systems and indicate the power and information flow.
2. Explain the advancement of future grids compared to the present grids.
3. Discuss the similarities and advantages of the present grid compared to the past grid.
4. What is the SCADA system? Explain the role of SCADA, PMU and RTUs in transmission and distribution.
5. Draw the architecture of the future electric grid. Explain the advancement in future smart grid compared with the present smart grid.
6. Explain why the future grid is named as a smart electric grid?

7. What is the bulk grid? Draw the architecture and explain the advantages and disadvantages of the bulk grid.
8. What is microgrid? Classify the microgrid. State the advantages and disadvantages of microgrids.
9. Discuss the locations and components of the microgrids.
10. What are the locations and components of nano grids?
11. Compare the following:
    a) Bulk grid vs. microgrid
    b) Microgrid vs. nano grid
    c) Nano grid vs. smart grid
    d) Smart grid vs. bulk grid
12. What do you understand about the single-line diagram of the power system? Draw the single-line diagram of the power system and mention all the voltage levels.
13. What do you mean by DER? Write the locations and types of DERs.
14. Explain the operation of a thermal power plant with the help of a neat block diagram by indicating all the components.
15. Provide the locations of the thermal power plants in India and also state the installed capacity of each thermal power station.
16. Explain the operation of a hydropower plant with the help of a neat block diagram by indicating all the components.
17. Provide the locations of the hydropower plants in India and also state the installed capacity of each hydropower station.
18. Explain the operation of a solar PV power plant with the help of a neat block diagram by indicating all the components.
19. Provide the locations of the solar PV power plants in India and also state the installed capacity of each solar PV power station.
20. Explain the operation of a nuclear power plant with the help of a neat block diagram by indicating all the components.
21. Provide the locations of the nuclear power plants in India and also state the installed capacity of each nuclear power station.
22. Derive the relation between reactive power consumed by the load and the receiving end voltage of a feeder.
23. Derive the expression for apparent power supplied by the generator to the transmission line, having line resistance and line reactance R_L *and* X_L respectively.
24. Derive the relation between line current and phase current for delta-connected load.
25. Derive the relation between line voltage and phase voltage for star-connected load with a phasor diagram.
26. Why do we not design a single-phase motor for a higher rating?
27. What do you mean by power factor? Why it should have a high value.

28. What is the importance of reactive power in the power system? Explain the disadvantages of low power factor.
29. What do you mean by synchronous condenser? What is the use of it explain.
30. In European countries supply frequency is 60 Hz, in our country, it is 50 Hz. What are the advantages and disadvantages if the supply frequency is higher than 50 or 60 Hz?
31. Write the objectives of the transmission of power. Explain with appropriate mathematics.
32. Explain all the grid-level storage technologies in detail.
33. Derive the formula for reactive power supplied by the shunt capacitor to improve the power factor of the load.
34. Write the disadvantages of the radial feeder.
35. What is the doubling effect in a single-phase supply system?
36. What are the methods to reduce the line reactance?
37. What is the need for HVDC transmission? Write the locations and ratings of the major HVDC links.
38. Explain the pumped storage plant with a neat diagram.
39. What are the base and peak load power plants?
40. Write the procedure to find the minimal voltage node in the ring distribution.

Exercise

1. For the balanced Y-Y connected 3-phase circuit, the line-to-line voltage is 400 V rms and the total power consumed by the balanced three-phase load is 1000 W at a power factor of 0.85 lagging. Find the impedance of the load.

2. A 380 V, 50 Hz three-phase balanced supply is connected to the balanced three-phase star-connected load, and its rating is 20 kVA, 0.85 (lag). The rating (kVAr) of delta connected capacitor bank to improve the p.f. to unity is ______.

3. The phase voltage is 230 V, 50 Hz, three - phase four wire line has a phase sequence of A, B and C. A load of 4 kW at unity p.f. is connected between phase A and neutral. It is designed to achieve zero neutral currents by connecting a few inductors and capacitors in the other two phases. What will be the values of the inductor and capacitor?

4. A balanced delta-connected load of (3+4j) Ω is connected to a 380 V, 50 Hz, balanced three-phase power supply. If the load p.f. is improved to 0.95 lagging by connecting a star-connected capacitor bank, then the required kVAr will be__________.

5. The reactive power supplied by the shunt capacitor at a rated voltage is 25 MVAr. If the rated voltage is increased by 10% and the supply frequency is reduced by 5%, what will be the new value of the reactive power supplied by the shunt capacitor?

6. A load of (8+10j) Ω/phase is connected in star to the bus bar of 11 kV. Find the following:
   a. Reactive power required to improve the p.f to 0.95 lagging.
   b. Capacitance/phase required for delta-connected capacitor bank.
   c. Capacitance of each cell if the bank contains 5 cells/phase.

7. A balanced three-phase 4–wire 400 volts, 50 Hz supply is connected to the motor and load. A motor of 15 kW, 0.85 pf lagging, is connected to this supply. A single-phase load of 10 amp at 0.8 p.f lagging, 20 amp at 0.9 p.f. leading, and 30 amp at 0.8 p.f. lagging is connected to phases R, Y, and B respectively. Find the following:
   a. Find the line currents and neutral currents.
   b. The required capacitance per phase to obtain the unity power factor.
   c. Draw the phasor diagram for neutral current, phase currents and phase voltages.

8. A load of (8+10j) Ω per phase is connected in star across a balanced three-phase supply, line voltage of 400 Volts. Find the following:
   a. Phase current and line current in each phase.
   b. Three-phase power consumed by the load.
   c. How much of the capacitance should be connected in parallel with load per phase to make the unity power factor?

9. The resistance of the go and return conductor is (0.2+0.1j) Ω/km. Calculate the voltage drop between points A and C of the given network.

A B C
150 m 250 m
25 A, 0.85 lag. 50 A, 0.9 lag.

10. A three-phase load is given in the network, connected across a line voltage of 400 Volts. Find the following:
   a. Phase current and line current in each phase.
   b. Three-phase power consumed by the load.
   c. How much of the capacitance should be connected in parallel with load per phase to make the unity power factor?

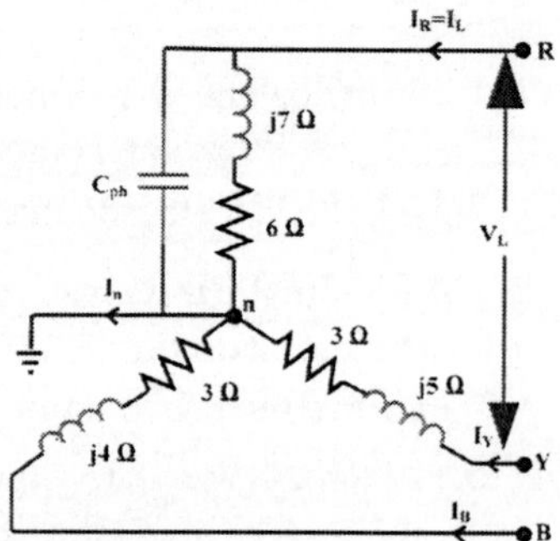

To know more about

Bulk Power Distribution
Grid Services and World's largest solar park

To know more about

Types of Microgrids with Real-Life Examples

To know more about

Smart Grid Pilot Projects under IPDS
Status of Ongoing NSGM Smart Grid Projects

To know more about

HVDC Products,
HVDC projects in the world
Milestones of DC Transmission system in India

To know more about

Status of EV's in India as on April 2024
Review of V2X Topologies and EV standards

To know more about

Grid Engineering Practices & IEEE Standards

02 OVERHEAD TRANSMISSION LINES AND CABLES

Unit specifics: In this unit, the following topics have been discussed for basic understanding of overhead transmission lines and cables:

- R, L and C Parameters of Transmission lines.
- Short, medium and long transmission lines.
- Generalised circuit constants of a transmission line.
- Sigle core cables and three core cables (belted, screened and pressured cables).
- Resistance and Capacitance of a Single-Core Cable.
- Dielectric Stress in a Single-Core Cable,
- Grading of Cables.
- Critical disruptive voltage, Visual critical voltage and Power loss of corona.
- Series and shunt compensation of transmission lines.

Rationale: This unit covers various topics related to transmission lines, including calculating resistance, inductance, and capacitance for different types of transmission lines. It also includes the analysis of short, medium, and long transmission lines, as well as the construction and characteristics of single-core and three-core cables. The unit further explores factors affecting corona, critical disruptive voltage, and power loss. The concepts are explained using diagrams, derivations, and examples.

Pre-Requisites: Basic knowledge of power system components.

Unit Outcomes: List of outcomes of this unit is as follows

U2-O1: To derive the expression for R, L and C Parameters of Transmission lines

U2-O2: To analyse short, medium and long transmission lines.

U2-O3: To derive the expression for generalised circuit constants of a transmission line

U2-O4: To understand construction of single core and three core cables, grading of cables.

U2-O5: To analyse critical disruptive voltage, visual critical voltage and power loss of corona

U2-O6: To understand series and shunt compensation of transmission lines

Unit-2 outcomes	**Expected mapping with course outcomes** (1: Weak, 2: medium, and 3: strong correlation)					
	CO-1	**CO-2**	**CO-3**	**CO-4**	**CO-5**	**CO-6**
U2-O1	2	3	-	-	1	3
U2-O2	2	2	-	-	1	3
U2-O3	2	-	-	1	1	3
U2-O4	2	-	3	2	1	3
U2-O5	2	2	-	-	1	3
U2-O6	2	2	-	-	3	-

2.1 Introduction:

When designing and operating a transmission line, it is crucial to calculate the voltage drop, line losses, and transmission efficiency. The values are greatly influenced by the line constants R, L, and C of the transmission line. For instance, the reduction in voltage in the line is dictated by the values of the three-line constants indicated earlier. The key component that contributes to power loss in transmission lines is the resistance of the conductors, which directly affects the efficiency of transmission.

A transmission line demonstrates homogeneous distribution of resistance, inductance, and capacitance along its whole length. Before delving into the methods for calculating these constants in a transmission line, it is necessary to possess a thorough comprehension of them.

2.2 Resistance of a Transmission Line:

Resistance refers to the inherent opposition that line conductors exhibit to the flow of electric current. The resistance is evenly spread over the whole length of the line, as shown in Figure 2.1(a). Nevertheless, the evaluation of a transmission line's efficiency can be streamlined by considering the scattered resistance as a consolidated parameter, as shown in Figure 2.1(b). The main determinant of power loss in a transmission line is the electrical resistance of the conductors.

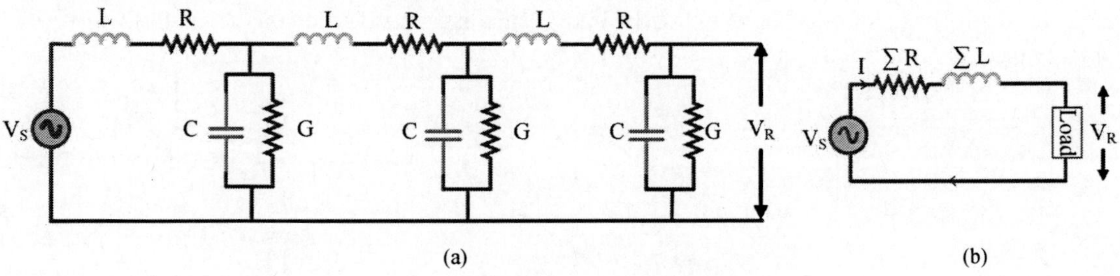

Fig. 2.1 Representation of transmission line

The resistance R of a line conductor can be calculated using the formula

$$\text{Resistance } R = \frac{\rho l}{a} \text{ Ohm} \quad \text{.... (2.1)}$$

Where ρ = resistivity, l = length and a= area of cross-section.

The resistance of metallic conductors exhibits a nearly linear relationship with temperature within their normal operating range. Let R_1 and R_2 represent the resistances of a conductor at temperatures T_1°C and T_2°C (where T_2 is greater than T_1) respectively. Let α_1 represent the temperature coefficient at T_1°C, then

$$R_2 = R_1\,[1 + \alpha_1(T_2 - T_1)] \quad \text{....(2.2)}$$

Where $\alpha_1 = \frac{\alpha_0}{1+\alpha_0 t_1}$ and α_0 = temperature coefficient at 0° C

In a single phase or 2-wire direct current line, the overall resistance, also known as loop resistance, is equal to twice the resistance of each individual conductor. In the context of a 3-phase transmission line, the resistance per phase represents the resistance of each individual conductor.

2.3 Inductance of a Transmission Line:

When an alternating current flows through a conductor, it generates a varying magnetic field that interacts with the conductor. The existence of these flux connections leads to the conductor possessing inductance. Inductance is a precise mathematical term that refers to the amount of magnetic flux that is associated with an electric circuit for each unit of electric current.

$$\text{Inductance, } L = \frac{\psi}{I} \text{ henry} \qquad \text{.... (2.3)}$$

Where ψ = flux linkages in weber-turns
I = current in amperes

Hence, to calculate the inductance of a circuit, it is crucial to first determine the flux linkages. The inductance is uniformly distributed along the entire length of the line, as depicted in Fig. 2.1(a). For the sake of analysis, it can be assumed that it is lumped (combined), as depicted in Fig. 2.1(b).

2.3.1 Flux linkages due to a single current carrying conductor:

Consider a cylindrical conductor that is long and straight, with a radius of r metres. This conductor is carrying a current of I amperes (root mean square), as shown in Fig. 2.2. This electric current creates a magnetic field. Magnetic field lines will exist both inside and outside the conductor. Both fluxes will increase the conductor's inductance.

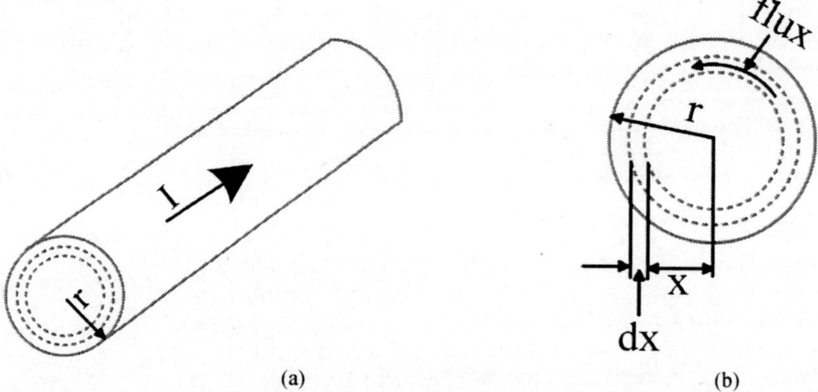

Fig. 2.2 Flux linkages due to internal flux

2.3.1.1 Flux linkages due to internal flux:

Refer to Fig. 2.2(b), which shows the conductors X-section magnified for clarity. Amperes law states that the magnetic field (m.m.f.) around a closed route is equal to the current contained within it. The current enclosed by the radial path at a distance x from centre is I_x, and the m.m.f. is $H_x * 2\pi x$. Where H_x represents magnetic field intensity at a distance x from the centre

$$\text{Therefore,} \quad I_x = H_x * 2\pi x \qquad \text{....(2.4)}$$

Assuming a uniform current density,

$$I_x = I * \frac{\pi x^2}{\pi r^2} = I * \frac{x^2}{r^2} \qquad \text{..... (2.5)}$$

The magnetic field intensity at a position 'x' metre from the centre is given by:

$$H_x = \frac{I_x}{2\pi x} = \frac{I * \frac{x^2}{r^2}}{2\pi x} = \frac{I\,x}{2\pi r^2} \quad \text{AT/m} \qquad .....(2.6)$$

The flux density at the considered point is

$$B_x = \mu H_x = \mu_0 \mu_r \frac{I\,x}{2\pi r^2} \quad \text{wb/m2} \qquad where\ \mu = \mu_0 \mu_r$$

$$= \frac{\mu_0\, I\, x}{2\pi r^2} \quad \text{wb/m}^2 \qquad where\ \mu_r = 1\ for\ non - magnetic\ material \qquad ......(2.7)$$

The flux $d\phi$ through a cylindrical shell with radial thickness dx and axial length 1m can be calculated as:

$$d\phi = B_x\, l\, dx = B_x * 1 * dx$$

$$= \frac{\mu_0\, I\, x}{2\pi r^2} dx \text{ weber} \qquad ......(2.8)$$

This flux $d\phi$ links with the current I_x. Thus, flux linkages per metre length of the conductor is

$$d\psi = \frac{\pi x^2}{\pi r^2} d\phi = \frac{x^2}{r^2} \frac{\mu_0\, I\, x}{2\pi r^2} dx = \frac{\mu_0\, I\, x^3}{2\pi r^4} dx \quad \text{weber-turns}$$

Total internal flux linkages from centre to the conductor surface is

$$\psi_{int} = \int_0^r d\psi = \int_0^r \frac{\mu_0\, I\, x^3}{2\pi r^4} dx = \frac{\mu_0\, I}{8\pi} \quad \text{weber-turns per metre length}$$

∴ Internal flux of a single current carrying conductor is $\psi_{int} = \frac{\mu_0\, I}{8\pi}$ .....(2.9)

2.3.1.2 Flux linkages due to external flux:

We will now compute the conductor's flux linkages caused by the external flux. External flux extends indefinitely from the conductor's surface. According to Fig. 2.3, the magnitude of the electric field at a distance of x metres from the centre of the conductor can be calculated using the following equation:

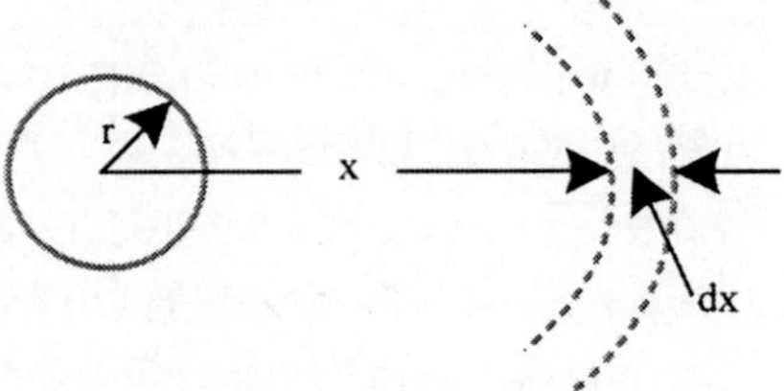

Fig. 2.3 Flux linkages due to external flux

$$H_x = \frac{I_x}{2\pi x} = \frac{I * \frac{x^2}{x^2}}{2\pi x} = \frac{I}{2\pi x} \quad AT/m \qquad .....(2.10)$$

Flux density $B_x = \mu H_x = \mu_0 \mu_r I \frac{1}{2\pi x} = \frac{\mu_0\, I}{2\pi x}$ wb/m² .....(2.11)

The flux dφ through a cylindrical shell with thickness dx and an axial length of one metre is

$$d\phi = B_x\, l\, dx = B_x * 1 * dx = \frac{\mu_0 I}{2\pi x} dx \text{ weber} \qquad .....(2.12)$$

The flux dφ links all current in the conductor just once.

$$d\psi = \frac{\pi x^2}{\pi x^2} d\phi = \frac{\mu_0 I}{2\pi x} dx \text{ weber-turns} \qquad .....(2.13)$$

Total flux linkages of the conductor from surface to infinity,

$$\psi_{ext} = \int_r^\infty d\psi = \int_r^\infty \frac{\mu_0 I}{2\pi x} dx \quad \text{weber-turns per metre length} \qquad .....(2.14)$$

Total flux linkages, $\psi = \psi_{int} + \psi_{ext}$

$$= \frac{\mu_0 I}{8\pi} + \int_r^\infty \frac{\mu_0 I}{2\pi x} dx$$

∴ Overall flux linkages, $\psi = \psi_{int} + \psi_{ext} = \frac{\mu_0 I}{2\pi}\left[\frac{1}{4} + \int_r^\infty \frac{1}{x} dx\right]$ .....(2.15)

2.3.2 Flux linkages in parallel current carrying conductors:

We will analyse the flux linkages between parallel current-carrying conductors. Fig. 2.4 depicts conductors A, B, and C that carry currents I_A, I_B, and I_C, respectively. Consider flux links using a single conductor, say A. As previously stated, conductor A's current will cause flux linkage to it. The mutual inductance effects of I_B, and I_C will also cause flux Linkages with this conductor. We shall calculate the total flow linkages to conductor A.

Fig. 2.4 Parallel current carrying conductors

Flux linkages with conductor *A* due to its own current is

$$= \frac{\mu_0 I_A}{2\pi}\left[\frac{1}{4} + \int_r^{\infty}\frac{dx}{x}\right] \qquad \text{......(2.16)}$$

Conductor B, which carries current I_B, is at a distance of d_1 from conductor A. Only the external flux from I_B links with conductor A. The external flux from I_B links with conductor A from d_1 to ∞, resulting in the term is

$$= \frac{\mu_0 I_B}{2\pi}\int_{d_1}^{\infty}\frac{dx}{x} \qquad \text{......(2.17)}$$

Similarly, the external flux from I_C links with conductor A from d_2 to ∞, resulting in the term

$$= \frac{\mu_0 I_C}{2\pi}\int_{d_2}^{\infty}\frac{dx}{x} \qquad \text{... ... (2.18)}$$

$$\therefore \text{Total flux linkages with conductor A, } \psi = \frac{\mu_0 I_A}{2\pi}\left[\frac{1}{4} + \int_r^{\infty}\frac{dx}{x}\right] + \frac{\mu_0 I_B}{2\pi}\int_{d_1}^{\infty}\frac{dx}{x} + \frac{\mu_0 I_C}{2\pi}\int_{d_2}^{\infty}\frac{dx}{x} \qquad \text{.....(2.19)}$$

Similarly, flux linkages to other conductors can be determined. The equation above can be used to determine the inductance of any circuit.

2.3.3 Inductance of a Single Phase Two-wire Line:

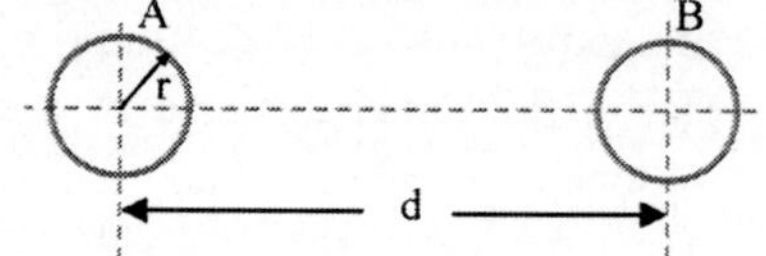

Fig. 2.5 Single Phase Two-wire Line

Two parallel wires combined to form a rectangular loop with one turn make up a single-phase line. Such a loop generates a fluctuating magnetic flux when an alternating current flows through it. The loop (or single-phase line) has inductance because of the fluctuating flux that links it. Because a single-phase wire has a flux path through high-reluctance air and is a one-turn loop, its inductance may seem low. Yet, even at low flux densities, the loop's X-sectional area is big, resulting in a relatively large total flux linking the loop and a noticeable inductance in the line.

An overhead single-phase line with two parallel wires, A and B, placed d meters apart is shown in Figure 2.5. The current flowing through conductors A and B is equal ($I_A = I_B$), but it flows in the opposite direction since one serve as the other's return circuit. So, $I_A + I_B = 0$. To calculate the inductance of conductor A (or conductor B), we must examine its flux linkages. Conductor A and conductor B will have flux linkages due to current I_A and mutual inductance.

Flux linkages with conductor *A* due to its own current $= \frac{\mu_0 I_A}{2\pi}\left[\frac{1}{4} + \int_r^{\infty}\frac{dx}{x}\right]$

Flux linkages with conductor *A* due to current I_B is $= \frac{\mu_0 I_B}{2\pi}\int_d^{\infty}\frac{dx}{x}$ ......(2.20)

∴ Total flux linkages with conductor A is

$$\psi = \frac{\mu_0 I_A}{2\pi}\left[\frac{1}{4}+\int_r^{\infty}\frac{dx}{x}\right]+\frac{\mu_0 I_B}{2\pi}\int_d^{\infty}\frac{dx}{x}$$

$$= \frac{\mu_0}{2\pi}\left[\left(\frac{1}{4}+\int_r^{\infty}\frac{dx}{x}\right)I_A + I_B\int_d^{\infty}\frac{dx}{x}\right]$$

$$= \frac{\mu_0}{2\pi}\left[\left(\frac{1}{4}+ log_e\infty - log_e r\right)I_A + I_B(log_e\infty - log_e d)\right]$$

$$= \frac{\mu_0}{2\pi}\left[\left(\frac{I_A}{4}+ log_e\infty(I_A+I_B) - I_A\, log_e r - I_B\, log_e d\right)\right] \qquad ......(2.21)$$

$$\psi_A = \frac{\mu_0}{2\pi}\left[\left(\frac{I_A}{4} - I_A\, log_e r + I_A\, log_e d\right)\right]$$ Where $I_A + I_B = 0$ and $I_B = - I_A$

$$= \frac{\mu_0}{2\pi}\left[\left(\frac{I_A}{4} + I_A\, log_e\frac{d}{r}\right)\right]$$

$$= \frac{\mu_0 I_A}{2\pi}\left[\left(\frac{1}{4} + log_e\frac{d}{r}\right)\right] \quad \text{weber-turns per metre} \qquad ......(2.22)$$

Inductance of conductor A is $L_A = \frac{\psi_A}{I_A}$

$$L_A = \frac{\mu_0}{2\pi}\left[\left(\frac{1}{4} + log_e\frac{d}{r}\right)\right] \quad \text{H/m}$$

$$= \frac{4\,\pi * 10^{-7}}{2\pi}\left[\left(\frac{1}{4} + log_e\frac{d}{r}\right)\right] \quad \text{H/m}$$

Inductance of conductor A is $L_A = 10^{-7}\left[\left(\frac{1}{2} + 2\, log_e\frac{d}{r}\right)\right]$ H/m ......(2.23)

Loop inductance = 2 L_A = $10^{-7}\left[\left(1 + 4\, log_e\frac{d}{r}\right)\right]$ H/m ......(2.24)

It should be noted that eq. (2.24) represents the inductance of a two-wire line, also known as loop inductance. Equation (2.23) calculates the inductance per conductor, which is equal to half the loop inductance.

Expression in another form: The inductance of a conductor can be expressed concisely.

$$L_A = 10^{-7}\left[\left(\frac{1}{2} + 2\, log_e\frac{d}{r}\right)\right] \quad \text{H/m}$$

$$= 2 * 10^{-7}\left[\left(\frac{1}{4} + log_e\frac{d}{r}\right)\right]$$

$$= 2 * 10^{-7}\left[\left(log_e e^{1/4} + log_e\frac{d}{r}\right)\right]$$

$$= 2 * 10^{-7}\, log_e\frac{d}{r\, e^{-1/4}}$$

$$= 2 * 10^{-7}\, log_e\frac{d}{r'} \qquad \text{where } r' = r\, e^{-1/4}$$

The radius r′ represents a fictitious conductor with no internal flux but the same inductance as the actual conductor with radius r.

The geometric mean radius (GMR) of a wire is denoted by $r' = r\, e^{-1/4} = 0.7788\, r$.

Note that r′ = 0·7788 r applies exclusively to solid round conductors.

Inductance of conductor A is $L_A = 2 * 10^{-7}\, log_e\frac{d}{r'} = 2 * 10^{-7}\, log_e\frac{d}{0.7788\, r}$ ......(2.25)

Loop inductance = 2 L_A = $2 * 2 * 10^{-7}\, log_e\frac{d}{r'}$ H/m ......(2.26)

2.3.4 Inductance of a 3-Phase Overhead Line:

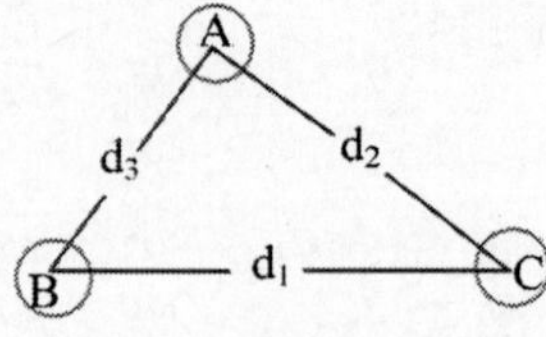

Fig. 2.6 Diagonally spaced conductors

Figure 2.6 depicts the three conductors (A, B, and C) of a three-phase line carrying currents (I_A, I_B, and I_C). Let d_1, d_2, and d_3 denote the conductor spacings mentioned. Assume a balanced load ($I_A + I_B + I_C = 0$). Consider the flux linkage with conductor A. Conductor A will experience flux linkages as a result of its own current as well as due to currents I_B and I_C.

Flux linkages with conductor *A* due to its own current is $= \frac{\mu_0 I_A}{2\pi}\left[\frac{1}{4} + \int_r^\infty \frac{dx}{x}\right]$

Flux linkages with conductor *A* due to current I_B is $= \frac{\mu_0 I_B}{2\pi}\int_{d_3}^\infty \frac{dx}{x}$

Flux linkages with conductor *A* due to current I_C is $= \frac{\mu_0 I_C}{2\pi}\int_{d_2}^\infty \frac{dx}{x}$

∴ Total flux linkages with conductor A,

$$\psi = \frac{\mu_0 I_A}{2\pi}\left[\frac{1}{4} + \int_r^\infty \frac{dx}{x}\right] + \frac{\mu_0 I_B}{2\pi}\int_{d_3}^\infty \frac{dx}{x} + \frac{\mu_0 I_C}{2\pi}\int_{d_2}^\infty \frac{dx}{x}$$

$$= \frac{\mu_0}{2\pi}\left[\left(\frac{1}{4} + \int_r^\infty \frac{dx}{x}\right) I_A + I_B\int_{d_3}^\infty \frac{dx}{x} + I_C\int_{d_2}^\infty \frac{dx}{x}\right]$$

$$= \frac{\mu_0}{2\pi}\left[\left(\frac{1}{4} + log_e\infty - log_e r\right) I_A + I_B(log_e\infty - log_e d_3) + I_C(log_e\infty - log_e d_2)\right]$$

$$= \frac{\mu_0}{2\pi}\left[\left(\frac{I_A}{4} + log_e\infty(I_A + I_B + I_C) - I_A\, log_e r - I_B\, log_e d_3 - I_C\, log_e d_2\right)\right] \qquad ......(2.27)$$

$$\psi_A = \frac{\mu_0}{2\pi}\left[\left(\frac{I_A}{4} - I_A\, log_e r - I_B\, log_e d_3 - I_C\, log_e d_2\right)\right] \qquad \text{Where } I_A + I_B + I_C = 0$$

$$= \frac{\mu_0}{2\pi}\left[\left(\frac{1}{4} - log_e r\right) I_A - I_B\, log_e d_3 - I_C\, log_e d_2\right] \qquad \text{weber-turns per metre} \qquad ......(2.28)$$

2.3.4.1 Symmetrical spacing:

If three conductors (A, B, and C) are placed symmetrically at the corners of an equilateral triangle with side d, then $d_1 = d_2 = d_3$. Under these conditions, flux linkages with conductor A become:

$$\psi_A = \frac{\mu_0}{2\pi}\left[\left(\frac{1}{4} - log_e r\right) I_A - I_B\, log_e d - I_C\, log_e d\right]$$

$$= \frac{\mu_0}{2\pi}\left[\left(\frac{1}{4} - log_e r\right) I_A - (I_B + I_C)\, log_e d\right]$$

$$= \frac{\mu_0}{2\pi}\left[\left(\frac{1}{4} - log_e r\right) I_A + I_A\, log_e d\right] \qquad \text{Where } I_A = -(I_B + I_C)$$

$$= \frac{\mu_0}{2\pi}\left[\left(\frac{1}{4} - log_e r + log_e d\right) I_A\right]$$

$$= \frac{\mu_0 I_A}{2\pi}\left[\left(\frac{1}{4} + log_e \frac{d}{r}\right)\right] \qquad \text{weber-turns per metre} \qquad ......(2.29)$$

Inductance of conductor *A* is $L_A = \frac{\psi_A}{I_A}$

$$L_A = \frac{\mu_0}{2\pi}\left[\left(\frac{1}{4} + log_e \frac{d}{r}\right)\right] = \frac{4\pi * 10^{-7}}{2\pi}\left[\left(\frac{1}{4} + log_e \frac{d}{r}\right)\right] \quad \text{H/m}$$

Inductance of conductor *A* is $L_A = 10^{-7}\left[\left(\frac{1}{2} + 2\, log_e \frac{d}{r}\right)\right]$ H/m ......(2.30)

Inductance formulas for conductors B and C are derived similarly.

2.3.4.2 Unsymmetrical spacing:

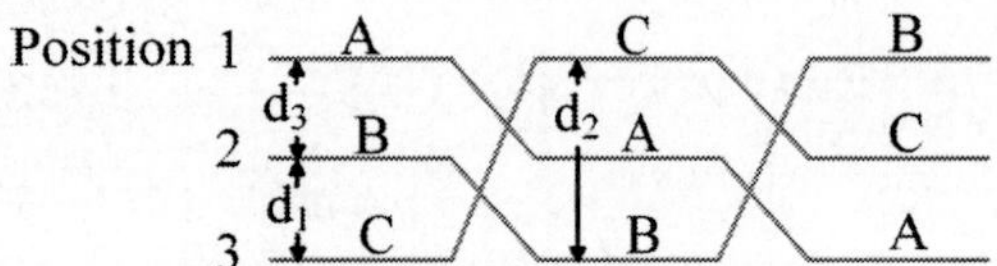

Fig. 2.7 Transposed transmission line

Unsymmetrical conductor spacing occurs when three-phase line conductors are not equidistant from one another. Under such circumstances, the flux linkages and inductance of each phase are not identical. Even if the currents in the conductors are balanced, each phase has a different inductance, resulting in disproportionate voltage drops. As a result, the voltage at the receiving end will vary for each phase. In order to achieve equal voltage drops across all conductors, it is normal to periodically exchange the positions of the conductors along the entire line. This ensures that each conductor occupies the original position of every other conductor for an equal distance. The act of swapping positions is known as transposition.

Figure 2.7 shows a transposed line. The phase conductors (A, B, and C) are assigned numbers 1, 2, and 3, respectively, for the three positions. Transposition results in each conductor having the same average inductance. Figure 2.7 illustrates a three-phase transposed line with an asymmetrical spacing. Assume each of the three sections is one metre long. Assume balanced conditions $I_A + I_B + I_C = 0$. Let the line currents be

$$I_A = I\,(1 + j\,0)$$

$$I_B = I\,(-0{\cdot}5 - j\,0{\cdot}866)$$

$$I_C = I\,(-0{\cdot}5 + j\,0{\cdot}866)$$

As seen previously, the total flux linkages per metre length of conductor A is

$$\psi_A = \frac{\mu_0}{2\pi}\left[\left(\frac{1}{4} - log_e r\right) I_A - I_B\ log_e d_3 - I_C\ log_e d_2\right]$$

$$= \frac{\mu_0}{2\pi}\left[\left(\frac{1}{4} - log_e r\right) I - I\,(-0.5 - j\,0.866)\ log_e d_3 - I\,(-0.5 + j\,0.866)\ log_e d_2\right]$$

$$= \frac{\mu_0 I}{2\pi}\left[\frac{1}{4} - log_e r + 0.5\,(\,log_e d_3 + log_e d_2) + j\,0.866\,(\,log_e d_3 - log_e d_2)\right]$$

$$= \frac{\mu_0 I}{2\pi}\left[\frac{1}{4} - log_e r + 0.5\,(\,log_e d_3 d_2) + j\,0.866\ log_e \frac{d_3}{d_2}\right]$$

$$= \frac{\mu_0 I}{2\pi}\left[\frac{1}{4} - log_e r + log_e (d_3 d_2)^{0.5} + j\,0.866\ log_e \frac{d_3}{d_2}\right]$$

$$= \frac{\mu_0 I}{2\pi}\left[\frac{1}{4} + log_e \frac{\sqrt{d_3 d_2}}{\mathrm{r}} + j\,0.866\ log_e \frac{d_3}{d_2}\right]$$

Inductance of conductor *A* is $L_A = \frac{\psi_A}{I_A} = \frac{\psi_A}{\mathrm{I}}$

$$L_A = \frac{\mu_0}{2\pi}\left[\frac{1}{4} + log_e \frac{\sqrt{d_3 d_2}}{\mathrm{r}} + j\,0.866\ log_e \frac{d_3}{d_2}\right]$$

$$= \frac{4\,\pi * 10^{-7}}{2\pi}\left[\frac{1}{4} + log_e \frac{\sqrt{d_3 d_2}}{\mathrm{r}} + j\,0.866\ log_e \frac{d_3}{d_2}\right]$$

$$= 10^{-7}\left[\frac{1}{2} + 2\ log_e \frac{\sqrt{d_3 d_2}}{\mathrm{r}} + j\,1.732\ log_e \frac{d_3}{d_2}\right]$$

Similarly inductance of conductors B and C will be :

$$L_B = 10^{-7}\left[\frac{1}{2} + 2\ log_e \frac{\sqrt{d_1 d_3}}{r} + j\ 1.732\ log_e \frac{d_1}{d_3}\right]$$

$$L_C = 10^{-7}\left[\frac{1}{2} + 2\ log_e \frac{\sqrt{d_2 d_1}}{r} + j\ 1.732\ log_e \frac{d_2}{d_1}\right]$$

Inductance of each line conductor $= \frac{1}{3}\ (L_A + L_B + L_C)$

$$= \frac{10^{-7}}{3}\left[\left(\frac{1}{2} + 2\ log_e \frac{\sqrt{d_3 d_2}}{r} + j\ 1.732\ log_e \frac{d_3}{d_2}\right) + \left(\frac{1}{2} + 2\ log_e \frac{\sqrt{d_1 d_3}}{r} + j\ 1.732\ log_e \frac{d_1}{d_3}\right) + \left(\frac{1}{2} + 2\ log_e \frac{\sqrt{d_2 d_1}}{r} + j\ 1.732\ log_e \frac{d_2}{d_1}\right)\right]$$

$$= \frac{10^{-7}}{3}\left[\frac{3}{2} + 2\ log_e \frac{\sqrt{d_3 d_2}}{r} + 2\ log_e \frac{\sqrt{d_1 d_3}}{r} + 2\ log_e \frac{\sqrt{d_2 d_1}}{r} + j\ 1.732\left(\ log_e \frac{d_3}{d_2} + \ log_e \frac{d_1}{d_3} + \ log_e \frac{d_2}{d_1}\right)\right]$$

$$= \frac{10^{-7}}{3}\left[\frac{3}{2} + 2\ log_e \frac{\sqrt{d_3 d_2}}{r} * \frac{\sqrt{d_1 d_3}}{r} * \frac{\sqrt{d_2 d_1}}{r} + j\ 1.732\left(\ log_e \frac{d_3}{d_2}\frac{d_1}{d_3}\frac{d_2}{d_1}\right)\right]$$

$$= \frac{10^{-7}}{3}\left[\frac{3}{2} + 2\ log_e \frac{d_1 d_2 d_3}{r^3} + j\ 1.732(\ log_e 1)\right]$$

$$= \frac{10^{-7}}{3}\left[\frac{3}{2} + 6\ log_e \frac{\sqrt[3]{d_1 d_2 d_3}}{r}\right] \qquad \text{where } log_e 1 = 0$$

$$= 10^{-7}\left[\frac{1}{2} + 2\ log_e \frac{\sqrt[3]{d_1 d_2 d_3}}{r}\right] \quad \text{H/m}$$

$$\text{Inductance of each line conductor} = 10^{-7}\left[\frac{1}{2} + 2\ log_e \frac{\sqrt[3]{d_1 d_2 d_3}}{r}\right] \quad \text{H/m} \qquad \text{......(2.31)}$$

When comparing an unsymmetrically spaced transposed line with a symmetrically spaced line, the inductance of each conductor is comparable if $d = \sqrt[3]{d_1 d_2 d_3}$. An unsymmetrically spaced transposed line's equivalent equilateral spacing is represented by the distance d.

2.3.5 Inductance of lines in-terms of Self GMD and Mutual GMD:

Using self-GMD and mutual-GMD simplifies inductance calculations for multiconductor configurations. These are denoted as D_s and D_m, respectively.

2.3.5.1 Self-GMD (D_s): To comprehend self-GMD (Geometrical Mean Radius or GMR), refer to the formula for inductance per conductor per metre in section 2.3.4. From eq. (2.25),

Inductance per conductor per meter is $= 2 * 10^{-7}\ log_e \frac{d}{r'} = 2 * 10^{-7}\ log_e \frac{d}{0.7788\ r}$

$$\text{Inductance per conductor per meter} = 2 * 10^{-7}\ log_e \frac{d}{D_S} \qquad \text{...... (2.32)}$$

where Ds = GMR or self-GMD = 0·7788 r

2.3.5.2 Mutual-GMD (D_m): The mutual geometric mean distance (mutual-GMD) is calculated as the average of the distances between two conductors. It is important to note that the mutual-GMD must always fall between the largest and smallest distance between the conductors. Mutual-GMD is a term that describes the equal spacing between geometrical elements.

(i). The mutual geometric mean distance (GMD) between two conductors is equivalent to the distance between their centres, given that the spacing between the conductors is greater than their diameter. $D_m = d$

(ii). The mutual geometric mean distance (GMD) of a single circuit 3-phase line is equivalent to the equilateral spacing between the conductors. $D_m = \sqrt[3]{d_1 d_2 d_3}$

The application of geometrical mean distances is most efficient when used for 3-φ double circuits. Examine the arrangement of the conductors in the double circuit, as depicted in Figure 2·8. Let's assume that each conductor has a radius of r.

Self-GMD of conductor = $0{\cdot}7788\, r$

Self-GMD of combination aa' is $D_{s1} = \sqrt[4]{D_{aa} * D_{aa'} * D_{a'a} * D_{a'a'}}$

Self-GMD of combination bb' is $D_{s2} = \sqrt[4]{D_{bb} * D_{bb'} * D_{b'b} * D_{b'b'}}$

Self-GMD of combination cc' is $D_{s3} = \sqrt[4]{D_{cc} * D_{cc'} * D_{c'c} * D_{c'c'}}$ ......(2.33)

Where D_{aa} or $D_{a'a'}$ refers to the conductor's self-GMD.

$D_{aa'}$ represents the distance between a and a'.

Equivalent self-GMD of one phase is $D_s = \sqrt[3]{D_{s1} D_{s2} D_{s3}}$ ......(2.34)

The value of self GMD (D_s) is the same for all phases since each conductor has the same radius.

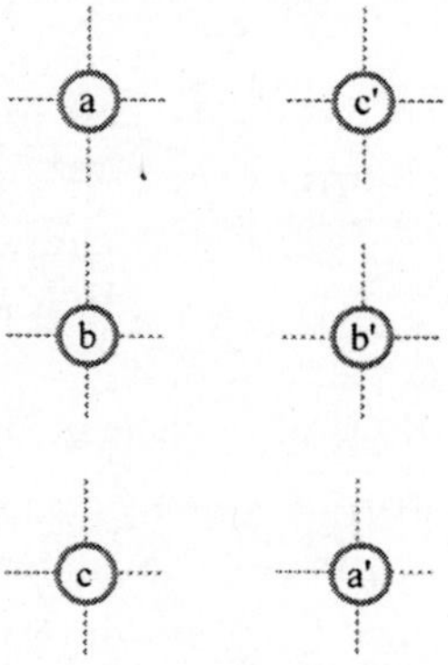

Fig. 2.8 Double circuit

Mutual-GMD between phases A and B is $D_{AB} = \sqrt[4]{D_{ab} * D_{ab'} * D_{a'b} * D_{a'b'}}$

Mutual-GMD between phases B and C is $D_{BC} = \sqrt[4]{D_{bc} * D_{bc'} * D_{b'c} * D_{b'c'}}$

Mutual-GMD between phases C and A is $D_{CA} = \sqrt[4]{D_{ca} * D_{ca'} * D_{c'a} * D_{c'a'}}$ ......(2.35)

Equivalent mutual-GMD, $D_m = \sqrt[3]{D_{AB} D_{BC} D_{CA}}$ ......(2.36)

Mutual GMD is not affected by the size, shape, or orientation of the conductor, but rather by its spacing alone. The inductance formulas from earlier parts can be stated in terms of geometrical mean distances as:

Inductance/conductor/metre of a single-phase line is

$$L = 2 * 10^{-7} \log_e \frac{D_m}{D_s} = 2 * 10^{-7} \log_e \frac{d}{0{\cdot}7788\, r} \qquad ......(2.37)$$

Inductance/conductor/metre of a single-circuit three-phase line is

$$L = 2 * 10^{-7} \log_e \frac{D_m}{D_s} = 2 * 10^{-7} \log_e \frac{\sqrt[3]{d_1 d_2 d_3}}{0{\cdot}7788\, r} \qquad ......(2.38)$$

Inductance/conductor/metre of a double-circuit three-phase line is

$$L = 2 * 10^{-7} \log_e \frac{D_m}{D_s} = 2 * 10^{-7} \log_e \frac{\sqrt[3]{D_{AB} D_{BC} D_{CA}}}{\sqrt[3]{D_{S1} D_{S2} D_{S3}}} \qquad ......(2.39)$$

2.3.6. Skin effect:

When a conductor carries continuous direct current (d.c.), the current is evenly distributed across its whole cross-section. Alternating current in a conductor does not distribute evenly, but rather concentrates toward the surface (Fig. 2.9). This is known as the skin effect. A solid conductor may be assumed to be devided into multiple concentric cylinders as shown in Fig. 2.10. When alternating current flows through a conductor, magnetic flux is produced. The magnetic flux linking with a cylindrical element near the center of the conductor is greater than the flux linking with cylindrical element near the surface of the conductor.

The central cylindrical element is surrounded by both internal and external flux, whereas the cylindrical element on the surface of the conductor is surrounded solely by external flux.

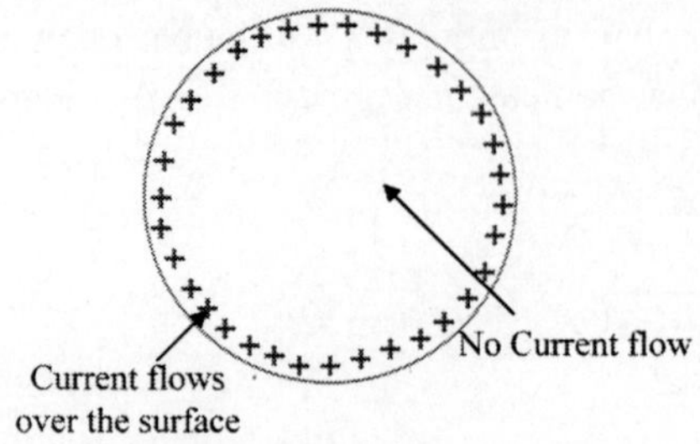

Fig 2.9 Skin effect

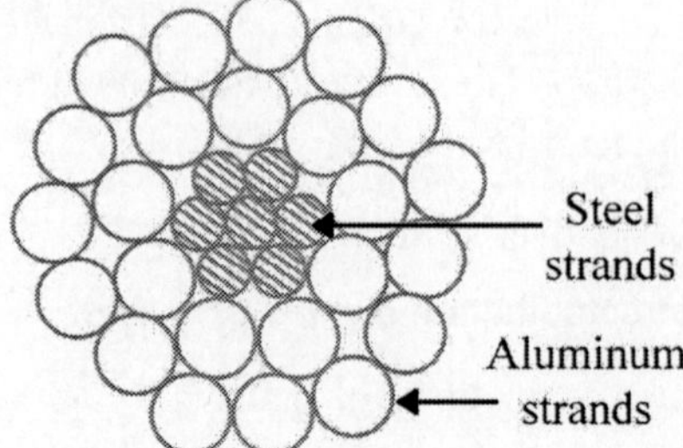

Fig. 2.10 ACSR conductors (07 steel and 24 Aluminium strands)

The self-inductance of the inner cylindrical element is greater, therefore providing a higher inductive reactance compared to the outer cylindrical element. The difference in inductive reactance causes the current to predominantly concentrate at the surface or skin of the conductor. The current density is higher near the conductor's surface and lower at its center. The effect corresponds to a decrease in the effective cross-sectional area of the conductor, hence increasing the effective resistance of the conductor.

Factors Affecting the Skin Effect:

- Frequency: The skin effect increases with an increase in frequency.
- Diameter: The diameter increases with the enlargement of the conductor's diameter.
- The shape of the conductor: The skin effect is more noticeable in solid conductors than in stranded conductors due to the greater surface area of the solid conductor.
- Type of material: The skin effect increases with an increase in the material's permeability, which is the capacity of a material to allow the formation of a magnetic field.

The skin effect is insignificant if the frequency is below 50 Hz and the conductor's diameter is less than 1 cm. In stranded conductors such as ACSR (Aluminium Conductor Steel Reinforced), the current predominantly flows through the outer aluminum layer, while the steel core, located centrally, conducts no current and provides significant tensile strength to the conductor. The current density at the surface enabled the utilization of ACSR conductor.

2.3.7. Proximity Effect: When two conductors are in close proximity, the flux produced by second conductor also links with inner half of the first conductor and flux produced by first conductor also links with inner half of the second conductor as shown in Fig. 2.11. When two conductors carry currents in same direction, the two fluxes oppose each other as shown in Fig. 2.11(a). This leads to lower flux linkage to inner halves of two conductors compared to outer halves. This leads to lower inductive reactance for inner halves of two conductors compared to outer halves resulting in higher current density in inner halves. When two conductors carry currents in opposite direction, the two fluxes add to each other as illustrated in Fig. 2.11(b). Therefore, inner halves of the two conductors get higher flux linkages compared to outer halves resulting in higher inductive reactance for inner halves. This leads to lower current density in inner halves of two conductors. This non-uniformity of current due to proximity of two conductors is called *proximity effect.* Proximity effect leads to decrease in effective cross-sectional area of conductor similar to skin effect resulting in increase in effective resistance of conductors. The proximity effect may be reduced by increase in spacing between the conductors. Almost all the factors influencing the Skin Effect also influence the Proximity Effect.

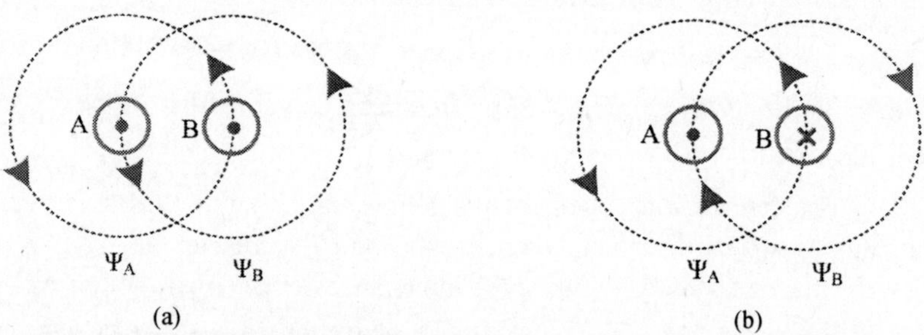

Fig. 2.11 Two conductors with currents in (a) same direction (b) opposite direction

Example 2.1. A single-phase line comprises two parallel wires separated by a distance of 3 meters. The diameter of each conductor measures 1.5cm. Calculate the inductance of the loop per km.

Ans : Spacing between the contacts d=300cm

Radius of the conductor $r = \frac{1.5}{2} = 0.75cm$

Loop inductance = 2 L_A

$$= 10^{-7}\left[\left(1 + 4\ log_e \frac{d}{r}\right)\right]$$

$$= 10^{-7}\left[\left(1 + 4\ log_e \frac{300}{0.75}\right)\right] = 2.49 * 10^{-6}\ H/m$$

Loop inductance per Km = $2.49 * 10^{-6} * 1000 = 2.49 * 10^{-3}\ H$ *(or)*

Loop inductance per Km =$0.4\ log_e \frac{d}{r'} = 0.4\ log_e \frac{d}{0.7788\ r} = 0.4\ log_e \frac{300}{0.7788*0.75} = 2.49mH$

Example 2.2. Calculate the inductance per kilometre of a 3-phase transmission line consisting of conductors with a diameter of 1.6cm, arranged at the corners of an equilateral triangle with a side length of 4m.

Ans : Spacing between the contacts d=400cm

Radius of the conductor $r = \frac{1.6}{2} = 0.8cm$

Inductance = $10^{-7}\left[\left(\frac{1}{2} + 2\ log_e \frac{d}{r}\right)\right] = 10^{-7}\left[\left(\frac{1}{2} + 2\ log_e \frac{400}{0.8}\right)\right] = 1.29 * 10^{-6}\ H/m$ *(or)*

Inductance =$2 * 10^{-7}\ log_e \frac{d}{0.7788\ r} = 2 * 10^{-7} log_e \frac{400}{0.7788*0.8} = 1.29 * 10^{-6}\ H/m$

Inductance per Km = $1.29 * 10^{-6} * 1000 = 1.29mH$

Example 2.3. The three conductors of a three-phase line are positioned at the vertices of a triangle with side lengths of 3m, 4m, and 5m. Calculate the inductance per kilometre of the line when the conductors are regularly interchanged. Each conductor has a diameter of 2.5 cm.

Ans : Spacing b/w conductors D_{AB}= 3m, D_{BC}= 4m, and D_{CA}= 5m

Equivalent spacing, $D_{eq} = \sqrt[3]{D_{AB} \times D_{BC} \times D_{CA}} = \sqrt[3]{3 \times 4 \times 5} = 3.91\ m$

Inductance =$2 * 10^{-7}\ log_e \frac{d}{0.7788\ r} = 2 * 10^{-7} log_e \frac{391}{0.7788*1.25} = 1.199 * 10^{-6}\ H/m$

Inductance per Km = $1.199 * 10^{-6} * 1000 = 1.199\ mH$

Example 2.4. Calculate the inductance of each conductor in a three-phase, three-wire system when they are arranged horizontally with a spacing of D_{AB} = 10m, D_{BC} = D_{CA} = 5m. The conductors possess a diameter of 4cm and are arranged in a transposed manner.

Ans : Equivalent spacing, $D_{eq} = \sqrt[3]{D_{AB} \times D_{BC} \times D_{CA}} = \sqrt[3]{10 \times 5 \times 5} = 6.29\, m$

$$\text{Inductance} = 2 * 10^{-7}\, log_e \frac{d}{0.7788\, r} = 2 * 10^{-7} log_e \frac{629}{0.7788 * 2} = 1.2 * 10^{-6}\, H/m$$

Inductance per Km = $1.2 * 10^{-6} * 1000 = 1.2\, mH$

Example 2.5. Figure depicts the arrangement of a double circuit, three-phase overhead wire with specific spacings. The line has been completely rearranged and adheres to the ABC phase sequence. The conductor has a radius of 1.5cm. Determine the inductance per kilometre for each phase.

Ans: GMR of conductor $D_{aa} = D_{a'a'} = 0 \cdot 7788\, r = 0.7788 * 1.5 = 1.168\, cm = 1.168 * 10^{-2} m$

$$D_{ab'} = \sqrt{7.5^2 + 2.5^2} = 7.905\, m\, ;\; D_{aa'} = \sqrt{7.5^2 + 5^2} = 9.013\, m$$

$$D_{s1} = \sqrt[4]{D_{aa} * D_{aa'} * D_{a'a} * D_{a'a'}} = \sqrt[4]{(1.168 * 10^{-2}) * 9.013 * 9.013 * (1.168 * 10^{-2})} = 0.323\, m$$

$$D_{s2} = \sqrt[4]{D_{bb} * D_{bb'} * D_{b'b} * D_{b'b'}} = \sqrt[4]{(1.168 * 10^{-2}) * 7.5 * 7.5 * (1.168 * 10^{-2})} = 0.295\, m$$

$$D_{s3} = \sqrt[4]{D_{cc} * D_{cc'} * D_{c'c} * D_{c'c'}} = \sqrt[4]{(1.168 * 10^{-2}) * 9.013 * 9.013 * (1.168 * 10^{-2})} = 0.323\, m$$

$$D_s = \sqrt[3]{D_{s1} D_{s2} D_{s3}} = \sqrt[3]{0.323 * 0.295 * 0.323} = 0.3133\, m$$

$$D_{AB} = \sqrt[4]{D_{ab} * D_{ab'} * D_{a'b} * D_{a'b'}} = \sqrt[4]{2.5 * 7.905 * 7.905 * 2.5} = 4.445\, m$$

$$D_{BC} = \sqrt[4]{D_{bc} * D_{bc'} * D_{b'c} * D_{b'c'}} = \sqrt[4]{2.5 * 7.905 * 7.905 * 2.5} = 4.445 m$$

$$D_{CA} = \sqrt[4]{D_{ca} * D_{ca'} * D_{c'a} * D_{c'a'}} = \sqrt[4]{5 * 7.5 * 7.5 * 5} = 6.12\, m$$

$$D_m = \sqrt[3]{D_{AB} D_{BC} D_{CA}} = \sqrt[3]{4.445 * 4.445 * 6.12} = 4.94\, m$$

Inductance/conductor/metre of a double-circuit three-phase line is

$$= 2 * 10^{-7}\, log_e \frac{\sqrt[3]{D_{AB} D_{BC} D_{CA}}}{\sqrt[3]{D_{s1} D_{s2} D_{s3}}} = 2 * 10^{-7}\, log_e \frac{4.94}{0.3133} = 0.5516 * 10^{-6}\, H/m$$

Inductance/km=$0.5516 * 10^{-6} * 1000 = 0.5516\, mH$

Example 2.6. Calculate the inductance per phase per kilometre of the double circuit three-phase line shown in the figure. The conductors are transposed and each possess a radius of 1.2cm. The pattern of phases is ABC.

Ans: GMR of conductor $D_{aa} = D_{a'a'} = 0 \cdot 7788\, r = 0.7788 * 1.2 = 0.9345 cm = 0.9345 * 10^{-2} m$

$$D_{ab} = \sqrt{2^2 + 1^2} = 2.236\, m\, ;$$

$$D_{ab'} = \sqrt{2^2 + 7^2} = 7.28\, m \;\; and \;\; D_{aa'} = \sqrt{4^2 + 6^2} = 7.211\, m$$

$$D_{s1} = \sqrt[4]{D_{aa} * D_{aa'} * D_{a'a} * D_{a'a'}}$$

$$= \sqrt[4]{(0.9345 * 10^{-2}) * 7.211 * 7.211 * (0.9345 * 10^{-2})} = 0.259\, m$$

$$D_{s2} = \sqrt[4]{D_{bb} * D_{bb'} * D_{b'b} * D_{b'b'}}$$

$$= \sqrt[4]{(0.9345 * 10^{-2}) * 8 * 8 * (0.9345 * 10^{-2})} = 0.2734\, m$$

$$D_{s3} = \sqrt[4]{D_{cc} * D_{cc'} * D_{c'c} * D_{c'c'}}$$

$$= \sqrt[4]{(0.9345 * 10^{-2}) * 7.211 * 7.211 * (0.9345 * 10^{-2})} = 0.259\, m$$

$D_s = \sqrt[3]{D_{s1}D_{s2}D_{s3}} = \sqrt[3]{0.259 * 0.2734 * 0.259} = 0.2637\ m$

$D_{AB} = \sqrt[4]{D_{ab} * D_{ab'} * D_{a'b} * D_{a'b'}} = \sqrt[4]{2.236 * 7.28 * 7.28 * 2.236} = 4.0346\ m$

$D_{BC} = \sqrt[4]{D_{bc} * D_{bc'} * D_{b'c} * D_{b'c'}} = \sqrt[4]{2.236 * 7.28 * 7.28 * 2.236} = 4.0346\ m$

$D_{CA} = \sqrt[4]{D_{ca} * D_{ca'} * D_{c'a} * D_{c'a'}} = \sqrt[4]{4 * 6 * 6 * 4} = 4.89\ m$

$D_m = \sqrt[3]{D_{AB}D_{BC}D_{CA}} = \sqrt[3]{4.0346 * 4.0346 * 4.89} = 4.3\ m$

Inductance/conductor/metre of a double-circuit three-phase line is $L = 2 * 10^{-7}\ log_e \frac{D_m}{D_s}$

$$= 2 * 10^{-7}\ log_e \frac{\sqrt[3]{D_{AB}D_{BC}D_{CA}}}{\sqrt[3]{D_{s1}D_{s2}D_{s3}}} = 2 * 10^{-7}\ log_e \frac{4.3}{0.2637} = 0.5583 * 10^{-6}\ H/m$$

Inductance/km$=0.5583 * 10^{-6} * 1000 = 0.5583\ mH$

Example 2.7. In a single-phase line (see Figure), the return path is created by the parallel conductors b and b', while the parallel conductors a and a' form go conductor. Calculate, the total inductance per km of the line, assuming the current to be uniformly distributed between the two parallel conductors. The conductor has a diameter of 1.3 cm.

Ans: $D_{aa} = D_{a'a'} = 0 \cdot 7788\ r = 0.7788 * 1.3 = 1.0124cm = 1.0124 * 10^{-2}m$

$D_s = \sqrt[4]{D_{aa}D_{aa'}D_{a'a}D_{a'a'}} = \sqrt[4]{(1.0124 * 10^{-2}) * (15 * 10^{-2}) * (15 * 10^{-2}) * (1.0124 * 10^{-2})} = 0.0389\ m$

$D_m = \sqrt[4]{D_{ab} * D_{ab'} * D_{a'b} * D_{a'b'}} = \sqrt[4]{1.15 * 1.3 * 1 * 1.15} = 1.145\ m$

Loop Inductance $L = 0.4\ log_e \frac{D_m}{D_s} = 0.4\ log_e \frac{1.145}{0.0389} = 1.295\ mH$

2.4 Capacitance of a Transmission Line:

It is widely accepted that the presence of an insulating substance between two conductors leads to the formation of a capacitor. The presence of air as an insulator results in capacitance between any two conductors of an overhead transmission line. The capacitance between the conductors is defined as the quotient of the charge stored on one of the conductors divided by the potential difference across them.

Capacitance, $C = \frac{q}{V}$ farad

Where q = Charge on the line in coulomb and
V = Potential difference between the conductors in volts

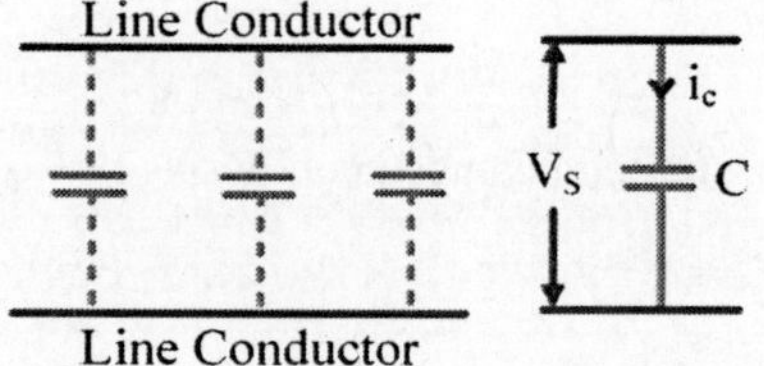

Fig. 2.12 Representation of capacitance in line

In Fig. 2.12, the capacitance is distributed uniformly throughout the line, similar to a series of capacitors connected by wires. When an alternating voltage is provided to a transmission line, the charge on the wires varies in response to the voltage between the conductors at that place. A charging current flows between the conductors. Even when the line is open-circuited and there is no load, charging current flows. It affects the voltage drop down the line, as well as the efficiency and power factor.

2.4.1 Electric Potential: The electric potential at a point is the amount of work required to move a unit positive charge from infinity to that location. Understanding electric potential is crucial for calculating capacitance in a circuit, which is defined as charge per unit potential.

2.4.1.1 Potential at a charged single conductor:

Consider a long, straight cylindrical conductor with a radius of r metres (see Fig. 2.13). Allow the conductor to function at a potential (V_A) sufficient to generate a charge of Q_A coulombs per metre. It is desired to determine the phrase for V_A.

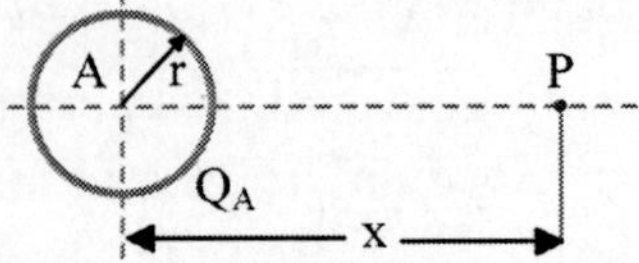

Fig. 2.13 Potential at a charged single conductor

The electric intensity E in air at a distance x from the centre of the conductor can be expressed as:

$$E = \frac{Q_A}{2\pi\varepsilon_0 x} \quad ......(2.40)$$

where Q_A = charge per metre length and ε_0 = permittivity of free space

As x approaches infinity, E approaches zero. The potential difference between conductor A and the distant neutral plane is represented as:

$$V_A = \int_r^{\infty} \frac{Q_A}{2\pi\varepsilon_0 x} dx = \frac{Q_A}{2\pi\varepsilon_0} \int_r^{\infty} \frac{dx}{x} \quad ......(2.41)$$

2.4.1.2 Potential at a conductor in a group of charged conductors: Consider a group of long straight conductors (A, B, C) running at potentials resulting in charges (Q_A, Q_B, Q_C) per metre length (see Fig. 2.14). Let us find the potential at *A* (*i.e.* V_A) in this arrangement.

Potential at *A* due to its own charge (*i.e.* Q_A) is $= \int_r^{\infty} \frac{Q_A}{2\pi\varepsilon_0 x} dx$

Potential at conductor *A* due to charge Q_B is $= \int_{d_1}^{\infty} \frac{Q_B}{2\pi\varepsilon_0 x} dx$

Potential at conductor *A* due to charge Q_C is $= \int_{d_2}^{\infty} \frac{Q_C}{2\pi\varepsilon_0 x} dx$

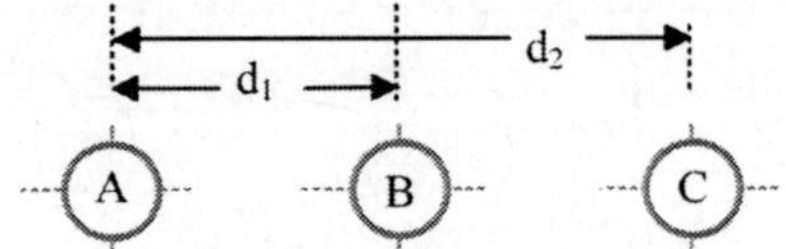

Fig. 2.14 Potential at a conductor in a group of charged conductors

Overall potential difference between conductor *A* and infinite neutral plane is

$$V_A = \int_r^{\infty} \frac{Q_A}{2\pi\varepsilon_0 x} dx + \int_{d_1}^{\infty} \frac{Q_B}{2\pi\varepsilon_0 x} dx + \int_{d_2}^{\infty} \frac{Q_C}{2\pi\varepsilon_0 x} dx$$

$$= \frac{1}{2\pi\varepsilon_0}[Q_A(\log_e\infty - \log_e r) + Q_B(\log_e\infty - \log_e d_1) + Q_c(\log_e\infty - \log_e d_2)]$$

$$= \frac{1}{2\pi\varepsilon_0}[-Q_A log_e r - Q_B log_e d_1 - Q_C log_e d_2 + log_e\infty(Q_A + Q_B + Q_c)]$$

$$= \frac{1}{2\pi\varepsilon_0}\left[Q_A log_e \frac{1}{r} + Q_B log_e \frac{1}{d_1} + Q_C log_e \frac{1}{d_2} + log_e\infty(Q_A + Q_B + Q_c)\right]$$

Assuming balanced conditions i.e., $Q_A + Q_B + Q_c = 0$, we have,

$$V_A = \frac{1}{2\pi\varepsilon_0}\left[Q_A log_e \frac{1}{r} + Q_B log_e \frac{1}{d_1} + Q_C log_e \frac{1}{d_2}\right] \quad ......(2.42)$$

2.4.2 Capacitance of a Single Phase Two-wire Line:

Let's examine a single-phase overhead transmission line consisting of two parallel wires, A and B, that are separated by a distance of ‘d’ meters in the air. Let's assume that each conductor has a radius of r meters. Consider charges of +Q and -Q coulombs per metre of length. The total potential difference between conductor A and the neutral "infinite" plane is

$$V_A = \int_r^{\infty} \frac{Q}{2\pi\varepsilon_0 x} dx + \int_d^{\infty} \frac{-Q}{2\pi\varepsilon_0 x} dx$$

$$= \frac{1}{2\pi\varepsilon_0}[Q(log_e\infty - log_e r) - Q(log_e\infty - log_e d)]$$

$$= \frac{Q}{2\pi\varepsilon_0}[log_e\infty - log_e r - log_e\infty + log_e d]$$

$$= \frac{Q}{2\pi\varepsilon_0}[log_e d - log_e r]$$

$$= \frac{Q}{2\pi\varepsilon_0}\left[log_e \frac{d}{r}\right] \text{ volts} \qquad ......(2.43)$$

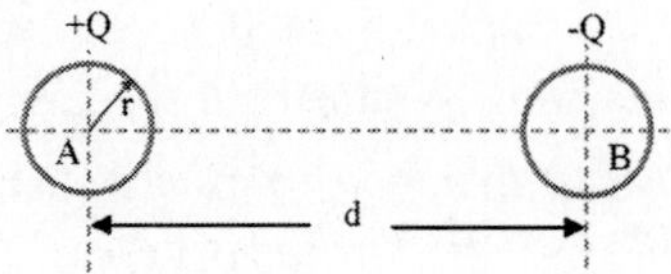

Fig. 2.15 Single Phase Two-wire Line

Therefore, capacitance C_{AN} between conductor A and neutral plane is

$$C_{AN} = \frac{Q}{V_A} = \frac{Q}{\frac{Q}{2\pi\varepsilon_0}\left[log_e\frac{d}{r}\right]} = \frac{2\pi\varepsilon_0}{log_e\frac{d}{r}} \text{ F/m} \qquad ......(2.44)$$

Similarly, the potential difference between conductor B and the neutral "infinite" plane is

$$V_B = \int_r^{\infty} \frac{-Q}{2\pi\varepsilon_0 x} dx + \int_d^{\infty} \frac{Q}{2\pi\varepsilon_0 x} dx$$

$$= \frac{1}{2\pi\varepsilon_0}[-Q(log_e\infty - log_e r) + Q(log_e\infty - log_e d)]$$

$$= \frac{-Q}{2\pi\varepsilon_0}[log_e\infty - log_e r - log_e\infty + log_e d]$$

$$= \frac{-Q}{2\pi\varepsilon_0}[log_e d - log_e r]$$

$$= \frac{-Q}{2\pi\varepsilon_0}\left[log_e \frac{d}{r}\right] \text{ volts} \qquad ......(2.45)$$

Therefore, capacitance C_{BN} between conductor B and neutral plane is

$$C_{BN} = \frac{-Q}{V_B} = \frac{-Q}{\frac{-Q}{2\pi\varepsilon_0}\left[log_e\frac{d}{r}\right]} = \frac{2\pi\varepsilon_0}{log_e\frac{d}{r}} \text{ F/m} \qquad ......(2.46)$$

Both of these potentials point to the same neutral plane. Because the opposite charges attract each other, the potential difference between the conductors

p.d. b/w the conductors $V_{AB} = 2V_A = \frac{2Q}{2\pi\varepsilon_0}\left[log_e \frac{d}{r}\right]$ volts ......(2.47)

Capacitance $C_{AB} = \frac{Q}{V_{AB}} = \frac{Q}{\frac{2Q}{2\pi\varepsilon_0}\left[log_e\frac{d}{r}\right]} = \frac{\pi\varepsilon_0}{log_e\frac{d}{r}}$ F/m ......(2.48)

The capacitance C_{AN} and C_{BN} are connected in series resulting in capacitance between two conductors (i.e. C_{AB}) to be half the capacitance between each conductor and neutral as illustrated in Fig. 2.16.

Fig. 2.16 Single Phase Two-wire Line

∴ Capacitance to neutral, $C_N = C_{AN} = C_{BN} = 2C_{AB}$

Capacitance $C_N = \frac{2\pi\varepsilon_0}{log_e\frac{d}{r}}$ F/m ......(2.49)

2.4.3 Capacitance of a 3-Phase Overhead Line: In a 3-phase transmission line, there are two possible scenarios: symmetrical spacing and unsymmetrical spacing.

2.4.3.1 Symmetrical Spacing: Figure 2.17 depicts the three conductors A, B, and C of the 3-phase overhead transmission line, with a charge of Q_A, Q_B, and Q_C per metre length. Let the conductors be equidistant (d metres) apart. In this symmetrically spaced line, we will determine the capacitance between the line conductor and the neutral. In Fig. 2.17, the potential difference between conductor A and the infinite neutral plane is given by equation (2.41) (see to section 2.4.1). The three conductors A, B, and C of the three-phase overhead transmission line have charges of Q_A, Q_B, and Q_C per metre length. Fig. 2.17 shows that the total potential difference between conductor A and the infinite neutral plane is

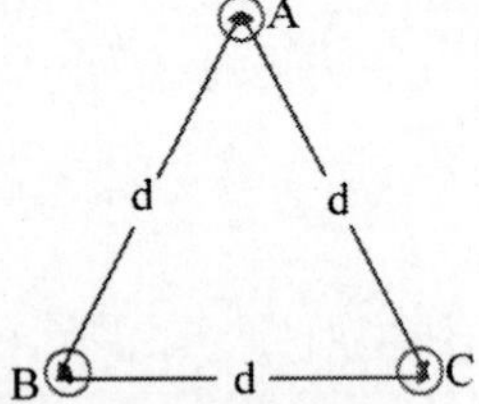

Fig. 2.17 Diagonally spaced conductors

$$V_A = \int_r^{\infty} \frac{Q_A}{2\pi\varepsilon_0 x} dx + \int_d^{\infty} \frac{Q_B}{2\pi\varepsilon_0 x} dx + \int_d^{\infty} \frac{Q_C}{2\pi\varepsilon_0 x} dx$$

$$= \frac{1}{2\pi\varepsilon_0}[Q_A(\log_e\infty - \log_e r) + Q_B(\log_e\infty - \log_e d) + Q_c(\log_e\infty - \log_e d)]$$

$$= \frac{1}{2\pi\varepsilon_0}[-Q_A\log_e r - Q_B\log_e d - Q_C\log_e d + \log_e\infty(Q_A + Q_B + Q_c)]$$

Assuming balanced conditions i.e., $Q_A + Q_B + Q_c = 0$, then $Q_B + Q_c = -Q_A$

$$V_A = \frac{1}{2\pi\varepsilon_0}[-Q_A\log_e r - (Q_B + Q_c)\log_e d]$$

$$= \frac{1}{2\pi\varepsilon_0}[-Q_A\log_e r + Q_A\log_e d]$$

$$= \frac{Q_A}{2\pi\varepsilon_0}\left[\log_e\frac{d}{r}\right] \text{ volts}$$

Capacitance of conductor A w.r.t neutral, $C_A = \frac{Q_A}{V_A} = \frac{Q_A}{\frac{Q_A}{2\pi\varepsilon_0}\left[\log_e\frac{d}{r}\right]} = \frac{2\pi\varepsilon_0}{\log_e\frac{d}{r}}$ F/m ……(2.50)

Note that this equation is the same as capacitance to neutral for a two-wire connection. The capacitance formulas for conductors B and C are derived similarly.

2.4.3.2 Unsymmetrical spacing: Fig. 2.18 depicts a three-phase transposed line with unsymmetrical spacing. Assume balanced conditions ($Q_A + Q_B + Q_C = 0$). Considering all the three sections of the transposed line for phase *A*,

Potential of conductor A at 1st position, $V_1 = \frac{1}{2\pi\varepsilon_0}\left[Q_A\log_e\frac{1}{r} + Q_B\log_e\frac{1}{d_3} + Q_C\log_e\frac{1}{d_2}\right]$

Potential of conductor A at 2nd position, $V_2 = \frac{1}{2\pi\varepsilon_0}\left[Q_A\log_e\frac{1}{r} + Q_B\log_e\frac{1}{d_1} + Q_C\log_e\frac{1}{d_3}\right]$

Potential of conductor A at 3rd position, $V_3 = \frac{1}{2\pi\varepsilon_0}\left[Q_A\log_e\frac{1}{r} + Q_B\log_e\frac{1}{d_2} + Q_C\log_e\frac{1}{d_1}\right]$

Average voltage on conductor *A* is $V_A = \frac{1}{3}(V_1 + V_2 + V_3)$

$$= \frac{1}{3*2\pi\varepsilon_0}\left\{\begin{aligned}&\left[Q_A\log_e\frac{1}{r} + Q_B\log_e\frac{1}{d_3} + Q_C\log_e\frac{1}{d_2}\right] + \left[Q_A\log_e\frac{1}{r} + Q_B\log_e\frac{1}{d_1} + Q_C\log_e\frac{1}{d_3}\right]\\ &+ \left[Q_A\log_e\frac{1}{r} + Q_B\log_e\frac{1}{d_2} + Q_C\log_e\frac{1}{d_1}\right]\end{aligned}\right\}$$

$$= \frac{1}{3*2\pi\varepsilon_0}\left[Q_A\log_e\frac{1}{r^3} + Q_B\log_e\frac{1}{d_1d_2d_3} + Q_C\log_e\frac{1}{d_1d_2d_3}\right]$$

$$= \frac{1}{3*2\pi\varepsilon_0}\left[Q_A log_e \frac{1}{r^3} + (Q_B + Q_C) log_e \frac{1}{d_1 d_2 d_3}\right]$$

Assuming balanced conditions i.e., $Q_A + Q_B + Q_c = 0$, then

$Q_B + Q_c = -Q_A$

$$V_A = \frac{1}{3*2\pi\varepsilon_0}\left[Q_A log_e \frac{1}{r^3} - Q_A log_e \frac{1}{d_1 d_2 d_3}\right]$$

$$= \frac{Q_A}{3*2\pi\varepsilon_0}\left[log_e \frac{d_1 d_2 d_3}{r^3}\right]$$

$$= \frac{Q_A}{2\pi\varepsilon_0}\left(log_e \frac{d_1 d_2 d_3}{r^3}\right)^{1/3} = \frac{Q_A}{2\pi\varepsilon_0}\left[log_e \frac{\sqrt[3]{d_1 d_2 d_3}}{r}\right]$$

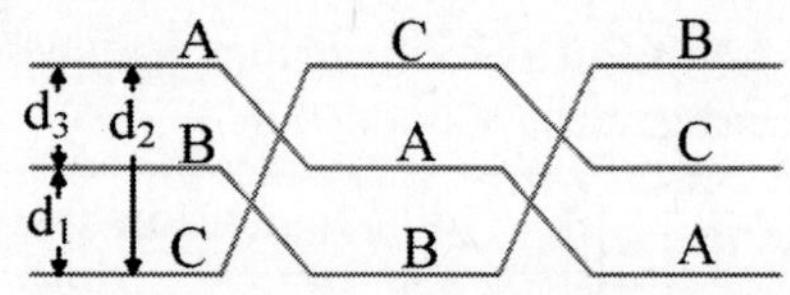

Fig. 2.18 Transposed transmission line

Capacitance of conductor A w.r.t neutral, $C_A = \frac{Q_A}{V_A} = \frac{Q_A}{\frac{Q_A}{2\pi\varepsilon_0}\left[log_e \frac{\sqrt[3]{d_1 d_2 d_3}}{r}\right]} = \frac{2\pi\varepsilon_0}{log_e \frac{\sqrt[3]{d_1 d_2 d_3}}{r}}$ F/m ......(2.51)

Example 2.8. The conductors of a three-phase overhead transmission line are placed at the corners of an equilateral triangle with a side length of three metres. Determine each line conductor's capacitance per kilometre. Considering that each conductor has a diameter of 1.5cm.

Ans: Capacitance $C = \frac{2\pi\varepsilon_0}{log_e \frac{d}{r}} = \frac{2\pi*8.854*10^{-12}}{log_e \frac{300}{0.75}} = 9.285 * 10^{-12}$ F/m $= 0.009285$ μF/km

Example 2.9. The conductor diameter of a 3-phase, 50 Hz, 110 kV overhead line is 16 cm, and it is arranged in a vertical plane as depicted in Fig. If the line length is 120 km, compute (i) capacitance per phase and (ii) charging current per phase, assuming full line transposition.

Ans: $D = \sqrt[3]{D_1 D_2 D_3} = \sqrt[3]{2.5 * 2.5 * 5} = 3.15\ m$

Capacitance per km $C = \frac{2\pi\varepsilon_0}{log_e \frac{d}{r}} = \frac{2\pi*8.854*10^{-12}}{log_e \frac{315}{0.8}}$

$= 9.309 * 10^{-12}$ F/m $= 0.009309$ μF/km

Total capacitance $C = 0.009309 * 120 = 1.117$ μF

Charging current $I_c = \frac{V_{ph}}{X_C} = V_{ph} * 2\pi f C$

$$= \frac{110000}{\sqrt{3}} * 2\pi * 50 * 1.117 * 10^{-6} = 22.28A$$

2.5 Classification of Overhead Transmission Lines:

A transmission line possesses three uniformly distributed constants (R, L, and C) over its length. The combination of resistance and inductance forms the series impedance. The presence of capacitance between conductors in a single-phase line or between a conductor and neutral in a three-phase line results in the formation of a shunt path down the line. Transmission line calculations are complicated by the presence of capacitance effects. Overhead transmission lines are categorised according to their capacitance, as outlined below:

(i). Short transmission lines

(ii). Medium transmission lines and

(iii). Long transmission lines.

When analysing the performance of a transmission line, it is important to calculate its voltage regulation and transmission efficiency.

Voltage regulation: The series parameters R & L leads to voltage drop when line is loaded at receiving end, whereas, shunt parameter C is responsible for voltage rise. The voltage regulation of a line represents amount of voltage rise at the receiving end when load is thrown out. Mathematically percentage voltage regulation of a transmission line is given by:

$$\% \text{ Voltage regulation} = \frac{V_{R,NL}-V_{R,FL}}{V_{R,FL}} * 100 \qquad .....(2.52)$$

where, $V_{R,NL}$ = Receiving end voltage at no load

$V_{R,FL}$ = Receiving end voltage at full load

It is desirable to have low voltage regulation so that available voltage to consumers doesn't get much affected as a result of switching. A high voltage regulation leads to significant voltage drop in the line under load resulting in low voltage at consumer end.

Transmission efficiency: A transmission line's resistance results in copper loss that reduces the power delivered to the load connected at receiving end. The ratio of power received at the receiving end of the line to the power transmitted at the sending end is known as transmission efficiency.

$$\% \text{ Transmission efficiency} = \frac{\text{Receiving end power}}{\text{Sending end power}} * 100 = \frac{V_R I_R \cos\phi_R}{V_S I_S \cos\phi_S} * 100 \qquad .....(2.53)$$

where V_R, I_R and $\cos\phi_R$ are the receiving end voltage, current and power factor while V_S, I_S and $\cos\phi_S$ are the corresponding values at the sending end.

2.6 Short transmission lines:

A transmission line is generally considered short if it is 50 kilometres or less long and carries a low voltage, often less than 20 kV. However, amount of capacitance present decides the actual length of the line to be considered short. Because of their shorter length and lower voltage, the effects of capacitance are small and may thus be ignored. As a result, while calculating the efficiency and regulation of a short transmission line, only the line's resistance and inductance are taken into account.

The impact of line capacitance is ignored for a short transmission line. Thus, when analysing the performance of such a line, only the resistance and inductance of the line are considered. Fig. 2.19(a) displays the schematic representation of the equivalent circuit for a single-phase short transmission line. Here, the total line resistance and inductance are represented as concentrated or lumped rather than being spread out.

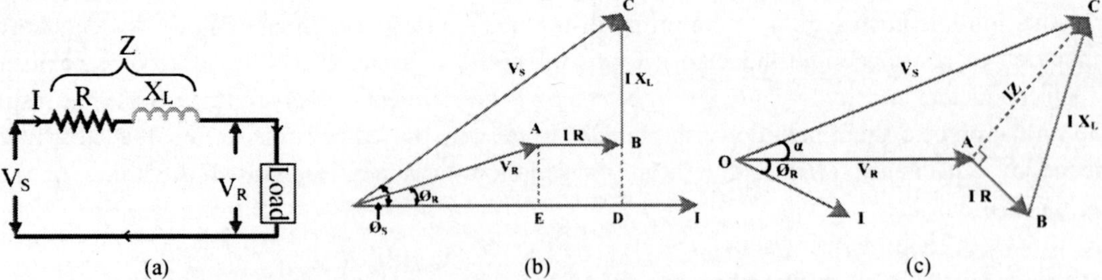

Fig. 2.19 (a) Equivalent circuit of short transmission line (b) and (c) Phasor diagrams

Figure 2.19(b) shows the phasor diagram of a short transmission line with current as the reference, whereas Figure 2.19(c) shows the phasor diagram of a short transmission line with voltage as the reference. Current 'I' has been chosen as the reference phasor. The receiving end phase voltage (V_R) is represented by OA and led by φ_R to 'I'. AB shows the drop IR in phase with the 'I'. BC represents the inductive drop IX_L and leads 'I' by 90°. OC represents the sending end phase voltage (V_S) and leads 'I' by φ_S.

$$V_S^2 = (V_R \cos\emptyset_R + IR)^2 + (V_R \sin\emptyset_R + IX_L)^2$$

$$V_S = \sqrt{(V_R \cos\emptyset_R + IR)^2 + (V_R \sin\emptyset_R + IX_L)^2} \quad \text{......(2.54)}$$

Where R = Resistance per phase
X_L = Reactance per phase
V_R = Receiving end phase voltage
V_S = Sending end phase voltage
$\cos\varphi_R$ = receiving end power factor (lagging)
$\cos\varphi_S$ = sending end power factor

Complex notation is often used to perform line calculations. It is evident that $\vec{V_S}$ is the phasor sum of $\vec{V_R}$ and $\vec{I}\,\vec{Z}$.

$$\vec{V_R} = V_R + j0$$
$$\vec{I} = I\angle{-\emptyset_R} = I(\cos\emptyset_R - j\sin\emptyset_R)$$
$$\vec{Z} = R + jX_L$$
$$\vec{V_S} = \vec{V_R} + \vec{I}\,\vec{Z}$$
$$= V_R + I(\cos\emptyset_R - j\sin\emptyset_R)(R + jX_L)$$
$$= V_R + IR\cos\emptyset_R + IX_L\sin\emptyset_R - j(IX_L\cos\emptyset_R - IR\sin\emptyset_R)$$
$$V_S = \sqrt{(V_R + IR\cos\emptyset_R + IX_L\sin\emptyset_R)^2 + (IX_L\cos\emptyset_R - IR\sin\emptyset_R)^2}$$

The second term under the root is relatively small and can be ignored. Hence, the approximate expression for V_S becomes:

Sending end voltage, $$V_S = V_R + IR\cos\emptyset_R + IX_L\sin\emptyset_R \quad \text{......(2.55)}$$

As capacitance is neglected, no load voltage at the receiving end is equal to sending end voltage.

% Voltage regulation, $$\%V_{reg} = \frac{V_S - V_R}{V_R} * 100$$

$$= \frac{IR\cos\emptyset_R + IX_L\sin\emptyset_R}{V_R} * 100 \quad \text{(for lagging p.f.)}$$

$$= \frac{IR\cos\emptyset_R - IX_L\sin\emptyset_R}{V_R} * 100 \quad \text{(for leading p.f.)} \quad \text{......(2.56)}$$

The expressions above can lead to the following conclusions:

- Positive voltage regulation occurs when the load p.f. is lagging or unity. This means that the receiving end voltage V_R is less than the sending end voltage V_S.
- Negative voltage regulation occurs when the load p.f. is leading and $IR\cos\emptyset_R < IX_L\sin\emptyset_R$. This means that the receiving end voltage V_R is more than the sending end voltage V_S.

The power factor determines the amount of power provided to the load.

For a single-phase line: $$P_R = V_R I_R \cos\emptyset_R \quad \text{......(2.57)}$$

$$I_R = \frac{P_R}{V_R\cos\emptyset_R} \quad \text{......(2.58)}$$

For a three-phase line: $$P_R = 3\, V_R I_R \cos\emptyset_R \quad \text{......(2.59)}$$

$$I_R = \frac{P_R}{3\, V_R\cos\emptyset_R} \quad \text{......(2.60)}$$

% Efficiency, $\eta = \frac{Receiving\ end\ power}{Sending\ end\ power} * 100 = \frac{V_R I_R \cos\emptyset_R}{V_S I_S \cos\emptyset_S} * 100 = \frac{V_R I_R \cos\emptyset_R}{V_R I_R \cos\emptyset_R + I^2 R} * 100$ ......(2.61)

For a given power transmission (P) and receiving end voltage (V_R), the load current (I) is inversely proportional to the load p.f. $\cos\emptyset_R$. As load p.f. decreases, the load current increases, leading to higher line losses. This suggests that a line's transmission efficiency decreases with lower load power factor, and vice versa.

2.7 Medium transmission lines:

A medium transmission line is one that is between 50 and 150 km long and transmits a moderately high voltage, precisely between 20 and 100 kV. The capacitance effects are taken into account due to the line's adequate length and voltage. The capacitance is equally spread along the entire length of the line. However, for the sake of ease in computation, it is assumed that the line capacitance is concentrated in the form of capacitors connected across the line at one or more specific locations. The method of localising line capacitance produces reasonably precise results. The end condenser method, nominal-T method, and nominal-π method are the most commonly used techniques for solving medium transmission lines, sometimes known as localised capacitance methods.

2.7.1 End Condenser Method: This approach, as shown in Fig. 2.20(a), lumps or concentrates the capacitance of the line at the receiving or load end. This method of determining line capacitance at the load end may overestimate its significance. The circuit's phasor diagram is depicted in Figure 2.20(b). Using the receiving end voltage as the reference phasor, we get

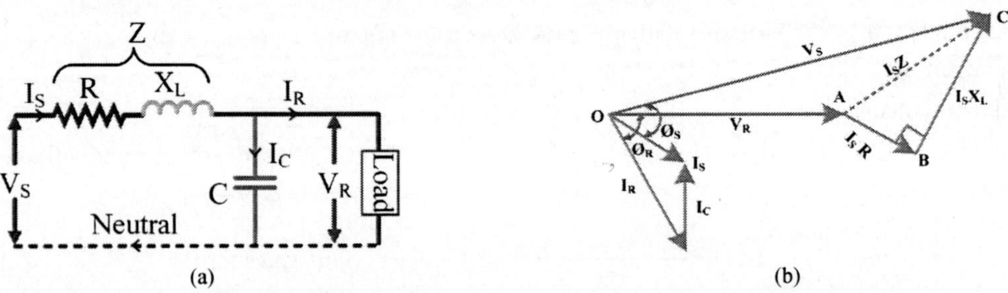

Fig. 2.20 (a) Equivalent circuit of End condenser method, (b) Phasor diagram

Receiving end voltage, $\overrightarrow{V_R} = V_R + j0$

Receiving end current, $\overrightarrow{I_R} = I_R\angle{-\emptyset_R} = I_R(\cos\emptyset_R - j\sin\emptyset_R)$

Capacitive current, $\overrightarrow{I_c} = j\,\overrightarrow{V_R}\,\omega\,C = j2\pi f C\,\overrightarrow{V_R}$ ......(2.62)

Sending end current, $\overrightarrow{I_s} = \overrightarrow{I_R} + \overrightarrow{I_c} = I_R(\cos\emptyset_R - j\sin\emptyset_R) + j2\pi f C\,\overrightarrow{V_R}$ ......(2.63)

Line Impedance/phase, $\vec{Z} = R + jX_L$

Sending end voltage, $\overrightarrow{V_s} = \overrightarrow{V_R} + \overrightarrow{I_s}\,\vec{Z} = \overrightarrow{V_R} + \overrightarrow{I_s}\,(R + jX_L)$ ......(2.64)

While the end condenser approach for medium line solutions is straightforward to calculate, it has several limitations.

- Calculations contain a significant inaccuracy (approximately 10%) since the dispersed capacitance was expected to be lumped or concentrated.
- This strategy overestimates the impact of line capacitance.

2.7.2 Nominal-T Method:

This technique considers that the capacitance of the entire line is concentrated at the midpoint of the line, while half of the line's resistance and reactance are lumped on either side of it, as illustrated in Fig. 2.21(a). Thus, with this configuration, the entire charging current passes through half of the line.

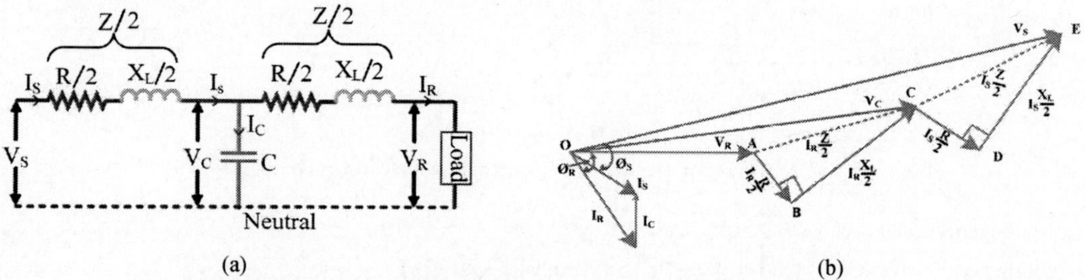

Fig. 2.21 (a) Equivalent circuit of Nominal-T method, (b) Phasor diagram

$\vec{V_R}$ is chosen as the reference phasor represented by OA. The load current ($\vec{I_R}$) lags behind $\vec{V_R}$ by φ_R. The drop AB = $I_R\,\frac{R}{2}$ is in phase with $\vec{I_R}$, while BC = $I_R\,\frac{X_L}{2}$ leads $\vec{I_R}$ by 90°. The phasor OC indicates the voltage $\vec{V_c}$ over condenser C. The capacitor current $\vec{I_c}$ leads $\vec{V_c}$ by 90°, as illustrated in Fig. 2.21(b). The phasor sum of $\vec{I_R}$ and $\vec{I_c}$ yields $\vec{I_s}$. CD = $I_S\,\frac{R}{2}$ is in phase with $\vec{I_s}$, while DE = $I_s\,\frac{X_L}{2}$ leads $\vec{I_s}$ by 90°. Then, OE denotes the sending end voltage $\vec{V_s}$.

Receiving end voltage, $\vec{V_R} = V_R + j0$

Receiving end current, $\vec{I_R} = I_R\angle{-\emptyset_R} = I_R(cos\emptyset_R - jsin\emptyset_R)$

Voltage across C, $\vec{V_c} = \vec{V_R} + \vec{I_R}\frac{\vec{Z}}{2} = V_R + I_R(cos\emptyset_R - jsin\emptyset_R)\,\frac{(R+jX_L)}{2}$ ......(2.65)

Charging current, $\vec{I_c} = j\,\vec{V_c}\,\omega\,C = j2\pi fC\,\vec{V_c}$ ......(2.66)

Sending end current, $\vec{I_s} = \vec{I_R} + \vec{I_c} = I_R(cos\emptyset_R - jsin\emptyset_R) + j2\pi fC\,\vec{V_c}$ ......(2.67)

Sending end voltage, $\vec{V_s} = \vec{V_c} + \vec{I_s}\frac{\vec{Z}}{2} = \vec{V_R} + \vec{I_R}\frac{\vec{Z}}{2} + \vec{I_s}\frac{\vec{Z}}{2}$ ......(2.68)

2.7.3 Nominal-π Method:

In this method, the capacitance of each conductor (specifically, the line to neutral capacitance) is divided into two equal portions. One component is concentrated at the sending end, while the other is at the receiving end, as shown in Fig. 2.22(a). The effect of capacitance at the sending end on line drop is insignificant. Nonetheless, the device's charging current must be coupled with the line current to obtain the total sending end current.

$\vec{V_R}$ is used as the reference phasor, which is represented by OA. IR now lags $\vec{V_R}$ by φ_R. The charging current ($\vec{I_{c1}}$) is 90° ahead of $\vec{V_R}$. $\vec{I_L}$ is the phasor sum of $\vec{I_R}$ and $\vec{I_{c1}}$. The drop AB = $I_L\,R$ is in phase with $\vec{I_L}$, while drop BC = $I_L\,X_L$ is 90° ahead of $\vec{I_L}$. The sending end voltage ($\vec{V_s}$) is therefore represented by OC. $\vec{I_{c2}}$ leads $\vec{V_s}$ by 90°. As a result, sending end current $\vec{I_s}$ is the phasor sum of $\vec{I_{c2}}$ and $\vec{I_L}$. The sending end p.f. cos φ_S is determined by the angle φ_S between sending end voltage ($\vec{V_s}$) and sending end current ($\vec{I_s}$).

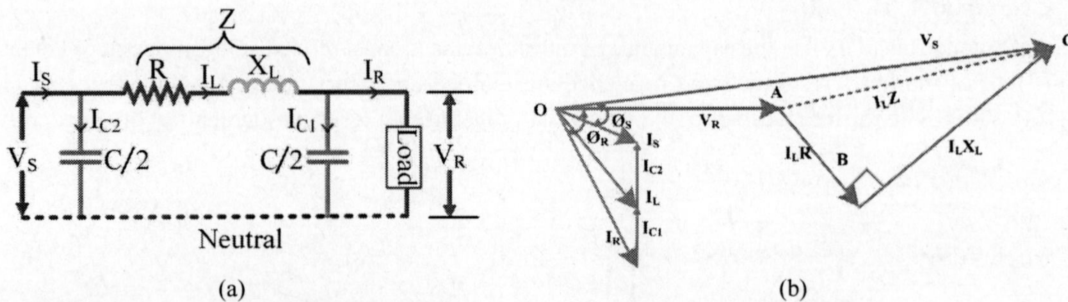

Fig. 2.22 (a) Equivalent circuit of Nominal- π method, (b) Phasor diagram

Receiving end voltage, $\vec{V_R} = V_R + j0$

Receiving end current, $\vec{I_R} = I_R\angle{-Ø_R} = I_R(cosØ_R - jsinØ_R)$

Charging current at load, $\vec{I_{c1}} = j\,\vec{V_R}\,\omega\,\frac{C}{2} = j\pi fC\,\vec{V_R}$ ......(2.69)

Line current, $\vec{I_L} = \vec{I_R} + \vec{I_{c1}} \;= I_R(cosØ_R - jsinØ_R) + j\pi fC\,\vec{V_R}$ ......(2.70)

Sending end voltage, $\vec{V_s} = \vec{V_R} + \vec{I_L}\,\vec{Z} \;= \vec{V_R} + \vec{I_L}\,(R + jX_L)$ ......(2.71)

Charging current at source, $\vec{I_{c2}} = j\,\vec{V_s}\,\omega\,\frac{C}{2} = j\pi fC\,\vec{V_s}$ ......(2.72)

Sending end current, $\vec{I_s} = \vec{I_L} + \vec{I_{c2}}$

$= \vec{I_R} + \vec{I_{c1}} + \vec{I_{c2}}$

$= I_R(cosØ_R - jsinØ_R) + +j\pi fC\,\vec{V_R} + j\pi fC\,\vec{V_s}$ ......(2.73)

2.8 Long transmission lines:

A transmission line is categorised as a long transmission line if its length surpasses 150 km and its voltage exceeds 100 kV. To address such a line, we assume that the line constants are uniformly distributed along its whole length and employ rigorous techniques to determine a solution. Figure 2.23 depicts the arrangement of a single phase and neutral connection in a 3-phase line. The line demonstrates impedance and admittance that are evenly distributed throughout as shown in Fig. 2.23. Consider a small element in the line of length *dx* situated at a distance *x* from the receiving end as shown in Fig. 2.24.

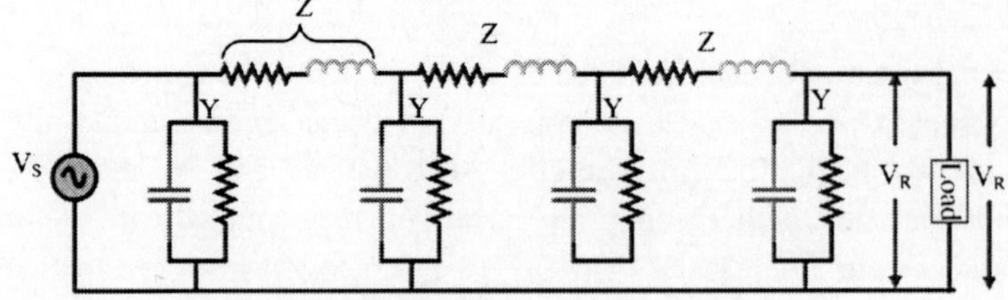

Fig. 2.23 Representation of long transmission line

Let

z = series Impedance per unit length of the line

y = shunt admittance per unit length of the line

V = voltage at the end of the element towards the receiving end

$V + dV$ = voltage at the end of the element towards the sending end

$I + dI$ = current entering the element dx

I = current leaving the element dx

For the small element dx,

zdx = series impedance

ydx = shunt admittance

Clearly, $dV = I\,zdx$

$$\frac{dV}{dx} = I\,z \quad \text{......(2.74)}$$

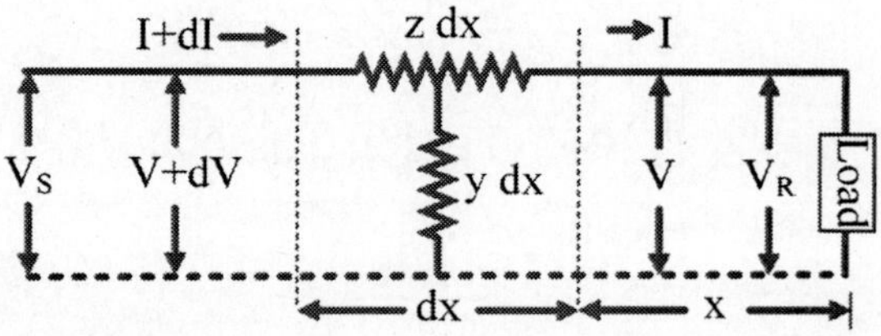

The current entering the element is $I + dI$,
while the current leaving the element is I.

I+dI→ z dx →I Vs V+dV y dx V VR Load dx X

Fig. 2.24 Representation of per unit length in long transmission line

The variation in current is determined by the shunt admittance of the element.

Current through shunt admittance of element $dI = V\,ydx \Rightarrow \frac{dI}{dx} = Vy$ ......(2.75)

Differentiating eq. (*2.74*) w.r.t. *x*, we get, $\frac{d^2V}{dx^2} = \frac{d}{dx}\frac{dV}{dx} = \frac{d}{dx}I\,z = z\,\frac{dI}{dx} = zVy$

Therefore, $\frac{d^2V}{dx^2} = Vyz$ ......(2.76)

The solution of this differential equation is $V = K_1\cosh(x\sqrt{yz}) + K_2\sinh(x\sqrt{yz})$ ......(2.77)

Differentiating exp. (*2.77*) *w.r.t. x*, we have,

$$\frac{dV}{dx} = K_1\sqrt{yz}\sinh(x\sqrt{yz}) + K_2\sqrt{yz}\cosh(x\sqrt{yz})$$

$$I\,z = K_1\sqrt{yz}\sinh(x\sqrt{yz}) + K_2\sqrt{yz}\cosh(x\sqrt{yz})$$

$$I = K_1\sqrt{\frac{y}{z}}\sinh(x\sqrt{yz}) + K_2\sqrt{\frac{y}{z}}\cosh(x\sqrt{yz})$$

$$I = \sqrt{\frac{y}{z}}\left(K_1\sinh(x\sqrt{yz}) + K_2\cosh(x\sqrt{yz})\right) \quad \text{......(2.78)}$$

Equations (2.77) and (2.78) provide the formulas for V and I using the variables K_1 and K_2 as unknown constants. The values of K_1 and K_2 can be determined by applying the specified end conditions:

At $x = 0, V = V_R$ *and* $I = I_R$; substituting these values into equation (2.77) and (2.78), we obtain:

$$V_R = K_1\cosh 0 + K_2\sinh 0$$

$$V_R = K_1 \quad \text{......(2.79)}$$

$$I_R = \sqrt{\frac{y}{z}}(K_1\sinh 0 + K_2\cosh 0)$$

$$I_R = K_2\sqrt{\frac{y}{z}} \Rightarrow K_2 = I_R\sqrt{\frac{z}{y}} \quad \text{......(2.80)}$$

Substituting the values of K_1 and K_2 in eqs. (*2.77*) and (*2.78*), we get,

$$V = V_R\cosh(x\sqrt{yz}) + I_R\sqrt{\frac{z}{y}}\sinh(x\sqrt{yz})$$

$$I = V_R\sqrt{\frac{y}{z}}\sinh(x\sqrt{yz}) + I_R\cosh(x\sqrt{yz})$$

Putting $x = l$ in the above equations yields the sending end voltage (V_S) and sending end current (I_S).

$$V_s = V_R\cosh(l\sqrt{yz}) + I_R\sqrt{\frac{z}{y}}\sinh(l\sqrt{yz}) = V_R\cosh(\sqrt{yl * zl}) + I_R\sqrt{\frac{zl}{yl}}\sinh(\sqrt{yl * zl})$$

$$V_s = V_R \cosh(\sqrt{YZ}) + I_R \sqrt{\frac{Z}{Y}} \sinh(\sqrt{YZ}) \quad \text{......(2.81)}$$

$$I_s = V_R \sqrt{\frac{Y}{Z}} \sinh(\sqrt{YZ}) + I_R \cosh(\sqrt{YZ}) \quad \text{......(2.82)}$$

where Y = total shunt admittance of the line

Z = total series impedance of the line

It is useful to express hyperbolic sine and cosine in terms of their power series.

$$coshx = 1 + \frac{x^2}{2!} + \frac{x^4}{4!} + \frac{x^6}{6!} \ldots\ldots$$

$$sinhx = x + \frac{x^3}{3!} + \frac{x^5}{5!} + \frac{x^7}{7!} \ldots\ldots = x\left(1 + \frac{x^2}{3!} + \frac{x^4}{5!} + \frac{x^6}{7!} \ldots\ldots\right)$$

$$cosh\sqrt{YZ} = 1 + \frac{YZ}{2} + \frac{Y^2Z^2}{24} + \frac{Y^3Z^3}{720} \ldots\ldots$$

$$sinh\sqrt{YZ} = \sqrt{YZ}\left(1 + \frac{YZ}{6} + \frac{Y^2Z^2}{120} + \frac{Y^3Z^3}{5040} \ldots\ldots\right) \quad \text{......(2.83)}$$

2.9 Generalised Circuit Constants of a Transmission Line:

In any network with four terminals, the relationship between input voltage and input current can be expressed in terms of output voltage and output current. A three-phase transmission line's input voltage $\vec{V_s}$ and input current $\vec{I_s}$ can be expressed mathematically as follows:

$$\vec{V_s} = \vec{A}\ \vec{V_R} + \vec{B}\ \vec{I_R} \text{ and } \vec{I_s} = \vec{C}\ \vec{V_R} + \vec{D}\ \vec{I_R} \quad \text{......(2.84)}$$

The transmission line's generalised circuit constants are $\vec{A}$, $\vec{B}$, $\vec{C}$, and $\vec{D}$, which are often complex numbers. The values of these constants depend on the method employed to solve a transmission line. Consider these points:

(i) The constants $\vec{A}$, $\vec{B}$, $\vec{C}$, and $\vec{D}$, are often complex numbers.

(ii) The constants $\vec{A}$ and $\vec{D}$ have no dimensions, while $\vec{B}$ has its unit in ohms and $\vec{C}$ has the dimension of siemens.

(iii) The value of $\vec{A}$ is equal to the value of $\vec{D}$ for a symmetrical circuit.

(iv) The value of $\vec{A}\,\vec{D} - \vec{B}\,\vec{C}$ for a reciprocal circuit is equal to 1.

2.9.1 Generalised Circuit Constants of Short Transmission Line:

Line capacitance is neglected in short transmission lines. Therefore, a short transmission line model represented by a series impedance is symmetrical as well as reciprocal that is quite obvious from Fig. 2.25.

$$\vec{I_s} = \vec{I_R} \text{ and } \vec{V_s} = \vec{V_R} + \vec{I_R}\ \vec{Z} \quad \text{......(2.85)}$$

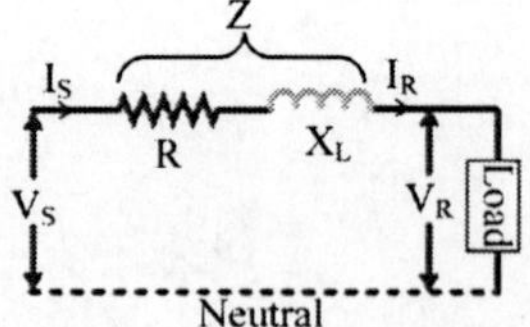

Fig. 2.25 Short Tx. line

From eq. (2.84) and (2.85):

Generalised Circuit Constants of a Short Transmission Line are $\vec{A} = \vec{D} = 1, \vec{B} = \vec{Z}, \text{ and } \vec{C} = 0$

$$\vec{A} = \vec{D}$$

$$\vec{A}\,\vec{D} - \vec{B}\,\vec{C} = 1 * 1 - Z * 0 = 1$$

Therefore, a short transmission line model represented by a series impedance is symmetrical as well as reciprocal that is quite obvious from Fig. 2.25.

2.9.2 Generalised Circuit Constants of Nominal-T Medium Tx. line: The Nominal-T approach considers the capacitance between the entire line and neutral to be concentrated at the line's midpoint. Therefore, Nominal-T representation of medium line represents symmetrical as well as reciprocal circuit as seen from Fig. 2.26.

Voltage across C, $\vec{V_c} = \vec{V_R} + \vec{I_R}\frac{\vec{Z}}{2}$

Charging current, $\vec{I_c} = \vec{V_c}\,Y = \left(\vec{V_R} + \vec{I_R}\frac{\vec{Z}}{2}\right)\vec{Y}$ Sending end current, $\vec{I_s} = \vec{I_R} + \vec{I_c} = \vec{I_R} + \left(\vec{V_R} + \vec{I_R}\frac{\vec{Z}}{2}\right)\vec{Y} = \vec{V_R}\,\vec{Y} + \vec{I_R}\left(1 + \frac{\vec{Y}\,\vec{Z}}{2}\right)$ ......(2.86)

Sending end voltage, $\vec{V_s} = \vec{V_c} + \vec{I_s}\frac{\vec{Z}}{2}$

$$= \vec{V_R} + \vec{I_R}\frac{\vec{Z}}{2} + \vec{I_s}\frac{\vec{Z}}{2}$$

$$= \vec{V_R} + \vec{I_R}\frac{\vec{Z}}{2} + \left(\vec{V_R}\,\vec{Y} + \vec{I_R}\left(1 + \frac{\vec{Y}\,\vec{Z}}{2}\right)\right)\frac{\vec{Z}}{2}$$

$$= \vec{V_R} + \vec{I_R}\frac{\vec{Z}}{2} + \left[\vec{V_R}\,\frac{\vec{Y}\,\vec{Z}}{2} + \vec{I_R}\left(\frac{\vec{Z}}{2} + \frac{\vec{Y}\,\vec{Z}^2}{4}\right)\right]$$

$$= \vec{V_R}\left(1 + \frac{\vec{Y}\,\vec{Z}}{2}\right) + \vec{I_R}\left(\vec{Z} + \frac{\vec{Y}\,\vec{Z}^2}{4}\right)$$

$$= \vec{V_R}\left(1 + \frac{\vec{Y}\,\vec{Z}}{2}\right) + \vec{I_R}\,\vec{Z}\left(1 + \frac{\vec{Y}\,\vec{Z}}{4}\right) \qquad ......(2.87)$$

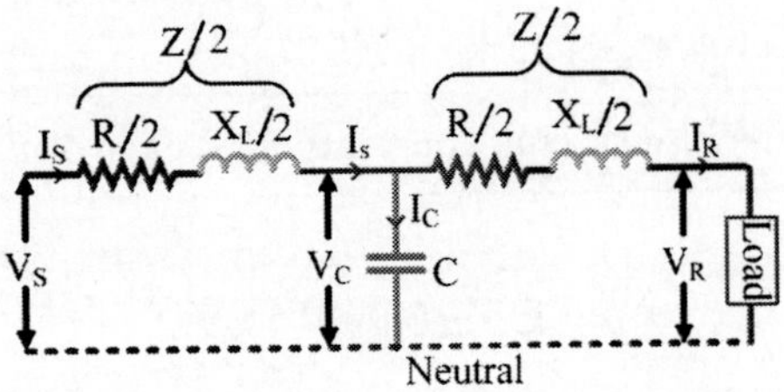

Fig. 2.26 Nominal-T Medium Tx. Line

From eq. (2.84), (2.86) and (2.87):

Generalised Circuit Constants of a Nominal – T Tx. Line are $\vec{A} = \vec{D} = 1 + \frac{\vec{Y}\,\vec{Z}}{2}$, $\vec{B} = \vec{Z}\left(1 + \frac{\vec{Y}\,\vec{Z}}{4}\right)$, and $\vec{C} = \vec{Y}$

$$\vec{A} = \vec{D}$$

$$\vec{A}\,\vec{D} - \vec{B}\,\vec{C} = \left(1 + \frac{\vec{Y}\,\vec{Z}}{2}\right) * \left(1 + \frac{\vec{Y}\,\vec{Z}}{2}\right) - \vec{Z}\left(1 + \frac{\vec{Y}\,\vec{Z}}{4}\right) * \vec{Y}$$

$$= 1 + \frac{\vec{Y}\,\vec{Z}}{2} + \frac{\vec{Y}\,\vec{Z}}{2} + \frac{(\vec{Y}\,\vec{Z})^2}{4} - \vec{Z}\,\vec{Y} - \frac{(\vec{Y}\,\vec{Z})^2}{4}$$

$$= 1$$

Therefore, Nominal-T representation of medium line represents symmetrical as well as reciprocal that is quite obvious from Fig. 2.26.

2.9.3 Generalised Circuit Constants of Nominal- π Medium Tx. line: This approach divides line-to-neutral capacitance into two halves, with one half concentrated at the load end and the other at the sending end. Therefore, Nominal-π representation of medium line represents symmetrical as well as reciprocal circuit as seen from Fig. 2.27.

Charging current at load, $\vec{I_{c1}} = \vec{V_R}\,\frac{\vec{Y}}{2}$

Line current, $\vec{I_L} = \vec{I_R} + \vec{I_{c1}} = \vec{I_R} + \vec{V_R}\,\frac{\vec{Y}}{2}$

Sending end voltage, $\vec{V_s} = \vec{V_R} + \vec{I_L}\,\vec{Z}$

$$= \vec{V_R} + \left(\vec{I_R} + \vec{V_R}\,\frac{\vec{Y}}{2}\right)\vec{Z}$$

$$= \vec{V_R}\left(1 + \frac{\vec{Y}\,\vec{Z}}{2}\right) + \vec{I_R}\,\vec{Z} \qquad ...(2.88)$$

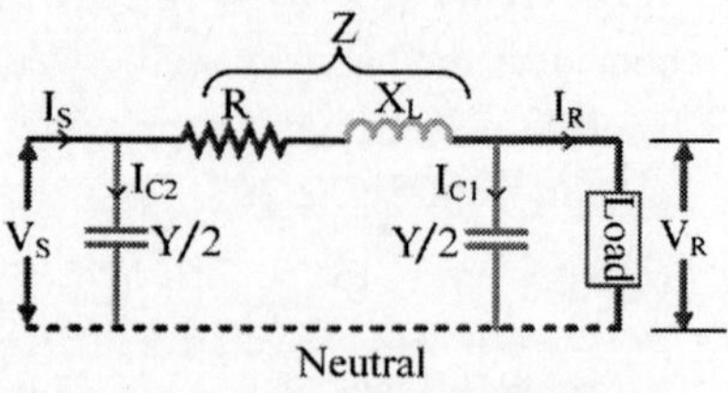

Fig. 2.27 Nominal- π Medium Tx. line

Charging current at source, $\vec{I_{c2}} = \vec{V_s}\,\frac{\vec{Y}}{2}$

$$= \left[\vec{V_R}\left(1+\frac{\vec{Y}\vec{Z}}{2}\right)+\vec{I_R}\,\vec{Z}\right]\frac{\vec{Y}}{2}$$

$$= \vec{V_R}\left(\frac{\vec{Y}}{2}+\frac{\vec{Y}^2\vec{Z}}{4}\right)+\vec{I_R}\,\frac{\vec{Y}\vec{Z}}{2}$$

Sending end current, $\vec{I_s} = \vec{I_L} + \vec{I_{c2}}$

$$= \vec{I_R} + \vec{V_R}\,\frac{\vec{Y}}{2} + \vec{V_R}\left(\frac{\vec{Y}}{2}+\frac{\vec{Y}^2\vec{Z}}{4}\right)+\vec{I_R}\,\frac{\vec{Y}\vec{Z}}{2}$$

$$= \vec{V_R}\left(\vec{Y}+\frac{\vec{Y}^2\vec{Z}}{4}\right)+\vec{I_R}\left(1+\frac{\vec{Y}\vec{Z}}{2}\right)$$

$$= \vec{V_R}\,\vec{Y}\left(1+\frac{\vec{Y}\vec{Z}}{4}\right)+\vec{I_R}\left(1+\frac{\vec{Y}\vec{Z}}{2}\right) \qquad \text{......(2.89)}$$

From eq. (2.82), (2.86) and (2.87) :

Generalised Circuit Constants of Nominal – π Tx. Line are $\vec{A} = \vec{D} = 1+\frac{\vec{Y}\vec{Z}}{2}$, $\vec{B} = \vec{Z}$, and $\vec{C} = \vec{Y}\left(1+\frac{\vec{Y}\vec{Z}}{4}\right)$

$$\vec{A} = \vec{D}$$

$$\vec{A}\,\vec{D} - \vec{B}\,\vec{C} = \left(1+\frac{\vec{Y}\vec{Z}}{2}\right)*\left(1+\frac{\vec{Y}\vec{Z}}{2}\right) - \vec{Z}*\vec{Y}\left(1+\frac{\vec{Y}\vec{Z}}{4}\right) = 1$$

Therefore, Nominal-π representation of medium line represents symmetrical as well as reciprocal that is quite obvious from Fig. 2.27.

2.9.4 Generalised Circuit Constants of a long transmission lines: Long line model is also symmetrical and reciprocal. From eq. (2.81) and (2.82), the sending end voltage and current of a long transmission line are: $V_s = V_R\cosh(\sqrt{YZ}) + I_R\sqrt{\frac{Z}{Y}}\sinh(\sqrt{YZ})$

$$I_s = V_R\sqrt{\frac{Y}{Z}}\,\sinh(\sqrt{YZ}) + I_R\cosh(\sqrt{YZ})$$

From eq. (2.82), (2.79) and (2.80): Generalised Circuit Constants of long Tx. Line are

$\vec{A} = \vec{D} = \cosh(\sqrt{YZ})$, $\vec{B} = \sqrt{\frac{Z}{Y}}\sinh(\sqrt{YZ})$, and $\vec{C} = \sqrt{\frac{Y}{Z}}\,\sinh(\sqrt{YZ})$

$$\vec{A}\,\vec{D} - \vec{B}\,\vec{C} = \cosh(\sqrt{YZ})*\cosh(\sqrt{YZ}) - \sqrt{\frac{Z}{Y}}\sinh(\sqrt{YZ})*\sqrt{\frac{Y}{Z}}\,\sinh(\sqrt{YZ})$$

$$= cosh^2\sqrt{YZ} - sinh^2\sqrt{YZ}$$

$$= 1$$

Therefore, long transmission line model is also symmetrical and reciprocal.

Example 2.10. What is the maximum length in km that a single-phase transmission line with a copper conductor with a cross-section of 085 cm2 can be used to supply 500kW at 0.8 power factor and 6600V? There is a 95% transmission efficiency. Select 1.62μΩ cm as the specific resistance.

Ans: a=0·85cm², P_R=500kW, V=6.6kV

$$\text{Efficiency} = \eta = \frac{P_R}{P_R+losses} \Rightarrow 0.95 = \frac{500*10^3}{500*10^3+losses} \Rightarrow losses = 26.31kW$$

$$I = \frac{P_R}{V_R * cos\,\emptyset_R} = \frac{500*10^3}{6600*0.8} = 94.69\ A$$

Line losses $= 2\ I^2R$

$26.31 * 10^3 = 2 * 94.69^2 * R \quad \Rightarrow R = \frac{26.31*10^3}{2*94.69^2} = 1.467\Omega$

$R = \frac{\rho l}{A} \Rightarrow l = \frac{RA}{\rho} = \frac{1.467*0.85}{1.62*10^{-6}} = 0.769 * 10^6 = 7.69\ km$

Example 2.11 A 3-phase overhead short transmission line can deliver 10MW at 66kV with 0.8 p.f. lagging. The conductors have an 8Ω resistance and a 10Ω reactance each. Calculate the (i) sending end voltage, (ii) percentage regulation, and (iii) transmission efficiency.

Ans: $P_R = 10MW, V_L = 66kV, cos\ \emptyset_R = 0.8\ lag$

$V_{ph} = \frac{66000}{\sqrt{3}} = 38105\ V$

$$I_R = \frac{P_R}{\sqrt{3}\ V_L\ cos\emptyset_R} = \frac{10 * 10^6}{\sqrt{3} * 66 * 10^3 * 0.8} = 109.34\ A$$

$\vec{I_R} = I_R(cos\emptyset_R - jsin\emptyset_R) = 109.34(0.8 - j0.6) = 87.47 - j65.6 = 109.34\angle - 36.86^0$ A

$\vec{V_s} = \vec{V_R} + \vec{I_s}\ (R + jX_L)$

$= 38105 + (87.47 - j65.6)(8 + j10) = 39460.76 + j349.9 = 39462\angle 0.508^0\ V$

$Line\ voltage\ V_L = \sqrt{3} * 39462 = 68.35kV$

In Short transmission line: $A = 1$

Receiving end voltage at no load, $\overrightarrow{V_{R\ NL}} = \frac{\vec{V_s}}{A} = \frac{\vec{V_s}}{1} = 39462\angle 0.508^0\ V$

% of Voltage Regulation, $\%V_{reg} = \frac{V_{R\ NL} - V_{R\ FL}}{V_{R\ FL}} * 100 = \frac{68350 - 66000}{66000} * 100 = 3.56\%$

Line losses $= 3\ I^2R = 3 * 109.34^2 * 8 = 286kW$

Efficiency $= \eta = \frac{P_R}{P_R + losses} = \frac{10 * 10^6}{10 * 10^6 + 286 * 10^3} = 97.21\%$

Example 2.12 Determine the maximum distance at which a 10MW load at 0·9p.f. lagging may be delivered by a 3-phase short transmission line with conductors that each have a resistance of 2Ω per km. It has a 3% transmission loss and 110kV of voltage at the receiving end.

Ans: $P_R = 10MW, V_L = 110kV, cos\ \emptyset_R = 0.9\ lag$

$V_{ph} = \frac{110000}{\sqrt{3}} = 63.508\ kV$

$$I_R = \frac{P_R}{\sqrt{3}\ V_L\ cos\emptyset_R} = \frac{10 * 10^6}{\sqrt{3} * 110 * 10^3 * 0.9} = 58.31\ A$$

Losses $=$ 3% of $P_R = 0.03 * 10 * 10^6 = 300kW$

Line losses $= 3\ I^2R$

$300 * 10^3 = 3 * 58.31^2 * R \quad \Rightarrow R = \frac{300*10^3}{3*58.31^2} = 29.41\Omega$

R = 2 Ω/km, therefore length $= \frac{29.41}{2} = 14.7\ km$

Example 2.13 A short 3-φ transmission line with an impedance of (5+j10) Ω per phase results in sending and receiving end voltages of 132kV and 123kV at 0.9 pf lagging, respectively. Calculate the power output and sending end power factor.

Ans: $Z = (5 + j10)\Omega,\ V_s = 132kV, V_R = 123kV, cos\ \emptyset_R = 0.9\ lag$

$V_{ph}^{s} = \frac{132000}{\sqrt{3}} = 76.21\ kV,\ V_{ph}^{R} = \frac{123000}{\sqrt{3}} = 71.014\ kV$

$V_s = V_R + IR\ cos\emptyset_R + IX_L\ sin\emptyset_R$

$76210 = 71014 + I * 5 * 0.86 + I * 10 * 0.51029$

$5196 = 4.3 * I + 5.102 * I$

$I = \frac{5196}{9.4029} = 552.59\ A$

$P_R = 3\ V_R\ I_R\ cos\emptyset_R = 3 * 71014 * 552.5 * 0.86 = 101.24MW$

$cos\emptyset_s = \frac{V_R\ cos\emptyset_R + I\ R}{V_s} = \frac{71014*0.9+552.59*5}{76210} = 0.837$

Example 2.14 The constants for a 100 km medium single-phase transmission line are as follows: Resistance per km is 0.6Ω, while reactance per km is 1.2Ω. Susceptibility per km = 18×10^{-6} siemens. The receiving end-line voltage is 66,000 volts. Assume that the line's total capacitance is only at the receiving end. Calculate (i) the sending end current, (ii) the sending end voltage, (iii) regulation, and (iv) the supply power factor. The line generates 12,000 kW at a trailing power factor of 0.86. Create a phasor diagram to illustrate your calculations.

Ans: Total resistance, $R = 0{\cdot}6 \times 100 = 60\ \Omega$

Total reactance, $X_L = 1.2 \times 100 = 120\ \Omega$

Total susceptance, $Y = 18 \times 10^{-6} \times 100 = 18 \times 10^{-4}\ S$

$cos\emptyset_R = 0.86\ lag \Rightarrow sin\emptyset_R = 0.5102$

Receiving end voltage, $V_R = 66{,}000$ V

Receiving end current $I_R = \frac{P_R}{V_R\ cos\emptyset_R} = \frac{12*10^6}{66*10^3*0.86} = 211.41\ A$

$\vec{I_R} = I_R(cos\emptyset_R - jsin\emptyset_R) = 211.41(0.86 - j0.5102) = 181.81 - j107.86 = 211.41\angle-30.68^0$ A

$\vec{I_c} = j\,\vec{V_R}\,Y = j66000 * 18 * 10^{-4} = j118.8$

$\vec{I_s} = \vec{I_R} + \vec{I_c} = 181.81 - j107.86 + j118.8 = 181.81 + j10.94 = 182.13\angle3.44^0$ A

$\vec{V_s} = \vec{V_R} + \vec{I_s}\,\vec{Z} = 66000 + (181.81 + j10.94)(60 + j120) = 75595.8 + j22473.6 = 78865\angle16.55^0 V$

In End condenser method: $A = 1 + \vec{Y}\vec{Z}$

Receiving end voltage at no load, $\overrightarrow{V_{R\,NL}} = \frac{\vec{V_s}}{A} = \frac{\vec{V_s}}{1+\vec{Y}\vec{Z}} = \frac{78865\angle16.55^0}{1+(j18*10^{-4})*(60+j120)} = 99652\angle8.70^0\ V$

% of Voltage Regulation, $\%V_{reg} = \frac{V_{R\,NL} - V_{R\,FL}}{V_{R\,FL}} * 100 = \frac{99652 - 66000}{66000} * 100 = 50.98\%$

Line losses $= {I_s}^2 R = 182.13^2 * 60 = 1.99MW$

Efficiency $= \eta = \frac{P_R}{P_R + losses} = \frac{12 * 10^6}{12 * 10^6 + 1.99 * 10^6} = 85.77\%$

Sendind end p.f. $= cos\emptyset_s = cos(\emptyset_{V_s} - \emptyset_{I_s}) = cos(16.55 - 3.44) = 0.973$

Sending end Power $P_s = P_R + losses = 12 * 10^6 + 1.99 * 10^6 = 13.99MW$

$P_s = V_s\,I_s\,cos\emptyset_s \Rightarrow cos\emptyset_s = \frac{P_s}{V_s\,I_s} = \frac{13.99*10^6}{78865.63*182.13} = 0.973$

Example 2.15 A 100-kilometer-long 3-phase, 50-Hz transmission line delivers 40 megawatts at 085 p.f. lag and 132 kV. The line resistance and reactance per km are 0·4Ω and 0·8Ω, respectively. The capacitance admittance is 5×10^{-6} Siemen/km/phase. Determine: (i) the current and voltage at the sending end, and (ii) the transmission line efficiency. Implement the nominal-T method.

Ans: Total resistance, R = 0·4 × 100 = 40 Ω

Total reactance, X_L = 0.8 × 100 = 80 Ω

Total susceptance, Y = 5 × 10^{-6} × 100 = 5×10^{-4} S

$cos\emptyset_R = 0.85\ lag \Rightarrow sin\emptyset_R = 0.526$

$$V_{ph}^R = \frac{132000}{\sqrt{3}} = 76.210\ kV$$

$$I_R = \frac{P_R}{\sqrt{3}\,V_L\,cos\emptyset_R} = \frac{40*10^6}{\sqrt{3}*132*10^3*0.85} = 205.82\ A$$

$$\vec{I_R} = I_R(cos\emptyset_R - jsin\emptyset_R) = 205.82\ (0.85 - j0.526) = 174.94 - j108.26 = 205.73\angle - 31.75^0\ A$$

$$\vec{V_c} = \vec{V_R} + \vec{I_R}\frac{\vec{Z}}{2} = 76210 + (174.94 - j108.26)\left(\frac{40+j80}{2}\right) = 84039.2 + j4832.4 = 84178\angle 3.29^0\ V$$

$$\vec{I_c} = jY\,\vec{V_c} = j\,5*10^{-4}*\ 84178\angle 3.29^0 = -2.41 + j42.01 = 42.08\angle 93.29^0\ \text{A}$$

$$\vec{I_s} = \vec{I_R} + \vec{I_c} = (174.94 - j108.26) + (-2.41 + j42.01) = 172.53 - j66.25 = 184.81\angle - 21^0\ A$$

$$\vec{V_s} = \vec{V_c} + \vec{I_s}\frac{\vec{Z}}{2} = (84039.2 + j4832.4) + (172.53 - j66.25)\left(\frac{40+j80}{2}\right)$$

$$= 90139.2 + j10407.4 = 90738.2\angle 6.58^0\ V$$

Line voltage $\vec{V_s} = \sqrt{3} * 90738\angle 6.58^0 = 157.162\angle 6.58^0\ kV$

$cos\emptyset_s = cos(\emptyset_{V_S} - \emptyset_{I_S}) = cos(6.58 + 21) = 0.8863$

Line losses $= 3\left(\frac{{I_s}^2R}{2} + \frac{{I_R}^2R}{2}\right) = 3\,[(184.81)^2 * 20 + (205.73)^2 * 20] = 4.588MW$

Sending end power $P_s = P_R + losses = 40*10^6 + 4.578*10^6 = 44.58\ MW$ ***(Or)***

Sending end power $P_s = 3V_sI_scos\emptyset_s = 3*90738.2*184.81*0.8863 = 44.587MW$

In Nominal-T method: $A = 1 + \frac{\vec{Y}\vec{Z}}{2}$

Receiving end voltage at no load, $\overrightarrow{V_{R\,NL}} = \frac{\vec{V_s}}{A} = \frac{\vec{V_s}}{1+\frac{\vec{Y}\vec{Z}}{2}} = \frac{90738\angle 6.58^0}{1+\frac{(j5*10^{-4})*(40+j80)}{2}} = 92584\angle 6^0\ V$

% of Voltage Regulation, $\%V_{reg} = \frac{V_{R\,NL} - V_{R\,FL}}{V_{R\,FL}} * 100 = \frac{92584 - 76210}{76210} * 100 = 21.48\%$

Efficiency $= \eta = \frac{P_R}{P_s} = \frac{40*10^6}{44.58*10^6} = 89.72\%$

Example 2.16 Line constants for a 150-kilometer, three-phase, 50-Hz transmission line are as follows: Resistance/phase/km = 0·2 Ω, Reactance/phase/km = 0.4 Ω, Susceptance/phase/km = 8 × 10^{-6} siemens. For a load of 25 MW at 0·9 p.f. and 66 kV at the receiving end, use the nominal-π method to determine: (i) power factor at the sending end (ii) regulation (iii) transmission efficiency

Ans: Total resistance, R = 0·2 × 150 = 30 Ω

Total reactance, X_L = 0.4 × 150 = 60 Ω

Total susceptance, Y = 8 × 10^{-6} × 150 = 1.2×10^{-3} siemens

$cos\emptyset_R = 0.9\ lag \Rightarrow sin\emptyset_R = 0.435;\ V_{ph}^R = \frac{66000}{\sqrt{3}} = 38.105\ kV$

$$I_R = \frac{P_R}{\sqrt{3}\,V_L\,cos\emptyset_R} = \frac{25*10^6}{\sqrt{3}*66*10^3*0.9} = 242.99\ A$$

$$\vec{I_R} = I_R(cos\emptyset_R - jsin\emptyset_R) = 242.99\ \ (0.9 - j0.435) = 218.69 - j105.7 = 242.99\angle - 25.79^0\ A$$

$\overrightarrow{I_{c1}} = j\frac{\vec{Y}}{2}\overrightarrow{V_R} = j\frac{1.2*10^{-3}}{2} * 38105 = j22.863$

$\overrightarrow{I_L} = \overrightarrow{I_R} + \overrightarrow{I_{c1}} = 218.69 - j105.7 + j22.863 = 218.69 - j82.83 = 233.85\angle -20.74^0\ A$

$\overrightarrow{V_s} = \overrightarrow{V_R} + \overrightarrow{I_L}\vec{Z} = 38105 + (218.69 - j82.83) * (30 + j60) = 49635 + j10636 = 50762.36\angle 12.09^0\ V$

$Line\ Voltage\ \overrightarrow{V_s} = \sqrt{3} * 50762.36\angle 12.09^0 = 87.922\angle 12.09^0\ kV$

$\overrightarrow{I_{c2}} = j\frac{\vec{Y}}{2}\overrightarrow{V_s} = j\frac{1.2*10^{-3}}{2} * (49635 + j10636) = -6.38 + j29.78 = 30.45\angle 102.09^0\ A$

$\overrightarrow{I_s} = \overrightarrow{I_L} + \overrightarrow{I_{c2}} = (218.69 - j82.83) + (-6.38 + j29.78) = 212 - j53.05 = 218.83\angle -14.02^0\ A$

$cos\emptyset_s = cos(\emptyset_{V_s} - \emptyset_{I_s}) = cos(12.09 + 14.02) = 0.8978$

Line losses $= 3{I_L}^2R = 3 * 233.85^2 * 30 = 4.921MW$

Sending end power $P_s = P_R + losses = 25 * 10^6 + 4.921 * 10^6 = 29.921\ MW$ ***(Or)***

Sending end power $P_s = 3V_sI_scos\emptyset_s = 3 * 50762.36 * 218.83 * 0.8978 = 29.92MW$

In Nominal- π method: $A = 1 + \frac{\vec{Y}\vec{Z}}{2}$

Receiving end voltage at no load, $\overrightarrow{V_{R\,NL}} = \frac{\overrightarrow{V_s}}{A} = \frac{\overrightarrow{V_s}}{1+\frac{\vec{Y}\vec{Z}}{2}} = \frac{50762.36\angle 12.09^0}{1+\frac{(j1.2*10^{-3})*(30+j60)}{2}} = 52648\angle 11.02^0\ V$

% of Voltage Regulation, $\%V_{reg} = \frac{V_{R\,NL} - V_{R\,FL}}{V_{R\,FL}} * 100 = \frac{52648-38105}{38105} * 100 = 38.16\%$

Efficiency $= \eta = \frac{P_R}{P_s} = \frac{25*10^6}{29.92*10^6} = 83.55\%$

Example 2.17 The constants for a 160 km long three-phase transmission line are as follows: Resistance/ph/km = 0.2Ω, reactance/ph/km = 0.1172Ω, and shunt admittance/ph/km=1.875x10^{-6} Siemens. Using a rigorous method, calculate the sending end voltage and current when providing a 25 MW load with 0·8 p.f. lag. The voltage at the receiving end remains constant at 110 kV.

Ans: Total resistance, $R = 0{\cdot}2 \times 160 = 32\ \Omega$

Total reactance, $X_L = 0.1172 \times 160 = 18.762\ \Omega$

Total susceptance, $Y = j1.875 \times 10^{-6} \times 160 = j\ 3\times10^{-4}$ siemens$=3 * 10^{-4}\angle 90^0$ siemens

Total impedance, $Z = R + jX_L = 32 + j18.762 = 37.0946\angle 30.38^0\ \Omega$

$cos\emptyset_R = 0.8\ lag \Rightarrow sin\emptyset_R = 0.6$

$$V_{ph}^R = \frac{110000}{\sqrt{3}} = 63.508\ kV$$

$$I_R = \frac{P_R}{\sqrt{3}\,V_L\,cos\emptyset_R} = \frac{25 * 10^6}{\sqrt{3} * 110 * 10^3 * 0.8} = 164.01\ A$$

$$\sqrt{YZ} = \sqrt{3 * 10^{-4}\angle 90 * 37.0946\angle 30.38^0} = 0.1054\angle 60.19^0$$

$$YZ = 3 * 10^{-4}\angle 90 * 37.0946\angle 30.38^0 = 0.011128\angle 120.38^0$$

$$Y^2Z^2 = (3 * 10^{-4}\angle 90)^2(37.0946\angle 30.38^0)^2 = 1.238 * 10^{-4}\angle 240.76^0$$

$$\sqrt{\frac{Z}{Y}} = \sqrt{\frac{37.0946\angle 30.38^0}{3 * 10^{-4}\angle 90}} = 343.42\angle -28.99^0$$

$$\sqrt{\frac{Y}{Z}} = \sqrt{\frac{3 * 10^{-4}\angle 90}{37.0946\angle 30.38^0}} = 2.911 * 10^{-3}\angle 28.99^0$$

$$cosh\sqrt{YZ} \approx 1 + \frac{YZ}{2} + \frac{Y^2Z^2}{24} + \cdots$$

$$= 1 + \frac{0.011128\angle 120.38^0}{2} + \frac{1.238 * 10^{-4}\angle 240.76^0}{24}$$

$$= 0.9971 + j4.795 * 10^{-3}$$

$$= 0.997\angle 0.2755^0$$

$$sinh\sqrt{YZ} \approx \sqrt{YZ} + \frac{(0.011128\angle 120.38^0)^{3/2}}{6} + \cdots\ldots$$

$$= 0.1054\angle 60.19^0 + \frac{1.1738 * 10^{-3}\angle 180.37^0}{6}$$

$$= 0.1054\angle 60.19^0 + 1.956 * 10^{-4}\angle 180.37^0$$

$$= 0.1053\angle 60.28^0$$

$$V_s = V_R \cosh(\sqrt{YZ}) + I_R\sqrt{\frac{Z}{Y}}\sinh(\sqrt{YZ})$$

$$= 63508 * 0.997\angle 0.2755^0 + 164.51 * 343.42\angle -28.99^0 * 0.1053\angle 60.28^0$$

$$= 63317.47\angle 0.2755^0 + 5930.95\angle 31.29^0$$

$$= 68468.74\angle 2.83^0\ V$$

$$I_s = V_R\sqrt{\frac{Y}{Z}}\ \sinh(\sqrt{YZ}) + I_R\cosh(\sqrt{YZ})$$

$$= 63508 * 2.911 * 10^{-3}\angle 28.99^0 * 0.1053\angle 60.28^0 + 164.51 * 0.997\angle 0.2755^0$$

$$= 19.46\angle 89.27^0 + 163.51\angle 0.2755^0$$

$$= 165\angle 7.047^0\ A$$

Example 2.18 A transmission line transmits a 25 MW balanced three-phase load at 132 kV, 50 Hz, and 0.9 p.f. lagging. A single conductor's series impedance is (30+j60) ohms, with a total phase-neutral admittance of 315×10^{-6} siemens. Use the nominal-T approach to determine the following: (i) line A, B, C, and D constants; (ii) sending end voltage; and (iii) line regulation.

Ans: **Total susceptance,** $Y = j\ 315\times10^{-6}$ **siemens** $= 315 * 10^{-6}\angle 90^0$ **siemens**

Total impedance, $Z = R + jX_L = 30 + j60\ \Omega$

$$cos\emptyset_R = 0.9\ lag \Rightarrow sin\emptyset_R = 0.435$$

$$V_{ph}^R = \frac{132000}{\sqrt{3}} = 76.21\ kV$$

$$\vec{A} = \vec{D} = 1 + \frac{\vec{Y}\,\vec{Z}}{2} = 1 + \frac{j315 * 10^{-6} * (30 + j60)}{2} = 0.9921 + j4.72 * 10^{-3} = 0.9921\angle 0.272^0$$

$$\vec{B} = \vec{Z}\left(1 + \frac{\vec{Y}\,\vec{Z}}{4}\right) = (30 + j60)\left(1 + \frac{j315 * 10^{-6} * (30 + j60)}{4}\right) = 29.76 + j48.87 = 58.08\angle 59.17^0\ \Omega$$

$$\vec{C} = \vec{Y} = 315 * 10^{-6}\angle 90^0\ \text{siemens}$$

$$I_R = \frac{P_R}{\sqrt{3}\,V_L\,cos\emptyset_R} = \frac{25 * 10^6}{\sqrt{3} * 132 * 10^3 * 0.9} = 121.49\ A$$

$$\vec{I_R} = I_R(cos\emptyset_R - jsin\emptyset_R) = 121.49(0.9 - j0.435) = 109.34 - j52.84 = 121.44\angle -25.79^0\ A$$

$$\vec{V_s} = \vec{A}\,\vec{V_R} + \vec{B}\,\vec{I_R}$$

$$= (0.9921 + j4.72 * 10^{-3}) * 76210 + (29.76 + j48.87)(109.34 - j52.84)$$

$= 81489.4 + j4239.97$

$= 81.599\angle 2.978^0\ kV$

$\vec{I} = \vec{C}\ \vec{V_R} + \vec{D}\ \vec{I_R}$

$= (j315 * 10^{-6}) * 76210 + (0.9921 + j4.72 * 10^{-3})(109.34 - j52.84)$

$= 108.714 - j27.89$

$= 112.23\angle -14.39^0\ A$

$Line\ Voltage\ \vec{V_s} = \sqrt{3} * 81599\angle 2.978^0 = 141.33\angle 2.978^0\ kV$

$cos\emptyset_s = cos(\emptyset_{V_s} - \emptyset_{I_s}) = cos(2.978 + 14.39) = 0.954$

Sending end power $P_s = 3V_sI_scos\emptyset_s = 3 * 81599 * 112.33 * 0.954 = 26.23MW$

Receiving end voltage at no load, $\overrightarrow{V_{R\ NL}} = \frac{\vec{V_s}}{A} = \frac{81599\angle 2.978^0}{0.9921\angle 0.272^0} = 82248.76\angle 2.706^0\ V$

% of Voltage Regulation, $\%V_{reg} = \frac{V_{R\ NL} - V_{R\ FL}}{V_{R\ FL}} * 100 = \frac{82248.76 - 76210}{76210} * 100 = 7.923\%$

Efficiency $= \eta = \frac{P_R}{P_s} = \frac{25*10^6}{26.23*10^6} = 95.31\%$

Example 2.19 A 132 kV, 50 Hz, 3-phase transmission line can provide a 50 MVA load with 086 p.f. lagging at the receiving end. The transmission line's generalised constants are A = D = 0·95 ∠4°; B = 105 ∠80°; C = 0·0005 ∠85°. Determine line regulation and charging current.

Ans: $V_{ph}^R = \frac{132000}{\sqrt{3}} = 76.21\ kV$

$P_R = 50\ MVA = 50 * 10^6 * 0.86 = 43\ MW$

$I_R = \frac{P_R}{\sqrt{3}\ V_L\ cos\emptyset_R} = \frac{50 * 10^6}{\sqrt{3} * 132 * 10^3} = 262.4\ A$

$\vec{I_R} = I_R(cos\emptyset_R - jsin\emptyset_R) = 262.4(0.86 - j0.51) = 225.66 - j133.82 = 262.36\angle -30.66^0\ A$

$\vec{V_s} = \vec{A}\ \vec{V_R} + \vec{B}\ \vec{I_R} = (0.95\angle 4^0) * 76210 + (105\angle 80^0) * 262.36\angle -30.66^0 = 82.061\angle 17.81^0\ kV$

$\vec{I} = \vec{C}\ \vec{V_R} + \vec{D}\ \vec{I_R} = (0.0005\angle 85) * 76210 + (0.95\angle 4) * 262.36\angle -30.66 = 239.34\angle -19.57^0\ A$

$cos\emptyset_s = cos(\emptyset_{V_s} - \emptyset_{I_s}) = cos(17.81 + 19.57) = 0.794$

Sending end power $P_s = 3V_sI_scos\emptyset_s = 3 * 82061 * 239.34 * 0.794 = 46.78MW$

Receiving end voltage at no load, $\overrightarrow{V_{R\ NL}} = \frac{\vec{V_s}}{A} = \frac{82061\angle 17.81^0}{0.95\angle 4^0} = 86380\angle 13.81^0\ V$

% of Voltage Regulation, $\%V_{reg} = \frac{V_{R\ NL} - V_{R\ FL}}{V_{R\ FL}} * 100 = \frac{86380 - 76210}{76210} * 100 = 13.34\%$

Efficiency $= \eta = \frac{P_R}{P_s} = \frac{43*10^6}{46.78*10^6} = 91.91\%$

2.10 Ferranti Effect:

The Ferranti effect refers to the phenomenon in which the voltage at the receiving end of a transmission line exceeds the sending voltage. This effect is mostly observed under no load or light load condition. The flow of charging current through distributed capacitance is the root cause of Ferranti effect. When line is fully loaded, load current which is in general inductive in nature is quite high compared to charging current resulting in receiving end voltage to be less than sending end voltage as a result of voltage drop in series impedance of line. However, under no load and light load conditions, charging current is quite significant leading to voltage rise at the receiving end.

2.10.1 Factors influencing Ferranti Effect:

Several factors influence the current in the transmission line. However, the Ferranti effect happens mainly for the following three reasons:

- Transmission line capacitance,
- Receiving end load, and
- Frequency of supply.

Transmission Line Capacitance: As the length of the transmission line increases, so as the distributed capacitance. Series impedance of line consumes lagging VAR causing voltage drop, whereas shunt capacitance consumes leading VAR causing voltage rise. Under no load and light conditions, leading VAR consumption by line capacitance may be higher than lagging VAR consumption of series inductance resulting in voltage rise in the receiving end.

Load at the Receiving End: The connected load at the receiving end is another component that influences the Ferranti effect. The load can be in either of three states:

- No load
- Light load
- Full load

In the no-load state, the transmission line only receives charging current and no-load current. The shunt capacitors draw the charging current from the line. It consumes leading reactive power, resulting in a voltage drop across the inductor being overcompensated that leads to increase of the voltage at the receiving end.

When a light load is connected, the load current is very low compared to the charging current flowing through the line. Because of the low load current, the capacitor generates reactive power that passes through the inductors, which overcompensates the reactive power consumed by the inductors. The Ferranti effect occurs when the charging current exceeds the load current.

In full load conditions, the load current exceeds the capacitor's charging current. Because a significant load current flows through the series inductors, the inductor consumes more reactive power than the capacitor produces, and voltage reduces at the receiving end.

Supply Frequency: As we know that the Ferranti effect happens because of the reactive power produced in the shunt capacitance of the line. Since DC has zero frequency, there is no Ferranti effect. The Ferranti effect is more prominent in the transmission lines that operate at high frequency as charging current increases due to increase in capacitive susceptance.

2.10.2 Ferranti effect in long transmission line: Consider an equivalent circuit diagram for a long transmission line as shown in Fig. 2.28(a). Because a long transmission line is composed of high capacitance and inductance dispersed along its whole length, this diagram addresses the parameters per km. In this manner, capacitance and inductance are proportional to line length. Inductors are connected to power lines in series, whereas shunt capacitors are connected in parallel. Phasor diagram of the given circuit is shown in Fig. 2.28(b). Under no load condition, Receiving current $\vec{I_R} = 0$

The receiving voltage $\vec{V_R}$ is designated as reference OB, whereas the capacitive current $\vec{I_c}$ is shown by line OE, which leads $\vec{V_R}$ by 90°. The capacitive current $\vec{I_c}$ experiences a voltage drop between the line resistance R and the line inductance L, as represented by the equation:

Voltage drop across the resistor = $\vec{I_c}R$ and Voltage drop across the inductor = $\vec{I_c}\,X_L$

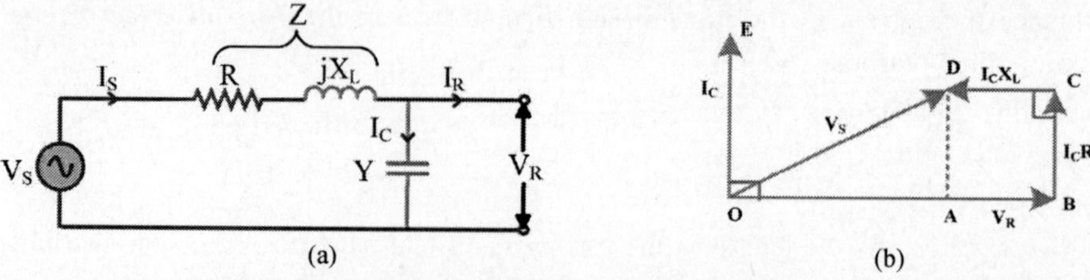

Fig. 2.28 (a) Transmission line under no-load (b) phasor diagram of Ferranti effect

The phasor graphic illustrates that the receiving voltage $\vec{V_R}$ at the load side exceeds the transmitting voltage $\vec{V_s}$ at the sending end side.

The inductive voltage drop $\vec{I_c}\,X_L$ is 90 degrees ahead of the resistive voltage drop $\vec{I_c}R$. The sending end voltage $\vec{V_s}$ equals the sum of all voltage drops plus the receiving voltage.

$$\vec{V_s} = \vec{V_R} + \text{Resistive voltage drop} + \text{Inductive voltage drop}$$

$$\vec{V_s} = \vec{V_R} + \vec{I_c}R + \vec{I_c}\,X_L$$

$$= \vec{V_R} + \vec{I_c}\,(R + jX_L)$$

$$= \vec{V_R} + j\omega C\vec{V_R}\,(R + jX_L) \qquad \text{Since capacitive current, } \vec{I_c} = j\omega C\vec{V_R}$$

$$= \vec{V_R} + j\omega C\vec{V_R}\,(R + j\omega L) \qquad \text{Since Inductive reactance, } X_L = j\omega L$$

$$= \vec{V_R} + j\omega CR\vec{V_R} + j^2\omega^2 CL\,\vec{V_R}$$

$$= \vec{V_R} + j\omega CR\vec{V_R} - \omega^2 CL\,\vec{V_R}$$

In long transmission lines, the line resistance is significantly smaller than the line reactance. Consequently, the resistance R and the associated resistive voltage loss are disregarded.

$$\vec{V_s} - \vec{V_R} = -\omega^2 CL\,\vec{V_R}$$

By considering the total length of the transmission l, the equation transforms into

$$\vec{V_s} - \vec{V_R} = -\omega^2 * C\,l * L\,l * \vec{V_R}$$

$$= -\omega^2 CL\,l^2\,\vec{V_R} \qquad \text{......(2.90)}$$

$$= -\frac{\omega^2 l^2}{v^2}\vec{V_R} \qquad \text{Since velocity of propagation, } v = \frac{1}{\sqrt{LC}}$$

The negative voltage difference between the sending and receiving voltages indicates an increase in voltage. Furthermore, it is directly proportional to the squares of the line length (l) and frequency (ω). This condition illustrates that the Ferranti effect amplifies with an increase in the length of the transmission line and supply frequency. As a result, small transmission lines and HVDC transmission are unaffected by the Ferranti effect.

2.10.3 Ferranti effect in medium transmission line: Capacitance and inductance are the primary characteristics of lines exceeding 200 kilometres in length. The capacitance in these transmission lines is not localized at certain points. It is uniformly distributed along the full length of the line. When voltage is applied at the sending end, the line's capacitance draws a greater current than the load. Consequently, under no or low load conditions, the voltage at the receiving end is considerably higher compared to the voltage at the sending end.

Consider the medium transmission line, where OB denotes the voltage at the receiving terminal and OD signifies the current transiting through the receiving end capacitor as shown in Fig. 2.29. The phasor BC represents the voltage drop across the resistance R. The phasor CD represents the voltage drop over the inductor (X). The phasor OD denotes the sending end voltage under no-load conditions. The phasor representation demonstrates that OB exceeds OD. In the absence of a load on the line, the voltage at the receiving end exceeds the value at the sending end.

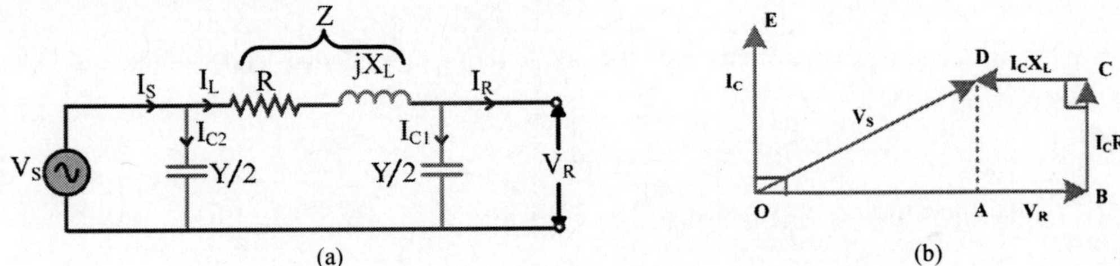

Fig. 2.29 (a) Nominal- π model of the line at no load (b) Phasor diagram

Sending end voltage, $\vec{V_S} = \vec{V_R}\left(1 + \frac{\vec{Y}\vec{Z}}{2}\right) + \vec{I_R}\vec{Z}$

At no load, $\vec{I_R} = 0$ then $\vec{V_S} = \vec{V_R}\left(1 + \frac{\vec{Y}\vec{Z}}{2}\right)$

$$\vec{V_S} - \vec{V_R} = \vec{V_R}\left(1 + \frac{\vec{Y}\vec{Z}}{2}\right) - \vec{V_R}$$

$$= \vec{V_R}\left(1 + \frac{\vec{Y}\vec{Z}}{2} - 1\right)$$

$$= \vec{V_R}\left(\frac{\vec{Y}\vec{Z}}{2}\right)$$

$$= \vec{V_R}\left(\frac{j\omega Cl * (R + j\omega L)l}{2}\right)$$

If the resistance of the line is neglected,

$$\vec{V_S} - \vec{V_R} = \vec{V_R}\left(\frac{j\omega C(j\omega L)}{2}\right)$$

$$= j^2\omega^2 l^2 CL\,\vec{V_R}$$

$$= -\omega^2 l^2 CL\,\vec{V_R}$$

$$= -\frac{\omega^2 l^2}{v^2}\vec{V_R} \qquad \text{......(2.91)}$$

2.10.4 How to reduce Ferranti effect: Electrical equipment is designed to operate at a specific voltage. Excessive voltages at the user terminals might result in equipment damage and the burning of windings due to excessive voltage exposure. The Ferranti effect on long transmission lines under low or no load elevates the receiving end voltage. The voltage can be regulated by positioning the shunt reactors at the receiving end of the lines. A shunt reactor is an inductive current component linked between the line and neutral to compensate the capacitive current from transmission lines. In long transmission lines, shunt reactors mitigate the capacitive VAR, thereby regulating the voltage within specified limits.

2.11 Power flow through transmission line: Consider a two-bus system as shown in Fig. 2.30. Let us take the receiving-end voltage as a reference phasor ($V_R = |V_R| \angle 0^\circ$) and let the sending-end voltage lead it by an angle δ ($V_S = |V_S| \angle \delta$). The complex power leaving the receiving end and entering the sending end of the transmission line can be given as

$$S_R = P_R + jQ_R = V_R I_R^*$$
$$S_S = P_S + jQ_S = V_S I_S^* \qquad ...(2.92)$$

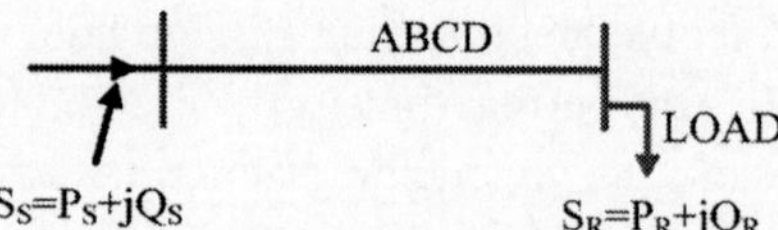

Fig. 2.30 A two bus system

A three-phase transmission line's input voltage $\vec{V_s}$ and input current $\vec{I_s}$ can be expressed mathematically as follows:

$$\vec{V_s} = \vec{A}\ \vec{V_R} + \vec{B}\ \vec{I_R}$$
$$\vec{I_s} = \vec{C}\ \vec{V_R} + \vec{D}\ \vec{I_R}$$

The receiving- and sending-end currents can be stated in terms of receiving- and sending-end voltages:

$$\vec{I_R} = \frac{1}{B}\vec{V_s} - \frac{A}{B}\vec{V_R} \qquad ......(2.93)$$
$$\vec{I_S} = \frac{D}{B}\vec{V_s} - \frac{1}{B}\vec{V_R} \qquad ......(2.94)$$

Let A, B, D, the transmission line constants, be written as $A = |A|\angle\alpha, B = |B|\angle\beta,$ and $D = |A|\angle\alpha$ as $A = D$.

Therefore,
$$\vec{I_R} = \left|\frac{1}{B}\right| |V_s|\angle(\delta - \beta) - \left|\frac{A}{B}\right| |V_R|\angle(\alpha - \beta) \qquad ......(2.95)$$
$$\vec{I_S} = \left|\frac{D}{B}\right| |V_s|\angle(\alpha + \delta - \beta) - \left|\frac{1}{B}\right| |V_R|\angle(-\beta) \qquad ......(2.96)$$
$$= \left|\frac{A}{B}\right| |V_s|\angle(\alpha + \delta - \beta) - \left|\frac{1}{B}\right| |V_R|\angle(-\beta)$$

Substituting for I_R in Eq. (2.91) we get

$$S_R = V_R I_R^* = |V_R|\angle 0\ \left[\left|\frac{1}{B}\right| |V_s|\angle(-\delta + \beta) - \left|\frac{A}{B}\right| |V_R|\angle(-\alpha + \beta)\right]$$
$$= \frac{|V_R|\,|V_S|}{|B|}\angle(\beta - \delta) - \left|\frac{A}{B}\right| |V_R|^2\angle(\beta - \alpha) \qquad ......(2.97)$$

Similarly,
$$S_S = V_S I_S^* = |V_S|\angle\delta\ \left[\left|\frac{A}{B}\right| |V_s|\angle(-\alpha - \delta + \beta) - \left|\frac{1}{B}\right| |V_R|\angle(\beta)\right]$$
$$= \left|\frac{A}{B}\right| |V_S|^2\angle(\beta - \alpha) - \frac{|V_R|\,|V_S|}{|B|}\angle(\beta + \delta) \qquad ......(2.98)$$

In the above equations, S_R and S_S are per phase complex voltamperes, whereas V_R and V_S are per phase volts. If V_R and V_S are expressed as line kVs, then the three-phase receiving-end complex power is given by

$$S_R = 3\left\{\frac{\left|\frac{V_R*10^3}{\sqrt{3}}\right|\left|\frac{V_S*10^3}{\sqrt{3}}\right|}{|B|}\angle(\beta - \delta) - \left|\frac{A}{B}\right| \left|\frac{V_R*10^3}{\sqrt{3}}\right|^2 \angle(\beta - \alpha)\right\}$$
$$= 3\left\{\frac{|V_R|\,|V_S|*10^6}{3*|B|}\angle(\beta - \delta) - \left|\frac{A}{B}\right| \frac{|V_R|^2}{3}\angle(\beta - \alpha)\right\}$$

3-ph complex power in MVA at receiving end is

$$S_R = \frac{|V_R|\,|V_S|}{|B|}\angle(\beta - \alpha) - \left|\frac{A}{B}\right| |V_R|^2\angle(\beta - \alpha) \qquad ......(2.99)$$

Similarly,
$$S_S = 3\left\{\left|\frac{A}{B}\right| \left|\frac{V_S*10^3}{\sqrt{3}}\right|^2 \angle(\beta - \alpha) - \frac{\left|\frac{V_R*10^3}{\sqrt{3}}\right|\left|\frac{V_S*10^3}{\sqrt{3}}\right|}{|B|}\angle(\beta + \delta)\right\}$$
$$= 3\left\{\left|\frac{A}{B}\right| \frac{|V_S|^2}{3}\angle(\beta - \alpha) - \frac{|V_R|\,|V_S|*10^6}{3*|B|}\angle(\beta + \delta)\right\}$$

3-ph complex power in MVA at sending end is $S_S = \left|\frac{A}{B}\right| |V_S|^2\angle(\beta - \alpha) - \frac{|V_R|\,|V_S|}{|B|}\angle(\beta + \delta)$ ......(2.100)

If Eq. (2.99) is expressed in real and imaginary parts, we can write the real and reactive powers at the receiving-end as

$$S_R = P_R + jQ_R = \frac{|V_R|\,|V_S|}{|B|}[cos(\beta - \delta) + jsin(\beta - \delta)] - \left|\frac{A}{B}\right| |V_R|^2[cos(\beta - \alpha) + jsin(\beta - \alpha)]$$

$$P_R = \frac{|V_R|\,|V_S|}{|B|} cos(\beta - \delta) - \left|\frac{A}{B}\right| |V_R|^2 cos(\beta - \alpha) \quad ......(2.101)$$

$$Q_R = \frac{|V_R|\,|V_S|}{|B|} sin(\beta - \delta) - \left|\frac{A}{B}\right| |V_R|^2 sin(\beta - \alpha) \quad ......(2.102)$$

Similarly, the real and reactive powers at sending-end are

$$S_S = P_S + jQ_S = \left|\frac{A}{B}\right| |V_S|^2[cos(\beta - \alpha) + jsin(\beta - \delta)] - \frac{|V_R|\,|V_S|}{|B|}[cos(\beta + \delta) + jsin(\beta + \delta)]$$

$$P_S = \left|\frac{A}{B}\right| |V_S|^2 cos(\beta - \alpha) - \frac{|V_R|\,|V_S|}{|B|} cos(\beta + \delta) \quad ......(2.103)$$

$$Q_S = \left|\frac{A}{B}\right| |V_S|^2 sin(\beta - \alpha) - \frac{|V_R|\,|V_S|}{|B|} sin(\beta + \delta) \quad ......(2.104)$$

Eq. (2.101) shows that the received power P_R will be maximum at $\delta = \beta$

$$P_{R\,max} = \frac{|V_R|\,|V_S|}{|B|} - \left|\frac{A}{B}\right| |V_R|^2 cos(\beta - \alpha) \quad ......(2.105)$$

The corresponding Q_R (at max P_R) is

$$Q_R = -\left|\frac{A}{B}\right| |V_R|^2 sin(\beta - \alpha) \quad ......(2.106)$$

Hence, the load must absorb this quantity of leading MVAR to attain the maximum real power.

Real and Reactive Power of Short Transmission lines:

Consider the specific scenario of a short line characterized by a series impedance Z. *Generalised Circuit Constants of a Short Transmission Line are* $A = D = 1\angle 0, B = |Z|\angle\theta,$ *and* $C = 0$

By substituting them into Equations (2.101), (2.102), (2.103) and (2.104), we obtain the simplified solutions for the short line as follows:

$$P_R = \frac{|V_R|\,|V_S|}{|Z|} cos(\theta - \delta) - \frac{|V_R|^2}{|Z|} cos\,\theta \quad ......(2.107)$$

$$Q_R = \frac{|V_R|\,|V_S|}{|Z|} sin(\theta - \delta) - \frac{|V_R|^2}{|Z|} sin\theta \quad ......(2.108)$$

$$P_S = \frac{|V_S|^2}{|Z|} cos\theta - \frac{|V_R|\,|V_S|}{|Z|} cos(\theta + \delta) \quad ......(2.109)$$

$$Q_S = \frac{|V_S|^2}{|Z|} sin\theta - \frac{|V_R|\,|V_S|}{|Z|} sin(\theta + \delta) \quad ......(2.110)$$

The aforementioned short line equation is applicable to a long line when the line is substituted with its equivalent π (or nominal-π) model, and the shunt admittances are consolidated with the receiving-end load and sending-end generation.

According to Eq. (2.105), the maximum power at the receiving end is attained when $\delta = \theta$.

$$P_{R\,max} = \frac{|V_R|\,|V_S|}{|Z|} - \frac{|V_R|^2}{|Z|} cos\theta \quad ......(2.111)$$

Substitute $cos\theta = \frac{R}{|Z|}$ in above eq.,

$$P_{R\,max} = \frac{|V_R|\,|V_S|}{|Z|} - \frac{|V_R|^2}{|Z|^2} R \quad ......(2.112)$$

Typically, the resistance of a transmission line is minimal relative to its reactance to ensure high transmission efficiency, resulting in $\theta = \tan^{-1}(X/R) \approx 90°$; where $Z = R+jX$. The equations at the receiving end, (2.107) and (2.108), can thereafter be approximated as

$$P_R = \frac{|V_R|\,|V_s|}{|X|} cos(90 - \delta) - \frac{|V_R|^2}{|X|} cos\ 90$$

$$P_R = \frac{|V_R|\,|V_s|}{|X|} sin\delta \qquad ......(2.113)$$

$$Q_R = \frac{|V_R|\,|V_s|}{|X|} cos\delta - \frac{|V_R|^2}{|X|} \qquad ......(2.114)$$

2.12 Underground Cables:

Electricity can be transmitted or distributed using either an overhead system or underground cables. Underground cables offer several benefits, including less storm and lightning damage, lower maintenance costs, fewer faults, lower voltage drop, and a better overall appearance. The main disadvantage of these systems is higher installation costs and insulation issues at higher voltages compared to overhead systems. Underground cables are used where overhead wires are impractical. Overhead wires may not be permitted in densely populated regions, around factories or substations, or due to maintenance requirements.

Underground cables have traditionally been used to distribute electricity in densely populated urban areas at low or moderate voltages. Recent advances in design and manufacturing have resulted in high-voltage cables. Underground cables can now be used to transmit power over short or moderate distances.

An underground cable primarily comprises one or more conductors that are encased in appropriate insulation and shielded by a protective covering. The selection of the appropriate cable depends on the operating voltage and specific service requirements, despite the availability of various cable types. Typically, a cable must meet the following essential criteria:

- Cable conductors should be made of high conductivity tinned stranded copper or aluminium. Stranding enhances a conductor's flexibility and current capacity.
- The conductor size should allow for adequate load current without overheating and voltage drop within acceptable limits.
- To ensure safety and dependability at the specified voltage, the cable insulation thickness must be sufficient.
- The cable needs mechanical protection to survive harsh laying.
- Cable manufacturing materials should be chemically and physically stable throughout.

2.13 Construction of Cables:

Fig. 2.31 illustrates the overall structure of a 3-conductor wire. The components are as follows:

- A cable may possess either a single or multiple cores (conductors) depending on the intended service type. The conductors consist of tinned copper or aluminium and are typically stranded to enhance the cable's flexibility.
- The voltage rating of the cable determines the insulation thickness of each core or conductor. Impregnated paper, varnished cambric, and rubber mineral compounds are common insulation materials.

- A lead or aluminium sheath is put over the insulation to protect the cable from moisture, gases, and hazardous elements (acids or alkalies) in the soil and atmosphere.
- Fiber bedding, such as jute or hessian tape, is applied over the metallic sheath. Bedding shields the metallic sheath from corrosion and mechanical damage caused by armouring.
- To protect armouring against atmospheric conditions, a layer of fibrous material (e.g. jute) similar to bedding is laid over it. This is referred to as serving.

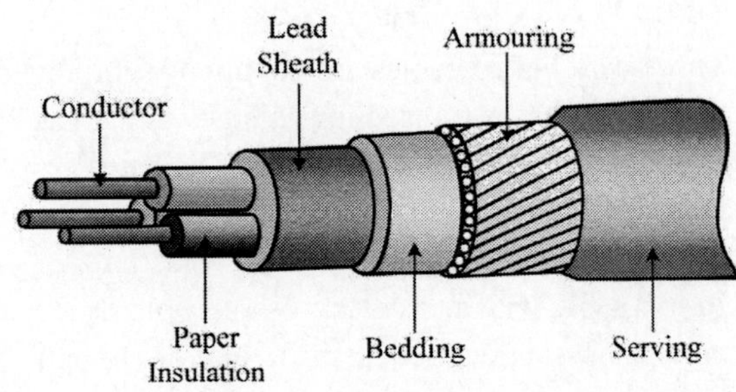

Fig. 2.31 Construction of a cable

2.14 Classification of Cables: Cables for underground service can be classified in two ways: (i) by the insulating substance used in their production, and (ii) by the voltage for which they are designed. However, the more typical way of classification is to divide cables into distinct types based on their voltage capacity. Low-tension (L.T.) cables can withstand voltages of up to 1000 V, but high-tension (H.T.) cables can manage voltages of up to 11,000 V. Super-tension (S.T.) cables have a voltage range of 22 kV to 33 kV, E.H.T. cables range from 33 kV to 66 kV, and Extra super-voltage cables can handle voltages greater than 132 kV. A cable may have one or several cores, depending on the intended service. It can be single-core, dual-core, triple-core, or quad-core. For a 3-phase service, either 3-single-core or three-core cables can be used based on the operating voltage and load requirement. Based on core it can also be categorised as Sigle core cables and three core cables (belted, screened and pressured cables).

2.14.1 Single Core cables:

Figure 2.32 depicts the specific characteristics of a single-core low voltage cable. The cable's structure is conventional because it can handle modest amounts of stress at low voltages (up to 6600 V). The device consists of a circular core composed of tinned stranded copper or aluminium, together with impregnated paper layers for insulation. The insulation is safeguarded by a lead enclosure, thereby preventing the ingress of moisture into the inside. In order to inhibit corrosion, a coating of fibrous material, such as jute, is put onto the lead sheath. Unprotected single-core cables often experience significant sheath losses.

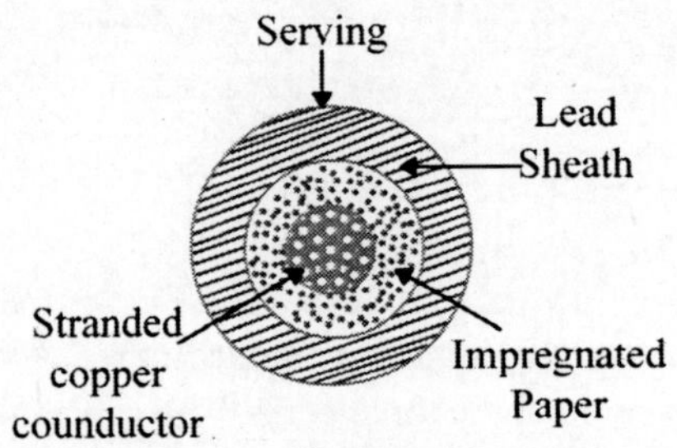

Fig. 2.32 Single core cable

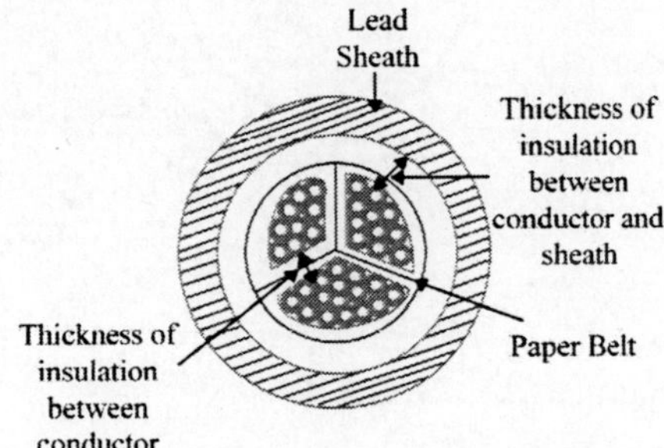

Fig. 2.33 Three core belted cable

2.14.2 Three Core Belted cables:

Three-core belted cables are commonly employed at voltage levels of up to 11kV, although in certain cases they can be utilised for voltages as high as 22kV. Figure 2.33 depicts the specific components and structure of a 3-core belted cable. The cores are divided by interleaved sheets of impregnated paper. A paper belt refers to a strip of paper tape that is saturated with a substance and wound around insulated cores. Fibre insulating material, such as jute, is used to fill the gap between insulated cores in order to form a cable with a circular cross-section. Cores are frequently arranged in a stranded configuration and may possess a noncircular shape in order to optimise the utilisation of available space.

The belt's lead sheath provides protection to the cable against moisture and mechanical damage. The lead sheath is equipped with many layers of armour and an exterior covering, which is not depicted in the illustration. Belted cables are appropriate only for low and medium voltages because of the radial distribution of electrostatic forces throughout the insulation. Tangential strains, however, reach a critical point when the voltage exceeds 22 kV. These forces exert influence on the paper insulating layers. Low insulation resistance in the paper insulation layers may result in leakage current when tangential loads are applied. The flow of current that escapes generate localised heat, which presents a risk of insulating breakdown. In order to resolve this problem, screened cables are utilised, which direct leakage currents to the ground through metallic screens.

2.14.3 Three Core Screened cables: Screened cables are designed for use up to 33 kV, but can be expanded to 66 kV in certain instances. There are two main types of screened cables: H-type cables and S.L. type cables.

2.14.3.1 H-type three Core Screened cables: H. Hochstadter was the pioneer in designing this type of cable, thus giving it its name. Figure 2.34(a) illustrates the structure of a standard three-core, H-type cable. Each core is insulated by layers of impregnated paper. The insulation of each core is encased with a metallic screen, typically composed of perforated aluminium foil.

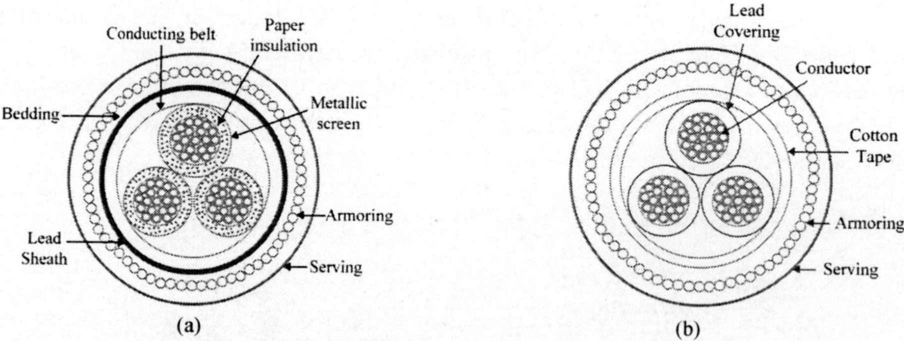

Fig. 2.34(a) H-type (b) S.L type three Core Screened cables

The cores are placed in a manner where metallic screens come into contact with each other. A copper woven fabric tape is employed as an additional conductor to encircle the three cores. The cable does not have an insulating belt, but it still has the normal lead sheath, bedding, armouring, and service. The central screens establish an electrical connection with the belt and sheath. Because all four screens (consisting of 3 core screens and one conducting belt) and the lead sheath are connected to the ground, the electrical stresses are exclusively directed towards the centre, resulting in a decrease in dielectric losses. H-type cables are reported to provide two unique benefits. The perforations of the metallic screens fully saturate the cable with the compound, ensuring the absence of any air pockets or voids in the dielectric. Presence of voids in the cable can diminish its ability to withstand breakdown and result in substantial harm to the paper insulation. Furthermore, the metallic screens aid in the dispersion of heat from the cable.

2.14.3.2 S.L. type three Core Screened cables: Figure 2.34(b) illustrates the structure of a 3-core separate lead (S.L.) type cable. Essentially, the cable is similar to an H-type cable, but with the distinction that each core insulation is enclosed by its own lead sheath. There is no overall lead covering, but rather just armour and protective layers are there. S.L. type cables offer two significant advantages compared to H-type cables. To begin with, the separate coverings decrease the likelihood of the core disintegrating and separating from each other. Furthermore, the lack of a comprehensive lead sheath allows for easier cable bending. However, a disadvantage of the S.L. cable is that its three lead sheaths are considerably thinner than the single sheath of the H-cable. This means that extra caution is required during production.

2.14.4 Pressure cables:

Cables of solid construction are not reliable for voltages exceeding 66 kV due to the potential failure of insulation caused by voids. Pressure cables are used when the operational voltage surpasses 66 kV. Pressure cables are wires that decrease voids by increasing compound pressure. The two most commonly used types are oil-filled cables and gas pressure cables.

2.14.4.1 Oil filled cables: These cables are equipped with channels or ducts to facilitate the circulation of oil. Oil is supplied under pressure from external reservoirs located along the cable path to the channel at regular intervals. This oil is the same oil used for impregnation. Applying pressure to oil causes it to compress the paper insulation layers and fill any empty spaces between them. Oil-filled cables have the capacity to endure greater voltages, usually ranging from 66 kV to 230 kV, by eliminating any voids. There are three types of oil-filled cables: single-core conductor channel, single-core sheath channel, and three-core filler-space channels. Figure 2.35(a) illustrates the configuration of a cable with a single core sheath channel that is filled with oil. The conductor of this cable is insulated with paper and is solid, just like a solid wire. Oil ducts are present within the metallic casing as indicated.

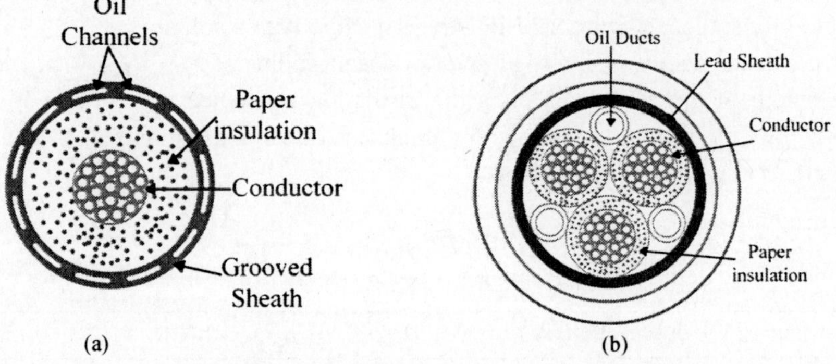

Fig. 2.35 (a) Single core (b) three core pressure cables

Figure 2.35(b) illustrates a 3-core oil-filled cable with oil ducts located in the spaces between the filler materials. The conduits, constructed from perforated metal-ribbon tubing, are grounded. Oil-filled wires possess three significant benefits. To begin with, the occurrence of void formation and ionisation is prevented. Furthermore, there is an increase in both the temperature range and dielectric strength. Furthermore, leaking serves as an indication of a flaw in the lead sheath and diminishes the likelihood of earth faults. The primary disadvantages of them are high initial expenses and a complicated installation process.

2.14.4.2 Gas pressure cables:

As the pressure increases, the voltage necessary for ionisation within a void also increases. By subjecting ordinary cable to high pressure, it is possible to totally prevent ionisation. Elevated pressure induces radial compression, effectively closing any gaps. The essential principle of gas pressure cables is being described here. Hochstadter, Vogal, and Bowden were the creators of the external pressure cable segment shown in Figure 2.36. The cable's structure is the same as a typical solid cable, except it has a triangle shape and a lead sheath that is 75% thicker. The triangular shape reduces weight and offers low thermal resistance. The primary purpose of the shape is for the lead sheath to act as a pressure membrane. A slender metallic strip provides protection for the sheath. The cable is contained within a hermetically sealed steel conduit. The pipe is pressurised with dry nitrogen gas at a pressure of 12 to 15 atmospheres. The gas pressure applies compression to the paper insulation layers, thereby sealing any voids that may have formed. These cables have a better capacity for carrying current and can operate at higher voltage levels compared to conventional cables. In addition, nitrogen gas is highly effective in extinguishing flames and requires less maintenance, resulting in inexpensive expenses. Nevertheless, it is characterised by the drawback of being quite costly.

Fig. 2.36 Gas pressure cables

2.15 Insulation Resistance of a Single-Core Cable:

To prevent current leakage, the cable conductor has a proper thickness of insulating material. The insulation creates a radial pathway for the flow of leakage current. Insulation resistance is the measure of a cable's ability to resist leakage currents. The cable's insulation resistance must be exceptionally high to ensure satisfactory operation. Figure 2.37 illustrates a cable with a solitary core, including a conductor radius of r_1 and an interior sheath radius of r_2. Define 'l' as the length of the cable and 'ρ' as the insulation resistance.

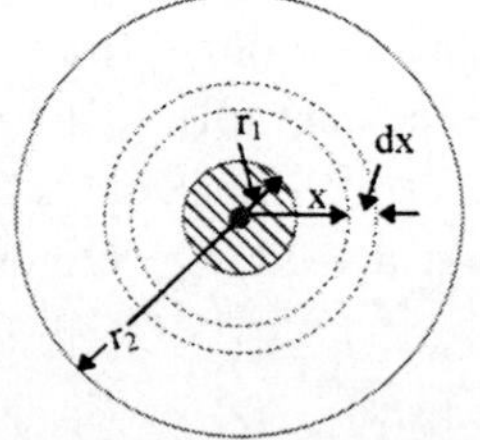

Fig. 2.37 Single-Core Cable

Consider an extremely thin insulation layer with a thickness of dx and a radius of x. Leakage current flows along a length (dx) with an area of $2\pi xl$.

Insulation resistance of the considered layer $= \rho \frac{dx}{2\pi xl}$

Insulation resistance of the entire cable is $R = \int_{r_1}^{r_2} \rho \frac{dx}{2\pi xl} = \frac{\rho}{2\pi l} \int_{r_1}^{r_2} \frac{dx}{x} = \frac{\rho}{2\pi l} \log_e \frac{r_2}{r_1}$

∴ Insulation resistance of the entire cable is $R = \frac{\rho}{2\pi l} \log_e \frac{r_2}{r_1}$ ......(2.115)

This demonstrates that a cable's insulation resistance varies inversely with length. Insulation resistance decreases as cable length rises, and vice versa.

2.16 Capacitance of a Single-Core Cable:

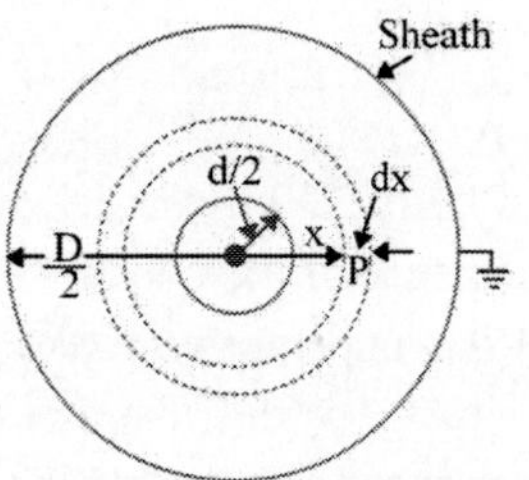

Fig. 2.38 Single-Core Cable

A single-core wire can be visualised as two elongated coaxial cylinders. The inside cylinder of the cable functions as the conductor, while the outer cylinder serves as a lead sheath that is connected to the earth potential. Let's examine a cable that has only one core. The cable has a conductor with a diameter of d and an inner sheath with a diameter of D (as shown in Figure 2.38). Let Q be the charge per meter of the cable's axial length, and represent the permittivity of the insulating medium between the core and lead sheath. Consider a cylinder with radius x metres and axial length one metre.

The surface area of the cylinder is $2\pi x * 1 = 2\pi x \quad m^2$

The electric flux density at any point P of the considered cylinder is $D_x = \frac{Q}{2\pi x}$ C/m^2 ...... (2.116)

The electrical intensity at point P is $E_x = \frac{D_x}{\varepsilon} = \frac{Q}{2\pi x \varepsilon} = \frac{Q}{2\pi x\, \varepsilon_0 \varepsilon_r} \quad V/m$ ...... (2.117)

The work done to move a unit positive charge from point P to dx in the direction of an electric field is denoted as $E_x\, dx$. The work required to transfer a positive charge from the conductor to the sheath, represented by the potential difference V, is:

$$V = \int_{\frac{d}{2}}^{\frac{D}{2}} E_x\, dx = \int_{\frac{d}{2}}^{\frac{D}{2}} \frac{Q}{2\pi x\, \varepsilon_0 \varepsilon_r}\, dx = \frac{Q}{2\pi x\, \varepsilon_0 \varepsilon_r} \log_e \frac{D}{d} \qquad ...... (2.118)$$

$$\text{Capacitance of the cable } C = \frac{Q}{V} = \frac{Q}{\frac{Q}{2\pi x\, \varepsilon_0 \varepsilon_r} \log_e \frac{D}{d}} = \frac{2\pi x\, \varepsilon_0 \varepsilon_r}{\log_e \frac{D}{d}} \quad F/m \qquad ...... (2.119)$$

2.17 Dielectric Stress in a Single-Core Cable:

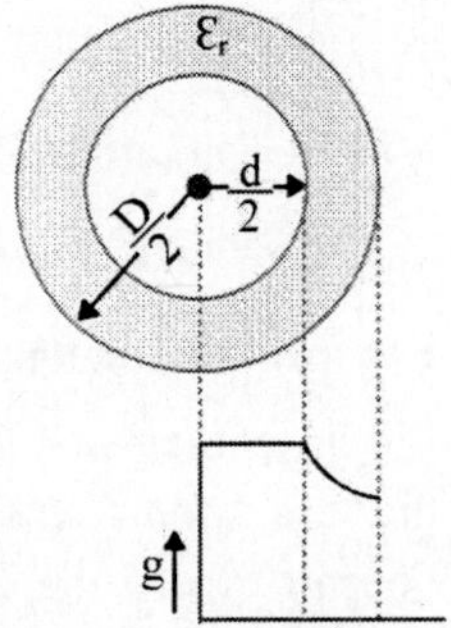

Fig. 2.39 Single-Core Cable

During cable operation, electrostatic forces exert an influence on the insulation. The term used to describe this occurrence is dielectric stress. The dielectric stress in a cable is determined by the electric intensity (potential gradient) at that specific location. Let's examine a wire that has only one core and is characterised by a diameter of d and an interior sheath with a diameter of D. Electric intensity is the measure of the change in electric potential per unit distance. The potential gradient, denoted as g, at a distance of x meters from the middle of the wire can be calculated as $g = E_x = \frac{Q}{2\pi x\, \varepsilon_0 \varepsilon_r} \quad V/m$ ...... (2.120)

From eq. 2.118, $Q = \frac{2\pi x\, \varepsilon_0 \varepsilon_r V}{\log_e \frac{D}{d}}$ ...... (2.121)

From eq. (2.119) and (2.120), $g = \frac{\frac{2\pi x\, \varepsilon_0 \varepsilon_r V}{\log_e \frac{D}{d}}}{2\pi x\, \varepsilon_0 \varepsilon_r} = \frac{V}{x \log_e \frac{D}{d}}$ ......(2.122)

Equation (2.119) clearly shows that the potential gradient varies inversely with distance x. The highest potential gradient occurs when x is minimum (x = d/2) or at the conductor's surface. On the other hand, at x = D/2 or at the sheath surface, the potential gradient is at its lowest.

∴ Maximum potential gradient is $g_{max} = \frac{2V}{d \log_e \frac{D}{d}} \quad Volts/m$ ......(2.123)

Minimum potential gradient is $g_{min} = \frac{2V}{D \log_e \frac{D}{d}} \quad Volts/m$ ......(2.124)

$$\frac{g_{max}}{g_{min}} = \frac{\frac{2V}{d\, log_e \frac{D}{d}}}{\frac{2V}{D\, log_e \frac{D}{d}}} = \frac{D}{d} \qquad \text{......(2.125)}$$

Figure 2.39 illustrates a variation in dielectric stress. The dielectric stress has its maximum value at the surface of the conductor and drops as the distance from the surface increases. When designing a cable, it is imperative to take into account the maximum stress. In order to maintain a cable operating at a voltage with a maximum stress of 5 kV/mm, it is necessary for the insulation to possess a dielectric strength of at least 5 kV/mm to avoid cable failure.

2.18 Most Economical Conductor Size in a Cable:

The maximum stress in a cable occurs at the conductor's surface, as previously demonstrated. To ensure cable safety, the insulation's dielectric strength should exceed the maximum stress limit. Rewriting the expression for maximal stress, we obtain, $g_{max} = \frac{2V}{d\, log_e \frac{D}{d}} \quad Volts/m$

For design purposes, it is imperative to keep the working voltage V and the internal sheath width D at a consistent value. The expression (2.98) only has one variable, which is the conductor diameter, represented by the symbol d. The conductor diameter that minimises the highest gradient, g_{max}, is the most cost-effective, given values of V and D. The value of g_{max} will be minimised when $d\, log_e \frac{D}{d}$ is maximised.

$$\frac{d}{dd}\left[d\, log_e \frac{D}{d}\right] = 0$$

$$log_e \frac{D}{d} + d\frac{d}{D}\left(\frac{-D}{d^2}\right) = 0$$

$$log_e \frac{D}{d} = 1$$

$$\frac{D}{d} = e = 2.718 \qquad \text{...... (2.126)}$$

∴ Most economical conductor diameter is $d = \frac{D}{2.718}$

The value of g_{max} under this condition is $g_{max} = \frac{2V}{d} \quad Volts/m \qquad \left(where\ log_e \frac{D}{d} = 1\right)$ ...... (2.127)

For low and medium voltage cables, this method's conductor diameter $\left(d = \frac{2V}{g_{max}}\right)$ may be insufficient for current density. Consequently, the conductor diameter of these cables is determined by calculating the safe current density. The use of this theory in high voltage cable designs leads to a value of d which doesn't exceed the acceptable limit for current carrying capacity. Thus, it is advisable to increase the diameter of the conductor by this magnitude.

2.19 Grading of Cables:

Grading is the procedure of attaining consistent electrostatic stress in the dielectric of cable. The electrostatic stress (g_{max}) in a single core cable is highest at the surface of the conductor and lowers as it moves towards the sheath. The highest voltage that can be safely delivered to a cable is defined by the electrostatic stress at the surface of the conductor, also known as g_{max}. In order for a cable with a uniform dielectric to operate without risk, the dielectric strength must exceed g_{max}. The effectiveness of high-strength dielectrics is greatest in close proximity to the conductor, where the stress is at its peak. However, as we move further away from the conductor, the electrostatic tension decreases, causing the dielectric to become too robust.

An uneven distribution of stress in a cable is detrimental for two reasons. Firstly, it is necessary to have a thicker layer of insulation, which in turn necessitates the use of larger cable sizes. Furthermore, it has the potential to result in insulation failure. Ensuring a uniform distribution of stress in cables is crucial for addressing the aforementioned challenges. In order to achieve this, it is necessary to evenly distribute the stress across the outer layers of the dielectric material, hence augmenting its value. This practice is known as cable grading. There are two primary ways for grading cables :(i) Capacitance Grading (ii) Intersheath Grading

2.19.1 Capacitance Grading: Capacitance grading is the technique of employing various dielectric layers to ensure a uniform distribution of dielectric stress. Capacitance grading entails the utilisation of a composite dielectric rather than a uniform one. The composite dielectric consists of many layers of different dielectrics, with the relative permittivity (ε_r) of each layer being inversely proportional to its distance from the centre. Under these circumstances, the potential gradient remains constant at every given place within the dielectric, regardless of its distance from the centre. The dielectric stress in the cable is uniformly distributed, leading to a high-quality grading. Optimal grading requires an unlimited quantity of dielectrics, a condition that is unattainable. Usually, two or three dielectrics are used in a certain order based on their permittivity, with the dielectric that has the highest permittivity placed closest to the core. Referring to Fig. 2.40, the capacitance grading is clearly explained. There are three dielectrics with different outer diameters (d_1, d_2) and relative permittivity (ε_1, ε_2, ε_3). When $\varepsilon_1 > \varepsilon_2 > \varepsilon_3$ and all three dielectrics are subjected to the same maximum stress,

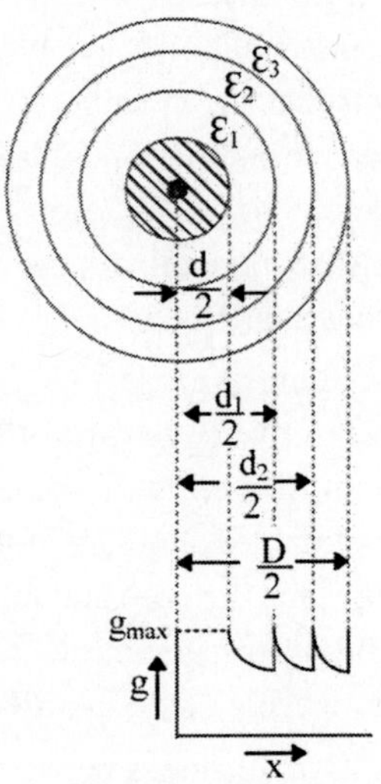

Fig. 2.40 Capacitance Grading

$$g_{1\,max} = g_{2\,max} = g_{3\,max}$$

$$\frac{Q}{2\pi\varepsilon_0\varepsilon_1 d/2} = \frac{Q}{2\pi\varepsilon_0\varepsilon_2 d_1/2} = \frac{Q}{2\pi\varepsilon_0\varepsilon_3 d_2/2}$$

$$\frac{Q}{\pi\varepsilon_0\varepsilon_1 d} = \frac{Q}{\pi\varepsilon_0\varepsilon_2 d_1} = \frac{Q}{\pi\varepsilon_0\varepsilon_3 d_2} \qquad \ldots\ldots (2.128)$$

$$\frac{1}{\varepsilon_1 d} = \frac{1}{\varepsilon_2 d_1} = \frac{1}{\varepsilon_3 d_2}$$

$$\varepsilon_1 d = \varepsilon_2 d_1 = \varepsilon_3 d_2 \qquad \ldots\ldots (2.129)$$

From eq. (2.118), $V_1 = \int_{\frac{d}{2}}^{\frac{d_1}{2}} g\, dx = \int_{\frac{d}{2}}^{\frac{d_1}{2}} \frac{Q}{2\pi\,\varepsilon_0\varepsilon_1 x}\, dx = \frac{Q}{2\pi\,\varepsilon_0\varepsilon_1} \log_e \frac{d_1}{d}$

$V_1 = \frac{g_{max}}{2} d \log_e \frac{d_1}{d}$ where $\frac{Q}{2\pi\,\varepsilon_0\varepsilon_1} = \frac{g_{max}}{2} d$

Similarly, $V_2 = \frac{g_{max}}{2} d_1 \log_e \frac{d_2}{d_1}$

$$V_3 = \frac{g_{max}}{2} d_2 \log_e \frac{D}{d_2} \qquad \ldots\ldots (2.130)$$

Total potential difference between core and earthed sheath is $V = V_1 + V_2 + V_3$

$$V = \frac{g_{max}}{2} d \log_e \frac{d_1}{d} + \frac{g_{max}}{2} d_1 \log_e \frac{d_2}{d_1} + \frac{g_{max}}{2} d_2 \log_e \frac{D}{d_2}$$

$$V = \frac{g_{max}}{2}\left(d \log_e \frac{d_1}{d} + d_1 \log_e \frac{d_2}{d_1} + d_2 \log_e \frac{D}{d_2}\right) \qquad \ldots\ldots (2.131)$$

If the cable had homogenous dielectric, the allowed potential difference between core and earthed sheath would have been

$$V = \frac{g_{max}}{2} d \log_e \frac{D}{d} \quad \text{...... (2.132)}$$

If the maximum stress in the three dielectrics is not the same, then,

$$V = \frac{g_{1\,max}}{2} d \log_e \frac{d_1}{d} + \frac{g_{2\,max}}{2} d_1 \log_e \frac{d_2}{d_1} + \frac{g_{3\,max}}{2} d_2 \log_e \frac{D}{d_2} \quad \text{...... (2.133)}$$

2.19.2 Intersheath Grading: The cable grading method involves the segmentation of a uniform dielectric material into many layers by introducing metallic intersheaths between the core and lead sheath. The intersheaths are maintained at potentials that are positioned between the potentials of the core and the earth. This arrangement improves the distribution of voltage in the dielectric of the cable, leading to a more consistent potential gradient.

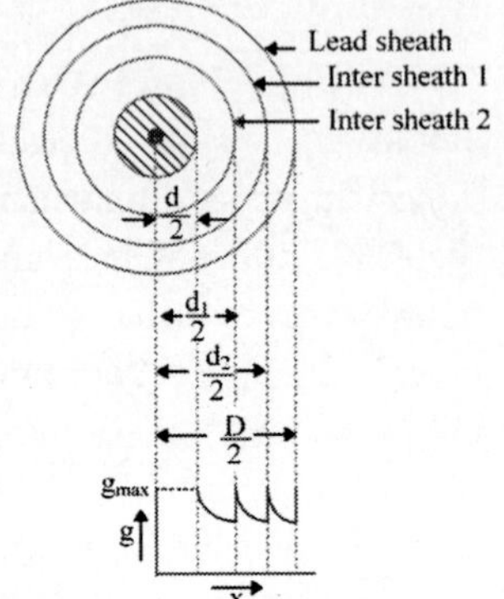

Fig. 2.41 Intersheath Grading

Imagine a cable that has a core dimension called "d" and an outside protective covering with a diameter called "D." Consider two intersheaths with diameters d_1 and d_2 that are placed within a uniform dielectric and maintained at certain potentials. Let V_1, V_2, and V_3 represent the voltages between the core and intersheath 1, intersheath 1 and 2, and intersheath 2 and outer lead sheath, respectively. Each intersheath can be considered as a single cable because of the noticeable potential difference between its inner and outer layers.

Maximum stress between core and intersheath 1 is $g_{1\,max} = \frac{V_1}{\frac{d}{2} \log_e \frac{d_1}{d}}$

Maximum stress between intersheath 1 and 2 is $g_{2\,max} = \frac{V_2}{\frac{d_1}{2} \log_e \frac{d_2}{d_1}}$

Maximum stress between intersheath 2 and outer sheath is $g_{3\,max} = \frac{V_3}{\frac{d_2}{2} \log_e \frac{D}{d_2}}$

The dielectric is homogenous, resulting in the same maximum stress across all layers.

$$g_{1\,max} = g_{2\,max} = g_{3\,max} \text{ then } \frac{V_1}{\frac{d}{2} \log_e \frac{d_1}{d}} = \frac{V_2}{\frac{d_1}{2} \log_e \frac{d_2}{d_1}} = \frac{V_3}{\frac{d_2}{2} \log_e \frac{D}{d_2}}$$

Because the cable functions like three capacitors in series, all potentials are in phase, resulting in a voltage between the conductor and earthed lead sheath. $V = V_1 + V_2 + V_3$

Intersheath grading features three significant drawbacks. Fixing sheath potentials can be a complex task. Furthermore, intersheaths may become damaged during the process of shipping and installation, resulting in the possibility of concentrated gradients occurring in some sites. Furthermore, the intersheaths experience substantial losses as a result of charging currents. Because of these factors, intersheath grading is hardly used.

Example 2.20 The conductor diameter and insulation thickness of a single-core cable are 2.4 cm and 0.6cm, respectively. Using a specific resistance of 6×10^{12} Ω-m, calculate the insulation resistance of a 3km cable.

Ans: Conductor radius, $r_1 = \frac{2.4}{2} = 1.2$ cm

Length of cable, $l = 3000\ m$

Resistivity of insulation, $\rho = 6 * 10^{12}\ \Omega - m$

Internal sheath radius, $r_2 = r_1 + t = 1.2 + 0.6 = 1.8$ cm

Insulation resistance of the entire cable is $R = \frac{\rho}{2\pi l} \log_e \frac{r_2}{r_1} = \frac{6*10^{12}}{2\pi*3000} \log_e \frac{1.8}{1.2} = 129\ M\Omega$

Example 2.21 Three single core cables make up a 5km underground cable that is 11kV, 50 Hz, and three phases. Each conductor has a diameter of 4 cm and 0.6 cm of radial insulation. Determine the following: (i) cable capacitance per phase, (ii) charging current per phase, and (iii) total charging kVAR. Insulation's relative permittivity is 3.

Ans: Total sheath diameter, $D = d + 2t = 4 + 2 * 0.6 = 5.2$ cm

Capacitance of the cable $C = \frac{2\pi\varepsilon_0\varepsilon_r l}{\log_e \frac{D}{d}} = \frac{2\pi * 8.854*10^{-12}*3*5000}{\log_e \frac{5.2}{4}} = 3.18\ \mu F$

$V_{ph} = \frac{110000}{\sqrt{3}} = 63.508\ kV$

Charging current $= I_c = 2\pi f C V_{ph} = 2\pi * 50 * 3.18 * 10^{-6} * 63508 = 6.34A$

Total charging kVAR $= 3V_{ph}I_c = 3 * 63508 * 6.34 = 120.77\ kVAR$

Example 2.22 or a single core cable, the maximum and minimum stresses on the dielectric are 50 and 20 kV/cm (r.m.s.), respectively. Calculate (i) the insulation thickness (ii) Operating voltage, if the conductor's diameter is 3 cm.

Ans: $g_{max} = 50\ kV/cm,\ g_{min} = 20\ kV/cm\ and\ d = 3cm$

We know that $\frac{g_{max}}{g_{min}} = \frac{D}{d} \Rightarrow D = d * \frac{g_{max}}{g_{min}} = 3 * \frac{50}{20} = 7.5cm$

We know that, the total sheath diameter, $D = d + 2t = 7.5$ cm

So, insulation thickness $t = \frac{D-d}{2} = \frac{7.5-3}{2} = 2.25\ cm$

$$V = \frac{g_{max}}{2} d \log_e \frac{D}{d} = \frac{50 * 10^6}{2} * 3 * \log_e \frac{7.5}{3} = 68.72\ kV$$

Example 2.23 An internal diameter of 3.5 cm and a conductor area of 0.666 cm² characterise a single phase, single core cable for a 33kV, 50Hz system. The dielectric in the cable has a permittivity of 4. Determine the cable's (i) maximum electrostatic stress, (ii) minimum electrostatic stress, (iii) capacitance per km of cable length, and (iv) charging current.

Ans: Area of cross – section of conductor, $a = 0.666\ cm^2$

Diameter of the conductor, $d = \sqrt{\frac{4a}{\pi}} = \sqrt{\frac{4*0.666}{\pi}} = 0.9208\ cm$

Internal diameter of sheath, D = 3.5 cm

Maximum electrostatic stress in the cable $g_{max} = \frac{2V}{d \log_e \frac{D}{d}} = \frac{2*33*10^3}{0.9208 \log_e \frac{3.5}{0.9208}} = 53.67\ kV/cm$

Minimum electrostatic stress in the cable $g_{min} = \frac{2V}{D \log_e \frac{D}{d}} = \frac{2*33*10^3}{3.5 \log_e \frac{3.5}{0.9208}} = 14.12\ kV/cm$

Capacitance of the cable $C = \frac{2\pi\varepsilon_0\varepsilon_r l}{\log_e \frac{D}{d}} = \frac{2\pi * 8.854*10^{-12}*4}{\log_e \frac{3.5}{0.9208}} = 0.166\ \mu F$

Charging current $= I_c = 2\pi f C V_{ph} = 2\pi * 50 * 0.166 * 10^{-6} * 33 * 10^3 = 1.727A$

Example 2.24 The conductor diameter and sheath inner diameter of a single-core, 66 kV cable used in a three-phase system are 2 and 6 cm, respectively. Determine the following: (i) intersheath placements, (ii) voltage on the intersheaths, and (iii) maximum and minimum stress, if two intersheaths are introduced so that the stress fluctuates between the same maximum and minimum in the three layers.

Ans: Diameter of the conductor, $d = 2\ cm$

Internal diameter of sheath, D = 6cm

$$V_{rms} = \frac{66000 * \sqrt{2}}{\sqrt{3}} = 53.9\ kV$$

As the maximum stress in the three layers is the same,

$$g_{1\,max} = g_{2\,max} = g_{3\,max}$$

$$\frac{V_1}{\frac{d}{2} log_e \frac{d_1}{d}} = \frac{V_2}{\frac{d_1}{2} log_e \frac{d_2}{d_1}} = \frac{V_3}{\frac{d_2}{2} log_e \frac{D}{d_2}}$$

In order that stress may vary between the same maximum and minimum in the three layers, we have,

$$\frac{d_1}{d} = \frac{d_2}{d_1} = \frac{D}{d_2} \text{ and } \frac{V_1}{d} = \frac{V_2}{d_1} = \frac{V_3}{d_2}$$

$$d_1 d_2 = Dd \Rightarrow d_1 d_2 = 6 * 2 = 12cm \qquad ....(1)$$

$$d_1^2 = dd_2 \Rightarrow d_1^2 = 2 * d_2 \Rightarrow \frac{d_1^2}{2} = d_2 \qquad ....(2)$$

$$From\ (1)\ and\ (2) \quad d_1 d_2 = 12cm \Rightarrow d_1 \frac{d_1^2}{2} = 12 \Rightarrow d_1^3 = 24 \Rightarrow d_1 = 2.88\ cm$$

$$d_2 = \frac{d_1^2}{2} = \frac{2.88^2}{2} = 4.147cm$$

Voltage on the intersheaths $\quad V = V_1 + V_2 + V_3$

$$53900 = V_1 + \frac{d_1}{d} V_1 + \frac{d_2}{d} V_1 = V_1 \left[1 + \frac{d_1}{d} + \frac{d_2}{d}\right] = V_1 \left[1 + \frac{2.88}{2} + \frac{4.147}{2}\right] = 4.513\ V_1$$

$$V_1 = \frac{53900}{4.513} = 11.94\ kV$$

$$V_2 = \frac{d_1}{d} * V_1 = \frac{2.88}{2} * 11940 = 17.19\ kV$$

Voltage on first intersheath $= V - V_1 = 53900 - 11940 = 41.96\ kV$

Voltage on second intersheath $= V - V_1 - V_2 = 53900 - 11940 - 17190 = 24.77\ kV$

Maximum electrostatic stress in the cable $g_{max} = \frac{V_1}{\frac{d}{2} log_e \frac{d_1}{d}} = \frac{11.94*10^3}{\frac{2}{2} log_e \frac{2.88}{2}} = 32.74\ kV/cm$

Minimum electrostatic stress in the cable $g_{min} = \frac{V_1}{\frac{d_1}{2} log_e \frac{d_1}{d}} = \frac{11.94*10^3}{\frac{2.88}{2} log_e \frac{2.88}{2}} = 22.74\ kV/cm$

Example 2.25 A 1.5 cm radius conductor is found in a single-core, lead encased cable with three layers of insulation and dielectric values of $\varepsilon_1 = 5, \varepsilon_2 = 4, \varepsilon_3 = 3$. Assume that the maximal potential gradient is being used by all three insulating materials. The sheath has a maximum potential gradient of 50 kV/cm and an inner radius of 3.5 cm. Determine the potential difference between the earthed sheath and core.

Ans: Diameter of the conductor, $d = 2 * r = 2 * 1.5 = 3\ cm$

Internal diameter of sheath, $D = 2 * R = 2 * 3.5 = 7$ cm

As the maximum stress in the three dielectrics is the same $\quad \varepsilon_1 d = \varepsilon_2 d_1 = \varepsilon_3 d_2$

$$d_1 = \frac{\varepsilon_1}{\varepsilon_2} d = \frac{5}{4} * 3 = 3.75cm$$

$$d_2 = \frac{\varepsilon_1}{\varepsilon_3} d = \frac{5}{3} * 3 = 5cm$$

$$V = \frac{g_{max}}{2} \left(d\ log_e \frac{d_1}{d} + d_1\ log_e \frac{d_2}{d_1} + d_2\ log_e \frac{D}{d_2}\right)$$

$$= \frac{50*10^3}{2} \left(3\ log_e \frac{3.75}{3} + 3.75\ log_e \frac{5}{3.75} + 5\ log_e \frac{7}{5}\right)$$

$$= 85.76\ kV$$

$$V_{rms} = \frac{85760}{\sqrt{2}} = 60.64\ kV$$

Example 2.26 A single-core lead sheathed cable with a voltage rating of 66kV to ground is to be designed. Its conductor radius is 0.75cm, and its three insulating materials, A, B, and C, have relative permittivities of 4, 3, and 2, with maximum permissible stresses of 50, 40, and 30kV/cm, respectively. Calculate the minimum internal diameter of the leadsheath.

Ans: Radius of the conductor, $r = 0.75cm$

Peak value of the voltage $V_{max} = 66000 * \sqrt{2} = 93.33\ kV$

$$g_{1\ max} = \frac{Q}{2\pi\varepsilon_0\varepsilon_1 d/2};\ g_{2\ max} = \frac{Q}{2\pi\varepsilon_0\varepsilon_2 d_1/2};\ g_{3\ max} = \frac{Q}{2\pi\varepsilon_0\varepsilon_3 d_2/2}$$

$$Q = 2\pi\varepsilon_0\varepsilon_1 g_{1\ max} d/2 = 2\pi\varepsilon_0\varepsilon_2 g_{2\ max} d_1/2 = 2\pi\varepsilon_0\varepsilon_3 g_{3\ max} d_2/2$$

$$\varepsilon_1 g_{1\ max} r = \varepsilon_2 g_{2\ max} r_1 = \varepsilon_3 g_{3\ max} r_2$$

$$r_1 = \frac{\varepsilon_1\, g_{1max}\, r}{\varepsilon_2\, g_{2max}} = \frac{4 * 50 * 10^3 * 0.75}{3 * 40 * 10^3} = 1.25cm$$

$$r_2 = \frac{\varepsilon_2\, g_{2max}\, r_1}{\varepsilon_3\, g_{3max}} = \frac{3 * 40 * 10^3 * 1.25}{2 * 30 * 10^3} = 2.5cm$$

$$V = \frac{g_{1\ max}}{2} d\ log_e \frac{d_1}{d} + \frac{g_{2\ max}}{2} d_1\ log_e \frac{d_2}{d_1} + \frac{g_{3\ max}}{2} d_2\ log_e \frac{D}{d_2}$$

$$= g_{1\ max}\, r\ log_e \frac{r_1}{r} + g_{2\ max}\, r_1\ log_e \frac{r_2}{r_1} + g_{3\ max} r_2\ log_e \frac{R}{r_2}$$

$$93330 = 50 * 10^3 * 0.75 * log_e \frac{1.25}{0.75} + 40 * 10^3 * 1.25 * log_e \frac{2.5}{1.25} + 30 * 10^3 * 2.5 * log_e \frac{R}{2.5}$$

$$93330 = 19155 + 34650 + 30 * 10^3 * 2.5 * log_e \frac{R}{2.5}$$

$$log_e \frac{R}{2.5} = \frac{39525}{30 * 10^3 * 2.5} = 0.527$$

$$R = 2.5 * e^{0.527} = 2.5 * 1.69 = 4.225cm$$

2.20 Corona: If a low voltage alternating potential difference is applied across two conductors with a substantial distance between them, the surrounding ambient air remains unaffected. Conductors emit a faint violet glow, called corona, when the applied voltage reaches the critical disruptive voltage. The term "corona" describes the occurrence of a violet glow, a hissing noise, and the production of ozone gas in an overhead transmission line. When the electric field strength at the surface of the conductor reaches a maximum value of 30 kV per cm, free electrons have enough energy to collide with a neutral molecule and remove one or more electrons from it. This process generates ions and free electrons, which are further accelerated and collide with neutral molecules, resulting in the formation of more ions. Therefore, the process of ionisation is additive. Ionisation generates a corona or spark between conductive materials.

2.20.1 Factors Affecting Corona: Corona formation is influenced by both atmospheric and line conditions. Corona relies on the following factors:

(i) The atmosphere: Corona is affected by the physical state of the environment since it is created as a result of the ionisation of the air surrounding conductors. Stormy weather produces more ions, resulting in lower corona voltage compared to fair weather.

(ii) Conductor size: The corona effect is determined by the shape and circumstances of the conductors. Uneven surfaces produce more corona due to lower breakdown voltage. As a result, stranded conductors have an uneven surface and generate more corona than solid conductors.

(iii) Spacing between conductors: Increasing the space between conductors beyond their diameters may eliminate the corona effect. Increasing the distance between conductors minimises electro-static forces at their surfaces, preventing corona formation.
(iv) The line voltage: Corona is significantly affected by line voltage. If it's low, the air around the conductors remains uncharged, resulting in no corona formation. Corona formation occurs when the line voltage is high enough to cause electrostatic tensions on the conductor surface, causing the air around it to conduct.

2.21 Critical disruptive voltage, Visual critical voltage and Power loss:

2.21.1 Critical disruptive voltage: The critical disruptive voltage refers to the minimal voltage between phases and neutral at which corona discharge takes place. Let us consider two conductors with radii r cm and a distance of d cm between them. The potential gradient at the conductor surface can be determined by using V as the phase-neutral potential. $g = \frac{V}{r\, log_e \frac{d}{r}}$ Volts/cm

To initiate the formation of corona, the value of g needs to be adjusted to match the breakdown strength of air. The electrical breakdown strength of air, measured at a pressure of 76 cm and a temperature of 25°C, is 30 kV per cm (maximum) or 21.2 kV per cm (root mean square). This value is represented by the symbol "g_o". If Vc represents the phase-neutral voltage necessary in these circumstances,

then, $g_0 = \frac{V_c}{r\, log_e \frac{d}{r}}$

The expression provided for the disruptive voltage pertains to standard conditions, specifically at a pressure of 76 cm of Hg and a temperature of 25°C. Nevertheless, as these conditions fluctuate, the air density likewise fluctuates, hence modifying the value of g_o. The magnitude of g_o is closely correlated with the density of the surrounding air. The breakdown strength of air at a barometric pressure of b cm of mercury and temperature of T°c is denoted as δg_o.

Air density factor $\delta = \frac{3.92 * b}{273+T}$

At standard conditions, the value of δ is equal to 1.

∴ Critical disruptive voltage, $V_c = g_0\, r\, \delta\, log_e \frac{d}{r}$

Additionally, the conductor's surface condition must be corrected. To account for this, multiply the above equation by the irregularity factor, m_o.

Critical disruptive voltage $V_C = g_0\, m_0\, \delta\, r\, log_e \frac{d}{r}$ Kv/ph ...... (2.134)

Where $m_0 = 1$ for polished conductors, 0·98 to 0·92 for dirty conductors, and 0·87 to 0·8 for stranded conductors.

2.21.2 Visual critical voltage: The visual critical voltage is the minimum phase-neutral voltage at which corona glow emerges throughout the line conductors. It has been observed that in parallel conductors, the corona light begins at a higher value V_v, known as the visual critical voltage, rather than at the disruptive voltage V_c. The empirical formula below provides the phase-neutral effective value of visual critical voltage:

Visual critical voltage $V_V = g_0\, m_v\, \delta\, r \left(1 + \frac{0.3}{\sqrt{\delta\, r}}\right) log_e \frac{d}{r}$ kV/ph ... ... (2.135)

The irregularity factor for polished conductors is 1·0, while rough conductors range from 0·72 to 0·82.

2.21.3 Power loss due to corona: Corona formation causes energy loss, which is through many means such as light, heat, sound, and chemical reactions. When the disruptive voltage is exceeded, the power loss caused by corona is calculated as:

$$Powerloss\ P = 242.2\left(\frac{f+25}{\delta}\right)\sqrt{\frac{r}{d}}\left(V_{ph}-V_C\right)^2 * 10^{-5}\ kW/km/ph \qquad \text{......(2.136)}$$

2.21.4 Advantages and Disadvantages of Corona: Corona exhibits both benefits and drawbacks. When designing a high voltage overhead line, it is crucial to achieve a harmonious equilibrium between the advantages and downsides of corona.

Benefits:

- ✓ Corona development causes the surrounding air to conduct, increasing the virtual diameter of the conductor. Increased diameter decreases electrostatic tensions between conductors.
- ✓ Corona mitigates the consequences of surge-related transients.

Drawbacks:

- × Corona causes a loss of energy. This affects the line's gearbox efficiency.
- × Corona produces ozone, which can induce conductor corrosion owing to chemical activity.
- × Corona causes a non-sinusoidal current draw, resulting in a non-sinusoidal voltage drop in the line. This can produce inductive interference with nearby communication lines.

Example 2.27 A 132 kV line with conductors that are 3 cm in diameter is intended to cause corona to arise when the line voltage surpasses 215 kV (r.m.s.). Determine the distance between the conductors if the potential gradient at which ionisation takes place is 30 kV per cm.

Ans: Given $m_0 = 1, \delta = 1, r = \frac{3}{2} = 1.5cm, V_{c\,ph} = \frac{215000}{\sqrt{3}} = 124\ kV$

$$Critical\ disruptive\ voltage\ \ V_C = g_0\ m_0\ \delta\, r\, log_e\frac{d}{r} \qquad Kv/ph$$

$$124 * 10^3 = 21.21 * 10^3 * 1 * 1 * 1.5 * log_e\frac{d}{1.5}$$

$$log_e\frac{d}{1.5} = \frac{124 * 10^3}{31.81 * 10^3} = 3.898$$

$$d = 1.5 * e^{3.898} = 1.5 * 49.303 = 73.95cm$$

Example 2.28 A 100km 3-phase, 220 kV, 50 Hz transmission line consists of equilateral triangle-shaped conductors with a radius of 1.5 cm that are spaced 3 meters apart. At 43°C and 78 cm of air pressure, find the line's corona loss per km. Assume that $m_o = 0.75$.

Ans: Given $l = 100km, m_0 = 0.75, b = 78cm,\ T = 43^0c, r = 1.5cm,\ \ V_{ph} = \frac{220000}{\sqrt{3}} = 127\ kV$

$$\text{Air density factor } \delta = \frac{3.92 * b}{273+T} = \frac{3.92 * 78}{273+43} = 0.9675$$

$$Critical\ disruptive\ voltage\ \ V_C = g_0\ m_0\ \delta\, r\, log_e\frac{d}{r} \qquad Kv/ph$$

$$= 21.21 * 10^3 * 0.75 * 0.9675 * 1.5 * log_e\frac{300}{1.5} = 122.2\ kV$$

$$Powerloss\ P = 242.2\left(\frac{f+25}{\delta}\right)\sqrt{\frac{r}{d}}\left(V_{ph}-V_C\right)^2 * 10^{-5}\ kW/km/ph$$

$$= 242.2\left(\frac{50+25}{0.9675}\right)\sqrt{\frac{1.5}{300}}\ (127 * 10^3 - 122.2 * 10^3)^2 * 10^{-5}$$

$$= 0.374\ kW/km/ph$$

$Three\ phase\ Powerloss = 3 * 0.374 = 1.1238\ kW/km$

$Total\ powerloss = 1.1238 * 100 = 112.38\ kW$

Example 2.29 The total corona loss of a three-phase equilateral transmission line is 66 kW at 105 kV and 99 kW at 115 kV. What precisely is the voltage that causes disruption? At 120 kV, how much corona loss is there?

Ans: $Powerloss\ P = 242.2\left(\frac{f+25}{\delta}\right)\sqrt{\frac{r}{d}}\left(V_{ph} - V_C\right)^2 * 10^{-5}\ \ kW/km/ph$

$$Powerloss\ P \propto \left(V_{ph} - V_C\right)^2$$

Transmission line has a total corona loss of 66kW at 105kV and a loss of 99kW at 115kV

$$Powerloss\ \ 66 * 10^3 \propto \left(\frac{105*10^3}{\sqrt{3}} - V_C\right)^2 \Rightarrow 66 * 10^3 \propto (60.62 * 10^3 - V_C)^2 \text{ and}$$

$$99 * 10^3 \propto \left(\frac{115*10^3}{\sqrt{3}} - V_C\right)^2 \Rightarrow 66 * 10^3 \propto (66.39 * 10^3 - V_C)^2$$

$$\frac{99*10^3}{66*10^3} = \frac{(66.39*10^3 - V_C)^2}{(60.62*10^3 - V_C)^2}$$

$$\sqrt{1.5} = \frac{66.39*10^3 - V_C}{60.62*10^3 - V_C}$$

$$1.224\,(60.62 * 10^3 - V_C) = 66.39 * 10^3 - V_C$$

$$74198 - 1.224 * V_C = 66390 - V_C$$

$$V_C = \frac{7808}{0.224} = 34.857\ kV$$

$$Powerloss\ \ W \propto \left(\frac{120 * 10^3}{\sqrt{3}} - V_C\right)^2 \Rightarrow W \propto (69.28 * 10^3 - V_C)^2$$

$$\frac{W}{99*10^3} = \frac{(69.28*10^3 - 34.857*10^3)^2}{(66.39*10^3 - 34.857*10^3)^2}$$

$$W = 1.1916 * 99 * 10^3 = 117.97kW$$

Example 2.30 The components of a 100km 3-phase, 220 kV, 50 Hz transmission line are conductors with a 0.5 cm radius spaced three meters apart in an equilateral triangle configuration. For a line at 21°C and 73.6cm of air pressure, find the visual disruptive voltage, critical disruptive voltage, and corona loss per km. Let $m_0 = 0.85, m_{v\ local} = 0.7\ and\ m_{v\ general} = 0.8$

Ans: Given $r = 0.5cm,\ b = 73.6cm,\ T = 21^0c$

$$m_0 = 0.85,\ \ m_{v\ local} = 0.7\ and\ \ m_{v\ general} = 0.8$$

Air density factor $\delta = \frac{3.92*b}{273+T} = \frac{3.92*73.6}{273+21} = 0.9813$

Critical disruptive voltage $V_C = g_0\, m_0\, \delta\, r \log_e \frac{d}{r}$ Kv/ph

$$= 21.21 * 10^3 * 0.85 * 0.9813 * 0.5 * \log_e \frac{200}{0.5} = 52.97\ kV$$

Local Visual critical voltage $V_{V\ local} = g_0\, m_{v\ local}\, \delta\, r \left(1 + \frac{0.3}{\sqrt{\delta r}}\right) \log_e \frac{d}{r}$ kV/ph

$$= 21.21 * 10^3 * 0.7 * 0.9813 * 0.5 * \left(1 + \frac{0.3}{\sqrt{0.9813*0.5}}\right) \log_e \frac{200}{0.5} = 63.2\ kV$$

General Visual critical voltage $V_{V\ general} = g_0\, m_{v\ general}\, \delta\, r \left(1 + \frac{0.3}{\sqrt{\delta r}}\right) \log_e \frac{d}{r}$ kV/ph

$$= 21.21 * 10^3 * 0.8 * 0.9813 * 0.5 * \left(1 + \frac{0.3}{\sqrt{0.9813*0.5}}\right) \log_e \frac{200}{0.5} = 71.2\ kV$$

(or) it can also be calculated as $V_{V\ general} = \frac{m_{v\ general}}{m_{v\ local}} * V_{V\ local} = \frac{0.8}{0.7} * 63.2 * 10^3 = 71.2kV$

2.22 Methods of Reducing Corona Effect:

Reports indicate that there is intense corona effects observed when the working voltage reaches 33 kV or higher. To prevent corona discharge in substations or bus bars intended for voltages of 33 kV or greater, meticulous design measures should be used. Failure to do so may lead to the ionisation of air, resulting in flash-over occurrences between insulators or phases, which can cause substantial equipment damage. The impact of the corona can be mitigated by employing the following methods:

- By increasing the size of the conductor, the voltage at which corona occurs increases, resulting in a significant decrease in corona effects. ACSR conductors with larger cross-sectional areas are used in transmission lines for this reason.
- By increasing the distance between conductors, the voltage at which corona occurs is raised, thus preventing corona effects. Nevertheless, it is important to note that increasing the spacing should be done cautiously, as it will lead to a substantial increase in the expenses associated with the supporting structure, such as larger cross arms and supports.
- As current is mostly concentrated towards outer layers of conductor due to skin effect, a hollow conductor of increased cross-section instead of solid conductor reduces corona off-course at the cost of lower mechanical strength.
- Use of bundled conductor is considered an effective means of reducing corona. Configuration of bundled conductors and its role in corona reduction is described in following section.

2.22.1 Bundled conductors:

As of now, we have examined three-phase systems including a singular conductor for each phase. However, for very high voltage lines, corona presents a significant issue when there is just one conductor per phase. Corona manifests when the surface potential gradient of a conductor exceeds the dielectric strength of the ambient air. This induces ionization in the vicinity of the conductor. Corona results in power loss. It also disrupts communication channels. Corona occurs as a hissing sound accompanied by ozone emission. Given that the majority of long-distance power lines in India operate at either 220 kV or 400 kV, it is preferable to prevent the occurrence of corona discharge.

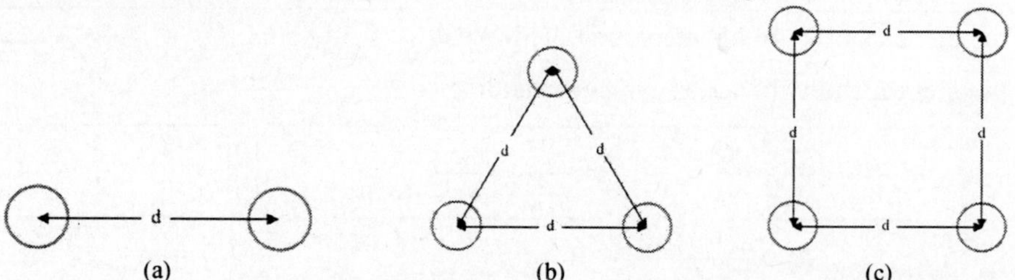

Fig. 2.42 Bundled conductors: (a) 2-conductor, (b) 3-conductor and (c) 4-conductor bundles

The high voltage surface gradient is significantly reduced by utilizing two or more conductors per phase in close proximity. This process is known as conductor bundling. The conductors are grouped in pairs, triplets, or quartets as illustrated in Fig. 2.42. The conductors within a bundle are spaced at regular intervals using spacer dampers, which inhibit conductor collision and mitigate swaying caused by wind. They additionally interconnect the lines in parallel. The geometric mean distance is computed under the assumption that the centre of a circular conductor equals with the centre of the bundle.

The geometric mean radius (GMR) of two, three and four- bundle conductors is given by

$$D_{S,2b} = \sqrt[4]{(D_S * d)^2} = \sqrt{D_S * d} \quad \text{......(2.137)}$$

$$D_{S,3b} = \sqrt[9]{(D_S * d * d)^2} = \sqrt[3]{D_S * d^2} \quad \text{......(2.138)}$$

$$D_{S,4b} = \sqrt[16]{\left(D_S * d * d * \sqrt{2}d\right)^4} = 1.09 * \sqrt[4]{D_S * d^3} \quad \text{......(2.139)}$$

It is apparent from equations (2.137), (2.138) and (2.139) that effective diameter of bundled conductor is higher than that of a single conductor. This reduces corona loss.

The inductance of the bundled conductor is then given by $L = 2 * 10^{-7}\, ln \frac{GMD}{D_{S,nb}}$ ......(2.140)

where D_s is the GMR of conductor and n=2,3,4 ....

2.23 Series compensation of transmission lines:

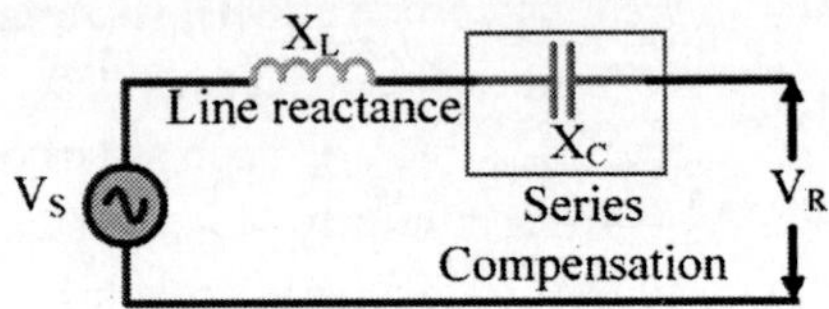

Fig. 2.43 Series compensation

A capacitor is connected in series with the line to reduce its net reactance as line's series impedance consists of resistance and inductive reactance. Reduction in line reactance results in enhancement in its power transfer capability leading to increased stability margin. An inductor is connected in series with the line to limit the fault current. This fault current limiter can be bypassed under normal operating conditions through a switch connected across it. Series capacitive compensation can be accomplished through the use of either fixed or switched capacitors, or by employing a Thyristor Controlled Series Capacitor (TCSC). Fixed capacitors offer a consistent amount of compensation, whereas switched capacitors can be activated or deactivated as required to modify the level of compensation. Thyristor Controlled Series Capacitor (TCSC) enable the provision of continuous and adjustable series compensation by regulating the current flow through the line. Series compensation not only enhances the power transfer capacity of the line but also reduces line losses and enhances the stability of the transmission system. It is frequently utilised in transmission lines that operate at extra high voltage and ultra-high voltage.

The power transfer capability of an uncompensated line is $= \frac{V_s V_r}{X} sin\delta$ ... ... (2.141)

The power transfer capability of series compensated line is $P_{series} = \frac{V_s V_r}{X - X_C} sin\delta$ ... ... (2.142)

$$\frac{P_{series}}{P_{UC}} = \frac{\frac{V_s V_r}{X-X_C} sin\delta}{\frac{V_s V_r}{X} sin\delta} = \frac{X}{X-X_c} = \frac{1}{1-\frac{X_c}{X}} = \frac{1}{1-K}$$

$$P_{series} = P_{UC}\left(\frac{1}{1-K}\right) \quad \text{... ... (2.143)}$$

where K is degree of compensation.

If k=0.2 then $P_{series} = P_{UC}\left(\frac{1}{1-0.2}\right) = 1.25\, P_{UC}$

k=0.3 then $P_{series} = P_{UC}\left(\frac{1}{1-0.3}\right) = 1.42\, P_{UC}$

k=0.4 then $P_{series} = P_{UC}\left(\frac{1}{1-0.4}\right) = 1.66\, P_{UC}$

k=0.5 then $P_{series} = P_{UC}\left(\frac{1}{1-0.5}\right) = 2\, P_{UC}$

k=0.6 then $P_{series} = P_{UC}\left(\frac{1}{1-0.6}\right) = 2.5\, P_{UC}$

k=0.7 then $P_{series} = P_{UC}\left(\frac{1}{1-0.7}\right) = 3.33\, P_{UC}$

k=0.8 then $P_{series} = P_{UC}\left(\frac{1}{1-0.8}\right) = 5\, P_{UC}$

k=0.9 then $P_{series} = P_{UC}\left(\frac{1}{1-0.9}\right) = 10\, P_{UC}$

Power transfer capability of the series compensator increases by increasing the value of K.

Benefits of Capacitive Series Compensation:

- ✓ Enhances voltage transmission: Series compensation involves injecting voltage into the transmission line, which leads to an improvement in the voltage profile at the load end and a reduction in voltage drop down the line. Implementing this can enhance equipment efficiency and minimise power loss.
- ✓ Minimises transmission losses: Series compensation helps mitigate transmission losses by enhancing power transmission efficiency over extended distances.
- ✓ Enhances system stability: Series compensation can enhance the stability of the transmission system by mitigating the voltage drop along the line and supplying a consistent source of reactive power.
- ✓ Enhances power transfer capacity: Series compensation reduces the impedance of the transmission line, so enabling it to transport a greater amount of power.
- ✓ Economical: Series compensation is typically more affordable than shunt compensation, making it a cost-effective choice for enhancing power transmission.
- ✓ Low maintenance: Series compensation devices, such as fixed capacitors, are typically self-regulating and require minimal or no control equipment, making them easy to maintain.

Drawbacks of Capacitive Series Compensation:

- × Insufficient efficacy: Series compensation is most efficient when there is a substantial voltage drop along the transmission line, particularly during periods of high load. Shunt compensation is more effective in boosting the power factor of the system during light load conditions.
- × Outage concerns: In the event of an outage on a transmission line equipped with series compensation, it is necessary to disable the series compensation to avoid overloading the other parallel lines. This process can be intricate and may necessitate the implementation of supplementary safeguards and regulatory measures.
- × Parallel line concerns: When incorporating series compensation into an existing transmission system, it may be imperative to have it applied to all lines in parallel in order to maintain system equilibrium. This process can be intricate and costly.
- × High voltage issues: High voltage problems may arise during system failures, potentially causing damage or failure to the series capacitors in the transmission line.
- × Sub-synchronous resonance: Series compensation has the potential to induce sub-synchronous resonance (SSR) in certain systems, resulting in instability and potential harm to equipment. Further expenditures may be required to resolve this matter.

2.24 Shunt compensation of transmission lines:

A capacitor is connected across the load terminals at the receiving end of transmission line to improve bus voltage magnitude and power factor. The capacitor consumes leading VAR compensating the lagging VAR consumed by inductive/motor loads. This reduces the VAR to be supplied to the load by the reactive power source through transmission network to maintain the voltage in desired range as capacitor effectively injects the VAR to the bus to meet reactive power demand. Apart from this as capacitor consumes leading VAR, the net reactive power consumption by capacitor-load combination reduces resulting in improved power factor. A capacitor bank may be switched either mechanically or through thyristor switches. A Thyristor Switched Capacitor (TSC) belongs to family of Static VAR compensator (SVC).

An inductor is connected across load terminals to overcome Ferranti effect under no load / light load condition of long transmission line. It has been discussed earlier that an unloaded / lightly loaded long transmission line may have voltage rise at the receiving end due to charging current in distributed capacitance of the line. This can be compensated by an inductor connected across load in series with a switch. The inductor connected in parallel with inductor increases reactive power consumption resulting in voltage drop compensating voltage rise due to Ferranti effect. The shunt inductor can be switched OFF once line is loaded. If the shunt inductor is controlled through thyristor, it is called Thyristor Controlled Reactor (TCR) which is a member of SVC family.

Benefits of Capacitive Shunt Compensation:

- ✓ Enhances voltage profile: Shunt compensation can enhance the voltage profile of a transmission system by supplying reactive power to the transmission line. Implementing this technique can effectively mitigate voltage drop and enhance the operational efficiency of equipment.
- ✓ Rapid regulation of excessive voltages: Over-voltages can be controlled quickly via shunt compensation. This can help to protect equipment and increase system stability.
- ✓ Effective for all load levels: Shunt compensation is effective at all load levels, unlike series compensation which is only effective during extreme load conditions. Regardless of the load on the system, this feature makes it a valuable tool for enhancing the power factor of a transmission system.
- ✓ Economical: Shunt compensation is typically more economical than series compensation, making it a financially efficient choice for enhancing power transmission and rectifying the power factor of a system.
- ✓ Low maintenance: Shunt compensation devices, such as fixed capacitors, are typically self-regulating and require minimal or no control equipment, making them easy to maintain.

Drawbacks of Capacitive Shunt Compensation:

- ✗ Increased expense: Shunt compensation typically incurs higher costs compared to series compensation, especially when there is a need for significant amounts of reactive power.
- ✗ Restricted overload capacity: The shunt compensation's ability to handle overload is limited, as the capacitors can only supply a specific amount of reactive power before reaching their overload limit.
- ✗ Complex control: Shunt compensation necessitates sophisticated control systems to guarantee the precise provision of reactive power to the transmission line. This can lead to an escalation in the expenses and maintenance demands of the system.
- ✗ Limited power transfer capability: Shunt compensation does not immediately enhance the power transfer capacity of a transmission line. In order to enhance the power transfer capability of the line, further procedures such as series compensation or line upgrading may be necessary.
- ✗ Risk of overvoltage: The improper coordination of shunt compensation with the rest of the transmission system might result in overvoltage and instability.

2.25 Comparisons between series and shunt compensation:

- Series compensation in electric power transmission refers to the incorporation of a capacitor or inductor in series with a transmission line to enhance its voltage transmission properties. Series compensation is employed to minimise transmission losses and enhance power transmission efficiency over extended distances.
- The primary distinction between series and shunt compensation lies in the placement of the compensation device inside the gearbox system. Series compensation is inserted in-line with the transmission line, whereas shunt compensation is across the load at the receiving end.
- Shunt compensation refers to the practice of enhancing the reactive power transmission properties of a transmission line by adding a capacitor or reactor across the load terminals at the receiving end of transmission line. Shunt compensation is employed to enhance the transfer of reactive power and rectify the power factor of the transmission system.
- Another distinction between the two is the nature of the remuneration offered. Series compensation enhances active power transmission by injecting voltage into the transmission line to offset voltage loss. Shunt compensation enhances the transfer of reactive power by supplying reactive power to the transmission line.
- Series compensation enhances the power transfer capability of the line, whereas, shunt compensation is beneficial in improving voltage profile and power factor.

2.26 Summary

- A transmission line's design and operation depend heavily on the line losses, voltage drop and transmission efficiency.
- Electricity can be transmitted or distributed using either an overhead system or underground cables.
- Inductance/conductor/metre of a

  single-phase line is $L = 2 * 10^{-7} \, log_e \frac{D_m}{D_s} = 2 * 10^{-7} \, log_e \frac{d}{0{\cdot}7788\, r}$

  single-circuit three-phase line is $L = 2 * 10^{-7} \, log_e \frac{\sqrt[3]{d_1 d_2 d_3}}{0{\cdot}7788\, r}$

  double-circuit three-phase line is $L = 2 * 10^{-7} \, log_e \frac{\sqrt[3]{D_{AB} D_{BC} D_{CA}}}{\sqrt[3]{D_{S1} D_{S2} D_{S3}}}$
- Capacitance of conductor A w.r.t neutral, $C_A = \dfrac{2\pi\varepsilon_0}{log_e \frac{\sqrt[3]{d_1 d_2 d_3}}{r}}$ F/m
- Overhead transmission lines are classified as short, medium and long transmission lines.
- A transmission line should ideally have low voltage regulation, meaning that an increase in load current should have little effect on the voltage at the receiving end.
- Generalised Circuit Constants of a

  Short Transmission Line are $\vec{A} = \vec{D} = 1, \vec{B} = \vec{Z}, \; and \; \vec{C} = 0$

  Nominal – T Tx. Line are $\vec{A} = \vec{D} = 1 + \frac{\vec{Y}\vec{Z}}{2}, \vec{B} = \vec{Z}\left(1 + \frac{\vec{Y}\vec{Z}}{4}\right)$, and $\vec{C} = \vec{Y}$

  Nominal – π Tx. Line are $\vec{A} = \vec{D} = 1 + \frac{\vec{Y}\vec{Z}}{2}, \vec{B} = \vec{Z}$, and $\vec{C} = \vec{Y}\left(1 + \frac{\vec{Y}\vec{Z}}{4}\right)$

  long Tx. Line are $\vec{A} = \vec{D} = \cosh(\sqrt{YZ}), \vec{B} = \sqrt{\frac{Z}{Y}}\sinh(\sqrt{YZ})$, and $\vec{C} = \sqrt{\frac{Y}{Z}} \sinh(\sqrt{YZ})$

- The type of insulating substance used in their construction and the voltage at which they are intended to be employed are the two ways that cables intended for Underground service cables can be classified: (i) based on the insulating material utilised in their production, and (ii) based on the voltage they are designed for.
- Based on core, cables can also be categorised as Sigle core cables and three core cables (belted, screened and pressured cables).
- Three-core belted cables are normally used for voltages of up to 11kV, but can occasionally reach 22kV.
- Screened cables are intended for use at up to 33 kV, but can be built up to 66 kV rating in specific circumstances.
- Maximum and minimum gradient of the cable are $g_{max} = \frac{2V}{d\ log_e \frac{D}{d}}$; $g_{min} = \frac{2V}{D\ log_e \frac{D}{d}}$ $Volts/m$
- Most economical conductor diameter is $d = \frac{D}{2.718}$
- If the cable had homogenous dielectric, the allowed potential difference between core and earthed sheath would have been $V = \frac{g_{max}}{2}\ d\ log_e \frac{D}{d}$
- If the maximum stress in the three dielectrics is not the same, then,

$$V = \frac{g_{1\,max}}{2}\ d\ log_e \frac{d_1}{d} + \frac{g_{2\,max}}{2}\ d_1\ log_e \frac{d_2}{d_1} + \frac{g_{3\,max}}{2}\ d_2\ log_e \frac{D}{d_2}$$

- The term "corona" describes the ozone gas production, violet hue, and hissing sound that occur in overhead transmission lines.
- The critical disruptive voltage refers to the minimal voltage between phases and neutral at which corona discharge takes place. $V_C = g_0\ m_0\ \delta\ r\ log_e \frac{d}{r}$ kV/ph
- The phase-neutral voltage that is required to produce corona glow throughout the line conductors is known as the visual critical voltage. $V_V = g_0\ m_v\ \delta\ r \left(1 + \frac{0.3}{\sqrt{\delta\, r}}\right) log_e \frac{d}{r}$ $\frac{kV}{ph}$
- The power loss caused by corona is $P = 242.2 \left(\frac{f+25}{\delta}\right) \sqrt{\frac{r}{d}}\ (V_{ph} - V_C)^2 * 10^{-5}\ kW/km/ph$

Short and Long Answer Questions

1. Determine the expression for the flux linkages caused by (i) a single conductor carrying current and (ii) parallel conductors carrying current.
2. In the following cases, find the equation for the inductance per phase of a three-phase overhead transmission line: (i) conductors organised symmetrically; (ii) conductors arranged asymmetrically with a fully transposed line
3. According to you, what does the term "electric potential" mean? Calculate the electric potential expression (i) at a single charged conductor and (ii) at a conductor that is a part of a larger group of charged conductors.
4. For a three-phase overhead transmission line with (i) symmetrically ordered conductors and (ii) unsymmetrically placed but transposed conductors, find the line-to-neutral capacitance expression.
5. What use does an overhead transmission line serve? How are these lines classified?
6. What effect does load power factor have on the regulation and efficiency of a transmission line?

7. Show how the End Condenser Method, Nominal-T Method, and Nominal-π Method are used to compute regulation and transmission efficiency for medium lines. To demonstrate your answer, use appropriate vector diagrams.
8. Derive expressions for the voltage and current at the sending end of a long transmission line using a rigorous method.
9. Regarding a transmission line, how would you interpret the generalised circuit constants? And why are they important?
10. Ascertain the generalised circuit parameters for the nominal-T and nominal-π methods, as well as for a short transmission line.
11. Analyse the advantages and disadvantages of an underground system compared to an overhead system.
12. Illustrate the different components of a high-voltage single-core cable with a neat diagram.
13. What is the most commonly used criterion for classifying cables? Create a sketch of a single-core low tension cable and label each component.
14. Draw an accurate cross-sectional drawing of the following:
    (i) Three-core belted cable (ii) Cables of the H and S.L. types
15. Derive an expression to represent a single-core cable's insulation resistance.
16. Determine the formula for a single-core cable's capacitance.
17. Prove g_{max}/g_{min} = D/d in a single-core cable.
18. Demonstrate the following cable grading techniques:
    (i) Intersheath grading (ii) Capacitance grading
19. Find the expression for the most cost-effective conductor size for a single core cable.
20. Explain the following terms in relation to corona:
    (i) Visually perceived critical voltage; (ii) Critical disruptive voltage; (iii) Corona-related power loss

Exercise

1. A 20 km single phase line consists of two parallel conductors 1.5 meters apart. The diameter of each conductor is 0.823 cm. If the resistance of the conductor is 0·311Ω/km, find the loop impedance of this line at 50 Hz.
2. A 150 km 3-phase, 50 Hz transmission line provides 45 MW at 110 kV and 0.85 pf. lag. A susceptibility of 3.5×10^{-6} Siemen/km/phase is observed, along with a line resistance and reactance per km of 1.5Ω and 2.5Ω, respectively. Compute (i) the sending end voltage and current; and (ii) the transmission efficiency. Apply the nominal-T approach.
3. The constants for a 100 km medium single-phase transmission line are as follows: Resistance/km 0.5Ω, reactance/km =0.7Ω. Susceptibility/km = 0.9×10^{-6} siemens. 66,000 volts is the line voltage at the receiving end. Compute (i) the sending end current, (ii) the sending end voltage, (iii) regulation, and (iv) the supply power factor, assuming that the line's entire capacitance is exclusively at the receiving end. With a lagging power factor of 0.8, the line delivers 10,000 kW. To illustrate your calculations, draw a phasor diagram.

4. The constants for a 250 km long three-phase transmission line are as follows: Resistance/ph/km =2Ω, Reactance/ph/km =4Ω, Shunt admittance/ph/km=6x10^{-6} Siemens. Apply a rigorous approach to determine the sending end voltage and current when delivering a 50 MW load at of 0·8 p.f. The receiving end voltage remains constant at 220 kV.
5. Fig. 2A1 shows the overhead line spacings for a double circuit with three phases. The ABC phase sequence is followed by the fully transposed line. The conductor's radius is 2 cm. Determine the inductance per km.
6. Find the double circuit 3-phase line's inductance per phase per km shown in Fig. 2A2. Each conductor has a radius of 2.5 cm and is transposed.

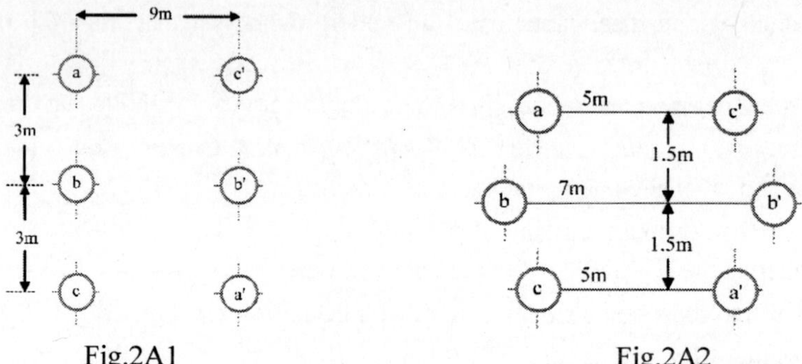

Fig.2A1 Fig.2A2

7. A 50 MW balanced three-phase load is delivered by a transmission line at 132 kV, 50 Hz, and 0.86 p.f. lagging. A single conductor has a phase-to-neutral admittance of 45×10^{-6} siemens and a series impedance of (25+j35) ohms. Using the nominal-π technique, find the following: (i) the A, B, C, and D constants of the line; (ii) the sending end voltage; and (iii) line regulation.
8. In a horizontal plane, the conductors of a 3-phase, 50 Hz, 132 kV overhead line are separated by 4.56 meters. The conductor has a diameter of 22·4 mm. If the line is 100 km long, find the charging current for each phase, assuming full transposition.
9. A 33 kV single-core cable with an inner diameter of 5 cm and a conductor diameter of 2.5 cm is utilised in a three-phase system. Assuming that the stress in the three layers changes between the same maximum and minimum ratio due to the introduction of two inter-sheaths, determine the following: Positions between sheaths, voltage between sheaths, and maximum and minimum stress.
10. A single-core lead sheathed cable with a voltage rating of 66kV to ground is to be designed. Its conductor radius is 1.5cm, and its three insulating materials, A, B, and C, have relative permittivities of 5, 4, and 3, with maximum permissible stresses of 60, 40, and 20kV/cm, respectively. Calculate the minimum internal diameter of the lead-sheath.
11. A 3 phase 220 kV 50 Hz transmission line is made up of conductors with a radius of 1.2 cm that are positioned 2 metres apart at the equilateral triangle's corners. Determine the corona loss for each kilometre along the line. The wire has a smooth weathered condition, and the temperature and barometric pressure are both fair at 20°C and 72·2 cm of Hg, respectively.

To know more about OH lines
Thermal Stress Technologies,
Machine learning methods for fault detection, and
Monitoring of Sag.

To know more about cables
Advancements & Challenges in Power Cable Laying,
Fault Detection & Location,
Optimization of Ampacity in HV UG Cables

To know more about corona
Inception Characteristics,
Dynamics of Surge Corona Growth and Corona on
Hybrid DC and AC OH Lines

To Model in MATLAB
Short Transmission Line,
Long Transmission Line, and
Unified Power Flow Controller

To Model in MATLAB
STATCOM,
SVC,
SVC & PSS

To Model in MATLAB
TCSC,
SSSC, and
Series Compensated Tx.

To know more about
Control techniques for compensating the fault current in distribution networks and
Power Quality Issues book

03 TRANSFORMERS

Unit specifics: In this unit, the following topics have been discussed for basic understating of transformers:

- Ideal transformer, practical transformer and different types of transformers
- Equivalent circuit diagram of a transformer referred to primary and secondary,
- Three-phase connections of transformers and phase-shifts,
- Three-winding transformers, construction and working of autotransformer,
- Neutral grounding transformers, and tap-changing transformers

Rationale: This unit describes the characteristics, advantages, and disadvantages of an ideal transformer, key characteristics of practical transformers and various types of transformers; Magnetically connected coils; the equivalent circuit of a transformer referred to primary and secondary; three-phase transformer connections and phase-shifts, including delta-delta (Δ-Δ), star-star (Y-Y), delta-star (Δ-Y), star-delta (Y-Δ), open delta or V-V, and Scott connections. Also addressed three-winding transformers, autotransformers, neutral grounding transformers, tap-changing transformers, and their construction, operation, functions, and applications. The concepts of transformers are explained thoroughly, with the aid of diagrams, derivations, and examples.

Pre-Requisites: Basic knowledge of power system and basic of electrical engineering.

Unit Outcomes: List of outcomes of this unit is as follows

U3-O1: To understand about ideal transformer, practical transformer and different types of transformers.
U3-O2: To analyse equivalent Circuit of a Transformer referred to Primary and Secondary.
U3-O3: To analyse three-phase connections of transformers and phase-shifts.
U3-O4: To analyse three-winding transformers and autotransformers.
U3-O5: To know more about Neutral grounding transformers and tap-changing in transformers
U3-O6: To understand and analyse transformers numerically.

Unit-3 outcomes	Expected mapping with course outcomes (1: Weak, 2: medium, and 3: strong correlation)					
	CO-1	CO-2	CO-3	CO-4	CO-5	CO-6
U3-O1	2	2	-	2	-	1
U3-O2	2	2	-	2	-	3
U3-O3	2	1	1	2	-	1
U3-O4	2	1	1	1	1	3
U3-O5	2	2	-	-	1	-
U3-O6	2	2	-	1	1	3

3.1 Introduction:

A transformer is a device that transfers energy from one electrical circuit to other through magnetic coupling. In an Alternating Current (AC) circuit, generation, transmission and distribution require different voltage levels. The need of voltage transformation from one level to other is achieved through transformer. This voltage transformation is done through change in flux linkage in mutually coupled windings with different number of turns. The winding connected to supply side is known as primary winding, whereas, the winding connected to load side is called secondary winding. A step-up transformer increases voltage level from primary to secondary side, whereas, step-down transformer reduces it. The configuration of a two winding transformer is shown in Fig. 3.1.

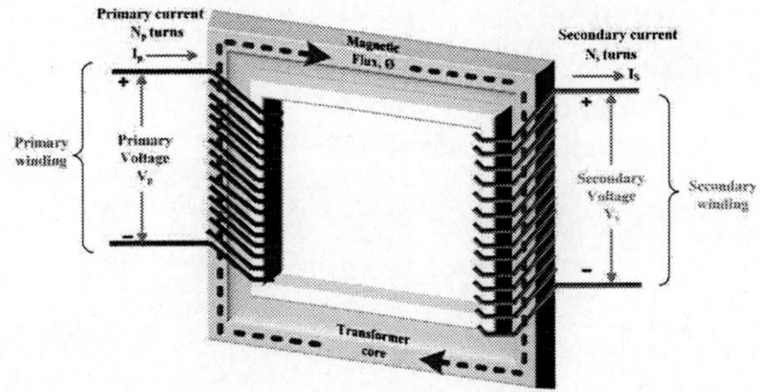

Fig. 3.1 Two winding Transformer

3.2 Types of Transformers:

The transformers are primarily categorized into different types based on their application, such as power transformers, distribution transformers, Isolation Transformers, instrument transformers (CT & PT), Step-Up Transformers, Step-Down Transformers, Auto Transformers, and Air-core transformers. These transformers are clearly depicted in Fig. 3.2 (a) to (h).

3.2.1 Power Transformers: Power transformers are used in transmission networks to step up or step-down voltage levels. They are designed specifically for exceptional performance and often have a large and heavy build, enabling them to handle high voltages and currents. The power transformer has a highly efficient layered metal design in its core, with windings composed of either copper or aluminium that have improved capabilities. Radiators are used to disperse heat, either through oil or air cooling.

3.2.2 Distribution Transformers: Distribution transformers serve for the ultimate voltage transition in distribution networks. They are smaller than power transformers.

3.2.3 Isolation Transformers: Isolation transformers isolate certain circuit components to provide protection and safety. The primary and secondary windings are not connected electrically. An isolation transformer is built up of a compact laminated steel core. The primary and secondary windings are controlled electrically via remote manipulation.

Application: Used in sensitive machinery and medical devices.

3.2.4 Instrument Transformers: Instrument transformers with measurement and safety capabilities are used in electric power infrastructure. The components include current and potential transformers (CT & PT).

3.2.5. Step-Up Transformers: Step-up transformers raise the voltage between the primary and secondary windings. They are used when a higher voltage is required for long-distance transmission. The secondary winding comprises more turns than the primary winding coil. When an alternating current flows through the primary winding, it produces a magnetic field in the core. The greater number of turns in the secondary winding induces a higher voltage from this magnetic field.

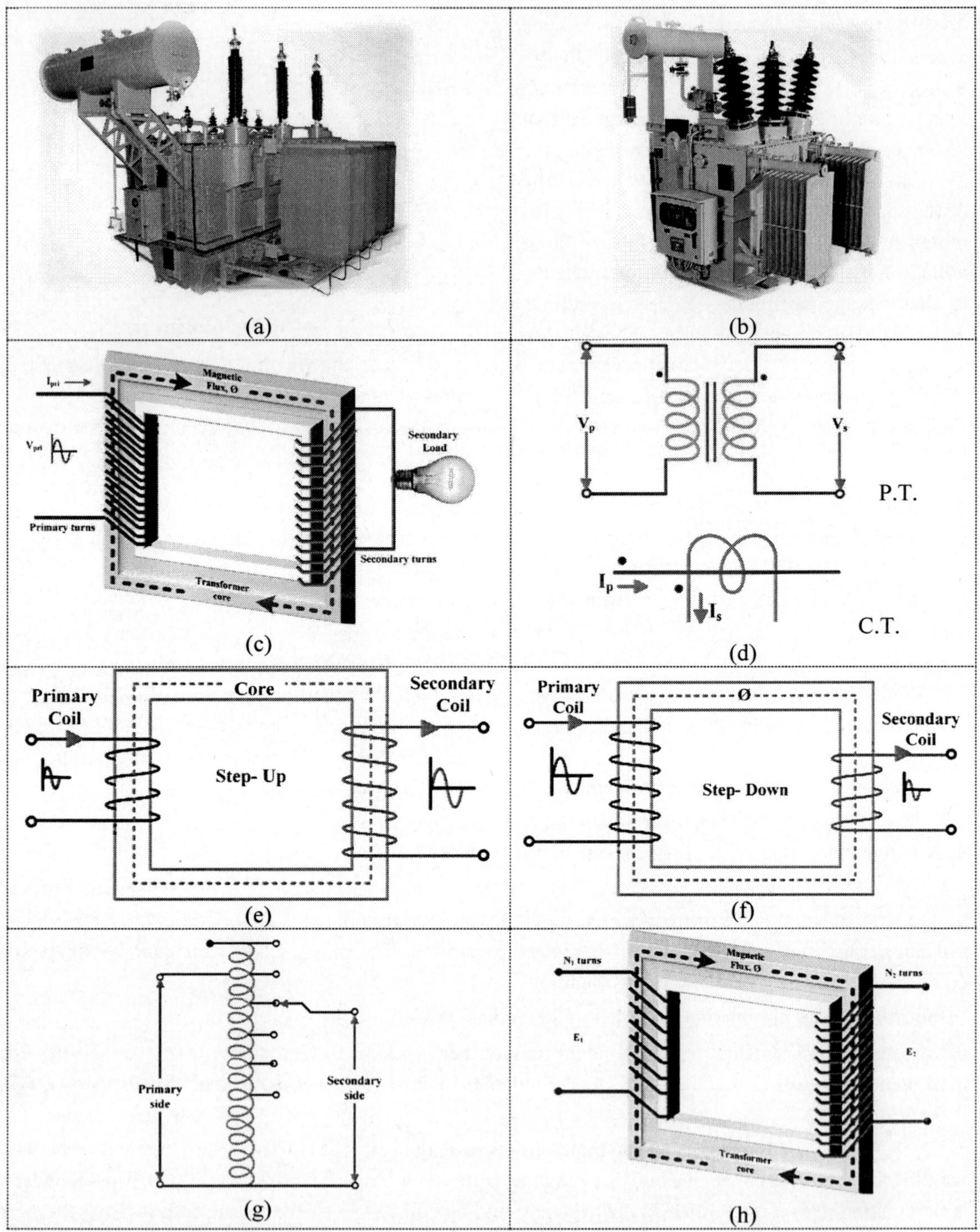

Fig. 3.2 (a) Power Transformer, (b) Distribution Transformer, (c) Isolation Transformer, (d) Instrument Transformers, (e) Step-Up Transformer, (f) Step-Down Transformer, (g) Auto Transformer and (h) Air-core Transformer

Applications: Utilized to increase the voltage of energy transmission in order to reduce power loss when transporting electricity via power lines.

3.2.6. Step-Down Transformers: Step-down transformers lower the voltage from the main to the secondary winding. They are used when it is necessary to lower the voltage for safe operation in residential and commercial buildings. The secondary winding comprises fewer turns than the primary winding. The magnetic field causes a lower voltage because the secondary winding has fewer turns.

Applications: Transformers are used in substations to convert high transmission voltages to lower distribution voltages, as well as in domestic appliances to provide the required voltage levels.

3.2.7. Autotransformers: Autotransformers may raise or lower voltage levels, and they contain a single winding that functions as both primary and secondary winding. Both access and output often require a portion of the winding. The transformer's voltage might change depending on the winding connections. The voltage is changed by tapping factors along the coil.

Applications: Used for voltage control, and to maintain voltage of sensitive equipment.

3.2.8. Air-Core Transformers: Air-core transformers use air rather than a magnetic core, lowering core losses and making them appropriate for high-frequency applications. Primary and secondary windings are wound on nonmagnetic materials, with no core to direct magnetic flow. The magnetic flux created by the primary winding flows through the air, activating a voltage in the secondary winding.

Applications: Used in radio frequency (RF) applications, wireless communication devices, and high-quality audio equipment that require extreme frequency and low loss.

3.3 Ideal transformer:

An ideal transformer is an imaginary transformer that is used as a theoretical tool to investigate the properties of real transformers. This transformer has no power loss, yielding a 100% efficiency. In an ideal transformer, there is no leakage flux, which means that the magnetic flux generated by the primary winding is completely linked with the secondary winding with no loss. Furthermore, a perfect transformer operates without magnetic saturation, keeping its linear magnetic properties regardless of the voltage or current applied.

The Ideal Transformer works on the idea of electromagnetic induction. The primary and secondary windings are coiled around the same magnetic core. When an electric current flows through the primary winding, it produces a magnetic field. The presence of this magnetic field causes the secondary winding to produce a voltage.

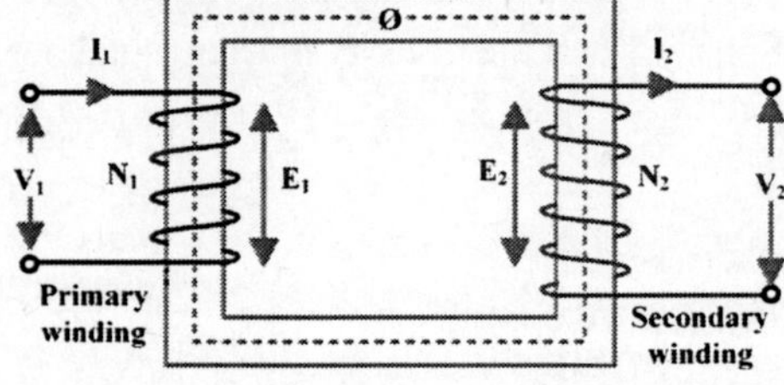

Fig. 3.3 (a) Two winding ideal transformer

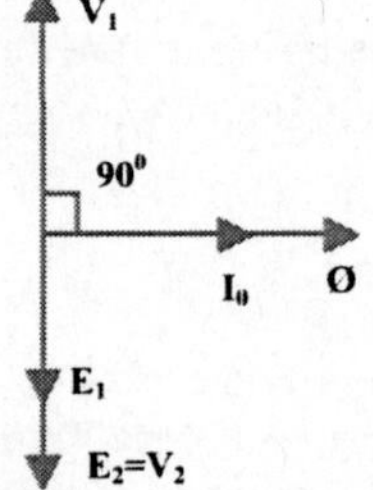

(b) Phasor diagram under no-load

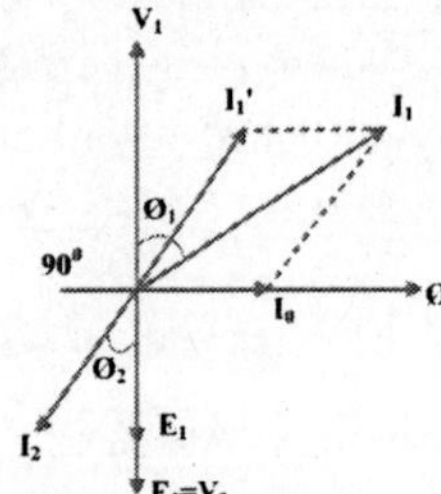

(c) Phasor diagram under load

Fig. 3.3 (a) shows a two winding ideal transformer with primary and secondary winding having N_1 and N_2 number of turns, respectively. An AC voltage V_1 is supplied to primary that circulates a small current, I_0 through it under no load condition that lags V_1 by 90° as circuit is purely inductive. This current produces flux, $\emptyset$ that flows through core and links both primary and secondary windings. As supply voltage is AC, the flux produced changes with time resulting in induced emf E_1 in primary and E_2 in secondary winding which is in phase opposition to supply voltage V_1 as illustrated by phasor diagram shown in Fig. 3.3 (b). When secondary winding is switched ON to a load, a current I_2 flows through secondary winding. This current produces flux that opposes main flux. As supply voltage is unchanged, a current I_1' in phase opposition to current I_2 appears in primary winding resulting in production of increased flux to nullify the flux decrease caused by load current, I_2. Therefore, when transformer is loaded, primary current is phasor sum of no-load current, I_0 and I_1', as illustrated by phasor diagram shown in Fig. 3.3 (c).

Power transformers use an iron core to contain the flux, ensuring that nearly all of the flux that links one coil links other coils as well. To form a single winding, multiple coils can be connected in series or parallel. These coils can be positioned on the core in alternating patterns with the coils from the other windings.

Fig. 3.4 (a) depicts how two windings on an iron core are arranged to make a single-phase shell transformer. A coil's number of windings can range from hundreds to thousands. Assume the flux in the transformer's core varies sinusoidally and the transformer is perfect. The core has infinite permeability (μ), allowing flux to travel through all turns of both windings with no losses or resistances. As a result, the voltages generated by the changing magnetic flux, E_1 and E_2, must be identical to the terminal voltages V_1 and V_2, respectively.

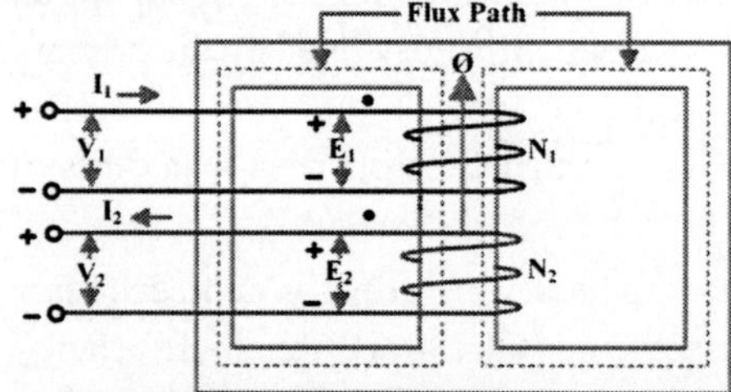

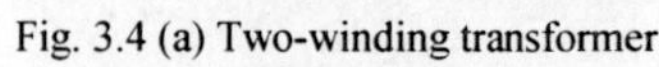
Fig. 3.4 (a) Two-winding transformer

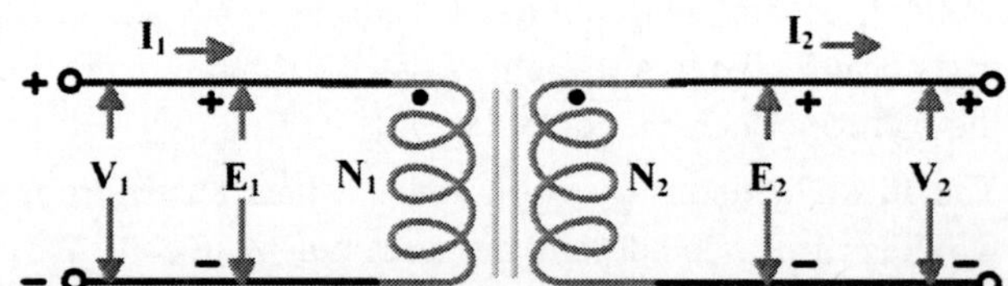

Fig.3.4 (b) Schematic representation of a two-winding transformer.

Examining the configuration of the windings represented in Fig. 3.4 (a) depicts that the voltages e_1 and e_2, which are generated by the changing magnetic flux, are in phase when the polarity markings denoted by the + and - symbols are applied. Then, by Faraday's law.

$$v_1 = e_1 = \sqrt{2}V_1 sin\omega t = N_1 \frac{d\emptyset}{dt} \quad \text{......... (3.1)}$$

$$v_2 = e_2 = \sqrt{2}V_2 sin\omega t = N_2 \frac{d\emptyset}{dt} \quad \text{......... (3.2)}$$

Where, v_1 = Instantaneous supply voltage to primary winding

e_1 = Instantaneous induced emf in primary winding

v_2 = Instantaneous terminal voltage of the secondary winding

e_2 = Instantaneous induced emf in the secondary winding

From equations (3.1) and (3.2), we get,

$$\frac{v_2}{v_1} = \frac{V_2}{V_1} = \frac{N_2}{N_1} = K \qquad \text{...... (3.3)}$$

Where, K represents the voltage transformation ratio which is equal to turns ratio $n = \frac{N_2}{N_1}$.

Typically, we have no idea which way the coils in a transformer are wound. A dot can be used to indicate winding information. By guaranteeing that all dotted endpoints of the windings have a positive voltage at the same time, the voltage from the dotted to the unmarked terminals of all windings are in phase. The two-winding transformer is depicted with dots in Fig. 3.4 (a) according to convention. Furthermore, the desired outcome can be achieved by positioning the dots in such a way that the current flowing from the terminal marked with dots to the terminal without any markings in each winding generates a magnetomotive force that acts in the same direction within the magnetic circuit. Fig. 3.4 (b) shows schematic representation of the two-winding transformer with core removed and dots representing sense of two windings wound on the core.

To find the relationship between currents I_1 and I_2 in the windings, we use Ampere's equation, which states that the magnetomotive force (mmf) in a closed channel can be determined using a line integral.

$$I = \oint H.ds \qquad \text{....... (3.4)}$$

Where I = net current that passes through the area bounded by the closed path

H = magnetic field intensit

H. ds = product of the tangential component of H and the incremental distance ds along the path

When evaluating the closed flux paths represented by the dotted lines in Figure 3.1, the law is used to determine that current I_1 passes through N_1 times and current I_2 flows through N_2 times. However, N_1I_1 and N_2I_2 produce mmfs in opposite directions, resulting in

$$\oint H.ds = N_1I_1 - N_2I_2 \qquad \text{....... (3.5)}$$

If we had chosen the opposite direction for the current I_2, the minus sign would be replaced with a plus sign. The line integral of the magnetic field intensity H along a closed path is zero when the permeability is infinite. If this statement is false, the flux density (corresponding to μH) would be infinite. However, the flux density must be finite in order to generate a finite electromotive force (e) in each winding when the flux changes.

$$N_1I_1 - N_2I_2 = 0 \Rightarrow \frac{I_1}{I_2} = \frac{N_2}{N_1}$$

$$\frac{V_2}{V_1} = \frac{E_2}{E_1} = \frac{N_2}{N_1} = \frac{I_1}{I_2} = K \qquad \text{...... (3.6)}$$

It should be noticed that I_1 and I_2 are in phase when the current is defined as positive when it enters the dotted terminal of one winding and exits the dotted terminal of another. If either current's direction is reversed, there will be a 180° phase displacement. The secondary winding of a transformer is the coil via which an impedance or load can be connected. Circuit elements connected to this winding are referred to as being on the transformer's secondary side. Similarly, the coil facing the energy source is known as the primary winding on the primary side. The difference between primary and secondary in the power system is irrelevant because power can flow bi-directionally through a transformer.

If an impedance Z_2 is connected across winding 2 of Fig. 3.4,

From eq. (3.2) and (3.6) $Z_2 = \frac{V_2}{I_2} = \frac{\left(\frac{N_2}{N_1}\right)V_1}{\left(\frac{N_1}{N_2}\right)I_1} = \left(\frac{N_2}{N_1}\right)^2 \left(\frac{V_1}{I_1}\right) = \left(\frac{N_2}{N_1}\right)^2 Z_1$ ……… (3.7)

The impedance connected to the secondary side referred to the primary $Z_2' = \frac{V_1}{I_1} = \left(\frac{N_1}{N_2}\right)^2 Z_2$ ……… (3.8)

3.4 Practical Transformer and its Equivalent Circuit:

The equivalent circuit diagram of any device can be quite useful in forecasting how it would perform under various operating conditions. It is basically the circuit representation of the device's performance equations. Figure 3.5 shows the corresponding circuit schematic for the transformer.

In a practical transformer :

- The flux produced by currents in primary and secondary windings are not completely linking with these windings, but small percentage of it leak through air. The flux that leaks through air is called leakage flux. The reactance X_1 and X_2 shown in Fig. 3.5 represent leakage reactance of primary and secondary winding, respectively, that account for leakage flux of corresponding winding.
- The primary and secondary windings have resistance R_1 and R_2, respectively, that result in voltage drop I_1R_1 and I_2R_2 and copper loss $I_1^2R_1$, and $I_2^2R_2$ in primary and secondary winding, respectively.
- The permeability of the core is finite that require finite current to magnetize the core.
- The magnetization and de-magnetization of core doesn't follow same path due to Hysteresis property of iron. This leads to Hysteresis loss in the core.
- The magnetic flux links not only to two windings, but also to iron core. The change in flux linking through core material leads to induced emf in it that result in flow of eddy currents through core. This causes eddy current loss in the core. Thus, apart from copper loss in the windings, power loss also occurs in the core. Since, core is made of iron, core loss is also known as iron loss that consists of two components -(i) Hysteresis loss and (ii) eddy current loss.
- The B-H curve is non-linear due to magnetic saturation at high magnetic field intensity.

In order to obtain the equivalent circuit of the transformer, it is necessary to take into account all the aforementioned factors. Fig. 3.5 represents a practical transformer that consists of a two winding ideal transformer connected to external circuits through the parameters that account for practical constraints mentioned above. Unlike ideal transformer, the no-load current, I_0 consists of two components- (i) The component I_m lags emf E_1 by 90° and is used to magnetize the core.

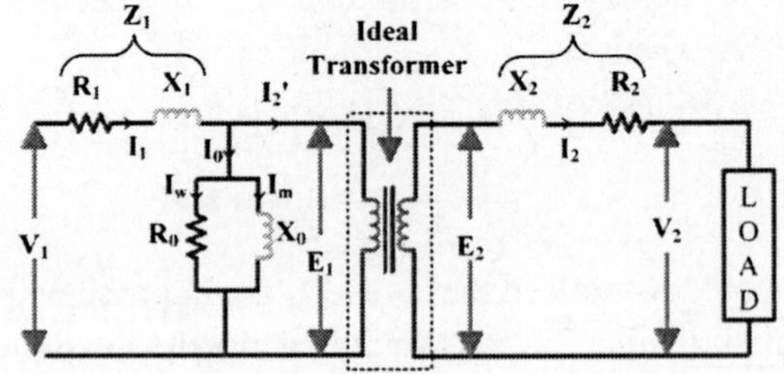

Fig. 3.5 Equivalent circuit diagram of a transformer

The current I_m is restricted by reactance X_0 known as magnetizing reactance. (ii) The component I_w that is in phase with E_1 accounts for iron loss. I_w is restricted by a fictitious resistance R_0 known as core loss resistance. X_0 and R_0 are connected across the primary winding. The resistance R_1 in series with X_1 is connected in series with primary winding to account for copper loss and leakage flux of primary winding. The resistance R_2 in series with X_2 is connected in series with secondary winding to account for copper loss and leakage flux of secondary winding. From the equivalent circuit of practical transformer shown in Fig. 3.5:

$$V_1 = E_1 + I_1 Z_1 \quad and \quad E_2 = V_2 + I_2 Z_2 \qquad \text{......... (3.9)}$$

The value of Ro and Xo can be calculated as, $R_0 = \frac{E_1}{I_w}$ and $X_0 = \frac{E_1}{I_m}$

3.4.1 Equivalent Resistance of the transformer:

When establishing a practical transformer, it is critical to account for winding resistance. The resistance of either of the two windings can be transferred to the other. To allow for simple computations, the transfer can be made to either the primary or secondary side. Figure 3.6 depicts the transformer's winding resistance as resistance in a series configuration.

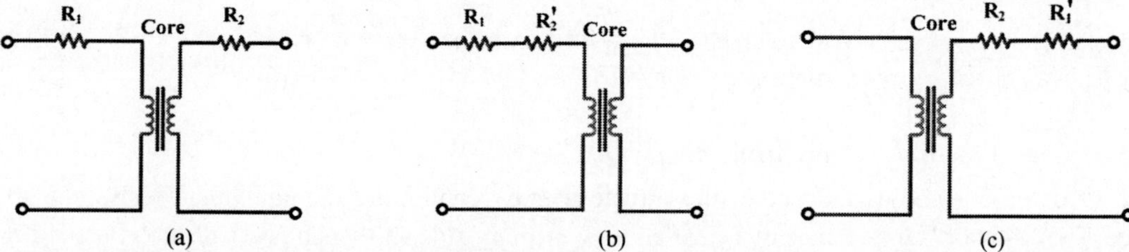

Fig. 3.6 (a) Equivalent Resistance of a T/F; equivalent resistance of T/F referred to (b) primary side (c) secondary side

Where, R_1 represent the resistance of the primary winding and R_2 represent the resistance of the secondary winding.

Case (i): Effective resistance of T/F referred to primary side: Fig. 3.6 (b) shows the effective resistance of a transformer referred to primary side.

$$I_1^2 R_2' = I_1^2 R_2 \;\Rightarrow\; R_2' = \left(\frac{I_2}{I_1}\right)^2 R_2 \;\Rightarrow\; R_2' = \frac{R_2}{K^2}$$

The equivalent secondary resistance R_2 referred to the primary side; $R_2' = \frac{R_2}{K^2}$ ... (3.10)

Where, K = Transformation ratio $= \frac{I_1}{I_2}$; R_2' = equivalent secondary resistance referred to the primary side

The total equivalent resistance of transformer referred to primary, $R_{01} = R_1 + R_2' = R_1 + \frac{R_2}{K^2}$ ... (3.11)

Case (ii): Effective resistance of T/F referred to secondary side: Fig. 3.6 (c) shows the effective resistance of a transformer referred to secondary side.

$$I_2^2 R_1' = I_1^2 R_1 \;\Rightarrow\; R_1' = \left(\frac{I_1}{I_2}\right)^2 R_1 \;\Rightarrow\; R_1' = K^2 R_1$$

The equivalent primary resistance R_1 referred to secondary side; $R_1' = K^2 R_1$ ... (3.12)

The total equivalent resistance of transformer referred to secondary, $R_{02} = R_1' + R_2 = K^2 R_1 + R_2$ ... (3.13)

3.4.2 Equivalent Leakage Reactance of the transformer:

The leakage reactance of both windings can be transferred to either of the windings, just like the resistance. The combined imaginary reactance of both windings is added, as depicted in Figure 3.7.

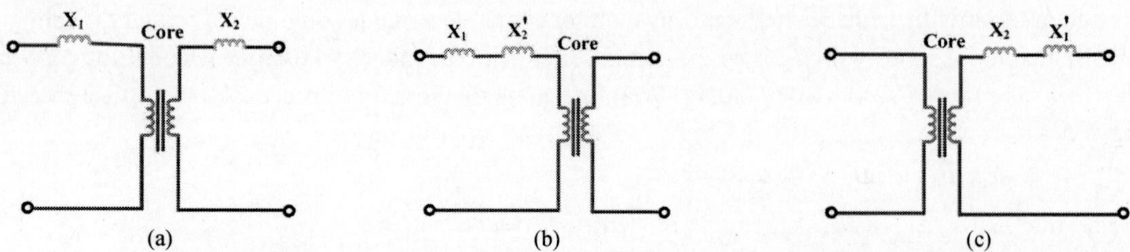

Fig. 3.7 (a) Equivalent Reactance of a T/F; equivalent reactance of T/F referred to (b) primary side (c) secondary side

Where, X1 represents the leakage reactance of primary winding and
X_2 represents the leakage reactance of secondary winding

The equivalent secondary reactance X_2 referred to the primary side; $X_2' = \frac{X_2}{K^2}$ ... (3.14)

The total equivalent reactance of transformer referred to primary, $X_{01} = X_1 + X_2' = X_1 + \frac{X_2}{K^2}$ ... (3.15)

The equivalent primary reactance X_1 referred to the secondary side; $X_1' = K^2 X_1$ ... (3.16)

The total equivalent reactance of transformer referred to secondary, $X_{02} = X_1' + X_2 = K^2 X_1 + X_2$ ... (3.17)

3.4.3 Equivalent impedance of the transformer:

Now, let us determine the impedance of the transformer by combining the equivalent resistance and reactance. The equivalent impedance of T/F referred to primary side and secondary side are given in Fig. 3.8.

Fig. 3.8 Equivalent impedance of transformer referred to (a) primary side and (b) secondary side

The total equivalent impedance of transformer referred to primary, $Z_{01} = R_{01} + jX_{01}$ ... (3.18)

The total equivalent impedance of transformer referred to secondary, $Z_{02} = R_{02} + jX_{02}$ ... (3.19)

3.4.4 Equivalent Circuit of Transformer referred to Primary:

When considering an equivalent circuit from the perspective of the primary side, we must transfer all components from the secondary side to the primary side. The diagram 3.9 (a) illustrates the simplified equivalent circuit of the transformer, referred to primary side. Assuming the negligible values of winding resistance and leakage reactance, we can consider V_1 and E_1 to be about identical. Thus, relocating the parallel combination of R_0 and X_0 to the input terminals, as depicted in Fig. 3.9 (b), would have minimal impact on the exciting current. By doing this action, a minute error will be introduced.

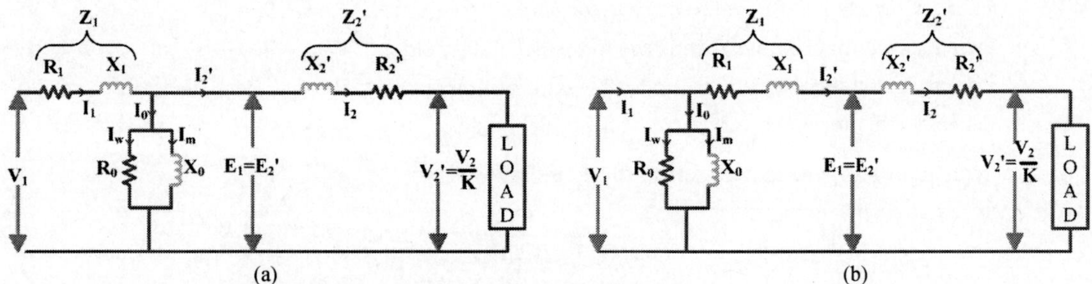

Fig. 3.9 (a) Equivalent circuit diagram of a transformer referring to the primary.
(b) Equivalent circuit diagram of a transformer by relocating the parallel combination of R_0 and X_0 to the input terminals.

Secondary induced emf E_2 referred to primary side; $E_2' = \frac{E_2}{K} = E_1$

Secondary terminal voltage V_2 referred to the primary side; $V_2' = \frac{V_2}{K}$

Secondary resistance R_2 referred to the primary side; $R_2' = \frac{R_2}{K^2}$

Secondary reactance X_2 referred to the primary side; $X_2' = \frac{X_2}{K^2}$

To simplify the process, we can add both primary and secondary resistance and reactance. The total resistance and reactance referred to primary side will be

$$R_{01} = R_1 + R_2' = R_1 + \frac{R_2}{K^2} \ and\ X_{01} = X_1 + X_2' = X_1 + \frac{X_2}{K^2}$$

By combining primary and secondary impedances, the equivalent circuit of transformer becomes as shown in the Fig. 3.10 (a). If the calculation mainly focuses on voltage regulation, then it is possible to disregard the entire excitation branch, which consists of the parallel combination of R_0 and X_0. Subsequently, the circuit can be represented by the equivalent circuit depicted in Fig. 3.10 (b). The circuit can be further simplified ignoring resistance R_{01} as shown in Fig. 3.10 (c).

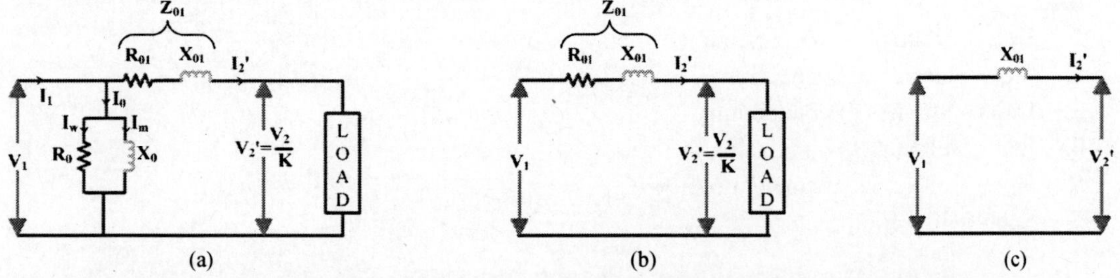

Fig. 3.10 (a) Equivalent circuit diagram of a transformer by combining primary and secondary impedance
(b) Equivalent impedance circuit diagram of a transformer by neglecting parallel combination of R_0 and X_0
(c) Equivalent reactance circuit diagram of a transformer by neglecting parallel combination of R_0 and X_0; and R_0.

3.4.5 Equivalent Circuit of Transformer referred to Secondary:

Similarly, we can find the approximate equivalent circuit referred to the secondary side and this circuit is shown in Fig. 3.11.

Where Primary induced emf E_1 referred to Secondary side; $E_1' = KE_1$

Primary terminal voltage V_1 referred to the Secondary side; $V_1' = KV_1$

Primary resistance R_1 referred to the Secondary side; $R_1' = K^2R_1$

Primary reactance X_1 referred to the Secondary side; $X_1' = K^2X_1$

The total resistance and reactance referred to secondary will be

$$R_{02} = R_1' + R_2 = K^2R_1 + R_2 \text{ and } X_{02} = X_1' + X_2 = K^2X_1 + X_2$$

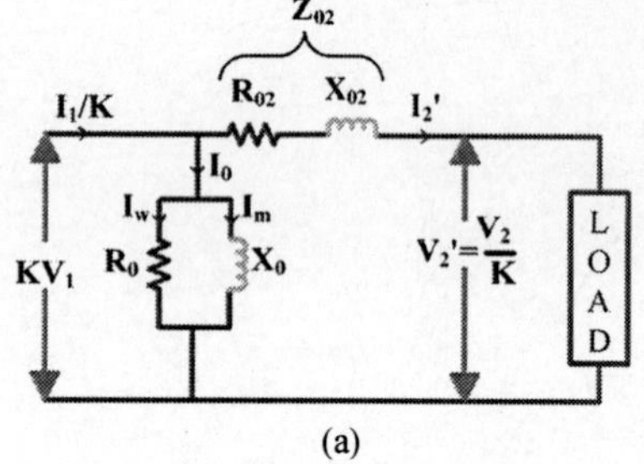

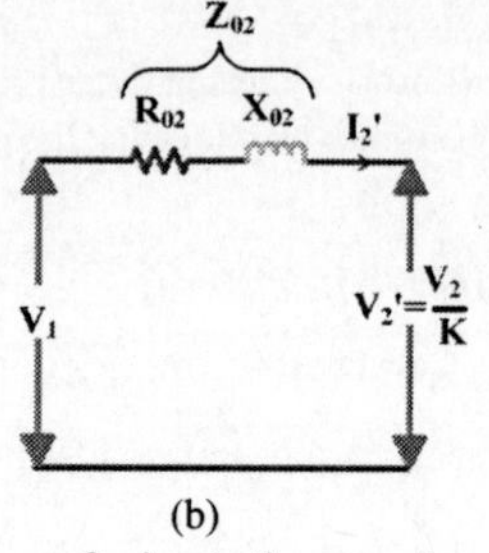

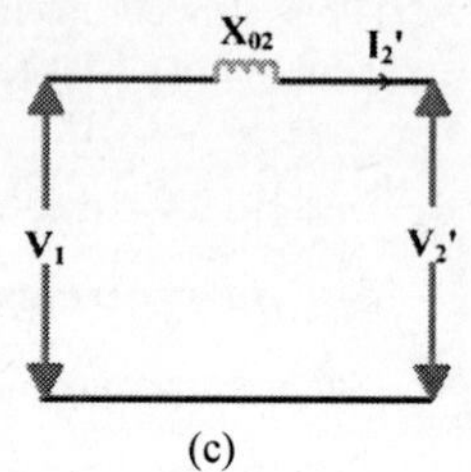

Fig. 3.11 (a) Equivalent circuit diagram of a transformer referring to the secondary.

(b) Equivalent circuit diagram of a transformer by relocating the parallel combination of R_0 and X_0 to the input terminals.

(c) Equivalent circuit diagram of a transformer referring to the secondary, ignoring the series resistance.

3.5 Three-phase connections of transformers and Phase-shifts:

In a three-phase electrical system, voltages are transformed using a three-phase transformer. A three-phase transformer consists of a three-phase primary winding that is magnetically coupled to a three-phase secondary winding. The primary and secondary windings can be connected in a number of ways, including star (wye), delta, or a combination of the two. The transformation ratio, phase shift between primary and secondary voltages, grounding availability, and other associated parameters are all determined by the type of connection. Various aspects determine which connections are suited for which applications. The most common types of three-phase transformer connections are as follows:

1. Delta - Delta (Δ - Δ) connection
2. Star - Star (Y-Y) connection
3. Delta - Star (Δ-Y) connection
4. Star - Delta (Y-Δ) connection
5. Open Delta or V-V connection
6. Scott Connection

3.5.1 Delta - Delta (Δ - Δ) connection of a 3-Phase transformer:

The delta-delta configuration of a three-phase transformer has both the primary and secondary windings connected in delta arrangement, as shown in Fig. 3.12 (a). The three primary windings are shown as A_1A_2, B_1B_2, C_1C_2 with line terminals marked as 1, 2, 3. The three secondary windings are shown as a_1a_2, b_1b_2, c_1c_2 with line terminals marked as $1', 2', 3'$. The phasor diagram for voltages and currents on primary side is shown in Fig. 3.12 (b), whereas, the phasor diagram for secondary side voltages and currents is shown in Fig. 3.12 (c). It is assumed that power factor is same on primary and secondary side with phase current lagging corresponding phase voltage by an angle θ.

It is observed from Fig. 3.12 that:

$$\frac{V_a}{V_A} = \frac{V_{1'2'}}{V_{12}} = K; \frac{V_b}{V_B} = \frac{V_{2'3'}}{V_{23}} = K; \frac{V_c}{V_C} = \frac{V_{3'1'}}{V_{31}} = K \qquad \text{... ... ... (3.20)}$$

Where, K represents line voltage transformation ratio (i.e. ratio of secondary line voltage to primary line voltage).

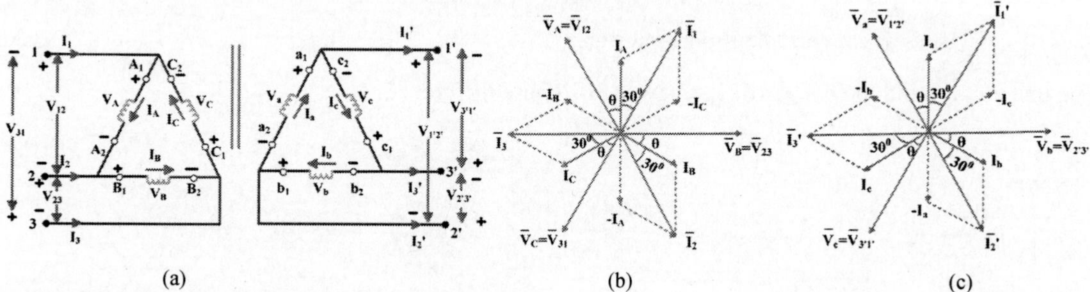

Fig. 3.12 (a) Delta-Delta configuration of a 3-Ph transformer
(b) Phasor Diagram of primary side voltages and currents
(c) Phasor Diagram of secondary side voltages and currents

Under balanced conditions, the current flowing through the line is √3 times the current flowing through each phase for both primary as well as secondary winding.

$$\text{In a delta connected 3 − phase system} \qquad V_L = V_{ph} \;\; and \;\; I_L = \sqrt{3}\, I_{ph} \qquad \text{... ... ... (3.21)}$$

Where, V_L and V_{ph} represent the magnitude of line voltage and phase voltage respectively, whereas, I_L and I_{ph} represent the magnitude of line current and phase current, respectively.

Neglecting the magnetizing current, the ratio of the currents can be expressed as:

$$\frac{I_A}{I_a} = \frac{I_B}{I_b} = \frac{I_C}{I_c} = K; \frac{I_1}{I_{1'}} = \frac{I_2}{I_{2'}} = \frac{I_3}{I_{3'}} = K \qquad \text{... ... ... (3.22)}$$

Important features:

- The phase voltage is equal to the line voltage.
- The transformation ratio is equivalent to the voltage ratio.
- Resilient against failures and capable to supply demand if one phase is out.
- Eliminates triple harmonics.

Applications: Typically employed where load needs higher voltage supply, such as in industrial operations.

Advantages of delta–delta connection of transformer:

✓ The delta-delta design is effective for both balanced and unbalanced load circumstances.

✓ The delta connection allows for uninterrupted three-phase power supply even if one phase fails.

✓ The delta arrangement allows third-harmonic currents to flow in a closed loop without affecting the output voltage waveform. The delta-delta configuration's main limitation is the absence of a neutral point.

✓ This setup is useful when there is no requirement for a neutral connection in either the primary or secondary side, and when working with low to moderate voltages.

3.5.2 Star - Star (Y - Y) connection of a 3-Phase transformer:

The three-phase transformer is configured in a star-star arrangement, with the primary and secondary windings in a star pattern. The configuration of the Y-Y transformer is shown in Fig. 3.13 (a) and its phasor diagram is shown in Fig. 3.13 (b). Under balanced conditions, the line voltage is $\sqrt{3}$ times the phase voltage, whereas, the line current is the same as the phase current.

$$\text{In a star connected } 3-\text{phase system} \qquad V_L = \sqrt{3}\, V_{ph} \;\; and \;\; I_L = I_{ph} \qquad \text{......... (3.23)}$$

The line voltage and the phase voltage have a 30° phase difference.

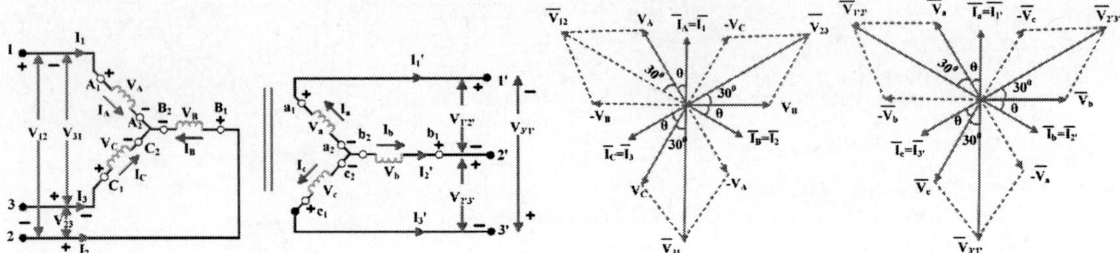

Fig. 3.13 (a) Star-Star Connection of a 3-Ph transformer

Fig. 3.13 (b) Phasor diagram of star-star 3-Ph transformer

Important features:

- The line voltage is $\sqrt{3}$ *times* the phase voltage.
- The neutral point is unstable under unbalanced conditions if load neutral is absent.

Applications: Typically employed in applications with balanced loads, particularly in the case of small-rated transformers.

The star-star (Y-Y) connection faces two major issues:

- The Y-Y configuration is not capable of effectively managing imbalanced loads with the absence of a load neutral. In the absence of a load neutral, the voltage across each phase might become imbalanced if the load is not evenly distributed.
- The Y-Y configuration also experiences third harmonics. When all elements are in equilibrium, these harmonics align precisely with the magnetizing current in terms of magnitude and orientation.

However, in the absence of a load neutral, the presence of third harmonics may distort the magnetic flux. This can result in voltage distortions in the transformers. In order to address the issue of presence of third harmonics under unbalanced loads in a Y-Y configuration, it is recommended to establish a direct connection between the neutral and the ground. Additionally, incorporating an additional set of windings known as tertiary windings can be beneficial.

3.5.3. Delta - Star (Δ - Y) connection of a 3-Phase transformer:

Fig. 3.14 (a) shows a Δ-Y Connection, where the primary winding is connected in a delta configuration and the secondary winding is connected in a star configuration. Fig. 3.14 (b) shows the phasor diagram for the Δ-Y connection of a three-phase transformer. It is observed from the phasor diagram that secondary line voltage leads the phase voltage by 30° and is $\sqrt{3}$ times the phase voltage. Thus, secondary winding turns may be made $1/\sqrt{3}$ times primary turns to get same line voltage on secondary side. This arrangement can be made to step-down phase voltage to $1/\sqrt{3}$ times the supply voltage.

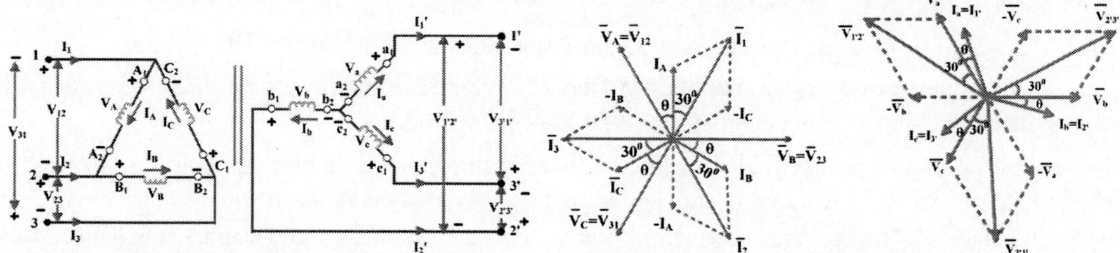

Fig. 3.14 (a) Delta-Star connection of a 3-Ph. transformer Fig. 3.14 (b) Phasor Diagram of Delta-Star 3-Ph. transformer

Important features:

- The ratio between line voltage and phase voltage on secondary side is $1/\sqrt{3}$. (i.e. $V_{L\,S} = \sqrt{3}\, V_{Ph\,S}$).
- The primary line voltage and secondary line voltage have a phase difference of 30°.
- Single-phase loads can be connected on secondary side due to availability of neutral.

Applications: Utilized to step-down voltage levels in distribution networks.

3.5.4. Star - Delta (Y - Δ) connection of a 3-phase Transformer:

The primary winding of a 3-phase transformer is connected in a star configuration, while the secondary winding is connected in a delta configuration, as depicted in Fig. 3.15 (a).

The phasor diagram of star-delta transformer is shown in Fig. 3.15 (b). It is observed from the phasor diagram that primary line voltage leads phase voltage by 30° and $\sqrt{3}$ times the phase voltage.

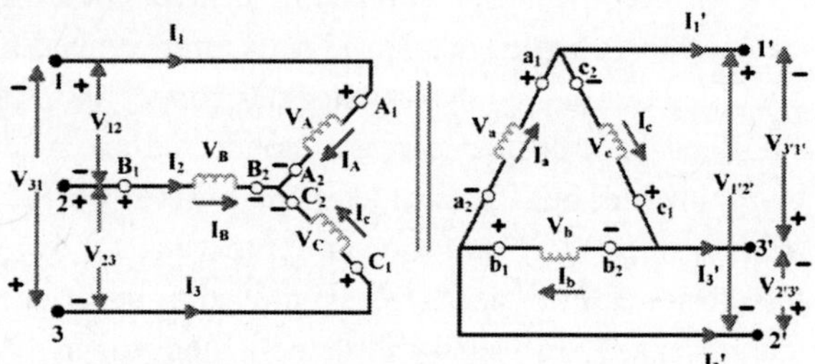

Fig. 3.15 (a) Star-Delta connection of 3-Ph transformer

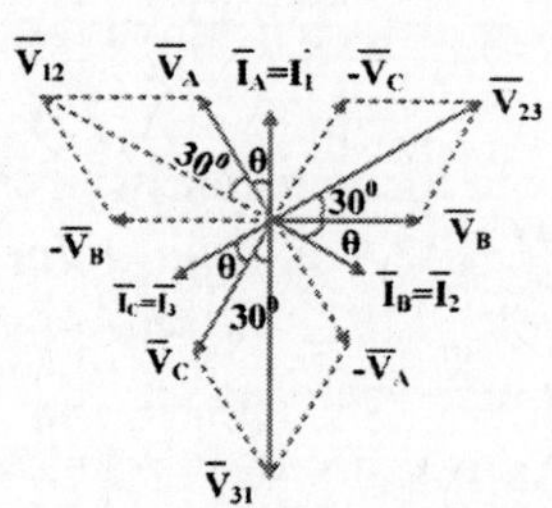

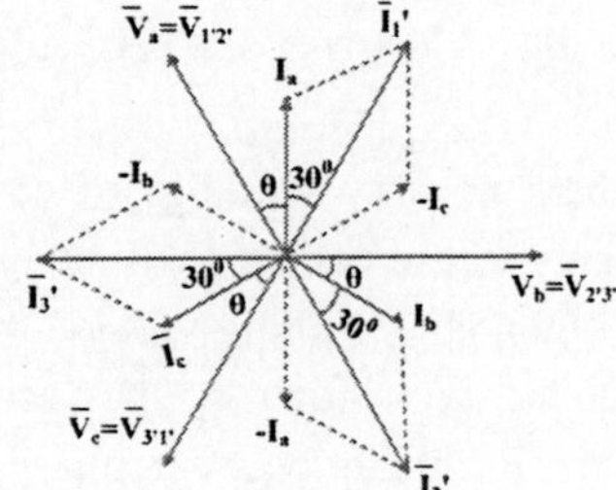

Fig. 3.15 (b) Phasor diagram of star-delta 3-phase transformer

The secondary phase voltage to primary phase voltage ratio can be given by:

$$\frac{V_{ph\,S}}{V_{ph\,P}} = \frac{N_2}{N_1} = n, \qquad \textit{where n reprents the turns ratio.}$$

The ratio of secondary line voltage to primary line voltage can be given by:

$$\frac{V_{LL\,S}}{V_{LL\,P}} = \frac{V_{ph\,S}}{\sqrt{3}V_{ph\,P}} = K = \frac{n}{\sqrt{3}} \qquad \text{......... (3.24)}$$

Thus, if the number of primary turns is $1/\sqrt{3}$ times the number of secondary turns, the secondary line voltage will be the same as the primary line voltage (i.e. $\sqrt{3}$ *times the primary phase voltage*). This arrangement is helpful in enhancing voltage on the secondary side.

Because of the closed path provided for the circulating third harmonic currents that pass through the winding phases, the delta winding guarantees automatic phase balancing even when a neutral wire is not used. The star-delta transformer connection, with a phase shift of 30°, can handle unbalanced loads without problem. This is because the delta winding construction naturally balances the currents.

Important features:

- The ratio between line voltage and phase voltage is $\sqrt{3}$ on the primary side.
- There is a phase shift of 30° between the line voltage and the phase voltage on the primary side.
- It is utilized to increase voltage in the transmission systems.

Applications: Utilized for voltage step-up in power transmission networks.

3.5.5. Open Delta or V-V connection of 3-phase Transformer:

Fig. 3.16 displays a 3-phase transformer's open Delta, or V-V connection, which is made up of two single-phase transformers coupled in an open delta arrangement. In an open delta or V-V delta transformer connection, three, single-phase transformers are connected in a delta configuration, but the third leg is purposefully left open or disconnected. This setup is appropriate for when a transformer fails or needs to be removed from service for maintenance. In such instances, the open delta connection allows the remaining transformers to function as a two-transformer bank, providing three-phase power to the load. However, the capacity of the three-transformer bank is reduced to 58% of the original rating.

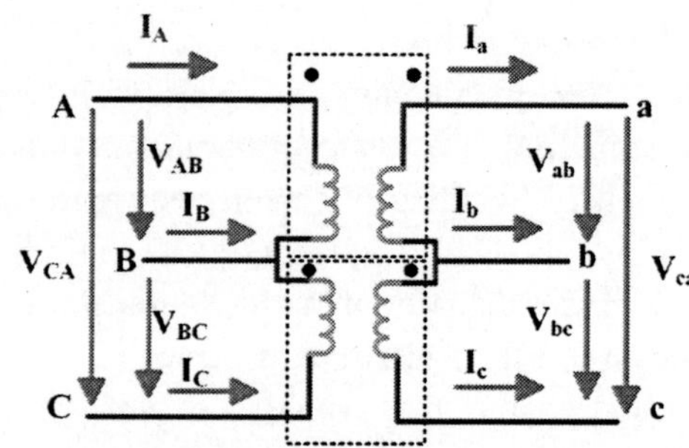

Fig. 3.16 V-V connection of a 3-phase transformer

The three transformers' primary windings are coupled in a delta configuration, with one leg unconnected. The primary windings' phase voltages will be indicated as V_{AB}, V_{BC}, and V_{CA}. As per Faraday's law the first transformer's secondary winding will get an induced voltage, V_{ab}. V_{bc} represents the phase voltage produced in the secondary winding of the second transformer. The open leg of the delta prevents an electrical connection or winding between phase terminals A and C in case of primary winding and a and c in case of secondary winding. The open delta, arrangement, enables three-phase power to be delivered using only two transformers rather than the usual three.

In an open delta transformer connection, the phase voltage V_{ca} in the third transformer's secondary winding is created by vectorially mixing the secondary voltages of the first two transformers. V_{ca} will have the same magnitude as the secondary voltages V_{ab} and V_{bc}, but will be out of phase by 120°.

When a balanced delta-delta configuration of three single-phase transformer switches over to open delta configuration due to disconnection of one single-phase transformer, the current flow through other two transformers fluctuate dramatically. In the V-V connection, the current flowing through each phase winding of the transformer increases by a factor of √3 when compared to the delta-delta. This is because the whole current that was supposed to pass through all three phase windings must suddenly be handled by just two windings.

The additional current causes each of the two transformers in the open delta arrangement to be overloaded by 73.2% of their rated capacity. To avoid transformer overheating and malfunction from high currents, the connected load is reduced by a factor of $\sqrt{3}$ while transitioning from a delta-delta to an open delta configuration during transformer maintenance.

Important features:

- The total capacity is 57.7% of the delta connected system.
- It is used in situations where one phase outage is anticipated.

Applications: Designed for uninterrupted operation in the event of a single-phase power failure that is expected to occur in the future.

3.5.6. Scott connection of a 3-phase transformer:

Scott connection consists of two single-phase transformers that converts three-phase voltage to two-phase voltgae and vice-versa. The primary winding of first transformer (called main transformer) has N_P number of turns with tapping at its centre, whereas, primary of second transformer (called teaser transformer) has 0.866 N_P ($\frac{\sqrt{3}}{2}N_P$) number of turns. One terminal of teaser transformer primary winding is connected to red phase, R of three-phase supply, while, its other terminal is connected to centre tap of main transformer. The two terminals of main transformer primary winding are connected to phase Y and phase B, respectively, of three-phase supply. The secondary of main as well as teaser transformer has N_s number of turms, and can supply a two-phase load. The Scott connection of two single-phase transformers is shown in Fig. 3.17 (a) and its phasor diagram is shown in Fig. 3.17 (b). It is clearly observed from the phasor diagrm that Scott connection of two single-phase transformers converts three-phase voltage to two-phase voltage. Alternately, if a two-phase supply is fed to the secondary winding, a three-phase voltage is generated on primary side.

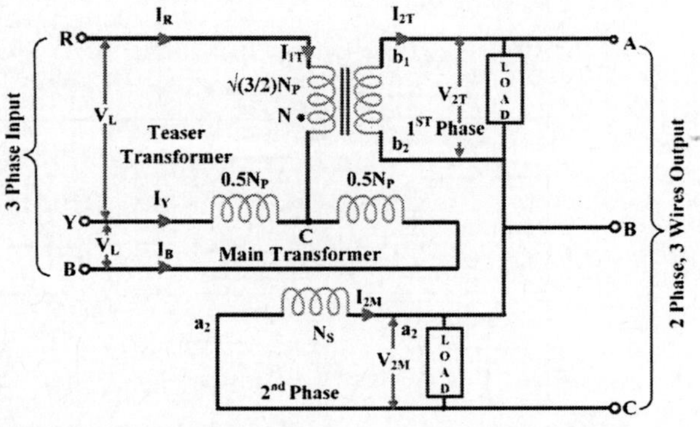

Fig. 3.17 (a) Scott connection of a 3-phase transformer.

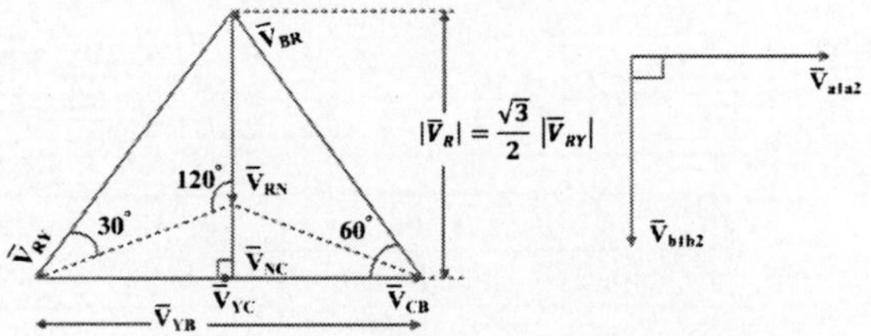

Fig. 3.17 (b) Phasor diagram of Scott connection.

It is observed from Fig. 3.17 (b) that ,

$$|\bar{V}_{RY}| = |\bar{V}_{YB}| = |\bar{V}_{BR}|$$

$$|\bar{V}_{YC}| = |\bar{V}_{CB}| = \frac{1}{2}|\bar{V}_{YB}|$$

$$|V_R| = \sqrt{|V_{RY}|^2 - |V_{YC}|^2} = \sqrt{|V_{RY}|^2 - \frac{1}{2}|V_{RY}|^2} = \frac{\sqrt{3}}{2}|V_{RY}| = 0.866|V_{RY}|$$

Despite the fact that the transformer cores are electrically linked, magnetic separation prevents them from contacting or communicating. This design generates the necessary third-phase voltage for three-phase power with only two transformers, protecting them from overloading. The Scott connection is a low-cost option for applications requiring voltage transfer from 3-phase to 2-phase or 2-phase to 3-phase with the fewest components.

Applications: Utilized for the transformation of power from three-phase systems to two-phase systems.

Analysis of three-phase Voltages and Currents:

Table 3.1 provides various connections and turn ratio of 3-phase transformer.

Table 3.1 various connections and turn ratio of 3-Ph transformer.

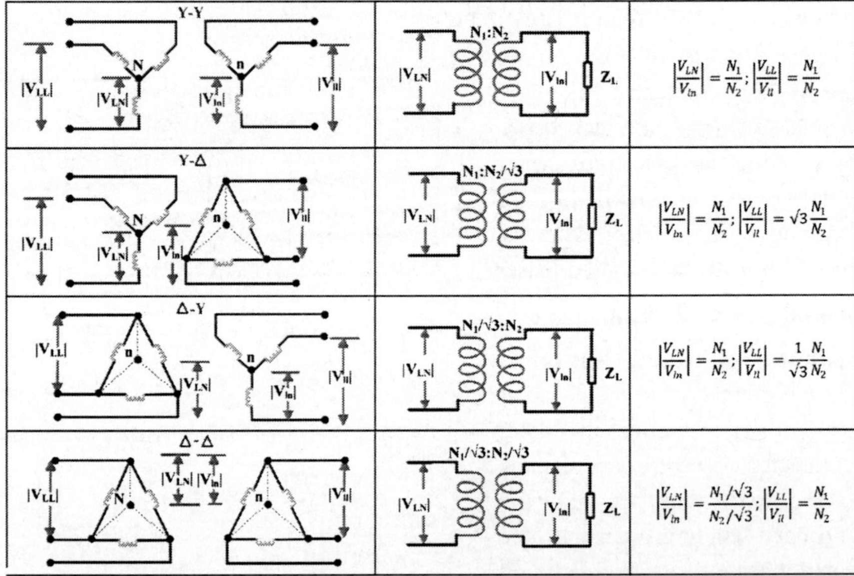

Configuration		Line Voltage		Line Current	
Primary	Secondary	Primary	Secondary	Primary	Secondary
Delta	Delta	V_L	nV_L	I_L	I_L/n
Delta	Star	V_L	$\sqrt{3}\,nV_L$	I_L	$I_L/\sqrt{3}n$
Star	Delta	V_L	$nV_L/\sqrt{3}$	I_L	$\sqrt{3}\,I_L/n$
Star	Star	V_L	nV_L	I_L	I_L/n

Where, $n = \frac{N_2}{N_1}$ which is also known as the transformer's "turns ratio".

V_L represents the line-to-line voltage, and V_P represents the phase-to-neutral voltage.

3.6 Three-winding transformers:

In addition to the primary and secondary windings, some high-rated transformers have a third winding known as the tertiary winding. Typically, a transformer's three windings have different voltage ratings. The primary winding has the greatest voltage rating, the tertiary winding the lowest, and the secondary winding intermediate. A two-winding transformer's primary and secondary windings have identical kilo-volt-ampere (kVA) values, whereas a three-winding transformer's windings may have different kVA ratings. The impedance of each winding in a three-winding transformer can be expressed in percentage or per unit, depending on the rating of its winding. These impedances can be determined by testing.

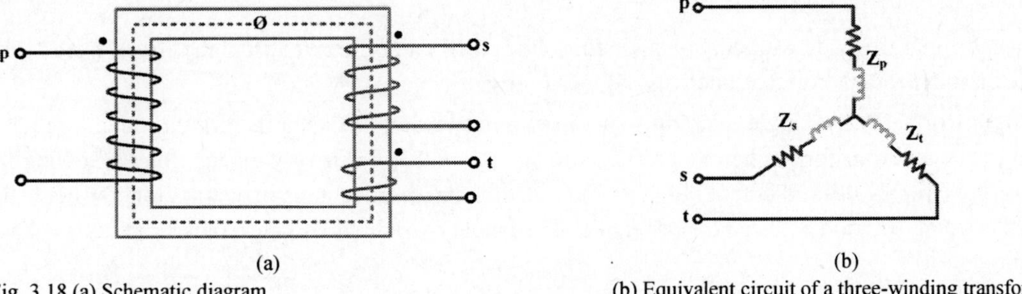

Fig. 3.18 (a) Schematic diagram (b) Equivalent circuit of a three-winding transformer.

Fig. 3.18 (a) shows a schematic of a single-phase three-winding transformer, with the three windings labelled primary, secondary, and tertiary. The points p, s and t represent identical polarity terminals (either all positive or all negative) as specified by dots for primary, secondary and tertiary winding, respectively.

Z_{ps} is the leakage impedance of the device, which is measured in the primary winding while the secondary winding is short-circuited and the tertiary winding is open.

Z_{pt} is the leakage impedance measured in the primary winding when the secondary winding is open and the tertiary winding is short-circuited.

Z_{st} is the leakage impedance measured in the secondary circuit while the tertiary circuit is short-circuited and the primary circuit is open.

The impedances of each individual winding, when referenced to the voltage of one of the windings, are related to the measured impedances as follows:

$$Z_{ps} = Z_p + Z_s; \;\; Z_{pt} = Z_p + Z_t \text{ and } Z_{st} = Z_s + Z_t \qquad \text{... ... ... (3.25)}$$

Where, Z_p, Z_s, and Z_t represent the impedance values at the primary, secondary, and tertiary windings, respectively. By simplifying the above equation, we obtain:

$$Z_p = \frac{1}{2}(Z_{ps} + Z_{pt} - Z_{st})$$

$$Z_s = \frac{1}{2}(Z_{ps} + Z_{st} - Z_{pt})$$

$$Z_t = \frac{1}{2}(Z_{pt} + Z_{st} - Z_{ps}) \qquad \text{... ... ... (3.26)}$$

The three windings' impedances are star connected to make the equivalent circuit of a single-phase three-winding transformer, ignoring the magnetizing current as shown in Fig. 3.18 (b). The star point is fictitious and has no connection to the system's neutral. As with two-winding transformers, converting to per-unit impedance necessitates using the same kilo-volt-ampere base for all three circuits, as well as voltage bases in the three circuits with the same ratio as the transformer's rated line-to-line voltages. When connecting three transformers for three-phase operation, the primary and secondary windings are usually Y-connected, while the tertiary windings are delta-connected to accommodate the third harmonic of the exciting current.

3.7 Autotransformer:

Autotransformers are made with only one winding. The prefix "auto" derived from the Greek word “self” refers to an independent coil that operates independently.

An autotransformer has a single winding, with a tap point located between the primary and secondary windings. The autotransformer has a significant advantage in that it can vary the tap point, allowing for exact control over the desired output voltage. The fundamental disadvantage of an autotransformer is the lack of electrical isolation between the primary and secondary windings. An autotransformer is a voltage regulator.

3.7.1 Construction and Working of Autotransformer:

Fig. 3.19 (a) depicts a standard transformer with electrically isolated primary and secondary windings that are magnetically coupled. The magnetic and electrical connection between an auto transformer's primary and secondary windings is its distinguishing feature. Auto transformers can be divided into two types based on their construction. The first method achieves the appropriate secondary voltage by using a continuous winding with carefully placed taps. In contrast, the second type is made up of at least two distinct coils that are connected to produce a continuous winding.

Fig. 3.19 (b) shows a schematic depiction of how to create an auto transformer. The primary winding AB has tapping at C which is adjustable. The section CB that is a part of primary winding, forms the secondary winding. The AB terminals receive the supply voltage, while the CB terminals are connected to the load. Applying an alternating current voltage V_1 across AB causes an alternating flux in the core, inducing an emf E_1 in the winding AB. A portion of the induced emf is directed to the secondary circuit.

Let, V_1 is Primary applied voltage
V_2 is Secondary voltage across the load
I_1 is Primary current
I_2 is load current
N_1 is number of turns between A and B
N_2 is number of turns between C and B

Neglecting no-load current, leakage reactance and losses, $V_1 = E_1 \text{ and } V_2 = E_2$

Therefore, the transformation ratio: $K = \frac{V_2}{V_1} = \frac{N_2}{N_1} = \frac{I_1}{I_2}$

Because the secondary ampere-turns have the opposite direction as the primary ampere-turns, the current I_2 is in phase opposition to I_1. The secondary voltage is less than the primary voltage. As a result, the current I_2 exceeds the existing I_1. The current flowing through section BC is (I_2–I_1) with direction as shown in Fig. 3.19 (b).

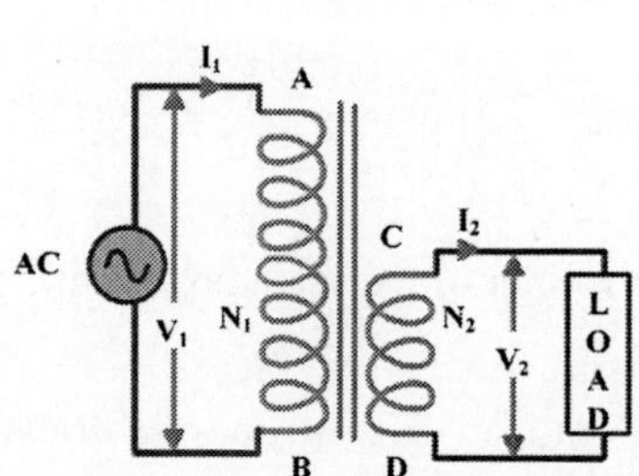

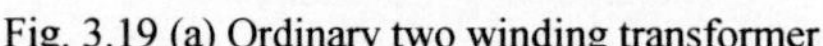

Fig. 3.19 (a) Ordinary two winding transformer

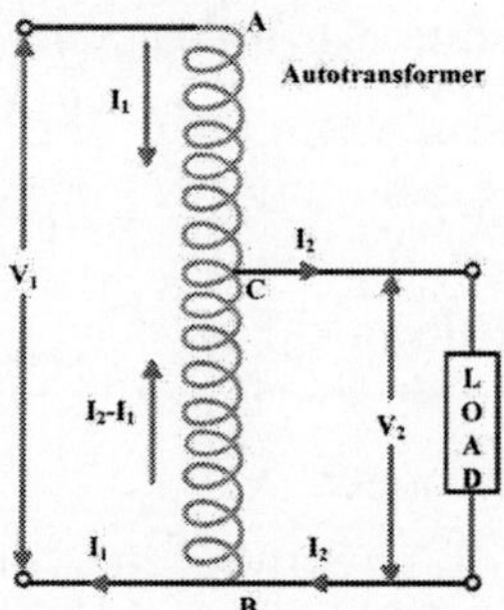

Fig. 3.19 (b) Auto transformer

The ampere-turns because of section BC = current * turns

The ampere – turns due to section BC, $= (I_2 - I_1)N_2 = \left(\frac{I_1}{K} - I_1\right)N_1K = I_1N_1(1-K)$ .... (3.27)

The ampere – turns due to section AC, $= I_1(N_1 - N_2) = I_1(N_1 - N_1K) = I_1N_1(1-K)$ .... (3.28)

Equations (3.27) and (3.28) demonstrate that the ampere-turns resulting from section BC and AC balance each other, which is a distinctive feature of transformer operation.

Case1: Autotransformer on No-Load

Fig. 3.20 (a) depicts a schematic diagram of an unloaded autotransformer that can perform step-down and step-up voltage transformations. The primary winding, 'ab', has N_1 turns, whereas the secondary winding, 'bc', has N_2 turns. The primary and secondary windings are clearly magnetically and electrically linked.

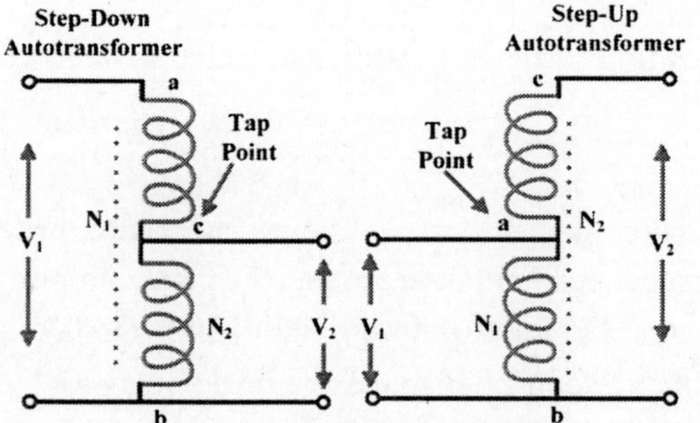

Fig. 3.20 (a) Step-down, step-up autotransformer on no-load

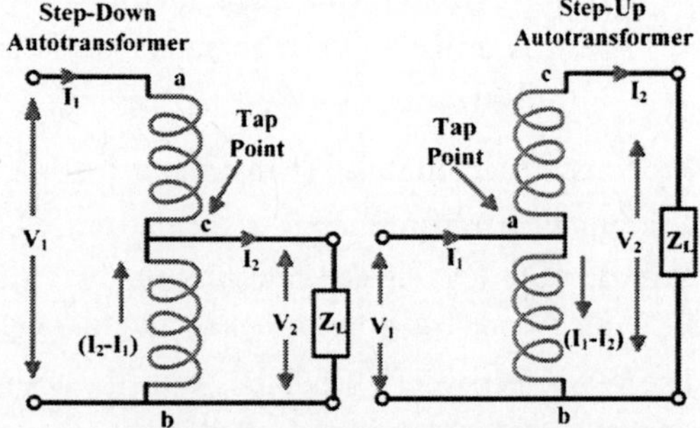

Fig. 3.20 (b) Step-down, step-up autotransformer on load

Case 2: Autotransformer on Load:

Fig. 3.20 (b) depicts the circuit layout of an autotransformer that is loaded. The current I_1 is the primary input current, whereas I_2 is the secondary output current or load current. Whether the autotransformer is stepdown or step up, the current flowing through the common portion of the primary and secondary windings equals the difference between the currents I_1 and I_2.

For the step – down autotransformer, the current $I_2 > I_1$,
thus current in common portion $I_2 - I_1$ *flowing from b to c.*

For the step – up autotransformer, the current $I_2 < I_1$,
thus current in common portion $I_1 - I_2$ *flowing from a to b.*

Benefits of Auto Transformer:

- ✓ Less costly
- ✓ Improved regulation
- ✓ Reduced losses compared to a conventional two-winding transformer of the same power rating.

Drawbacks of Auto Transformer:

The auto transformer has several benefits, but it is not frequently utilized due to one significant drawback.

- × The secondary winding is not isolated from the primary winding. When an autotransformer is used to convert high voltage to low voltage and the secondary winding fails, the entire primary voltage appears across the secondary terminals, creating a dangerous condition for both the user and the equipment. Auto transformers are not appropriate for connecting high and low voltage systems.
- × Only utilized in select situations where a minor difference in output voltage from input voltage is necessary.

Applications of Auto Transformer:

- It is used to supply around 50 to 60% of the rated voltage to the stator of a squirrel cage induction motor during its starting.
- It is utilized to provide a small voltage boost to a distribution cable in order to compensate the voltage drop.
- It serves as a voltage regulator.
- It is utilized in power transmission and distribution systems, as well as in audio systems and railways.

3.8 Neutral Grounding Transformer:

A grounding transformer is a specialized electrical device used in power systems essentially for the purpose of protecting the neutral point by establishing a connection to the ground. Typically, in power systems, the neutral point is connected to the ground in order to improve the reliability of the system.

The key objective of a grounding transformer is to establish a connection to the neutral point, divert fault current to the ground, and effectively safeguard personnel from electric shock or potential harm to electrical equipment. The quality of grounding transformer has a significant impact on the safety, stability, reliability, and efficiency of electrical systems.

Grounding transformers are often categorized as indoor grounding transformers or outdoor grounding transformers based on their intended locations of use. This can be further categorized into hanging grounding transformers and bracket grounding transformers based on their installation position. Outdoor grounding transformers closely resemble dry-type transformers in their physical appearance.

The outdoor grounding transformers exhibit superior performance in terms of pollution resistance, insulation deformation, and partial discharge performance. Earthing or neutral grounding transformers can be two-winding with a zig-zag linked primary and a star connected secondary, or single-winding three-phase auto-transformers with interconnected star or zig-zag windings. An earthing transformer is a transformer having a three-limbed core, each with two equally balanced windings. To form the neutral point, one set of windings is connected in a star arrangement. Fig. 3.21 shows how the remaining terminals of this series of windings are joined to the second set of windings.

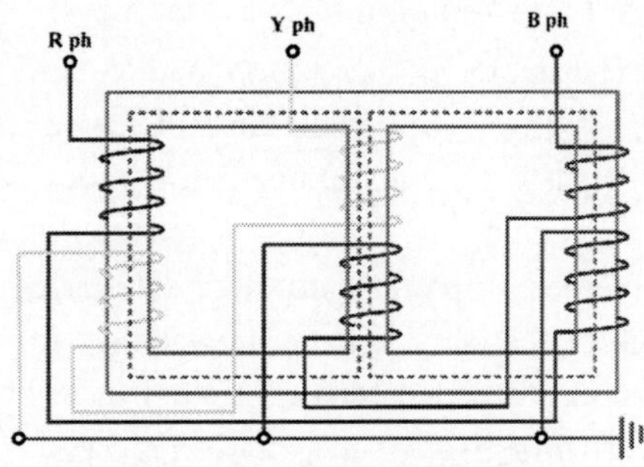

Fig. 3.21 Neutral grounding transformer

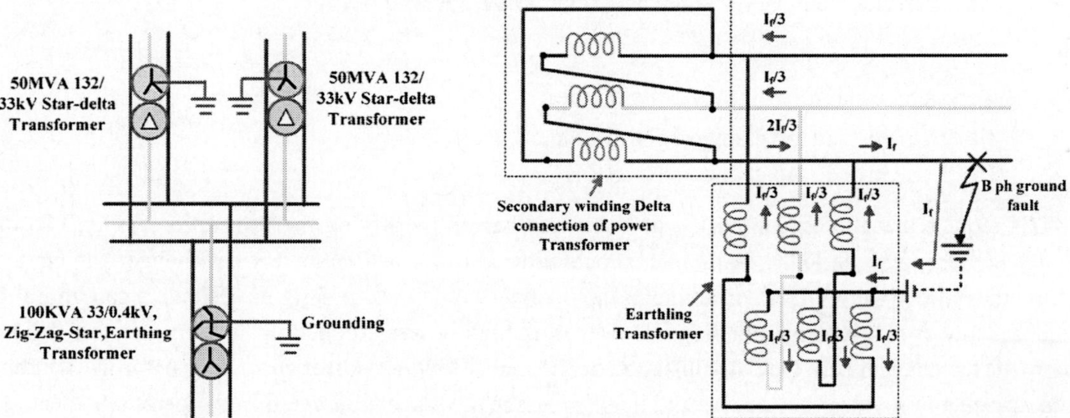

Fig. 3.22 (a) Representation of NGT in power system   Fig. 3.22 (b) Schematic diagram of NGT during ground fault on phase B

Fig. 3.22 (a) shows connection of Neutral Grounding Transformer (NGT) in power system. Fig. 3.22 (b) depicts the distribution of electric currents in the different windings of the earthing transformer during a ground fault on phase B. The earth fault current is routed back to the power system via the earth star point on the earthing transformer. It's uniformly dispersed across all three phases. The diagram clearly shows that the currents in two windings on the same limb run in opposite directions. As a result, the magnetic flux generated by the currents in the two windings cancels out. There is no impediment that prevents the flow of fault current. Zig-Zag earthing transformers are designed to handle the rated normal current flow in the case of a solid single-line to ground fault at the transformer's terminals. It is suggested that the earthing transformer's current rating match the full load current rating of the largest generator or transformer unit. The kVA rating of a three-phase earthing transformer is calculated by multiplying the normal line to neutral voltage by the neutral current that the transformer is designed to manage during fault conditions for a given time period.

3.9 Tap-Changing in Transformers:

It is usual for an increase in load to cause a drop in supply voltage. As a result, it is critical to ensure that the voltage supplied by the transformer to the load remains within the prescribed limits. One method is to change the transformer's turns ratio. Taps are terminals or connections that are positioned at different points along the winding. The turns ratio varies between taps, allowing for a range of voltages to be received. Taps fall into three categories: principle, positive, and negative. A principle tap delivers the rated secondary voltage in exchange for the rated primary voltage. Positive and negative taps cause an increase and decrease in primary turns being used, respectively.

System voltage control is essential for:

- Regulating the terminal voltage of consumers within the specified limits.
- Voltage adjustment in response to load variations.
- To regulate both the active and reactive power.
- To adjust the secondary voltage according to the specific needs.

3.9.1 Types of Tap Changers:

Tap changers can be classified into two main types:

(i) No-load tap changers (NLTC) and

(ii) On-load tap changers (OLTC).

NLTCs require that the equipment be de-energized before modifying the turn ratio, but OLTCs can change the turn ratio while the equipment is operational. Tap selection on any tap changer can be done by an automatic system OLTC or a human tap changer NLTC. Automatic tap changers can be put on either the low or high-voltage windings. However, in high-power production and transmission settings, automatic tap changers are typically installed on the higher voltage (lower current) transformer winding. This placement provides easy access and helps to lessen the current burden during operation.

3.9.1.1 No-load tap changer (NLTC/OCTC/DETC):

An NLTC also known as an off-circuit tap changer (OCTC) or de-energized tap changer (DETC), is a device used when the turn ratio does not need to be modified often and the transformer system can be disconnected safely. This transformer is widely used in applications that require minimal power and voltage. In these transformers, the tap point is frequently configured as a terminal for connecting the transformer. This requires that the input line be manually disconnected and connected to the new terminal. A rotary or slider switch can be used in some systems to facilitate tap changes. No-load tap changers are used on the primary winding of high-voltage distribution transformers. This enables the transformer to respond to fluctuations in the transmission system within a narrow range of the nominal rating. In these systems, the tap changer is usually adjusted only once during installation, but it can be modified later to accommodate a long-term change in the system's voltage profile.

3.9.1.2 On-Load Tap Changers (OLTC):

OLTCs, are specifically intended to adjust the tap settings of a transformer while it is energized and operating under load. OLTCs employ several techniques, such as diverter switches and selector switches, to modify the winding connections of the transformer. This enables voltage modifications to counterbalance oscillations in the power grid, guaranteeing a consistent provision of electricity. OLTCs are widely installed in substation transformers to regularly adjust the voltage in order to meet fluctuations in loads and voltage profiles.

3.9.2. Applications of Tap Changers:

Voltage Regulation: A tap changer's main purpose is to control and adjust the voltage output of transformers. Fluctuations in voltage within the power grid can have an adverse effect on the quality of electricity that is delivered to consumers. Tap changers allow transformers to adjust and maintain a constant output voltage, guaranteeing a steady and reliable power supply.

Grid Stability: Tap changers are essential for ensuring the stability of the electrical system. They provide rapid attentiveness to changes in voltage and load, hence mitigating the potential for voltage sags or surges which can disrupt the power supply.

Distribution Transformers: Tap changers play a crucial role in providing power to residential, commercial, and industrial buildings. They ensure that voltage levels stay within acceptable thresholds, regardless of fluctuations of load or disturbances in the power grid.

Substation Transformers: Tap changers are commonly employed in substations to control voltage levels in high-voltage transmission networks. Efficient long-distance power transmission and grid synchronization depend on this.

Renewable Energy Integration: Tap changers play a crucial role in handling the fluctuating power production of renewable energy sources such as wind and solar, as these sources become increasingly integrated. They help with maintaining grid stability during fluctuations in renewable generation.

Industrial Applications: Industries that have sensitive machinery depend on tap changers to maintain consistent voltage levels, protecting their operations from disturbances caused by voltage fluctuations.

Example 3.1. Given that $N_1 = 1500$ and $N_2 = 500$ in the two winding transformer. Also $V_1 = 1000\angle 0^0$ V and $I_1 = 5\angle -30^0 A$, with an impedance Z_2 connected across winding 2. Determine V_2, I_2, Z_2 and the impedance Z_2', which is defined as the value of Z_2 referred to the primary side of the transformer.

Ans: Given $N_1 = 1500$, $N_2 = 500$, $V_1 = 1000\angle 0^0$ V and $I_1 = 5\angle -30^0 A$

$$V_2 = \frac{N_2}{N_1} V_1 = \frac{500}{1500} * 1000\angle 0^0 = 333.33\angle 0^0\ V$$

$$I_2 = \frac{N_1}{N_2} I_1 = \frac{1500}{500} * 10\angle -30^0 = 15\angle -30^0\ A$$

$$Z_2 = \frac{V_2}{I_2} = \frac{333.33\angle 0^0}{15\angle -30^0} = 22.22\angle 30^0\ \Omega$$

$$Z_2' = \left(\frac{N_1}{N_2}\right)^2 Z_2 = \left(\frac{1500}{500}\right)^2 * 22.22\angle 30^0 = 199.99\angle 30^0\ \Omega \qquad (or)$$

$$Z_2' = \frac{V_1}{I_1} = \frac{1000\angle 0^0}{5\angle -30^0} = 200\angle 30^0\ \Omega$$

Example 3.2. For a 100V/400V, 50kVA, single-phase transformer primary and secondary leakage reactance are 0.08Ω and 0.2Ω respectively. Show that net per unit reactance of the transformer to LV side is same as the referred to HV side.

Ans: Transformation ratio $K = \frac{V_2}{V_1} = \frac{400}{100} = 4$

LV side base voltage, $kV_{base}^{LV} = 0.1kV$

LV side base current, $I_{base}^{LV} = \frac{kVA_{base}}{kV_{base}^{LV}} = \frac{50}{0.1} = 500A$

LV side base impedance, $Z_{base}^{LV} = \frac{kV_{base}^{LV}}{I_{base}^{LV}} * 1000 = \frac{0.1}{500} * 1000 = 0.2\Omega$

LV side actual impedance, $Z_{actual}^{LV} = Z_1 + \frac{Z_2}{K^2} = 0.08 + \frac{0.1}{4^2} = 0.0862\Omega$

LV side p.u. impedance, $Z_{p.u}^{LV} = \frac{Z_{actual}^{LV}}{Z_{base}^{LV}} = \frac{0.0862}{0.2} = 0.431\ p.u.$

HV side base voltage, $kV_{base}^{HV} = 0.4kV$

HV side base current, $I_{base}^{HV} = \frac{kVA_{base}}{kV_{base}^{HV}} = \frac{50}{0.4} = 125A$

HV side base impedance, $Z_{base}^{HV} = \frac{kV_{base}^{HV}}{I_{base}^{HV}} * 1000 = \frac{0.4}{125} * 1000 = 3.2\Omega$

HV side actual impedance, $Z_{actual}^{HV} = Z_1K^2 + Z_2 = 0.08 * 4^2 + 0.1 = 1.38\Omega$

HV side p.u. impedance, $Z_{p.u}^{HV} = \frac{Z_{actual}^{HV}}{Z_{base}^{HV}} = \frac{1.38}{3.2} = 0.431\ p.u.$

Example 3.3. The single-phase transformer has a power rating of 10MVA and a voltage rating of 11kV/33kV. The leakage reactance, measured from the low-tension side, is 10Ω. Calculate the per unit value of the leakage reactance.

Ans: LV side base impedance, $X_{base}^{LV} = \frac{(11*10^3)^2}{10*10^6} = 12.1\Omega$

LV side p.u. impedance, $X_{p.u}^{LV} = \frac{X_{actual}^{LV}}{X_{base}^{LV}} = \frac{10}{12.1} = 0.826\ p.u.$

HV side actual impedance, $X_{actual}^{HV} = 10 * \left(\frac{33*10^3}{11*10^3}\right)^2 = 90\Omega$

HV side base impedance, $X_{base}^{HV} = \frac{(33*10^3)^2}{10*10^6} = 108.9\Omega$

HV side p.u. impedance, $X_{p.u}^{HV} = \frac{X_{actual}^{HV}}{X_{base}^{HV}} = \frac{90}{108.9} = 0.826\ p.u.$

Example 3.4. The three-phase ratings of a three-winding transformer are, neglecting resistance, the leakage impedances are:

Primary : star – connected, 33kV, 10MVA
Secondary : star – connected, 11kV, 7.5MVA
Tertiary : delta – connected, 2.2kV, 5MVA
Z_{ps} = 10% on 10MVA, 33kV base
Z_{pt} = 9% on 10MVA, 33kV base
Z_{st} = 8% on 7.5MVA, 11kV base

Determine the per-unit impedances of the star-connected equivalent circuit using a base of 10MVA and 33kV in the primary circuit.

Ans: Consider a base of 10MVA, 33kV in the primary circuit.

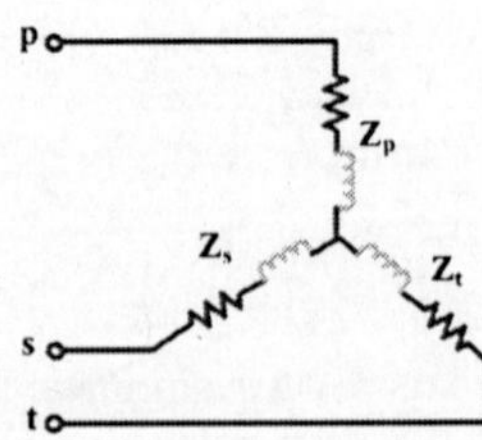

Z_{ps} = 10% on 10MVA, 33kV base = j0.1 pu

Z_{pt} = 9% on 10MVA, 33kV base = j0.09 pu

$$Z_{st} = 8\% \text{ on } 7.5\text{MVA}, 11\text{kV base} = 0.08 * \frac{10}{7.5} = j0.1066\text{pu}$$

$$Z_p = \frac{1}{2}(Z_{ps} + Z_{pt} - Z_{st}) = \frac{1}{2}(j0.1 + j0.09 - j0.1066) = j0.0417pu$$

$$Z_s = \frac{1}{2}(Z_{ps} + Z_{st} - Z_{pt}) = \frac{1}{2}(j0.1 + j0.1066 - j0.09) = j0.0583pu$$

$$Z_t = \frac{1}{2}(Z_{pt} + Z_{st} - Z_{ps}) = \frac{1}{2}(j0.09 + j0.1066 - j0.1) = j0.0483pu$$

Example 3.5. An infinite bus, which is a constant voltage source, provides power to a fully resistive load of 5MW and 2.2kV, as well as a synchronous motor with a power rating of 7.5MVA and voltage rating of 11kV. The motor has a transient reactance of $X'' = 25\%$. The source is linked to the primary coil of the three-winding transformer mentioned in example 3.4. The motor and resistive load are linked to the secondary and tertiary terminals of the transformer. Construct the impedance diagram of the system and indicate the per-unit impedances for a base of 33kV and 10MVA in the primary.

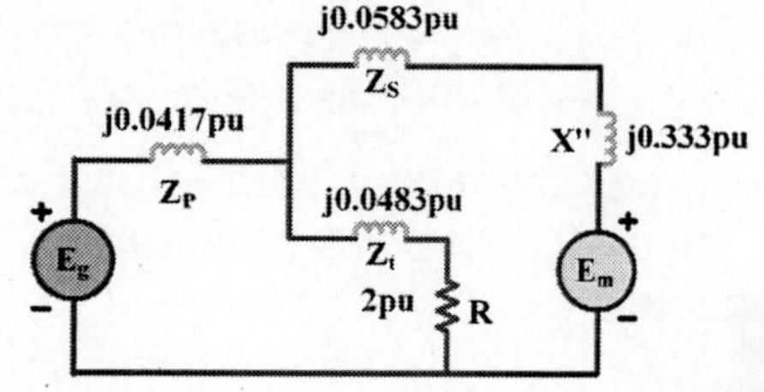

Ans: The resistance of the load is 1pu on a base of 5MVA, 2.2kV in the tertiary.

Expressed on a 10MVA, 2.2kV base the load resistance is $R = 1 * \frac{10}{5} = 2pu$

Changing the reactance of the motor to a base of 10MVA, 11kV yields

$$X'' = 0.25 * \frac{10}{7.5} = j0.333pu.$$

Example 3.6. As illustrated in the figure, a 25kVA single phase transformer rated 220V/110V is used as an auto-transformer. The low-tension winding of the transformer receives the rated voltage. Assume the transformer is ideal and the load is such that the windings carry the rated currents I_1 and I_2. Find the auto transformer's secondary voltage (V_2) and kVA rating.

Ans: From the given auto-transformer

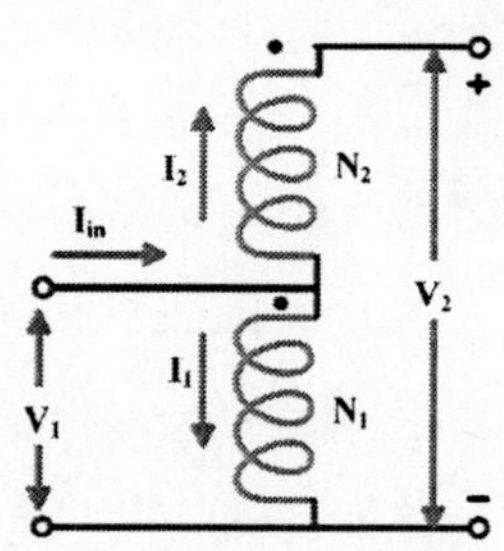

$$I_1 = \frac{25000}{110} = 227.27\ A$$

$$I_2 = \frac{25000}{220} = 113.63\ A$$

$$V_2 = 220 + 110 = 330\ V$$

$$I_{in} = I_1 + I_2 = 227.27 + 113.63 = 340.9\ A$$

$$Input\ kVA = I_{in} * V_{sec} = 340.9 * 110 = 37.49\ kVA$$

$$Output\ kVA = I_2 * V_2 = 113.63 * 330 = 37.49\ kVA$$

Example 3.7. The primary winding of the single-phase transformer consists of 2400 turns, whereas the secondary winding consists of 800 turns. The values of the winding resistances are $R_1 = 5\Omega\ and\ R_2 = 0.25\Omega$. The leakage reactances are $X_1 = 10\Omega\ and\ X_2 = 0.75\Omega$. The impedance of the load $Z_2 = 12\Omega$. Given an applied voltage of 1000V at the primary winding terminals, determine the value of V_2 and calculate the voltage regulation. Ignore the magnetizing current and core loss.

Ans: Given $N_1 = 2400,\ N_2 = 800,\ V_1 = 1000\angle 0^0\ V$

$$R_1 = 5\Omega, R_2 = 0.25\ \Omega, X_1 = 10\Omega, X_2 = 0.75\ \Omega\ and\ \text{load}\ Z_2 = 12\Omega$$

$$\text{Turns ratio}\ \ K = \frac{N_2}{N_1} = \frac{800}{2400} = 0.333$$

The total equivalent resistance and reactances of transformer referred to primary are,

$$R_{01} = R_1 + R_2' = R_1 + \frac{R_2}{K^2} = 5 + \frac{0.25}{0.333^2} = 7.25\ \Omega$$

$$X_{01} = X_1 + X_2' = X_1 + \frac{X_2}{K^2} = 10 + \frac{0.75}{0.333^2} = 16.76\ \Omega$$

The equivalent secondary impedance Z_2 referred to the primary side; Z_2' *is*

$$Z_2' = \frac{Z_2}{K^2} = \frac{12}{0.333^2} = 108.21\ \Omega$$

$$I_1 = \frac{V_1}{Z_2'+Z_{01}} = \frac{V_1}{Z_2'+R_{01}+jX_{01}} = \frac{1000}{108.21+7.25+j16.76} = 8.57\angle{-8.25^0}\ A$$

Secondary terminal current I_2 referred to the primary side; $I_2' = I_1$

Secondary terminal voltage V_2 referred to the primary side; $V_2' = \frac{V_2}{K}$

$$V_2' = Z_2' I_2' = Z_2' I_1 = 108.21 * 8.57\angle{-8.25^0} = 923.37\angle{-8.25^0}\ V$$

$$V_2 = K\ V_2' = 0.333 * 923.37\angle{-8.25^0} = 308.81\angle{-8.25^0}\ V$$

At No load : $V_2 = KV_1 = 0.333 * 1000 = 333\Omega$

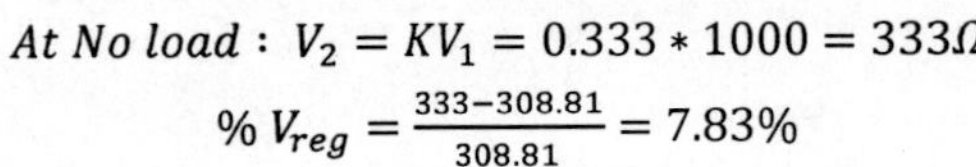

$$\%\ V_{reg} = \frac{333-308.81}{308.81} = 7.83\%$$

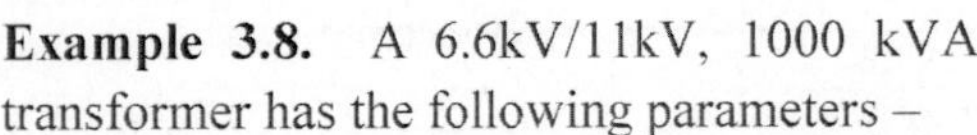

Example 3.8. A 6.6kV/11kV, 1000 kVA transformer has the following parameters –

$R_1 = 0.055\Omega, R_2 = 0.66\Omega$ *and* $R_0 = 1500\Omega$

$X_1 = 0.077\Omega, X_2 = 1.44\Omega$ *and* $X_0 = 500\Omega$

The transformer is supplying full load with a power factor of 0.8 at the rated voltage. Determine the input current by using an exact equivalent circuit.

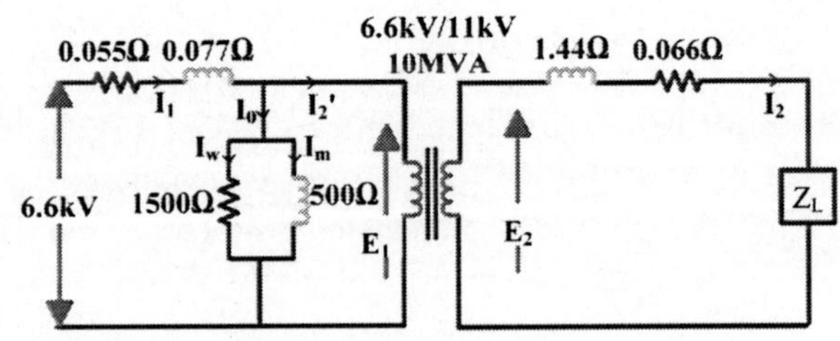

Ans: Turns ratio $K = \frac{V_2}{V_1} = \frac{11000}{6600} = 1.666$

$$I_2 = \frac{kVA}{kV} = \frac{1000*10^3}{11*10^3} = 90.909\angle{-36.86^0}\ A$$

$$I_2' = KI_2 = 1.666 * 90.909\angle{-36.86^0} = 151.45\angle{-36.86^0}\ A$$

$$Z_2 = R_2 + jX_2 = 0.66 + j1.44 = 1.584\angle 65.37^0\ A$$

$$E_2 = V_2 + I_2 Z_2 = 11000 + 90.909\angle{-36.86^0} * 1.584\angle 65.37^0 = 11126.74\angle 0.353^0\ V$$

$$E_1 = \frac{E_2}{K} = \frac{11126.74\angle 0.353^0}{1.666} = 6678.71\angle 0.353^0\ V$$

$$I_m = \frac{E_1}{X_0} = \frac{6678.71\angle 0.353^0}{j500} = 13.357\angle{-89.64^0}\ A$$

$$I_w = \frac{E_1}{R_0} = \frac{6678.71\angle 0.353^0}{1500} = 4.452\angle 0.353^0\ A$$

$$I_0 = I_m + I_w = 13.357\angle{-89.64^0} + 4.452\angle 0.353^0 = 14.079\angle{-89.64^0}\ A$$

$$I_1 = I_2' + I_0 = 151.45\angle{-36.86^0} + 14.079\angle{-89.64^0} = 163.26\angle - 39.64^0 A$$

3.10 Summary

- The transformers are primarily categorized into different types based on their application, such as Power Transformers, Distribution Transformers, Isolation Transformers, Instrument Transformers, Step-Up Transformers, Step-Down Transformers, Auto Transformers, and Air-core Transformers.
- Power transformers are used in transmission networks to increase or reduce voltage levels. They are built for outstanding performance and generally have a massive and heavy build, allowing them to manage high voltages and currents.
- Distribution transformers are used in distribution networks to provide the ultimate voltage transition within the energy-distributing equipment.
- Isolation transformers are used to isolate specific circuit components for protection and safety.
- It is evident from the voltage transformation ratio that the ratio of E_2 to E_1 is equal to the ratio of N_2 to N_1, which is denoted as K. $K = \frac{V_2}{V_1} = \frac{E_2}{E_1} = \frac{N_2}{N_1} = \frac{I_1}{I_2}$
- The total resistance and reactance referred to primary side and secondary are

$$R_{01} = R_1 + R_2' = R_1 + \frac{R_2}{K^2} \text{ and } X_{01} = X_1 + X_2' = X_1 + \frac{X_2}{K^2}$$

$$R_{02} = R_1' + R_2 = K^2R_1 + R_2 \text{ and } X_{02} = X_1' + X_2 = K^2X_1 + X_2$$

- Three-phase transformer connections include Y-Y, Y-Δ, Δ-Y, and Δ - Δ, Open Delta (V-V), and Scott connections.
- In certain high-rated transformers, a tertiary winding is included in addition to the primary and secondary windings. The tertiary winding is the third winding in the transformer, which gives it the name "three winding transformer" due to the presence of three windings.
- An autotransformer is a one winding transformer with an adjustable tap on the winding. The full winding forms one side of the transformer, whereas, the part of winding between tapping position and one fixed terminal forms other side of the transformer.
- A grounding transformer is a specialized electrical device used in power systems essentially for the purpose of protecting the neutral point by establishing a connection to the ground.
- A no-load tap changer, often referred to as an off-circuit tap changer or de-energized tap changer is a device used in transformers when the turn ratio does not need to be adjusted frequently and it is acceptable to disconnect the transformer system.
- On-load tap changers are specifically intended to adjust the tap settings of a transformer while it is energized and operating under load.

Short and Long Answer Questions

1. Describe the features, benefits, and drawbacks of an ideal transformer.
2. Differentiate between power and distribution transformers.
3. What is the transformation ratio for the step-up, step-down, and isolation transformer?
4. Give equivalent resistance, reactance and impedance referred to primary and secondary circuits.
5. Why to compute equivalent impedance based on the primary or secondary side?
6. Comment the characteristics of a two-winding transformer with those of an auto transformer.
7. Why isn't an autotransformer used for distribution?
8. What are the benefits of a three-phase transformer over three single-phase transformers?

9. Explain the equivalent circuit of a transformer, referred to as primary and secondary, with the relevant derivations and illustrations.
10. What's the distinction between "insulating", "isolating", and "shielded winding" transformers?
11. What are the applications of step-up & step-down transformer?
12. What is the phase relationship of a transformer's core flux, magnetizing current, and induced emfs in its primary and secondary windings? Sketch the phasor diagram.
13. What is the maximum flux value in a transformer core when excited from the primary side? Is the value of flux much altered when the secondary is loaded? Explain the reason for this.
14. Explain the three-phase transformer connections and phase-shifts of Star-Star, Star-Delta, Delta-Star, Delta - Delta, Open Delta, and Scott connections, with derivations and illustrations.
15. Explain the construction, operation, functions, and applications of tap-changing transformers, including derivations and illustrations.

Exercise

1. The single-phase transformer has a power rating of 20MVA and a voltage rating of 22kV/66kV. The leakage reactance, measured from the low-tension side, is 20Ω. Calculate the per unit value of the leakage reactance.
2. Given that $N_1 = 2400$ and $N_2 = 800$ in the two winding transformer. Also $V_1 = 1100\angle 0^0$ V and $I_1 = 15\angle -30^0 A$, with an impedance Z_2 connected across winding 2. Determine V_2, I_2, Z_2 and the impedance Z_2', which is defined as the value of Z_2 referred to the primary side of the transformer.
3. For a 220V/440V, 50kVA, single-phase transformer primary and secondary leakage reactances are 0.1Ω and 0.2Ω respectively. Show that net per unit reactance of the transformer to LV side is same as the referred to HV side.
4. The primary winding of the single-phase transformer consists of 1200 turns, whereas the secondary winding consists of 300 turns. The values of the winding resistances are $R_1 = 3\Omega$ and $R_2 = 0.15\Omega$. The leakage reactances are $X_1 = 5\Omega$ and $X_2 = 0.35\Omega$. The impedance of the load $Z_2 = 7\Omega$. Given an applied voltage of 1100V at the primary winding terminals, determine the value of V_2 and calculate the voltage regulation. Ignore the magnetizing current and core loss.
5. A 11kV/22kV, 1500 kVA transformer has the following parameters –

   $R_1 = 0.045\Omega, R_2 = 0.56\Omega \text{ and } R_0 = 1450\Omega$

   $X_1 = 0.078\Omega, X_2 = 1.56\Omega \text{ and } X_0 = 350\Omega$

   The transformer is supplying full load with a power factor of 0.9. Determine the input current by using an exact equivalent circuit.
6. The three-phase ratings of a three-winding transformer are:

   Primary : star – connected, 66kV, 20MVA

   Secondary : star – connected, 33kV, 15MVA

   Tertiary : delta – connected, 11kV, 5MVA

   $Z_{ps} = 16\%$ on 20MVA, 66kV base

   $Z_{pt} = 12\%$ on 20MVA, 33kV base

   $Z_{st} = 8\%$ on 5MVA, 11kV base

   Determine the per-unit impedances of the star-connected equivalent circuit using a base of 20MVA and 66kV in the primary circuit.

To know more about

Transformers: Advanced Topologies, Implementation Issues, Recent Progress and Improvements

To know more about

Why do we need transformer, ABB UZ Tap Changer and Motor Drive Unit Operation and Maintenance

To know more about

Development of a New Equivalent Circuit of DGS with Ideal Transformer, active cell balancing circuit using multi-winding T/f

To know more about

Modelling of tap-changing T/f and Studies on performance parameters of a practical transformer for various utilizations

To know more about

A review of the applications of machine learning in the condition monitoring of transformers

To know more about

Review of Projects with Relevant Prototypes and Demonstrators, Twelve-step voltage source inverter

04 SYNCHRONOUS MACHINES

Unit specifics: In this unit, the following topics have been discussed for basic understating of Synchronous machines:

- Equivalent circuits of synchronous generator and motor during steady-state,
- Real and reactive power output of synchronous machine connected to an infinite bus,
- Capability curve of synchronous generators,
- Equivalent circuits of a synchronous generator during short-circuit,
- Electrical loads with their voltage and frequency dependency,
- Per-unit system and per-unit calculations.

Rationale: This unit covers the topics of synchronous reactance and equivalent circuits of synchronous generators and motors in steady state conditions. It also includes the study of synchronous machines connected to an infinite bus, as well as the real and reactive power output of synchronous machines in this system. The capability curve of synchronous generators is explained. The steady state, transient, and sub-transient equivalent circuits of a three-phase synchronous generator during a short-circuit occurrence are presented. Additionally, the unit explores electrical loads along with their voltage and frequency dependency. The concepts of per-unit system and per-unit calculations are explained thoroughly, with the aid of diagrams, derivations, and examples.

Pre-Requisites: Basic knowledge of power system components.

Unit Outcomes: List of outcomes of this unit is as follows

U4-O1: To analyse the equivalent circuits of synchronous machines during steady-state.

U4-O2: To analyse the real and reactive power o/p of synchronous machine connected to an infinite bus.

U4-O3: To analyse capability curve of synchronous generators.

U4-O4: To analyse the equivalent circuits of synchronous generator and motor during short-circuit.

U4-O5: To know about various electric loads along with their voltage and frequency dependency.

U4-O6: To understand and analyse per-unit system numerically.

Unit-4 outcomes	Expected mapping with course outcomes (1: Weak, 2: medium, and 3: strong correlation)					
	CO-1	**CO-2**	**CO-3**	**CO-4**	**CO-5**	**CO-6**
U4-O1	2	2	-	2	-	1
U4-O2	2	2	-	2	-	1
U4-O3	2	1	1	2	-	1
U4-O4	2	1	1	3	1	2
U4-O5	2	2	-	-	1	-
U4-O6	2	2	-	3	1	3

4.1 Introduction to Synchronous machine:

The electrical power system (EPS) has various components, including a generator, transformer, transmission line, motor, circuit breaker, busbar, feeder, and relays. These components have their minimum and maximum capabilities for safe operation of power system, and to withstand excessive rise in voltage and currents under fault conditions.

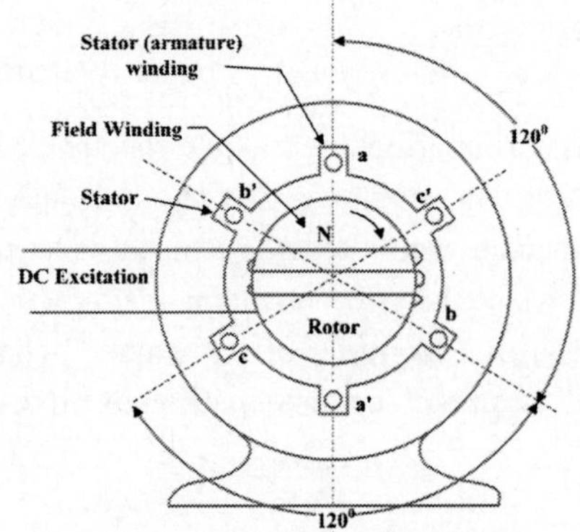

Fig. 4.1 synchronous generator

The synchronous machine, functioning as an alternating current generator, is powered by a turbine to transform mechanical energy into electrical energy. It is the primary source of electric power generation worldwide. The machine functions as a motor by transforming electrical energy into mechanical energy. Figure 4.1 depicts a schematic cross-sectional diagram of a 3-phase synchronous generator (alternator) with a two-pole configuration. The stator is equipped with a symmetrical three-phase winding, labelled aa', bb', and cc'. The depicted winding is a concentrated one, whereas the winding in a real machine is dispersed throughout the stator perimeter. The depicted rotor is a cylindrical one, also known as a round rotor or non-salient pole rotor, which contains a concentric winding known as field winding energized by a DC source. This winding is known as field winding as current through it sets up electromagnetic field surrounding it. The rotor winding is positioned on the outside edge of the rotor in such a way that the field excitation generates a flux/pole (ϕ_f) that is spread in a nearly sinusoidal manner in the air gap. As the rotor rotates, the field flux linking through stator winding (known as armature winding) changes resulting in induction of three-phase electromotive force (emf) in armature winding since three phases of armature winding are space displaced by 120° with respect to each other. Under balanced steady-state condition, the three-phase armature quantities can be analysed on per phase basis. In a machine with a pole count above two, the aforementioned configuration is electrically replicated for each pair of poles.

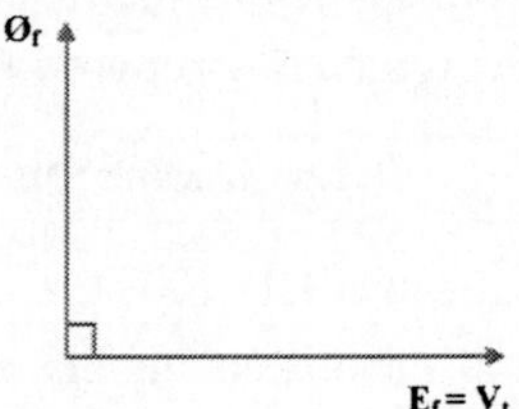

Fig. 4.2 Phasor of Φ_f Vs E_f at No-load

The frequency of induced emf $f = \frac{P}{2} * \frac{N}{60} = \frac{P}{2} f_m$ .....(4.1)

where
f = electrical frequency in Hz
P = number of poles
N = rotor speed in revolutions per minute (rpm)
$f_m = \frac{N}{60}$; mechanical frequency in revolutions per second

As per Faraday's law of electromagnetic induction and Lenz's law, the induced emf in phase-a is equal to negated rate of change of flux linkage. Therefore, the induced emf E_f in phase-a lags the field flux ϕ_f by 90°, as shown in Fig. 4.2. This voltage is proportional to ϕ_f if the magnetic circuit is unsaturated.

At no load : terminal voltage = electromotive force $\Rightarrow V_t = E_f$ .....(4.2)

When a balanced and steady load is connected to the three-phase stator winding, the stator currents generate a synchronously rotating magnetic field that aligns with the direction of the rotor's rotation. The flux ϕ_a generated by the armature current I_a, known as armature reaction flux, remains fixed with respect to the field flux ϕ_f. Considering the magnetic circuit to be unsaturated, we can apply the superposition principle to get the resultant air gap flux by calculating the phasor sum.

$$\text{Resultant air gap flux} = \text{Field flux} + \text{armature reaction flux} \Rightarrow \phi_r = \phi_f + \phi_a \qquad \text{.....(4.3)}$$

Ignoring armature leakage reactance X_l and armature resistance R_a, the resultant magnetic flux ϕ_r generates an electromotive force (emf) in the armature that is equal to terminal voltage V_t. As per Faraday's and Lenz's laws, V_t lags the resultant magnetic flux ϕ_r by 90°. Fig. 4.3 shows the phasor diagram of a loaded (balanced) synchronous generator, illustrating the representation of fluxes, currents, and voltages as phasors.

Here $\theta =$ *power factor angle*

$\delta =$ *angle by which* E_f *leads* V_t *called load angle or torque angle*

Fig. 4.3 Phasor of Syn. m/c when loaded

The power output of the generator is primarily determined by δ, while the VAr output is determined by the magnitude of E_f, which represents the excitation. Due to the assumed linearity of the magnetic circuit, the voltage phasors E_f, E_a *and* V_t are directly proportional to the flux phasors ϕ_f, ϕ_a *and* ϕ_r respectively. Additionally, the voltage phasors lag behind the flux phasors by 90°. From Fig. 4.3, it can be observed that phasor AB, denoted as E_a, is directly proportional to ϕ_a (and consequently I_a), and it leads ϕ_a (or I_a) by 90°.

$$\text{During loading condition and when } R_a \text{ and } X_l \text{ are neglected: } V_t = E_f - jI_aX_a \qquad \text{.....(4.4)}$$

In equation (4.4), E_f is the voltage that is generated solely by the magnetic field flux ϕ_f, and is commonly referred to as the no-load electromotive force (emf). Fig. 4.4 represents circuit diagram corresponding to equation (4.4). In this diagram, X_a represents the inductive reactance, which takes into account the impact of armature reaction.

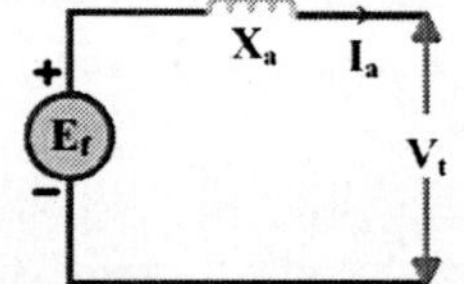

Fig. 4.4 Circuit model of Syn. m/c when R_a and X_l neglected

4.2 Synchronous reactance and equivalent circuits of synchronous generator and motor:

The circuit depicted in Fig. 4.4 can be modified to include armature leakage reactance, X_l and armature resistance, R_a, as shown in Fig. 4.5. The sum of the reactance values $(X_l + X_a)$ is referred to as the synchronous reactance (X_d) of the machine.

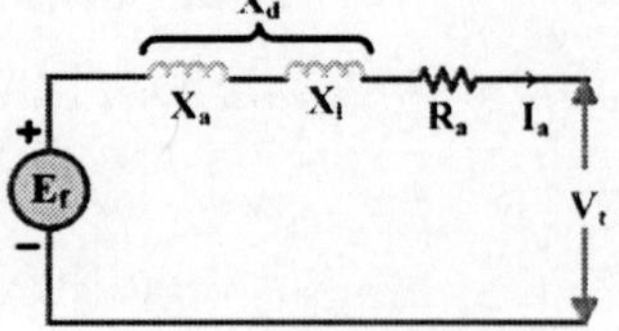

Fig. 4.5 Circuit model of Syn. m/c when R_a and X_l considered

$$\text{Synchronous reactance} \quad X_d = X_l + X_a \qquad \text{.....(4.5)}$$

$$\text{Synchronous impedance} \quad Z_d = R_a + jX_d \qquad \text{.....(4.6)}$$

During loading condition when R_a and X_l are considered:

$$\text{For generator: } V_t = E_f - I_aZ_d = E_f - I_aR_a - jI_aX_d \qquad \text{.....(4.7)}$$

Fig. 4.6 shows the equivalent circuits of a synchronous generator and motor when R_a and X_l are taken into account, and Fig. 4.7 shows corresponding phasor diagrams. In case of synchronous motor, the current I_a direction is reversed, as depicted in Fig. 4.6. As a result, equation (4.7) is modified as:

$$\text{For motor}: V_t = E_f + I_a Z_d = E_f + I_a R_a + jI_a X_d \qquad \ldots\ldots (4.8)$$

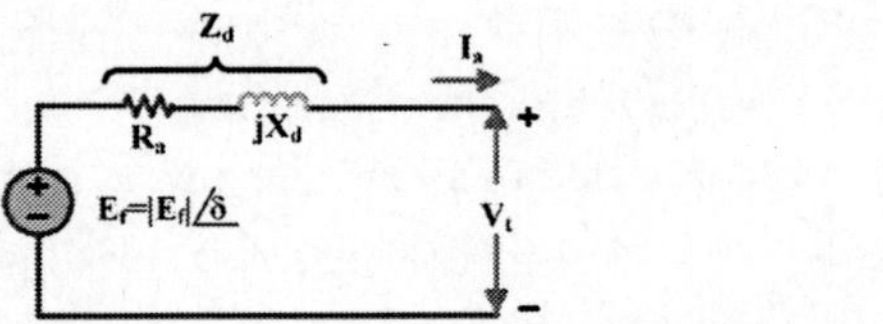

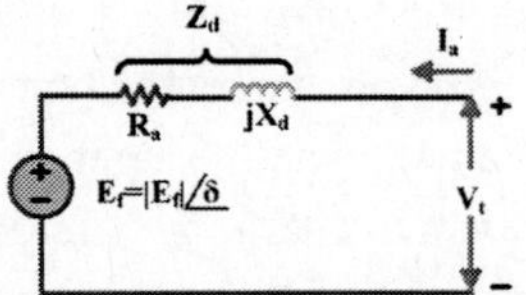

Fig. 4.6 Equivalent circuits for (a) Synchronous generator (b) Synchronous motor when R_a and X_l are considered

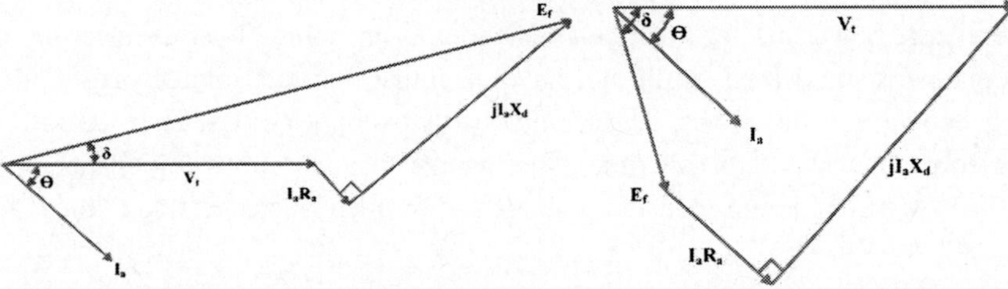

Fig. 4.7 Phasor diagrams of (a) Synchronous generator (b) Synchronous motor when R_a and X_l are considered

Fig. 4.8 shows the equivalent circuits of the synchronous generator and motor when R_a and X_l are ignored, and fig. 4.9 shows corresponding phasor diagrams. Resistance of the armature R_a is usually neglected in power system assessments. The induced electromotive force (emf) E_f in the armature of synchronous generator leads the terminal voltage by the torque angle δ. Indeed, this is the essential requirement for the generator to produce and deliver active power. The amount of power delivered is determined by $sin\delta$.

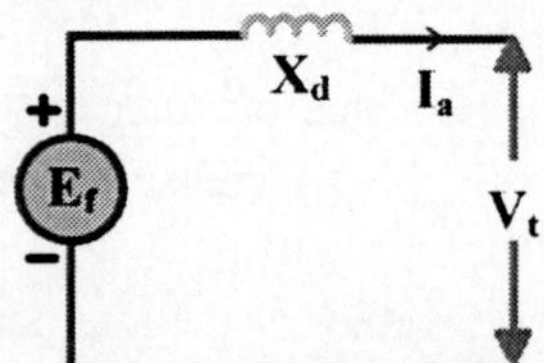

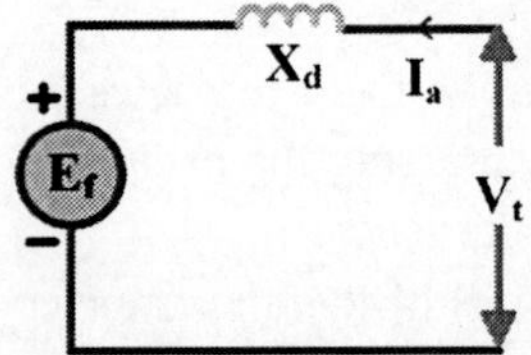

Fig. 4.8 Equivalent circuits for (a) Synchronous generator (b) Synchronous motor when R_a and X_l are neglected

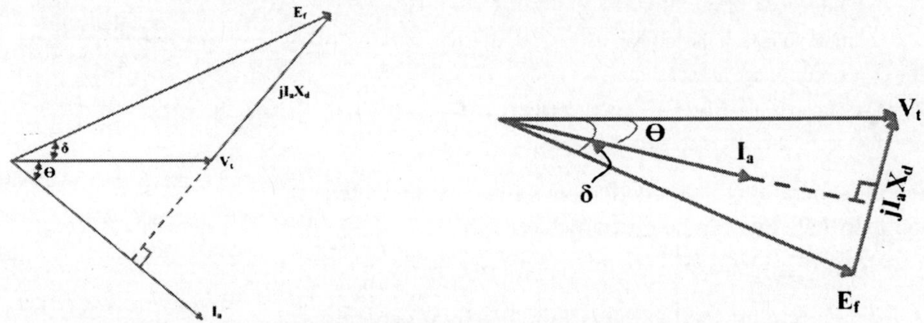

Fig. 4.9 Phasor diagrams of (a) Synchronous generator (b) Synchronous motor when R_a and X_l are neglected

During loading condition when R_a and X_l are neglected:

For generator : $V_t = E_f - jI_aX_d$ ..... (4.9)

For motor : $V_t = E_f + jI_aX_d$ ..... (4.10)

4.3 Synchronous machine connected to an Infinite Bus:

Typically, a synchronous generator functions in parallel with other generators that are connected to the power system. We will examine a generator that is connected to an infinite bus, as depicted in Fig. 4.10. An infinite bus refers to a large network where the voltage and frequency stay constant regardless of the power exchange between the synchronous machine and the bus as well as the excitation of the synchronous machine.

A bus that injects power to a large power network can be considered as an infinite bus. Connecting or disconnecting a single alternator or small load would not have an impact on the voltage magnitude, voltage phase angle, or frequency at this bus. Thus, the infinite bus that supplies power to large power network can be considered equivalent to a large alternator that has zero synchronous impedance and infinite rotating inertia. An infinite bus maintains constant voltage and frequency irrespective of the load connected to it.

The characteristics of an infinite bus are as follows:

- The frequency of the infinite bus remains constant due to infinite rotating inertia. The dynamics of rotating machines connected to it doesn't change the frequency.
- The connection/disconnection of generators and/or loads doesn't change voltage magnitude at the infinite bus.

An alternator that is linked to an infinite bus exhibits the following operational characteristics:

- The power system to which it is attached regulates the terminal voltage and frequency of the alternator.
- The excitation of the alternator regulates the amount of reactive power that the alternator supplies to the infinite bus. Increasing the excitation of the alternator results in an increase in the reactive power output of the alternator.
- The governor set points of the alternator control the real power delivered by the alternator to the infinite bus.

Fig. 4.10 shows n alternators connected to an infinite bus.

Let V = Terminal voltage of the bus bar

E = Generated voltage in each alternator

Z_d = Synchronous impedance of each alternator

n = Number of alternators connected in parallel

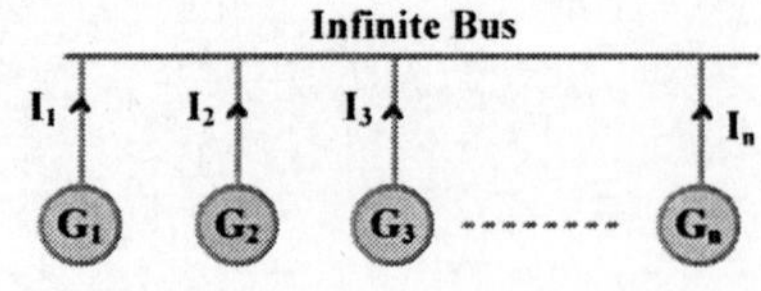

Fig. 4.10 Representation of an Infinite bus

Considering all generators to be identical, the currents injected by these to the infinite bus can be given by:

$$I_1 = I_2 = \dots = I_n = \frac{I}{n}$$

Where, I = Total current injected to the infinite bus by n generators.

Thus, the voltage of the infinite bus would be

$$V = E - \frac{I}{n} Z_d = E - I\frac{Z_d}{n} = E - IZ_{d\,eq} \qquad \text{...... (4.11)}$$

From equation (4.11), when n is very large, $Z_{d\,eq.} = \frac{Z_d}{n} = 0 \Rightarrow I * Z_{d\,eq.} = 0$

Therefore, $V = E$ (constant)

Let J = moment of inertia of each alternator

Then, the total moment of inertia of all n alternator operating in parallel is

$$\Sigma J = J + J + \cdots + J = nJ$$

$$\text{Acceleration of alternator} = \frac{\text{Accelerating torque}}{\text{Moment of inertia}} = \frac{\tau_a}{nJ} \qquad \text{...... (4.12)}$$

From equation (4.12), if n is very large, then nJ is very large, thus Acceleration of alternator $= 0$

Therefore, the speed of the machine remains constant irrespective of the load/generation change at the bus. Therefore, the frequency remains constant. Thus, to achieve constant voltage and frequency at a bus, it is to be connected to as many alternators in parallel as possible.

4.4 Real and Reactive Power of synchronous machine connected to an Infinite Bus:

When a synchronous machine is connected to infinite bus, the bus voltage and frequency remains unaffected by the rotational dynamics of the machine. The real power injected to the bus (in case of synchronous generator) / drawn from the bus (in case of synchronous motor) can be regulated through control of input mechanical torque in case of generator and load torque in case of motor. The reactive power injected to the bus / drawn from the bus can be regulated through control of field current known as excitation control that regulates generated voltage E_f in the armature winding.

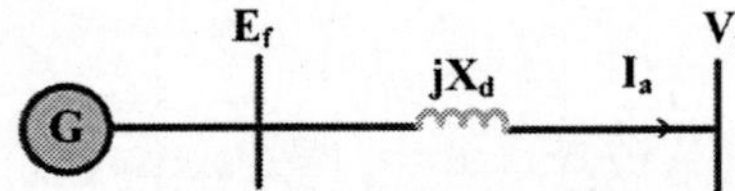

Fig. 4.11 Syn. m/c connected to an Infinite bus

While considering the control of reactive power in a round-rotor generator, it is convenient to disregard resistance. Let's assume that the generator is supplying electricity in such a way that there is a specific angle (δ) between the terminal voltage (V_t) and the generated voltage (E_f) of the machine. The complex power delivered to the system by the generator can be expressed as:

$$S = P + jQ = V_t I_a^* = |V_t||I_a|(cos\theta - jsin\theta) \qquad \text{...... (4.13)}$$

Equating real and imaginary quantities in equation (4.13), we obtain

$$P = |V_t||I_a|cos\theta \text{ and } Q = -|V_t||I_a|sin\theta \qquad \text{...... (4.14)}$$

It is observed from equation (4.14) that synchronous generator injects real power to the bus whether power factor is lagging or leading. However, the generator injects reactive power to the bus if power factor is lagging, whereas, it draws reactive power from the bus if power factor is leading. The generator neither injects reactive power to the bus nor draws it from the bus in case of unity power factor, and the machine generates zero reactive power under this condition. A synchronous motor draws real power from the bus irrespective of bus power factor. The motor draws reactive power from the bus if bus power factor is lagging and injects reactive power to the bus if power factor is leading. The motor neither injects reactive power to the bus nor draws it in case of unity power factor.

Fig. 4.12 (a) shows phasor diagrams of a synchronous generator for lagging, unity and leading power factors. The phasor diagrams of a synchronous motor for lagging, leading and unity power factors are shown in Fig. 4.12 (b). It is observed from the phasor diagrams that:

For normal excitation $|E_f|cos\delta = |V_t|$,

For over-excitation $|E_f|cos\delta > |V_t|$ and

For under-excitation $|E_f|cos\delta < |V_t|$ ... ... (4.15)

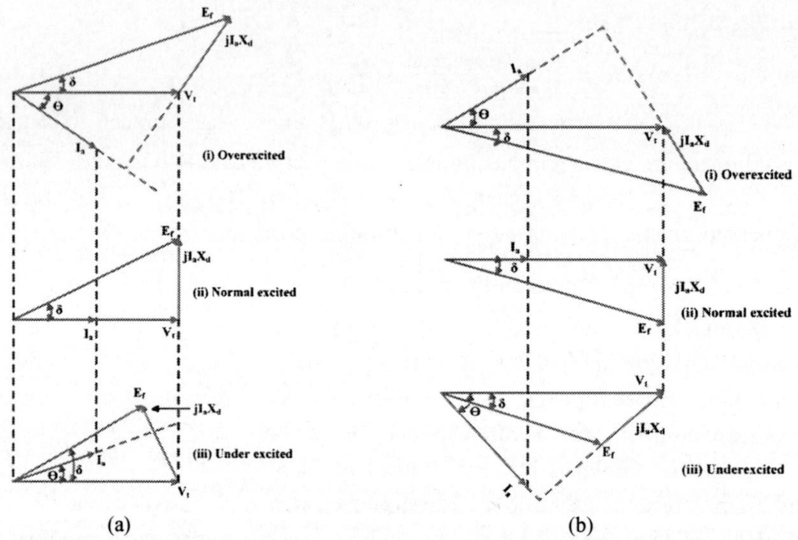

Fig. 4.12 Phasor diagrams of (a) synchronous generator feeding constant power as excitation is varied (b) motor drawing constant power as excitation is varied

A synchronous generator always supplies real power to the bus, whereas, a synchronous motor always draws real power from the bus. However, a synchronous machine can either supply reactive power to the bus or draws it from the bus depending upon amount of excitation. The overexcited synchronous generator injects reactive power to the infinite bus. Thus, infinite bus acts as an inductive load that consumes reactive power, whereas, generator acts as a source of reactive power. The power factor at the bus is lagging under this condition. An under excited synchronous generator draws reactive power from the bus. Thus, infinite bus acts as a capacitor that consumes leading VAr that is equivalent to supply of lagging VAr to the generator, and the generator acts as an inductor that consumes reactive power. The bus power factor is leading under this condition. A synchronous generator with normal excitation neither delivers nor consumes reactive power. The bus power factor is unity under this condition.

An overexcited synchronous motor draws leading VAr from the bus similar to capacitor. The bus power factor is leading under this condition. Thus, an overexcited synchronous motor without any mechanical load can inject reactive power to the bus, and the device is called synchronous condenser as its function is similar to a condenser. An under excited synchronous motor operates at lagging power factor and draws reactive power from the bus. A synchronous motor with normal excitation neither consumes nor delivers reactive power to the bus, and bus power factor is unity under this condition.

It is clear from above discussion that change in excitation (via control of generated voltage through field current variation) controls reactive power flow between synchronous machine and infinite bus. The armature current variation controls the angle δ, the phase-shift between generated voltage and terminal voltage. When armature current I_a is zero, generated emf E_f coincides with terminal voltage V_t as I_aX_d drop is zero. Increase in armature current increases phase-shift δ between E_f and V_t. Armature current of synchronous generator increases with increase in demand. Therefore, δ represents load angle that increases with increase in demand. If demand increases, mechanical power input to the generator from prime mover also increases to keep power balance between generation and demand as bus frequency is constant.

The dependence of P on the power angle δ is also shown as follows. If $V_t = |V_t|\angle 0^0$ *and* $E_f = |E_f|\angle\delta$

$$I_a = \frac{|E_f|\angle\delta - |V_t|}{jX_d} \text{ and } I_a^* = \frac{|E_f|\angle -\delta - |V_t|}{-jX_d}$$

As a result, the complex power delivered to the system by the generator is given by:

$$S = P + jQ = V_t I_a^*$$

$$= |V_t|\left(\frac{|E_f|\angle -\delta - |V_t|}{-jX_d}\right)$$

$$= \frac{|V_t||E_f|\angle -\delta - |V_t|^2}{-jX_d}$$

$$= \frac{|V_t||E_f|(\cos\delta - j\sin\delta) - |V_t|^2}{-jX_d}$$

$$= \frac{|V_t||E_f|\sin\delta}{X_d} + j\frac{|V_t||E_f|\cos\delta - |V_t|^2}{X_d}$$

$$\text{Active Power } P = \frac{|V_t||E_f|\sin\delta}{X_d} \quad and \quad \text{Reactive Power } Q = \frac{|V_t||E_f|\cos\delta - |V_t|^2}{X_d} \qquad \ldots\ldots(4.16)$$

If $|E_f|$ and $|V_t|$ are constant, equation (4.16) very clearly illustrates how P depends on the power angle δ. However, eq. (4.16) demonstrates that δ must drop if $|E_f|$ is increased by enhancing the dc field excitation if P and $|V_t|$ remain unchanged. In equation (4.16), when P is constant, an increase in $|E_f|$ and a decrease in δ indicate that Q will either rise if it is already positive or will fall in magnitude and may even become positive in case it is negative prior to the field excitation being increased.

4.5 Capability Curve of synchronous generators:

When selecting a large generator, it is important to consider not only the rated MVA and power factor, but also the maximum permissible stator and rotor currents. These currents have an impact on mechanical stress and temperature increase. The operation's restrictive limitations are illustrated through the use of an operating chart or performance chart. A Synchronous Generator's Capability Curves establish its safe operating limits. The allowed region of operation is limited to the points listed below:

- The MVA loading should not exceed the generators rated capacity. This limit is established by the stator's armature heating caused by the armature current.
- The MW loading should not exceed the prime mover rated capacity.

- The field current should not be allowed to exceed a certain value determined by the heating of the field.
- The load angle δ must be smaller than 90º to ensure steady-state operation. At δ = 90º, the stable situation reaches its theoretical stability limit.

The construction process of generator's capability curves commences by creating a phasor diagram for the machine, with terminal voltage V_t as the reference phasor, as depicted in Fig. 4.12 (a). Fig. 4.13 (a) is drawn from Fig. 4.12 (a) by rotating all the phasors by 90º in counter clockwise direction, and considering V_t as reference phasor along vertical axis all other phasors drawn in their mirror image position. Fig. 4.13 (b) is drawn by multiplying all the phasors drawn in Figure 4.13 (a) by $\frac{V_t}{X_d}$. This figure shows five loci passing through point m corresponding to: (a) constant real power P; (b) constant reactive power Q; (c) constant internal voltage $|E_f|$; (d) constant armature current $|I_a|$; and (e) constant power factor angle θ.

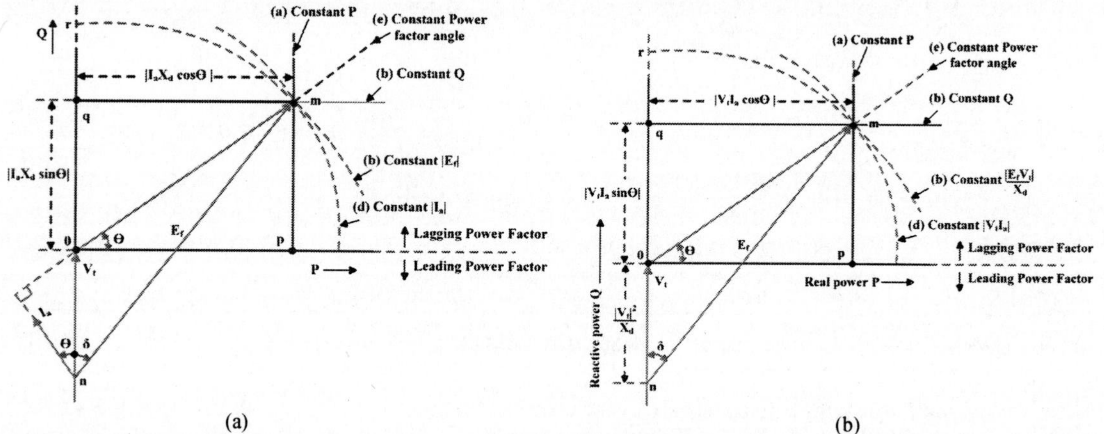

Fig. 4.13 Phasor obtained (a) from mirror image of Fig. 4.12 (a) showing five loci through point *m* corresponding to constant P, Q, E_f, I_a, and θ. (b) by multiplying (rescaling) the distances in Fig. 4.13 (a) by $\frac{V_t}{X_d}$

Constant excitation: The constant excitation circle is centred at point n and has a radius equal to the difference between n and m. This radius is equal to the magnitude of the internal voltage $|E_f|$, which may be kept constant by maintaining a constant dc current I_f, in the field winding.

Constant armature current: The circle representing constant armature current has point *o* as its centre and a radius of length *o-m* that is directly proportional to a fixed value of $|I_a|$. Since the value of V_t is constant, the operational points on this locus represent a consistent output of mega-volt-amperes ($|V_t||I_a|$) from the generator.

Constant active power: The active power output of the machine is represented by the equation P = $|V_t||I_a|cos\theta$. The constant $|V_t|$ results in a fixed distance $X_d|I_a|cos\theta$ between the vertical line *m-p* and the vertical axis *n-o*. This provides a locus of operational points for constant P. The generator always produces a positive megawatt output, regardless of the power factor of the output.

Constant reactive power: The machines reactive power output is calculated as $Q = |V_t||I_a|sin\theta$ in per unit when the angle θ is positive for lagging power factors. When $|V_t|$ is constant, the horizontal line q-m at a set distance $X_d|I_a|sin\theta$ from the horizontal axis represents the locus of operational points for constant Q. For unity power-factor operation, the generators Q output is zero, corresponding to an operating point on the horizontal axis *o-p*. For lagging power factor, the operating point is in the first quadrant, whereas, for leading power factor, the operating point is in fourth quadrant.

Constant power factor: The radial line *o-m* represents a fixed power-factor angle θ between the armature current I_a and terminal voltage V_t. In Fig. 4.13, angle θ represents a lagging power factor load. When θ = 0°, the power factor is unity, and the operating point is on the horizontal axis (*o-p*). The points below the horizontal axis correspond to leading power factors.

Example 4.1. A synchronous generator is rated at 500 MVA, 22 kV, and 0.9 pf lagging. Its synchronous reactance is 1.5 Ω. The generator is supplying full load having a pf of 0.9 lagging at rated voltage. Calculate excitation emf, power angle δ, and reactive power drawn by the load, in p.u. and actual values.

Ans: Consider base values as 500MVA and 22 kV. Load voltage $= \frac{22*10^3}{22*10^3} = 1\ p.u.$

$$\text{p.u. synchronous reactance}, X_d = \frac{X_{actual}}{kV_{base}^2} * MVA_{base} = \frac{1.5}{(22*10^3)^2} * 500 * 10^6 = 1.549\ p.u.$$

Full load MVA = 1 p.u., 0.9 pf lagging

Load current = generator currrent

$$I_a = 1\ \text{pu}, 0.9\ \text{pf lagging} = I_a(cos\emptyset_a - jsin\emptyset_a) = 0.9 - j0.436\ p.u.$$

pu Excitation emf, $E_{f\ pu} = V_t + jI_aX_d$

$$= 1\angle 0^0 + j(0.9 - j0.436)1.549$$

$$= 1.6753 + j1.3941$$

$$= 2.179\angle 39.76^0\ p.u.$$

Actual Excitation emf, $E_{f\ actual} = E_{f\ pu} * V_{base} = 2.179 * 22 * 10^3 = 47.938\ kV$

Power angle $\delta = 39.76^0$ leading

Reactive power drawn by the load $Q = V_tI_a\sin\theta$

$$Q_{pu} = 1 * 1 * 0.436 = 0.436\ \text{p.u.}$$

$$Q_{actual} = 0.436 * 500 * 10^6 = 218\ \text{MVAR}$$

Example 4.2. A synchronous generator is rated at 150 MVA, 10 kV, and 0.8 pf lagging. It has a synchronous reactance of 1.2 Ω. If its excitation emf is (i) increased by 25% and (ii) reduced by 15%, calculate in each case load pf, power angle δ, and reactive power drawn by the load. Carry out calculations in pu form and convert the results to actual values.

Ans : Consider base values as 150 MVA and 10 kV. Load voltage $= \frac{10*10^3}{10*10^3} = 1\ p.u.$

$$\text{p.u. synchronous reactance}, X_d = \frac{X_{actual}}{kV_{base}^2} * MVA_{base} = \frac{1.2}{(10*10^3)^2} * 150 * 10^6 = 1.8\ p.u.$$

Full load MVA = 1 p.u., 0.8 pf lagging = 1 * 0.8 = 0.8 p.u.

Load current = generator currrent

$$I_a = 1\ \text{pu}, 0.8\ \text{pf lagging} = I_a(cos\emptyset_a - jsin\emptyset_a) = 0.8 - j0.6\ p.u.$$

pu Excitation emf, $E_{f\,pu} = V_t + jI_aX_d$

$$= 1\angle 0^0 + j(0.8 - j0.6)1.8 = 2.08 + j1.44 = 2.52\angle 34.69^0\ p.u.$$

Actual Excitation emf, $E_{f\,actual} = E_{f\,pu} * V_{base} = 2.52 * 10 * 10^3 = 25.2\ kV$

Power angle $\delta = 34.69^0$ leading

Case (i): Excitation emf is increased by 25%

$$E_{f\,pu} = 2.52 * 1.25 = 3.15$$

Active Power $P = \dfrac{|V_t||E_f|sin\delta}{X_d}$

$$0.8 = \frac{1 * 3.15}{1.8} sin\delta$$

$$sin\delta = 0.457 \Rightarrow \delta = sin^{-1}(0.457) = 27.19^0$$

$$I_a = \frac{|E_f|\angle\delta - |V_t|}{jX_d} = \frac{3.15\angle 27.19^0 - 1}{j1.8} = 1.28\angle -51.38^0$$

$$p.f = \cos(-51.38) = 0.624$$

Reactive power drawn by load $Q = V_tI_a\sin\theta = 1 * 1 * \sin(51.38) = 1 * 1 * 0.781 = 0.781$ p.u.

$$Q_{actual} = 0.781 * 150 * 10^6 = 117.19\ \text{MVAr}$$

Case (ii): Excitation emf is decreased by 15%

$$E_{f\,pu} = 2.52 * 0.85 = 2.142$$

Active Power $P = \dfrac{|V_t||E_f|sin\delta}{X_d}$

$$0.8 = \frac{1 * 2.142}{1.8} sin\delta$$

$$sin\delta = 0.672 \Rightarrow \delta = sin^{-1}(0.672) = 42.22^0$$

$$I_a = \frac{|E_f|\angle\delta - |V_t|}{jX_d} = \frac{2.142\angle 42.22^0 - 1}{j1.8} = 0.863\angle -22.16^0$$

$$p.f = \cos(-22.16°) = 0.926$$

Reactive power drawn by load $Q = V_tI_a\sin\theta = 1 * 1 * \sin(22.16°) = 1 * 1 * 0.377 = 0.377$ p.u.

$$Q_{actual} = 0.377 * 150 * 10^6 = 56.57\ \text{MVAr}$$

4.6 Steady state, transient and sub-transient equivalent circuits of a synchronous generator during Short-Circuit:

When a synchronous generator's terminals are shorted, the current flowing in the armature is identical to the current flowing when a sinusoidal voltage is quickly supplied to a series R-L circuit. However, there is one significant difference: in a series R-L circuit, inductive reactance is constant, whereas in a synchronous generator, reactance varies with time. In practice, three discrete values are assigned and thus we have three reactances direct-axis sub-transient reactance X_d'', direct-axis transient reactance X_d', and direct-axis synchronous reactance X_d. When the machine's terminals are suddenly short-circuited, the armature current surges and the power factor approaches zero.

A short-circuit at the generator terminals results in mismatch between mechanical power input and electrical power output causing the machine to accelerate. This creates slip between synchronously rotating magnetic field and machine rotor that causes induced emf in rotor circuits (i. e. in field winding and in short-circuited damper bars). The induced emf increases field current and causes current flow through damper windings. The increased current in field winding and current in damper windings produce flux that link with armature winding. As flux linking with armature increases, the generated voltage increases that increases armature current. However, for analysis of synchronous machine under transient, machine is represented by a constant voltage source, and the impact of increased voltage is considered by a reduced reactance (if voltage is considered constant a reduction in reactance will result in increased armature current). The reduced reactance is represented by reactances X_{dw} (corresponding to damper winding) and X_f (corresponding to field winding) in parallel with X_a as shown in Fig. 4.14 (a). The current in damper windings decay to zero after 2-3 cycles as there is no source to sustain this current. The initial 2-3 cycle period following short-circuit is known as sub-transient period. The reactance X_{dw} corresponding to sub-transient period is made open after first 2-3 cycles and only X_f remains in parallel with X_a as shown in Fig. 4.14 (b). The period following sub-transient period is known as transient period. When fault is cleared and steady-state is recovered, reactance X_f is removed and only X_a remains in the circuit as shown in Fig. 4.14 (c).

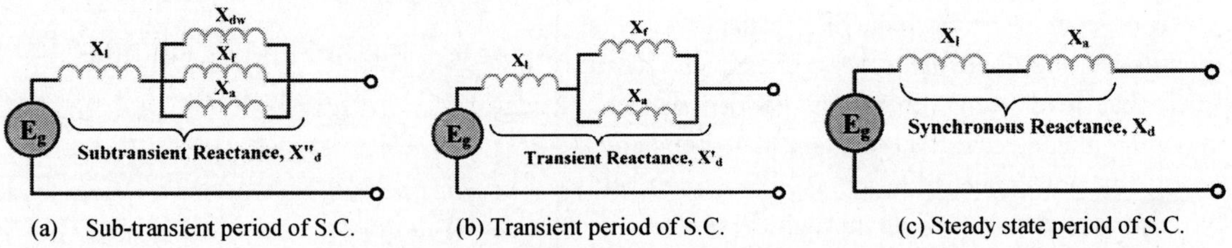

Fig. 4.14 Approximate model of an alternator

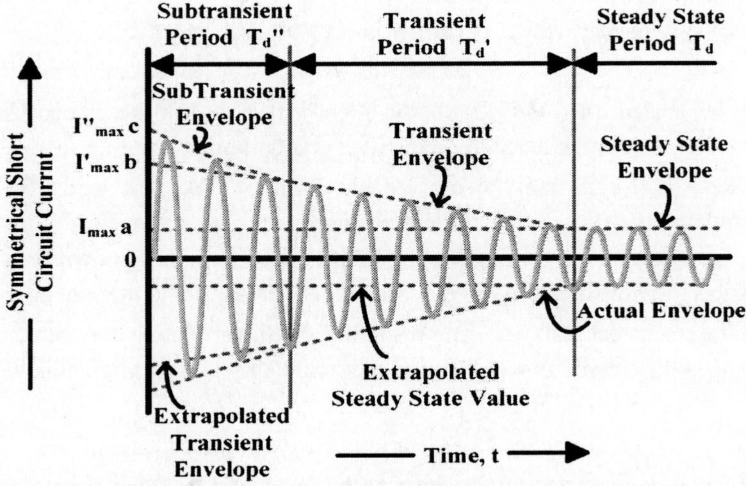

Fig. 4.15 Symmetrical short-circuit current of a synchronous generator

Sub – transient reactance $X''_d = X_l + \frac{1}{\frac{1}{X_a}+\frac{1}{X_f}+\frac{1}{X_{dw}}}$ ……… (4.17)

Transient reactance $X'_d = X_l + \frac{1}{\frac{1}{X_a}+\frac{1}{X_f}}$ ……… (4.18)

Steady – state reactance $X_d = X_l + X_a$ ……… (4.19)

sub-transient short-circuit current $I''_g = \frac{E_g}{X''_d}$ ……… (4.20)

Transient short-circuit current $I'_g = \frac{E_g}{X'_d}$ ……… (4.21)

Steady-state short-circuit current $I_g = \frac{E_g}{X_d}$ ……… (4.22)

From the above equations it can be observed that reactance of a 3-phase synchronous generator during short-circuit will be $X''_d < X'_d < X_d$ ……… (4.23)

and the respective current will be $I''_g > I'_g > I_g$ ……… (4.24)

4.7 Electrical Loads:

An electrical load is a device that converts electrical energy, carried by a current, into different forms such as heat, light, or mechanical work. The electrical load can consist of resistive, inductive, capacitive, or a mixture of these elements. In general, the term "electrical load" can encompass various aspects, including the electrical equipment that consumes energy, the power needed from a specific circuit, and the current or power flowing through the line.

4.7.1 Classification of Loads:

Electrical loads can be classified based on nature of the load and power system load which is shown in Fig. 4.16.

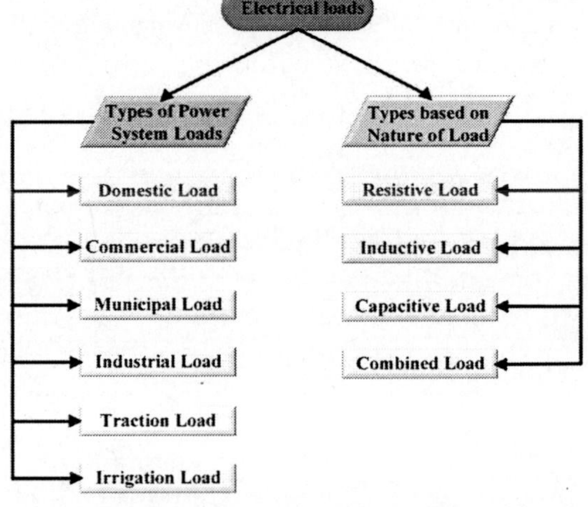

Fig. 4.16 Classification of loads.

4.7.1.1 Classification based on nature of loads: Electrical loads can be categorized based on their characteristics as either resistive, capacitive, inductive, or a combination of these.

Resistive Load: Resistive loads are the most basic form of electrical load. They have a power consumption behaviour that does not include any energy storage and instead converts all electrical energy into heat. A load that includes a heating element is commonly known as a resistive load. A resistive load restricts the flow of energy in the circuit, converting it into thermal energy.

These loads exhibit a consistent resistance and result in a direct correlation between current and voltage. Resistive loads consume electrical power in a way that maintains a phase alignment between the current wave and the voltage wave. Therefore, the power factor for a resistive load is equal to unity.

Some examples of resistive load are incandescent light bulbs, electric heaters, ovens etc.

Inductive Load: Inductive loads accumulate energy in a magnetic field when an electric current flows through them. These loads can induce a phase lag between current and voltage, resulting in power factor problems that need to be addressed in order to uphold system efficiency. Some examples of inductive loads include transformers, motors, coils etc.

Capacitive Load: Capacitive loads have the ability to store energy within an electric field. Capacitive loads can induce a phase shift; however, it occurs in the opposite direction compared to inductive loads. A capacitive load results in a phase shift where the current wave leads the voltage wave. Therefore, the power factor of a capacitive load is in a leading phase. Effectively handling these loads is essential for preserving system stability and optimizing efficiency. Capacitive loads encompass different examples, including capacitor banks, buried cables, capacitors utilized in circuits of motors starters etc.

Combination Loads: The majority of loads exhibit a combination of resistive, capacitive, and inductive characteristics. Several practical applications utilize different combinations of resistors, capacitors, and inductors. The power factor of these loads is less than unity and can either be lagging or leading.

Examples of combined loads include the employment of capacitors in single phase motors to assist with motor starting and running, as well as in tuning circuits or filter circuits.

4.7.1.2 Classification of power system loads:

Power system loads are mainly classified as domestic load / residential load, commercial load, industrial load, municipal load, irrigation load and traction load.

Domestic load / residential load: A domestic load refers to the aggregate energy consumed by electrical appliances in a residential building. Indeed, the extent of this variation is contingent upon individual households and exhibits notable disparities across different nations.

Domestic load encompasses many electrical devices used in households, such as lights, fans, home appliances (including TV, AC, refrigerators, heaters, etc.), as well as small motors for water pumping etc. Most of these devices are connected only for a few hours in a day and use a moderate amount of electricity.

Commercial load: A commercial load refers to electrical loads that are utilized for business purposes, such as shop lighting, office appliances, and restaurant equipments. These appliances are often connected for extended periods of time compared to household loads.

Industrial load: An industrial load refers to the quantity of energy consumed by factories and manufacturing plants to power their robust industrial machinery. This load encompasses a wide range of sectors, from small-scale to big industries. The load primarily consist of robust machinery and other systems that utilize induction motors. Industrial power loads are often deployed in conjunction with utility feeders, which are electrical transmission cables that connect the industrial load to the electric grid. It is probable that they will be continuously connected.

Examples of machinery and equipment that operate utilizing an industrial load include: Conveyor belts, forklifts, and excavators.

Municipal load: A municipal load refers to the aggregate power consumption required to sustain the optimal functioning of residents, businesses, and infrastructure within a town or city. Municipal loads are commonly quantified using either kilowatt-hour (kWh) or megawatt-hour (MWh).

Some examples of municipal load include: sewage systems, traffic signals, power plants, water distribution systems, and drainage systems. Certain systems, such as street lights or pumps that refill overhead storage tanks, function during nighttime hours.

Irrigation load: This category encompasses motors and pumps utilized in irrigation systems to provide water for agricultural purposes. Typically, irrigation loads are fed during non-peak or nighttime hours.

Traction load: Traction loads encompass electric trains, tram cars, and similar transportation systems. These loads have their maximum intensity throughout the morning and evening periods.

4.7.2 Voltage and Frequency Dependence of Loads:

The real power demand (P_D) and reactive power demand (Q_D) are given by:

$$P_D = VI\cos\theta = \frac{V^2}{Z}\cos\theta\,;\quad Q_D = VI\sin\theta = \frac{V^2}{Z}\sin\theta \qquad \text{......... (4.25)}$$

where,

V = voltage across the load
I = current drawn by the load
Z = load impedance.

Based on dependency on voltage, electrical loads can be classified in three categories as mentioned below:

i. **Constant impedance load:** The impedance of such type of loads is constant irrespective of the voltage. Therefore, for a load with constant power factor, real and reactive power drawn are proportional to square of the voltage. A normal heater is an example of such type of load.
ii. **Constant current load:** The current drawn by the load is constant in such type of loads. Therefore, for a load with constant power factor, real and reactive power drawn are proportional to voltage. Power electronic devices come under this category of loads.
iii. **Constant power load:** Such type of loads draws constant real and reactive power irrespective of the load voltage. Thermostatically controlled devices come under this category of loads.

In general, a load may have all the three components mentioned above and can be represented as:

$$P_D = P_{D0}(\alpha_2 V^2 + \alpha_1 V + \alpha_0) \qquad \text{......... (4.26)}$$

$$Q_D = Q_{D0}(\beta_2 V^2 + \beta_1 V + \beta_0) \qquad \text{......... (4.27)}$$

where,

P_{D0} = base case real power demand
Q_{D0} = base case reactive power demand
$\alpha_2, \alpha_1, \alpha_0, \beta_2, \beta_1, \beta_0$ are constants.

Above representation of voltage dependent load is called ZIP model as its first, second and third components represent constant impedance (Z) type, constant current (I) type and constant power (P) type, respectively.

Another way of representation of voltage dependent loads is exponential model as per following:

$$P_D = P_{D0}\left(\frac{V}{V_0}\right)^a \qquad \text{......... (4.28)}$$

$$Q_D = Q_{D0}\left(\frac{V}{V_0}\right)^b \qquad \text{......... (4.29)}$$

Where, V_0 is the rated voltage and exponents a and b are constants whose value depend upon amount of constant impedance, constant current and constant power components present in the load.

The frequency sensitivity of load is represented by adding a factor of load proportional to change of frequency in ZIP model as well as in exponential model as per following:

$$P_D = P_{D0}(\alpha_2 V^2 + \alpha_1 V + \alpha_0)(1 + c\Delta f) \quad \text{.........(4.30)}$$

$$Q_D = Q_{D0}(\beta_2 V^2 + \beta_1 V + \beta_0)(1 + d\Delta f) \quad \text{.........(4.31)}$$

$$P_D = P_{D0}\left(\frac{V}{V_0}\right)^a (1 + c\Delta f) \quad \text{.........(4.32)}$$

$$Q_D = Q_{D0}\left(\frac{V}{V_0}\right)^b (1 + d\Delta f) \quad \text{.........(4.33)}$$

where, c and d are constants.

4.8 The Per-Unit System:

When analysing power networks, it is a common practice to represent quantities as fractions of their reference values such as rated or full-load values, rather than using actual values. The fractions are referred to as per unit (abbreviated as p.u.) and the p.u. value of any quantity is defined as

$$p.u.value = \frac{actual\ value\ in\ any\ unit}{base\ value\ in\ the\ same\ unit} \quad \text{......(4.34)}$$

Power transmission lines are operated at a specific voltage level, with kilovolt (kV) being the preferred unit for expressing voltage. The standard convention used to describe the large amount of power transmitted are kilowatts or megawatts for real power and kilo volt-amperes or mega volt-amperes for apparent power. Nevertheless, many measurements, including amperes and ohms, are frequently represented as a percentage or as a fraction of predetermined base or reference value. For example, when a base voltage of 150kV is selected, the voltages of 110kV, 150kV, and 220kV correspond to 0.733, 1, and 1.46 per unit, or 73.33%, 100%, and 146%, respectively. The ratio expressed as a percentage is equal to 100 times the value expressed as a per unit. Both the percentage and per-unit methods of calculation are simple.

4.8.1. Selection of base values:

The relationship between voltage, current, kilo-volt-amperes, and impedance is such that the selection of base values for any two of these variables dictates the base values of the remaining two. By specifying the base values of current and voltage, we may determine the base impedance and base kilo-volt-amperes. The base impedance refers to the impedance that will result in a voltage drop equal to the base voltage when the current flowing through the impedance is equal to the base current value. The base kilovolt-amperes in single-phase systems can be calculated by multiplying the base voltage in kilo-volts with the base current in amperes. The base is typically specified by selecting the base mega-volt-amperes and base voltage in kilo-volts. In case of three phase system, base voltage and base current correspond to line quantities, whereas, the base MVA correspond to three phase MVA.

The following equations establish the relationship between the different base quantities:

In a 1-phase system:

Apparent power $S_{base} = V_{base} * I_{base}$

$$\text{Base Impedance} = \frac{base\ voltage}{base\ current} \Rightarrow Z_{base} = \frac{kV_{base} \times 1000}{I_{base}} \quad \ldots\ldots\ldots (4.35)$$

$$Z_{base} = \frac{V_{base}}{I_{base}} = \frac{kV_{base}/1000}{kVA_{base}/kV_{base}} = \frac{kV_{base}^2}{kVA_{base} \cdot 1000} = \frac{kV_{base}^2}{MVA_{base}} \quad \ldots\ldots\ldots (4.36)$$

$$\text{Per unit impedance}, Z_{p.u.} = \frac{\text{Actual impedance}}{\text{Base impedance}} \Rightarrow Z_{p.u.} = \frac{Z_{actual}}{Z_{base}} \quad \ldots\ldots\ldots (4.37)$$

$$\text{Per unit impedance}, Z_{p.u.} = \frac{Z_{actual}}{\frac{kV_{base}^2}{MVA_{base}}} = \frac{Z_{actual}}{kV_{base}^2} * MVA_{base} \quad \ldots\ldots\ldots (4.38)$$

$$\text{Per unit voltage} = \frac{\text{Actual voltage}}{\text{Base voltage}} \Rightarrow kV_{p.u.} = \frac{kV_{actual}}{kV_{base}} \quad \ldots\ldots\ldots (4.39)$$

$$\text{Base current} = \frac{\text{Base kVA}}{\text{Base kV}} = \frac{kVA_{base}}{kV_{base}} \quad \ldots\ldots\ldots (4.40)$$

$$\text{Per unit current}, I_{p.u.} = \frac{\text{Actual current}}{\text{Base current}} = \frac{\text{Actual current}}{\frac{\text{Base kVA}}{\text{Base kV}}} = \frac{I_{actual}}{kVA_{base}} * kV_{base} \quad \ldots\ldots\ldots (4.41)$$

$$kW_{1-\emptyset,base} = kVAr_{1-\emptyset,base} = kVA_{1-\emptyset,base} \quad \ldots\ldots\ldots (4.42)$$

$$MW_{1-\emptyset,base} = MVAr_{1-\emptyset,base} = MVA_{1-\emptyset,base} \quad \ldots\ldots\ldots (4.43)$$

In a 3-phase system:

$$\text{Base Current}, I_{base} = \frac{kVA_{base}}{\sqrt{3} * kV_{base}} \quad \ldots\ldots\ldots (4.44)$$

$$\text{Base Impedance}, Z_{base} = \frac{\left(\frac{kV_{LL,base}}{\sqrt{3}}\right)^2}{\frac{MVA_{3\emptyset,base}}{3}} = \frac{kV_{LL,base}^2}{MVA_{3\emptyset,base}} \quad \ldots\ldots\ldots (4.45)$$

Where, $kV_{LL,base} = Base\ line\ to\ line\ voltage$

$$\text{Per unit impedance}, Z_{p.u.} = \frac{Z_{actual}}{\frac{kV_{LL,base}^2}{MVA_{3\emptyset,base}}} = \frac{Z_{actual}}{kV_{LL,base}^2} * MVA_{3\emptyset,base} \quad \ldots\ldots\ldots (4.46)$$

4.8.2. Change of base:

Generally, rated voltage and rated MVA of a power system component (such as generator, motor, transformer, transmission line, and load) are taken as corresponding base quantities. However, all these components of power system that are interconnected are not having same rated voltage and MVA. Therefore, in order to perform power system computations on per unit basis, it is required to have a common system base for MVA. However, a common voltage base is not considered for the whole system, rather these are taken as per voltage transformation ratio of transformers in different parts of power system (the reason behind this is presented in the next sub-section). As system MVA base is different from the base MVA for an individual power system component, and also base voltages in different parts of the power system are not uniform, a formula that converts per unit impedance expressed based on a particular voltage and MVA base to that expressed based on another MVA and voltage base is required. The conversion formula is derived below:

$$\text{Per unit impedance}, \ Z_{p.u.}^{given} = Z_{actual} * \frac{MVA_{base}^{given}}{kV_{given\ base}^2}$$

$$Z_{actual} = Z_{p.u.}^{given} * \frac{kV_{given\ base}^2}{MVA_{base}^{given}} \qquad \text{……… (4.47)}$$

The per-unit impedance is directly proportional to the base MVA and inversely proportional to the square of the base voltage. The per-unit impedance on a new base is obtained as per following:

Per unit impedance on new base, $Z_{p.u.}^{new} = Z_{actual} * \frac{MVA_{base}^{new}}{kV_{new\ base}^2} = Z_{p.u.}^{given} * \frac{kV_{given\ base}^2}{MVA_{base}^{given}} * \frac{MVA_{base}^{new}}{kV_{new\ base}^2}$

Per unit impedance on new base, $Z_{p.u.}^{new} = Z_{p.u.}^{given} * \left(\frac{kV_{given\ base}}{kV_{new\ base}}\right)^2 * \left(\frac{MVA_{base}^{new}}{MVA_{base}^{given}}\right)$ ……… (4.48)

Certain scholars represent the p.u. number as a percentage. While the usage of p.u. values may initially appear to be an indirect form of representation, it actually offers significant advantages. These advantages can be summarised as follows:

1. The size of the apparatus being considered might vary significantly, as can the losses and voltage drops. The per unit voltage drops and losses for apparatus of the same general type are comparable, irrespective of their size.
2. The selection of suitable voltage bases simplifies the solution of networks that include several transformers as explained in next sub-section.
3. The utilisation of $\sqrt{3}$ in three-phase computations is minimised.

4.8.3 Per unit impedance of transformer:

Let us consider three-phase rated MVA ($MVA_{3-\emptyset,rated}$) of the transformer as base MVA common to both primary and secondary sides. However, let us consider different base voltages on primary and secondary sides with rated primary voltage $kV_{1,rated}$ as base kV for the primary side and rated secondary voltage $kV_{2,rated}$ as base kV for the secondary side.

Neglecting shunt branch,

The equivalent impedance referred to primary side $\qquad Z_{1,eq} = Z_1 + Z_2'$ ……… (4.49)

Where, Z_1 = Leakage impedance of primary side

Z_2' = Leakage impedance of secondary side referred to primary side

Therefore, the per unit equivalent impedance referred to primary side

$$\frac{Z_1+Z_2'}{\frac{kV_{1,rated}^2}{MVA_{3-\emptyset,rated}}} = \frac{Z_1}{\frac{kV_{1,rated}^2}{MVA_{3-\emptyset,rated}}} + \frac{Z_2\frac{kV_{1,rated}^2}{kV_{2,rated}^2}}{\frac{kV_{1,rated}^2}{MVA_{3-\emptyset,rated}}}$$

$$= \frac{Z_1}{\frac{kV_{1,rated}^2}{MVA_{3-\emptyset,rated}}} + \frac{Z_2}{\frac{kV_{2,rated}^2}{MVA_{3-\emptyset,rated}}}$$

$$= Z_1(p.u) + Z_2(p.u) \qquad \text{……… (4.50)}$$

Where, Z_2 = Leakage impedance of secondary side

The equivalent impedance referred to secondary side $\qquad Z_{2,eq} = Z_2 + Z_1'$ ……… (4.51)

Where, Z_1' = Leakage impedance of primary side referred to secondary side

Therefore, the per unit equivalent impedance referred to secondary side

$$\frac{Z_2+Z_1'}{\frac{kV^2_{2,rated}}{MVA_{3-\emptyset,rated}}} = \frac{Z_2}{\frac{kV^2_{2,rated}}{MVA_{3-\emptyset,rated}}} + \frac{Z_1\frac{kV^2_{2,rated}}{kV^2_{1,rated}}}{\frac{kV^2_{2,rated}}{MVA_{3-\emptyset,rated}}}$$

$$= \frac{Z_2}{\frac{kV^2_{2,rated}}{MVA_{3-\emptyset,rated}}} + \frac{Z_1}{\frac{kV^2_{1,rated}}{MVA_{3-\emptyset,rated}}} = Z_2(p.u) + Z_1(p.u) \quad \ldots\ldots\ldots (4.52)$$

It is observed from above that whereas, actual impedance of the transformer referred to primary/secondary side depends upon voltage transformation ratio, the per unit impedance referred to either of the two sides of the transformer is same. A large interconnected power system contains a large number of transformers to step-up / step-down voltages at different levels. So, if power system computations are performed using absolute impedance of transformers, the calculation becomes very tedious and time consuming as impedances of all the elements connected to one of the sides of the transformer are to be converted to equivalent impedance referred to other side as per voltage transformation ratio for every transformer present in the system. However, if calculations are performed in terms of per unit quantities no such transformation is required, which simplifies power system computations. This is very important advantage of using per unit system.

4.8.4. Per unit impedance diagram of a power system:

The per unit impedance diagram can be derived directly from the single line diagram of a power system by following the prescribed procedure:

Select a common kVA or MVA base for the entire system. Divide the system into multiple portions using transformers. Select a suitable kV base for one of the sections and calculate the kV bases for the remaining sections as per voltage transformation ratio of different transformers. Now compute the values of voltages and impedances in per unit for each segment and link them according to the topology of the single line diagram.

Example 4.3. For the following power system, draw the impedance and reactance diagrams.

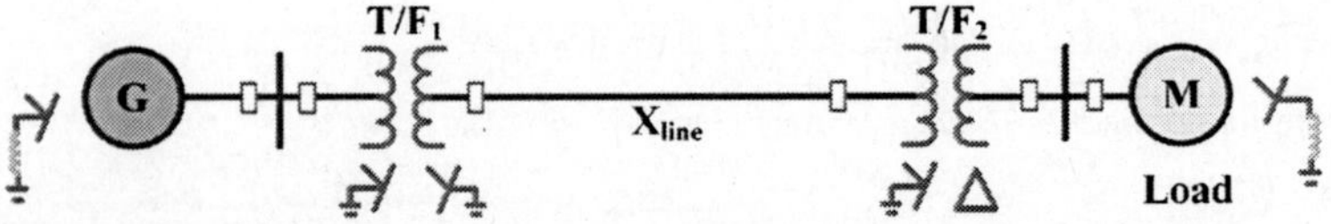

Ans: The impedance and reactance diagrams of the above single line diagram are given below.

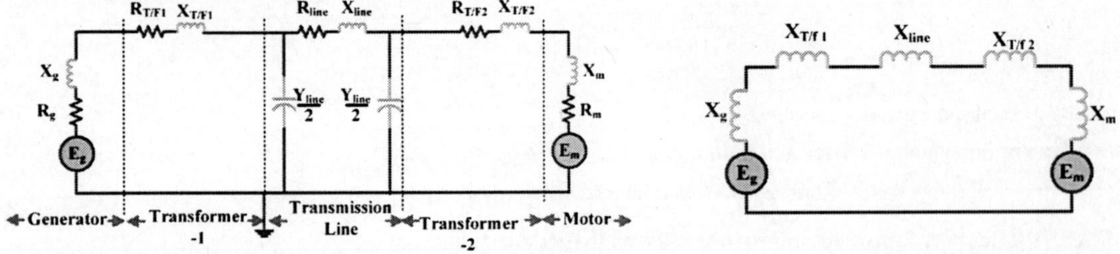

Example 4.4. For the following power system, draw the impedance and reactance diagrams.

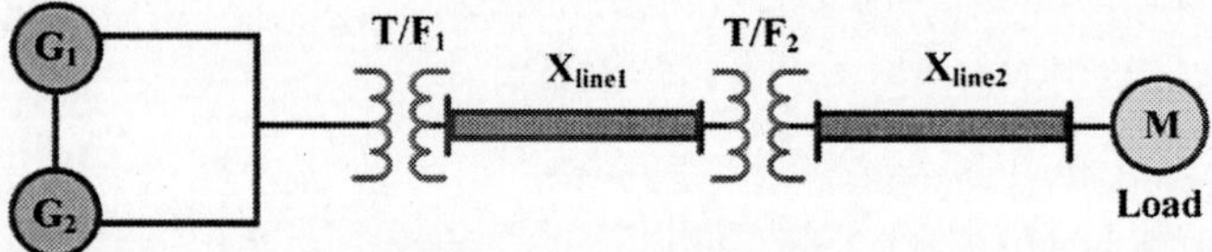

Ans: The impedance and reactance diagrams of the above single line diagram are given below.

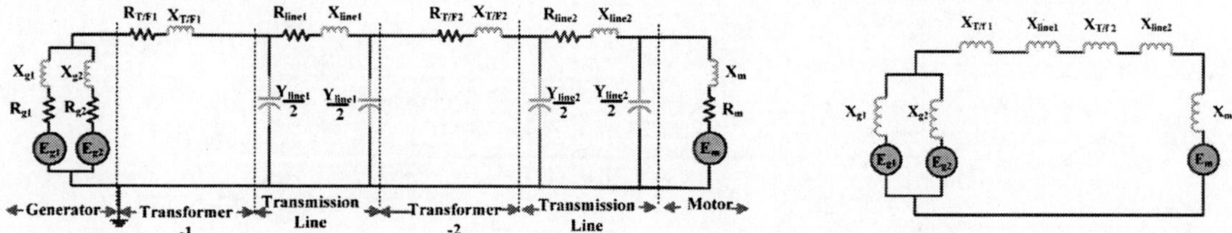

Example 4.5. For the following power system, draw the impedance and reactance diagrams.

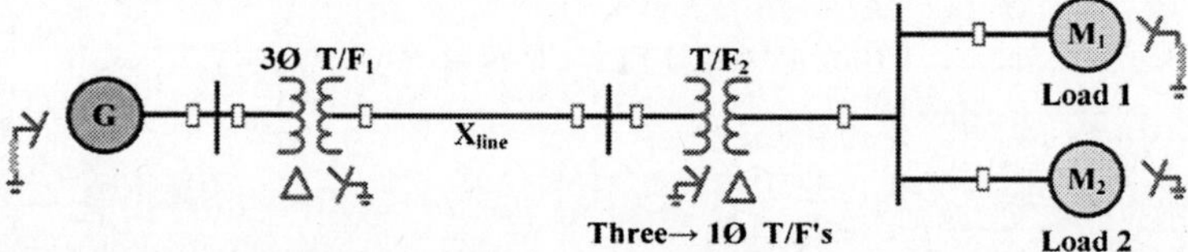

Ans: The impedance and reactance diagrams of the above single line diagram are given below.

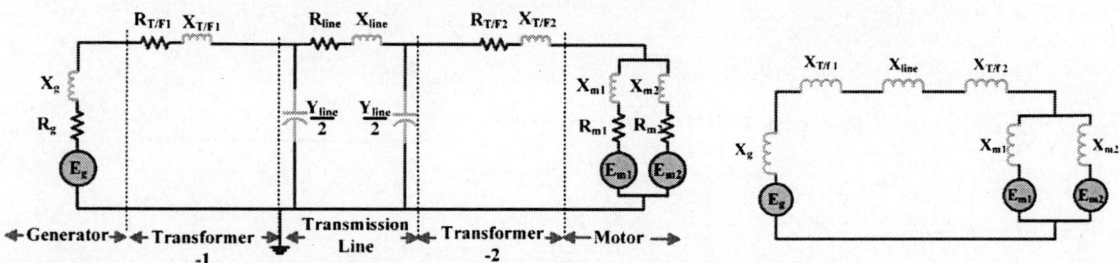

Example 4.6. For the following power system, draw the reactance diagram on 30 MVA base with base voltage of 11 kV on the motor side.

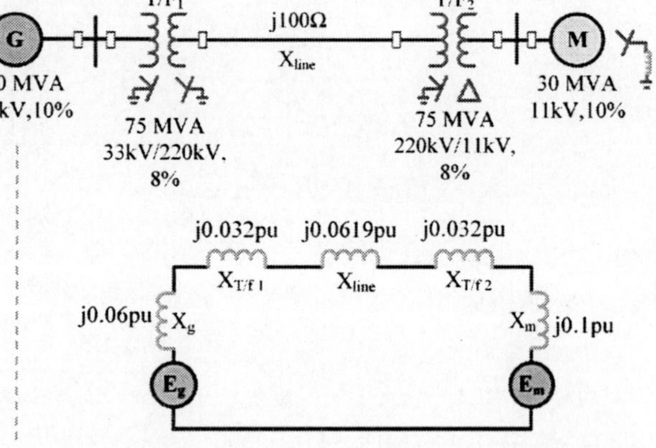

Ans. Base values of 30 MVA and 11kV at M

$$X^{M}_{p.u.} = j0.10 * \left(\frac{11*10^3}{11*10^3}\right)^2 * \frac{30*10^6}{30*10^6} = 0.10\ pu$$

$$X^{T/f2}_{p.u.} = j0.08 * \left(\frac{11*10^3}{11*10^3}\right)^2 * \frac{30*10^6}{75*10^6} = j0.032\ p.u.$$

$$X^{line1}_{p.u.} = \frac{100*30*10^6}{(220*10^3)^2} = j0.0619\ p.u.$$

$$X^{T/f2}_{p.u.} = j0.08 * \left(\frac{33*10^3}{33*10^3}\right)^2 * \frac{30*10^6}{75*10^6} = j0.032\ p.u.$$

$$X^{G}_{p.u.} = j0.10 * \left(\frac{33*10^3}{33*10^3}\right)^2 * \frac{30*10^6}{50*10^6} = j0.06\ p.u.$$

Example 4.7. A 100 MVA, 11 kV, 3φ generator has a sub-transient reactance of 10%. The generator supplies two synchronous motors via a 25 km transmission line with transformers on both sides. Transformer-1 is a 3φ (Δ-Y) transformer with a rating of 150 MVA, 11kV/110kV, and 15% reactance, whereas Transformer-2 is made up of three identical 1φ transformers each with ratings of 60 MVA, 63.5kV/13.2kV(Y-Δ), and 10% reactance. The transmission line has a series reactance of 2Ω per km. The rating of motor M_1 is 120 MVA, 6.9 kV, and 5% reactance. Motor M_2 is rated at 150 MVA, 6.9 kV, and 10% reactance. Draw the reactance diagram with each reactance marked in per unit values. Consider generator ratings as base values.

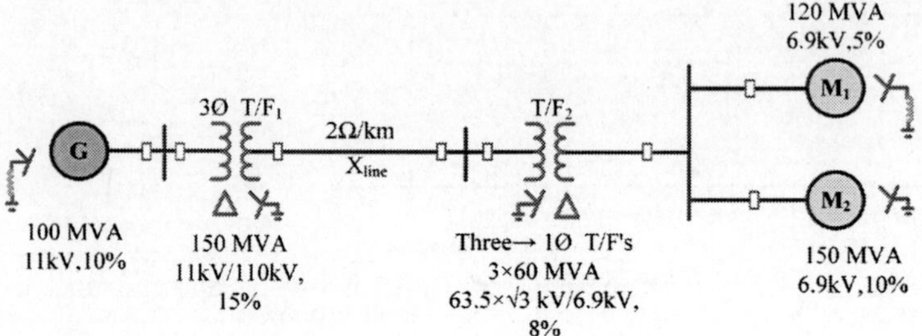

Ans. Consider the base values of 100 MVA and 11 kV at generator G.

$$X_{p.u.}^{G} = j0.10 * \left(\frac{11*10^3}{11*10^3}\right)^2 * \frac{100*10^6}{100*10^6} = j0.1\ p.u.$$

$$X_{p.u.}^{T/f1} = j0.15 * \left(\frac{11*10^3}{11*10^3}\right)^2 * \frac{100*10^6}{150*10^6} = j0.1\ p.u.$$

$$X_{p.u.}^{line} = \frac{2*25*100*10^6}{(110*10^3)^2} = j0.4132\ p.u.$$

At T/f 2: $kV_{LT} = kV_{HT} * \left(\frac{LT}{HT}\right)_{T/f}$

$$= 110 * 10^3 * \left(\frac{6.9*10^3}{\sqrt{3}*63.5*10^3}\right) = 6.9\ kV$$

$$X_{p.u.}^{T/f2} = j0.10 * \left(\frac{6.9*10^3}{6.9*10^3}\right)^2 * \frac{100*10^6}{3*60*10^6} = j0.055\ p.u.$$

$$X_{p.u.}^{M_1} = j0.05 * \left(\frac{6.9*10^3}{6.9*10^3}\right)^2 * \frac{100*10^6}{120*10^6} = j0.0416\ p.u.$$

$$X_{p.u.}^{M_2} = j0.10 * \left(\frac{6.9*10^3}{6.9*10^3}\right)^2 * \frac{100*10^6}{150*10^6} = j0.066\ p.u.$$

j0.1pu $X_{T/f1}$ j0.4132pu X_{line} j0.055pu $X_{T/f2}$
j0.1pu X_g E_g j0.0416pu X_{m1} E_{m1} X_{m2} j0.066pu E_{m2}

Example 4.8. The single-line diagram of an unloaded power network is shown below. Draw the reactance diagram on the basis of:

a. 15 MVA & 13.8 kV on generator G_2 b. 15 MVA & 13.8 kV on generator G_1

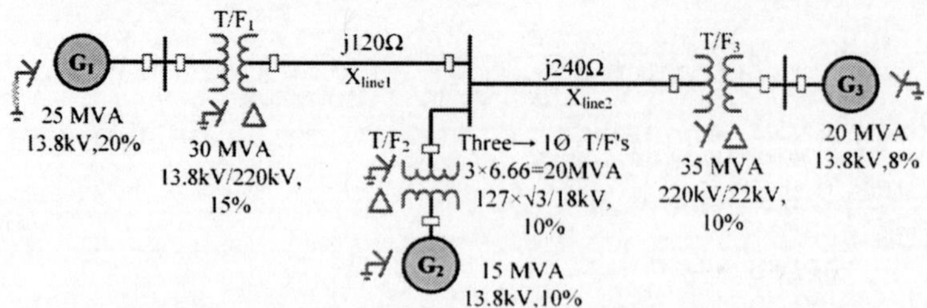

Ans. Case i: Consider the base values of 15 MVA and 13.8 kV at the generator G_2

Per unit reactance value of generator-2 $X_{p.u.}^{G_2} = j0.10 * \left(\frac{13.8*10^3}{13.8*10^3}\right)^2 * \frac{15*10^6}{15*10^6} = j0.1\ p.u.$

Per unit reactance value of T/f-2 $X_{p.u.}^{T/f2} = j0.10 * \left(\frac{18*10^3}{13.8*10^3}\right)^2 * \frac{15*10^6}{20*10^6} = j0.12759\, p.u.$

At T/f-2: $kV_{HT} = kV_{LT} * \left(\frac{HT}{LT}\right)_{T/f} = 13.8 * 10^3 * \left(\frac{\sqrt{3}*127*10^3}{18*10^3}\right) = 168.644\, kV$

Per unit reactance value of the line-2 $X_{p.u.}^{line2} = \frac{240*\,15*10^6}{(168.644*10^3)^2} = j0.12657\, p.u.$

At T/f 3: $kV_{LT} = kV_{HT} * \left(\frac{LT}{HT}\right)_{T/f} = 168.644 * 10^3 * \left(\frac{22*10^3}{220*10^3}\right) = 16.86\, kV$

Per unit reactance value of T/f-3 $X_{p.u.}^{T/f3} = j0.10 * \left(\frac{22*10^3}{16.86*10^3}\right)^2 * \frac{15*10^6}{35*10^6} = j0.055\, p.u.$

Per unit reactance value of generator-3 $X_{p.u.}^{G_3} = j0.08 * \left(\frac{13.8*10^3}{16.86*10^3}\right)^2 * \frac{15*10^6}{20*10^6} = j0.04\, p.u.$

Per unit reactance value of the line-1 $X_{p.u.}^{line1} = \frac{120*\,15*10^6}{(168.644*10^3)^2} = j0.06328\, p.u.$

At T/f 1: $kV_{LT} = kV_{HT} * \left(\frac{LT}{HT}\right)_{T/f} = 168.644 * 10^3 * \left(\frac{13.8*10^3}{220*10^3}\right) = 10.578\, kV$

Per unit reactance value of T/f-1 $X_{p.u.}^{T/f1} = j0.15 * \left(\frac{13.8*10^3}{10.578*10^3}\right)^2 * \frac{15*10^6}{30*10^6} = j0.1276\, p.u.$

Per unit reactance value of generator-1 $X_{p.u.}^{G_1} = j0.20 * \left(\frac{13.8*10^3}{10.578*10^3}\right)^2 * \frac{15*10^6}{25*10^6} = j0.2\, p.u.$

Case ii: Consider the base values of 15 MVA and 13.8 kV at the generator G_1

Per unit reactance value of generator-1 $X_{p.u.}^{G_1} = j0.20 * \left(\frac{13.8*10^3}{13.8*10^3}\right)^2 * \frac{15*10^6}{25*10^6} = j0.12\, p.u.$

Per unit reactance value of T/f-1 $X_{p.u.}^{T/f1} = j0.15 * \left(\frac{13.8*10^3}{13.8*10^3}\right)^2 * \frac{15*10^6}{30*10^6} = j0.075\, p.u.$

Per unit reactance value of the line-1 $X_{p.u.}^{line1} = \frac{120*\,15*10^6}{(220*10^3)^2} = j0.03719\, p.u.$

At T/f 2: $kV_{LT} = kV_{HT} * \left(\frac{LT}{HT}\right)_{T/f} = 220 * 10^3 * \left(\frac{18*10^3}{127*10^3*\sqrt{3}}\right) = 18\, kV$

Per unit reactance value of T/f-2 $X_{p.u.}^{T/f2} = j0.10 * \left(\frac{18*10^3}{18*10^3}\right)^2 * \frac{15*10^6}{20*10^6} = j0.075\, p.u.$

Per unit reactance value of generator-2 $X_{p.u.}^{G_2} = j0.10 * \left(\frac{13.8*10^3}{18*10^3}\right)^2 * \frac{15*10^6}{15*10^6} = j0.0587\, p.u.$

Per unit reactance value of the line-2 $X_{p.u.}^{line2} = \frac{240*\,15*10^6}{(220*10^3)^2} = j0.07438\, p.u.$

At T/f 3: $kV_{LT} = kV_{HT} * \left(\frac{LT}{HT}\right)_{T/f} = 220 * 10^3 * \left(\frac{22*10^3}{220*10^3}\right) = 22\, kV$

Per unit reactance value of T/f-3 $X_{p.u.}^{T/f3} = j0.10 * \left(\frac{22*10^3}{22*10^3}\right)^2 * \frac{15*10^6}{35*10^6} = j0.04285\, p.u.$

Per unit reactance value of generator-3 $X_{p.u.}^{G_3} = j0.08 * \left(\frac{13.8*10^3}{22*10^3}\right)^2 * \frac{15*10^6}{20*10^6} = j0.0236\, p.u.$

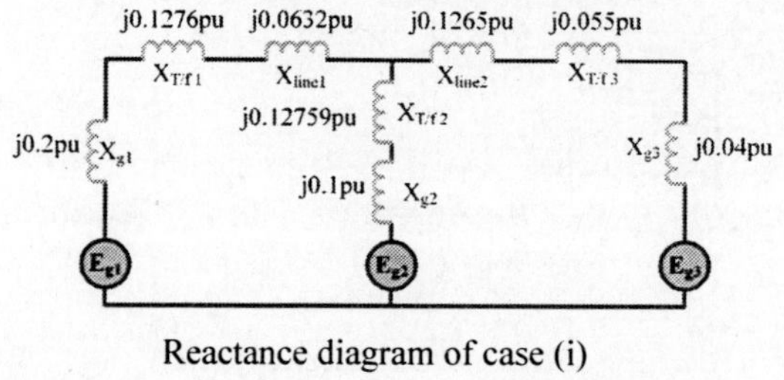

Reactance diagram of case (i)

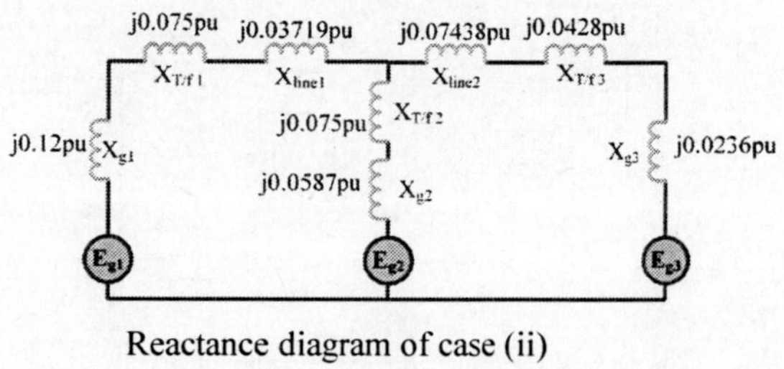

Reactance diagram of case (ii)

Example 4.9. Draw the p.u. reactance diagram of the system shown in Figure. The ratings are as follows:

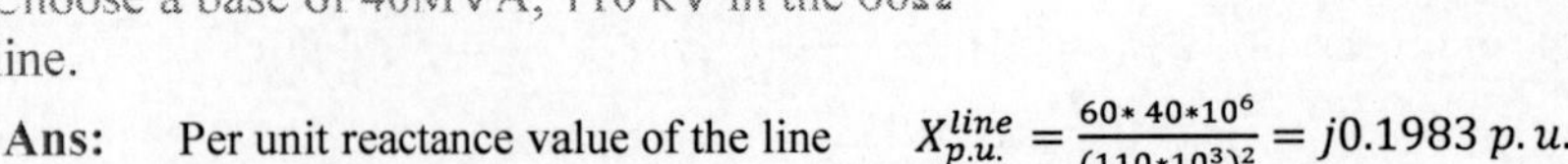

G_1: 10 MVA, 22 kV, X" = 8%
G_2: 20 MVA, 22 kV, X = 12%
M: 30 MVA, 22 kV, X" = 14%
3-ph Y-Y T/F: 15MVA, 22/110 kV, X = 10%
3-ph Y-Δ T/F: 20MVA, 110/22 kV, X = 8%
Choose a base of 40MVA, 110 kV in the 60Ω line.

Ans: Per unit reactance value of the line $X_{p.u.}^{line} = \frac{60*40*10^6}{(110*10^3)^2} = j0.1983\ p.u.$

Per unit reactance value of T/f (Y-Y) $X_{p.u.}^{T/f} = j0.10*\left(\frac{110*10^3}{110*10^3}\right)^2 * \frac{40*10^6}{15*10^6} = j0.266\ p.u.$

Per unit reactance value of generator-1 $X_{p.u.}^{G_1} = j0.08*\left(\frac{22*10^3}{22*10^3}\right)^2 * \frac{40*10^6}{10*10^6} = j0.32\ p.u.$

Per unit reactance value of generator-2 $X_{p.u.}^{G_2} = j0.12*\left(\frac{22*10^3}{22*10^3}\right)^2 * \frac{40*10^6}{20*10^6} = j0.24\ p.u.$

Per unit reactance value of T/f (Y-Δ) $X_{p.u.}^{T/f} = j0.08*\left(\frac{22*10^3}{22*10^3}\right)^2 * \frac{40*10^6}{20*10^6} = j0.16\ p.u.$

Per unit reactance value of motor $X_{p.u.}^{M} = j0.14*\left(\frac{22*10^3}{22*10^3}\right)^2 * \frac{40*10^6}{30*10^6} = j0.186\ p.u.$

Example 4.10. Draw the p.u. reactance diagram of the system shown in Example 4.9 by considering following data

G_1: 10 MVA, 22 kV, X = 8%
G_2: 20 MVA, 18 kV, X" = 12%
M: 30 MVA, 13.8 kV, X" = 14%
3-ph Y-Y T/F: 15MVA, 22/110 kV, X = 10%
3-ph Y-Δ T/f: 20MVA, 110/13.8 kV, X = 8%
Base: 40MVA, 110 kV in the 60 Ω line.

Ans: Per unit reactance value of line $X_{p.u.}^{line} = \frac{60*40*10^6}{(110*10^3)^2} = j0.19834\ p.u.$

Per unit reactance value of T/f (Y-Y) $X_{p.u.}^{T/f} = j0.10*\left(\frac{110*10^3}{110*10^3}\right)^2 * \frac{40*10^6}{15*10^6} = j0.266\ p.u.$

Per unit reactance value of generator-1 $X_{p.u.}^{G_1} = j0.08*\left(\frac{22*10^3}{22*10^3}\right)^2 * \frac{40*10^6}{10*10^6} = j0.32\ p.u.$

Per unit reactance value of generator-2 $X_{p.u.}^{G_2} = j0.12*\left(\frac{18*10^3}{22*10^3}\right)^2 * \frac{40*10^6}{20*10^6} = j0.1606\ p.u.$

Per unit reactance value of T/f (Y-Δ) $X_{p.u.}^{T/f} = j0.08*\left(\frac{22*10^3}{22*10^3}\right)^2 * \frac{40*10^6}{20*10^6} = j0.16\ p.u.$

Per unit reactance value of motor $X_{p.u.}^{M} = j0.14*\left(\frac{13.8*10^3}{13.8*10^3}\right)^2 * \frac{40*10^6}{30*10^6} = j0.186\ p.u.$

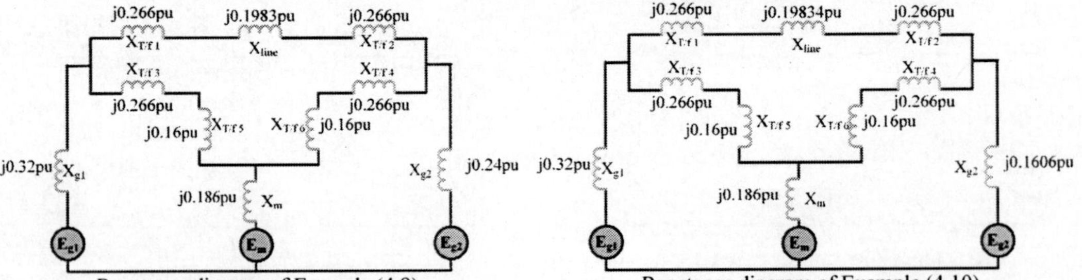

Reactance diagram of Example (4.9) Reactance diagram of Example (4.10)

Example 4.11. Draw the pu reactance diagram choosing a base of 30 MVA, 11kV in the G_1 circuit. The data is as follows:

G_1-30 MVA, 11kV, X"=5% T_1-25 MVA, 11kV/110 kV (Δ-Y), X=15%

G_2-20MVA,11kV, X"=10% T_2-15 MVA, 11kV/110 kV (Δ-Y), X=10%

G_3-10MVA,11kV, X"=15% T_3-3 single-phase T/fs 10 MVA each, 63.5 kV/6.35 kV (Y-Y) each, X=5%

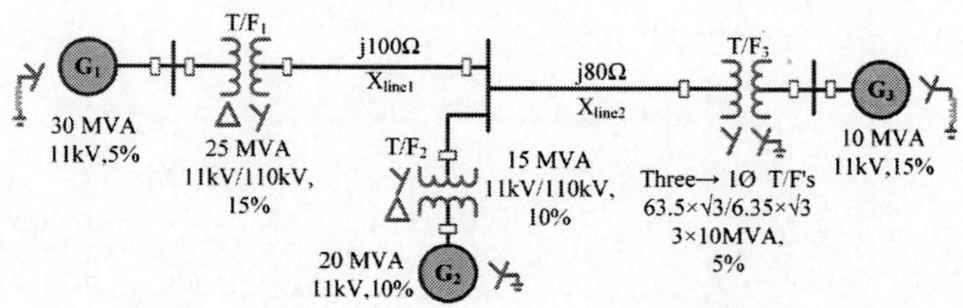

Ans:

$$X_{p.u.}^{G_1} = j0.05 * \left(\frac{11*10^3}{11*10^3}\right)^2 * \frac{30*10^6}{30*10^6} = j0.05\ p.u.$$

$$X_{p.u.}^{T/f1} = j0.15 * \left(\frac{11*10^3}{11*10^3}\right)^2 * \frac{30*10^6}{25*10^6} = j0.18\ p.u.$$

$$X_{p.u.}^{line1} = \frac{100*30*10^6}{(110*10^3)^2} = j0.2479\ p.u.$$

At T/f 3: $kV_{LT} = kV_{HT} * \left(\frac{LT}{HT}\right)_{T/f}$

$$= 110 * 10^3 * \left(\frac{\sqrt{3}*6.35*10^3}{\sqrt{3}*63.5*10^3}\right) = 11\ kV$$

$$X_{p.u.}^{T/f3} = j0.05 * \left(\frac{11*10^3}{11*10^3}\right)^2 * \frac{30*10^6}{3*10*10^6} = j0.05\ p.u.$$

$$X_{p.u.}^{G_3} = j0.15 * \left(\frac{11*10^3}{11*10^3}\right)^2 * \frac{30*10^6}{10*10^6} = j0.45\ p.u.$$

$$X_{p.u.}^{T/f2} = j0.10 * \left(\frac{11*10^3}{11*10^3}\right)^2 * \frac{30*10^6}{15*10^6} = j0.2\ p.u.$$

$$X_{p.u.}^{G_2} = j0.10 * \left(\frac{11*10^3}{11*10^3}\right)^2 * \frac{30*10^6}{20*10^6} = j0.15\ p.u.$$

$$X_{p.u.}^{line2} = \frac{80*30*10^6}{(110*10^3)^2} = j0.1983\ p.u.$$

Example 4.12. Draw the p.u. reactance diagram choosing a base of 30 MVA, 11kV in the G1 circuit for the system shown in Example 4.11 by considering following data.

G_1-30MVA, 11kV, X"=5% T_1-25 MVA, 11kV (delta) - 110 (star) kV, X=15%

G_2-20MVA, 13.8kV, X"=10% T_2-15 MVA, 11kV (delta) -110 (star) kV, X=10%

G_3-10MVA, 6.9kV, X"=15% T_3-10 MVA, 110kV (star)-11 (star) kV, X=5%

Ans.:

$$X_{p.u.}^{G_1} = j0.05 * \left(\frac{11*10^3}{11*10^3}\right)^2 * \frac{30*10^6}{30*10^6} = j0.05\ p.u.$$

$$X_{p.u.}^{T/f1} = j0.15 * \left(\frac{11*10^3}{11*10^3}\right)^2 * \frac{30*10^6}{25*10^6} = j0.18\ p.u.$$

$$X_{p.u.}^{line1} = \frac{100*30*10^6}{(110*10^3)^2} = j0.2479\ p.u.$$

At T/f 2: $kV_{LT} = kV_{HT} * \left(\frac{LT}{HT}\right)_{T/f}$

$$= 110 * 10^3 * \left(\frac{\sqrt{3}*110*10^3}{\sqrt{3}*110*10^3}\right) = 11\ kV$$

$$X_{p.u.}^{T/f2} = j0.10 * \left(\frac{11*10^3}{11*10^3}\right)^2 * \frac{30*10^6}{15*10^6} = j0.2\ p.u.$$

$$X_{p.u.}^{G_2} = j0.10 * \left(\frac{13.8*10^3}{11*10^3}\right)^2 * \frac{30*10^6}{20*10^6} = j0.0166\ p.u.$$

$$X_{p.u.}^{line2} = \frac{80*30*10^6}{(110*10^3)^2} = j0.1983\ p.u.$$

$$X_{p.u.}^{T/f3} = j0.05 * \left(\frac{11*10^3}{11*10^3}\right)^2 * \frac{30*10^6}{3*10*10^6} = j0.05\ p.u.$$

$$X_{p.u.}^{G_3} = j0.15 * \left(\frac{6.9*10^3}{11*10^3}\right)^2 * \frac{30*10^6}{10*10^6} = j0.177\ p.u.$$

Example 4.13. The Single line diagram of a 3-phase power system is shown in Fig. Select a common base of 100MVA and 22 kV on the generation side. Draw the reactance diagram with all reactances marked in per unit.

G: 90 MVA, 22 kV, X-18% — T_3: 40 MVA, 22/110 kV, X-6.4%

T_1: 50 MVA, 22/220 kV, X-10% — T_4: 40 MVA, 110/11 kV, X-8%

T_2: 40 MVA, 220/11 kV, X-6% — M: 66.5 MVA, 10.45 kV, X-18.5%

The 3-phase load absorbs 57 MVA, 0.6 p.f. lagging at 10.45 kV. Line1 and Line2 have reactances of 48.4Ω and 65.43Ω respectively.

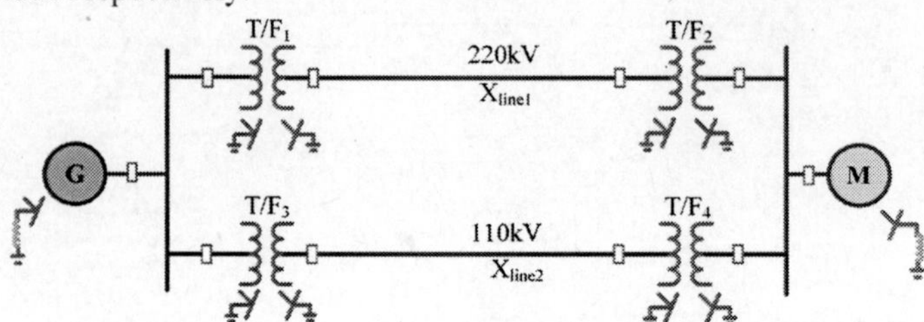

Ans.:

$$X^{G}_{p.u.} = j0.18 * \left(\frac{22*10^3}{22*10^3}\right)^2 * \frac{100*10^6}{90*10^6} = j0.2\ p.u.$$

$$X^{T/f1}_{p.u.} = j0.10 * \left(\frac{22*10^3}{22*10^3}\right)^2 * \frac{100*10^6}{50*10^6} = j0.2\ p.u.$$

$$X^{line1}_{p.u.} = \frac{48.4*100*10^6}{(220*10^3)^2} = j0.1\ p.u.$$

$$X^{T/f2}_{p.u.} = j0.06 * \left(\frac{11*10^3}{11*10^3}\right)^2 * \frac{100*10^6}{40*10^6} = j0.15\ p.u.$$

$$X^{T/f3}_{p.u.} = j0.064 * \left(\frac{22*10^3}{22*10^3}\right)^2 * \frac{100*10^6}{40*10^6} = j0.16\ p.u.$$

$$X^{line2}_{p.u.} = \frac{65.43*100*10^6}{(110*10^3)^2} = j0.54\ p.u.$$

$$X^{T/f4}_{p.u.} = j0.08 * \left(\frac{110*10^3}{110*10^3}\right)^2 * \frac{100*10^6}{40*10^6} = j0.2\ p.u.$$

$$X^{M}_{p.u.} = j0.185 * \left(\frac{10.45*10^3}{11*10^3}\right)^2 * \frac{100*10^6}{66.5*10^6} = 0.251pu.$$

j0.2pu $X_{T/f1}$, j0.1pu X_{line1}, j0.15pu $X_{T/f2}$, $X_{T/f3}$ j0.16pu, X_{line2} j0.54pu, $X_{T/f4}$ j0.2pu, j0.2pu X_g, X_m j0.251pu, E_g, E_m

4.9 Summary

- The synchronous machine operates as an AC generator, converting mechanical energy into electrical energy via a turbine. It also serves as a motor, converting electrical energy into mechanical energy.
- The frequency of induced emf $f = \frac{P}{2} * \frac{N}{60} = \frac{P}{2} f_m$
- At no load : terminal voltage = electromotive force ⇒ $V_t = E_f$
- During loading condition when R_a and X_l neglected: $V_t = E_f - jI_aX_a$
- Resultant air gap flux = Field flux + armature reaction flux ⇒ $\phi_r = \phi_f + \phi_a$
- Synchronous reactance $X_d = X_l + X_a$
- Synchronous impedance $Z_d = R_a + jX_d$
- During loading condition when R_a and X_l are considered:

  For generator: $V_t = E_f - I_aR_a - jI_aX_d$

  For motor: $V_t = E_f + I_aR_a + jI_aX_d$
- During loading condition when R_a and X_l are neglected:

  For generator : $V_t = E_f - jI_aX_d$

  For motor : $V_t = E_f + jI_aX_d$

- An infinite bus refers to a large network where voltage and frequency stay constant regardless of the amount of excitation of the machine and the power exchange between the machine and the bus.
- When the synchronous machine is connected to an infinite bus, bus voltage and frequency remain constant irrespective of the rotational dynamics of the machine.
- Active Power output of the generator $P = \frac{|V_t||E_f|\sin\delta}{X_d}$ and --

  Reactive Power output of the generator $Q = \frac{|V_t||E_f|\cos\delta - |V_t|^2}{X_d}$
- It can be observed that reactance of a 3-phase synchronous generator during short-circuit will be $X_d'' < X_d' < X_d$ and the respective current will be $I_g'' > I_g' > I_g$.
- The electrical load can consist of resistive, inductive, capacitive, or a mixture of these elements.
- Power system loads are mainly classified as domestic load, commercial load, industrial load, municipal load, irrigation load and traction load.
- The voltage and frequency sensitivity of loads may be represented by ZIP model /exponential model.
- The p.u. value of any quantity is defined as $\text{p.u. value} = \frac{\text{actual value in any unit}}{\text{base value in the same unit}}$
- Per unit impedance, $Z_{p.u.} = \frac{Z_{actual}}{kV_{base}^2} * MVA_{base}$
- Per unit impedance on new base, $Z_{p.u.}^{new} = Z_{p.u.}^{given} * \left(\frac{kV_{given\ base}}{kV_{new\ base}}\right)^2 * \left(\frac{MVA_{base}^{new}}{MVA_{base}^{given}}\right)$

Short and Long Answer Questions

1. Draw the equivalent circuits and explain the operation of synchronous generator and motor during steady-state condition.
2. Derive an expression for real and reactive power output of a synchronous machine when it is connected to an infinite bus.
3. With neat illustrations explain in detail about capability curve of synchronous generators.
4. Draw the equivalent circuits and explain the operation of synchronous generator during short-circuit condition.
5. Discuss and explain the sub-transient, transient, and steady-state reactances of a synchronous generator under 3-phase short-circuit at its terminals.
6. Define and classify electrical loads.
7. Classify electrical loads based on the nature of the load and power system loads. Provide the examples of each load.
8. What do you understand by single line diagram? What is the difference between single line diagram and impedance diagram? Explain with the help of example.
9. Why the per unit system is used in Power System Analysis? Which of the electrical quantities are chosen as base values?
10. Explain voltage and frequency dependency of electrical loads.

Exercise

1. Convert an impedance of (4+j24) Ω into per unit impedance with base voltage of 11kV and base kVA of 500.

2. For the given system, draw the p.u. reactance diagram choosing a base of 50 MVA, 11kV in the G_2 circuit.

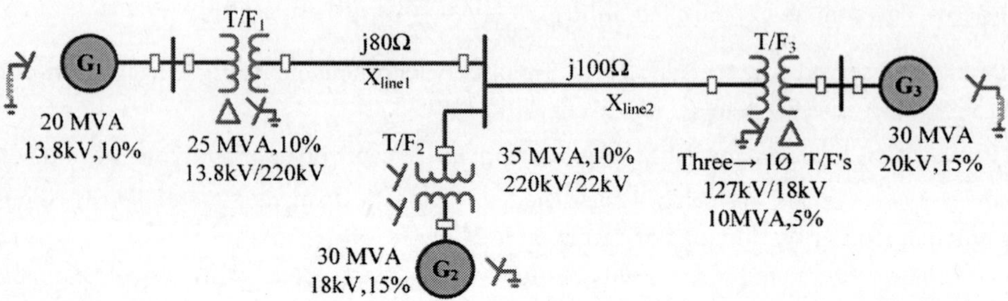

3. Draw pu reactance diagram of the given system by considering base values as 11 kV, 100 MVA at G_2.

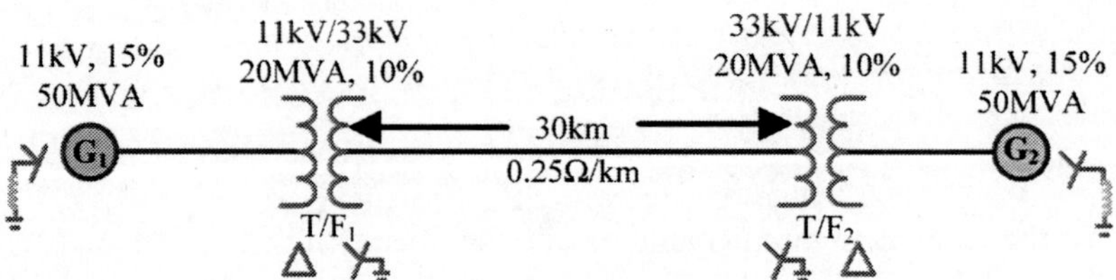

4. Draw pu reactance diagram of the given system by considering base values as 6.6 kV, 50 MVA at G_1.

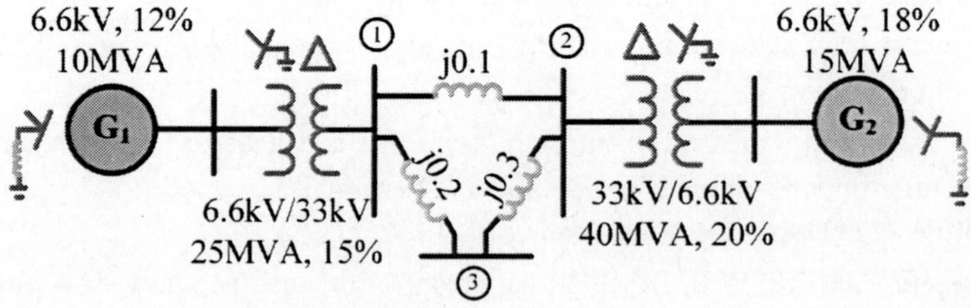

To know more about

Synchronous Machines: Topologies, Design and Analysis; Generator Capability Curve

To know more about

Per-Unit Modelling via Similarity Transformation and Hybrid Excitation PM SM: Per Unit Analysis

To design

To study the Synchronization of the alternator with infinite bus bar, and to determine $X_{d''}$, $X_{d'}$ and X_d of a SM

To Model & Simulate

Synchronous Machine, 3-Ph & 5-Ph PMSM, and 6-Phase SM with Post-Fault Operating Strategy

05 OVER-VOLTAGES AND INSULATION REQUIREMENTS

Unit specifics: In this unit, the following topics have been discussed for basic understating related to Over-voltages and Insulation Requirements

- Generation of Over-voltages,
- Mechanism of Lightning,
- Propagation of Surges,
- Protection against Over-voltages,
- Insulation Coordination,
- Voltages produced by traveling wave and
- Bewley's Lattice Diagram

Rationale: In this unit students will be introduced to over-voltages and Insulation Requirements. Causes of over-voltages, Mechanism of lightning, Propagation of travelling wave, Protection Schemes against over-voltages, Insulation Coordination (BIL & SIL, V-T curves), various line terminations and Bewley's Lattice Diagram are clearly described with the help of necessary diagrams, derivations and examples.

Pre-Requisites: Basic Knowledge of Power System components.

Unit-5 Outcomes: List of outcomes of this unit is as follows

U5-O1: To understand and classify over-voltages
U5-O2: To understand the mechanism of lightning and switching surges
U5-O3: To know the protection schemes used against over-voltages.
U5-O4: To follow the best practices for Insulation coordination levels
U5-O5: To analyse the travelling wave when the line is terminated
U5-O6: To analyse the travelling wave with Bewley's Lattice Diagram

Unit-5 outcomes	**Expected mapping with course outcomes** (1: Weak, 2: medium, and 3: strong correlation)					
	CO-1	**CO-2**	**CO-3**	**CO-4**	**CO-5**	**CO-6**
U5-O1	2	1	2	-	-	-
U5-O2	2	-	3	-	-	-
U5-O3	2	-	2	-	3	-
U5-O4	2	-	3	-	-	-
U5-O5	2	2	2	-	1	3
U5-O6	2	2	2	-	-	3

5.1 Generation of Over-Voltages:

The rise in the voltage which exceeds the maximum value of the operating voltage is called *over-voltage*. Over-voltages mainly occur between the phases and Phase to earth. Over-voltages may be caused due to lightning, switching and etc.

5.1.1. Classification of Over-voltages: All voltage surges in high-voltage installations are broadly classified as

(i). Internal Over-voltages
(ii). External Over-voltages

5.1.1.1 Internal Over-voltages: Internal Over voltages originate in the system itself and it can be further divided into

- Dynamic Over-voltages and
- stationary Over-voltages.

Dynamic Over-voltages occur at normal system frequency and persist for a short duration, may be few seconds. They may be caused, when a large portion of the load is suddenly thrown off or by the withdrawal of a generator which over-speeds.

Stationary Over-voltages occur at normal system frequency and persist for some time, may be hours. This situation may arise when an earth fault on one line is sustained and there by leading to over-voltages on healthy phases.

5.1.1.2 External Over-voltages: External Over-voltages originate from atmospheric disturbances, mainly due to lightning. The surge produced by external over-voltages (lightning surges) are very severe and may increase the system voltage to several times of the normal value. External over-voltages may occur due to any of the following

- Direct lightning stroke
- Electromagnetically induced over voltages
- Electrostatically induced voltages due to presence of charged clouds nearby
- Voltages induced due to atmospheric changes along the length of the line

5.1.2 Causes of Over-voltages:

The over-voltages on a power system mainly occur due to Internal causes and External causes.

5.1.2.1 Internal Causes of Over-voltages: The main causes of Internal Over-voltages are due to Switching surges, Arcing ground, Resonance, Insulation failure and etc.

Switching surges: The over-voltages produced on the power system due to switching operations are known as switching surges. The causes for production of switching surges are as follows

- Switching operations on unloaded line
- Sudden opening of a loaded line
- Current chopping
- Short Circuits and line faut interruptions
- Auto-Reclosing of C.B. in long EHV lines

Insulation failure: The insulation failure may take place in various ways such as

- the cores of an insulated cable,
- b/w the lines of an O.H. line or
- b/w line and earth

Arcing ground: If a line to ground fault occurs on a three-phase un-grounded system then the voltage of the faulty phase fluctuates due to capacitive charging current, which is called arcing ground. The transients produced by arcing ground are cumulative and may cause serious damage to the equipment.

Resonance: It occurs when $X_L = X_C$. Resonance also causes high voltages in the electrical system.

5.1.2.2 External Causes of Over-voltages: External Over-voltages instigate from atmospheric disturbances, mainly due to lightning. The discharge of electricity take place when a cloud attains an excess electrical charge, either "+ve" or "-ve", that is sufficient to break down the resistance of air, which is called lightning.

5.2 Mechanism of Lightning Discharge:

Benjamin Franklin performed his famous experiment of flying kite in thunder cloud in 1752. Before his discovery the lightning used to be considered as "Act of God". He also proved that the lightning stroke is because of the discharge of electricity. He also invented lightning rods for fixation on tall buildings and grounding for their protection from lightning strokes. Lightning phenomenon is clearly illustrated in Fig. 5.1. An electric discharge can be mainly observed

- between cloud and earth,
- between the clouds (Inter-cloud) or
- within the cloud between the "+ve" and "-ve" charges (Intra-cloud).

Various theories exist about how the potentials required to produce lightning are buildup. A cloud consists of many positive and negative charges. During foul weather conditions, the pressure and temperature will vary in the atmosphere. If the potential of the charges increases then the potential between the cloud and earth increases. This potential gradient is not uniformly distributed; it is usually more intense at the charge centre in the cloud.

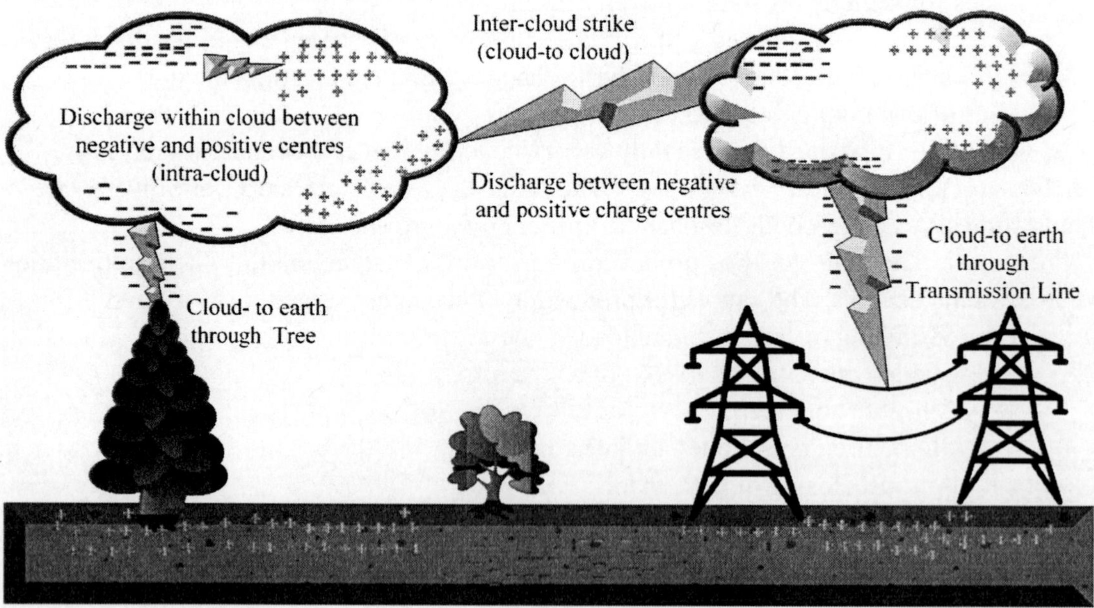

Fig. 5.1. Lightning phenomenon

The process of lightning discharge is clearly depicted in Fig. 5.2 (a-f). When the gradient exceeds the strength of the portion of the air, the air brakes down and a *streamer* starts from the cloud towards earth which is shown. The gradient required to start the streamer is about 5 to 10 kV/cm. It tries to discharge the surge on the highest peak on the earth which is nearby (i.e. longest trees, on EHV line, on tall building etc.). So other than the streamer side streamers may also exist, it is clearly illustrated in Fig 5.2 (a).

The streamer will be called as a leader when it identifies and reaches nearer to the earth, it can be seen in Fig. 5.2 (b). As the leader approaches the earth, the electrostatic field increases and the gradient at the earth surface becomes great. When the downstream leader reaches the earth, the complete breakdown will occur b/w the cloud and earth.

Due to complete breakdown b/w the cloud and earth, the upward streamer or return streamer will start from earth and it will reach the cloud in μs, which can be seen in Fig. 5.2 (c and d). When complete disruption happens b/w the cloud to earth and earth to cloud, the spark will be generated. If another charge point is present in the same cloud, then it tries to ionise the medium between the charged point and the other charge points in the same cloud with in fraction of seconds, which can be seen in Fig. 5.2 (e and f).

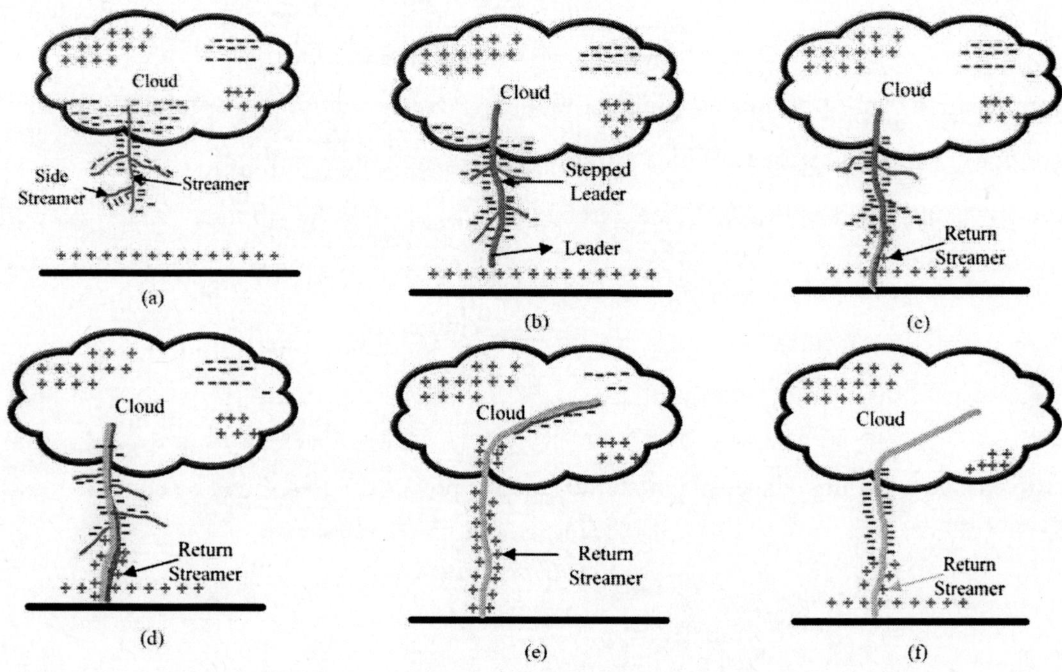

Fig. 5.2. Process of Lightning Discharge

If multiple charge points are available in the same cloud, then we can observe multiple or repetitive sparks in the cloud when the complete disruption of the other charged particles is accomplished.

5.3 Switching Surges: A voltage surge/transient voltage is the sudden increase in voltage for a very short duration. Even though Transients/surges exist for a very short duration (few ms or μs) but it causes over voltages on the power system.

5.3.1 Surge Impedance: A line possesses characteristic impedance under normal condition which is given by square root of ratio of its series impedance and shunt admittance. The characteristic impedance is complex in general due to presence of series resistance and shunt conductance. When high frequency surge falls on the line, series resistance becomes negligible compared to series reactance, same is the case with shunt elements with conductance becoming negligible compared to shunt susceptance. This makes the line lossless with characteristic impedance becoming equal to square root of L and C. The characteristic impedance of the line which is purely resistive in nature under the fall of high frequency surge on it is called 'Surge impedance'. So, the surge Impedance can be defined as the square root of the ratio of Series Impedance (Z) and Shunt Admittance (Y). The representation of the long transmission line is given in Fig. 5.3.

Surge Impedance or characteristic impedance is $Z_C = \sqrt{\frac{Z}{Y}}$ ....(5.1)

Where $Z = R + jX$ and $Y = G + j\,B$

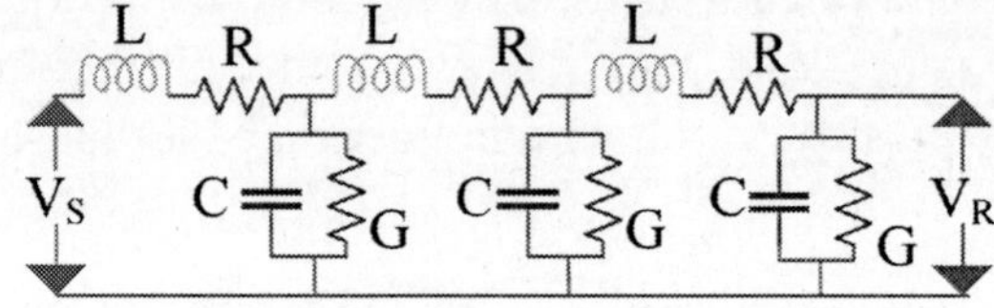

Fig. 5.3 Representation of Overhead Transmission line

The representation of the transmission line during high-frequency surge is given in Fig. 5.4.

Reactive power supplied by the O.H. line will be, $Q_{generated} = \frac{V^2}{X_C} = V^2\omega C$ ....(5.2)

Reactive power absorbed by the O.H. line will be $Q_{absorbed} = I^2X_L = I^2\omega L$ ....(5.3)

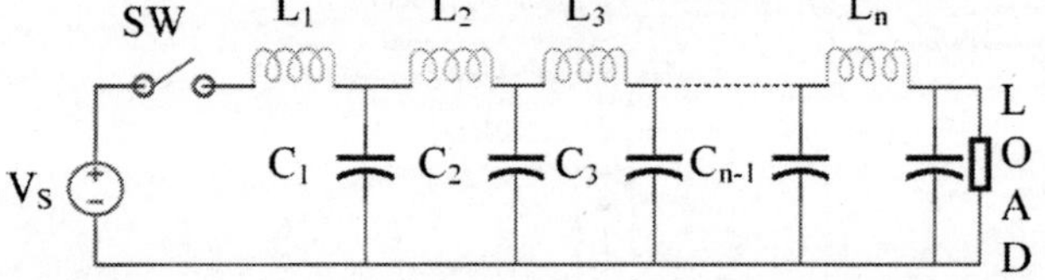

Fig. 5.4. Representation of Transmission line during high-frequency surge

Surge Impedance is the impedance at which the reactive power generated will be equal to reactive power absorbed.

$$Q_{generated} = Q_{absorbed}$$

$$V^2\omega C = I^2\omega L$$

$$\frac{V^2}{I^2} = \frac{L}{C}$$

The characteristic impedance (Z_c) of a loss-less transmission line is $Z_C = \sqrt{\frac{L}{C}}$ ...(5.4)

The characteristic impedance (Z_c) varies from 400Ω to 600Ω in case of overhead transmission lines and 40Ω to 60Ω in case of underground cables. Surge Impedance can also be evaluated by measuring the line impedance when

(i). Receiving end is O.C and

(ii). Receiving end is S.C.

Sending end voltage and current (V_s and I_s) of the O.H. line can be written as

$$V_s = AV_R + BI_R$$
$$I_s = CV_R + DI_R \quad \text{.........(5.5)}$$

When the Line at receiving end is open-circuited $I_R = 0$, so Sending end voltage and current (V_s and I_s) of the O.H. line will be $V_s = AV_R \ and \ I_s = CV_R$

$$Z_{OC} = \frac{V_s}{I_s} = \frac{AV_R}{CV_R} = \frac{A}{C} \quad \text{.........(5.6)}$$

When the Line at receiving end is short-circuited $V_R = 0$, so the sending-end Voltage and current (V_s and I_s) of the transmission line will be $V_s = BI_R$ and $I_s = DI_R$

$$Z_{SC} = \frac{V_s}{I_s} = \frac{BI_R}{DI_R} = \frac{B}{D} \quad \text{......(5.7)}$$

From eq. 5.6 and 5.7 $$Z_{OC} * Z_{SC} = \frac{A}{C} * \frac{B}{D} = \frac{B}{C} \quad \text{.........(5.8)}$$

In a long transmission $$B = Z_C * Sinh\sqrt{YZ} \text{ and } C = \frac{1}{Z_C} * Sinh\sqrt{YZ}$$

Eq. 5.8 will become $$Z_{OC} * Z_{SC} = \frac{B}{C} = \frac{Z_C * Sinh\sqrt{YZ}}{\frac{1}{Z_C} * Sinh\sqrt{YZ}} = {Z_C}^2$$

Surge Impedance of a long transmission will be $Z_C = \sqrt{Z_{OC} * Z_{SC}}$ (5.9)

5.3.2 Surge Impedance Loading:

A transmission line under surge may be represented by Thevenin's equivalent with Thevenin's impedance equal to its surge impedance. When load impedance matches with surge impedance of the line, the power delivered to the load is maximum. The connected load under this condition is called Surge Impedance Loading (SIL). i.e. The loading being at unity power factor.

Power transmitted, $$P_R = \frac{{V_{RL}}^2}{Z_C} = \frac{{V_{RL}}^2}{\sqrt{\frac{L}{C}}} \quad \text{.........(5.10)}$$

Where P_R is the surge impedance loading
V_{RL} is the receiving-end voltage
Z_C is the surge impedance

The power transmitted through a O.H. line can be increased either by increasing the voltage or by reducing the surge impedance. To increase the power limit of heavily loaded long transmission lines, the most commonly adopted method is employing higher voltage for transmission. But there is a limit beyond which it is neither economical nor practical to increase the value of the receiving-end voltage.

Inductance of a transmission line $L = 2 * 10^{-7} \ln\frac{d}{r} \ H/m$ ....(5.11)

Capacitance of a transmission line $C = \frac{2\pi\varepsilon}{\ln\frac{d}{r}} \frac{F}{m}$ ....(5.12)

Inductance and capacitance of a transmission line are mainly dependent on spacing between the conductors. The spacing b/w the conductors is dictated by the operating voltage, it cannot be reduced much. So, the value of the surge impedance cannot be varied much. However, some artificial means, such as series reactance or shunt capacitance can be used to reduce the surge impedance.

5.4 Propagation of Surges:

Travelling wave is a temporary wave which occurs for a very short duration (few μs) but cause a much disturbance in the line. The transient wave is set up in the O.H. line mainly due to faults, switching and lightning. A travelling wave is characterised by following specifications: Crest, front, tail and polarity. Characteristics of travelling wave are given in Fig. 5.5.

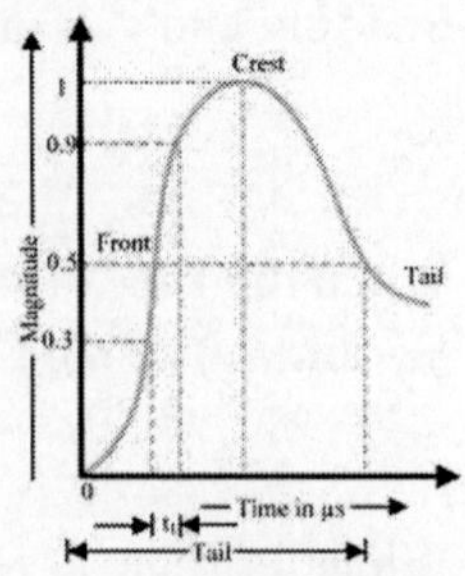

Fig. 5.5 Characteristics of travelling wave

5.4.1 Velocity of Propagation of travelling wave: Consider a transmission line as lossless transmission line loaded with SIL, so the voltage and current are same throughout the transmission line. Suppose that the wave travelled a distance ***x*** in a time ***t*** then the velocity of the propagation is the ratio of change in distance to the change in time.

Velocity of propagation $$v = \frac{dx}{dt} \qquad \text{.........(5.13)}$$

The electromagnetic flux is associated with the current wave and electrostatic flux is associated with the voltage wave. The electrostatic flux of the line up to a distance ***x*** is given by

$$q = VCx \qquad \text{.........(5.14)}$$

The current will become $$I = \frac{dq}{dt} = \frac{d(VCx)}{dt} = VC\frac{dx}{dt} = VCv \qquad \text{...(5.15)}$$

Here $\frac{dx}{dt}$ is the velocity of the travelling wave.

Similarly, the electromagnetic flux of the line up to a distance ***x*** is $\psi = ILx$ ....(5.16)

The voltage will become $$V = \frac{d\psi}{dt} = \frac{d(ILx)}{dt} = IL\frac{dx}{dt} = ILv \qquad \text{.........(5.17)}$$

Multiplying eq.5.15 with eq.5.17, we get

$$I * V = VCv * ILv \Rightarrow 1 = LCv^2 \Rightarrow v^2 = \frac{1}{LC} \Rightarrow v = \frac{1}{\sqrt{LC}} \qquad \text{.........(5.18)}$$

By Substituting L and C values in eq.5.18,

$$v = \frac{1}{\sqrt{\left(2*10^{-7}\, ln\frac{d}{r}\right)\left(\frac{2\pi\varepsilon}{ln\frac{d}{r}}\right)}} = \frac{1}{\sqrt{4\pi\varepsilon*10^{-7}}} = \frac{1}{\sqrt{4\pi\varepsilon_0\varepsilon_r*10^{-7}}} = 3*10^8 \; metres/sec \qquad \text{.........(5.19)}$$

Where $\varepsilon = \varepsilon_0\varepsilon_r$
$\varepsilon_0 = 8.854 * 10^{-12}$
$\varepsilon_r = 1$ for overhead transmission lines
$\varepsilon_r > 1$ for cables

i.e. the velocity of travelling wave over the O.H. lines is equal to the velocity of light. In actual practice, the velocity of the travelling wave is slightly less than the velocity of the light due to the resistance and leakage reactance of the lines. In overhead transmission lines $\varepsilon_r = 1$ and in cables $\varepsilon_r > 1$ due to presence of dielectric material across the conductor, therefore the velocity of the wave over the cables is smaller than over the transmission lines.

5.4.2 Wavelength (λ) : Wavelength of the travelling wave is defined as the ratio of the velocity of the travelling wave and the frequency. $\lambda = \frac{v}{f}$ ....(5.20)

5.5 Protection against Over-voltages:

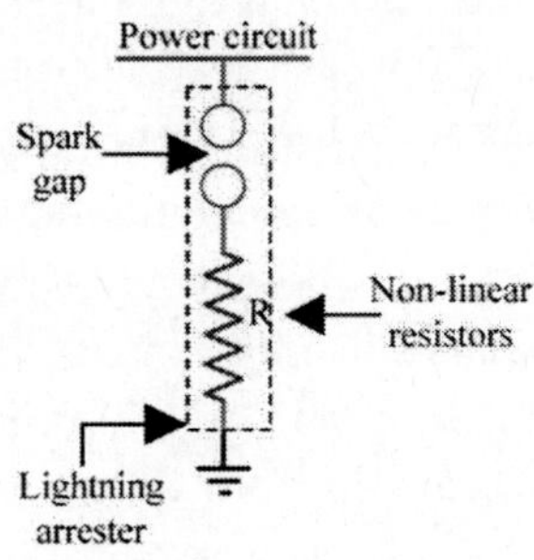

Fig. 5.6 lightning arrester

To protect the power system against over-voltages, lightning arresters/surge diverters are used. A lightning arrester (L.A.) is a protective device which diverts the surges to ground. L.A. consists of a sphere gap in series with a non-linear resistor, which is shown in Fig. 5.6. One end is connected to the system to be protected and the other end is grounded. The length of the sphere-gap is so set, when the line is operating at rated voltage, this low voltage will be insufficient to conduct the gap. So, the gap remains open and there will not be any discharge through the arrester. In this situation, non-linear resistor will act as high resistance.

But, when an internal/external fault occurs, the high voltage surge will be applied across the gap. When the high voltage surge reaches to a particular defined voltage then the gap starts conducting and non-linear resistor will act as low resistance. So, the high voltage surge will be discharged through the arrester. Lightning arrester/Surge diverter are divided into following types they are:

(i). Rod gap L.A.
(ii). Horn gap L.A.
(iii). Multiple gap L.A.
(iv). Expulsion type L.A.
(v). Valve type L.A.
(vi). Zinc-oxide gapless L.A.

5.5.1 Rod gap lightning arrester/Surge diverter:

This is the simplest form of lightning arrester, consists of two rods with ends facing each other, one connected to the transmission line and the other connected to the earth. The distance between insulator and rod is 'd', which is shown in Fig. 5.7. The rod-gap should be set to breakdown at about 80% of the impulse spark over-voltage to avoid the cascading across the insulator surface of very steep fronted waves. The distance between the gap and the insulator should be 30% of the rod gap length in order to prevent the arc from being blown on the insulator. When the line is operating at rated voltage, this low voltage will be insufficient to conduct the rod-gap. So, the rod gap remains open and there will not be any discharge through the arrester. When an internal/external fault occurs, the high voltage surge will be generated. When the high voltage surge reaches to a particular defined voltage then the rod-gap starts conducting and it will discharge the surge through the arrester.

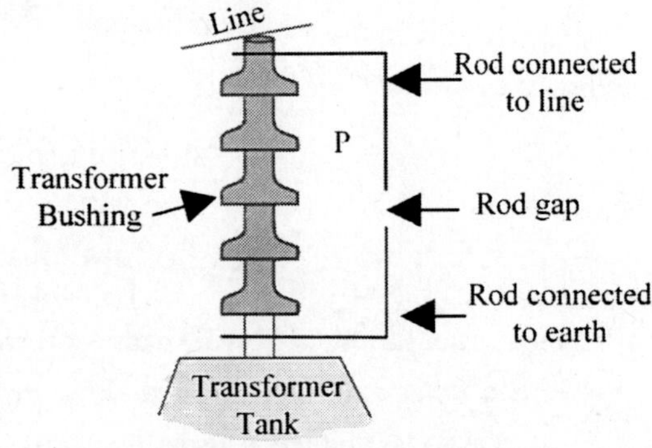

Fig. 5. 7 Rod gap lightning arrester

Disadvantages of Rod gap lightning arrester:

- Once the spark has taken place, it may continue for some time even at low supply voltage.
- The presence of high temperature on the arc may cause the rods to melt.

- × The performance of the rod gaps is badly affected due to climatic variations and also the polarity of the surge.

Thus, the rod-gap cannot be relied upon as main protection in high voltage power systems where the continuity of supply and also the protection of equipment is given priority. However, it can be employed as an auxiliary line due to its lower cost.

5.5.2 Horn gap lightning arrester/Surge diverter:

This was the oldest lightning arrester. Due to its great simplicity, still it is being used to a certain extent on low-voltage lines. It consists of two horn shaped metal pieces separated by a minor air gap and connected in b/w the line and earth, which is shown in Fig. 5.8. The gap b/w the horns is less at the bottom and large at the top. The distance b/w two electrodes are such that the normal operating voltage is insufficient to breakdown the horn-gap medium, but abnormally high voltages will breakdown the gap and so find a path to earth. The arc thus formed by reason of heated air and electromagnetic action will rise up the horn and extinguish itself, thus preventing a follow-on arc. The time taken for the complete operation is usually 3 to 5 seconds.

Usually, a choking coil consisting of several turns of a bare copper wire, is connected in the line between the arrester and the apparatus to be protected. Its reactance is negligible at normal power frequency. The horn gap cannot rupture the arc currents much in excess of 10A, and as the arc is a dead short circuit it is necessary to limit the current to a small value. This is accomplished by inserting a non-inductive resistance. When the line is operating at rated voltage, this low voltage will be insufficient to conduct the horn-gap. So, the horn gap remains open and there will not be any discharge through the arrester. When an internal/external fault occurs, the high voltage surge will be generated. When the high voltage surge reaches to a particular defined voltage then the horn-gap starts conducting and it will discharge the surge through the arrester. For the utmost protection to the terminal apparatus the arrester should be located as close to the end of transmission line as possible.

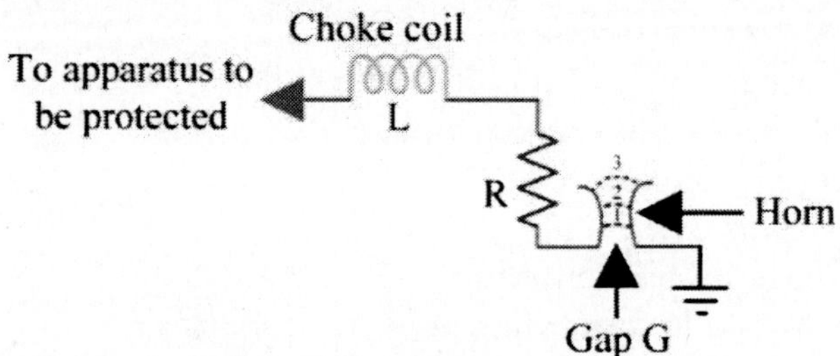

Fig. 5.8 Horn gap lightning arrester

Horn gap arrester has the following characteristics:

- The voltage at which the breakdown occurs during transients, depends upon the impulse ratio of the gap.
- Breakdown voltage is also affected by atmospheric conditions such as temperature and pressure of air.
- The performance is also affected by any roughness of the horn gap and the frequent settings are required to be made for the gap
- At heights a longer air gap is required as it depends upon the air density and is inversely proportional to it.

Disadvantages of Horn gap lightning arrester:

- × Operation time of the gap is relatively large as compared to the modern protective schemes
- × The bridging of gap by birds etc. can render the device useless.

5.5.3 Multiple gap lightning arrester:

Multiple gap lightning arrester consists of a series of small metal cylinders insulated from one another. These are separated by an air gap of about 1mm width. The first one of the series connected to the line and the last to the ground. From Fig. 5.9, it is obvious that the point 'B' is at ground potential under normal conditions and therefore a discharge will take place when the voltage is sufficient to break down the series gaps between A and B.

When fault occurs, the high voltage surge will be generated. When the transient surge reaches to a particular defined voltage then the series gap starts conducting and it will discharge the surge to the ground via small series resistance through the shunted gaps between B and C, instead of passing through the parallel path i.e. shunt resistance.

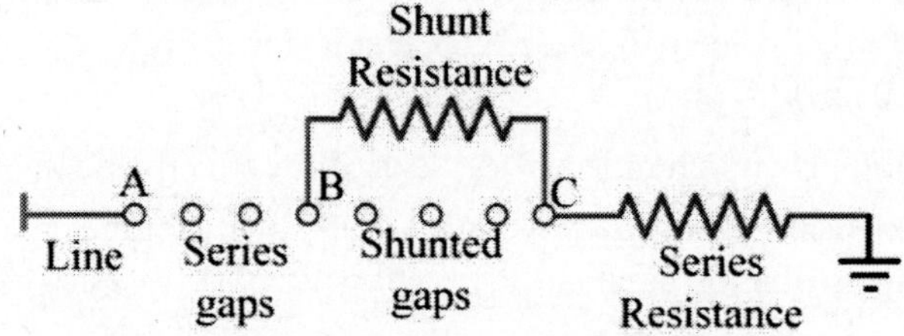

Fig. 5.9 Multiple gap lightning arrester

As soon as the impulsive rush of current is over, the arcs in the shunted gaps die out, with the result that any dynamic current attempting to flow through both the resistances, which are now in series. The effect of shunting of some of the gaps is therefore to provide the arrester a certain amount of selective action, since it interposes a low resistance in the path of high-frequency disturbances, and a high resistance in the low-frequency discharges.

5.5.4 Expulsion type or Protector tube lightning arrester:

It is an improved version of a rod-gap arrester. In this arrester, a rod gap is connected in series with the protector tube, which is shown in Fig. 5.10 (a). Similar to Fig. 5.10(a), a slight modified arrester is also existing, only modification is a horn shaped metal piece is considered in the upper portion, which is shown in Fig. 5.10(b).

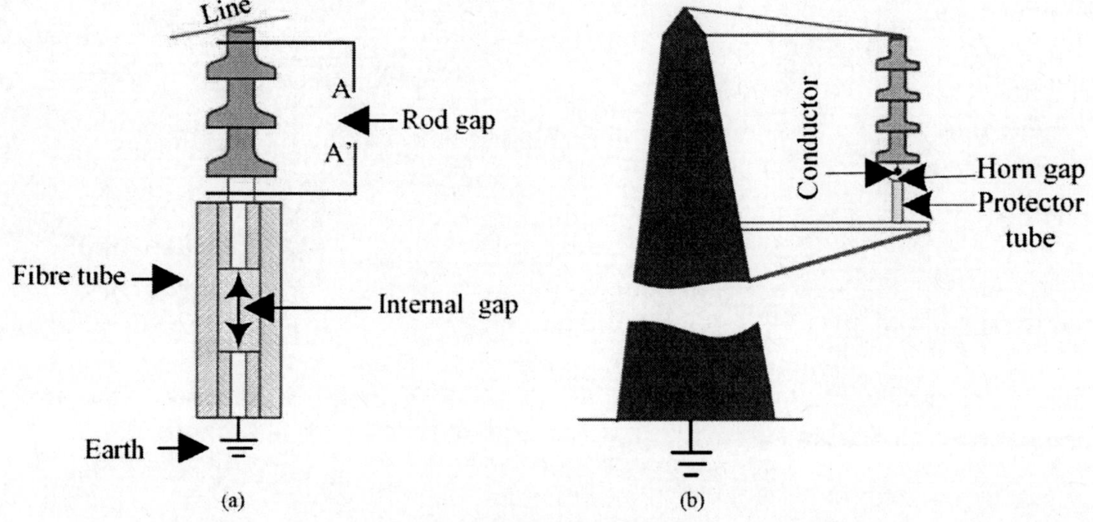

Fig. 5.10 Expulsion type L.A.

It consists of (i) an isolating rod-gap

(ii) a protecter tube made of fibre

(iii) an interrupting spark gap inside the protecter tube

When the line is operating at rated voltage, this low voltage will be insufficient to conduct the rod gap. So, the rod gap remains open and there will not be any discharge through the protector tube. When an internal/external fault occurs, the high voltage surge will be generated. When the high voltage surge reaches to a particular defined voltage then the breakdown of rod-gap takes place and the surge current will flow into the protector tube. During this faulty condition, arc due to the impulse spark-over inside the protector tube causes some fibrous material of the tube vapourised in the form of gas, which is expelled through a vent from the bottom of the tube. The extinction of the arc in this protector tube arrester is same as that of circuit breakers. Since the gases generated have to be ejected, one of the electrodes is hollow and diverter is open at its lower end.

Disadvantages of Expulsion type arrester:

- Transmission type Expulsion lightning arrester is used for the protection of transmission lines
- Distribution type Expulsion lightning arresters are used largely for the protection of distribution t/f's and similar apparatus on distribution circuits.

5.5.5 Valve type lightning arrester:

Valve type lightning arrester is also called as non-linear lightning arrester. It is an advanced arrester but more expensive. It consists of a multiple spark gap asssembly in series with set of non-linear resistors. Valve type lightning arrester is clearly illustrated in Fig. 5.11 (a) and its simplified representation is given in Fig. 5.11 (b).

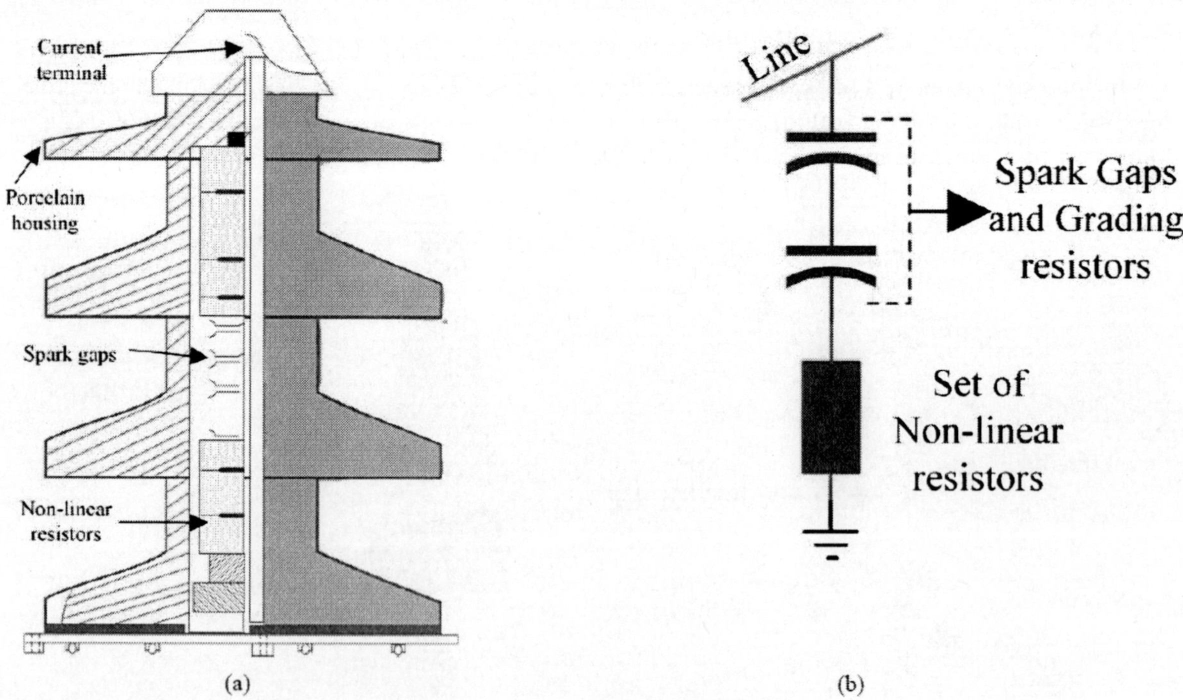

Fig. 5.11 Valve type lightning arrester

The spark gap assembly consists of a number of identical elements united in series. Each spark gap consists of two electrodes. A grading resistor of high ohmic value is connected in parallel between each spark gap and it ensures uniform grading of the voltage between different elements. This structural design is similar to that of a number of capacitors connected in series with grading resistors connected across each capacitor. The non-linear resistors are made-up of an inorganic compound such as thyrite or metrosil. When the line is operating at rated voltage, this low voltage will be insufficient to breakdown the airgap assembly. When an internal/external fault occurs, the high voltage surge will be generated. When the high voltage surge reaches to a particular defined voltage then the breakdown of series spark gap takes place and the surge current will be discharged through the non-linear resistor.

Advantages:

- ✓ It provides very effective protection, especially for t/f's and cables
- ✓ It operates very rapidly

Disadvantages:

- ✗ Its performance is badly affected by the entrance of moisture into the enclosure.
- ✗ It is more costly

5.5.6 Zinc-oxide/Metal-oxide gapless lightning arrester:

As the name indicate it is gapless arrester. It resembles valve type arrester spark gaps. Zinc-oxide materials have good non-linear characteristics, due to which it is possible to construct Zinc-oxide surge arresters without series connected spark gaps. Zinc-oxide/Metal-oxide gapless lightning arrester is clearly depicted in Fig. 5.12.

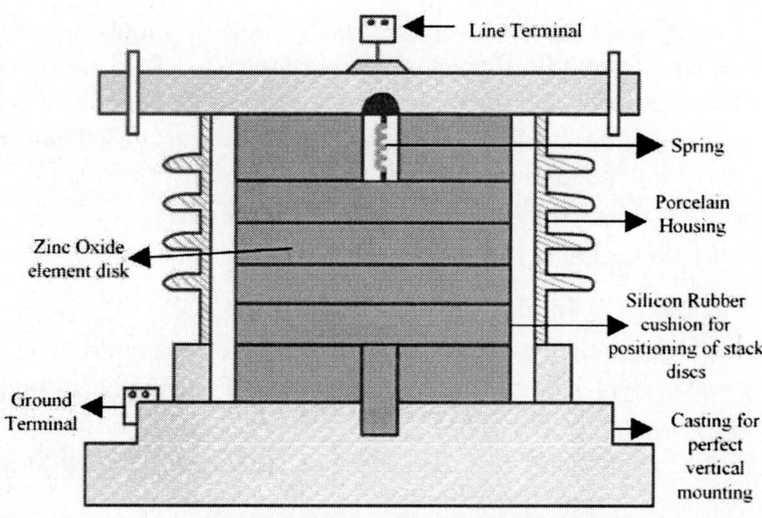

Fig. 5.12 Zinc-oxide gapless lightning arrester

When certain elements are slightly heated, they will change from a good conductor to almost ideal insulator. This arrester also works on the same principle. During normal operating conditions zinc-oxide element discs offer high resistance and during fault conditions it offer a very low resistance through which it discharges currents to ground. The zinc-oxide surge arresters consist of series connected stacks of zinc-oxide elements. ZnO arresters have excellent V-I characteristics and higher energy absorption level. This arrester is preferable for protection of EHV and HVDC systems.

5.6 Insulation Co-ordination:

It is the process of selecting suitable values for the insulation levels of the various components in any electrical system. A proper insulation co-ordination of insulation seeks to ensure:

- that it shall withstand all working stresses and majority of abnormal ones
- that the breakdown may occur only due to external flashover
- the effective discharge of over-voltages,
- breakdown may cause no or comparatively little damage.

The insulation problem in a power system shall, therefore, involve the following:

- Selection of Basic Impulse insulation level (BIL)
- Selection of Standard Impulse insulation level (SIL)
- Selection of lightning arrestor
- Determination of line insulation

5.6.1 Basic Impulse insulation level (BIL):

BIL can be defined as reference level expressed in kV crest (peak) voltage value with standard $1.2/50\mu s$ lightning impulse wave. Apparatus should be capable of withstanding test waves above BIL. The significance of BIL in power system studies is to get the desired insulation level for the system to be protected such that below BIL, any insulation should not breakdown or flashover to provide efficient protective device for the power system. The gap between the BIL and the protective device should be kept such that it is not only economical but also give necessary protection. Volt-Time curves of BIL is given in Fig. 5.15.

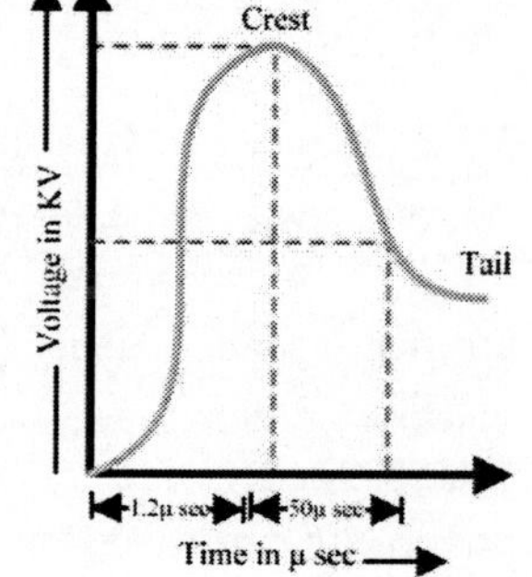

Fig.5.13 Volt-Time curves of BIL

Protective margin = (withstand level of the system) – (Protective level of the surge arrester)

5.6.2 Standard Impulse insulation level (SIL):

Impulse wave shape gives the time in μsec for the impulse to reach crest followed by the time in μsec for wave to reach half of the crest magnitude. For Lightning Impulse Test wave : $T_1 = 1.5\mu s$ and $T_2 = 50\mu s$

For Switching Surge Test wave : $T_1 = 250\mu s$ and $T_2 = 2500\mu s$

5.6.3 Volt-Time curves:

Volt-Time curves give the relation between peak flashover voltage and the time taken to flash over of a particular wave shape. Fig. 5.14. provides the basic V-T characteristics of the protective device and device to be protected. Curve 'X' is the volt-time curve of the protective device and Curve 'Y' is the volt-time curve of the system to be protected. Voltage-Time Characteristics are given in Fig. 5.15. The volt-time curve is constructed depending upon impulse voltage of the applied waves having similar wave shapes but different peak values to the insulation of the equipment whose volt-time curve is desired. If the

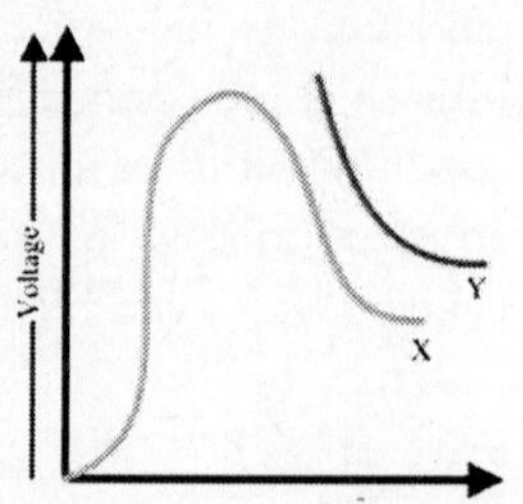

Fig. 5.14 Volt-Time curves

adjustment is made in the polarity and shape of all impulse waves such that the breakdown occurs on the front of the wave then the breakdown point is called *front flashover*. It gives starting point of the *V-T* curve. If the breakdown or flashover occurs at the crest value of the wave, the point is the second point on the wave and it is called as *crest flashover*. The next possibility is that the flashover occurs on the tail end of the wave, the point is the third point on the wave and it is called as *tail flashover*. The insulation characteristics of various equipment are defined for rated voltage, power frequency voltage withstand, switching impulse, lightning impulse. These are correlated with characteristics of the surge arresters.

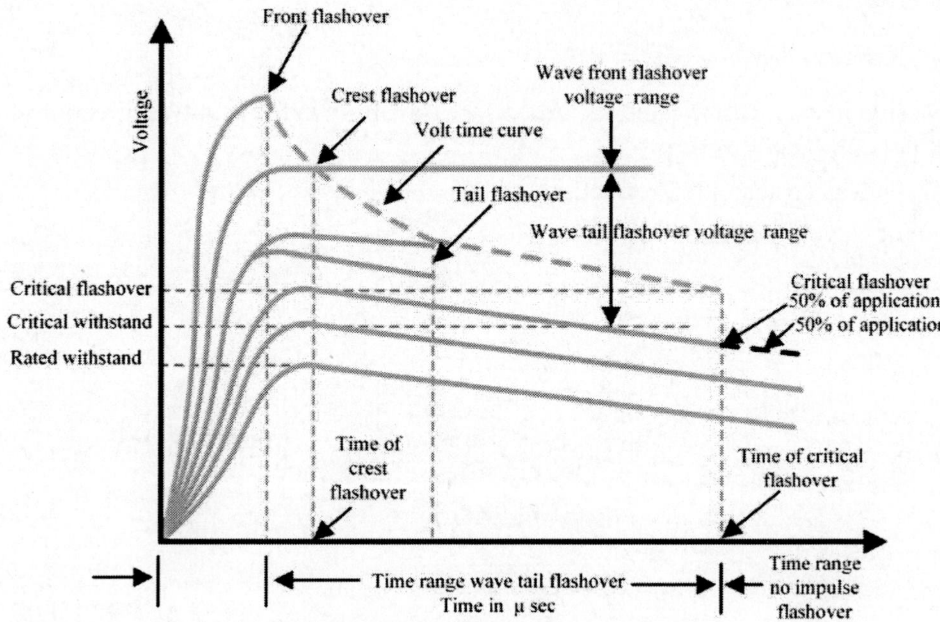

Fig. 5.15 Voltage-Time Characteristics

5.7 Voltages produced by traveling wave:

Voltages produced by travelling wave can be calculated with the help of Bewley's lattice diagram. Bewley's lattice diagram theory, necessary mathematical expressions and examples (numerical) are clearly given in section 5.8.

Travelling wave consists of three voltages, i.e. incident, reflected and refracted voltages. To measure the reflected and refracted voltages, we need to have Reflection and Refraction coefficients. Reflection and Refraction coefficients of a travelling wave are given in section 5.7.1, necessary mathematical expressions are also provided. There are various line terminations possible when the travelling wave reach the other end of the lines. Typical cases of Line Terminations are clearly discussed in section 5.7.2, necessary mathematical expressions and examples (numerical) are also provided.

5.7.1 Reflection and Refraction coefficients of a travelling wave:

The reflected voltage wave and its associated current wave travel backdown the line, and it is superimposed on the initial or forward or incident wave. The refracted or transmitted wave penetrates beyond the discontinuity. To develop general formulae for reflection and refraction coefficients we consider a line with surge impedance Z_c terminated by an impedance Z

e_i, i_i are incident (or) forward voltage and current

e_r, i_r are reflected voltage and current

e_t, i_t are refracted (or) transmitted voltage and current

For incident wave $\frac{e_i}{i_i} = Z_c$ ......(5.21)

For reflected wave $\frac{e_r}{i_r} = -Z_c$ ....(5.22)

For refracted wave $\frac{e_t}{i_t} = Z$ ....(5.23)

The negative sign in eq.5.22 indicates, e_r *and* i_r are travelling in the negative direction (or) backward direction of the same line e_i *and* i_i .

Voltage wave transmitted/refracted $e_t = e_i + e_r$ ....(5.24)

Current wave transmitted/refracted $i_t = i_i + i_r$ ....(5.25)

$$\frac{e_t}{Z} = \frac{e_i}{Z_C} - \frac{e_r}{Z_C}$$

$$= \frac{1}{Z_C}(e_i - e_r)$$

$$= \frac{1}{Z_C}\{e_i - (e_t - e_i)\} \quad \text{from (5.24))}$$

$$= \frac{1}{Z_C}(2e_i - e_t)$$

$$\frac{e_t}{Z} = \frac{2e_i}{Z_C} - \frac{e_t}{Z_C}$$

$$\frac{2e_i}{Z_C} = e_t\left(\frac{Z + Z_C}{Z * Z_C}\right)$$

$$e_t = e_i\left(\frac{2Z}{Z + Z_C}\right) \quad ....(5.26)$$

Voltage wave transmitted/refracted $e_t = e_i + e_r$

$$e_r = e_t - e_i$$

$$= e_i\left(\frac{2Z}{Z + Z_C}\right) - e_i$$

$$e_r = e_i\left(\frac{Z - Z_C}{Z + Z_C}\right) \quad ....(5.27)$$

The coefficients $\frac{2Z}{Z+Z_C}, \frac{Z-Z_C}{Z+Z_C}$ are called refraction and reflection coefficients.

From eq.5.23 and eq. 5.26, $i_t = \frac{e_t}{Z} = \frac{e_i\left(\frac{2Z}{Z+Z_C}\right)}{Z}$

$$i_t = \frac{2e_i}{Z + Z_C} \quad ....(5.28)$$

From eq.5.22 and eq. 5.27, $$i_r = \frac{-e_r}{Z_C} = \frac{-e_i}{Z_C}\left(\frac{Z-Z_C}{Z+Z_C}\right)$$

$$i_r = -i_i\left(\frac{Z-Z_C}{Z_C+Z}\right) \quad \text{where } \frac{e_i}{Z_C} = i_i$$

$$i_r = i_i\left(\frac{Z_C-Z}{Z_C+Z}\right) \qquad \text{.........(5.29)}$$

5.7.2 Termination of Lines:

There are various line terminations possible when the travelling wave reach the other end of the lines. Incident, reflected, transmitted Voltages and currents are shown in Fig. 5.16. Typical cases of Line Terminations are as follows:

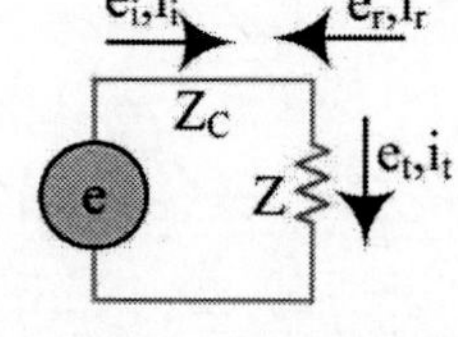

Fig. 5.16 Incident, Reflected, Refracted V and I

- Load Resistance equal to Surge Impedance
- Open-circuited line
- Short-circuited line
- Forked Line/Bifurcated line
- Line terminated by an Inductance
- Line terminated by Capacitance

Case (i): Line terminated by resistance equal to Surge Impedance ($Z = Z_C$)

Fig. 5.17 gives the line terminated by Resistance. If the line is terminated by an impedance equal to Surge Impedance i.e. $Z = Z_C$ $e_t = e_i\left(\frac{2Z}{Z+Z_C}\right) = e_i\left(\frac{2Z_C}{Z_C+Z_C}\right) = e_i\left(\frac{2Z_C}{2Z_C}\right) = e_i$ ....(5.30)

$$e_r = e_i\left(\frac{Z-Z_C}{Z+Z_C}\right) = e_i\left(\frac{Z_C-Z_C}{Z_C+Z_C}\right) = 0 \qquad \text{.........(5.31)}$$

$$i_t = \frac{2e_i}{Z+Z_C} = \frac{2e_i}{Z_C+Z_C} = \frac{2e_i}{2Z_C} = \frac{e_i}{Z_C} = i_i \qquad \text{.........(5.32)}$$

$$i_r = i_i\left(\frac{Z_C-Z}{Z_C+Z}\right) = i_i\left(\frac{Z_C-Z_C}{Z_C+Z_C}\right) = 0 \qquad \text{.........(5.33)}$$

If the line is terminated by *an* impedance equal to Surge Impedance i.e. $Z = Z_C$

Then we can observe that $e_t = e_i$; $e_r = 0$;

$$i_t = i_i;\ i_r = 0$$

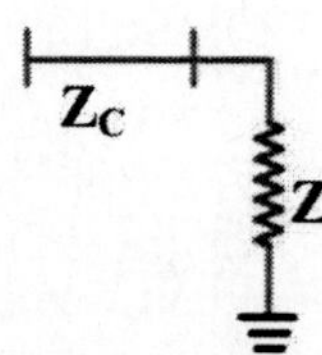

Fig. 5.17 Line terminated by Resistance

Case (ii): Line is terminated by an Open-Circuit ($Z = \infty$)

If the Line is terminated by an Open-Circuit, i.e. $I = 0$ *then*

$$Z = \frac{V}{I} = \frac{V}{0} = \infty$$

$$e_t = e_i\left(\frac{2Z}{Z+Z_C}\right) = \frac{2e_i Z}{Z\left(1+\frac{Z_C}{Z}\right)} = \frac{2e_i}{\left(1+\frac{Z_C}{\infty}\right)} = \frac{2e_i}{(1+0)} = 2e_i \qquad \text{.........(5.34)}$$

$$e_r = e_i\left(\frac{Z-Z_C}{Z+Z_C}\right) = \frac{e_i Z\left(1-\frac{Z_C}{Z}\right)}{Z\left(1+\frac{Z_C}{Z}\right)} = \frac{e_i\left(1-\frac{Z_C}{\infty}\right)}{\left(1+\frac{Z_C}{\infty}\right)} = \frac{e_i(1-0)}{(1+0)} = e_i \qquad \text{.........(5.35)}$$

$$i_t = \frac{2e_i}{Z+Z_C} = \frac{2e_i}{\infty+Z_C} = 0 \qquad \text{.........(5.36)}$$

$$i_r = i_i\left(\frac{Z_C - Z}{Z_C + Z}\right) = i_i\frac{Z\left(\frac{Z_C}{\infty} - 1\right)}{Z\left(\frac{Z_C}{\infty} + 1\right)} = i_i\frac{(0-1)}{(0+1)} = -i_i \qquad \text{.........(5.37)}$$

If the Line is terminated by an Open-Circuit i.e. $Z = \infty$

Then we can observe that $e_t = 2e_i;\ e_r = e_i;\ i_t = 0;\ i_r = -i_i$

Case (iii): Line is terminated by Short-Circuit ($Z = 0$)

If the Line is terminated by Short-Circuit $i.e.\ V = 0\ then\ Z = \frac{V}{I} = \frac{0}{I} = 0$

$$e_t = e_i\left(\frac{2Z}{Z + Z_C}\right) = \frac{2e_i Z}{Z\left(1 + \frac{Z_C}{Z}\right)} = \frac{2e_i}{\left(1 + \frac{Z_C}{0}\right)} = \frac{2e_i}{(1+\infty)} = 0 \qquad \text{.........(5.38)}$$

$$e_r = e_i\left(\frac{Z - Z_C}{Z + Z_C}\right) = e_i\left(\frac{0 - Z_C}{0 + Z_C}\right) = -e_i \qquad \text{.........(5.39)}$$

$$i_t = \frac{2e_i}{Z + Z_C} = \frac{2e_i}{0 + Z_C} = 2i_i \qquad \text{.........(5.40)}$$

$$i_r = i_i\left(\frac{Z_C - Z}{Z_C + Z}\right) = i_i\left(\frac{Z_C - 0}{Z_C + 0}\right) = i_i \qquad \text{.........(5.41)}$$

If the Line is terminated by Short-Circuit i.e. $Z = 0$

Then we can observe that $e_t = 0;\ e_r = -e_i;\ i_t = 2i_i;\ i_r = i_i$

Case (iv): Line terminated by Forked /Bifurcated Line

Fig.5.18 gives a line terminated by Bifurcated Line and its Equivalent circuit is given in Fig. 5.19. Let a line of natural impedance Z_C bifurcated into two branches of natural impedances Z_1 and Z_2. Transmitted voltage across both the branches will be same as they are in parallel but the transmitted currents will be different as $Z_1 \neq Z_2$.

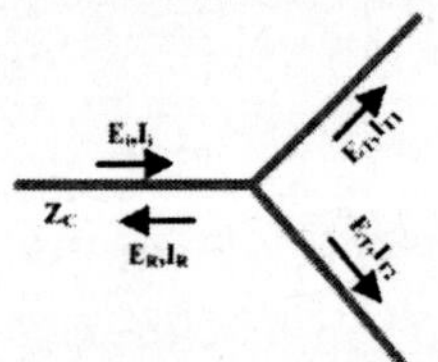

Fig.5.18 Line terminated by Bifurcated Line

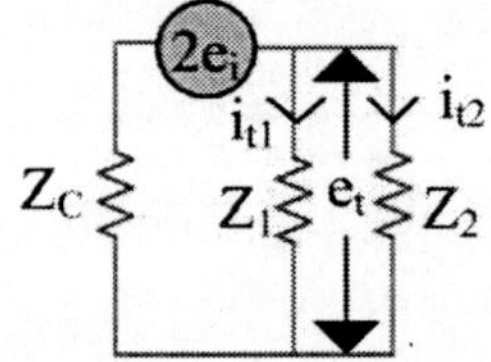

Fig. 5.19 Equivalent circuit for Fig. 5.18

Let incident wave be (E_i and I_i) travelling to the right, the reflected wave (E_R and I_R) be travelling to the left and transmitted waves (E_T and I_{T1}) (E_T and I_{T2}) travelling towards right as shown in Fig. 5.17.

From equivalent circuit $\quad z = \dfrac{Z_1 * Z_2}{Z_1 + Z_2}$ ....(5.42)

Transmitted voltage $\quad E_t = \dfrac{2 * E_i}{\frac{Z_1 * Z_2}{Z_1 + Z_2} + Z_C} * \dfrac{Z_1 * Z_2}{Z_1 + Z_2} = \dfrac{2 * E_i * \frac{1}{Z_C}}{\frac{1}{Z_C} + \frac{1}{Z_1} + \frac{1}{Z_2}}$ ....(5.43)

Transmitted current in line 1 is $I_{t1} = \dfrac{E_t}{Z_1}$ ....(5.44)

Transmitted current in line 2 is $I_{t2} = \dfrac{E_t}{Z_2}$ ....(5.45)

Reflected voltage $E_r = E_t - E_i = \frac{2 * E_i * \frac{1}{Z_C}}{\frac{1}{Z_C} + \frac{1}{Z_1} + \frac{1}{Z_2}} - E_i$

$$= E_i \left\{ \frac{2Z_1Z_2}{Z_1Z_2 + Z_C(Z_1+Z_2)} - 1 \right\}$$

$$= E_i \left\{ \frac{Z_1Z_2 - Z_C(Z_1+Z_2)}{Z_1Z_2 + Z_C(Z_1+Z_2)} \right\} \quad \text{.........(5.46)}$$

Reflected current $I_r = I_{t1} + I_{t2} - I_i$ ....(5.47)

Case (v): Line terminated by an Inductance

Let the line be terminated by an inductance L, as illustrated in Fig. 5.20 and equivalent circuit diagram is shown in Fig. 5.21 for the DC surge.

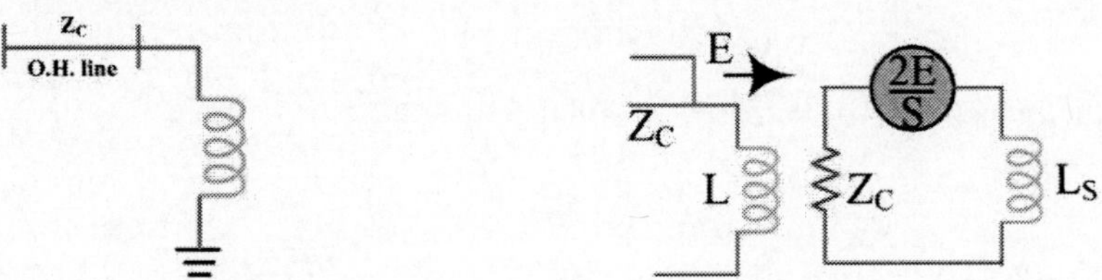

Fig. 5.20 Line terminated by an Inductance

Fig. 5.21 Equivalent circuit for Fig. 5.20

Transmitted current through the inductor is $i_t = i_L = i_i + i_r = \frac{e_i}{Z_C} - \frac{e_R}{Z_C}$ ......(5.48)

The voltage across the inductor is $e_L = L\frac{di_L}{dt} = L\frac{d\left(\frac{e_i}{Z_C} - \frac{e_R}{Z_C}\right)}{dt}$ ....(5.49)

$$e_t = e_L = e_i + e_R$$

$$e_i + e_r = L\frac{d\left(\frac{e_i}{Z_C} - \frac{e_R}{Z_C}\right)}{dt}$$

$$e_r + \frac{L}{Z_C}\frac{de_R}{dt} = -e_i + \frac{L}{Z_C}\frac{de_i}{dt} \quad \text{.........(5.50)}$$

Let the incident wave be of constant value ‘E’, so that its time derivative is zero. So, the above eq. becomes $e_r + \frac{L}{Z_C}\frac{de_R}{dt} = -e_i$

$$\frac{de_R}{(E + e_R)} = \frac{-Z_C}{L}dt \quad \text{.........(5.51)}$$

Integrating both the sides of above eq., we have

$$log_e(E + e_r) = \frac{-Z_C}{L}t + A \quad \text{... ... .(5.52)}$$

Where A is a constant of integration which can be evaluated from initial condition, i.e. $t = 0, i_L = 0$; because when the wave arrives at the inductance terminal the inductance L does not carry any current, the whole of it is reflected. It means when $t = 0, e_r = E$

Therefore $log_e(E + E) = \frac{-Z_C}{L} * 0 + A$

$$A = log_e(2E) \quad \text{........(5.53)}$$

Substituting 'A' value in the above eq.5.52, we have

$$log_e(\mathrm{E} + e_r) = \frac{-Z_C}{L}t + log_e(2\mathrm{E})$$

$$log_e\frac{(\mathrm{E} + e_r)}{2\mathrm{E}} = \frac{-Z_C}{L}t$$

$$\frac{(\mathrm{E} + e_r)}{2\mathrm{E}} = e^{\frac{-Z_C}{L}t}$$

$$\mathrm{E} + e_r = 2Ee^{\frac{-Z_C}{L}t}$$

$$e_r = \mathrm{E}\left(2 * e^{\frac{-Z_C}{L}t} - 1\right)$$

By replacing E=e_i, then e_R will become

Reflected Voltage $e_r = e_i\left(2 * e^{\frac{-Z_C}{L}t} - 1\right)$ ....(5.54)

The voltage across the inductor is $e_t = e_L = e_i + e_R$

$$= e_i + e_i\left(2 * e^{\frac{-Z_C}{L}t} - 1\right)$$

Therefore, Voltage across the inductor is $E_t = E_L = 2 * E_i * e^{\frac{-Z_C}{L}t}$ ....(5.55)

Transmitted current through the inductor is $i_t = i_L = i_i + i_R = \frac{e_i}{Z_C} - \frac{e_R}{Z_C}$

$$i_t = \frac{e_i}{Z_C} - \frac{e_i\left(2 * e^{\frac{-Z_C}{L}t} - 1\right)}{Z_C}$$

$$= \frac{e_i}{Z_C} - \frac{2e_i\left(e^{\frac{-Z_C}{L}t}\right)}{Z_C} + \frac{e_i}{Z_C}$$

$$i_t = \frac{2e_i}{Z_C}\left(1 - e^{\frac{-Z_C}{L}t}\right)$$ ....(5.56)

Case (vi): Line terminated by a Capacitance

Let the line be terminated by a capacitance 'C', as illustrated in Fig. 5.22 and equivalent circuit diagram is shown in Fig. 5.23 for the DC surge. The reflected voltage varies exponentially and its mathematical expression can be arrived in the same way as in case of terminal inductance.

Voltage across the Capacitor is $e_t = e_C = 2 * e_i * \left(1 - e^{\frac{-t}{C * Z_C}}\right)$ ....(5.57)

Transmitted current through the Capacitor is $i_t = i_C = \frac{2e_i}{Z_C}e^{\frac{-t}{C * Z_C}}$ ....(5.58)

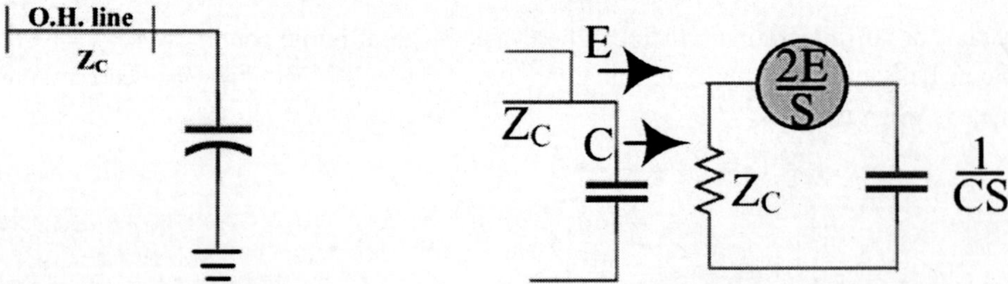

Fig. 5.22 Line terminated by capacitance

Fig. 5.23 Equivalent circuit for Fig. 5.22

Reflected Voltage $e_r = e_i\left(1 - 2 * e^{\frac{-t}{C*Z_C}}\right)$ ………(5.59)

Reflected current $i_r = i_T - i_i$

$$= \frac{2e_i}{Z_C} e^{\frac{-t}{C*Z_C}} - \frac{e_i}{Z_C}$$

$$= \frac{e_i}{Z_C}\left(e^{\frac{-t}{C*Z_C}} - 1\right)$$ ………(5.60)

Example 5.1. A Surge of 12kV is incident on a cable having an impedance 100Ω. A cable meets the transmission line having an impedance 400Ω. Determine the surge voltage transmitted into the overhead line.

Ans. Incident Voltage $E_i = 12\ kV$

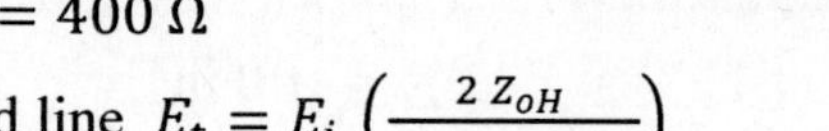

Surge impedance of the cable $Z_{cable} = 100\ \Omega$

Surge impedance of the O.H. line $Z_{OH} = 400\ \Omega$

Surge voltage transmitted into the overhead line $E_t = E_i\left(\frac{2\ Z_{OH}}{Z_{cable} + Z_{OH}}\right)$

$$E_t = 12 * 10^3\left(\frac{2*400}{100+400}\right)$$

$$= 19.2\ kV$$

Example 5.2 A Surge of 50kV is incident on transmission line having an impedance of 500Ω. The transmission meets a cable having an impedance of 50Ω. Determine the transmitted voltage, reflected voltage and energy transmitted into a cable during a period of 5µs.

Ans. Incident Voltage $E_i = 50\ kV$

Surge impedance of the O.H. line $Z_{OH} = 500\ \Omega$

Surge impedance of the cable $Z_{cable} = 50\ \Omega$

Transmitted voltage $E_t = E_i\left(\frac{2\ Z_{cable}}{Z_{OH} + Z_{cable}}\right) = 50 * 10^3\left(\frac{2*50}{500+50}\right) = 9.09\ kV$

Reflected voltage $E_r = E_i\left(\frac{Z_{cable} - Z_{OH}}{Z_{OH} + Z_{cable}}\right) = 50 * 10^3\left(\frac{50-500}{500+50}\right) = -40.9kV$

Transmitted current $I_t = \frac{E_t}{Z_2} = \frac{9.09 * 10^3}{50} = 181.8\ A$

Energy transmitted into the cable during a period of 5µs is $= E_t * I_t * t$

$$= (9.09 * 10^3) * 181.8 * (5 * 10^{-6})$$

$= 8.26\ J$

Example 5.3 A Surge of 20kV is incident on the cable having an inductance and capacitance of 0.2mH and 0.4µF. A cable meets the transmission line with an inductance and capacitance of 1.2mH and 0.024µF, respectively. Determine the surge voltage transmitted into the overhead line.

Ans. Incident Voltage $E_i = 20\ kV$

Natural impedance of the cable

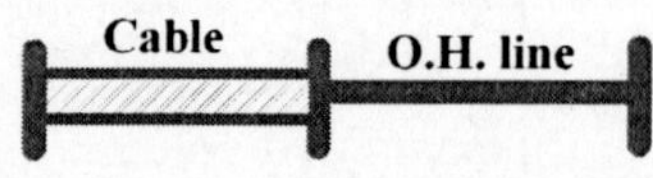

$$Z_{cable} = \sqrt{\frac{L}{C}} = \sqrt{\frac{0.2 * 10^{-3}}{0.4 * 10^{-6}}} = 22.36\ \Omega$$

Natural impedance of the O.H. line $Z_{OH} = \sqrt{\frac{L}{C}} = \sqrt{\frac{1.2 * 10^{-3}}{0.024 * 10^{-6}}} = 223.6\ \Omega$

Voltage transmitted into the O.H. line $E_t = E_i \left(\frac{2\,Z_{cable}}{Z_{OH} + Z_{cable}}\right)$

$$E_t = 20 * 10^3 \left(\frac{2*223.6}{22.36 + 223.6}\right)$$

$$= 36.36\ kV$$

Example 5.4. Two O.H. lines 'A' and 'B' having surge impedance of 500Ω and 400Ω respectively are connected by a short cable 'C' of surge impedance 50Ω. A rectangular surge of 100kV and of infinite length travels along 'A' towards the cable 'C. Determine

(i). Reflected voltage at junction B and C

(ii). Reflected voltage at junction A and C

(iii). Total Surge Voltage transmitted to cable

Ans. Incident Voltage $E_i = 100\ kV$

Surge impedance of the O.H. line A $Z_A = 500\ \Omega$

Surge impedance of the O.H. line B $Z_B = 400\ \Omega$

Surge impedance of the cable C $Z_C = 50\ \Omega$

Transmitted voltage at junction A and C is $E_t = E_i \left(\frac{2\,Z_C}{Z_A + Z_C}\right)$

$$= 100 * 10^3 \left(\frac{2*50}{500 + 50}\right) = 18.18\ kV$$

Reflected Coefficient at junction B and C is $r_{BC} = \frac{Z_B - Z_C}{Z_B + Z_C} = \frac{400-50}{400+50} = 0.777$

Reflected voltage at junction B and C is $E_{r_{BC}} = E_t * r_{BC}$

$$= (18.18 * 10^3) * 0.777 = 14.14\ kV$$

Reflected Coefficient at junction A and C is $r_{AC} = \frac{Z_A - Z_C}{Z_A + Z_C} = \frac{500-50}{500+50} = 0.818$

Reflected voltage at junction A and C is $E_{r_{AC}} = E_{r_{BC}} * r_{AC}$

$$= (14.14 * 10^3) * 0.818 = 11.56\ kV$$

Total Surge Voltage transmitted to cable $= E_t + E_{r_{AC}} = 18.18 + 11.56 = 29.74\ kV$

Example 5.5. A Surge of 50kV is incident on transmission line having natural impedance of 500Ω arrives at a junction of 750Ω and 250Ω respectively. Find the surge voltage and currents transmitted into each of the branch lines.

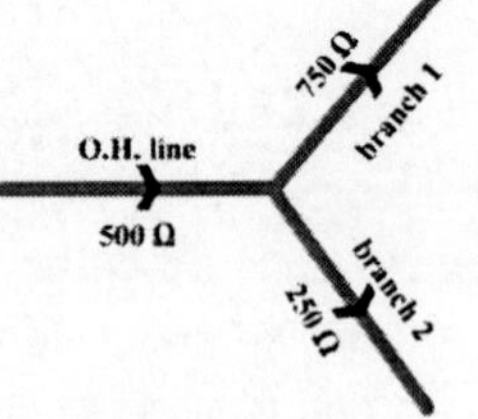

Ans. Incident Voltage $E_i = 50\ kV$

Natural impedance of the OH, $Z_C = 500\ \Omega$

Natural impedance of branch 1, $Z_1 = 750\ \Omega$

Natural impedance of branch 2, $Z_2 = 250\ \Omega$

Transmitted voltage $E_t = \frac{2*E_i*\frac{1}{Z_C}}{\frac{1}{Z_C}+\frac{1}{Z_1}+\frac{1}{Z_2}} = \frac{2*(50*10^3)*\left(\frac{1}{500}\right)}{\frac{1}{500}+\frac{1}{750}+\frac{1}{250}} = 27.27\ kV$

Transmitted current in line 1 is $I_{t1} = \frac{E_t}{Z_1} = \frac{27.27*10^3}{750} = 36.36\ A$

Transmitted current in line 2 is $I_{t2} = \frac{E_t}{Z_2} = \frac{27.27*10^3}{250} = 109.08\ A$

Example 5.6. A rectangular Surge of 50kV, 2μs is incident on transmission line having impedance of 500Ω, terminated by an inductance of 4mH. Determine the reflected voltage and voltage across the inductance.

Ans. Incident Voltage $E_i = 50\ kV$

Inductance of the overhead line $L = 4\ mH = 4000\ \mu H$

Surge impedance of the overhead line $Z_c = 500\ \Omega$

Voltage across the inductor is $E_L = 2 * E_i * e^{\frac{-Z_C}{L}t}$

$$= 2 * (50 * 10^3) * e^{\frac{-500}{4000}*2}$$
$$= (100 * 10^3) * e^{-0.25}$$
$$= (100 * 10^3) * 0.7788$$
$$= 77.88\ kV$$

Reflected Voltage $E_r = E_i\left(2 * e^{\frac{-Z_C}{L}t} - 1\right)$

$$= (50 * 10^3)\left(2 * e^{\frac{-500}{4000}*2} - 1\right)$$
$$= (50 * 10^3) * (2 * e^{-0.25} - 1)$$
$$= 27.88\ kV$$

Example 5.7. A rectangular Surge of 400kV, 2μs is incident on transmission line having impedance of 500Ω, terminated by a capacitance of 2500pF. Determine the reflected voltage and voltage across the capacitance.

Ans. Incident Voltage $E_i = 400\ kV$

Capacitance connected at the receiving end of the O.H. line is $C = 2000pF$

Surge impedance of the overhead line $Z_c = 500\ \Omega$

Voltage across the Capacitor is $E_C = 2 * E_i * \left(1 - e^{\frac{-t}{C*Z_C}}\right)$

$$= 2 * (50 * 10^3) * \left(1 - e^{\frac{-2}{0.0025*500}}\right)$$
$$= (100 * 10^3) * (1 - e^{-1.6})$$
$$= 798.1\ kV/Ph$$

Reflected Voltage $E_r = E_i\left(1 - 2 * e^{\frac{-t}{C*Z_C}}\right)$

$$= (50 * 10^3)\left(1 - 2 * e^{\frac{-2}{0.0025*500}}\right)$$
$$= (50 * 10^3)(1 - 2 * e^{-1.6})$$
$$= 298.1\ kV/Ph$$

Example 5.8. A 3-ph O.H. line has conductors 3cms in diameter and spaced 2metre apart in equilibrium formation. A surge of 11kV travels along the O.H. line. Calculate

(i). the impedance of the O.H. line

(ii). the line current

(iii). Voltage across the load, reflected voltage, Rate of Reflected Energy if the line is terminated through a star connected resistor of 1k Ω per phase

(iv). the value of terminated resistance of no reflection and

(v). Rate of refracted and reflected energy of a cable if the line is connected to a cable with an inductance and capacitance per phase per cm of 2*10⁻⁸H and 0.9*10⁻⁶ μF respectively.

Ans. Inductance $L = 2 * 10^{-7}\, ln\frac{d}{r}\; H/m \;=\; 2 * 10^{-7}\, ln\frac{200}{1.5} = 9.78 * 10^{-7}\; H/m$

Capacitance $C = \frac{2\pi\epsilon}{ln\frac{d}{r}}\; F/m \;=\; \frac{2*3.14*8.854*10^{-12}}{ln\frac{200}{1.5}} = 1.137 * 10^{-11}\; F/m$

(i) Natural impedance of the O.H. line $Z_1 \;=\; \sqrt{\frac{L}{C}} = \sqrt{\frac{9.78*10^{-7}}{1.137*10^{-11}}} = 293.41\;\Omega$

(ii) Line current $= \frac{11*10^3}{\sqrt{3}*293.41} = 21.64\;A$

(iii) Voltage across the terminating resistance $E = E_i\left(\frac{2\,Z_2}{Z_1+Z_2}\right)$

$$= \frac{11*10^3}{\sqrt{3}}\left(\frac{2*1000}{293.41+1000}\right) = 9.815\;kV$$

Reflected voltage $E_r = E_i\left(\frac{Z_2-Z_1}{Z_1+Z_2}\right) = \frac{11*10^3}{\sqrt{3}}\left(\frac{1000-293.41}{1000+293.41}\right) = 3.465\;kV$

Rate of Reflected Energy $= \frac{3*E_r{}^2}{R} = \frac{3*(3.465*10^3)^2}{293.41} = 121.8\;kw$

(iv) The terminating resistance for no reflection is 293.41 Ω

(v) Impedance of the Cable $Z_{cable} \;=\; \sqrt{\frac{L}{C}} = \sqrt{\frac{2*10^{-8}}{0.9*10^{-12}}} = 149\;\Omega$

Refracted Voltage of cable $= E_i\left(\frac{2\,Z_C}{Z_1+Z_C}\right) = \frac{11*10^3}{\sqrt{3}}\left(\frac{2*149}{293.41+149}\right) = 4.27\;kV$

Reflected voltage of cable $= E_i\left(\frac{Z_C-Z_1}{Z_C+Z_1}\right) = \frac{11*10^3}{\sqrt{3}}\left(\frac{149-293.41}{149+293.41}\right) = -2.07kV$

Rate of Refracted Energy of cable $= \frac{3*E_{t\,cable}{}^2}{R} = \frac{3*(4.27*10^3)^2}{149} = 367\;kw$

Rate of Reflected Energy of cable $= \frac{3*E_{r\,cable}{}^2}{R} = \frac{3*(2.078*10^3)^2}{149} = 44\;kw$

5.8 Bewley's Lattice Diagram:

This is a convenient diagram developed by Bewley, which shows at a glance the position and direction of motion of every incident, reflected, and transmitted wave on the system at every instant of time.

In an extensive network having many junctions and terminations, the number of transmitted and reflected waves, initiated by a single incident wave meets different junctions. Generally, it becomes difficult to keep track of the transmitted and reflected waves, but with the use of Bewley's Lattice Diagram it is possible. Bewley's lattice diagram is drawn, which is also called the zig-zag diagram. In this section Bewley's lattice diagram are drawn for voltage waves.

Case (i): Overhead transmission line is connected to a cable, at receiving end is terminated by open circuit

Consider a line with two sections having $Z_{line,}$ and Z_{cable} as shown in Fig. 5.24. These 02 sections have different surge impedances and different velocity of wave propagation, shown in Fig.5.25.

Reflection coefficient from line to cable $a_1 = \frac{Z_{line} - Z_{cable}}{Z_{line} + Z_{cable}}$ ....(5.61)

Reflection coefficient from cable to line $a_1' = \frac{Z_{cable} - Z_{line}}{Z_{cable} + Z_{line}}$ ....(5.62)

Refraction coefficient from cable to far end $b_1 = \frac{2\, Z_{cable}}{Z_{cable} + Z_{line}}$ ....(5.63)

Refraction coefficient from cable to line is summation of the Incident coefficient at that junction and reflected coefficient from line to cable $b_1' = 1 + a_1$ ......(5.64)

Fig.5.24 O.H. line is connected to a cable

At Receiving end: Open Circuited line

Impedance $Z_R = \infty$

Reflection coefficient at open circuit end $= \frac{Z_R - Z_{cable}}{Z_R + Z_{cable}}$

$$= \frac{Z_R\left(1 - \frac{Z_{cable}}{Z_R}\right)}{Z_R\left(1 + \frac{Z_{cable}}{Z_R}\right)}$$

$$= \frac{\left(1 - \frac{Z_C}{\infty}\right)}{\left(1 + \frac{Z_C}{\infty}\right)}$$

$$= 1 \qquad \text{.........(5.65)}$$

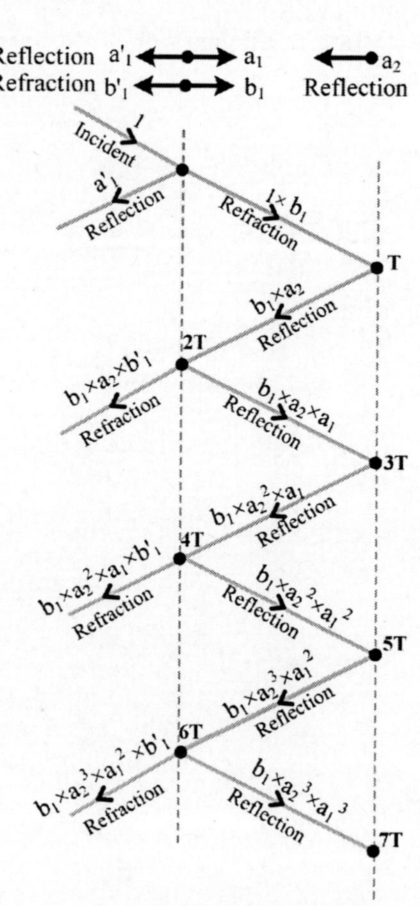

Fig. 5.25 Bewley's lattice diagram

Case (ii): Overhead transmission line is connected to a transformer through a cable

Consider a line with two sections having Z_{line}, Z_{cable} and $Z_{T/F}$ as shown in Fig. 5.26. These 03 sections have different surge impedances and different velocity of wave propagation, shown in Fig.5.27.

Fig. 5.26 O.H. line is connected to a T/F through a cable

Reflection coefficient from line to cable $a_1 = \frac{Z_{line}-Z_{cable}}{Z_{line}+Z_{cable}}$ ……...(5.66)

Reflection coefficient from cable to line $a_1' = \frac{Z_{cable}-Z_{line}}{Z_{cable}+Z_{line}}$ ……...(5.67)

Refraction coefficient from cable toT/f $b_1 = \frac{2\,Z_{cable}}{Z_{cable}+Z_{line}}$ ……...(5.68)

Refraction coefficient from cable to line $b_1' = 1 + a_1$ ……...(5.69)

Reflection coefficient from T/f to cable $a_2 = \frac{Z_{T/F}-Z_{cable}}{Z_{T/F}+Z_{cable}}$ ……...(5.70)

Refraction coefficient from T/f to far end $b_2' = 1 + a_2$ ……...(5.71)

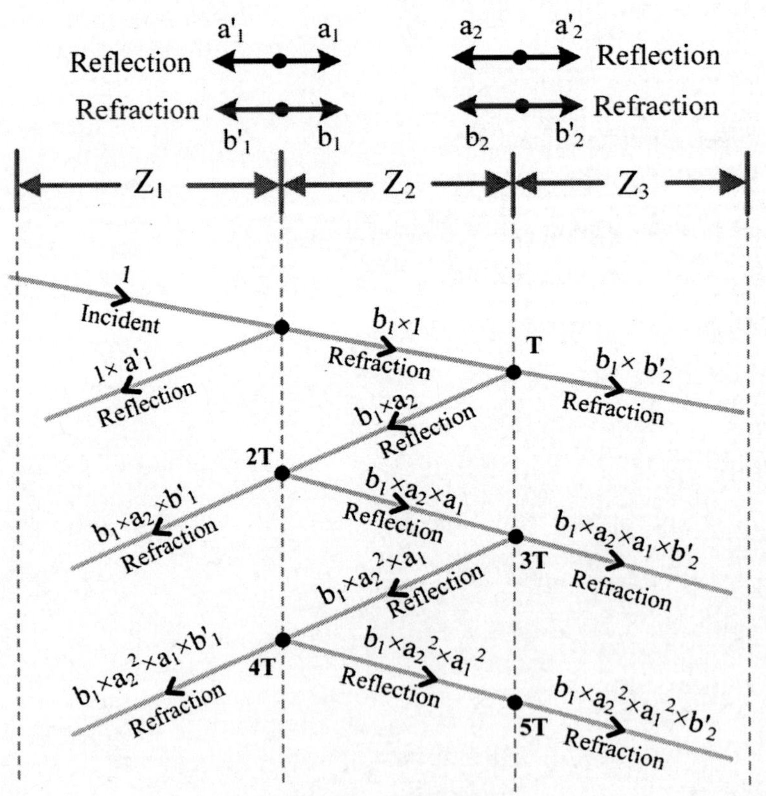

Fig. 5.27 Bewley's lattice diagram

Example 5.9 A line of surge impedance 400Ω is charged from a battery of constant voltage 100V. The line is 300metres long and is terminated in a resistance 200Ω. Draw the Bewley's lattice diagram and calculate the voltage at time intervals from T to 6T.

Ans. Incident Voltage $E_i = 100\ V$

Surge Impedance $Z_c = 400\Omega$

Terminal resistance $Z = 200\Omega$

At Sending end $Z_s = 0\Omega,\ Z_c = 400\Omega$

Reflection coefficient at sending end $a_1 = \frac{Z_s - Z_c}{Z_s + Z_c} = \frac{0 - 400}{0 + 400} = -1$

At Receiving end $Z_R = 200\Omega,\ Z_c = 400\Omega$

Reflection coefficient at receiving end $a_2 = \frac{Z_R - Z_c}{Z_R + Z_c} = \frac{200 - 400}{200 + 400} = -\frac{1}{3}$

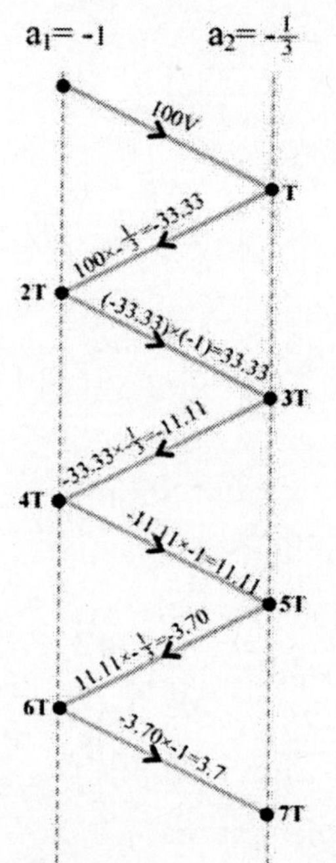

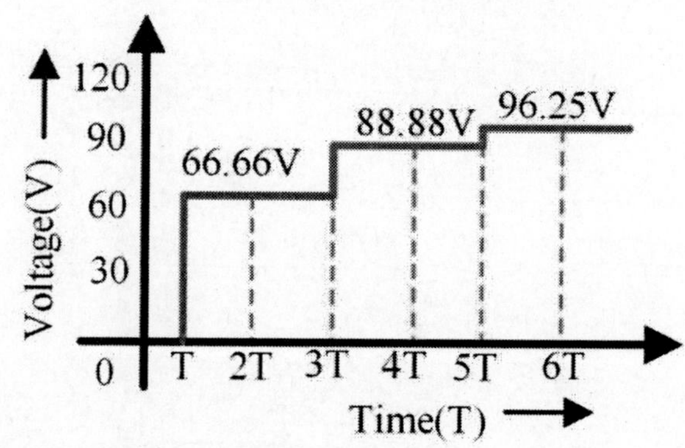

$$at\ T,\ V_T = 100 - 33.33 = 66.66V$$
$$at\ 2T, V_{2T} = V_T - 33.33 - 33.33 = 66.66V$$
$$at\ 3T, V_{3T} = V_{2T} + 33.33 - 11.11 = 88.88V$$
$$at\ 4T, V_{4T} = V_{3T} - 11.11 + 11.11 = 88.88V$$
$$at\ 5T, V_{5T} = V_{4T} + 11.11 - 3.7 = 96.25V$$
$$at\ 6T, V_{6T} = V_{5T} - 3.7 + 3.7 = 96.25V$$

Example 5.10. The ends of two long transmission lines A and B are connected by a cable C, 1.5km long. Each line has the capacitance of $10pF/m$ and inductance of $1.6 * 10^{-6} H/m$ and cable has the capacitance of $89pF/m$ and inductance of $5 * 10^{-7} H/m$. A rectangular surge 10kV travels along line 'A' towards the cable 'C' and line 'B'. What will be the voltage at the junction line 'A' and cable, $20\mu s$ after the initial surge reaches the point.

Ans.: Incident Voltage $E_i = 10\ kV$

Natural impedance of the cable $Z_{cable} = \sqrt{\frac{L}{C}} = \sqrt{\frac{5 * 10^{-7}}{89 * 10^{-12}}} = 75\ \Omega$

Natural impedance of the O.H. line $Z_{OH} = \sqrt{\frac{L}{C}} = \sqrt{\frac{1.6 * 10^{-6}}{10 * 10^{-12}}} = 400\ \Omega$

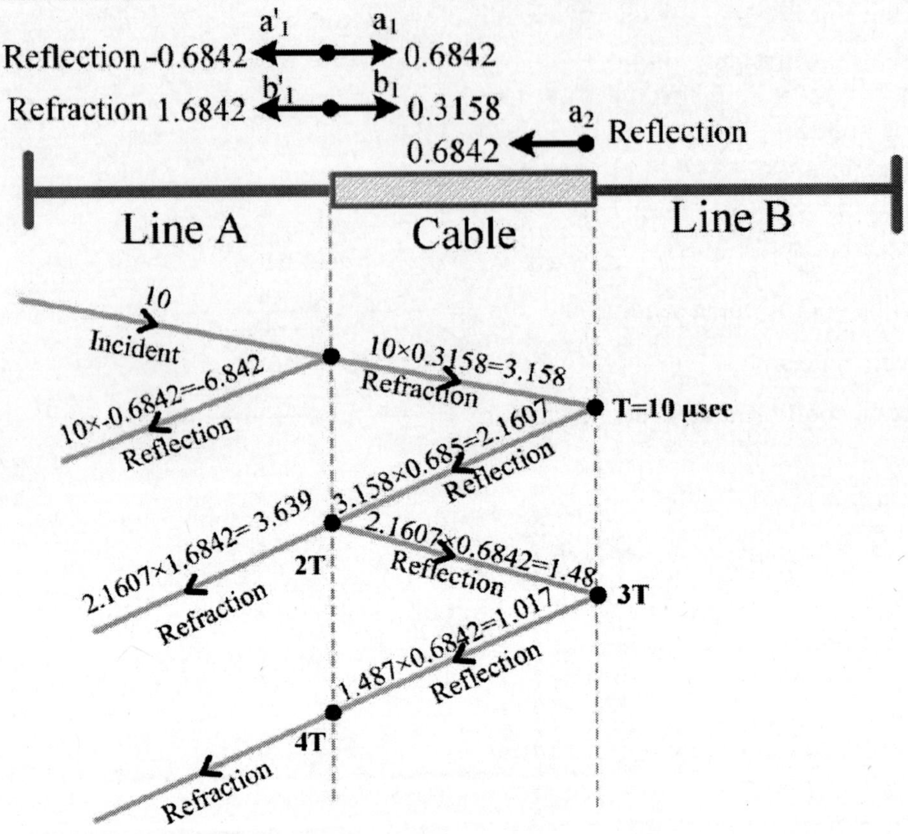

Reflection coefficient from line to cable $a_1 = \frac{Z_{line\ A} - Z_{cable}}{Z_{line\ A} + Z_{cable}} = \frac{400 - 75}{400 + 75} = 0.6842$

Reflection coefficient from cable to line $a_1' = \frac{Z_{cable} - Z_{line\ A}}{Z_{cable} + Z_{line\ A}} = \frac{75 - 400}{75 + 400} = -0.6842$

Refraction coefficient from cable to farend $b_1 = \frac{2\ Z_{cable}}{Z_{cable} + Z_{line\ A}} = \frac{2 * 75}{75 + 400} = 0.3158$

Refraction coefficient from cable to line is summation of the Incident coefficient at that junction and reflected coefficient from line to cable

$b_1' = 1 + a_1 = 1 + 0.6842 = 1.6842$ Reflection coefficient from T/f to cable $a_2 = \frac{Z_{Line\ B} - Z_{cable}}{Z_{Line\ B} + Z_{cable}} = \frac{400 - 75}{400 + 75} = 0.6842$

Surge velocity in cable $= \frac{1}{\sqrt{LC}} = \frac{1}{\sqrt{5 * 10^{-7} * 89 * 10^{-12}}} = 149906\ km/sec$

Time of surge travelling through the cable $t = \frac{length}{velocity} = \frac{1.5}{149906} = 10\mu s.$

At $20\mu s$ (i.e @ 2T); $V = 3.158 + 2.16 + 1.47 = 6.797\ kV$.

Example 5.11. A surge of 10kV is incident on a O.H. line which is joined to a cable having open circuit at its far end. The ratio of characteristic impedance of line to that of cable is 10. Draw the Bewley's lattice diagram if the wave originates in the line.

Ans.

Incident Voltaage $E_i = 10\ kV,$

$$\frac{Z_{line}}{Z_{cable}} = 10;\ Z_{line} = 10, Z_{cable} = 1$$

Reflection coefficient from line to cable

$$a_1 = \frac{Z_{line} - Z_{cable}}{Z_{line} + Z_{cable}} = \frac{10 - 1}{10 + 1} = 0.8182$$

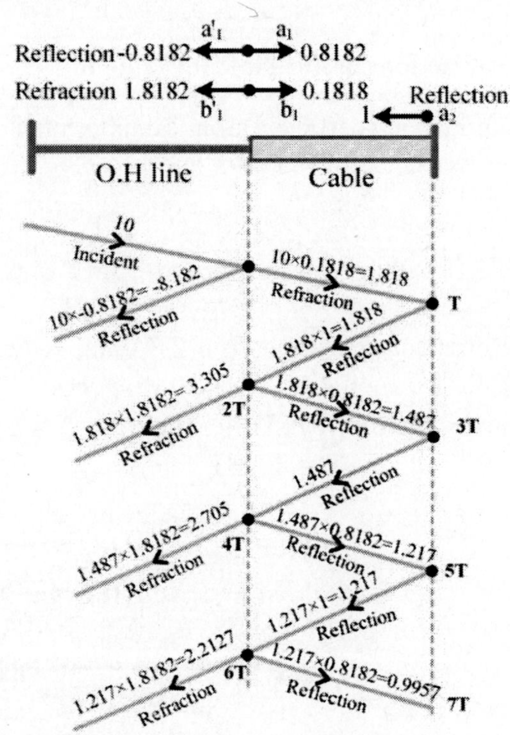

At Receiving end:

Open Circuit line; Impedance $Z_R = \infty$

Reflection coefficient at open circuit end will be

$$a_2 = \frac{Z_R - Z_{cable}}{Z_R + Z_{cable}} = \frac{Z_R\left(1 - \frac{Z_{cable}}{Z_R}\right)}{Z_R\left(1 + \frac{Z_{cable}}{Z_R}\right)}$$

$$= \frac{\left(1 - \frac{Z_C}{\infty}\right)}{\left(1 + \frac{Z_C}{\infty}\right)}$$

$$= \frac{(1 - 0)}{(1 + 0)}$$

$$= 1$$

Reflection coefficient from cable to line $\quad a_1' = \frac{Z_{cable} - Z_{line}}{Z_{cable} + Z_{line}} = \frac{1 - 10}{1 + 10} = -0.8182$

Refraction coefficient from cable to far end $\quad b_1 = \frac{2\,Z_{cable}}{Z_{cable} + Z_{line}} = \frac{2 * 1}{1 + 10} = 0.1818$

Refraction coefficient from cable to line is summation of the Incident coefficient at that junction and reflected coefficient from line to cable $\quad b_1' = 1 + a_1 = 1 + 0.8182 = 1.8182$

Example 5.12. A surge of 500kV is incident on a long OH line of characteristic impedance 400 Ω. It connects to a cable AB of length 1km having a total inductance of 264μH and a total capacitance of 0.165μF. At the far end of the cable, connection is made to a transformer of characteristic impedance 1000Ω. Draw the Bewley's lattice diagram at junction A of the cable for 26.4μs after the arrival at this junction of the original surge.

Ans. Incident Voltage $\quad E_i = 500\ kV$

Impedance of the Cable $\quad Z_{cable} = \sqrt{\frac{L}{C}} = \sqrt{\frac{264*10^{-6}}{0.165*10^{-6}}} = 40\ \Omega$

Velocity of the surge through cable $v = \frac{1}{\sqrt{LC}} = \frac{1}{\sqrt{264*10^{-6} * 0.165*10^{-6}}} = 151.51\ km/sec$

Time for surge to travel through cable $\quad t = \frac{length}{velocity} = \frac{1}{151.51} = 6.6\ \mu s$

Reflection coefficient from line to cable $a_1 = \dfrac{Z_{line} - Z_{cable}}{Z_{line} + Z_{cable}} = \dfrac{400 - 40}{400 + 40} = 0.8182$

Reflection coefficient from cable to line $a_1' = \dfrac{Z_{cable} - Z_{line}}{Z_{cable} + Z_{line}} = \dfrac{40 - 400}{40 + 400} = -0.8182$

Refraction coefficient from cable to T/f $b_1 = \dfrac{2\,Z_{cable}}{Z_{cable} + Z_{line}} = \dfrac{2 * 40}{40 + 400} = 0.1818$

Refraction coefficient from cable to line is summation of the Incident coefficient at that junction and reflected coefficient from line to cable

$$b_1' = 1 + a_1 = 1 + 0.8182 =$$

1.8182 Reflection coefficient from T/f to cable $a_2 = \dfrac{Z_{\frac{T}{f}} - Z_{cable}}{Z_{\frac{T}{f}} + Z_{cable}} = \dfrac{1000-40}{1000+40} = 0.9231$

Refraction coefficient from T/f to far end is summation of the Incident coefficient at that junction and reflected coefficient from T/f to cable $b_2' = 1 + a_2 = 1 + 0.9231 = 1.9231$

From the Bewley'slattice diagram $T = 6.6\mu s, 2T = 13.2\mu s, 3T = 19.8\mu s, 4T = 26.4\mu s, 5T = 33\mu s$

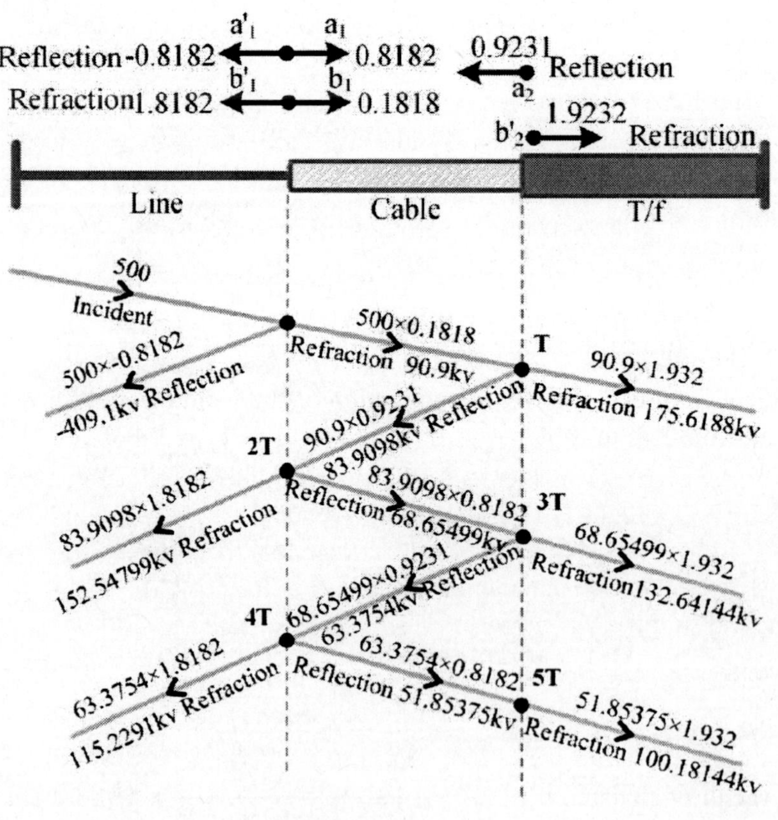

5.9 Unit Summary:

- Over-voltages are broadly classified into two major classes: Internal Over-voltages and External Over-voltages
- Over-voltages mainly caused due to lightning, switching and etc.
- Benjamin Franklin performed his famous experiment of flying kite in thunder cloud in 1745 and proved that the lightning stroke is because of the discharge of electricity.
- Surge Impedance is the impedance at which the reactive power generated will be equal to reactive power absorbed.
- When load impedance matches with surge impedance of the line, the power delivered to the load is maximum. The connected load under this condition is called Surge Impedance Loading (SIL).
- The velocity of propagation of travelling wave over the O.H. lines is equal to the velocity of light. In actual practice, the velocity of the travelling wave is slightly less than the velocity of the light due to the resistance and leakage reactance of the lines.
- To protect the power system against over-voltages, lightning arresters/surge diverters are used.
- To protect the buildings/tall objects from over-voltages surge absorbers are used.
- Lightning arrester consists of a sphere gap in series with a non-linear resistor
- There are different types of Lightning arrester/Surge diverter available they are Rod gap, Horn gap, Multi-gap, Expulsion type, Valve type, Zinc-oxide gapless lightning arrester and etc.
- BIL can be defined as reference level expressed in kV crest (peak) voltage value with standard $1.2/50\mu s$ lightning impulse wave.
- Protective margin = (withstand level of the system) - (Protective level of the surge arrester)
- There are various line terminations possible when the travelling wave reach the other end of the lines; they are Open-circuited line, Short-circuited line, Forked Line/Bifurcated line, Line terminated by an Inductance, Line terminated by Capacitance and etc.
- Voltages produced by travelling wave can be calculated with the help of Bewley's lattice diagram.

Short and Long Answer Questions

1. Define over-voltage. List the causes and consequences of over-voltages in Power Systems.
2. What is Volt-time Curves? Explain their significance in Power system studies.
3. What is the best location of lightning arrestor and why?
4. Which arrester is called as gap less arrester, why?
5. Which arrester is called as non-linear lightning arrester, why?
6. Is there a way to protect electronic instrument from a direct lightning strike?
7. Why are steep fronted surges more dangerous to power system equipment?
8. Explain how the rating of lightning arrester is selected.
9. Explain the significance of Surge Impedance and Surge Impedance Loading in Power System.
10. With neat diagrams, explain how the lightning occurs in (i) Intra Cloud (ii) Inter Clouds and (iii) Cloud and Earth.

11. Starting from the first principles derive the Velocity of Propagation of travelling wave and comment on it.
12. Explain the construction and working of Zinc-oxide gapless lightning arrester and comment how it is different from Valve Type arrester
13. Compare and contrast the advantages and disadvantages of Rod-gap, Horn-gap and multi-gap lightning arrester
14. Why insulation co-ordination is required in a large power system?
15. Explain the significance of Basic Impulse Insulation Level and give the Standard Impulse wave values of Lightning Impulse wave and Switching Impulse wave.
16. Derive the surge voltages and currents of reflected and refracted waves when the line is terminated by (i) Open-circuited line and (ii) Short-circuited line
17. Derive the reflection and refraction coefficients for the following cases
    a. When the O.H. line is connected to a cable
    b. When a cable is connected to the O.H. line
    c. When two O.H. lines are connected through a cable

Exercises

1. A Surge of 30kV is incident on a transmission line having an impedance 300Ω. A transmission line meets the cable having an impedance 600Ω. Determine the surge voltage transmitted into the cable.
2. A Surge of 100kV is incident on the cable having an impedance of 500Ω. The cable meets a transmission line having an impedance of 100Ω. Determine the transmitted voltage, reflected voltage and energy transmitted into transmission line during a period of 10μs.
3. A Surge of 20kV is incident on the cable having an inductance and capacitance of 0.3mH and 0.6μF. A cable meets the transmission line with an inductance and capacitance of 2.2mH and 0.044μF, respectively. Determine the surge voltage transmitted into the overhead line.
4. A Surge of 60kV incident on transmission line having natural impedance of 250Ω arrives at a junction of 350Ω and 450Ω, respectively. Find the surge voltage and currents transmitted into each of the branch line.
5. A rectangular Surge of 30kV, 3μs is incident on transmission line having impedance of 300Ω, terminated by an inductance of 9mH. Determine the reflected voltage and voltage across the inductance.
6. A rectangular Surge of 40kV, 4μs is incident on transmission line having impedance of 400Ω, terminated by a capacitance of 1400pF. Determine the reflected voltage and voltage across the capacitance.
7. A line of surge impedance 200Ω is charged from a battery of constant voltage 150V. The line is 250metres long and is terminated in a resistance 100Ω. Draw the Bewley's lattice diagram and calculate the voltage at time intervals from T to 5T.
8. A surge of 20kV is incident on a O.H. line which is joined to a cable having open circuit at its far end. The ratio of characteristic impedance of line to that of cable is 15. Draw the Bewley's lattice diagram if the wave originates in the line.

9. The ends of two long transmission lines A and B are connected by a cable C, 2km long. Each line has the capacitance of $15pF/m$ and inductance of $3.5 * 10^{-6}H/m$ and cable has the capacitance of $45pF/m$ and inductance of $5.5 * 10^{-7}H/m$. A rectangular surge 15kV travels along line 'A' towards the cable 'C' and line 'B'. What will be the voltage at the junction line 'A' and cable, $30\mu s$ after the initial surge reaches the point.
10. A surge of 300kV is incident on a long OH line of characteristic impedance 600 Ω. It connects to a cable AB of length 1km having a total inductance of 125μH and a total capacitance of 0.25μF. At the far end of the cable, connection is made to a transformer of characteristic impedance 900Ω. Draw the Bewley's lattice diagram at junction A of the cable for 27.95μs after the arrival at this junction of the original surge.

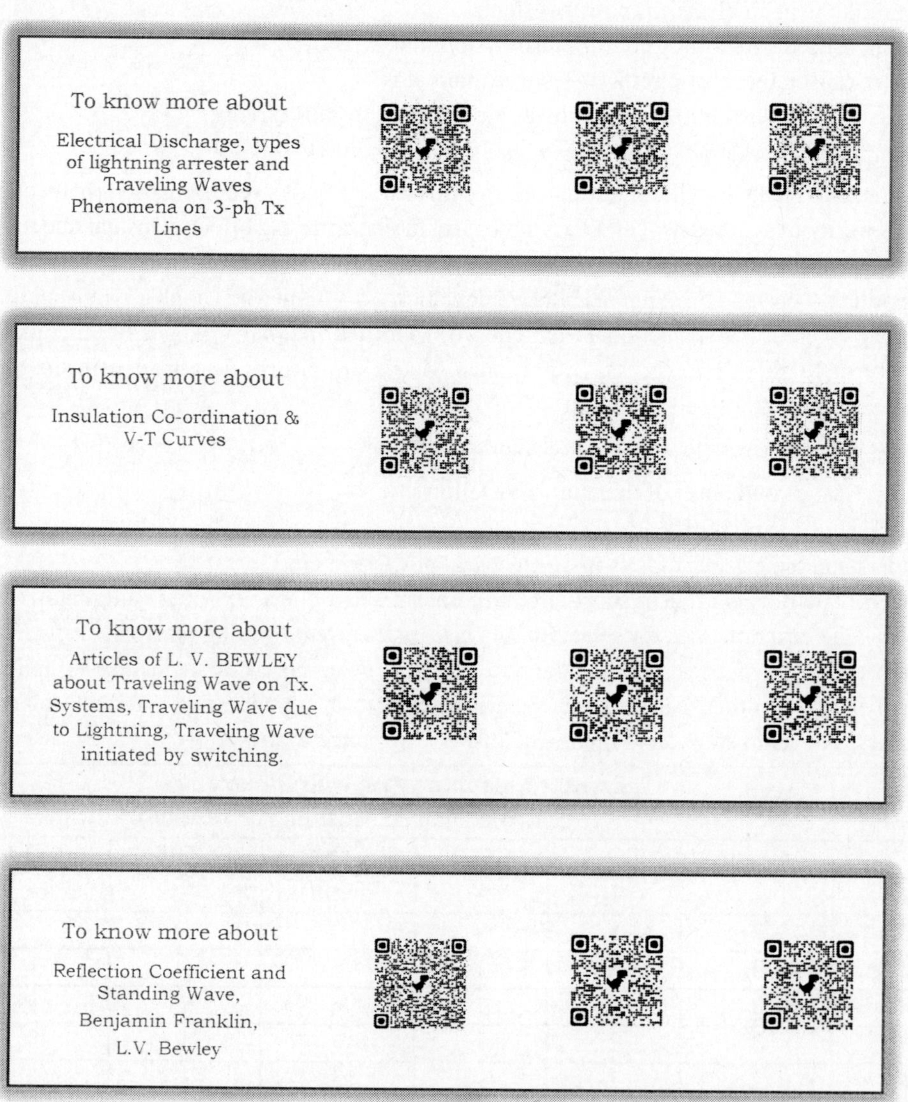

06 SYMMETRICAL FAULTS

Unit specifics:

In this unit, the following topics have been discussed for basic understating of Symmetrical faults:

- Classification of faults in an electrical power system.
- Effect of faults in an electrical power system.
- Transient caused by short-circuit in transmission line.
- Transient caused by short-circuit in 3-phase alternator.
- Analysis of electrical equipment in power system during short-circuit.
- Calculation of symmetrical fault current and short-circuit MVA.

Rationale: In this unit, students will be introduced to symmetrical faults, representation, percentage of occurrence and severity of symmetrical and unsymmetrical faults, doubling effect, transient due to short-circuit in transmission line and 3-phase alternator, per unit values of electrical equipment, selection of a circuit breaker ratings, current limiting reactors. Calculation of symmetrical fault current and short-circuit MVA using (i) Network Reduction Technique, (ii) Modified Internal Voltages of Machines, (iii) Thevenin Equivalent Method, and (iv) Bus Impedance Matrix are clearly described with the help of necessary diagrams, derivations and examples.

Pre-Requisites: Basic knowledge of power system components.

Unit Outcomes: List of outcomes of this unit is as follows

U6-O1: To understand and classify faults.

U6-O2: To understand the effect of faults in an electrical power system.

U6-O3: To analyse the transient due to short-circuit in transmission line and 3-phase alternator.

U6-O4: To follow the best practices for selection of a C.B. and current limiting reactors.

U6-O5: To analyse the symmetrical faults in the power system using Network Reduction Technique.

U6-O6: To analyse the symmetrical faults in the power system using Modified Internal Voltages of Machines, Thevenin equivalent method and Bus Impedance Matrix.

Unit-6 outcomes	Expected mapping with course outcomes (1: Weak, 2: medium, and 3: strong correlation)					
	CO-1	**CO-2**	**CO-3**	**CO-4**	**CO-5**	**CO-6**
U6-O1	2	1	-	2	-	-
U6-O2	2	-	-	2	-	-
U6-O3	2	-	1	2	-	1
U6-O4	2	-	1	3	1	3
U6-O5	2	2	-	3	1	3
U6-O6	2	2	-	3	1	3

6.1 Introduction:

The electrical power system (EPS) is made up of numerous components such as generator, transformer, transmission line, motor, circuit breaker, busbar, feeder, relays, and so on. All electrical components must be able to function normally as well as endure short-term over-currents and over-voltages under fault conditions.

A fault in electrical equipment is defined as a problem in the electrical circuit that causes current to be diverted away from its normal route. The fault in the system reduces the insulation strength between phases or between phase and earth. This reduces the system's effective impedance as the current rises. Transmission lines are more likely to fail or develop faults. Based on the cases of occurrence, reasons of failure can be characterized as

(i). Breakdown may occur at normal voltage,

(ii). Breakdown may occur at abnormal voltage.

Breakdown at normal voltage will occur due to

- Insulation degradation or aging and
- The damages resulting from unforeseen events, such as strong winds, trees falling onto transmission lines, birds causing line shorts, aircraft crashing with lines, cars impacting with poles/towers, and line breakage, are significant contributing factors.

Switching surges or lightning strikes will cause a breakdown at abnormal voltage.

6.2 Classification of Faults in an Electrical Power System:

Faults in an electrical power system are mainly classified as

(i). Symmetrical faults

(ii). Unsymmetrical faults

As the name suggests, in symmetrical faults an equal amount of fault current will flow in all three phases with 120^0 displacement and in unsymmetrical faults different currents flow in the phases depending on type of fault. Symmetrical faults are further classified into two types

a) L-L-L : All three phases experienced a short circuit.

b) L-L-L-G : All three phases are connected to the ground.

L-L-L fault mainly occur due to breakdown of insulation between all the three phases and L-L-L-G fault occur due to the breakdown of insulation between all three phases as well as to ground. Unsymmetrical Faults are further classified into four types

a) L-G : Single phase to ground

b) L-L : Phase to phase

c) L-L-G : Two-phases to ground

d) L-L & L-G : Phase to phase and third phase to ground

L-G faults occur when insulation between one of the phases and ground fails; L-L faults occur when insulation fails between either of the two phases; L-L-G faults occur when insulation fails between either of the two phases and ground; and similarly, L-L & L-G faults occur when insulation fails between two phases and simultaneously between the third phase and ground. Table 6.1 shows the representation, percentage of occurrence, and severity of symmetrical and unsymmetrical faults.

Table 6.1: Representation, percentage of occurrence and severity of symmetrical and unsymmetrical faults

S.No.	Type of S.C. Fault	Representation	Percentage of occurrence	Severity
1	L-G	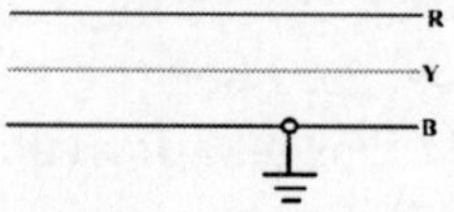 R Y B	70 to 80%	Less severe
2	L-L	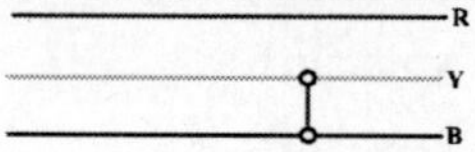 R Y B	10 to 15%	Severe than LG
3	L-L-G	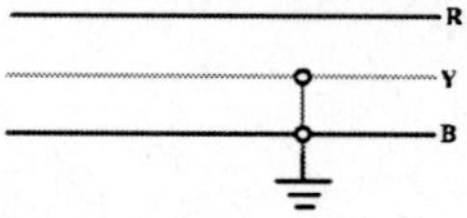 R Y B	5 to 10%	Severe than LL
4	L-L and L-G	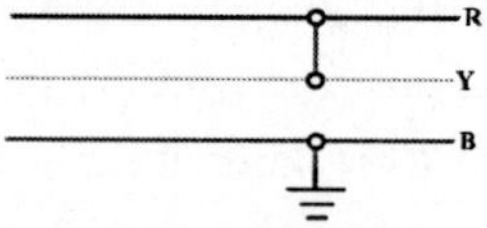 R Y B	1 to 2 %	Severe than LLG
5	L-L-L	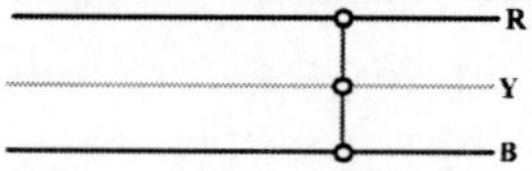 R Y B	2 to 3 %	Most severe
6	L-L-L-G	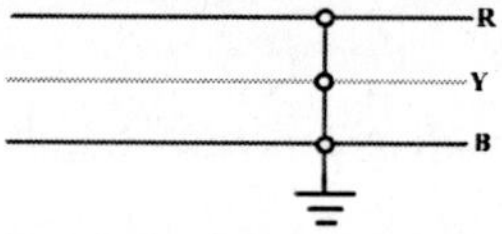 R Y B	2 to 3 %	Most severe

6.3 Effect of Faults in an Electrical Power System:

The effect of faults in an electrical power system may be varied depending upon voltage level, type of relay circuits employed, type of neutral grounding considered, and presence of regulating devices. The flow of heavy short circuit currents incident to the occurrence of interphase short circuits near the generating stations frequently result in substantial disturbance to normal operation of electrical power system. The effect of short circuits in an electrical power system may have any of the following

- Short-circuits generate high currents that lead to excessive heating, perhaps resulting in an explosion.
- If the fault is not promptly resolved, it has the potential to compromise the stability of the power system and result in a full shutdown of the system.
- Unbalanced symmetrical circuits can be introduced by unsymmetrical short-circuit faults.
- When a fault occurs in an interconnected system, it causes a decrease in voltage or frequency. As a consequence, other connected loads, such as motors, which typically draw power from the power supply, begin to feed the faults.

- If a short-circuit occurs, it can result in significant damage to the components of the power system if the fault is not promptly resolved.
- The conductors have the potential to fracture as a result of an excessive fault current within the conductor.
- Excessive overheating can result in harm to other equipment.
- There is a possibility of a disruption in the provision of electricity to end-users.

6.4 Transient due to Short-circuit in Transmission line:

Figure 6.1 illustrates the equivalent circuit of a transmission line. In the scenario where the line is not subjected to any load, the line capacitance is disregarded in a short-circuit state. Consequently, the lines can be expressed as a lumped series RL circuit.

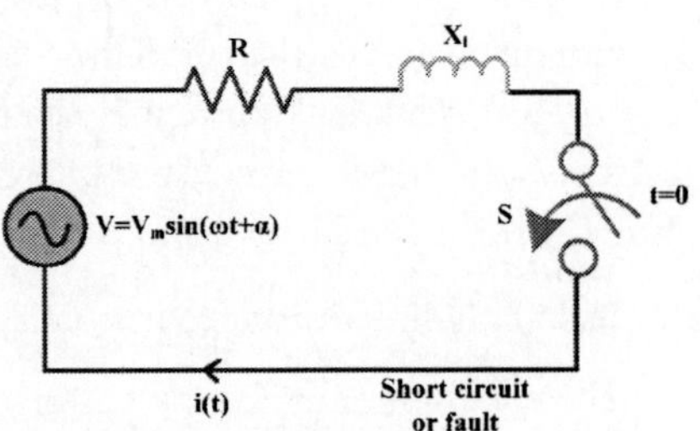

Fig. 6.1 Representation of transmission line under short-circuit Source voltage

$$Z = R + jX_l$$

$$|Z| = \sqrt{(R^2 + X_l^2)} \qquad \text{......... (6.1)}$$

$$V = V_m Sin\ (\omega t + \alpha) \qquad \text{......... (6.2)}$$

Fault may be represented by switch 'S' closed at t=0 in Fig. 6.1.

By Applying KVL $\quad V_m Sin\ (\omega t + \alpha) = R\ i(t) + L\dfrac{di(t)}{dt}$ .... (6.3)

The complete solution of the above equation consists two parts which are called the particular integral and the complementary function.

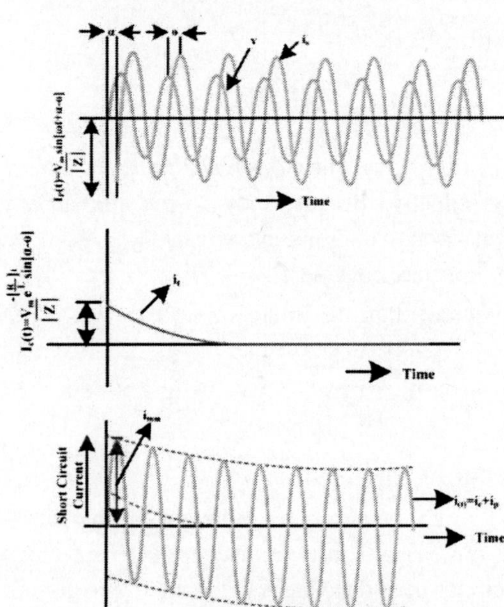

Fig. 6.2 Short-circuit current of a transmission line

The complementary function of eq. (6.3) is $i_c(t) = C * e^{-\left(\frac{R}{L}\right)t}$ .... (6.4)

The particular solution of eq. (6.3) is

$$i_p(t) = \frac{V_m}{\sqrt{(R^2+X_l^2)}}\ Sin\left[\omega t + \alpha - tan^{-\left(\frac{X_l}{R}\right)}\right] \text{... (6.5)}$$

$$i_p(t) = \frac{V_m}{|Z|}\ Sin\ [\omega t + \alpha - \theta] \qquad \text{......... (6.6)}$$

Where $\theta = tan^{-\left(\frac{X_l}{R}\right)}$

The complete solution is $i(t) = i_c(t) + i_p(t)$

$$i(t) = C * e^{-\left(\frac{R}{L}\right)t} + \frac{V_m}{|Z|}\ Sin\ [\omega t + \alpha - \theta]$$

.... (6.7)

At initial conditions, $i.e., at\ t = 0, i(t) = 0$

$$0 = C * e^{-\left(\frac{R}{L}\right)*0} + \frac{V_m}{|Z|}\ Sin\ [\alpha - \theta]$$

$$C = -\frac{V_m}{|Z|}\ Sin\ [\alpha - \theta] \qquad \text{......... (6.8)}$$

Substituting these values in eq. (6.7)

$$i(t) = -\frac{V_m}{|Z|} e^{-\left(\frac{R}{L}\right)t} \, Sin\,[\alpha - \theta] + \frac{V_m}{|Z|} Sin\,[\omega t + \alpha - \theta] \qquad \text{... ... ... (6.9)}$$

$$i(t) = I_{max}\; Sin\,[\omega t + \alpha - \theta] - I_{max}\, e^{-\left(\frac{R}{L}\right)t} \, Sin\,[\alpha - \theta] \qquad \text{... ... ... (6.10)}$$

Where I_{max} is the maximum symmetrical short-circuit current, $I_{max} = \frac{V_m}{|Z|}$

In the foregoing equation (6.9), the first term is the "DC off-set current," that makes the total short-circuit current unsymmetrical until the transient decays. The second term is known as "symmetrical short-circuit current". This fault current is at its peak when t=π/ω, when the fault occurs. Figure 6.2 clearly depicts the short-circuit current in a transmission line. It has sub-figures for symmetrical short-circuit current, DC off-set current, and short-circuit current.

Assuming line resistance is small, $i.e., \theta = 90^0$ *and at* $\omega t = \pi$

The maximum momentary current $i_{mm}(t) = \frac{V_m}{|Z|} cos(\alpha) + \frac{V_m}{|Z|} cos(\alpha) = \frac{2V_m}{|Z|} cos(\alpha)$ ... ... (6.11)

If $\alpha = 0$, *then* the maximum momentary current will be $i_{mm}(t) = 2\frac{V_m}{|Z|}$ ... ... (6.12)

$$i_{mm}(t) = 2I_{max} \qquad \text{... ... ... (6.13)}$$

Therefore, the maximum momentary current $i_{mm}(t)$ is twice the maximum symmetrical short-circuit current, known as "*Doubling Effect*".

6.5 Transient due to Short-circuit in 3-Phase Alternator:

The transients may occur due to

(i) Sudden changes of load,

(ii) Switching,

(iii) Sudden short-circuit (either symmetrical or unsymmetrical faults).

A synchronous generator generates alternating voltage. In case of a synchronous machine, the armature and field windings can be assumed to be almost purely inductive because they do not contain any capacitance and their resistance is almost negligible in comparison to their inductive reactance. A sudden short-circuit occurs on the terminals, the following assumptions are considered:

(i) Synchronous machine is under no load condition before short-circuit.

(ii) Armature resistance is negligible.

(iii) All three-phases are short-circuited.

6.5.1 Current of a 3-Ph Synchronous generator during Short-Circuit:

Steady state, transient and sub-transient equivalent circuits of a synchronous generator during Short-circuit are clearly explained previous unit section 4.6. Symmetrical short-circuit current of a 3-phase synchronous generator was given in Fig. 4.15. The short-circuit period can be divided into three periods - sub-transient period T_d'', transient period T_d' and steady state period T_d and the corresponding reactance of three periods are X_d'', X_d' and X_d respectively.

Base current $I_{base} = \frac{kVA_{base}}{\sqrt{3}*kV_{base}}$ ……… (6.14)

The sustained, transient and sub-transient short-circuit current of a 3-ph synchronous generator are I''_g, I'_g and I_g respectively.

Per unit value of sub-transient short-circuit current $I''_{g\,p.u.} = \frac{E_g}{X''_d}$ ……… (6.15)

Actual value of sub-transient short-circuit current $I''_{g\,actual} = I''_{g\,p.u.} * I_{base}$ ……… (6.16)

Per unit value of transient short-circuit current $I'_{g\,p.u.} = \frac{E_g}{X'_d}$ ……… (6.17)

Actual value of transient short-circuit current $I'_{g\,actual} = I'_{g\,p.u.} * I_{base}$ ……… (6.18)

Per unit value of steady-state short-circuit current $I_{g\,p.u.} = \frac{E_g}{X_d}$ ……… (6.19)

Actual value of steady-state short-circuit current $I_{g\,actual} = I_{g\,p.u.} * I_{base}$ ……… (6.20)

The short-circuit current in a system is constrained by the impedance of the system till the fault location. Figure 6.3 illustrates the occurrence of a failure on the feeder at point F. Subsequently, the short-circuit current originating from the generating station will be constrained by the impedance of the generator, transformer, and the line preceding the location of fault.

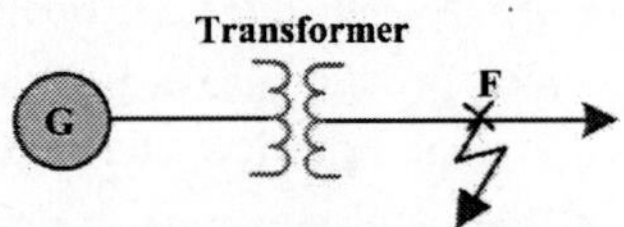

Fig. 6.3. Representation of power system with fault

6.6 Analysis of Electrical equipment in power system during short-circuit:

To determine short-circuit currents, it is required to have knowledge of the impedances of different equipment/components in the system's line.

6.6.1: Per unit reactance values of Electrical Equipment: Per unit reactance values of generator, transformer, transmission line and motor can be calculated as follows:

Per unit reactance value of generator $X^G_{p.u.} = X^G_{actual} * \left(\frac{kV_{actual}}{kV_{base}}\right)^2 * \left(\frac{MVA_{base}}{MVA_{actual}}\right)$ ……… (6.21)

Per unit reactance value of T/f $X^{T/f}_{p.u.} = X^{T/f}_{actual} * \left(\frac{kV_{actual}}{kV_{base}}\right)^2 * \left(\frac{MVA_{base}}{MVA_{actual}}\right)$ ……… (6.22)

Per unit reactance value of Tx. Line $X^{Tx.line}_{p.u.} = \frac{X_{actual}}{X_{base}} = \frac{X_{actual}}{\frac{kV^2_{base}}{MVA_{base}}}$ ……… (6.23)

Base current, base reactance, fault Current and short-circuit MVA can be calculated as follows: Base Current $I_{base} = \frac{kVA_{base}}{\sqrt{3}*kV_{base}}$ ……… (6.24)

Base Impedance $Z_{base} = \frac{kV_{base}}{I_{base}}$ (or) $\frac{kV^2_{base}}{MVA_{base}}$ ……… (6.25)

Per unit fault Current $I^F_{p.u.} = \frac{V^{p.f}_{p.u.}}{X^{p.u.}_{equivalent}}$ ……… (6.26)

where $V^{p.f}_{p.u.}$ is the pre-fault voltage at fault location

Actual fault Current $I^F_{actual} = I^F_{p.u.} * I_{base}$ ……… (6.27)

Short-circuit MVA $MVA_{SC} = \sqrt{3} * I^F_{actual} * kV_{base}$ ……… (6.28)

Short-circuit MVA and fault Current can also be calculated as:

Assuming pre-fault voltage as 1 p.u., Short-circuit MVA is $MVA_{SC} = \frac{MVA_{base}}{X^{p.u.}_{equivalent}}$ ……… (6.29)

Actual fault Current $I^F_{actual} = \frac{MVA_{SC}}{\sqrt{3} * kV_{base}}$ ……… (6.30)

6.6.2 Selection of a circuit breaker:

The circuit breakers are protective devices which are used in electrical power system to isolate healthy section from the faulty section. In normal operating period, these can be used as a switch. The circuit breakers are normally used in places where the power level is very high. They are mainly used in transmission line, substations, generating stations and load side. During short-circuit conditions, it has to interrupt heavy fault currents. Since the power system is predominantly inductive, the interruption of current when the C.B. open its contact is associated with large voltages induced across the contacts, which produce arc across the contacts. So, in C.B. the magnitude of the current to be interrupted is an important criterion. The C.B.s are selected for a particular load, based on the following aspects:

- Breaking capacity of the circuit breaker (Interrupting current and Interrupting MVA)
- Making capacity of the circuit breaker (momentary current rating, initial symmetric current, maximum possible dc component)
- Short-time current rating of the circuit breaker
- Normal operating voltage and speed of the breaker

The initial symmetrical rms current in the C.B. is same as the sub-transient short-circuit current

$$I_{sym} = I''_g \quad \text{……… (6.31)}$$

The maximum possible dc component of the short-circuit current in the circuit breaker is the peak value of the sub-transient short-circuit current $I_{DC} = \sqrt{2} * I_{sym} = \sqrt{2} * I''_g$ ……… (6.32)

The momentary current rating of the breaker is 1.6 times the initial symmetrical rms current in the breaker

$$I_{momentary} = 1.6 * I_{sym} = 1.6 * I''_g \quad \text{……… (6.33)}$$

The current to be interrupted by a C.B. is $I_{interrupt} = M.F. * I_{sym}$ ……… (6.34)

The multiplication factor (M.F.) can be considered based on the position of the interruption done.

If the interruption is at 3 Cycles the M.F. is 1.2,

at 1 Cycle the M.F. is 1.5, at 5 Cycles the M.F. is 1.1, and

at 2 Cycles the M.F. is 1.4, at 8 Cycles the M.F. is 1.

The interrupting kVA will be $kVA^{interrupt}_{sc} = \sqrt{3} * I^{interrupt}_{actual} * kV_{base}$ ……… (6.35)

6.6.3 Reactors:

An increase in the per unit reactance in the power system clearly leads to a reduction in short-circuit currents. The shunt currents are dependent upon the generating capacity, voltage at the fault point, and the equivalent reactance, which refers to the reactance between the generators and the fault point. The circuit breaker (C.B.) should possess a sufficient breaking capacity that exceeds the anticipated fault currents. When the fault exceeds the capacity of the circuit breaker (C.B.), it is possible that the fault current will not be interrupted. In large interconnected systems, the overall rating of generators is quite high, and the p.u. reactance is low, so the fault current is very high. It is possible that sufficient breaking capacity C.B.s are not available. Furthermore, as the system is expanded with the installation of more generating units, the fault current to be halted by the C.B. will increase above the prior value. As a result, by putting reactors into the system, the fault current can be limited. There are various types of reactors in the electrical power system:

(i). Series reactors (current limiting reactors/protective reactors),
(ii). Shunt reactors (reactive power compensation reactors),
(iii). Neutral to earth reactors.

6.6.3.1: Current limiting reactors

The current limiting reactors are connected in series with the line to limit the fault current and hence the fault MVA. It is an inductive coil with a high inductive reactance that is used to limit the short-circuit current that will be interrupted by the circuit breaker.

Current limiting reactors can accomplish the following functions:

- These limit current flow in a short-circuit to prevent against mechanical stress and overheating.
- Voltage disturbances induced by short-circuits can be mitigated.
- In addition, they effectively confine the fault by restricting the influx of current from adjacent intact portions, so preventing the propagation of the fault.

Current limiting reactors must ensure that the reactance is unaffected by saturation when subjected to short-circuit situations. An iron core reactor with practically constant permeability would require a significantly larger cross-sectional area of core if the fault current exceeds around three times the rated full load current. Consequently, the reactor would be both expensive and bulky. Therefore, air-core coils are commonly employed in order to restrict fault currents.

6.6.3.2: Types of Air-cored Current limiting reactors : There are two types of air-cored reactors available

(i). Dry-type/bare/unshielded air-cored reactors,
(ii). Oil-immersed/shielded air-cored reactors.

Dry-type/bare/unshielded Air-cored Current limiting reactors: Typically, these systems are cooled using natural ventilation and occasionally incorporate forced air and heat exchanger auxiliary components.

Advantages of Dry-type/bare/unshielded air-cored reactors:

✓ From a constructional perspective, they exhibit simplicity.

✓ Resilient.

Disadvantages of Dry-type/bare/unshielded air-cored reactors:

× Significant area is necessary.
× Challenges arise when attempting to cool huge coils.
× Their application is restricted to 33kV.
× These coils are not appropriate for outdoor applications.

Oil-immersed/Shielded Current limiting reactors: The insulating and cooling characteristics of these reactors have resemblance to those observed in conventional transformers. In an air-cored structure, it is necessary to incorporate laminated-iron/copper shields around the conductor in order to mitigate the ingress of magnetic flux into the tank walls and mitigate the risk of excessive heating. In an iron-cored construction, the incorporation of air-gaps within the core serves the purpose of preventing saturation and achieving the appropriate magnetizing current.

Advantages of Oil-immersed/Shielded Current limiting reactors:

- ✓ Applicable across all voltage levels.
- ✓ Suitable for indoor or outdoor services.
- ✓ Smaller in size compared to dry-type air-cored reactors.
- ✓ They have high factor of safety against flash-over.
- ✓ They have higher thermal capacity.
- ✓ There is an absence of an external magnetic field that would induce heating in the tank.
- ✓ Air-cored oil-immersed reactor having no iron, has a constant reactance at all currents.

Disadvantages of Oil-immersed/Shielded Current limiting reactors:

- × Iron-cored oil-immersed/shielded current limiting reactors may drop by 10% reactance due to saturation during short-circuits.

6.6.3.3: Selection of Current limiting reactors

The aspects to be considered in the selection of a current limiting reactors are:

- Reactance in ohms or in percentage
- Rated voltage
- Rated through-put MVA
- Normal current rating
- Short-time current rating
- Type of reactor (dry/oil immersed)
- No. of phases (single/three)
- Indoor/Outdoor

6.6.3.4: Location of Current limiting reactors:

There are three possible locations for Current Limiting reactors: (i) in series with generators, (ii) in series with feeders, and (iii) in bus-bars.

Generator Reactors: Generator reactors are connected between generator terminals and bus-bar as shown in Fig. 6.4(a). These reactors provide individual protection for the generators. Contemporary power generators possess adequate transient reactance to safeguard themselves against symmetrical three-phase faults, hence obviating the need for external current limiting generator reactors. Nevertheless, the integration of modern machinery into an old substation may necessitate the incorporation of generator reactors to replace the outdated generators. During typical operating conditions, these reactors have a constant voltage drop and power loss due to the flow of a full load current. One primary drawback associated with this approach is the potential for a short-circuit to occur on one feeder, resulting in a significant decrease in voltage at the generator. Consequently, the synchronous machines linked to this same bus-bar may experience a loss of synchronization.

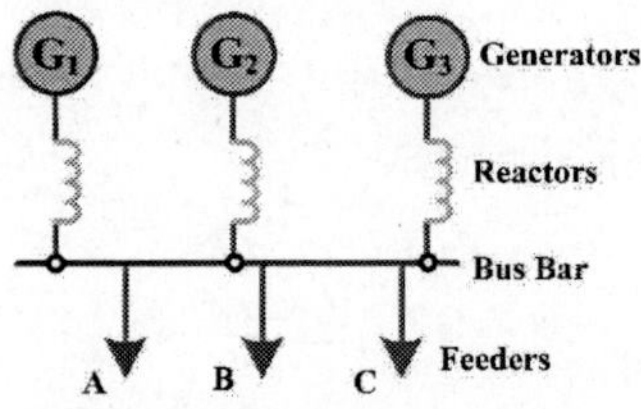

Fig. 6.4 (a) Generator Reactors

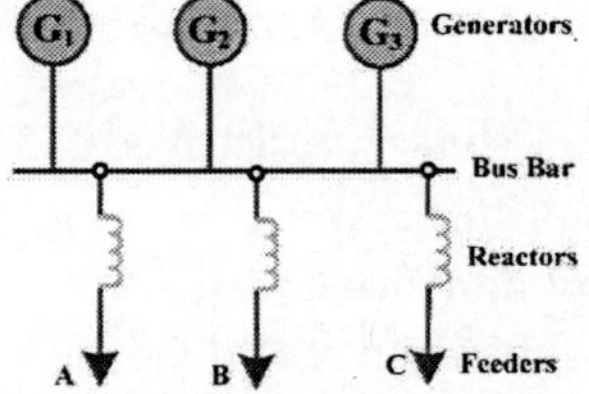

Fig. 6.4 (b) Feeder Reactors

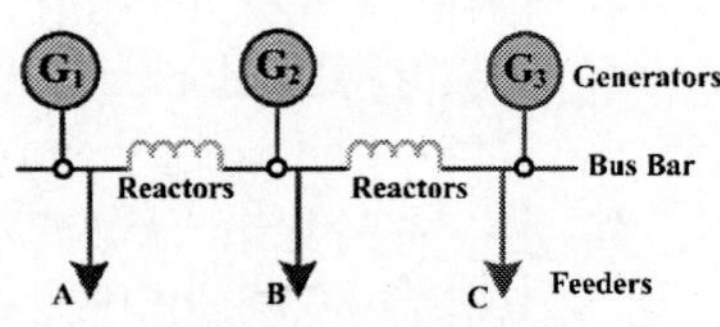

Fig. 6.4 (c) Bus-bar Reactors

Feeder Reactors: As depicted in Figure 6.4(b), the feeder reactors are interconnected in a series configuration with the feeders. If a symmetrical 3-phase fault occurs on a specific feeder, the primary voltage drop is limited to the reactor and has no impact on the bus-bar voltage. As a result, other machines are able to continue supplying the load. One primary drawback associated with situating reactors in this particular location is the absence of protection for generators against S.C. faults. This issue does not provide a challenge, as contemporary generators exhibit significant transient reactance.

Bus-bar Reactors: The substantial voltage drop and power loss in reactors can be decreased by putting them into the bus-bars, as shown in Fig. 6.4(c). Bus-bar reactors are used to connect individual bus sections. In this setup, one feeder is typically fed by a single generator. Under typical operating conditions, a minimal quantity of power travels through the reactors, resulting in modest voltage drop and power loss. In the event of a fault on any one feeder, only one generator feeds the fault, while the current from the other generators is reduced due to the existence of reactors.

6.6.3.5: Drawbacks of Current limiting reactors:

Current-limiting reactors have various drawbacks, such as by the introduction of the reactors, the total p.u. reactance of the system increases, resulting in

- Increased voltage drop.
- Reduced power factor owing to lag angle.
- Poor control.

6.7 Calculation of Symmetrical Fault current and Short-Circuit MVA in power system:

Symmetrical fault current and short-circuit MVA in an electrical power system can be calculated using following techniques/methods:

(i) Network Reduction Technique
(ii) Modified Internal Voltages of Machines
(iii) Thevenin Equivalent Method
(iv) Bus Impedance Matrix.

6.7.1 Symmetrical Fault Analysis using Network Reduction Technique:

Because of the system's balanced nature, every condition that applies to one phase also applies to the other two. Thus, the problem is simplified to a single-phase problem, with a single supply source acting across the equivalent network impedance up to the fault. The equivalent network impedance up to the fault can be achieved through network reduction, which includes series parallel combinations and reactance conversion to star/delta or delta/star. Various steps involved in the short-circuit calculations are given below

- Create a single line diagram of the entire network.
- Identify each component's rating, voltage, resistance, and reactance.
- Choose the base kV and common base MVA.
- Calculate each component's per unit reactance value.
- Draw a single line reactance diagram and record the per unit reactance value.
- Use network reduction to simplify the reactance diagram while preserving fault point identification.
- Determine the system's reactance (Thevenin equivalent reactance).
- Calculate fault current and short-circuit MVA using the calculations provided in section 6.6.

6.7.2 Symmetrical Fault Analysis using Modified Internal Voltages of Machines:

Consider a synchronous generator supplying a three-phase balanced load before symmetrical fault with impedance Z_f occurs at its terminals. Steady-state equivalent circuit of the system is shown in Fig. 6.5(a). Pre-fault condition is represented by open switch status, whereas, closed switch status represents occurrence of fault at the terminals of the machine. As reactance of the machine decreases to X_g'' from its pre-fault steady-state value of X_g , consideration of pre-fault generated voltage E_g will lead to a different terminal voltage.

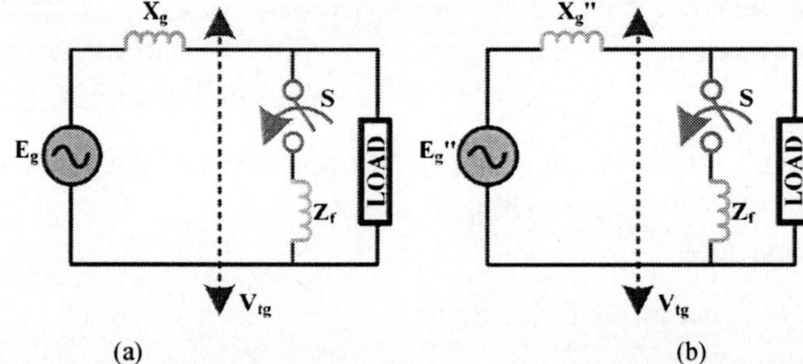

Fig. 6.5 Synchronous generator: (a) Steady-state equivalent circuit (b) Equivalent circuit for fault calculation

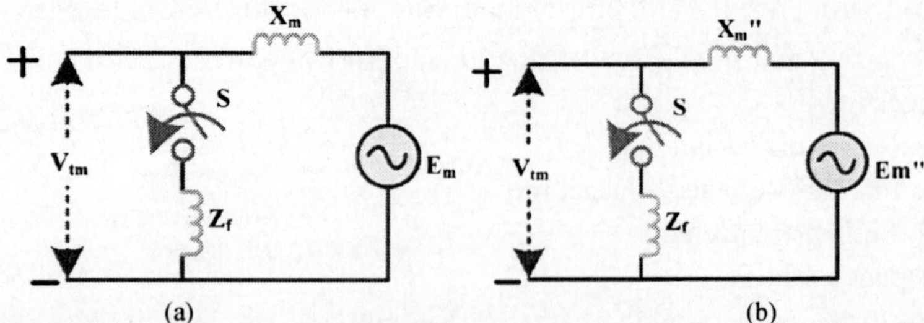

Fig. 6.6 Synchronous motor: (a) Steady-state equivalent circuit (b) Equivalent circuit for fault calculation

However, the terminal voltage of machine for pre-fault condition must be maintained to V_{tg} shown in Fig. 6.5(a) that reduces as a result of fault (becomes zero in the absence of fault impedance). Therefore, generated voltage of machine is modified as a result of change in machine reactance and is calculated as;

$$E_g'' = V_{tg} + jX_g''I_0 \qquad \text{... ... ... (6.36)}$$

Where, I_0 represents pre-fault current through machine terminals.

The modified equivalent circuit meant for symmetrical fault calculations is shown in Fig. 6.5(b). The fault current obtained from Fig. 6.5(b) as;

$$I_g'' = \frac{E_g''}{X_g''+Z_f} \qquad \text{... ... ... (6.37)}$$

Steady state equivalent circuit and equivalent circuit used for fault calculation for a synchronous motor are shown in Fig. 6.6 (a) and 6.6(b), respectively. The internal voltage of synchronous motor is changed from pre-fault value of E_m to E_m'' as per following;

$$E_m'' = V_{tm} - jX_m''I_0 \qquad \text{... ... ... (6.38)}$$

If three phase fault occurs at motor terminals, the motor acts as synchronous generator and supplies the fault current given by;

$$I_m'' = \frac{E_m''}{X_m''+Z_f} \qquad \text{... ... ... (6.39)}$$

In case of multi-machine network internal voltages of all the machines are modified as per above and modified network is analysed using methods available for circuit analysis.

6.7.3 Symmetrical Fault Analysis using Thevenin Equivalent Method

This method obtains Thevenin's equivalent across fault terminals shown in Fig. 6.7 where V_{th} represents pre-fault voltage at the fault terminals and Z_{th} represents input impedance seen from the fault terminals.

The power system network connected across the fault terminals is replaced by a passive network with all sources dead. This network is excited by Thevenin's voltage source V_{th} and currents in different branches and bus voltages are calculated. Computed currents and voltages are superimposed to pre-fault values to determine voltages and currents during fault condition. The fault is represented by closed switch status.

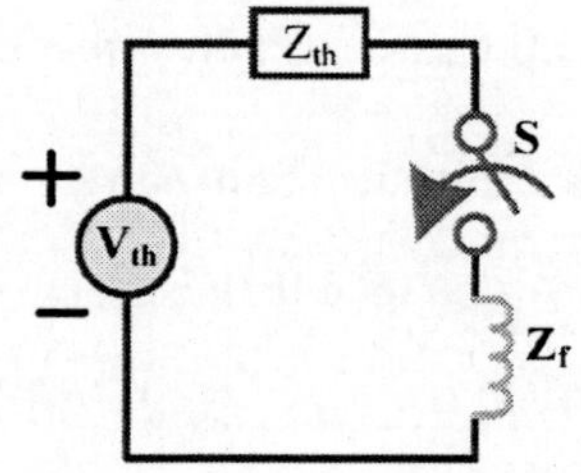

Fig. 6.7 Thevenin's equivalent circuit at fault location

The fault current I_f is obtained from Fig.6.7 as;

$$I_f = \frac{V_{th}}{Z_{th}+Z_f} \qquad \text{... ... ... (6.40)}$$

In this section fault calculations for a two-machine radial network (a synchronous generator supplying to a synchronous motor through a transmission line) are illustrated using (i) modified internal voltages of machines and (ii) by Thevenin equivalent method for six different cases – symmetrical fault occurs at the terminals of motor/generator/half-way of transmission line when the pre-fault voltage is specified at motor/generator terminals.

Case(i): Symmetrical fault occurs at the terminals of motor when the pre-fault terminal voltage is specified at the terminals of the motor itself

In case(i), pre-fault terminal voltage is specified at the motor terminals and symmetrical fault also considered at the terminals of motor, it is clearly given in Fig. 6.8(a) and (b) respectively. Thevenin equivalent circuit for case(i) is shown in Fig. 6.8(c).

Pre-fault condition:

Voltage behind sub-transient reactance of the generator $E_g'' = V_{tm} + jI_0(X_g'' + X_{line})$ .... (6.41)

Voltage behind sub-transient reactance of the motor $E_m'' = V_{tm} - jI_0(X_m'')$ .... (6.42)

Where I_0 is pre-fault output current of the generator

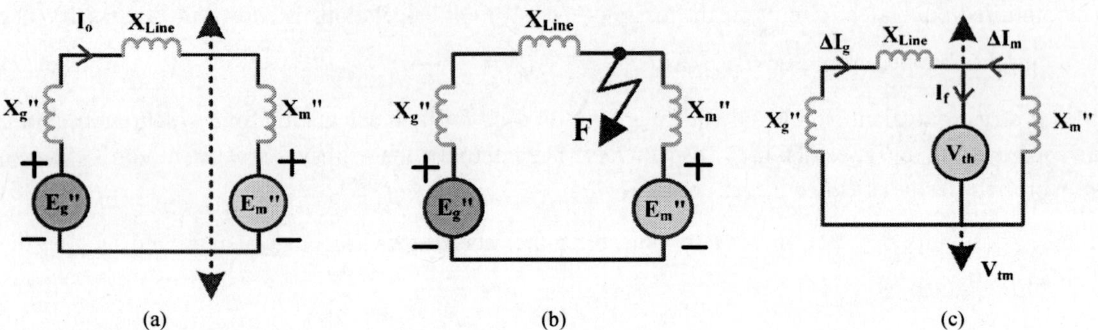

Fig. 6.8. (a) terminal voltage at motor (b) symmetrical fault at the terminals of motor (c) Thevenin equivalent circuit

Fault condition:

Sub-transient fault current of the generator $I_g'' = \frac{E_g''}{X_g''+X_{line}}$ .... (6.43)

Sub-transient fault current of the motor $I_m'' = \frac{E_m''}{X_m''}$ .... (6.44)

Per unit total fault current $I_f^{p.u.} = I_g''+I_m''$ .... (6.45)

Using Thevenin Equivalent method:

Terminal voltage is specified at the terminals of the motor and fault has also occurred at the terminals of the motor itself, so $V_{th} = V_{tm}$ .... (6.46)

Equivalent impedance $Z_{p.u.}^{eq} = (X_{p.u.}^{G} + X_{p.u.}^{line}) \parallel X_{p.u.}^{M}$ .... (6.47)

$$Z_{p.u.}^{eq} = \frac{(X_{p.u.}^{G} + X_{p.u.}^{line}) * X_{p.u.}^{M}}{(X_{p.u.}^{G} + X_{p.u.}^{line} + X_{p.u.}^{M})} \quad \text{......... (6.48)}$$

Total fault current $I_f = \frac{V_{th}}{Z_{p.u.}^{eq}}$ .... (6.49)

Change in generator current due to fault $\Delta I_g = I_f * \frac{X_{p.u.}^{M}}{(X_{p.u.}^{G}+X_{p.u.}^{line}+X_{p.u.}^{M})}$ .... (6.50)

Change in motor current due to fault $\Delta I_m = I_f * \frac{(X_{p.u.}^{G}+X_{p.u.}^{line})}{(X_{p.u.}^{G}+X_{p.u.}^{line}+X_{p.u.}^{M})}$ .... (6.51)

Sub-transient fault current of the generator $I_g''^{(p.u.)} = I_0 + \Delta I_g$ .... (6.52)

Sub-transient fault current of the motor $I_m''^{(p.u.)} = -I_0 + \Delta I_m$ .... (6.53)

Per unit total fault current can also be calculated by adding eq. 6.52 and 6.53.

Case(ii): Symmetrical fault occurs at the terminals of generator when the pre-fault terminal voltage is specified at the terminals of the motor:

In case(ii), pre-fault terminal voltage is specified at the motor terminals and symmetrical fault is considered at the terminals of generator, it is clearly given in Fig. 6.9(a) and (b) respectively. Thevenin equivalent circuit for case(ii) is shown in Fig. 6.9(c).

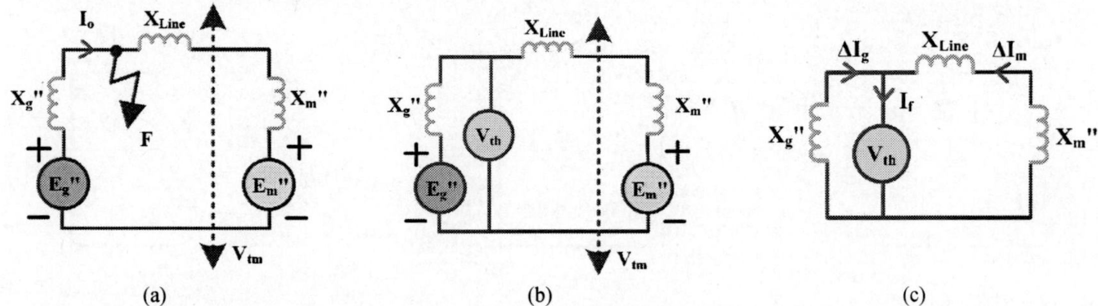

Fig. 6.9. (a) symmetrical fault at the terminals of generator (b) terminal voltage at motor (c) Thevenin equivalent circuit

Pre-fault condition: Voltage behind sub-transient reactance of the generator and the motor will be same as in case(i).

Fault condition:

Sub-transient fault current of the generator $I_g'' = \frac{E_g''}{X_g''}$ ... ... (6.54)

Sub-transient fault current of the motor $I_m'' = \frac{E_m''}{X_{m+X_{line}}''}$ ... ... (6.55)

Total fault current can be calculated using equation 6.50

Using Thevenin Equivalent method:

Terminal voltage is specified at the terminals of the motor and fault occurs at the terminals of the generator, so $V_{th} = V_{tm} + I_0 * X_{p.u.}^{line}$ ... ... (6.56)

Equivalent impedance $Z_{p.u.}^{eq} = X_{p.u.}^{G} \parallel (X_{p.u.}^{M} + X_{p.u.}^{line})$ ... ... (6.57)

$$Z_{p.u.}^{eq} = \frac{X_{p.u.}^{G} * (X_{p.u.}^{M} + X_{p.u.}^{line})}{(X_{p.u.}^{G} + X_{p.u.}^{line} + X_{p.u.}^{M})} \quad \text{... ... ... (6.58)}$$

Total fault current can be calculated using eq. 6.54

Change in generator current due to fault $\Delta I_g = I_f * \frac{(X_{p.u.}^{M} + X_{p.u.}^{line})}{(X_{p.u.}^{G} + X_{p.u.}^{line} + X_{p.u.}^{M})}$ ... ... (6.59)

Change in motor current due to fault $\Delta I_m = I_f * \frac{(X_{p.u.}^{G})}{(X_{p.u.}^{G} + X_{p.u.}^{line} + X_{p.u.}^{M})}$ ... ... (6.60)

To calculate the sub-transient fault current of the generator, motor and per unit total fault current; case(i) formulas can be used.

Case(iii): Symmetrical fault occurs at the half-way point of the transmission line when the pre-fault terminal voltage is specified at the terminals of the motor:

In case(iii), pre-fault terminal voltage is specified at the motor terminals and symmetrical fault is considered at the half-way point of the transmission line, it is clearly given in Fig. 6.10(a) and (b) respectively. Thevenin equivalent circuit for case(iii) is shown in Fig. 6.10(c).

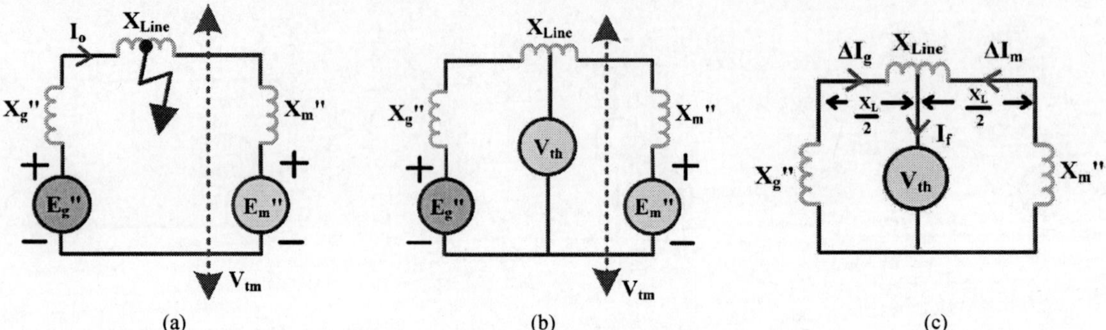

Fig. 6.10. (a) Symmetrical fault at half-way of the transmission line (b) terminal voltage at motor (c) Thevenin equivalent circuit

Pre-fault condition:

Voltage behind sub-transient reactance of the generator and the motor will be same as in case(i).

Fault condition:

Sub-transient fault current of the generator $I_g'' = \frac{E_g''}{X_g'' + \frac{X_{line}}{2}}$ ... ... (6.61)

Sub-transient fault current of the motor $I_m'' = \frac{E_m''}{X_m'' + \frac{X_{line}}{2}}$ ... ... (6.62)

Total fault current can be calculated using equation 6.45

Using Thevenin Equivalent method:

Terminal voltage is specified at the terminals of the motor and fault has also occurred at the half-way of the transmission line so $V_{th} = V_{tm} + I_0 * \frac{X_{line}}{2}$ ... ... (6.63)

Equivalent impedance $Z_{p.u.}^{eq} = \left(X_g'' + \frac{X_{line}}{2}\right) \parallel \left(X_m'' + \frac{X_{line}}{2}\right)$ ... ... (6.64)

$$Z_{p.u.}^{eq} = \frac{\left(X_g'' + \frac{X_{line}}{2}\right) * \left(X_m'' + \frac{X_{line}}{2}\right)}{\left(X_{p.u.}^{G} + X_{p.u.}^{line} + X_{p.u.}^{M}\right)} \quad \text{... ... ... (6.65)}$$

Total Fault current can be calculated using eq. 6.54

Change in generator current due to fault $\Delta I_g = I_f * \frac{\left(X_m'' + \frac{X_{line}}{2}\right)}{\left(X_{p.u.}^{G} + X_{p.u.}^{line} + X_{p.u.}^{M}\right)}$ ... ... (6.66)

Change in motor current due to fault $\Delta I_m = I_f * \frac{\left(X_g'' + \frac{X_{line}}{2}\right)}{\left(X_{p.u.}^{G} + X_{p.u.}^{line} + X_{p.u.}^{M}\right)}$ ... ... (6.67)

To calculate the sub-transient fault current of the generator, motor and per unit total fault current; case(i) formulas can be used.

Case(iv): Symmetrical fault occurs at the terminals of motor when the pre-fault terminal voltage is specified at the terminals of the generator:

In case(iv), pre-fault terminal voltage is specified at the generator terminals and symmetrical fault is considered at the terminals of motor, it is clearly given in Fig. 6.11(a) and (b) respectively. Thevenin equivalent circuit for case(iv) is shown in Fig. 6.11(c).

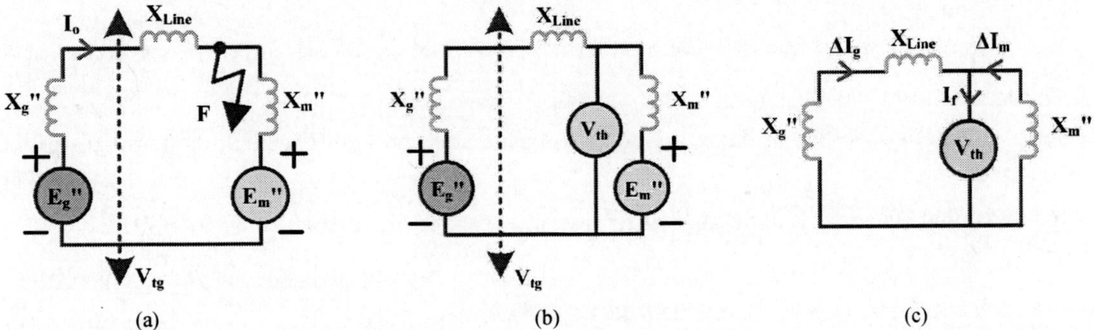

Fig. 6.11. (a) Symmetrical fault at the terminals of motor (b) terminal voltage at the generator (c) Thevenin equivalent circuit

Pre-fault condition: Voltage behind sub-transient reactance of the generator and motor are

$$E_g'' = V_{tg} + jI_0(X_g'') \qquad \text{......... (6.68)}$$

$$E_m'' = V_{tg} - jI_0(X_m'' + X_{line}) \qquad \text{......... (6.69)}$$

Fault condition: To calculate the sub-transient fault current of the generator, motor and Per unit total fault current; case(i) formulas can be used.

Using Thevenin Equivalent method: Terminal voltage is specified at the terminals of the generator and fault occurs at the terminals of the motor, so $V_{th} = V_{tg} - I_0 * X_{p.u.}^{line}$ .... (6.70)

All the remaining equations will be same as that of case(i) except eq. 6.46 replaced by eq. 6.70.

Case(v): Symmetrical fault occurs at the terminals of generator when the pre-fault terminal voltage is specified at the terminals of the generator itself:

In case(v), pre-fault terminal voltage is specified at the generator terminals and symmetrical fault also considered at the terminals of generator, it is clearly given in Fig. 6.12(a) and (b) respectively. Thevenin equivalent circuit for case(v) is shown in Fig. 6.12(c).

Pre-fault condition:

Voltage behind sub-transient reactance of the generator and the motor will be same as in case (iv).

Fault condition:

To calculate the sub-transient fault current of the generator, motor and per unit total fault current; case(ii) formulas can be used.

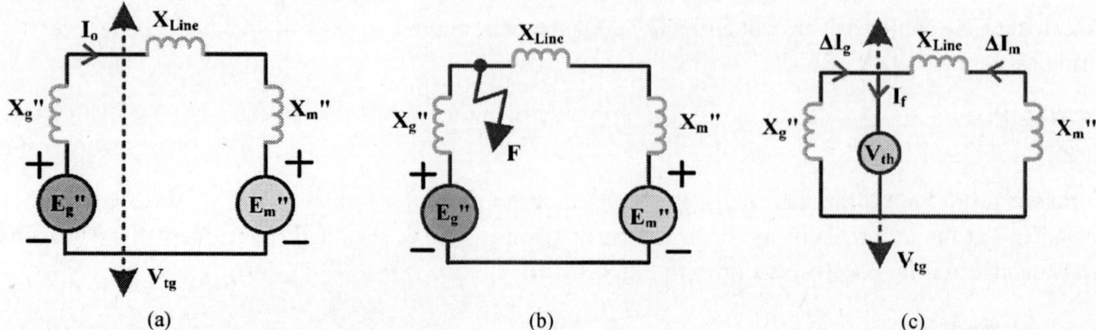

Fig. 6.12. (a) Terminal voltage at generator (b) symmetrical fault at generator terminal and (c) Thevenin equivalent circuit

Using Thevenin Equivalent method:

Terminal voltage is specified at the terminals of the generator and fault occurs at the terminals of the generator so $V_{th} = V_{tg}$ ... ... (6.71)

All the remaining equations will be same as of case(ii) except eq. 6.56 replaced by eq. 6.71.

Case(vi): Symmetrical fault occurs at the half-way point of the transmission line when the pre-fault terminal voltage is specified at the terminals of the generator:

In case(vi), pre-fault terminal voltage is specified at the generator terminals and Symmetrical fault is considered at the half-way point of the transmission line, it is clearly given in Fig. 6.13(a) and (b) respectively. Thevenin equivalent circuit for case(vi) is shown in Fig. 6.13(c).

Pre-fault condition:

Voltage behind sub-transient reactance of the generator and the motor will be same as in case (iv).

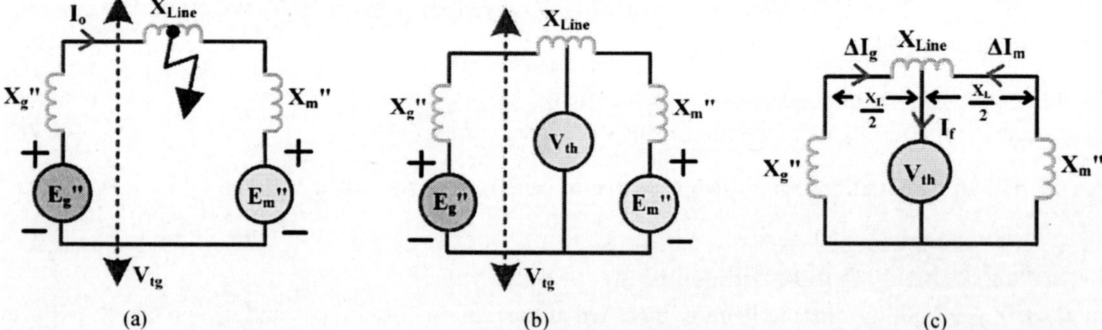

Fig. 6.13. (a) Symmetrical fault at half-way of the tx. line (b) terminal voltage at generator (c) thevenin equivalent circuit

Fault condition: To calculate the sub-transient fault current of the generator, motor and Per unit total fault current; case(iii) formulas can be used.

Using Thevenin Equivalent method:

Terminal voltage is specified at the terminals of the generator and fault is occurred at the half-way of the transmission line, so $V_{th} = V_{tg} - I_0 * \frac{X_{line}}{2}$ ... ... (6.72)

All the remaining equations will be same as of case(iii) except eq. 6.63 replaced by eq. 6.72.

6.7.4 Symmetrical Fault Analysis using Bus Impedance Matrix:

Consider a system with balanced phase impedance in generators, transformers, lines, motors, and balanced voltages in each generator. For short-circuit analysis, the following assumptions are made to simplify computations without much compromise on accuracy in the results. The assumptions are

- All shunt connections to ground are neglected.
- All the transformers are considered to be at their normal tap settings.
- For transmission systems, usually line resistances are much smaller than the line reactance and hence resistances are neglected.
- The generators are represented by a constant voltage source behind the machine sub-transient or transient reactance.

The bus impedance matrix is a useful approach to calculate short-circuit currents and voltages when ground is used as the reference bus. The advantage of this method is that it considers a bus impedance matrix, which can then be used to directly calculate the short-circuit currents and voltages associated with various sorts of defects and fault locations. The pre-fault bus voltages are considered as $V_1^{pf}, V_2^{pf}, \dots\dots\dots\dots V_n^{pf}$ and fault bus voltages are $V_1^{f}, V_2^{f}, \dots\dots\dots\dots V_n^{f}$ respectively.

The bus voltage vector under fault $$V_{bus}^{f} = \begin{bmatrix} V_1^f \\ V_2^f \\ \vdots \\ . \\ . \\ V_n^f \end{bmatrix} \quad \dots\dots\dots (6.73)$$

The Pre-fault bus voltage vector $$V_{bus}^{pf} = \begin{bmatrix} V_1^{pf} \\ V_2^{pf} \\ \vdots \\ . \\ . \\ V_n^{pf} \end{bmatrix} \quad \dots\dots\dots (6.74)$$

Pre-fault voltages at various buses are considered as $V_1^{pf} = V_2^{pf} = \dots\dots\dots V_n^{pf} = 1\angle 0$ .... (6.75)

Let fault occurs at bus-k. Bus voltages changes due to occurrence of fault can be calculated by considering a current source equal to negated fault current at bus-k with all other sources considered dead as per following:

$$I_{bus}^{f} = \begin{bmatrix} 0 \\ . \\ -I_K^f \\ . \\ 0 \end{bmatrix} \quad \dots\dots\dots (6.76)$$

Bus Impedance matrix

$$Z_{bus} = \begin{bmatrix} Z_{11} & Z_{12} & \ldots\ldots\ldots\ldots & Z_{1n} \\ & & \vdots & \\ Z_{k1} & Z_{k2} & \ldots\ldots\ldots\ldots & Z_{kn} \\ & & \vdots & \\ Z_{n1} & Z_{n2} & \ldots\ldots\ldots\ldots & Z_{nn} \end{bmatrix} \quad \ldots\ldots\ldots (6.77)$$

The voltage equation during the fault is $V^{f}_{bus} = V^{pf}_{bus} + \Delta V^{f}_{bus} = V^{pf}_{bus} + Z_{bus} * I^{f}_{bus}$ ……… (6.78)

Where, ΔV^{f}_{bus} represents bus voltage changes under fault given by

$$\begin{bmatrix} \Delta V^{f}_{1} \\ \cdot \\ \cdot \\ \cdot \\ \Delta V^{f}_{k} \\ \cdot \\ \cdot \\ \Delta V^{f}_{n} \end{bmatrix} = \begin{bmatrix} Z_{11} & Z_{12} \ldots Z_{1k} .. Z_{1n} \\ \cdot \\ \cdot \\ \cdot \\ Z_{k1} & Z_{k2} \ldots Z_{kk} .. Z_{kn} \\ \cdot \\ \cdot \\ Z_{n1} & Z_{n2} \ldots Z_{nk} .. Z_{nn} \end{bmatrix} \begin{bmatrix} 0 \\ \cdot \\ \cdot \\ \cdot \\ -I^{f}_{K} \\ \cdot \\ \cdot \\ 0 \end{bmatrix} \quad \ldots\ldots\ldots (6.79)$$

Post-fault voltages are obtained by superimposing pre-fault voltages to voltage changes under fault as per equation 6.78, where pre-fault voltages are given by equation 6.74 and voltage changes under fault are given by equation 6.79. Combining two equations, bus voltages under fault become

$$\begin{aligned} V^{f}_{1} &= V^{pf}_{1} - Z_{1k} * I^{f}_{K} \\ V^{f}_{k} &= V^{pf}_{k} - Z_{kk} * I^{f}_{K} \\ V^{f}_{n} &= V^{pf}_{n} - Z_{nk} * I^{f}_{K} \end{aligned} \quad \ldots\ldots\ldots (6.80)$$

The voltage at fault bus 'k' is also given by $V^{f}_{k} = Z_{f} * I^{f}_{K}$ ……… (6.81)

Where Z_f is the fault impedance.

From Eq. 6.80 & 6.81, the kth bus voltage will be $Z_f * I^{f}_{K} = V^{pf}_{k} - Z_{kk} * I^{f}_{K}$

$$V^{pf}_{k} = I^{f}_{K}(Z_{kk} + Z_f)$$

The fault current at k[th] bus will be $I^{f}_{k} = \frac{V^{pf}_{k}}{Z_{kk}+Z_f}$ ……… (6.82)

Substituting I^{f}_{k} *in eq.* (6.80) $V^{f}_{k} = Z_f * \frac{V^{pf}_{k}}{Z_{kk}+Z_f}$ ……… (6.83)

From equations 6.80 and 6.82 $V^{f}_{i} = V^{pf}_{i} - Z_{ik} * I^{f}_{K} = V^{pf}_{i} - Z_{ik} * \frac{V^{pf}_{k}}{Z_{kk}+Z_f}$ ……… (6.84)

The post Fault current flowing in tx. lines between i and j buses is $I^{f}_{ij} = \frac{V^{f}_{i} - V^{f}_{j}}{Z_{ij}}$ ……… (6.85)

Example 6.1. The sub-transient reactance of a 3-ph, 50MVA, 33kV alternator is measured to be 12%. There is a short-circuit defect at the terminals of a 3-ph system. Determine the fault current and short-circuit MVA. **Ans.** Consider 50 MVA and 33kV as base values.

Base Current $I_{base} = \frac{MVA_{base}}{\sqrt{3}*kV_{base}} = \frac{50*10^6}{\sqrt{3}*33*10^3} = 874\ A$

Per unit fault Current $I^F_{p.u.} = \frac{V_{p.u.}}{X_{p.u.}} = \frac{1}{0.12} = 8.33\ p.u.$

Actual fault Current $I^F_{actual} = I^F_{p.u.} * I_{base} = 8.33 * 874 = 7.28\ kA$

Short-circuit MVA $MVA_{SC} = \sqrt{3} * I^F_{actual} * kV_{base} = \sqrt{3} * 7280 * 33000 = 416\ MVA$

It can also be calculated as follows:

Short-circuit MVA $MVA_{SC} = \frac{MVA_{base}}{X_{p.u.}} = \frac{50*10^6}{0.12} = 416\ MVA$

Actual fault Current $I^F_{actual} = \frac{MVA_{SC}}{\sqrt{3}*kV_{base}} = \frac{416*10^6}{\sqrt{3}*33*10^3} = 7.28\ kA$

Example 6.2. A 3-phase, 10MVA, 11kV alternator with 12% reactance is coupled to a feeder with series impedance (0.12+j0.24)Ω/km. The transformer has a 5MVA rating, an 11kV/33kV voltage range, and a 5% reactance. When a symmetrical fault occurs at a location 20km along the feeder, calculate the fault current delivered by the generator operating at no load and with a voltage of 11.5kV.

Ans. Consider the base values of 10 MVA and 11kV at generator.

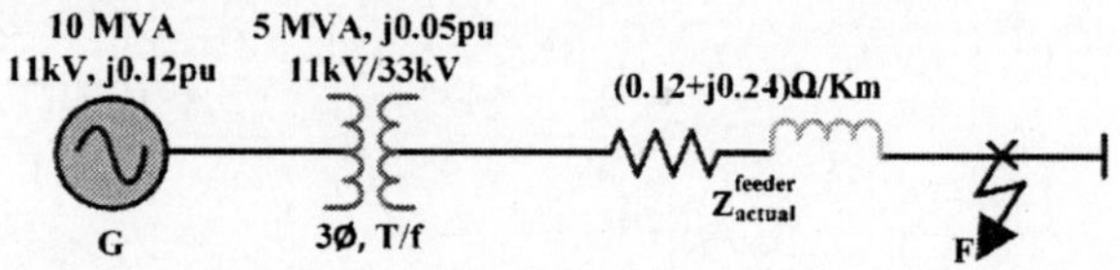

10 MVA
11kV, j0.12pu
G
5 MVA, j0.05pu
11kV/33kV
3Ø, T/f
(0.12+j0.24)Ω/Km
Z^{feeder}_{actual}
F

$$X^G_{p.u.} = j0.12 * \left(\frac{11}{11}\right)^2 * \frac{10}{10} = j0.12\ p.u.$$

$$X^{T/f}_{p.u.} = j0.05 * \left(\frac{11}{11}\right)^2 * \frac{10}{5} = j0.1\ p.u.$$

$$Z^{feeder}_{actual} = 20 * (0.12 + j0.24) = (2.4 + j4.8)\Omega$$

$$Z^{feeder}_{p.u.} = \frac{Z_{actual}}{Z_{base}} = \frac{Z_{actual}}{\frac{kV^2_{base}}{MVA_{base}}} = \frac{(2.4+j4.8)}{(33*10^3)^2} * (10 * 10^6) = (0.022 + j0.044)\ p.u.$$

$$Z^{eq}_{p.u.} = X^G_{p.u.} + X^{T/f}_{p.u.} + Z^{feeder}_{p.u.} = j0.12 + j0.1 + (0.022 + j0.044) = (0.022 + j0.264)\ p.u.$$

$$I_{base} = \frac{MVA_{base}}{\sqrt{3}*kV_{base}} = \frac{10*10^6}{\sqrt{3}*33*10^3} = 174.95A$$

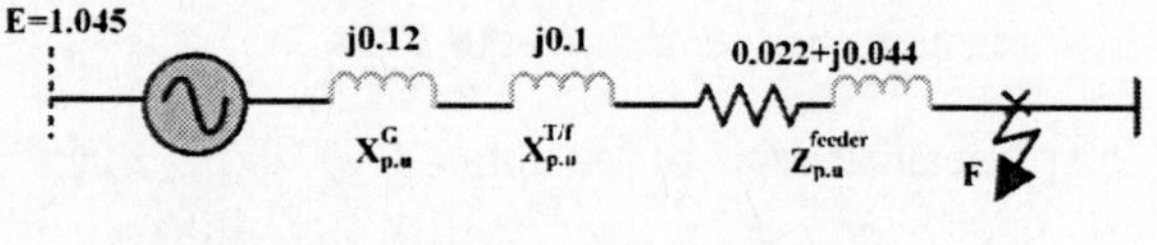

E=1.045
j0.12
$X^G_{p.u}$
j0.1
$X^{T/f}_{p.u}$
0.022+j0.044
$Z^{feeder}_{p.u}$
F

$$I^F_{p.u.} = \frac{V_{p.u.}}{Z^{eq}_{p.u.}} = \frac{\frac{11.5}{11}}{0.022+j0.264}$$

$$= \frac{1.045}{0.022 + j0.264} = 3.94\ p.u.$$

Actual fault Current $I^F_{actual} = I^F_{p.u.} * I_{base} = 3.94 * 174.95 = 689\ A$

Example 6.3. Two generating stations with short-circuit capacities of 2500MVA and 2000MVA operate at 22kV and are joined by a 1.2Ω interconnection cable. Determine each station's new short-circuit capacity. Consider 100MVA to be the base MVA.

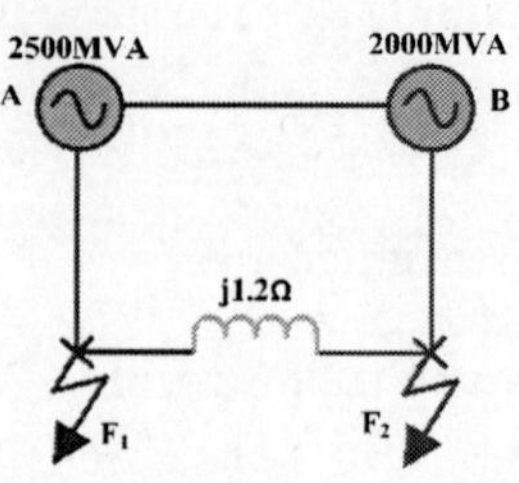

Ans. $X^{A}_{p.u.} = \frac{100}{2500} = j0.04\,p.u.$ and $X^{B}_{p.u.} = \frac{100}{2000} = j0.05\,p.u.$

Per unit reactance value of interconnector

$$X^{I}_{p.u.} = \frac{X_{actual}}{X_{base}} = \frac{X_{actual}}{\frac{kV^2_{base}}{MVA_{base}}} = \frac{j1.2}{(22*10^3)^2} * (100 * 10^6) = j0.2479\,p.u.$$

(i) Fault on Generator A at F_1:

$$X^{F1}_{p.u.} = \frac{j0.04 * (j0.05 + j0.2479)}{(j0.04 + j0.2479 + j0.05)} = j0.0352\,p.u.$$

$$MVA_{SC} = \frac{MVA_{base}}{X^{F1}_{p.u.}} = \frac{100*10^6}{0.0352} = 2840\ MVA$$

(ii) Fault on Generator B at F_2:

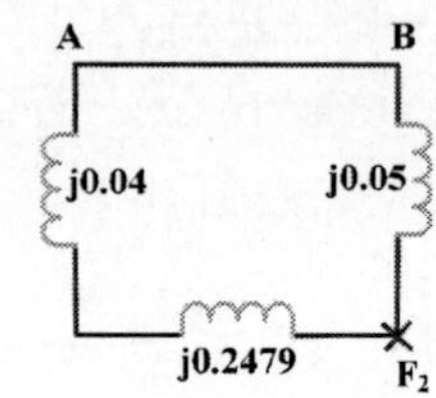

$$X^{F2}_{p.u.} = \frac{(j0.04+j0.2479)*j0.05}{(j0.04+j0.2479+j0.05)} = j0.0426\,p.u.$$

$$MVA_{SC} = \frac{MVA_{base}}{X^{F2}_{p.u.}} = \frac{100 * 10^6}{0.0426} = 2347\ MVA$$

Example 6.4. Generators A and B are identical, rated 11kV, 20MVA, and have a transient reactance of 25% at their respective MVAs. Two transformers have equal ratings of 11kV/66kV, 10MVA, and 15% reactance. The tie-line is 50km long, with each conductor having a reactance of 0.35Ω/km. When the system is in no-load mode but at rated voltage, a symmetrical fault occurs 20 kilometers from one end of the line. Determine the fault MVA and fault current. Consider 20MVA to be the base MVA.

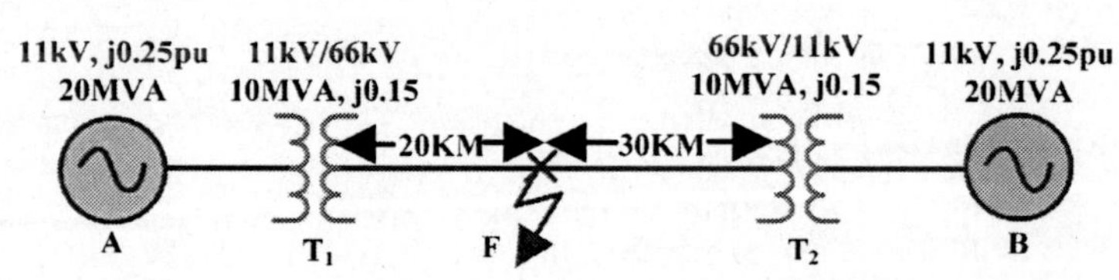

Ans. Consider the base values of 20 MVA and 11kV at generator A.

Per unit reactance value of Generators $\quad X^{A}_{p.u.} = X^{B}_{p.u.} = j0.25 * \left(\frac{11}{11}\right)^2 * \frac{20}{20} = j0.25\,p.u.$

Per unit reactance value of T/f's $\quad X^{T/f1}_{p.u.} = X^{T/f2}_{p.u.} = j0.15 * \left(\frac{11}{11}\right)^2 * \frac{20}{10} = j0.3\,p.u.$

p.u. reactance value of line upto 20km $\quad X^{line1}_{p.u.} = \frac{20* j0.35}{(66*10^3)^2} * (20 * 10^6) = j0.0321\,p.u.$

p.u. reactance value of remaining line 30km $\quad X^{line2}_{p.u.} = \frac{30* j0.35}{(66*10^3)^2} * (20 * 10^6) = j0.0482\,p.u.$

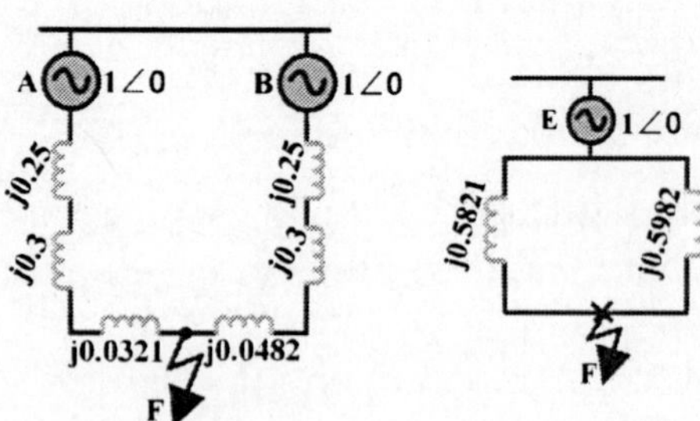

$$X^{eq}_{p.u.} = \frac{j0.5821 * j0.5982}{j0.5821 + j0.5982} = j0.295\,p.u.$$

$$MVA_{SC} = \frac{MVA_{base}}{X^{eq}_{actual}} = \frac{20 * 10^6}{0.295} = 67.79MVA$$

$$I_{SC} = \frac{MVA_{SC}}{\sqrt{3} * kV_{base}} = \frac{67.79 * 10^6}{\sqrt{3} * 66 * 10^3} = 593\ A$$

Example 6.5. Figure below depicts a generating station that provides power to a 110kV grid. Calculate the total fault current, short-circuit MVA, and fault current delivered by each alternator for a three-phase fault at the receiving end. The line is 100 km long.

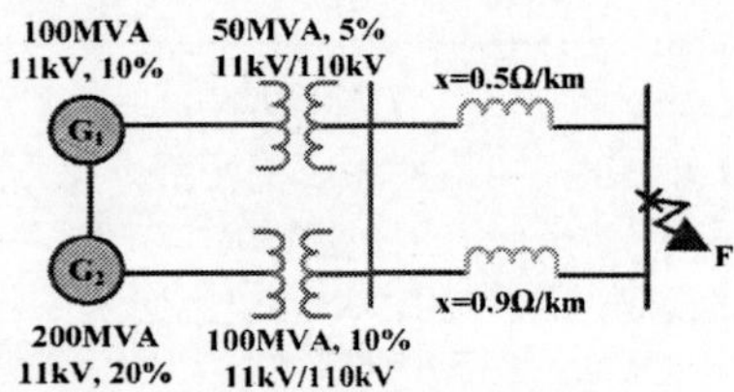

Ans Consider base values100MVA, 11kV at G_1

$X^{G1}_{p.u.} = j0.1 * \left(\frac{11}{11}\right)^2 * \frac{100}{100} = j0.1\, p.u.$ $\quad X^{T/f2}_{p.u.} = j0.1 * \left(\frac{11}{11}\right)^2 * \frac{100}{100} = j0.1\, p.u.$

$X^{G2}_{p.u.} = j0.2 * \left(\frac{11}{11}\right)^2 * \frac{100}{200} = j0.1\, p.u.$ $\quad X^{line1}_{p.u.} = \frac{100 * j0.5}{110^2} * 100 = j0.4132\, p.u.$

$X^{T/f1}_{p.u.} = j0.05 * \left(\frac{11}{11}\right)^2 * \frac{100}{50} = j0.1\, p.u.$ $\quad X^{line2}_{p.u.} = \frac{100 * j0.9}{110^2} * 100 = j0.7438\, p.u.$

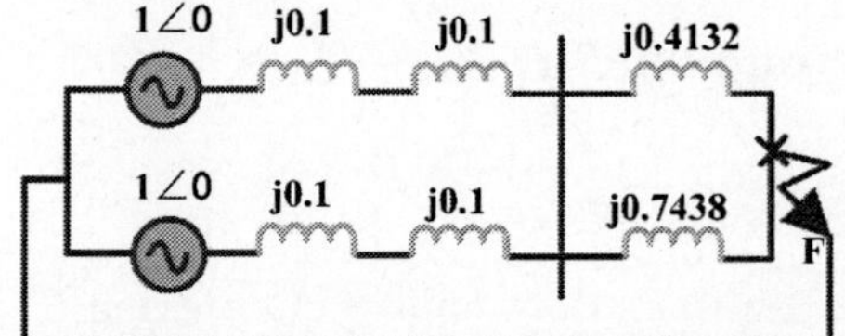

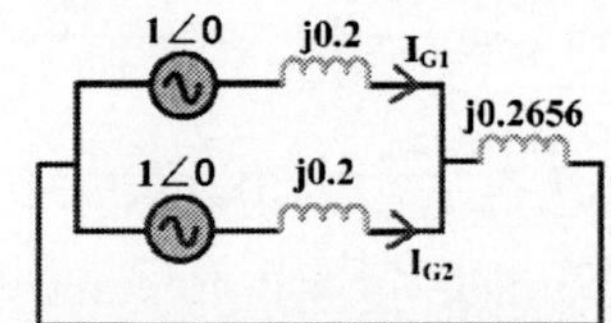

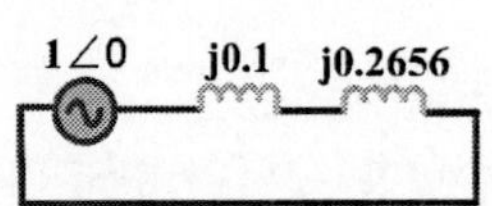

Short-circuit MVA $\quad MVA_{SC} = \frac{MVA_{base}}{X^{eq}_{p.u.}} = \frac{100}{0.3656} = 273.52\, MVA$

Total fault current on feeder side $\quad I^{feeder}_{SC} = \frac{MVA_{SC}}{\sqrt{3} * kV_{base}} = \frac{273 * 10^6}{\sqrt{3} * 110 * 10^3} = 1.432\, kA$

Total fault current supplied by two generators $= I^{feeder}_{SC} * \frac{t}{f} ratio = 1432 * \frac{110}{11} = 14.320\, kA$

Fault current supplied G_1 $= I^{feeder}_{SC} * \frac{X_{G2}}{X_{G2} + X_{G1}} = 14320 * \frac{j0.2}{j0.2 + j0.2} = 7.160\, kA$

Fault current supplied G_2 $= I^{feeder}_{SC} * \frac{X_{G1}}{X_{G1} + X_{G2}} = 14320 * \frac{j0.2}{j0.2 + j0.2} = 7.160\, kA$

Example 6.6. The values stated in th2e figure are per unit reactances with 20MVA and 11kV as the generator circuit's base voltages. Both transformers are rated at 11kV and 110kV. At bus 3, there is a three-phase to ground fault with a fault impedance of 0.9p.u. Determine the fault current.

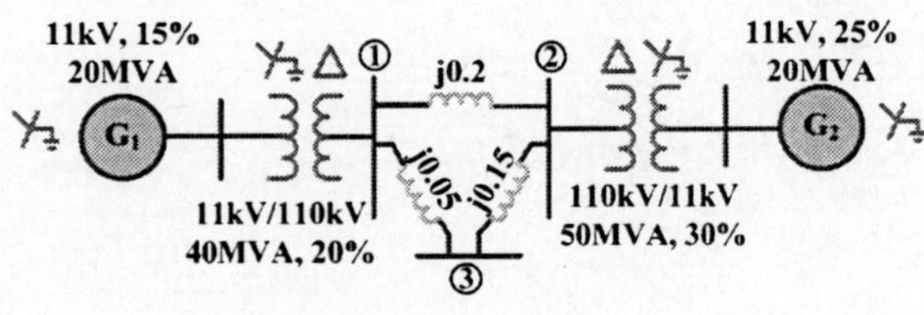

Ans. Consider the base values of 20 MVA and 11kV at G1.

$X^{G1}_{p.u.} = j0.15 * \left(\frac{11}{11}\right)^2 * \frac{20}{20} = j0.15\, p.u.$ $\quad X^{T/f1}_{p.u.} = j0.2 * \left(\frac{11}{11}\right)^2 * \frac{20}{40} = j0.1\, p.u.$

$X^{G2}_{p.u.} = j0.25 * \left(\frac{11}{11}\right)^2 * \frac{20}{20} = j0.25\, p.u.$ $\quad X^{T}_{\frac{f2}{p.u.}} = j0.3 * \left(\frac{110}{110}\right)^2 * \frac{20}{50} = j0.12\, p.u.$

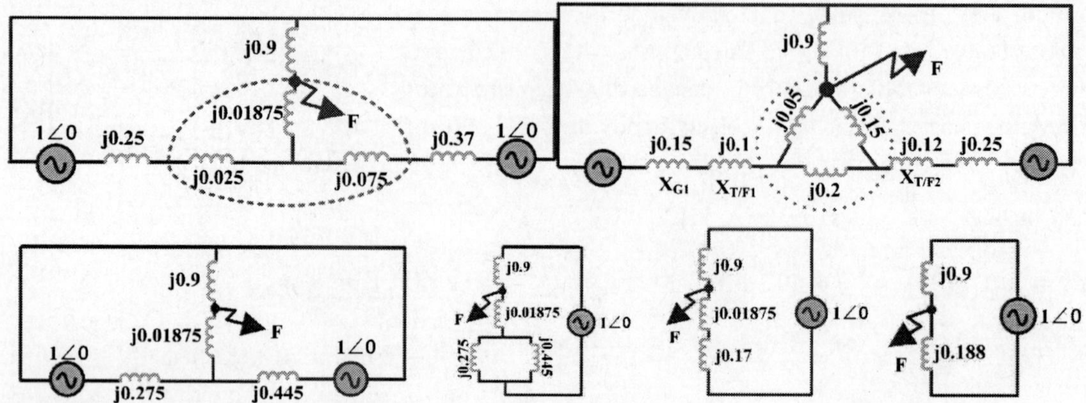

The line reactances of Δ are converted to Y and the network is reduced as per the figures given.

Equivalent impedance is $X^{eq}_{p.u.} = j0.9 + j0.188 = j0.1088\ p.u$

Short-circuit MVA is $MVA_{SC} = \frac{MVA_{base}}{X^{eq}_{p.u.}} = \frac{20*10^6}{1.088} = 18.38\ MVA$

Fault current is $I_{SC} = \frac{MVA_{SC}}{\sqrt{3}*kV_{base}} = \frac{18.38*10^6}{\sqrt{3}*110*10^3} = 96.46\ A$

Example 6.7. Figure depicts a single-line representation of a 3-Ø system. Determine the three-phase short-circuit MVA at point F using the per-unit analysis approach. Consider 50MVA as base MVA.

Ans.

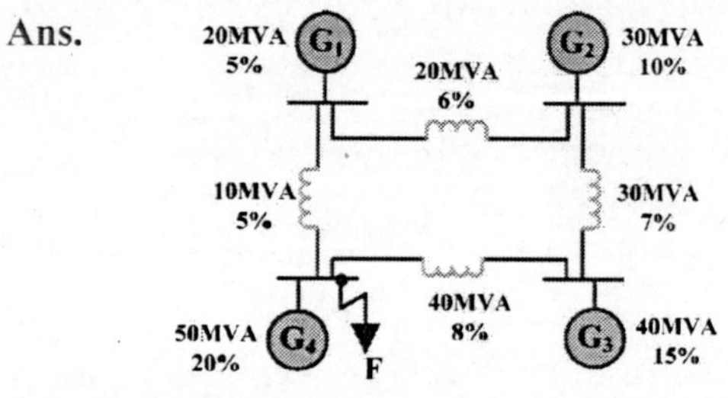

G1 G2 G3 G4 j0.125 j0.166 j0.187 j0.2 j0.15 j0.1166 j0.1 F j0.25

$$X^{G1}_{p.u.} = j0.05 * \frac{50}{20} = j0.125\ p.u. \qquad X^{line12}_{p.u.} = j0.06 * \frac{50}{20} = j0.15\ p.u.$$

$$X^{G2}_{p.u.} = j0.1 * \frac{50}{30} = j0.166\ p.u. \qquad X^{line34}_{p.u.} = j0.08 * \frac{50}{40} = j0.1\ p.u.$$

$$X^{G3}_{p.u.} = j0.15 * \frac{50}{40} = j0.1875\ p.u. \qquad X^{line41}_{p.u.} = j0.05 * \frac{50}{10} = j0.25\ p.u.$$

$$X^{G4}_{p.u.} = j0.2 * \frac{50}{50} = j0.2\ p.u. \qquad X^{line23}_{p.u.} = j0.07 * \frac{50}{30} = j0.116\ p.u.$$

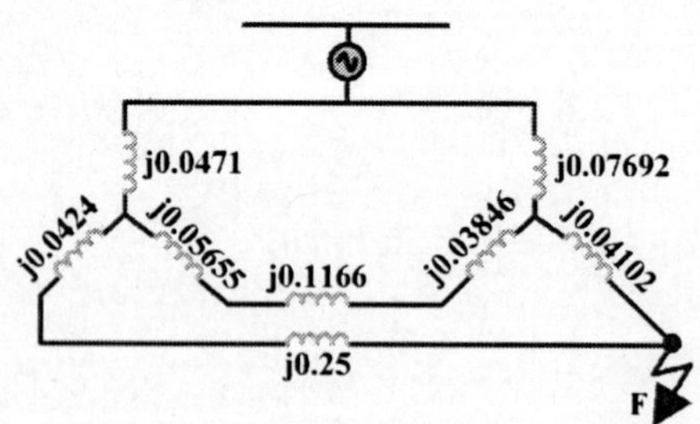

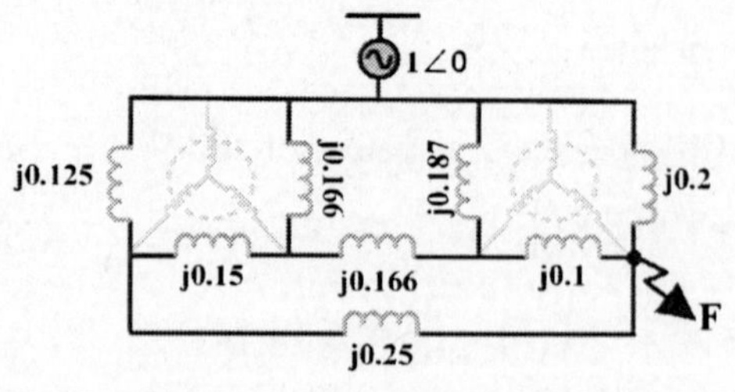

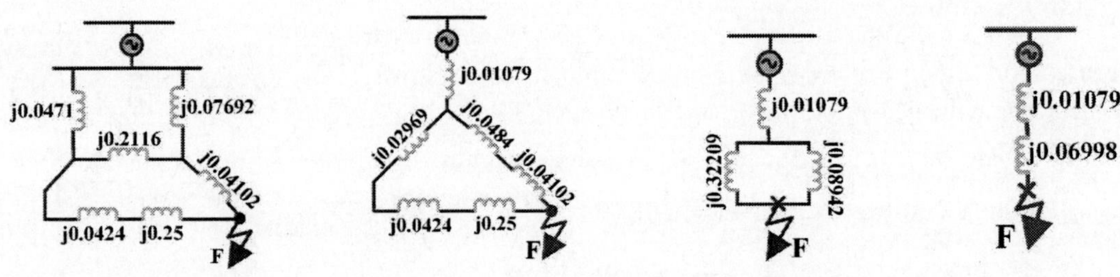

The line reactances of Δ are converted to Y and the network is reduced as per the figures given.

Equivalent impedance $\quad X^{eq}_{p.u.} = j0.01079 + j0.06998 = j0.08077\ p.u$

Short-circuit MVA is $\quad MVA_{SC} = \frac{MVA_{base}}{X^{eq}_{p.u.}} = \frac{50**10^6}{0.08077} = 619.041\ MVA$

Example 6.8. A three-phase fault occurs at F in the radial network depicted in Figure. Determine the fault current of a 11kV bus under fault conditions. Assume 100MVA and 11kV as base values.

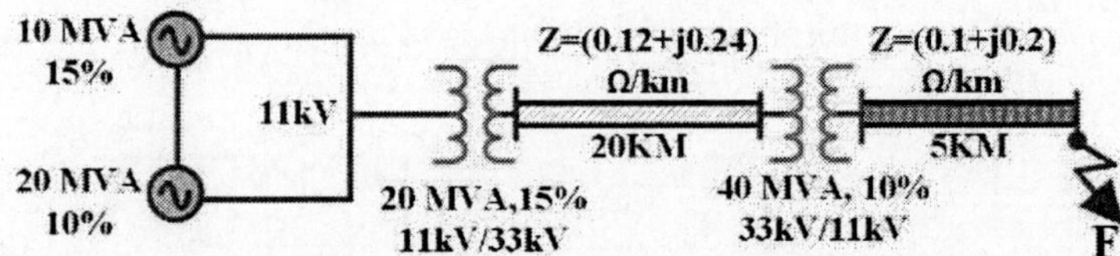

Ans: Consider the base values of 100MVA and 11kV at generator.

$X^{G1}_{p.u.} = j0.15 * \left(\frac{11}{11}\right)^2 * \frac{100}{10} = j1.5\ p.u.$ $\qquad X^{T}_{\substack{f1\\p.u.}} = j0.15 * \left(\frac{11}{11}\right)^2 * \frac{100}{20} = j0.75\ p.u.$

$X^{G2}_{p.u.} = j0.1 * \left(\frac{11}{11}\right)^2 * \frac{100}{20} = j0.5\ p.u.$ $\qquad X^{T/f2}_{p.u.} = j0.1 * \left(\frac{33}{33}\right)^2 * \frac{100}{40} = j0.25\ p.u.$

$Z^{line1}_{p.u.} = \frac{20*(0.12+j0.24)}{33^2} * 100$

$= 0.2203 + j0.4407\ p.u.$

$Z^{line2}_{p.u.} = \frac{5*(0.1+j0.2)}{11^2} * 100$

$= 0.4132 + j0.8264\ p.u.$

1∠0 j1.5
1∠0 j0.5
j0.75 0.2203+ j0.4407 j0.25 0.4132+ j0.8264
F

Equivalent impedance is $Z^{eq}_{p.u.} = (X^{G1}_{p.u.} \parallel X^{G2}_{p.u.}) + X^{T/f1}_{p.u.} + Z^{line1}_{p.u.} + X^{T/f2}_{p.u.} + Z^{line2}_{p.u.}$

$$Z^{eq}_{p.u.} = (j1.5 \parallel j0.5) + j0.75 + (0.2203 + j0.4407) + j0.25 + (0.4132 + j0.8264)$$

$$= 0.6335 + j2.642\ p.u = 2.716\angle 76.51^0\ p.u.$$

Base Current $\quad I_{base} = \frac{kVA_{base}}{\sqrt{3}*kV_{base}} = \frac{100*10^6}{\sqrt{3}*11*10^3} = 5.248\ kA$

Per unit fault Current $\quad I^F_{p.u.} = \frac{V_{p.u.}}{Z^{eq}_{p.u.}} = \frac{1+j0}{2.716} = 0.368\ p.u.$

Actual fault Current $I^F_{actual} = I^F_{p.u.} * I_{base} = 0.368 * 5248 = 1.931\ kA$

Example 6.9. A synchronous generator and motor, each rated at 50MVA, 11kV, and 15% reactance. The line connecting the two has a reactance of 10% on a base of 11 kV, 50 MVA. The motor is receiving

40MW at 0.85 pf leading and a terminal voltage of 10.8kV. a symmetrical three-phase short circuit occurs at the motor's terminals. Determine sub-transient current in the generator, motor, and fault.

Ans. Consider base values of 11kV and 50MVA at the generator.

<u>*Pre-fault:*</u> Pre-fault Voltage $V_{tm} = \frac{10.8}{11} = 0.9818\ p.u.$

Load = 40 MW at 0.85 pf = $\frac{40}{50*0.85} = 0.9411\ p.u.$

Pre-fault current I_0 is

$$I_0 = \frac{0.9411}{0.9818}\angle cos^{-1}0.85 = 0.9585\angle 31.78^0\ p.u.$$

$$E_g'' = V_{tm} + jI_0(X_g'' + X_{line})$$

$$= 0.9818 + 0.9585\angle 31.78^0 * (j0.15 + j0.1)\ 0.879\angle 13.39^0\ p.u.$$

$$E_m'' = V_{tm} - jI_0(X_m'') = 0.9818 - 0.9585\angle 31.78^0 * (j0.15) = 1.064\angle -6.593^0\ p.u.$$

<u>*fault:*</u> A 3-phase fault occurs at the terminals of motor Pre-fault terminal voltage is specified at the terminals of the motor and fault has occurred at the terminals of the motor so $V_{th} = V_{tm}$

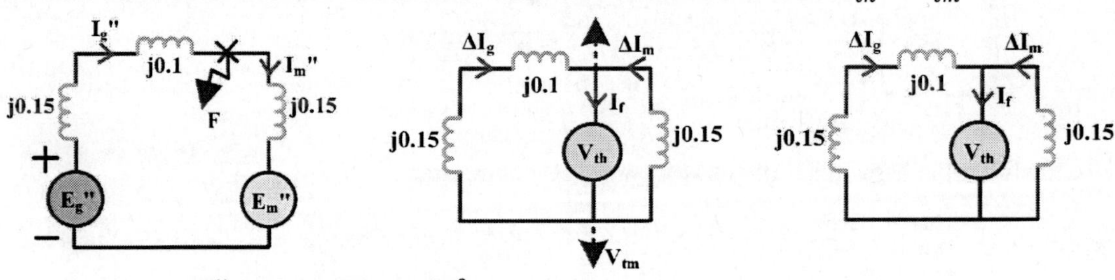

$$I_g'' = \frac{E_g''}{X_g'' + X_{line}} = \frac{0.879\angle 13.39^0}{j0.15 + j0.1} = 0.8148 - j3.42 = 3.515\angle -76.59^0\ p.u.$$

$$I_m'' = \frac{E_m''}{X_m''} = \frac{1.064\angle -6.593^0}{j0.15} = -0.8144 - j7.046 = 7.093\angle -96.59^0\ p.u.$$

$$I_f^{p.u.} = I_g'' + I_m'' = 3.515\angle -76.59^0 + 7.093\angle -96.59^0 = 10.41\angle -89.96^0\ p.u.$$

$$I_{base} = \frac{kVA_{base}}{\sqrt{3}*kV_{base}} = \frac{50*10^6}{\sqrt{3}*11*10^3} = 2.624\ kA$$

$$I_g''^{(actual)} = I_g''^{(p.u.)} * I_{base} = 3.515\angle -76.59^0 * 2624 = 9.223\angle -76.59^0\ kA$$

$$I_m''^{(actual)} = I_m''^{(p.u.)} * I_{base} = 7.093\angle -96.59^0 * 2624 = 18.612\angle -96.59^0\ kA$$

$$I_f''^{(actual)} = I_f''^{(p.u.)} * I_{base} = 10.41\angle -89.96^0 * 2624 = 27.315\angle -89.96^0\ kA$$

<u>*By using Thevenin Equivalent method:*</u>

$$Z_{p.u.}^{eq} = (X_{p.u.}^{G} + X_{p.u.}^{line}) \parallel X_{p.u.}^{M} = (j0.15 + j0.1) \parallel j0.15 = j0.09375\ p.u.$$

$$I_f = \frac{V_{th}}{Z_{p.u.}^{eq}} = \frac{0.9818}{J0.09375} = -j10.47\ p.u.$$

Change in generator current due to fault $\Delta I_g = I_f * \frac{j0.15}{j0.15+j0.1+j0.15} = -j3.92\ p.u.$

Change in motor current due to fault $\Delta I_m = I_f * \frac{j0.15+j0.1}{j0.15+j0.1+j0.15} = -j6.54\ p.u.$

$$I_g''^{(p.u.)} = I_0 + \Delta I_g = 0.9585\angle 31.78^0 + (-j3.92) = 3.511\angle -76.58^0\ p.u.$$

$$I_m''^{(p.u.)} = -I_0 + \Delta I_m = -0.9585\angle 31.78^0 + (-j6.54) = 7.091\angle -96.59^0\ p.u.$$

$$I_f^{p.u.} = I_g'' + I_m'' = 3.511\angle -76.58^0 + 7.091\angle -96.59^0 = 10.45\angle -89.99^0\ p.u.$$

$$I_g''^{(actual)} = I_g''^{(p.u.)} * I_{base} = 3.511\angle -76.58^0 * 2624 = 9.212\angle -76.58^0\ kA$$

$$I_m''^{(actual)} = I_m''^{(p.u.)} * I_{base} = 7.091\angle -96.59^0 * 2624 = 18.606\angle -96.59^0\ kA$$

$$I_f''^{(actual)} = I_f''^{(p.u.)} * I_{base} = 10.45\angle -89.99^0 * 2624 = 27.423\angle -89.99^0\ kA$$

Example 6.10. A generator is connected to a synchronous motor via a transformer. The sub-transient reactances of the generator and motor are 0.15 and 0.35 p.u., respectively. The leakage reactance of the transformer is j0.1 p.u., and all reactances are determined using a common base value. When the generator's terminal voltage is 0.9p.u., a three-phase fault develops at the motor terminals. The generator's pre-fault output current is 0.98 p.u. with 0.8 pf leading. Find the sub-transient fault current. Assume 50MVA and 11kV as base values.

Ans. Pre-fault terminal voltage of generator is $V_{tg} = 0.9\ p.u.$

The pre-fault output current of the generator is $I_0 = 0.98\angle 36.86^0\ p.u.$

<u>*Pre-fault*</u>: $E_g'' = V_{tg} + jI_0(X_g'') = 0.9 + 0.98\angle 36.86^0 * (j0.15) = 0.8202\angle 8.24^0\ p.u.$

$$E_m'' = V_{tg} - jI_0(X_{T/f} + X_m'') = 0.9 - 0.98\angle 36.86^0 * (j0.1 + j0.35) = 1.21\angle -16.85^0\ p.u.$$

<u>*fault:*</u> A 3-phase fault occurs at the terminals of motor

$$I_g'' = \frac{E_g''}{X_g'' + X_{T/f}} = \frac{0.8202\angle 8.24^0}{j0.15 + j0.1} = 3.28\angle -81.76^0\ p.u.$$

$$I_m'' = \frac{E_m''}{X_m''} = \frac{1.21\angle -16.85^0}{j0.35} = 3.457\angle -106.85^0\ p.u.$$

$$I_f^{p.u.} = I_g'' + I_m'' = 3.28\angle -81.76^0 + 3.457\angle -106.85^0 = 6.576\angle -94.63^0\ p.u.$$

Base Current $I_{base} = \frac{kVA_{base}}{\sqrt{3}*kV_{base}} = \frac{50*10^6}{\sqrt{3}*11*10^3} = 2624\ A$

$$I_g''^{(actual)} = I_g''^{(p.u.)} * I_{base} = 3.28\angle -81.76^0 * 2624 = 8.606\angle -81.76^0\ kA$$

$$I_m''^{(actual)} = I_m''^{(p.u.)} * I_{base} = 3.457\angle -106.85^0 * 2624 = 9.071\angle -106.85^0\ kA$$

$$I_f''^{(actual)} = 6.576\angle -94.63^0 * 2624 = 17.255\angle -94.63^0\ kA$$

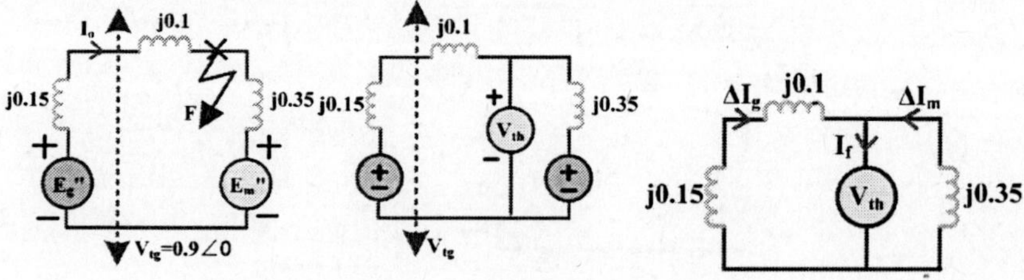

By using Thevenin Equivalent method:

Pre-fault terminal voltage is specified at the terminals of the generator and fault has occurred at the terminals of the motor so

$$V_{th} = V_{tg} - I_0(j0.1) = 0.9 - 0.98\angle 36.86^0 * j0.1 = 0.9619\angle -4.675^0\ p.u.$$

$$Z_{p.u.}^{th} = (X_{p.u.}^{G} + X_{p.u.}^{T/f}) \parallel (X_{p.u.}^{M}) = (j0.15 + j0.1) \parallel j0.35 = j0.1458\ p.u.$$

$$I_f = \frac{V_{th}}{Z_{p.u.}^{th}} = \frac{0.9619\angle -4.675^0}{j0.1458} = 6.597\angle -94.67^0\ p.u.$$

Change in generator current due to fault ΔI_g is $\Delta I_g = I_f * \frac{j0.35}{j0.35+j0.1+j0.15}$

$$= 6.597\angle -94.67^0 * \frac{j0.35}{j0.6} = 3.848\angle -94.67^0\ p.u.$$

Change in motor current due to fault ΔI_m is $\Delta I_m = I_f * \frac{j0.15+j0.1}{j0.15+j0.1+j0.35}$

$$= 6.597\angle -94.67^0 * \frac{j0.25}{j0.6} = 2.748\angle -94.67^0\ p.u.$$

$$I_g''^{(p.u.)} = I_0 + \Delta I_g = 0.9585\angle 31.78^0 + 3.848\angle -94.67^0 = 3.28\angle -81.76^0\ p.u.$$

$$I_m''^{(p.u.)} = -I_0 + \Delta I_m = -0.9585\angle 31.78^0 + 2.742\angle -94.67^0 = 3.476\angle -106.865^0\ p.u.$$

$$I_f^{p.u.} = I_g'' + I_m'' = 3.28\angle -81.76^0 + 3.476\angle -106.865^0 = 6.594\angle -94.68^0\ p.u.$$

$$I_g''^{(actual)} = I_g''^{(p.u.)} * I_{base} = 3.28\angle -81.76^0 * 2624 = 8.606\angle -81.76^0\ kA$$

$$I_m''^{(actual)} = I_m''^{(p.u.)} * I_{base} = 3.476\angle -106.865^0 * 2624 = 9.121\angle -106.865^0\ kA$$

$$I_f''^{(actual)} = I_f''^{(p.u.)} * I_{base} = 6.594\angle -94.68^0 * 2624 = 17.302\angle -94.68^0\ kA$$

Example 6.11. A generator is linked to a synchronous motor via a transmission line. The sub-transient reactances of generator and motor are j0.2 and j0.4 p.u., respectively. The reactance of the transmission line is j0.3 p.u.; all reactances are calculated using the same base value. The generator's output current is 0.96 p.u. at 0.8 pf leading. When the generator pre-fault terminal voltage is 0.9p.u., a three-phase fault develops halfway down the transmission line. Consider the following base values: 50MVA and 11kV. Determine the sub-transient current in the generator, motor, and fault.

Ans. Pre-fault terminal voltage of generator is $V_{tg} = 0.9\ p.u.$

The pre-fault output current of the generator is $I_0 = 0.96\angle 36.86^0\ p.u.$

Pre-fault: $E_g'' = V_{tg} + jI_0(X_g'') = 0.9 + 0.96\angle 36.86^0 * (j0.2) = 0.7997\angle 11.07^0\ p.u.$

$$E_m'' = V_{tg} - jI_0(X_{line} + X_m'') = 0.9 - 0.96\angle 36.86^0 * (j0.3 + j0.4) = 1.409\angle -22.42^0 p.u.$$

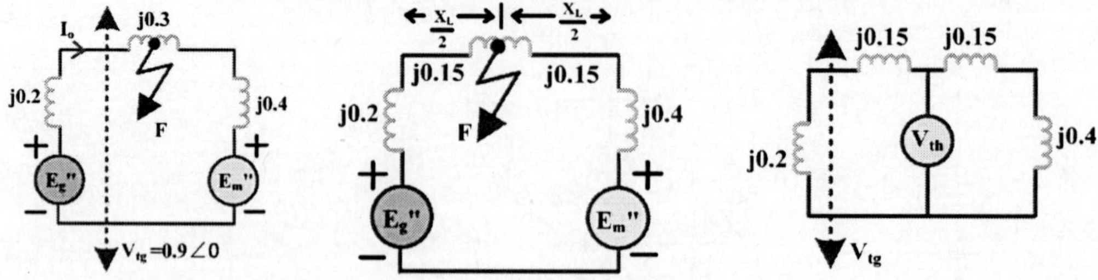

<u>fault:</u> A 3-phase fault occurs at the half-way of the transmission line

Base Current $I_{base} = \frac{kVA_{base}}{\sqrt{3}*kV_{base}} = \frac{50*10^6}{\sqrt{3}*11*10^3} = 2.624\ kA$

$$I_g'' = \frac{E_g''}{X_g'' + \frac{X_{line}}{2}} = \frac{0.7997\angle 11.07^0}{j0.2 + j0.15} = 2.284\angle -78.81^0\ p.u.$$

$$I_m'' = \frac{E_m''}{X_m'' + X_{line}/2} = \frac{1.4096\angle -22.42^0}{j0.4 + j0.15} = 2.562\angle -112.42^0\ p.u.$$

$$I_f^{p.u.} = I_g'' + I_m'' = 2.284\angle -78.81^0 + 2.562\angle -112.42^0 = 4.64\angle -96.65^0\ p.u.$$

$$I_g''^{(actual)} = I_g''^{(p.u.)} * I_{base} = 2.284\angle -78.81^0 * 2624 = 5.993\angle -78.81^0\ kA$$

$$I_m''^{(actual)} = I_m''^{(p.u.)} * I_{base} = 2.562\angle -112.42^0 * 2624 = 6.722\angle -112.42^0\ kA$$

$$I_f''^{(actual)} = I_f''^{(p.u.)} * I_{base} = 4.64\angle -96.65^0 * 2624 = 12.17\angle -96.65^0\ kA$$

<u>By using Thevenin Equivalent method</u>:

Pre-fault terminal voltage is specified at the terminals of the generator and fault occurs in the middle of the transmission line.

$$V_{th} = V_{tg} - I_0(j0.15) = 0.9 - 0.96\angle 36.86^0 * j0.15 = 0.9930\angle -6.66^0\ p.u.$$

$$Z_{p.u.}^{th} = (X_{p.u.}^{G} + X_{p.u.}^{line}/2) \parallel (X_{p.u.}^{M} + X_{p.u.}^{line}/2)$$

$$= (j0.2 + j0.15) \parallel (j0.4 + j0.15) = j0.2138\ p.u.$$

$$I_f = \frac{V_{th}}{Z_{p.u.}^{th}} = \frac{0.9930\angle -6.66^0}{j0.2138} = 4.644\angle -96.66^0\ p.u.$$

Change in generator current due to fault is $\Delta I_g = I_f * \frac{(X_{p.u.}^{M} + X_{p.u.}^{line}/2)}{(X_{p.u.}^{G} + X_{p.u.}^{line} + X_{p.u.}^{M})}$

$$= 4.644\angle -96.66^0 * \frac{(j0.4 + j0.15)}{(j0.2 + j0.3 + j0.4)} = 2.838\angle -96.66^0\ p.u.$$

Change in motor current due to fault is $\Delta I_m = I_f * \frac{(X_{p.u.}^{G} + X_{p.u.}^{line}/2)}{(X_{p.u.}^{G} + X_{p.u.}^{line} + X_{p.u.}^{M})}$

$$= 4.644\angle -96.66^0 * \frac{(j0.2 + j0.15)}{(j0.2 + j0.3 + j0.4)} = 1.806\angle -96.66^0\ p.u.$$

$$I_g''^{(p.u.)} = I_0 + \Delta I_g = 0.96\angle 36.86^0 + 2.838\angle -96.66^0 = 2.285\angle -78.92^0\ p.u.$$

$$I_m''^{(p.u.)} = -I_0 + \Delta I_m = -0.96\angle 36.86^0 + 1.806\angle -96.66^0 = 2.563\angle -112.41^0\ p.u.$$

$$I_f^{p.u.} = I_g'' + I_m'' = 2.285\angle -78.92^0 + 2.563\angle -112.41^0 = 4.643\angle -96.65^0\ p.u.$$

$$I_g''^{(actual)} = I_g''^{(p.u.)} * I_{base} = 2.285\angle -78.92^0 * 2624 = 5.995\angle -78.92^0\ kA$$

$$I_m''^{(actual)} = I_m''^{(p.u.)} * I_{base} = 2.563\angle -112.41^0 * 2624 = 6.725\angle -112.41^0\ kA$$

$$I_f''^{(actual)} = I_f''^{(p.u.)} * I_{base} = 4.643\angle -96.65^0 * 2624 = 12.183\angle -96.65^0\ kA$$

Example 6.12. A generating station has three alternators: 3MVA, 5MVA, and 7MVA, with reactances of j0.06, j0.04, and j0.02 p.u. The circuit breaker shown in the figure is rated at 100 MVA. The system will be extended by a grid supply via fourth generator with a rating of 10MVA and an 8% reactance. Determine the reactance required to safeguard the switchgear when the busbar voltage is 6.6kV.

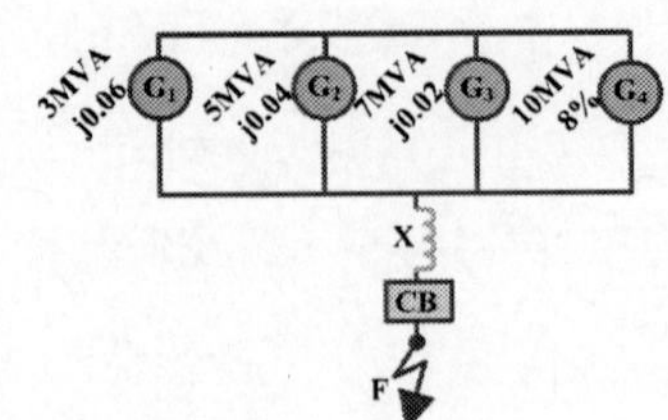

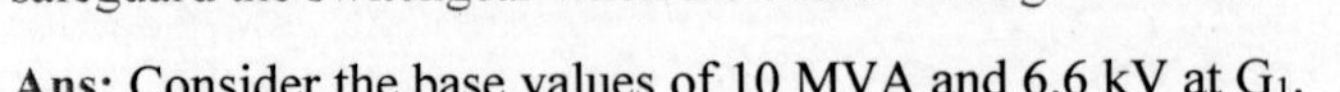

Ans: Consider the base values of 10 MVA and 6.6 kV at G_1.

$X^{G1}_{p.u.} = j0.06 * \frac{10}{3} = j0.2\,p.u.$ $\quad X^{G3}_{p.u.} = j0.02 * \frac{10}{7} = j0.02857\,p.u.$

$X^{G2}_{p.u.} = j0.04 * \frac{10}{5} = j0.08\,p.u.$ $\quad X^{G4}_{p.u.} = j0.08 * \frac{10}{10} = j0.08\,p.u.$

Four alternators are connected in parallel $X^{G1}_{p.u.} \parallel X^{G2}_{p.u.} \parallel X^{G3}_{p.u.} \parallel X^{G4}_{p.u.}$

$$= \frac{1}{\frac{1}{j0.2}+\frac{1}{j0.08}+\frac{1}{j0.02857}+\frac{1}{j0.08}} = j0.01538\,p.u.$$

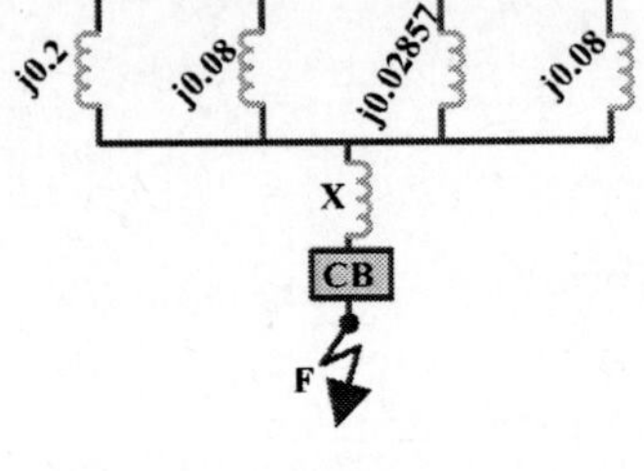

then $\quad X^{eq}_{p.u.} = j(0.01538 + X)$

MVA_{SC} is given as 100MVA

$MVA_{SC} = \frac{MVA_{base}}{X^{eq}_{p.u.}} \Rightarrow 100 = \frac{10}{0.01538+X}$ then $X^{eq}_{p.u.} = 0.0846\,pu$

$$X_{actual} = X^{eq}_{p.u.} * X_{base} = \frac{0.0846 * (6.6 * 10^3)^2}{10 * 10^6} = 0.3685\,\Omega$$

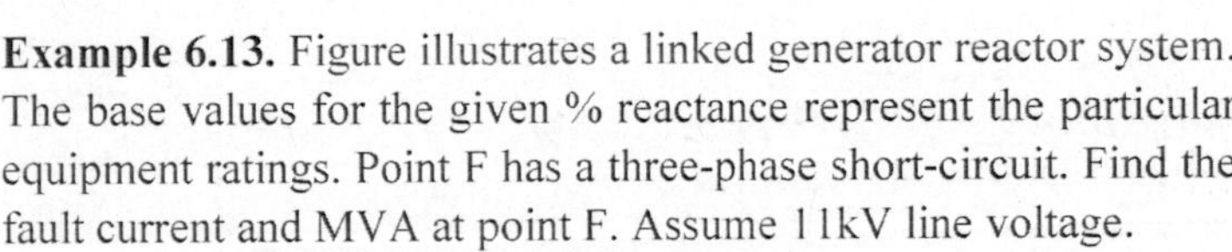

Example 6.13. Figure illustrates a linked generator reactor system. The base values for the given % reactance represent the particular equipment ratings. Point F has a three-phase short-circuit. Find the fault current and MVA at point F. Assume 11kV line voltage.

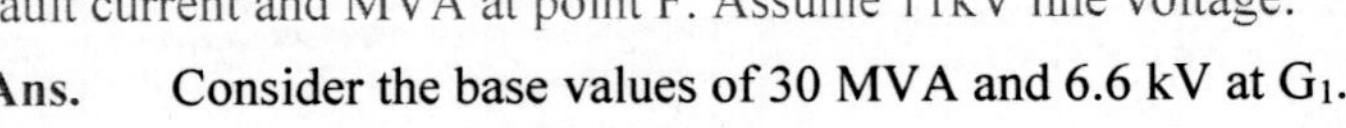

Ans. Consider the base values of 30 MVA and 6.6 kV at G_1.

$X^{G1}_{p.u.} = j0.1 * \frac{30}{10} = j0.3\,p.u.$

$X^{G2}_{p.u.} = j0.15 * \frac{30}{20} = j0.225\,p.u.$

$X^{G3}_{p.u.} = j0.2 * \frac{30}{30} = j0.2\,p.u.$

$X^{line1}_{p.u.} = j0.05 * \frac{30}{4} = j0.375\,p.u.$

$X^{line2}_{p.u.} = j0.07 * \frac{30}{8} = j0.2625\,p.u.$

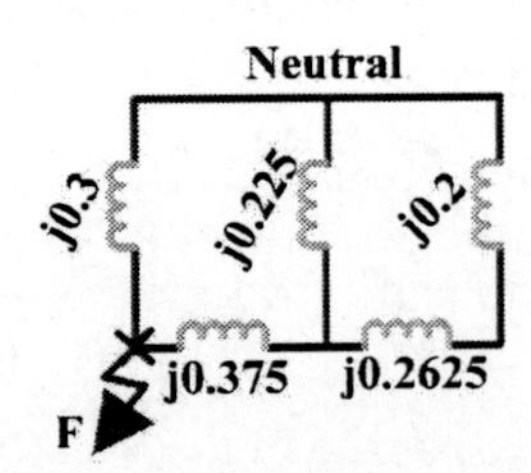

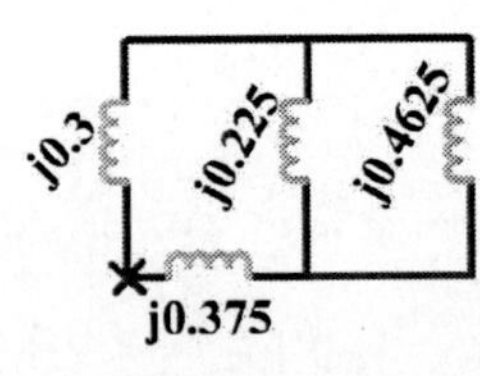

In Fig. $j0.2$ is in series with $j0.2625$ so both are added $j0.2 + j0.2625 = j0.4625$

Now $j0.4625$ is in parallel with $j0.225$ so

$$j0.4625 \parallel j0.225 = \frac{j0.4625 * j0.225}{j0.4625+j0.225} = j0.1513p.u.$$

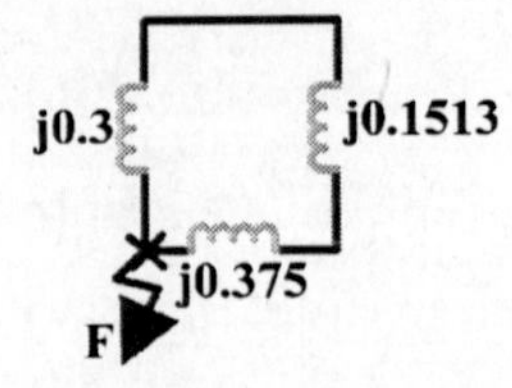

$$X^{eq}_{p.u.} = j0.3 \parallel (j0.375 + j0.1513) = \frac{j0.3*(j0.375+j0.1513)}{j0.3+(j0.375+j0.1513)} = j0.19108\,p.u.$$

$$MVA_{SC} = \frac{MVA_{base}}{X^{eq}_{p.u.}} = \frac{30*10^6}{0.19108} = 157MVA$$

$$I_{SC} = \frac{MVA_{SC}}{\sqrt{3} * kV_{base}} = \frac{157 * 10^6}{\sqrt{3} * 11 * 10^3} = 8.24\,kA$$

Example 6.14. Figure depicts a system with four alternators, each rated at 11kV and 50MVA, with a sub-transient reactance of 25%. Determine

(a) Fault level for a fault at the terminals of generators.

(b) the reactance 'X' of the current-limiting reactor shown in figure to limit the fault level to 200MVA.

Ans. Consider base values as 11kV, 50MVA

Case (a) : Consider X=0 , All four alternators of same value are connected in parallel

$$X_{p.u.}^{eq} = \frac{j0.25}{4} = j0.0625p.u.$$

$$MVA_{SC} = \frac{MVA_{base}}{X_{p.u.}^{eq}} = \frac{50}{0.0625} = 800MVA$$

Case (b) : Considering p.u. reactance 'X'

$$X_{p.u.}^{eq} = j(0.0625 + X)$$

MVA_{SC} is given as 200MVA

$$MVA_{SC} = \frac{MVA_{base}}{X_{p.u.}^{eq}} \Rightarrow 200 = \frac{50}{0.0625 + X}$$

$$X_{p.u.}^{eq} = 0.1875\ p.u.$$

$$X_{actual} = X_{p.u.}^{eq} * X_{base} = \frac{0.1875*(11*10^3)^2}{50*10^6} = 0.4537\ \Omega$$

Example 6.15. Two three-phase alternators running in parallel, each of 10MVA and having 25% reactance feed directly 11kV busbars. These busbars feed 33kV busbars through two transformers in parallel each of 5MVA with 10% reactance. Two overhead lines in parallel are connected to 33kV busbars. The line reactance's are (2+j4) Ω. If a three-phase fault occurs at the end of the transmission line, find short-circuit MVA.

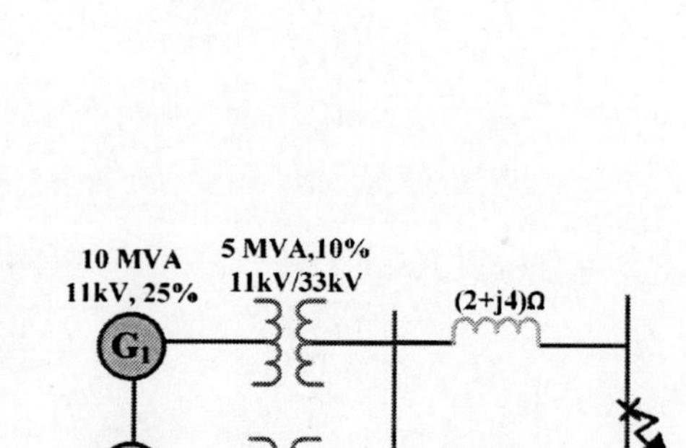

Ans : Base =10MVA

$$X_{p.u.}^{G1} = X_{p.u.}^{G2} = j0.25 * \left(\frac{11}{11}\right)^2 * \frac{10}{10} = j0.25\ p.u.$$

$$X_{p.u.}^{T/f1} = X_{p.u.}^{T/f2} = j0.1 * \left(\frac{11}{11}\right)^2 * \frac{10}{5} = j0.2\ p.u.$$

$$Z_{p.u.}^{line1} = Z_{p.u.}^{line2} = \frac{(2 + j4)}{33^2} * 10 = (0.01836 + j0.0367)\ p.u.$$

Equivalent reactance is $Z_{p.u.}^{eq} = (X_{p.u.}^{G1} \parallel X_{p.u.}^{G2}) + (X_{p.u.}^{\frac{T}{f1}} \parallel X_{p.u.}^{\frac{T}{f2}}) + (Z_{p.u.}^{line1} \parallel Z_{p.u.}^{line2})$

$$Z_{p.u.}^{eq} = (j0.25 \parallel j0.25) + (j0.2 \parallel j0.2) + [(0.01836 + j0.0367) \parallel (0.01836 + j0.0367)]$$

$$= j0.125 + j0.1 + (0.00918 + j0.01835) = 0.00918 + j0.24335\ p.u = 0.2435\angle 87.83^0\ p.u.$$

Base Current $I_{base} = \frac{kVA_{base}}{\sqrt{3}*kV_{base}} = \frac{10*10^6}{\sqrt{3}*33*10^3} = 174.95\ A$

Per unit fault Current $I^{F}_{p.u.} = \frac{V_{p.u.}}{Z^{eq}_{p.u.}} = \frac{1+j0}{0.2435} = 4.105\ p.u.$

Actual fault Current $I^{F}_{actual} = I^{F}_{p.u.} * I_{base} = 4.105 * 174.95 = 718\ A$

Short-circuit MVA $MVA_{SC} = \frac{MVA_{base}}{Z^{eq}_{p.u.}} = \frac{10*10^6}{0.2435} = 41.06\ MVA$

Example 6.16. A three-phase generating station's plant capacity is made up of two 10MVA generating stations with 12% reactance each and one 5MVA generating station with 8% reactance connected to a common 11kV busbar from which loads are drawn via a 20MVA, 6% reactance step-up transformer. Find the MVA rating of the circuit breaker on (a) the L.V. side and (b) the H.V. side, using the base value of 100 MVA.

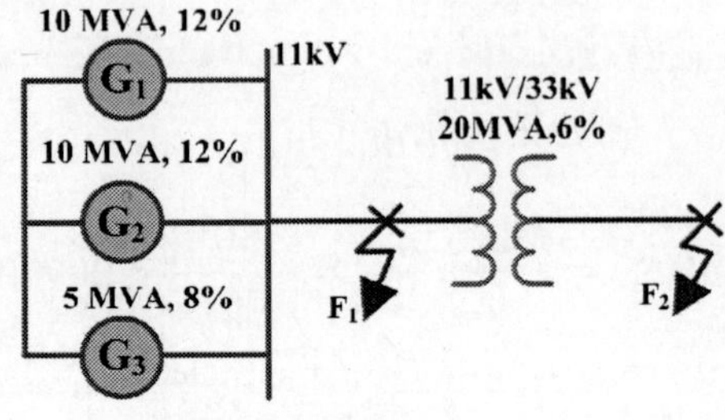

Ans : $X^{G1}_{p.u.} = X^{G2}_{p.u.} = j0.12 * \left(\frac{11}{11}\right)^2 * \frac{100}{10} = j1.2\ p.u.$

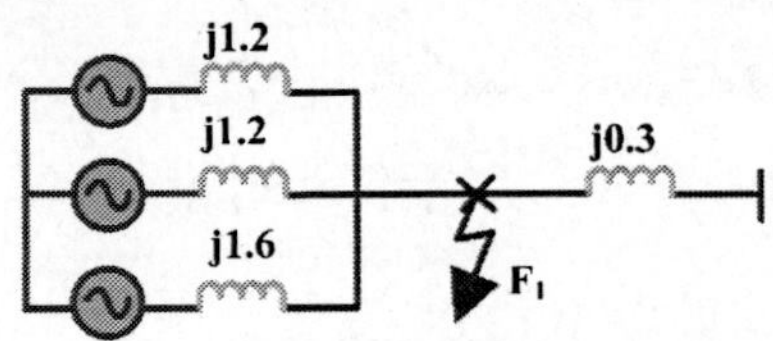

$X^{G3}_{p.u.} = j0.08 * \left(\frac{11}{11}\right)^2 * \frac{100}{5} = j1.6\ p.u.$

$X^{T/f}_{p.u.} = j0.06 * \left(\frac{11}{11}\right)^2 * \frac{100}{20} = j0.3\ p.u.$

(a) When fault is occurred on LV side at F1

$$X^{F1}_{p.u.} = (j1.2 \parallel j1.2 \parallel j1.6) = \frac{1}{\frac{1}{j1.2} + \frac{1}{j1.2} + \frac{1}{j1.6}} = j0.436\ p.u$$

Base Current $I_{base} = \frac{kVA_{base}}{\sqrt{3} * kV_{base}} = \frac{100*10^6}{\sqrt{3}*11*10^3} = 5248\ A$

Per unit fault Current $I^{F1}_{p.u.} = \frac{V_{p.u.}}{X^{eq}_{p.u.}} = \frac{1+j0}{0.436} = 2.293\ p.u.$

Actual fault Current $I^{F1}_{actual} = I^{F1}_{p.u.} * I_{base} = 2.293 * 5248 = 12.036\ kA$

Short-circuit MVA $MVA^{F1}_{SC} = \frac{MVA_{base}}{X^{F1}_{p.u.}} = \frac{100}{0.436} = 229.36\ MVA$

(b) When fault is occurred on HV side at F2

$X^{F2}_{p.u.} = (X^{G1}_{p.u.} \parallel X^{G2}_{p.u.} \parallel X^{G3}_{p.u.}) + X^{T/f}_{p.u.}$

$X^{F2}_{p.u.} = (j1.2 \parallel j1.2 \parallel j1.6) + j0.3 = j0.436 + j0.3 = j0.736$

Base Current on HV side $I_{base} = \frac{100*10^6}{\sqrt{3}*33*10^3} = 1.749\ kA$

Per unit fault Current $I^{F2}_{p.u.} = \frac{V_{p.u.}}{X^{eq}_{p.u.}} = \frac{1+j0}{0.736} = 1.358\ p.u.$

Actual fault Current $I^{F2}_{actual} = I^{F2}_{p.u.} * I_{base} = 1.358 * 1749 = 2.375\ kA$

Short-circuit MVA $MVA^{F2}_{SC} = \frac{MVA_{base}}{X^{F2}_{p.u.}} = \frac{100*10^6}{0.736} = 135.86\ MVA$

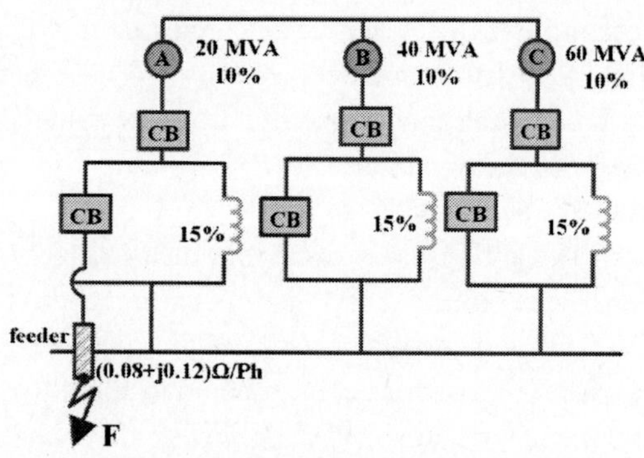

Example 6.17. Three 11kV generators A, B, and C, each with a 10% leakage reactance and MVA ratings of 20, 40, and 60MVA, are electrically coupled by a tie-line/tie-bar to a current limiting reactor with a 15% reactance dependent on the machine rating to which it is connected. A 3-phase feeder is powered by generator 'A's busbar at an 11kV line voltage. Its reactance is (0.08+j0.12) Ω/ph. Find the Short-circuit MVA at the feeder's terminus. Consider a 50MVA/11kV base.

Ans: Per unit reactance values of Generator's are $X^{GA}_{p.u.}$, $X^{GB}_{p.u.}$, $X^{GC}_{p.u.}$ and Per unit reactance of current limiting reactors are $X^{LA}_{p.u.}$, $X^{LB}_{p.u.}$, $X^{LC}_{p.u.}$ respectively.

$$X^{GA}_{p.u.} = j0.1 * \left(\frac{11}{11}\right)^2 * \frac{50}{20} = j0.25\ p.u. \qquad X^{LA}_{p.u.} = j0.15 * \left(\frac{11}{11}\right)^2 * \frac{50}{20} = j0.375\ p.u.$$

$$X^{GB}_{p.u.} = j0.1 * \left(\frac{11}{11}\right)^2 * \frac{50}{40} = j0.125\ p.u. \qquad X^{LB}_{p.u.} = j0.15 * \left(\frac{11}{11}\right)^2 * \frac{50}{40} = j0.1875\ p.u.$$

$$X^{GC}_{p.u.} = j0.1 * \left(\frac{11}{11}\right)^2 * \frac{50}{60} = j0.0833\ p.u. \qquad X^{LC}_{p.u.} = j0.15 * \left(\frac{11}{11}\right)^2 * \frac{50}{60} = j0.125\ p.u.$$

$$X^{feeder}_{p.u.} = \frac{(0.08 + j0.12)}{(11 * 10^3)^2} * (50 * 10^3) = 0.03305 + j0.0495\ p.u. = 0.05959\angle 56.30^0\ p.u.$$

$$X^{eq}_{p.u.} = \{[(X^{GC}_{p.u.} + X^{LC}_{p.u.}) \parallel (X^{GB}_{p.u.} + X^{LB}_{p.u.}) + X^{LA}_{p.u.}] \parallel X^{GA}_{p.u.}\} + X^{feeder}_{p.u.}$$

$$= \{[(j0.0833 + j0.125) \parallel (j0.125 + j0.1875) + j0.375] \parallel j0.25\} + (0.03305 + j0.0495)$$

$$= \{[(j0.2083 \parallel j0.3125) + j0.375] \parallel j0.25\} + (0.03305 + j0.0495)$$

$$= [(j0.1249 + j0.375) \parallel j0.25] + (0.03305 + j0.0495)$$

$$= (j0.499 \parallel j0.25) + (0.03305 + j0.0495)$$

$$= (j0.166) + (0.03305 + j0.0495)$$

$$= 0.03305 + j0.2155$$

$$= 0.218\angle 81.28^0\ p.u.$$

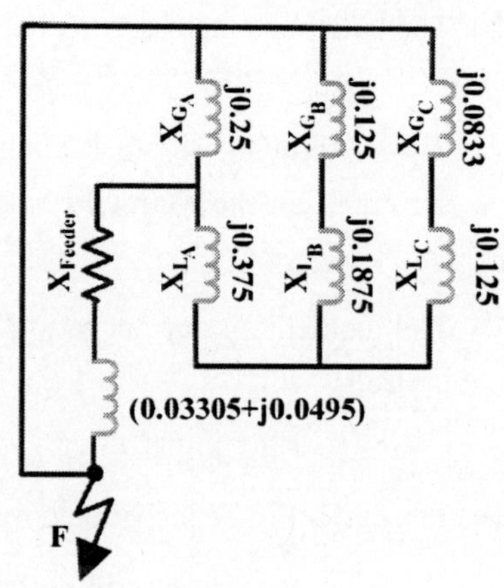

Base Current $I_{base} = \frac{50*10^6}{\sqrt{3}*11*10^3} = 2624\ A$

Per unit fault Current $I^F_{p.u.} = \frac{V_{p.u.}}{X^{eq}_{p.u.}} = \frac{1+j0}{0.218} = 4.587\ p.u.$

Actual fault Current $I^F_{actual} = I^F_{p.u.} * I_{base} = 4.587 * 2624$

$= 12.036kA$

Short-circuit MVA $MVA^F_{sc} = \frac{MVA_{base}}{X^{eq}_{p.u.}} = \frac{50*10^6}{0.218} = 229.35\ MVA$

Example 6.18. A generator connected through a 3-cycle Circuit Breaker to a transformer is rated 100 MVA, 11 kV with reactance's of $X_d'' = 10\%, X_d' = 20\%, and\ X_d = 100\%$ respectively. The system is functioning under load and at its designated voltage when a 3-phase short-circuit event takes place between the circuit breaker and the transformer. The following parameters need to be determined:
(a) The sustained, transient, and sub-transient short-circuit current in the breaker;
(b) The initial symmetrical root mean square (rms) current in the breaker;
(c) The maximum possible direct current (dc) component of the short-circuit current in the breaker;
(d) The momentary current rating of the breaker;
(e) The current that needs to be interrupted by the breaker; and
(f) The interrupting kVA.

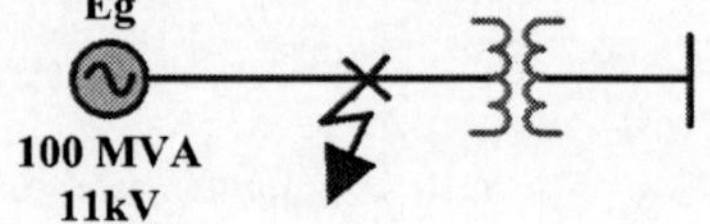

Ans: Base values are 11kV, 100MVA

$$I_{base} = \frac{kVA_{base}}{\sqrt{3} * kV_{base}} = \frac{100 * 10^6}{\sqrt{3} * 11 * 10^3} = 5.248\ kA$$

(a) The sustained, transient and sub-transient short-circuit current in breaker

Sub-transient short-circuit current $I''_{g\,p.u.} = \frac{E_g}{X_d''} = \frac{1}{j0.1} = -j10\ p.u.$

$$I''_{g\,actual} = I''_{g\,p.u.} * I_{base} = -j10 * 5248 = 52.48\angle - 90^0 kA$$

Transient short-circuit current $I'_{g\,p.u.} = \frac{E_g}{X_d'} = \frac{1}{j0.2} = -j5\ p.u.$

$$I'_{g\,actual} = I'_{g\,p.u.} * I_{base} = -j5 * 5248 = 26.24\angle - 90^0 kA$$

Sustained short-circuit current $I_{g\,p.u.} = \frac{E_g}{X_d} = \frac{1}{j1} = -j1\ p.u.$

$$I_{g\,actual} = I_{g\,p.u.} * I_{base} = -j1 * 5248 = 5.248\angle - 90^0 kA$$

(b) The initial symmetrical rms current in the breaker is same as of sub-transient short-circuit current

$$I_{sym} = I_g'' = -j10\ p.u. = 52.48\angle - 90^0 kA$$

(c) The maximum possible dc component of the short-circuit current in the breaker is the peak value of the sub-transient short-circuit current $I^{dc}_{p.u.} = \sqrt{2} * I_{sym} = \sqrt{2} * I_g'' = \sqrt{2} * (-j10) = -j14.14\ p.u.$

$$I^{dc}_{actual} = I^{dc}_{p.u.} * I_{base} = -j14.14 * 5248 = 74.217\angle - 90^0 kA$$

(d) The momentary current rating of the breaker is 1.6 times the initial symmetrical rms current in the breaker $I^{momentary}_{p.u.} = 1.6 * I_{sym} = 1.6 * I_g'' = 1.6 * (-j10)\ p.u. = -j16\ p.u.$

$$I^{momentary}_{actual} = I^{momentary}_{p.u.} * I_{base} = -j16 * 5248 = 83.96\angle - 90^0 kA$$

(e) The current to be interrupted by a 3-cycle C.B.: $I^{interrupt}_{p.u.} = M.F. * I_{sym}$

We need to consider multiplication factor as 1.2 since the current is to be interrupted by a 3-cycle Circuit Breaker $I^{interrupt}_{p.u.} = 1.2 * I_{sym} = 1.2 * I_g'' = 1.2 * (-j10)\ p.u. = -j12\ p.u.$

$$I^{interrupt}_{actual} = I^{interrupt}_{p.u.} * I_{base} = -j12 * 5248 = 62.976\angle - 90^0 kA$$

(f) The interrupting kVA

$$MVA^{interrupt}_{sc} = \sqrt{3} * I^{interrupt}_{actual} * kV_{base} = \sqrt{3} * 62976 * 11 * 10^3 = 1199\ MVA$$

Example 6.19. A generator-transformer unit is connected to a line through a 3-cycle Circuit Breaker. Generator ratings are 10MVA, 6.6kV with reactance's of $X_d'' = 10\%, X_d' = 20\%, and\ X_d = 100\%$ respectively. Transformer ratings are 10 MVA, 6.6kV/33kV, 8.5% reactance. The system is functioning under load and at its designated voltage when a 3-phase short-circuit event takes place between the circuit breaker and the transformer. The following parameters need to be determined:

(a) The sustained, transient, and sub-transient short-circuit current in the breaker;
(b) The initial symmetrical root mean square (rms) current in the breaker;
(c) The maximum possible direct current (dc) component of the short-circuit current in the breaker;
(d) The momentary current rating of the breaker;
(e) The current that needs to be interrupted by the breaker; and
(f) The interrupting kVA.

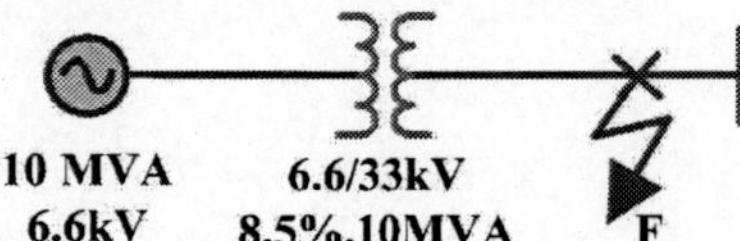

Ans: Base values are 6.6kV, 10MVA

$$I_{base} = \frac{kVA_{base}}{\sqrt{3} * kV_{base}} = \frac{10 * 10^6}{\sqrt{3} * 6.6 * 10^3} = 174.95\ A$$

(a) Sub-transient short-circuit current $I''_{g\,p.u.} = \frac{E_g}{X_d'' + X_{T/f}} = \frac{1}{j0.1 + j0.085} = -j5.405\ p.u.$

$$I''_{g\,actual} = I''_{g\,p.u.} * I_{base} = -j5.405 * 174.95 = 945.67\angle - 90^0\ A$$

Transient short-circuit current $I'_{g\,p.u.} = \frac{E_g}{X_d' + X_{T/f}} = \frac{1}{j0.2 + j0.085} = -j3.508\ p.u.$

$$I'_{g\,actual} = I'_{g\,p.u.} * I_{base} = -j3.508 * 174.95 = 613.85\angle - 90^0\ A$$

Sustained short-circuit current $I_{g\,p.u.} = \frac{E_g}{X_d + X_{T/f}} = \frac{1}{j1 + j0.085} = -j0.921\ p.u.$

$$I_{g\,actual} = I_{g\,p.u.} * I_{base} = -j0.921 * 174.95 = 161.24\angle - 90^0\ A$$

(b) The initial symmetrical rms current in the breaker is same as of sub-transient short-circuit current

$$I_{sym} = I_g'' = -j5.405\ p.u. = 935.48\angle - 90^0\ A$$

(c) The maximum possible dc component of the short-circuit current in the breaker is the peak value of the sub-transient SC current $I^{dc}_{p.u.} = \sqrt{2} * I_{sym} = \sqrt{2} * I_g'' = \sqrt{2} * (-j5.405) = -j7.64\ p.u.$

$$I^{dc}_{actual} = I^{dc}_{p.u.} * I_{base} = -j7.64 * 174.95 = 1.337\angle - 90^0\ kA$$

(d) The momentary current rating of the breaker is 1.6 times the initial symmetrical rms current in the breaker $I^{momentary}_{p.u.} = 1.6 * I_{sym} = 1.6 * I_g'' = 1.6 * (-j5.405)\ p.u. = -j8.648\ p.u.$

$$I^{momentary}_{actual} = I^{momentary}_{p.u.} * I_{base} = -j8.648 * 174.95 = 1.512\angle - 90^0\ kA$$

(e) The current to be interrupted by a 5-cycle C.B.: $I^{interrupt}_{p.u.} = M.F. * I_{sym}$

We need to consider multiplication factor as 1.1 since the current is to be interrupted by a 5-cycle Circuit Breaker $I^{interrupt}_{p.u.} = 1.1 * I_{sym} = 1.1 * I_g'' = 1.1 * (-j5.405)\ p.u. = -j5.945\ p.u.$

$$I^{interrupt}_{actual} = I^{interrupt}_{p.u.} * I_{base} = -j5.945 * 174.95 = 1.04\angle - 90^0\ kA$$

(f) $MVA^{interrupt}_{SC} = \sqrt{3} * I^{interrupt}_{actual} * kV_{base} = \sqrt{3} * 1040 * 11 * 10^3 = 59.45\ MVA$

Example 6.20. Consider $Z_{bus} = j\begin{bmatrix} 0.0745 & 0.0355 & 0.0615 \\ 0.0355 & 0.0745 & 0.0485 \\ 0.0615 & 0.0485 & 0.2175 \end{bmatrix}$ $p.u.$; a three-phase fault at bus 1, calculate

(a) fault current I_1^f

(b) The bus voltages during the fault (V_2^f and V_3^f)

(c) Post fault-current in lines (I_{12}^f, I_{13}^f and I_{23}^f)

If a fault is through fault impedance of $Z_f = j0.05\ p.u.$ then calculate fault current, bus voltages and branch currents

Ans: Pre-fault voltages at buses 1,2 and 3 are considered as $V_1^{pf} = V_2^{pf} = V_3^{pf} = 1\angle 0^0$

Case (i): A 3-phase fault occurs at bus 1 and without fault impedance $Z_f = 0$

(a) $I_1^f = \frac{V_1^{pf}}{Z_{11}+Z_f} = \frac{1}{j0.0745+0} = -j13.42 = 13.42\angle -90^0\ p.u.$

(b) Voltage at i[th] bus when fault on k[th] bus is $V_i^f = V_i^{pf} - \frac{Z_{ik}}{Z_{kk}+Z_f} * V_i^{pf} = V_i^{pf} - Z_{ik} * I_k^f$

A three-phase fault occurs at bus 1, so $V_1^f = 0$

$$V_2^f = V_2^{pf} - \frac{Z_{21}}{Z_{11}+Z_f} * V_2^{pf} = 1\angle 0^0 - \frac{j0.0355}{j0.0745+0} * 1\angle 0^0 = 0.5234\ p.u.$$

$$V_3^f = V_3^{pf} - \frac{Z_{31}}{Z_{11}+Z_f} * V_3^{pf} = 1\angle 0^0 - \frac{j0.0615}{j0.0745+0} * 1\angle 0^0 = 0.1746\ p.u.$$

(c) Fault current between i and j buses is $I_{ij}^f = \frac{V_i^f - V_j^f}{Z_{ij}}$

$$I_{12}^f = \frac{V_1^f - V_2^f}{Z_{12}} = \frac{0-0.5234}{j0.0355} = j14.743\ p.u. = 14.743\angle 90^0\ p.u.$$

$$I_{13}^f = \frac{V_1^f - V_3^f}{Z_{13}} = \frac{0-0.1746}{j0.0615} = j2.839\ p.u. = 2.839\angle 90^0\ p.u.$$

$$I_{23}^f = \frac{V_2^f - V_3^f}{Z_{23}} = \frac{0.5234-0.1746}{j0.0485} = -j7.191\ p.u. = 7.191\angle -90^0\ p.u.$$

Case (ii) : A 3-phase fault occurs at bus 1, through fault impedance $Z_f = j0.05\ p.u.$

$$I_1^f = \frac{V}{Z_{11}+Z_f} = \frac{1\angle 0^0}{j0.0745+j0.05} = -j8.032\ p.u. = 8.032\angle -90^0\ p.u.$$

$$V_2^f = V_2^{pf} - \frac{Z_{21}}{Z_{11}+Z_f} * V_2^{pf} = 1\angle 0^0 - \frac{j0.0355}{j0.0745+j0.05} * 1\angle 0^0 = 0.7148\ p.u.$$

$$V_3^f = V_1^{pf} - \frac{Z_{31}}{Z_{11}+Z_f} * V_1^{pf} = 1\angle 0^0 - \frac{j0.0615}{j0.0745+j0.05} * 1\angle 0^0 = 0.506\ p.u.$$

$$I_{12}^f = \frac{V_1^f - V_2^f}{Z_{12}} = \frac{0-0.7148}{j0.0355} = j20.135\ p.u. = 20.135\angle 90^0\ p.u.$$

$$I_{13}^f = \frac{V_1^f - V_3^f}{Z_{13}} = \frac{0 - 0.506}{j0.0615} = j8.227\ p.u. = 8.227\angle 90^0\ p.u.$$

$$I_{23}^f = \frac{V_2^f - V_3^f}{Z_{23}} = \frac{0.7148 - 0.506}{j0.0485} = j4.305\ p.u. = 4.305\angle 90^0\ p.u.$$

Example 6.2. Consider $Z_{bus} = j\begin{bmatrix} 0.7166 & 0.6099 & 0.5334 & 0.5805 \\ 0.6099 & 0.7319 & 0.6401 & 0.6966 \\ 0.5334 & 0.6401 & 0.7166 & 0.6695 \\ 0.5805 & 0.6966 & 0.6695 & 0.7631 \end{bmatrix}\ p.u.;$

a three-phase fault at bus 4, calculate

(a) fault current I_4^f

(b) The bus voltages during the fault (V_1^f, V_2^f and V_3^f)

(c) Post fault-current in lines ($I_{12}^f, I_{13}^f, I_{14}^f, I_{23}^f, I_{24}^f$ and I_{34}^f)

Ans: Pre-fault voltages at buses 1,2,3 and 4 are considered as $V_1^{pf} = V_2^{pf} = V_3^{pf} = V_4^{pf} = 1\angle 0^0$

Case (i): A 3-phase fault occurs at bus 4 and without fault impedance $Z_f = 0$

(a) $I_4^f = \frac{V_4^{pf}}{Z_{44}+Z_f} = \frac{1\angle 0^0}{j0.7631+0} = -j1.3104 = 1.3104\angle -90^0\ p.u.$

(b) Voltage at ith bus when fault on kth bus is $V_i^f = V_i^{pf} - \frac{Z_{ik}}{Z_{kk}+Z_f} * V_i^{pf} = V_i^{pf} - Z_{ik} * I_k^f$

$$V_1^f = V_1^{pf} - Z_{14} * I_4^f = 1\angle 0^0 - j0.5805 * (-j1.3104) = 0.2393\ p.u.$$

$$V_2^f = V_2^{pf} - Z_{24} * I_4^f = 1\angle 0^0 - j0.6966 * (-j1.3104) = 0.0871\ p.u.$$

$$V_3^f = V_3^{pf} - Z_{34} * I_4^f = 1\angle 0^0 - j0.6695 * (-j1.3104) = 0.1226\ p.u.$$

(c) Post fault-current in lines ($I_{12}^f, I_{13}^f, I_{14}^f, I_{23}^f, I_{24}^f$ and I_{34}^f)

$$I_{12}^f = \frac{V_1^f - V_2^f}{Z_{12}} = \frac{0.2393 - 0.0871}{j0.6099} = -j0.2495\ p.u. = 0.2495\angle -90^0\ p.u$$

$$I_{13}^f = \frac{V_1^f - V_3^f}{Z_{13}} = \frac{0.2393 - 0.1226}{j0.5334} = -j0.2175\ p.u. = 0.2175\angle -90^0\ p.u.$$

$$I_{14}^f = \frac{V_1^f - V_4^f}{Z_{14}} = \frac{0.2393 - 0}{j0.5805} = -j0.4122\ p.u. = 0.4122\angle -90^0\ p.u$$

$$I_{23}^f = \frac{V_2^f - V_3^f}{Z_{23}} = \frac{0.0871 - 0.1226}{j0.6401} = j0.0554\ p.u. = 0.0554\angle 90^0\ p.u$$

$$I_{24}^f = \frac{V_2^f - V_4^f}{Z_{24}} = \frac{0.0871 - 0}{j0.6966} = -j0.1250\ p.u. = 0.1250\angle -90^0\ p.u$$

$$I_{34}^f = \frac{V_3^f - V_4^f}{Z_{34}} = \frac{0.1226 - 0}{j0.6695} = -j0.1831\ p.u. = 0.1831\angle -90^0\ p.u$$

6.8 Unit Summary:

- The classification of faults in an Electrical Power System encompasses two main categories: symmetrical faults and unsymmetrical faults.
- The impact of faults in an electrical power system can vary based on factors such as voltage level, the type of relay circuits used, the type of neutral grounding considered, and the existence of regulating devices.
- Short-circuits generate high currents that lead to excessive heating, perhaps resulting in an explosion.
- If the fault is not promptly resolved, it might potentially disrupt the stability of the power system and lead to a total shutdown.
- The maximum momentary current $i_{mm}(t)$ is twice the maximum symmetrical short-circuit current is known as "Doubling Effect".
- This unit provides a description of the sub-transient, transient, and steady state reactance and current of a 3-phase alternator in the context of a short-circuit.
- The reactance of a 3-Ph Synchronous generator during Short-Circuit will be $X_d'' < X_d' < X_d$ and the respective current will be $I_g'' > I_g' > I_g$.
- The essential requirement of a current limiting reactors is that the reactance should not reduce to saturation level under short-circuit conditions
- Current Limiting reactors can be located in different ways (i) in series with generators, (ii) in series with feeders and (ii) in bus-bars
- Symmetrical Fault current and Short-Circuit MVA in an electrical power system can be calculated using following techniques/methods.

  (i). Network Reduction Technique
  (ii). Modification of machine internal voltages
  (iii). Thevenin Equivalent Method
  (iv). Bus Impedance Matrix

Short and Long Answer Questions

1. What exactly does it mean to have a single line diagram? What are the distinctions between a single line diagram and an impedance diagram? Explain by an example.
2. What exactly do you mean by a short circuit? Discuss the probable causes of power system short-circuits.
3. Classify the electrical power system's faults. Briefly address the percentage of occurrence and the severity of symmetrical and unsymmetrical faults.
4. Discuss the harmful effects of short-circuit fault on the power system.
5. What is doubling effect? Prove that the maximum momentary current $i_{mm}(t)$ is twice the maximum symmetrical short-circuit current.
6. What do you mean by "instantaneous maximum momentary current for line"? Explain it with the help of an appropriate diagram.
7. Discuss and explain the sub-transient, transient, and steady-state reactances that occur at the generator terminals during a 3-phase short-circuit.

8. Why is the per-unit system employed in power system analysis? Which of the electrical quantities are used as base values?
9. Discuss the significance of a circuit breaker in power system during a short-circuit. Provide the ratings and specifications of a circuit breaker.
10. What are current limiting reactors? Mention different type of reactors used for reducing fault current.
11. Discuss the use and location of current limiting reactors in a power system.
12. List the methods used for computing short-circuit current.
13. With necessary diagrams and equations explain how the Symmetrical fault current and short-circuit MVA are calculated in Thevenin Equivalent Method
14. Discuss various steps involved in Network Reduction Technique for measuring Symmetrical fault current and short-circuit MVA
15. With necessary diagrams and equations explain how the Symmetrical fault current and short-circuit MVA calculated in Thevenin Equivalent Method when the symmetrical fault occurs at the terminals of motor/generator/half-way of the transmission line with pre-specified voltage at the terminals of the motor.
16. What is the significance of Bus Impedance Matrix in power system? With necessary diagrams and equations explain how the Symmetrical fault current and short-circuit MVA are calculated with Bus Impedance Matrix.

Exercises

1. _A 50MVA, 33kV alternator with 10% reactance is coupled to a feeder with a series impedance of (0.2+j0.3) Ω/km through a 250MVA, 11kV/33kV transformer having 5% leakage reactance. When a 3-phase symmetrical fault occurs at a position 15km down the feeder, calculate the fault current delivered by the generator operating at 32.5kV with no load through a transformer.
2. The values highlighted in Fig. are per unit reactance, with 10MVA and 6.6kV as the generator circuit's base voltages. Both transformers are rated at 6.6kV and 33kV. A three-phase to ground fault with a fault impedance of j0.6p.u. occurs at bus 3. Determine the fault current at bus 3 and the output current of each generator under fault.

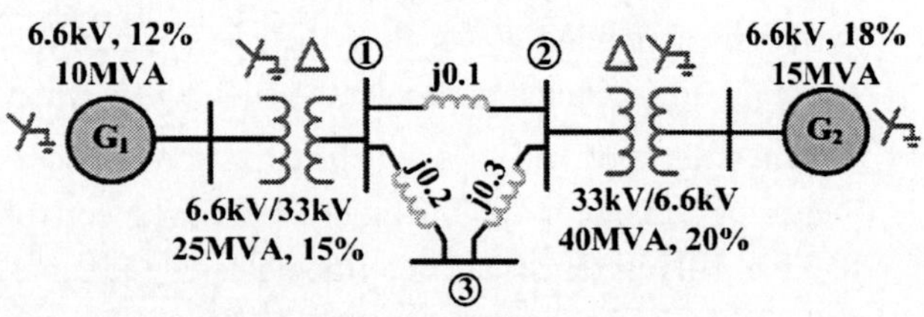

3. Generators A and B are identical, rated at 11kV and 50MVA, with a transient reactance of 15% at their own MVA. Two transformers have equal ratings of 11kV/33kV, 20MVA, and 10% reactance. The tie-line is 30km long, with each conductor having a reactance of 0.25Ω/km. A 3-phase fault occurs 15 kilometers from one end of the line while the system is not loaded but at rated voltage. Determine the fault MVA and fault current. Consider the base MVA to be 50MVA.

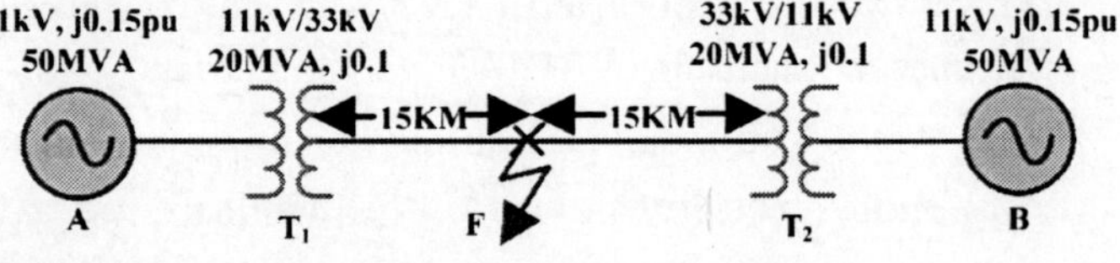

4. A synchronous generator is connected to a motor with ratings of 30MVA, 11kV, and 12% reactance each. The line that connects the two has an 18% reactance on a base of 11 kV, 30 MVA. The motor receives 25MW at 0.8 pf leading and a terminal voltage of 10.6 kV. If there is a symmetrical three-phase short-circuit at the generator terminals, determine the sub-transient current in the generator, motor, and fault.

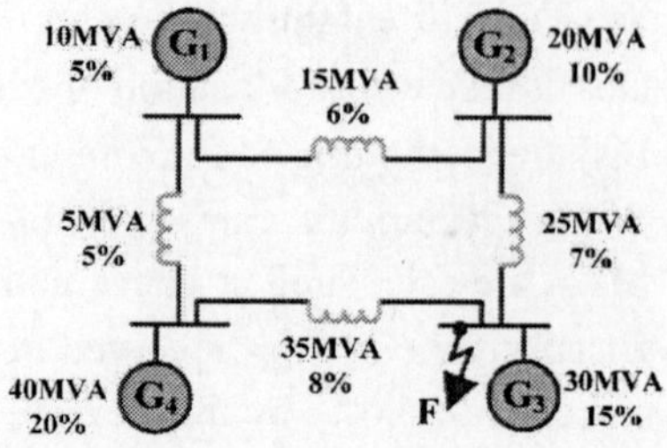

5. Single line diagram of 3-phase power system is shown in Fig. Using per unit method of analysis calculate the three-phase short-circuit MVA at point F. Consider base MVA as 50MVA.

6. A generator-transformer unit is connected to a line through a 3-cycle Circuit Breaker. Generator ratings are 50MVA, 11kV with reactances of $X''_d = 9\%, X'_d = 18\%, and\ X_d = 100\%$ respectively. Transformer ratings are 100 MVA, 11kV/110kV, 10% reactance. The system is functioning under load and at its designated voltage when a 3-phase short-circuit event takes place between the circuit breaker and the transformer. Determine the following:
   (a) The sustained, transient, and sub-transient short-circuit current in the breaker;
   (b) The initial symmetrical root mean square (rms) current in the breaker;
   (c) The maximum possible dc component of the short-circuit current in the breaker;
   (d) The momentary current rating of the breaker;
   (e) The current that needs to be interrupted by the breaker; and
   (f) The interrupting kVA.

7. A generating station has three alternators: 5MVA, 7.5MVA, and 10MVA, each with a reactance of j0.12, j0.24, and j0.36 p.u. The circuit breakers have a rating of 50 MVA. It is planned to extend the system by supplying power from the grid via a 20MVA generator with a 15% reactance. Determine the required reactance to protect the switchgear when the busbar voltage is 6.6kV.

8. A generator is linked to a synchronous motor via a transformer. The sub-transient reactances of generator and motor are j0.2 and j0.4 p.u., respectively. The leakage reactance of the transformer is j0.3 p.u.; all reactances are determined using the same base value. When the motor's terminal voltage is 0.925p.u., a three-phase fault develops at its terminals. The generator's output current under pre-fault condition is 0.96 p.u. at 0.8 pf leading. Determine the sub-transient fault current. Consider 20MVA and 6.6kV as a base.

9. A 33kV 3-phase transmission line with 5Ω resistance and 20Ω reactance is connected to the generating station via a 15 MVA step-up transformer with 5% reactance. The generating station features two alternators, one 10 MVA with 10% reactance and another 5 MVA with 7.5% reactance. Calculate the short-circuit MVA for a symmetrical fault between phases
   (a) At the load end of the transmission line
   (b) At the transformer's high voltage terminals, use 20 MVA as the base MVA.

10. Two 3-phase 11kV, 10 MVA generators with a sub-transient reactance of 12% operate in parallel. The generators deliver power to a transmission line via a 15MVA transformer with an 11kV/33kV ratio and a leakage reactance of 6%. Calculate the fault current and MVA for a symmetrical fault on the (i) LV and (ii) HV sides of the T/F.

11. The figure depicts a three-phase electrical system. The reactances shown are p.u. values on a common base. For a three-phase fault at bus 3, compute.

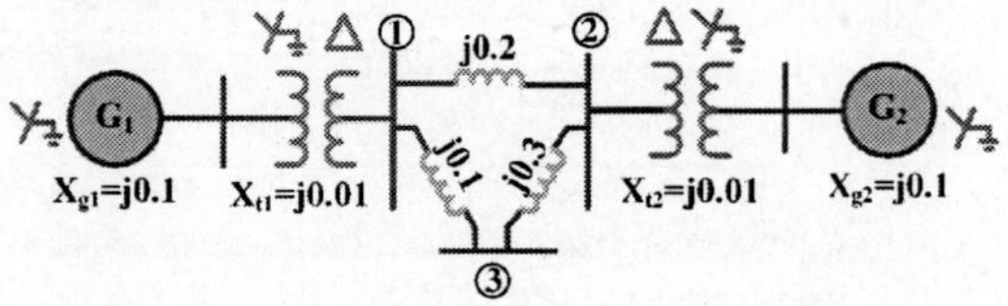

(a) Bus Impedance Matrix Z_{bus}

(b) The fault current I_3^f

(c) The bus voltages during the fault (V_1^f, V_2^f, V_3^f)

(d) The line currents (I_{12}^f, I_{13}^f and I_{23}^f) and

(e) The generator currents (I_{G1}^f and I_{G2}^f)

use Bus Impedance matrix method

To know more about

Types of Faults in Power System, Fault Current & Short circuit MVA and Current Limiting Reactors

To know more about

Fault Analysis using Thevenin's Method, Network Reduction Technique and Bus Impedance Matrix

To know more about

Synchronous Generator Parameter Identification, 3-PhCircuits and Power Measurements and Top 10 Famous Scientists

To Design (MATLAB)

Long Tx Line Model, Load Flow Analysis and Symmetrical Fault Analysis

07 SYMMETRICAL COMPONENTS, UNSYMMETRICAL FAULTS AND SEQUENCE NETWORKS

Unit specifics: In this unit, the following topics have been discussed for basic understating of symmetrical components, unsymmetrical faults and sequence networks:

- Symmetrical components of an unbalanced 3-phase system.
- Relation between unbalanced voltages and symmetrical components.
- Relation between unbalanced currents and symmetrical components.
- Sequence impedances of a synchronous generator, transmission lines, and transformers.
- Analysis of unsymmetrical faults (LG, LL and LLG) in power system during short-circuit.
- Calculation of symmetrical components, sequence impedances and fault currents.

Rationale: In this unit, students will be introduced to symmetrical components, unsymmetrical faults and sequence networks. Significance of symmetrical components in an unbalanced 3-phase system, significance of sequence operator 'a', relation between unbalanced voltages, currents with symmetrical components, three-phase power in terms of symmetrical currents and voltages, sequence impedances of a synchronous generator, transmission lines, and transformers, analysis of unsymmetrical faults (LG, LL and LLG) in power system during short-circuit, sequence networks of unsymmetrical faults when the synchronous machine is solidly grounded and grounded through impedance during direct short circuit and short circuit through fault impedance, are clearly described with the help of necessary diagrams, derivations and examples.

Pre-Requisites: Basic knowledge of power system components.

Unit Outcomes: List of outcomes of this unit is as follows:

U7-O1: To understand symmetrical components.

U7-O2: To derive the relation between unbalanced voltages, currents with symmetrical components.

U7-O3: To analyse the sequence impedances of a synchronous generator, Transmission lines, and Transformers.

U7-O4: To originate the relation between three-phase power in terms of symmetrical currents and voltages.

U7-O5: To analyse the unsymmetrical faults (LG, LL and LLG) in power system when the synchronous machine is solidly grounded and grounded through impedance during direct short circuit and short circuit through fault impedance.

U7-O6: To analyse the sequence networks of unsymmetrical faults when the synchronous machine is solidly grounded and grounded through impedance during direct short circuit and short circuit through fault impedance.

Unit-7 outcomes	Expected mapping with course outcomes (1: Weak, 2: medium, and 3: strong correlation)					
	CO-1	CO-2	CO-3	CO-4	CO-5	CO-6
U7-O1	2	1	-	2	-	-
U7-O2	2	-	-	3	-	3
U7-O3	2	-	1	2	-	3
U7-O4	2	-	1	2	1	1
U7-O5	2	2	-	3	1	3
U7-O6	2	2	-	3	1	3

7.1 Introduction:

There are two main classifications of faults in an electrical power system: symmetrical faults and unsymmetrical faults. The previous unit provides a comprehensive analysis of symmetrical faults. The effects of faults in an electrical power system may be varied depending upon voltage level, type of relay circuits employed, type of neutral grounding considered, and presence of regulating devices. The chances of occurrence of symmetrical fault are rare but severe but asymmetrical faults are frequent, especially line to ground fault is most frequent.

The analysis of balanced three-phase system with balanced loads is much easier to analyse with the help of Network Reduction Technique, Thevenin Equivalent Method and Bus Impedance Matrix, which we have discussed in last unit. But when the system is unbalanced, it is very difficult to analyse the network with the above-mentioned methods. Analysis of unbalanced system is made simpler with the method "symmetrical components". C L Fortesque introduced the method of symmetrical components in 1918. The symmetrical components are used to analyse both balanced and unbalanced systems.

7.2 Symmetrical components of an unbalanced 3-phase system:

As per the CL Fortesque theorem, it is possible to resolve any unbalanced three-phase system into three balanced systems of phasors, referred to as symmetrical components of the given unbalanced system.

(i). Positive sequence components

(ii). Negative sequence components

(iii). Zero sequence components

7.2.1 Positive sequence components:

Figure 7.1 displays the positive phase sequence components. A positive sequence system refers to a system where the phase or line currents/voltages reach their maximum values in the same cyclic order as those in a normal supply. The components of the positive sequence have the same magnitude and are with phase sequence same as that of normal balanced system. The components of the positive sequence are denoted by the subscript '1'.

Positive sequence components of voltages are represented as $V_{a1}, V_{b1},$ and V_{c1} respectively.

Positive sequence components of currents are represented as $I_{a1}, I_{b1},$ and I_{c1} respectively.

Positive sequence components of reactance are represented as $X_{a1}, X_{b1},$ and X_{c1} respectively.

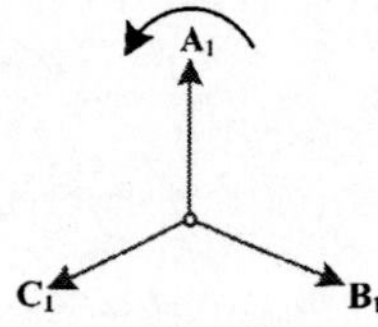

Fig. 7.1 Positive phase sequence components

7.2.2 Negative sequence components: Figure 7.2 displays the components of the negative phase sequence. The negative sequence system refers to a system where the phase or line currents/voltages reach their maximum values in the reverse cyclic order compared to normal balanced system. The magnitudes of the negative sequence components are equal, and they are equally spaced at intervals of 120^0 in reverse order compared to normal balanced system. The components of the negative sequence are denoted by the subscript '2'.

Negative sequence components of voltages are represented as $V_{a2}, V_{b2},$ and V_{c2} respectively.

Negative sequence components of currents are represented as $I_{a2}, I_{b2},$ and I_{c2} respectively.

Negative sequence components of reactance are represented as $X_{a2}, X_{b2},$ and X_{c2} respectively.

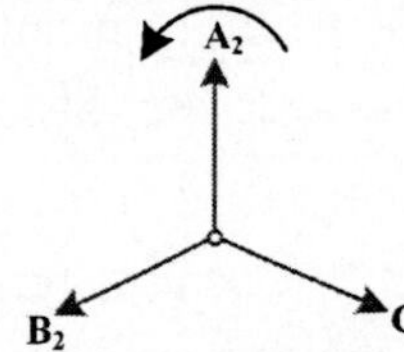

Fig. 7.2 Negative phase sequence components

7.2.3 Zero sequence components:

Figure 7.2 displays the components of the zero-phase sequence. The magnitude of the zero sequence components is equal, and they exhibit zero phase displacement. The subscript '0' is used to denote the zero sequence components. Zero sequence components of voltages are represented as $V_{ao}, V_{bo},$ and V_{co} respectively.

Zero sequence components of currents are represented as $I_{ao}, I_{bo},$ and I_{co} respectively.

Zero sequence components of reactance are represented as $X_{ao}, X_{bo},$ and X_{co} respectively.

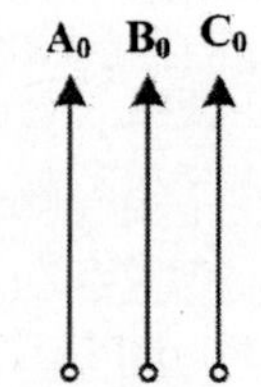

Fig. 7.3 Zero phase sequence components

7.3 Significance of Operator 'a':

In order to address the issue of an unbalanced system in a broader sense and to facilitate the determination of the relationship between phase voltages and/or phase currents, the sequence operator 'a' is employed. Due to the presence of 120^0 displacement in both the positive and negative sequence components inside the symmetrical component theory, it is advantageous to develop an operator capable of inducing 120^0 displacement. To achieve this objective, the operator 'a' is employed. The next section presents the fundamental properties of phasors.

$$a = 1\angle 120^0 = e^{j\frac{2\pi}{3}} = cos\frac{2\pi}{3} + j\ sin\frac{2\pi}{3} = -0.5 + j0.8666 \quad \text{...... (7.1)}$$

$$a^2 = 1\angle 240^0 = e^{j\frac{4\pi}{3}} = cos\frac{4\pi}{3} + j\ sin\frac{4\pi}{3} = -0.5 - j0.8666 \quad \text{...... (7.2)}$$

$$a^3 = 1\angle 360^0 = e^0 = 1\angle 0^0 \quad \text{...... (7.3)}$$

$$a^4 = 1\angle 480^0 = 1\angle 120^0 = a \quad \text{...... (7.4)}$$

$$a^5 = 1\angle 600^0 = 1\angle 240^0 = a^2 \quad \text{...... (7.5)}$$

$$1 + a + a^2 = 1 + -0.5 + j0.8666 - 0.5 - j0.8666 = 0 \quad \text{...... (7.6)}$$

$$a + a^2 = -0.5 + j0.8666 - 0.5 - j0.8666 = -1 \quad \text{...... (7.7)}$$

$$a - a^2 = -0.5 + j0.8666 + 0.5 + j0.8666 = j1.732 = j\sqrt{3} \quad \text{...... (7.8)}$$

$$a^2 - a = -0.5 - j0.8666 + 0.5 - j0.8666 = -j1.732 = -j\sqrt{3} \quad \text{...... (7.9)}$$

The three phasors 1, a^2 *and* a exhibits a balanced and symmetrical. This is due to the fact that the phasors possess identical lengths and are displaced by equal angles of 120^0 degrees relative to each other.

7.4 Relation between unbalanced voltages and symmetrical components:

Let $V_{a,}$ V_b *and* V_c represent an unbalanced set of voltage phasors. Each of the unbalanced phasors is sum of the its symmetrical components.

In Fig. 7.1, by considering V_{a1} as a reference V_{b1} *and* V_{c1} will become

$$V_{b1} = a^2 V_{a1} = 1\angle 240^0 * V_{a1} \quad \text{...... (7.10)}$$

$$V_{c1} = a\ V_{a1} = 1\angle 120^0 * V_{a1} \quad \text{...... (7.11)}$$

In Fig. 7.2, by considering V_{a2} as a reference V_{b2} *and* V_{c2} will become

$$V_{b2} = aV_{a2} = 1\angle 120^0 * V_{a2} \quad \text{...... (7.12)}$$

$$V_{c2} = a^2 V_{a2} = 1\angle 240^0 * V_{a2} \quad \text{...... (7.13)}$$

From Fig. 7.3, the zero sequence components

$$V_{ao} = V_{bo} = V_{co} \quad \text{...... (7.14)}$$

Voltage in phase 'a' is $\quad V_a = V_{ao} + V_{a1} + V_{a2}$ ...... (7.15)

Voltage in phase 'b' is $\quad V_b = V_{bo} + V_{b1} + V_{b2}$

Voltage in phase 'c' is $\quad V_c = V_{co} + V_{c1} + V_{c2}$

From eq. (7.10), (7.12) and (7.14) $\quad V_b = V_{ao} + a^2\ V_{a1} + a\ V_{a2}$ ...... (7.16)

From eq. (7.11), (7.13) and (7.14) $\quad V_c = V_{ao} + a\ V_{a1} + a^2\ V_{a2}$ ...... (7.17)

From eq. (7.15), (7.16) and (7.17)

Phase voltages of a, b, c are $\begin{bmatrix} V_a \\ V_b \\ V_c \end{bmatrix} = \begin{bmatrix} 1 & 1 & 1 \\ 1 & a^2 & a \\ 1 & a & a^2 \end{bmatrix} \begin{bmatrix} V_{ao} \\ V_{a1} \\ V_{a2} \end{bmatrix}$ ...... (7.18)

Consider a transformation matrix $[A] = \begin{bmatrix} 1 & 1 & 1 \\ 1 & a^2 & a \\ 1 & a & a^2 \end{bmatrix}$

$$[A]^{-1} = \begin{bmatrix} 1 & 1 & 1 \\ 1 & a^2 & a \\ 1 & a & a^2 \end{bmatrix}^{-1} = \frac{1}{3}\begin{bmatrix} 1 & 1 & 1 \\ 1 & a & a^2 \\ 1 & a^2 & a \end{bmatrix}$$

The symmetrical voltage components of phase 'a' are $\begin{bmatrix} V_{a0} \\ V_{a1} \\ V_{a2} \end{bmatrix} = [A]^{-1} \begin{bmatrix} V_a \\ V_b \\ V_c \end{bmatrix}$

The symmetrical voltage components of phase 'a' are $\begin{bmatrix} V_{a0} \\ V_{a1} \\ V_{a2} \end{bmatrix} = \frac{1}{3}\begin{bmatrix} 1 & 1 & 1 \\ 1 & a & a^2 \\ 1 & a^2 & a \end{bmatrix}\begin{bmatrix} V_a \\ V_b \\ V_c \end{bmatrix}$ ... ... (7.19)

Zero sequence voltage of phase 'a' is $V_{a0} = \frac{1}{3}(V_a + V_b + V_c)$ ... ... (7.20)

Positive sequence voltage of phase 'a' is $V_{a1} = \frac{1}{3}(V_a + a\,V_b + a^2\,V_c)$ ... ... (7.21)

Negative sequence voltage of phase 'a' is $V_{a2} = \frac{1}{3}(V_a + a^2\,V_b + a\,V_c)$ ... ... (7.22)

7.5 Relation between unbalanced currents and symmetrical components:

Let I_a, I_b *and* I_c represent an unbalanced set of current phasors. Each of the unbalanced phasors is sum of its components. Similarly to section 7.4, the currents in phase a, b and c can be calculated as follows:

Current in phase 'a' is $I_a = I_{a0} + I_{a1} + I_{a2}$ ... ... (7.23)

Current in phase 'b' is $I_b = I_{b0} + I_{b1} + I_{b2} = I_{a0} + a^2\,I_{a1} + a\,I_{a2}$ ... ... (7.24)

Current in phase 'c' is $I_c = I_{c0} + I_{c1} + I_{c2} = I_{a0} + a\,I_{a1} + a^2\,I_{a2}$ ... ... (7.25)

The currents in phase a, b and c are $\begin{bmatrix} I_a \\ I_b \\ I_c \end{bmatrix} = \begin{bmatrix} 1 & 1 & 1 \\ 1 & a^2 & a \\ 1 & a & a^2 \end{bmatrix}\begin{bmatrix} I_{a0} \\ I_{a1} \\ I_{a2} \end{bmatrix}$ ... ... (7.26)

The symmetrical current components of phase 'a' are $\begin{bmatrix} I_{a0} \\ I_{a1} \\ I_{a2} \end{bmatrix} = \frac{1}{3}\begin{bmatrix} 1 & 1 & 1 \\ 1 & a & a^2 \\ 1 & a^2 & a \end{bmatrix}\begin{bmatrix} I_a \\ I_b \\ I_c \end{bmatrix}$ ... ... (7.27)

Zero sequence current of phase 'a' is $I_{a0} = \frac{1}{3}(I_a + I_b + I_c)$ ... ... (7.28)

Positive sequence current of phase 'a' is $I_{a1} = \frac{1}{3}(I_a + a\,I_b + a^2\,I_c)$ ... ... (7.29)

Negative sequence current of phase 'a' is $I_{a2} = \frac{1}{3}(I_a + a^2\,I_b + a\,I_c)$ ... ... (7.30)

Example 7.1. The phase voltages of a 3-phase unbalanced system are $V_a = 100\angle 0^0\,V, V_b = 200\angle 90^0\,V, V_c = 300\angle{-135^0}\,V$. Determine the symmetrical components of voltages.

Ans: The symmetrical voltage components of phase 'a' are $\begin{bmatrix} V_{a0} \\ V_{a1} \\ V_{a2} \end{bmatrix} = \frac{1}{3}\begin{bmatrix} 1 & 1 & 1 \\ 1 & a & a^2 \\ 1 & a^2 & a \end{bmatrix}\begin{bmatrix} V_a \\ V_b \\ V_c \end{bmatrix}$

Zero sequence voltage of phase 'a' is $V_{a0} = \frac{1}{3}(V_a + V_b + V_c)$

$$= \frac{1}{3}(100\angle 0^0 + 200\angle 90^0 + 300\angle{-135^0})$$

$$= -37.37 - j4.04 = 37.59\angle - 173.82^0\ V$$

Positive sequence voltage of phase 'a' is $V_{a1} = \frac{1}{3}(V_a + a\,V_b + a^2\,V_c)$

$$= \frac{1}{3}(100\angle 0^0 + 1\angle 120^0 * 200\angle 90^0 + 1\angle 240^0 * 300\angle -135^0)$$

$$= -50.28 + j63.25$$

$$= 80.8\angle 128.48^0\ V$$

Negative sequence voltage of phase 'a' is $V_{a2} = \frac{1}{3}(V_a + a^2\,V_b + a\,V_c)$

$$= \frac{1}{3}(100\angle 0^0 + 1\angle 240^0 * 200\angle 90^0 + 1\angle 120^0 * 300\angle -135^0)$$

$$= 186.91 - j58.97$$

$$= 196\angle -17.51^0\ V$$

Zero sequence voltage of all three phases are $V_{ao} = V_{bo} = V_{co} = 37.59\angle -173.82^0\ V$

Positive sequence voltage of all three phases is V_{a1}, V_{b1} *and* V_{c1}

$$V_{a1} = 80.8\angle 128.48^0\ V$$

$$V_{b1} = a^2 V_{a1} = 1\angle 240^0 * 80.8\angle 128.48^0 = 80.8\angle 8.48^0\ V$$

$$V_{c1} = a\,V_{a1} = 1\angle 120^0 * 80.8\angle 128.48^0 = 80.8\angle -111.52^0\ V$$

Negative sequence voltage of all three phases is V_{a2}, V_{b2} *and* V_{c2}

$$V_{a2} = 196\angle -17.51^0\ V$$

$$V_{b2} = a\,V_{a2} = 1\angle 120^0 * 196\angle -17.51^0 = 196\angle 102.49^0\ V$$

$$V_{c2} = a^2\,V_{a2} = 1\angle 240^0 * 196\angle -17.51^0 = 196\angle -137.51^0 V$$

Example 7.2. Symmetrical voltage components of phase 'a' in an unbalanced system are $V_{a0} = 50\angle 140^0\ V, V_{a1} = 150\angle 0^0\ V, V_{a2} = 250\angle -100^0 V$. Calculate the phase voltages of a, b and c.

Ans: Phase voltages of a, b and c are $\begin{bmatrix} V_a \\ V_b \\ V_c \end{bmatrix} = \begin{bmatrix} 1 & 1 & 1 \\ 1 & a^2 & a \\ 1 & a & a^2 \end{bmatrix} \begin{bmatrix} V_{a0} \\ V_{a1} \\ V_{a2} \end{bmatrix}$

Voltage in phase 'a' is $V_a = V_{ao} + V_{a1} + V_{a2}$

$$= 50\angle 140^0 + 150\angle 0^0 + 250\angle -100^0$$

$$= 68.28 - j214.06$$

$$= 224.69\angle -72.3^0\ V$$

Voltage in phase 'b' is $V_b = V_{ao} + a^2\,V_{a1} + a\,V_{a2}$

$$= 50\angle 140^0 + 1\angle 240^0 * 150\angle 0^0 + 1\angle 120^0 * 250\angle -100^0$$

$$= 121.62 - j12.25$$

$$= 122.23\angle -5.755^0\ V$$

Voltage in phase 'c' is $V_c = V_{ao} + a\,V_{a1} + a^2\,V_{a2}$

$$= 50\angle 140^0 + 1\angle 120^0 * 150\angle 0^0 + 1\angle 240^0 * 250\angle -100^0$$

$$= -304.81 + j322.74$$

$$= 443.92\angle 133.36^0\ V$$

Example 7.3. The phase currents of a 3-phase unbalanced system are $I_a = 10\angle 25^0 A$, $I_b = 20\angle 80^0 A, I_c = 30\angle -100^0 A$. Calculate the symmetrical components of currents.

Ans: The symmetrical current components of phase 'a' are $\begin{bmatrix} I_{a0} \\ I_{a1} \\ I_{a2} \end{bmatrix} = \frac{1}{3}\begin{bmatrix} 1 & 1 & 1 \\ 1 & a & a^2 \\ 1 & a^2 & a \end{bmatrix}\begin{bmatrix} I_a \\ I_b \\ I_c \end{bmatrix}$

Zero sequence current of phase 'a' is $I_{a0} = \frac{1}{3}(I_a + I_b + I_c)$

$$= \frac{1}{3}(10\angle 25^0 + 20\angle 80^0 + 30\angle -100^0)$$
$$= 2.442 - j1.8739$$
$$= 3.0783\angle -37.5^0 A$$

Positive sequence current of phase 'a' is $I_{a1} = \frac{1}{3}(I_a + a\,I_b + a^2\,I_c)$

$$= \frac{1}{3}(10\angle 25^0 + 1\angle 120^0 * 20\angle 80^0 + 1\angle 240^0 * 30\angle -100^0)$$
$$= -10.90 + j5.556$$
$$= 12.238\angle 152.99^0\ A$$

Negative sequence current of phase 'a' is $I_{a2} = \frac{1}{3}(I_a + a^2\,I_b + a\,I_c)$

$$= \frac{1}{3}(10\angle 25^0 + 1\angle 240^0 * 20\angle 80^0 + 1\angle 120^0 * 30\angle -100^0)$$
$$= 17.52 + j0.5436$$
$$= 17.53\angle 1.77^0\ A$$

Zero sequence current of all three phases is $I_{a0} = I_{b0} = I_{c0} = 3.0783\angle -37.5^0\ A$

Positive sequence current of all three phases is $I_{a1}, I_{b1}\ and\ I_{c1}$

$$I_{a1} = 12.238\angle 152.99^0\ A$$
$$I_{b1} = a^2\,I_{a1} = 1\angle 240^0 * 12.238\angle 152.99^0 = 12.238\angle 32.99^0\ A$$
$$I_{c1} = a\ I_{a1} = 1\angle 120^0 * 12.238\angle 152.99^0 = 12.238\angle -87.01^0\ A$$

Negative sequence current of all three phases is $I_{a2}, I_{b2}\ and\ I_{c2}$

$$I_{a2} = 17.53\angle 1.77^0\ A$$
$$I_{b2} = a\,I_{a2} = 1\angle 120^0 * 17.53\angle 1.77^0 = 17.53\angle 121.77^0 A$$
$$I_{c2} = a^2\,I_{a2} = 1\angle 240^0 * 17.53\angle 1.77^0 = 17.53\angle -118.23^0\ A$$

Example 7.4. Symmetrical current components of phase 'a' in an unbalanced system are $I_{a0} = 25\angle 80^0 A, I_{a1} = 50\angle -40^0\ A, I_{a2} = 75\angle 0^0 A$. Calculate the phase currents of a, b and c.

Ans: The currents in Phase a, b and c are $\begin{bmatrix} I_a \\ I_b \\ I_c \end{bmatrix} = \begin{bmatrix} 1 & 1 & 1 \\ 1 & a^2 & a \\ 1 & a & a^2 \end{bmatrix}\begin{bmatrix} I_{a0} \\ I_{a1} \\ I_{a2} \end{bmatrix}$

Current in phase 'a' is $I_a = I_{a0} + I_{a1} + I_{a2}$

$$= 25\angle 80^0 + 50\angle -40^0 + 75\angle 0^0$$
$$= 117.64 - j7.519 = 117.88\angle -3.657^0\ A$$

Current in phase 'b' is $I_b = I_{a0} + a^2 I_{a1} + a I_{a2}$

$$= 25\angle 80^0 + 1\angle 240^0 * 50\angle -40^0 + 1\angle 120^0 * 75\angle 0^0$$
$$= -80.143 + j72.471$$
$$= 108.05\angle 137.87^0\ A$$

Current in phase 'c' is $I_c = I_{a0} + a\, I_{a1} + a^2 I_{a2}$

$$= 25\angle 80^0 + 1\angle 120^0 * 50\angle -40^0 + 1\angle 240^0 * 75\angle 0^0$$
$$= 24.47 + j8.908$$
$$= 26.04\ \angle 160^0\ A$$

Example 7.5. There is an open conductor in a 3-phase line. Line A is experiencing a current of 25A. Given that line 'C' is open, determine the symmetrical components of the line currents assuming neutral ungrounded.

Ans: Consider I_a as reference $I_a = 25\angle 0^0$

Since line 'C' is open, $I_c = 0$, So $I_b\, must\ be\ 25\angle 180^0$

Zero sequence current of phase 'a' is $I_{a0} = \frac{1}{3}\,(I_a + I_b + I_c)$

$$= \frac{1}{3}\,(25\angle 0^0 + 25\angle 180^0 + 0)$$
$$= 0$$

Positive sequence current of phase 'a' is $I_{a1} = \frac{1}{3}\,(I_a + a\, I_b + a^2\, I_c)$

$$= \frac{1}{3}\,(25\angle 0^0 + 1\angle 120^0 * 25\angle 180^0 + 1\angle 240^0 * 0)$$
$$= 14.43\angle -30^0\ A$$

Negative sequence current of phase 'a' is $I_{a2} = \frac{1}{3}\,(I_a + a^2\, I_b + a\, I_c)$

$$= \frac{1}{3}\,(25\angle 0^0 + 1\angle 240^0 * 25\angle 180^0 + 1\angle 120^0 * 0)$$
$$= 14.43\angle 30^0\ A$$

Zero sequence current of all three phases is $I_{a0} = I_{b0} = I_{c0} = 0$

Positive sequence current of all three phases is $I_{a1}, I_{b1}\ and\ I_{c1}$

$$I_{a1} = 14.43\angle -30^0\ A$$
$$I_{b1} = a^2\, I_{a1} = 1\angle 240^0 * 14.43\angle -30^0 = 14.43\angle -150^0\ A$$
$$I_{c1} = a\ I_{a1} = 1\angle 120^0 * 14.43\angle -30^0 = 14.43\angle 90^0\ A$$

Negative sequence current of all three phases is $I_{a2}, I_{b2}\ and\ I_{c2}$

$$I_{a2} = 14.43\angle 30^0\ A$$
$$I_{b2} = a\, I_{a2} = 1\angle 120^0 * 14.43\angle 30^0 = 14.43\angle 150^0\ A$$
$$I_{c2} = a^2\, I_{a2} = 1\angle 240^0 * 14.43\angle 30^0 = 14.43\angle -90^0\ A$$

To verify: Current in phase 'a' is $I_a = I_{a0} + I_{a1} + I_{a2}$

$= 0 + 14.43\angle{-30^0} + 14.43\angle 30^0$

$= 25\ A$

Current in phase 'b' is $I_b = I_{a0} + a^2\ I_{a1} + a\ I_{a2}$

$= 0 + 1\angle 240^0 * 14.43\angle{-30^0} + 1\angle 120^0 * 14.43\angle 30^0$

$= 25\angle 180^0\ A$

Current in phase 'c' is $I_c = I_{a0} + a\ I_{a1} + a^2\ I_{a2}$

$= 0 + 1\angle 120^0 * 14.43\angle{-30^0} + 1\angle 240^0 * 14.43\angle 30^0$

$= 0$

Example 7.6. A delta connected resistive load is connected across an unbalanced supply. The line currents are expressed as depicted in Fig. Identify the symmetrical components of the line currents.

Ans: $I_a + I_b + I_c = 0$

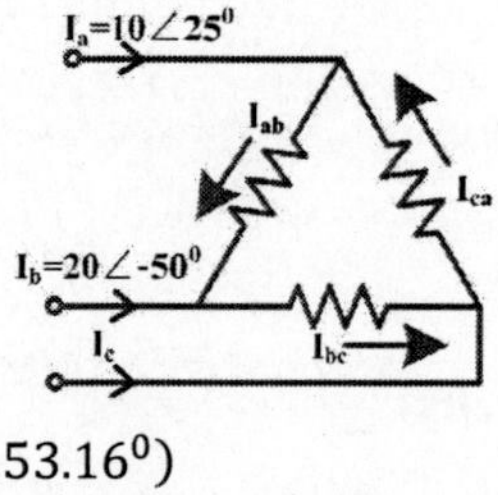

$I_c = -(I_a + I_b) = -(10\angle 25^0 + 20\angle{-50^0})$

$= 24.56\angle 153.16^0$

Zero sequence current of all three phases is $I_{a0} = I_{b0} = I_{c0} = 0$

Positive sequence current of phase 'a' is $I_{a1} = \frac{1}{3}\ (I_a + a\ I_b + a^2\ I_c)$

$= \frac{1}{3}\ (10\angle 25^0 + 1\angle 120^0 * 20\angle{-50^0} + 1\angle 240^0 * 24.56\angle 153.16^0)$

$= 17.18\angle 44.99^0\ A$

Negative sequence current of phase 'a' is $I_{a2} = \frac{1}{3}\ (I_a + a^2\ I_b + a\ I_c)$

$= \frac{1}{3}\ (10\angle 25^0 + 1\angle 240^0 * 20\angle{-50^0} + 1\angle 120^0 * 24.56\angle 153.16^0)$

$= 8.505\angle{-111.32^0}\ A$

Positive sequence current of all three phases is $I_{a1}, I_{b1}\ and\ I_{c1}$

$I_{a1} = 17.18\angle 44.99^0\ A$

$I_{b1} = a^2\ I_{a1} = 1\angle 240^0 * 17.18\angle 44.99^0 = 17.18\angle{-75^0}\ A$

$I_{c1} = a\ I_{a1} = 1\angle 120^0 * 17.18\angle 44.99^0 = 17.18\angle 165^0\ A$

Negative sequence current of all three phases is $I_{a2}, I_{b2}\ and\ I_{c2}$

$I_{a2} = 8.505\angle{-111.32^0}\ A$

$I_{b2} = a\ I_{a2} = 1\angle 120^0 * 8.505\angle{-111.32^0} = 8.505\angle 8.68^0\ A$

$I_{c2} = a^2\ I_{a2} = 1\angle 240^0 * 8.505\angle{-111.32^0} = 8.505\angle 128.68^0\ A$

Example 7.7. A resistive load linked in a delta configuration is connected across a balanced 3-phase power supply with a voltage of 415V. $R_{ab} = 50\Omega, R_{bc} = 100\Omega, and\ R_{ca} = 150\Omega$. Calculate the symmetrical components of the line currents and delta currents. (Note: Students should complete the solution by solving for symmetrical components of I_b, I_c, I_{bc} and I_{ca}).

Ans: $V_{ab} = 415\angle 0^0 V, V_{bc} = 415\angle - 120^0 V, \text{and } V_{ca} = 415\angle 120^0 V$

$$I_{ab} = \frac{V_{ab}}{R_{ab}} = \frac{415}{50} = 8.3\angle 0^0\ A$$

$$I_{bc} = \frac{V_{bc}}{R_{bc}} = \frac{415\angle - 120^0}{100} = 4.15\angle - 120^0 A$$

$$I_{ca} = \frac{V_{ca}}{R_{ca}} = \frac{415\angle 120^0}{150} = 2.766\angle 120^0\ A$$

Symmetrical components of delta currents are I_{ab0}, I_{ab1} *and* I_{ab2}

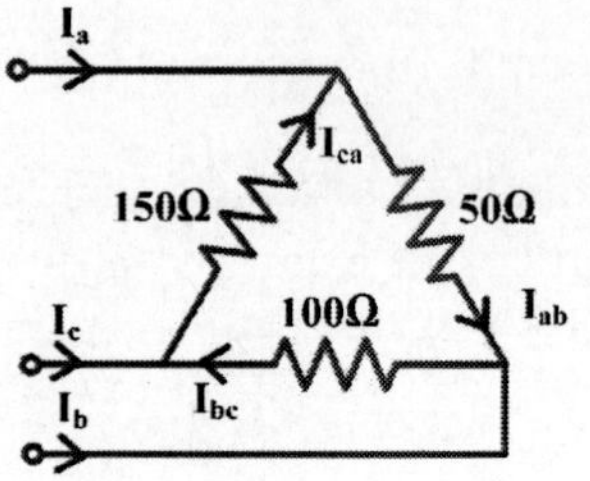

Ia, Ica, 150Ω, 50Ω, Ic, 100Ω, Iab, Ib, Ibc

$$I_{ab0} = \frac{1}{3}(I_{ab} + I_{bc} + I_{ca})$$

$$= \frac{1}{3}(8.3\angle 0^0 + 4.15\angle - 120^0 + 2.766\angle 120^0)$$

$$= 1.662\angle{-13.903^0}\ A$$

$$I_{ab1} = \frac{1}{3}(I_{ab} + a\ I_{bc} + a^2\ I_{ca})$$

$$= \frac{1}{3}(8.3\angle 0^0 + 1\angle 120^0 * 4.15\angle - 120^0 + 1\angle 240^0 * 2.766\angle 120^0)$$

$$= 5.072\ A$$

$$I_{ab2} = \frac{1}{3}(I_{ab} + a^2\ I_{bc} + a\ I_{ca})$$

$$= \frac{1}{3}(8.3\angle 0^0 + 1\angle 240^0 * 4.15\angle - 120^0 + 1\angle 120^0 * 2.766\angle 120^0)$$

$$= 1.662\angle 13.903^0\ A$$

The line currents are calculated as follows

$$I_A = I_{ab} - I_{ca} = 8.3\angle 0^0 - 2.766\angle 120^0 = 9.975\angle{-13.89^0}\ A$$

$$I_B = I_{bc} - I_{ab} = 4.15\angle - 120^0 - 8.3\angle 0^0 = 10.979\angle{-160.89^0}\ A$$

$$I_C = I_{ca} - I_{bc} = 2.766\angle 120^0 - 4.15\angle - 120^0 = 6.029\angle 83.4^0\ A$$

Zero sequence current of phase 'a' is $I_{a0} = \frac{1}{3}(I_a + I_b + I_c)$

$$= \frac{1}{3}(9.975\angle{-13.89^0} + 10.979\angle{-160.89^0} + 6.029\angle 83.4^0) = 0$$

Positive sequence current of phase 'a' is $I_{a1} = \frac{1}{3}(I_a + a\ I_b + a^2\ I_c)$

$$= \frac{1}{3}(9.975\angle{-13.89^0} + 1\angle 120^0 * 10.979\angle{-160.89^0} + 1\angle 240^0 * 6.029\angle 83.4^0)$$

$$= 8.784\angle{-30^0}\ A$$

Negative sequence current of phase 'a' is $I_{a2} = \frac{1}{3}(I_a + a^2\ I_b + a\ I_c)$

$$= \frac{1}{3}(9.975\angle{-13.89^0} + 1\angle 240^0 * 10.979\angle{-160.89^0} + 1\angle 120^0 * 6.029\angle 83.4^0)$$

$$= 2.88\angle 43.91^0\ A$$

7.6 Sequence impedances of a synchronous generator

An unloaded synchronous generator grounded through impedance Z_n is shown in Fig. 7.4. $E_a, E_b,$ and E_c are the induced emfs of the three phases. $I_a, I_b,$ and I_c are the fault currents flow in the lines and I_n is the neutral current which is flowing through the impedance Z_n during the fault.

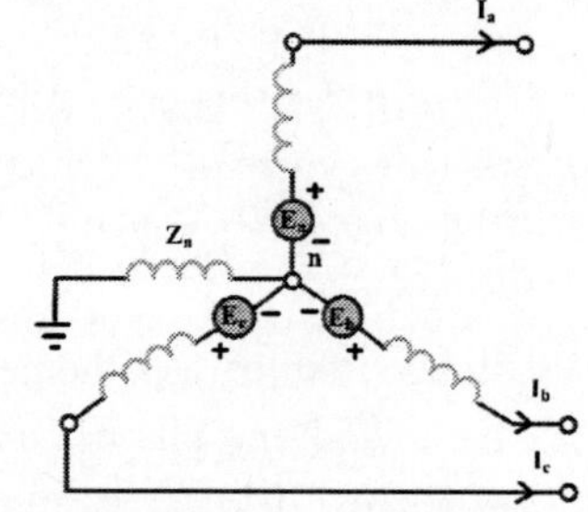

Fig. 7.4 Synchronous generator

Neutral current $I_n = (I_a + I_b + I_c)$ ……… (7.31)

7.6.1 Positive sequence impedance:

Positive sequence impedance refers to the impedance of the network that allows the flow of positive sequence current. Due to the symmetrical windings of the synchronous generator, it only induces electromotive forces (emfs) in the positive sequence. This means that the emf induced in the negative and zero sequence is zero. As mentioned in the preceding unit, the machine provides a direct axis reactance value that increases from sub-transient reactance X_d'' to transient reactance X_d', and finally to steady state reactance X_d. As the short circuit transient progresses in time, so $X_d'' < X_d' < X_d$.

A synchronous machine's positive sequence network is represented by the induced emf and the positive sequence impedance Z_1. Figure 7.5 shows the three-phase and single-phase positive sequence networks of a synchronous generator. Since all three phasors have the same length and are displaced by identical angles of 120^0 degrees from each other, the sum of the positive sequence currents$I_{a1}, I_{b1},$ and I_{c1} will be zero. It is not possible for neutral current to pass through the neutral impedance Z_n. In the positive sequence network, it may be concluded that the neutral impedance Z_n is absent.

Neutral current in positive sequence network $I_n = I_{a1} + I_{b1} + I_{c1} = 0$ ……… (7.32)

Positive sequence impedance during sub-transient, transient and steady-state period can be considered as below:

during sub-transient $Z_g^1 = jX_d''$ ……… (7.33)

during transient period $Z_g^1 = jX_d'$ ……… (7.34)

during steady-state period $Z_g^1 = jX_d$ ……… (7.35)

Positive sequence voltage of terminal 'a' $V_{a1} = E_a - I_{a1} * Z_1$ ……… (7.36)

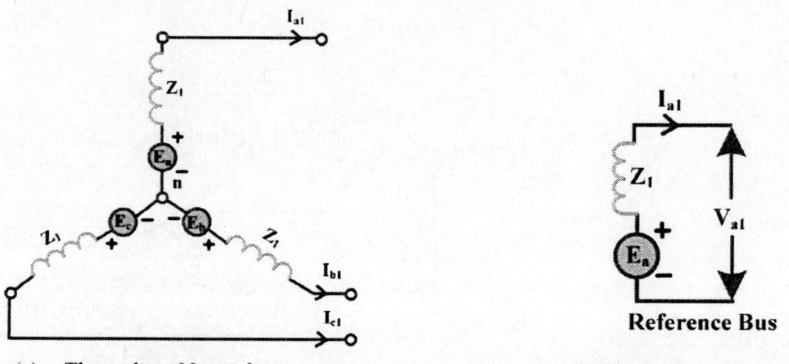

(a) Three phase Network (b) Single phase Network

Fig. 7.5 Synchronous generator's positive sequence network

7.6.2 Negative sequence impedance:

The negative sequence impedance refers to the impedance of the network that allows the flow of negative sequence currents. The negative-sequence network for a synchronous machine can be described solely by the negative sequence impedance Z_2, as the synchronous machine does not produce any negative sequence emf.

Figure 7.6 illustrates the negative sequence network of a synchronous generator, which consists of three phases and one phase. Given that all three phasors have the same length and are displaced by equal angles of 120^0 degrees in phase opposition, the sum of the positive sequence currents I_{a2}, I_{b2}, and I_{c2} will be zero. The neutral impedance Z_n does not exist in the negative sequence network because it prevents the flow of neutral current.

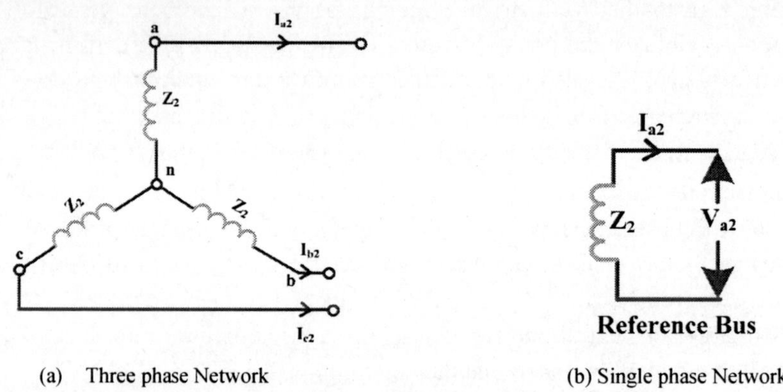

(a) Three phase Network (b) Single phase Network

Fig. 7.6 Negative sequence network of a synchronous generator

Neutral current in negative sequence network $I_n = I_{a2} + I_{b2} + I_{c2} = 0$ ... ... (7.37)

Negative-sequence voltage of terminal 'a' $V_{a2} = -I_{a2} * Z_2$ ... ... (7.38)

7.6.3 Zero sequence impedance:

Zero-sequence impedance refers to the impedance of the network that allows the flow of zero sequence currents. The zero sequence currents in all three phases contribute to the current flowing in the neutral impedance Z_n. Consequently, the voltage drop resulting from this current is $3I_{a0}Z_n$, as depicted in Figure 7.7 (b).

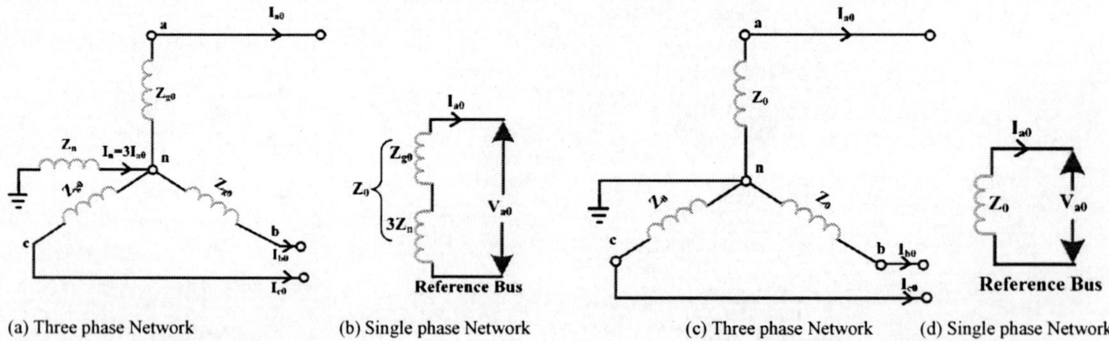

(a) Three phase Network (b) Single phase Network (c) Three phase Network (d) Single phase Network

Fig. 7.7 Zero sequence network of a synchronous generator: (a&b) with neutral impedance Z_n, (c&d) Solid grounding

Neutral current in zero sequence network $I_n = I_{a0} + I_{b0} + I_{c0} = 3I_{a0}$ ……… (7.39)

The voltage drop caused by neutral impedance $= 3I_{a0}Z_n$ ……… (7.40)

Due to the absence of zero sequence electromotive force (emf) generation in synchronous machines, the zero-sequence network associated with such machines can be mathematically represented by the zero-sequence impedance Z_0 and a neutral impedance $3Z_n$.

The voltage-drop of a zero-sequence from terminal ‘a’ to ground $= 3I_{a0}Z_n + I_{a0}Z_{g0}$

$$= I_{a0}(3Z_n + Z_{g0}) \quad \text{……… (7.41)}$$

Zero-sequence impedance of a synchronous machine with neutral impedance will be

$$Z_0 = Z_{g0} + 3Z_n \quad \text{……… (7.42)}$$

Zero-sequence voltage of a synchronous machine with neutral impedance will be

$$V_{a0} = -I_{a0} * (Z_{g0} + 3Z_n) \quad \text{……… (7.43)}$$

Whereas, zero-sequence impedance of a solidly grounded synchronous machine will be Z_0 only which is shown in Fig. 7.7 (c and d).

Zero-sequence voltage of a solidly grounded synchronous machine at terminal ‘a’

$$V_{a0} = -I_{a0} * Z_0 \quad \text{……… (7.44)}$$

7.7 Symmetrical component voltages of a synchronous generator:

Figure 7.8 depicts a three-phase synchronous generator that is grounded via impedance Z_n. The relationship between the voltages and currents in a sequence for a balanced voltage supply can be determined by applying Kirchhoff's rules to each phase. Let's consider the impedance of each phase winding as Z_s.

$$V_a = E_a - Z_sI_a - Z_nI_n$$

$$V_b = E_b - Z_sI_b - Z_nI_n$$

$$V_c = E_c - Z_sI_c - Z_nI_n \quad \text{……… (7.45)}$$

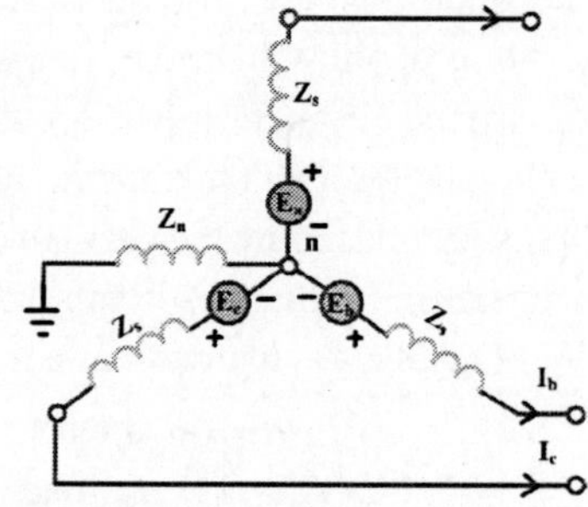

Fig. 7.8. A three-phase synchronous generator grounded through impedance Z_n

As discussed in previous sections, the sequence voltages can be written in terms of sequence currents

The sequence voltages at terminal ‘a’ are $\begin{bmatrix} V_{a0} \\ V_{a1} \\ V_{a2} \end{bmatrix} = \begin{bmatrix} 0 \\ E \\ 0 \end{bmatrix} - \begin{bmatrix} Z_{g0} + 3Z_n & 0 & 0 \\ 0 & Z_1 & 0 \\ 0 & 0 & Z_2 \end{bmatrix} \begin{bmatrix} I_{a0} \\ I_{a1} \\ I_{a2} \end{bmatrix}$ ……… (7.46)

Zero-sequence voltage of terminal ‘a’ $V_{a0} = -I_{a0} * (Z_{g0} + 3Z_n)$ ……… (7.47)

Positive sequence voltage of terminal ‘a’ $V_{a1} = E - I_{a1} * Z_1$ ……… (7.48)

Negative-sequence voltage of terminal ‘a’ $V_{a2} = -I_{a2} * Z_2$ ……… (7.49)

Symmetrical component voltages of a synchronous generator obtained in eq. (7.47), (7.48) and (7.49) are same as the equations obtained in eq. (7.43), (7.36) and (7.38) respectively.

When a three-phase synchronous generator is solidly grounded then the sequence voltages at terminal 'a' will become

$$\begin{bmatrix} V_{a0} \\ V_{a1} \\ V_{a2} \end{bmatrix} = \begin{bmatrix} 0 \\ E \\ 0 \end{bmatrix} - \begin{bmatrix} Z_0 & 0 & 0 \\ 0 & Z_1 & 0 \\ 0 & 0 & Z_2 \end{bmatrix} \begin{bmatrix} I_{a0} \\ I_{a1} \\ I_{a2} \end{bmatrix} \quad \ldots\ldots\ldots (7.50)$$

The positive and negative-sequence voltages at terminal 'a' remain same but the zero-sequence impedance of a solidly grounded synchronous machine will be Z_0.

Zero-sequence voltage of a solidly grounded synchronous machine at terminal 'a' is

$$V_{a0} = -I_{a0} * Z_0 \quad \ldots\ldots\ldots (7.51)$$

7.8 Sequence Impedances of Transmission lines:

A fully transposed three-phase line is inherently symmetrical, meaning that the positive and negative-sequence impedances of a transmission line are not affected by the order of the phases and are identical. When there is simply zero-sequence current flowing in a transmission line, the currents in each phase have the same amplitude and phase. A portion of these currents is redirected through the ground, while the remaining portion is rerouted through above ground wires. The magnetic field generated by the zero-sequence currents flowing through the transmission lines, ground wires, and ground differs significantly from the magnetic field produced by the passage of positive and negative sequence currents. The zero-sequence impedance of transmission lines is approximately 2 to 4 times greater than the positive-sequence impedance. Figure 7.9 displays a depiction of a three-phase transmission line.

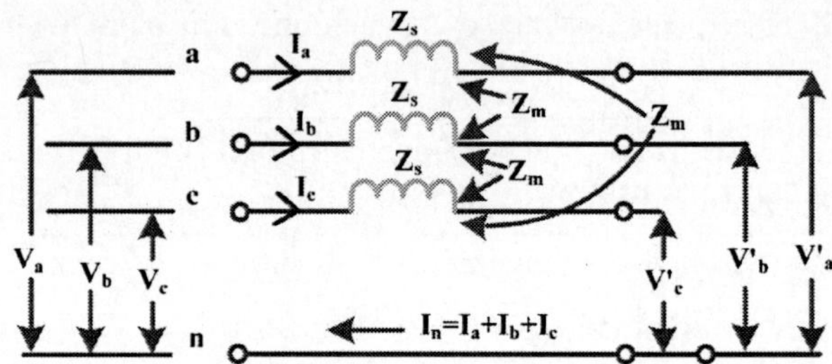

Fig. 7.9 Representation of a three-phase transmission line

Let I_a, I_b *and* I_c represent an unbalanced set of current phasors.

V_a, V_b *and* V_c are voltages of a transmission line on primary side.

$V_a', V_b',$ *and* V_c' are voltages of a transmission line on secondary side.

Z_s is the impedance of each phase winding.

Z_m is the impedance between phases a and b, b and c, c and a respectively.

Applying KVL for the above circuit,

$$V_a - V_a' = Z_s I_a + Z_m I_b + Z_m I_c$$
$$V_b - V_b' = Z_m I_a + Z_s I_b + Z_m I_c$$
$$V_c - V_c' = Z_m I_a + Z_m I_b + Z_s I_c \quad \ldots\ldots\ldots (7.52)$$

$$\begin{bmatrix} V_a \\ V_b \\ V_c \end{bmatrix} - \begin{bmatrix} V_a' \\ V_b' \\ V_c' \end{bmatrix} = \begin{bmatrix} Z_s & Z_m & Z_m \\ Z_m & Z_s & Z_m \\ Z_m & Z_m & Z_s \end{bmatrix} \begin{bmatrix} I_a \\ I_b \\ I_c \end{bmatrix} \quad \ldots\ldots\ldots (7.53)$$

Consider $[V_L] = \begin{bmatrix} V_a \\ V_b \\ V_c \end{bmatrix}, [V_L'] = \begin{bmatrix} V_a' \\ V_b' \\ V_c' \end{bmatrix}, [z] = \begin{bmatrix} Z_s & Z_m & Z_m \\ Z_m & Z_s & Z_m \\ Z_m & Z_m & Z_s \end{bmatrix}$ and $[I_L] = \begin{bmatrix} I_a \\ I_b \\ I_c \end{bmatrix}$

Sequence voltages and currents $[V_s] = \begin{bmatrix} V_{a0} \\ V_{a1} \\ V_{a2} \end{bmatrix}, [V_s'] = \begin{bmatrix} V'_{a0} \\ V'_{a1} \\ V'_{a2} \end{bmatrix}, [z] = \begin{bmatrix} Z_s & Z_m & Z_m \\ Z_m & Z_s & Z_m \\ Z_m & Z_m & Z_s \end{bmatrix}$ *and* $[I_s] = \begin{bmatrix} I_{a0} \\ I_{a1} \\ I_{a2} \end{bmatrix}$ ....(7.54)

By substituting above parameters in eq. (7.53) $[V_L] - [V_L'] = [z][I_L]$

Expressing phase voltages and currents in terms of sequence voltages and currents

$$[V_s] - [V_s'] = [z][I_s]$$

$$[A][V_s - V_s'] = [z][A][I_s]$$

$$[V_s - V_s'] = [A]^{-1}[z][A][I_s] \qquad \text{.........(7.55)}$$

Where $[A] = \begin{bmatrix} 1 & 1 & 1 \\ 1 & a^2 & a \\ 1 & a & a^2 \end{bmatrix}$ and $[A]^{-1} = \begin{bmatrix} 1 & 1 & 1 \\ 1 & a^2 & a \\ 1 & a & a^2 \end{bmatrix}^{-1} = \frac{1}{3}\begin{bmatrix} 1 & 1 & 1 \\ 1 & a & a^2 \\ 1 & a^2 & a \end{bmatrix}$

$$[A]^{-1}[z][A] = \frac{1}{3}\begin{bmatrix} 1 & 1 & 1 \\ 1 & a & a^2 \\ 1 & a^2 & a \end{bmatrix}\begin{bmatrix} Z_s & Z_m & Z_m \\ Z_m & Z_s & Z_m \\ Z_m & Z_m & Z_s \end{bmatrix}\begin{bmatrix} 1 & 1 & 1 \\ 1 & a^2 & a \\ 1 & a & a^2 \end{bmatrix}$$

$$= \frac{1}{3}\begin{bmatrix} 1 & 1 & 1 \\ 1 & a & a^2 \\ 1 & a^2 & a \end{bmatrix}\begin{bmatrix} Z_s + Z_m + Z_m & Z_s + a^2Z_m + aZ_m & Z_s + aZ_m + a^2Z_m \\ Z_m + Z_s + Z_m & Z_m + a^2Z_s + aZ_m & Z_m + aZ_s + a^2Z_m \\ Z_m + Z_m + Z_s & Z_m + a^2Z_m + aZ_s & Z_m + aZ_m + a^2Z_s \end{bmatrix}$$

$$= \frac{1}{3}\begin{bmatrix} 1 & 1 & 1 \\ 1 & a & a^2 \\ 1 & a^2 & a \end{bmatrix}\begin{bmatrix} Z_s + 2Z_m & Z_s + (a^2 + a)Z_m & Z_s + (a + a^2)Z_m \\ Z_s + 2Z_m & a^2Z_s + (1 + a)Z_m & aZ_s + (1 + a^2)Z_m \\ Z_s + 2Z_m & aZ_s + (1 + a^2)Z_m & a^2Z_s + (1 + a)Z_m \end{bmatrix}$$

$$= \frac{1}{3}\begin{bmatrix} 1 & 1 & 1 \\ 1 & a & a^2 \\ 1 & a^2 & a \end{bmatrix}\begin{bmatrix} Z_s + 2Z_m & Z_s - Z_m & Z_s - Z_m \\ Z_s + 2Z_m & a^2(Z_s - Z_m) & a(Z_s - Z_m) \\ Z_s + 2Z_m & a(Z_s - Z_m) & a^2(Z_s - Z_m) \end{bmatrix}$$

$$= \frac{1}{3}\begin{bmatrix} 3(Z_s + 2Z_m) & (Z_s - Z_m)(1 + a + a^2) & (Z_s - Z_m)(1 + a + a^2) \\ (Z_s + 2Z_m)(1 + a + a^2) & (Z_s - Z_m)(1 + a^3 + a^3) & (Z_s - Z_m)(1 + a^2 + a^4) \\ (Z_s + 2Z_m)(1 + a + a^2) & (Z_s - Z_m)(1 + a^4 + a^2) & (Z_s - Z_m)(1 + a^3 + a^3) \end{bmatrix}$$

$$= \frac{1}{3}\begin{bmatrix} 3(Z_s + 2Z_m) & (Z_s - Z_m)(0) & (Z_s - Z_m)(0) \\ (Z_s + 2Z_m)(0) & (Z_s - Z_m)(1 + 1 + 1) & (Z_s - Z_m)(1 + a^2 + a) \\ (Z_s + 2Z_m)(0) & (Z_s - Z_m)(1 + a + a^2) & (Z_s - Z_m)(1 + 1 + 1) \end{bmatrix}$$

$$= \frac{1}{3}\begin{bmatrix} 3(Z_s + 2Z_m) & 0 & 0 \\ 0 & 3(Z_s - Z_m) & 0 \\ 0 & 0 & 3(Z_s - Z_m) \end{bmatrix}$$

$$= \begin{bmatrix} (Z_s + 2Z_m) & 0 & 0 \\ 0 & (Z_s - Z_m) & 0 \\ 0 & 0 & (Z_s - Z_m) \end{bmatrix} \qquad \text{.........(7.56)}$$

From eq. (7.54), (7.55) and (7.56), $[V_s - V_s'] = [A]^{-1}[z][A][I_s]$

$$\begin{bmatrix} V_{a0} \\ V_{a1} \\ V_{a2} \end{bmatrix} - \begin{bmatrix} V'_{a0} \\ V'_{a1} \\ V'_{a2} \end{bmatrix} = \begin{bmatrix} (Z_s + 2Z_m) & 0 & 0 \\ 0 & (Z_s - Z_m) & 0 \\ 0 & 0 & (Z_s - Z_m) \end{bmatrix}\begin{bmatrix} I_{a0} \\ I_{a1} \\ I_{a2} \end{bmatrix} \qquad \text{.........(7.57)}$$

Zero-sequence impedance of a transmission line	$Z_{0\ line} = Z_s + 2Z_m$	
Positive-sequence impedance of a transmission line	$Z_{1\ line} = Z_s - Z_m$	
Negative-sequence impedance of a transmission line	$Z_{2\ line} = Z_s - Z_m$	... ... (7.58)

From the above equations we can conclude that

(i). The positive and negative-sequence impedances of a transposed transmission line are identical.

(ii). The zero-sequence impedance of a transposed transmission line is approximately two to four times higher than the positive-sequence impedance.

7.9 Sequence Impedances of Transformers:

The positive-sequence series impedance of a transformer is equivalent to its leakage reactance. Since transformers are static devices, their positive and negative-sequence impedances are the same. This is because impedance is not affected by phase order, as long as the applied voltages are balanced.

$$Z_{1\,T/F} = Z_{2\,T/F} \qquad \text{... ... ... (7.59)}$$

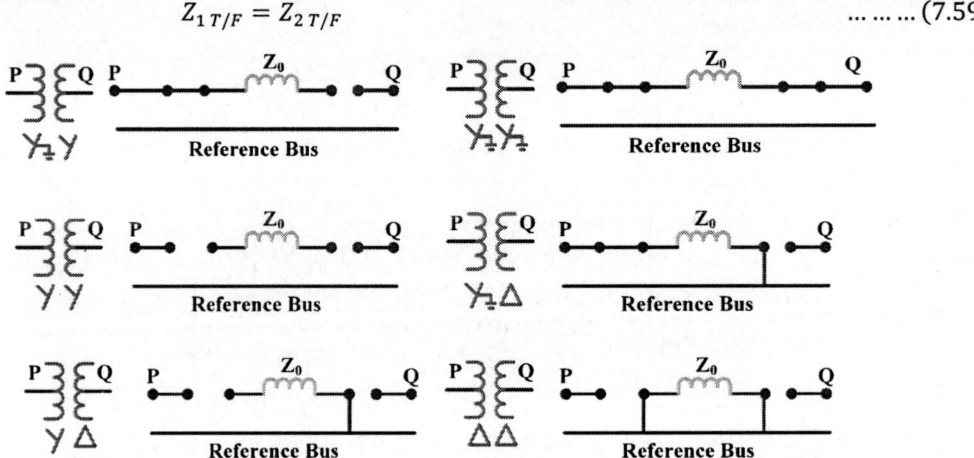

Fig. 7.10 Zero-sequence networks of different combinations of transformer connections

As you are aware, primary and the secondary windings of the three-phase transformers can be connected in different ways like star-star, star with ground – star, star-star with ground, star with ground- star with ground, star-delta, delta-star, star with ground-delta, delta-star with star ground, and delta-delta respectively. Due to possibility of different three-phase transformer connections, it is more complex to find zero-sequence impedance. Zero-sequence networks of different combinations of transformer connections are given in Fig. 7.10.

Zero-sequence currents can only pass through a star-connected winding if the star point is grounded. However, if the star point is isolated, the Zero-sequence currents are unable to pass through the windings. Zero currents cannot flow via the lines linked to a delta-connected winding because there is no suitable channel for these zero-sequence currents to return. If there are any zero-sequence voltages induced in the delta-connected windings, zero-sequence currents can flow through them.

(i). Star with ground to Star (or) Star to Star with ground: If either of the two neutrals of a star-star transformer is not connected to the ground, there will be no flow of zero-sequence currents in the ungrounded/isolated star point, and as a result, there will be no flow in the grounded star. Therefore, there is an open circuit in the zero-sequence network connecting points P and Q.

(ii). Star with ground to Star with ground: As both the neutrals of a star-star transformer are connected to the ground, there will be the presence of zero-sequence currents in the grounded star point. Therefore, there is a closed circuit in the zero-sequence network connecting points P and Q.

(iii). Star-Star: Due to the isolation of the neutrals in both windings, it is not possible for zero-sequence currents to flow.

(iv). Star with ground to Delta: The neutral of the star side is connected to the ground, allowing zero-sequence currents to pass via the grounded star point. Additionally, the balancing zero-sequence currents can flow through the delta connection.

(v). Star to Delta: Due to the star connection with an isolated neutral on the main side, it is necessary to keep the primary side open. On the other hand, the secondary side is delta connected, which means it should be closed. Zero-sequence currents are unable to pass through the windings in this context.

(vi). Delta to Delta: As both sides are connected in a delta configuration, the series switches are open and the shunt switches are closed on both sides. A delta circuit does not allow for the passage of zero-sequence currents in or out of a delta-delta transformer as it lacks a return path. Nevertheless, zero-sequence currents have the ability to flow through the delta windings.

Example 7.8. A single line diagram of a power system is shown in figure. Draw positive, negative and zero-sequence networks.

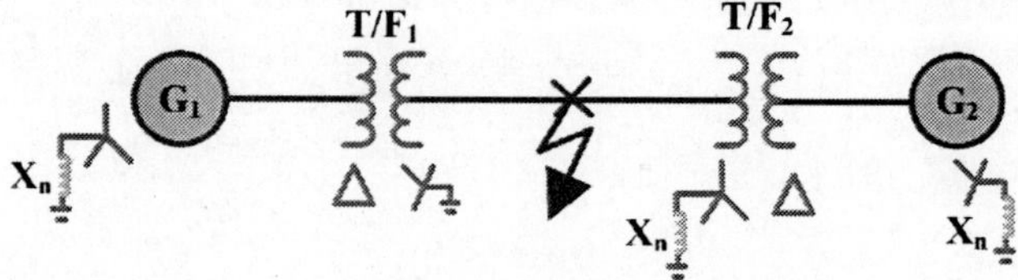

Ans: Positive, negative and zero-sequence networks of the given single line diagram are as follows

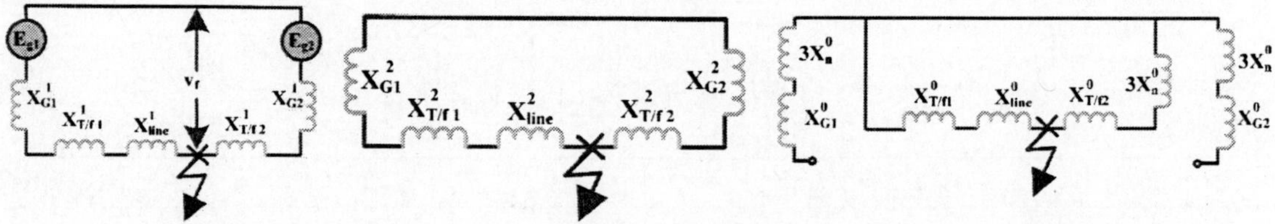

7.10 Three-phase power in terms of symmetrical components:

The power consumed in a 3-phase circuit can be calculated directly with the symmetrical voltage and current components. The aggregate complex power entering a three-phase circuit over all three lines, namely a, b, and c, is

$$S = P + jQ = V_a I_a^* + V_b I_b^* + V_c I_c^* \qquad \text{......... (7.60)}$$

Where V_a, V_b, V_c are phase voltages

I_a, I_b, I_c are phase currents

Power equation can also be expressed in matrix form as:

$$S = [V_a \;\; V_b \;\; V_c]\begin{bmatrix} I_a \\ I_b \\ I_c \end{bmatrix}^* = \begin{bmatrix} V_a \\ V_b \\ V_c \end{bmatrix}^T \begin{bmatrix} I_a \\ I_b \\ I_c \end{bmatrix}^* \qquad \text{......... (7.61)}$$

We know that $\begin{bmatrix} V_a \\ V_b \\ V_c \end{bmatrix} = \begin{bmatrix} 1 & 1 & 1 \\ 1 & a^2 & a \\ 1 & a & a^2 \end{bmatrix}\begin{bmatrix} V_{a0} \\ V_{a1} \\ V_{a2} \end{bmatrix}$ and $\begin{bmatrix} I_a \\ I_b \\ I_c \end{bmatrix} = \begin{bmatrix} 1 & 1 & 1 \\ 1 & a^2 & a \\ 1 & a & a^2 \end{bmatrix}\begin{bmatrix} I_{a0} \\ I_{a1} \\ I_{a2} \end{bmatrix}$

Consider $[V] = \begin{bmatrix} V_{a0} \\ V_{a1} \\ V_{a2} \end{bmatrix}$, $[I] = \begin{bmatrix} I_{a0} \\ I_{a1} \\ I_{a2} \end{bmatrix}$, $[A] = \begin{bmatrix} 1 & 1 & 1 \\ 1 & a^2 & a \\ 1 & a & a^2 \end{bmatrix}$ and $[A]^{-1} = \frac{1}{3}\begin{bmatrix} 1 & 1 & 1 \\ 1 & a & a^2 \\ 1 & a^2 & a \end{bmatrix}$

$$S = [AV]^T[AI]^*$$

$$S = [V]^T[A]^T[A]^*[I]^* = [V]^T[A][A]^*[I]^*$$

$$S = [V_{a0} \;\; V_{a1} \;\; V_{a2}]\,[A][A]^*\begin{bmatrix} I_{a0} \\ I_{a1} \\ I_{a2} \end{bmatrix}^* \quad \text{Where } [A]^T = [A] \qquad \text{......... (7.62)}$$

$$[A][A]^* = \begin{bmatrix} 1 & 1 & 1 \\ 1 & a^2 & a \\ 1 & a & a^2 \end{bmatrix}\begin{bmatrix} 1 & 1 & 1 \\ 1 & a & a^2 \\ 1 & a^2 & a \end{bmatrix}$$

$$= \begin{bmatrix} 1+1+1 & 1+a+a^2 & 1+a^2+a \\ 1+a^2+a & 1+a^3+a^3 & 1+a^4+a^2 \\ 1+a+a^2 & 1+a^2+a^4 & 1+a^3+a^3 \end{bmatrix}$$

$$= \begin{bmatrix} 3 & 0 & 0 \\ 0 & 3 & 0 \\ 0 & 0 & 3 \end{bmatrix} \qquad \text{......... (7.63)}$$

Where, $a^4 = a, a^3 = 1, 1 + a + a^2 = 0$

From (7.62) and (7.63) $S = [V_{a0} \;\; V_{a1} \;\; V_{a2}]\,[A][A]^*\begin{bmatrix} I_{a0} \\ I_{a1} \\ I_{a2} \end{bmatrix}^*$

$$S = [V_{a0} \;\; V_{a1} \;\; V_{a2}]\begin{bmatrix} 3 & 0 & 0 \\ 0 & 3 & 0 \\ 0 & 0 & 3 \end{bmatrix}\begin{bmatrix} I_{a0} \\ I_{a1} \\ I_{a2} \end{bmatrix}^* = 3[V_{a0} \;\; V_{a1} \;\; V_{a2}]\begin{bmatrix} I_{a0} \\ I_{a1} \\ I_{a2} \end{bmatrix}^*$$

$$S = 3[V_{a0}I_{a0}^* + V_{a1}I_{a1}^* + V_{a2}I_{a2}^*] \qquad \text{......... (7.64)}$$

7.11 Single line to ground (LG) fault: Consider a, b, and c as the terminals of an unloaded alternator. A fault arises between terminal 'a' and the ground. There are five cases that are taken into consideration during direct short circuit when neutral is solidly grounded through impedance/ungrounded.

- Dead short circuit when neutral is solidly grounded.
- Dead short circuit when neutral is grounded through an impedance Z_n.
- Short circuit with fault impedance Z_f when neutral is solidly grounded.
- Short circuit with fault impedance Z_f when neutral is grounded through an impedance Z_n.
- When neutral is isolated.

7.11.1 Direct short circuit (SLG) when neutral is solidly grounded:

Fig. 7.11 shows a direct short circuit (**SLG fault**) when neutral is solidly grounded. Under this fault condition the currents and voltages can be considered as

$$I_f = I_a \,, I_b = 0, I_c = 0,$$

$$V_a = 0 \quad \text{......... (7.65)}$$

The symmetrical current components of phase 'a' are $\begin{bmatrix} I_{a0} \\ I_{a1} \\ I_{a2} \end{bmatrix} = \frac{1}{3}\begin{bmatrix} 1 & 1 & 1 \\ 1 & a & a^2 \\ 1 & a^2 & a \end{bmatrix}\begin{bmatrix} I_a \\ 0 \\ 0 \end{bmatrix}$

$$I_{a0} = I_{a1} = I_{a2} = \frac{1}{3} I_a \quad \text{......... (7.66)}$$

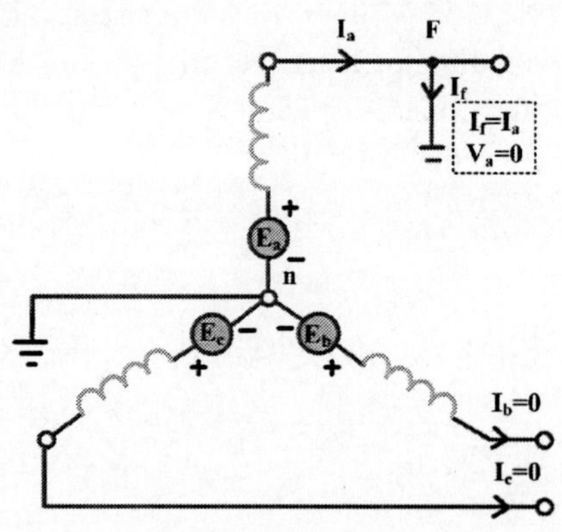

Fig. 7.11 **SLG fault:** Direct short circuit when neutral is solidly grounded

Symmetrical components of voltages at terminal "a" are V_{a0}, V_{a1} *and* V_{a2}

$$V_{a0} = -I_{a0} * Z_0$$

$$V_{a1} = E_a - I_{a1} * Z_1$$

$$V_{a2} = -I_{a2} * Z_2$$

Voltage in phase 'a' $\quad V_a = 0$

$$V_{a0} + V_{a1} + V_{a2} = 0$$

$$-I_{a0} * Z_0 + E_a - I_{a1} * Z_1 - I_{a2} * Z_2 = 0$$

$$E_a = I_{a0} * Z_0 + I_{a1} * Z_1 + I_{a2} * Z_2$$

$$E_a = I_{a1}(Z_0 + Z_1 + Z_2) \quad \text{......... (7.67)}$$

$$I_{a1} = \frac{E_a}{Z_0 + Z_1 + Z_2} \quad \text{......... (7.68)}$$

From eq. (7.66 &7.67) $\quad I_{a0} = I_{a1} = I_{a2} = \frac{E_a}{Z_0 + Z_1 + Z_2}$ .... (7.69)

From eq. (7.66) $\quad I_f^{LG} = I_a = 3 * I_{a1}$

Direct short circuit (SLG) when neutral is solidly grounded, the fault current $I_f^{LG} = \frac{3E_a}{Z_0 + Z_1 + Z_2}$ ..... (7.70)

7.11.2. Direct short circuit when neutral is grounded through an impedance Z_n:

Fig.7.12. Shows a direct short circuit (**SLG fault**) when neutral is grounded through an impedance. If the generator neutral is grounded through an impedance Z_n Then Z_0 Will become

$$Z_0 = Z_{g0} + 3 * Z_n \quad \text{......... (7.71)}$$

Direct short circuit when neutral is grounded through an impedance Z_n, *the fault current will be*

$$I_f^{LG} = \frac{3E_a}{Z_1 + Z_2 + (Z_{g0} + 3 * Z_n)} \quad \text{......... (7.72)}$$

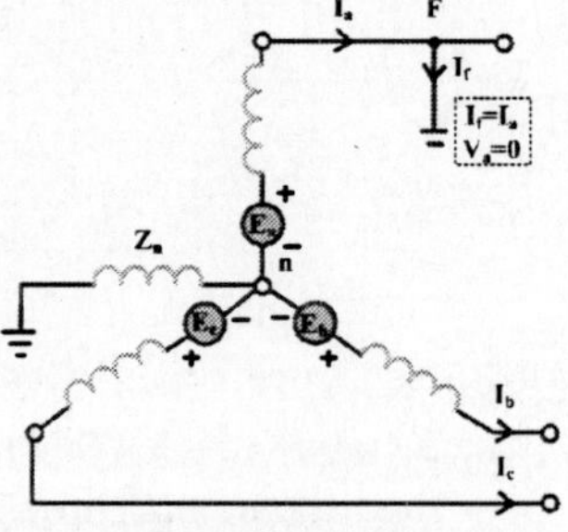

Fig. 7.12: **SLG Fault** - Direct short circuit when neutral is grounded through an impedance

7.11.3. Short circuit with fault impedance Z_f When neutral is solidly grounded:

Figure 7.13 depicts a short circuit (also known as an SLG fault) with a fault impedance, occurring when the neutral is firmly connected to the ground. During this fault situation, only V_a will deviate from the preceding condition.

$I_f = I_a, I_b = 0 \; and \; I_c = 0$

$$V_a = I_a Z_f \quad \text{......... (7.73)}$$

$$V_{a0} + V_{a1} + V_{a2} = I_a Z_f$$

$$-I_{a0} * Z_0 + E_a - I_{a1} * Z_1 - I_{a2} * Z_2 = I_a Z_f$$

$$E_a = I_a Z_f + I_{a1}(Z_0 + Z_1 + Z_2)$$

$$E_a = I_a Z_f + \frac{1}{3} I_a (Z_0 + Z_1 + Z_2)$$

$$E_a = I_a \left[Z_f + \frac{(Z_0+Z_1+Z_2)}{3}\right]; \quad E_a = I_a\left[\frac{Z_0+Z_1+Z_2+3Z_f}{3}\right]; I_a = \frac{3E_a}{Z_0+Z_1+Z_2+3Z_f}$$

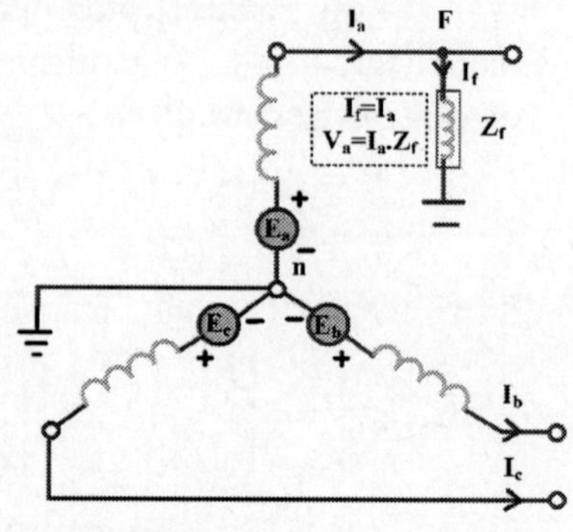

Fig.7.13. **SLG fault:** Short circuit with fault impedance Z_f When neutral is solidly grounded

Short circuit with fault impedance Z_f When neutral is solidly grounded, the fault current will be

$$I_f^{LG} = \frac{3E_a}{Z_0 + Z_1 + Z_2 + 3Z_f} \quad \text{......... (7.74)}$$

7.11.4. Short circuit with fault impedance Z_f when neutral is grounded through an impedance Z_n

Fig.7.14. shows a short circuit **(SLG fault)** with fault impedance Z_f when neutral is grounded through an impedance Z_n

If the generator neutral is grounded through an impedance Z_n Then Z_0 will become

$$Z_0 = Z_{g0} + 3 * Z_n$$

Short circuit with fault impedance Z_f when neutral is grounded through an impedance Z_n, the fault current will be

$$I_f^{LG} = \frac{3E_a}{Z_1 + Z_2 + (Z_{g0} + 3 * Z_n) + 3Z_f} \quad \text{...... (7.75)}$$

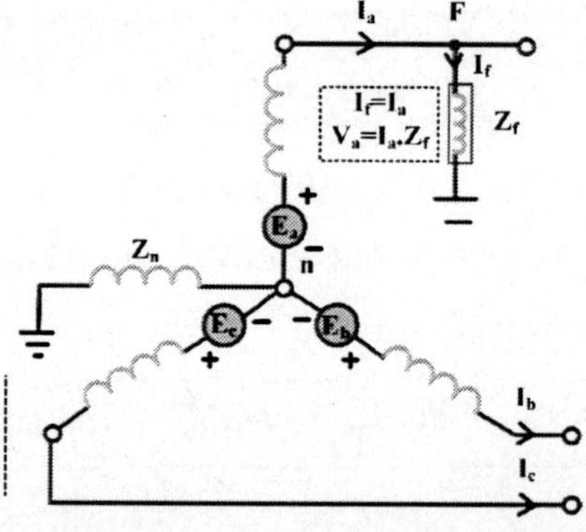

Fig.7.14. SLG fault: Short circuit with fault impedance when neutral is grounded through an impedance

7.11.5. Sequence networks of single line to ground (LG):

The equations (7.70), (7.72), (7.74), and (7.75) demonstrate that the positive, negative, and zero-sequence impedances should be linked in series when a single line to ground fault occurs. The sequence networks for single line to ground (LG) faults are shown in Figure 7.15, considering both solid grounding and grounding through impedance. The analysis includes direct short circuit and short circuit through fault impedance. Fig. 7.15 (a) represents the sequence network of a direct short circuit when the neutral is solidly grounded. Fig. 7.15 (b) represents the sequence network of a direct short circuit when the neutral is grounded through an impedance Z_n. Fig. 7.15 (c) represents the sequence network of a short circuit with fault impedance Z_f when the neutral is solidly grounded. Fig. 7.15 (d) represents the sequence network of a short circuit with fault impedance Z_f when the neutral is grounded through an impedance Z_n.

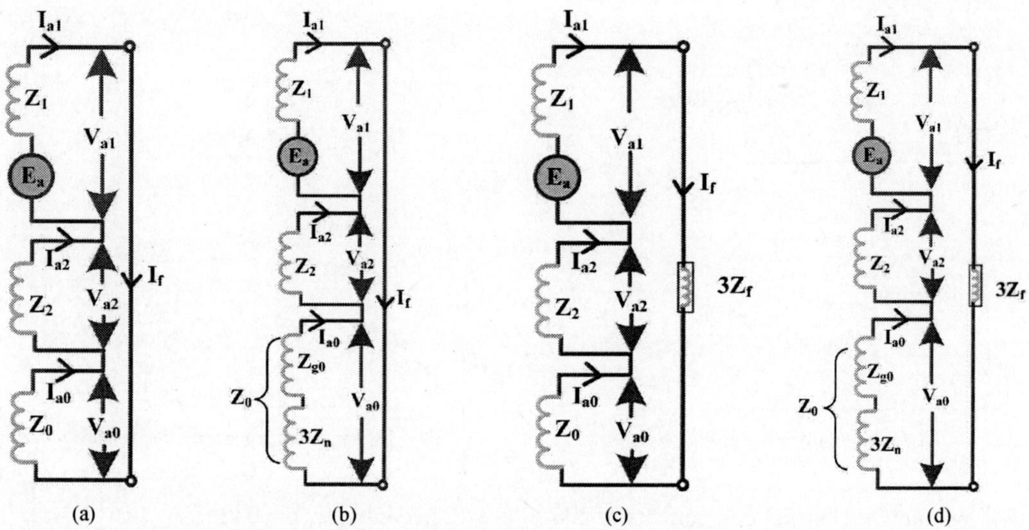

Fig.7.15. Sequence network of Single line to ground (LG) fault

7.11.6. When neutral is isolated :

If a generator has an isolated neutral, the zero-sequence impedance Z_0 will have an infinite value.

$$I_{a0} = I_{a1} = I_{a2} = \frac{E_a}{Z_0+Z_1+Z_2+Z_f}$$

$$= \frac{E_a}{\alpha + Z_1 + Z_2 + Z_f} = 0 \quad \dots\dots\dots (7.76)$$

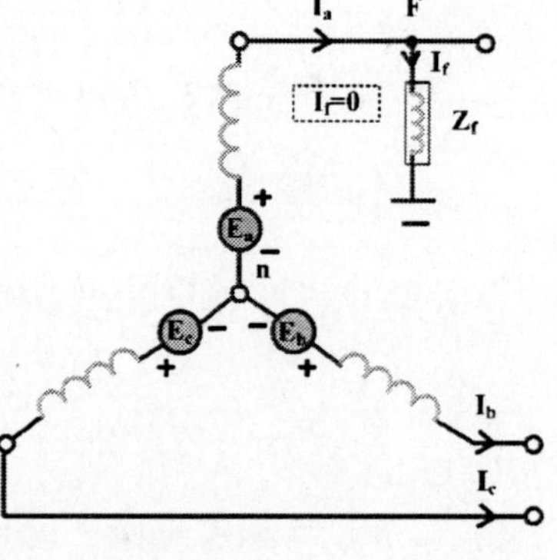

Fig.7.16. **SLG fault:** When neutral is isolated

When the neutral is isolated, fault current $I_f^{LG} = \frac{3E_a}{Z_0+Z_1+Z_2+3Z_f} = \frac{3E_a}{\alpha+Z_1+Z_2+Z_f} = 0$

$\dots\dots\dots (7.77)$

7.12 Double line/Line to Line (LL) fault:

Let a, b, c be the terminals of an unloaded alternator and fault occurs between terminals ‘b’ and ‘c’. Two cases are considered

- Direct short circuit between terminals ‘b’ and ‘c’.
- Short circuit between terminals ‘b’ and ‘c’ through fault impedance Z_f.

7.12.1 Direct short circuit between terminals 'b' and 'c':

Fig.7.17. shows double line/Line to Line (LL) fault when the neutral is solidly grounded and Fig.7.18. shows LL fault when the neutral is grounded through impedance. Under this fault condition the currents and voltages can be considered as

$$I_f = I_b = -I_c \quad \text{......... (7.78)}$$

$$I_b + I_c = 0, I_a = 0$$

$$V_b - V_c = 0 \text{ then } V_b = V_c$$

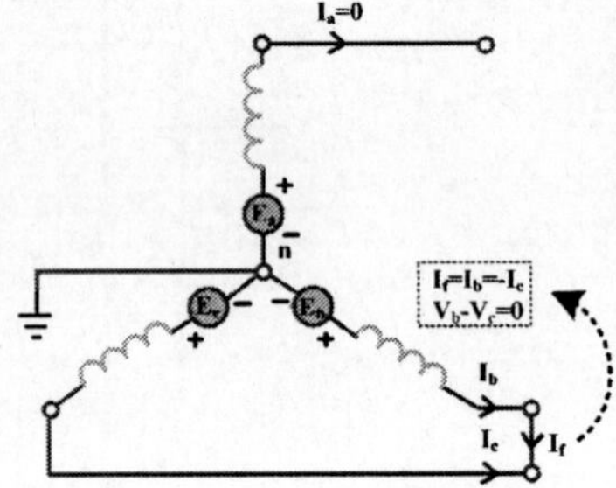

Fig.7.17. **LL fault:** solidly grounded

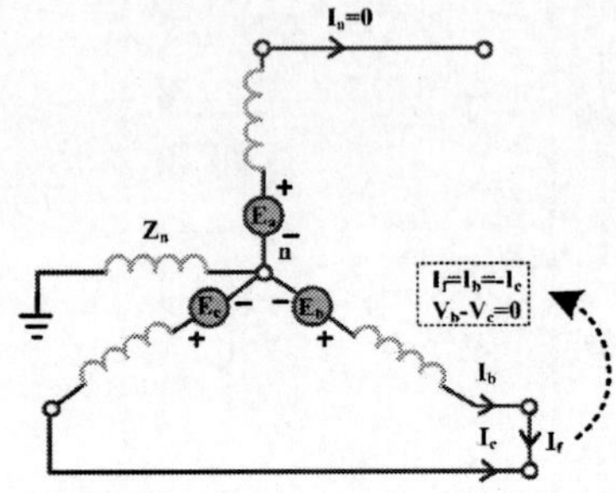

Fig.7.18. **LL fault:** grounded through impedance

The symmetrical current components of phase 'a' will become $\begin{bmatrix} I_{a0} \\ I_{a1} \\ I_{a2} \end{bmatrix} = \frac{1}{3}\begin{bmatrix} 1 & 1 & 1 \\ 1 & a & a^2 \\ 1 & a^2 & a \end{bmatrix}\begin{bmatrix} 0 \\ I_b \\ -I_b \end{bmatrix}$

Sequence currents of phase 'a' is $I_{a0} = 0$

$$I_{a1} = \frac{1}{3} I_b(a - a^2) = j\sqrt{3}\,\frac{I_b}{3} = j\,\frac{I_b}{\sqrt{3}} \qquad (From\ eq.7.8)$$

$$I_{a2} = \frac{1}{3} I_b(a^2 - a) = -j\sqrt{3}\,\frac{I_b}{3} = -j\,\frac{I_b}{\sqrt{3}} \quad \text{......... (7.79)}$$

$$I_{a1} = -I_{a2} \quad \text{......... (7.80)}$$

The symmetrical voltage components of phase 'a' will be $\begin{bmatrix} V_{a0} \\ V_{a1} \\ V_{a2} \end{bmatrix} = \frac{1}{3}\begin{bmatrix} 1 & 1 & 1 \\ 1 & a & a^2 \\ 1 & a^2 & a \end{bmatrix}\begin{bmatrix} V_a \\ V_b \\ V_b \end{bmatrix}$

$$V_{a0} = \frac{1}{3}(V_a + 2V_b)$$

$$V_{a1} = \frac{1}{3}[V_a + (a + a^2)V_b] = \frac{1}{3}[V_a - V_b]$$

$$V_{a2} = \frac{1}{3}[V_a + (a^2 + a)V_b] = \frac{1}{3}[V_a - V_b]$$

So, $V_{a1} = V_{a2}$ .... (7.81)

$$E_a - I_{a1} * Z_1 = -I_{a2} * Z_2$$

$$E_a = I_{a1}(Z_1 + Z_2)$$

$$I_{a1} = \frac{E_a}{Z_1 + Z_2} \quad \text{......... (7.82)}$$

$$I_{a2} = -I_{a1} = -\frac{E_a}{Z_1 + Z_2} \quad \text{......... (7.83)}$$

Current in phase 'b' is $I_b = I_{ao} + a^2\, I_{a1} + a\, I_{a2}$

$$= 0 + a^2\left(\frac{E_a}{Z_1 + Z_2}\right) + a\left(\frac{-E_a}{Z_1 + Z_2}\right)$$

$$= \frac{E_a}{Z_1 + Z_2}(a^2 - a)$$

$$= \frac{-j\sqrt{3}E_a}{Z_1 + Z_2} \quad \text{......... (7.84)}$$

From Eq. (7.78), $I_f = I_b = -I_c$

Direct short circuit between terminals 'b' and 'c', the fault current will be $I_f^{LL} = \frac{-j\sqrt{3}E_a}{Z_1 + Z_2}$ .... (7.85)

7.12.2 Short circuit between terminals 'b' and 'c' through fault impedance Z_f:

Fig.7.19. shows double line/Line to Line (LL) fault through fault impedance Z_f when the neutral is solidly grounded and Fig.7.20. shows LL fault when the neutral is grounded through impedance. Under this fault condition the currents and voltages can be considered as

$$I_f = I_b = -I_c$$

$$I_b + I_c = 0,\ I_a = 0,\ \ V_b - V_c = I_b Z_f \quad \text{......... (7.86)}$$

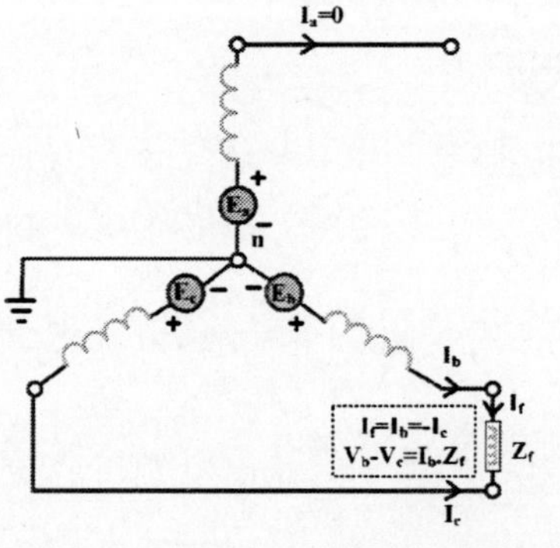

Fig.7.19. LL fault through Z_f: solidly grounded

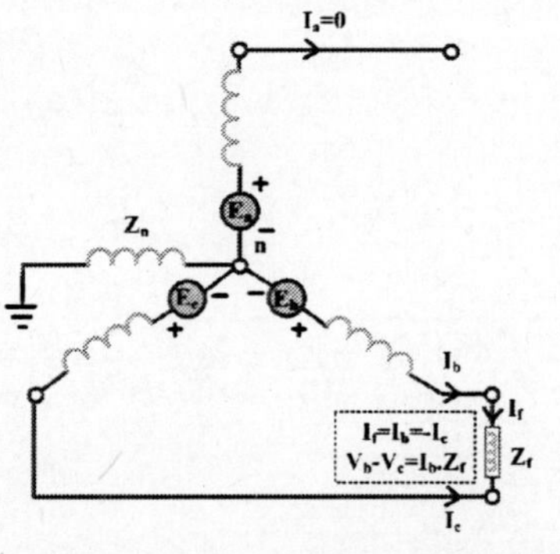

Fig.7.20. LL fault through Z_f: grounded through Z_n

The symmetrical voltage components of phase 'a' will be $\begin{bmatrix} V_{ao} \\ V_{a1} \\ V_{a2} \end{bmatrix} = \frac{1}{3}\begin{bmatrix} 1 & 1 & 1 \\ 1 & a & a^2 \\ 1 & a^2 & a \end{bmatrix}\begin{bmatrix} V_a \\ V_b \\ V_b - I_b Z_f \end{bmatrix}$

$$V_{ao} = \frac{1}{3}(V_a + 2V_b - I_b Z_f)$$

$$V_{a1} = \frac{1}{3}[V_a + aV_b + a^2(V_b - I_b Z_f)]$$

$$V_{a1} = \frac{1}{3}[V_a + (a + a^2)V_b - a^2 I_b Z_f] \quad \text{......... (7.87)}$$

$$V_{a2} = \frac{1}{3}[V_a + a^2V_b + a(V_b - I_bZ_f)]$$

$$V_{a2} = \frac{1}{3}[V_a + (a^2 + a)V_b - aI_bZ_f] \quad \text{......... (7.88)}$$

So, $$V_{a1} - V_{a2} = \frac{1}{3}[V_a + (a + a^2)V_b - a^2I_bZ_f] - \frac{1}{3}[V_a + (a^2 + a)V_b - aI_bZ_f]$$

$$E_a - I_{a1} * Z_1 + I_{a2} * Z_2 = \frac{1}{3}[(a - a^2)I_bZ_f]$$

$$E_a - I_{a1} * Z_1 - I_{a1} * Z_2 = \frac{1}{3}[(a - a^2)I_bZ_f] \qquad \textit{From eq.}\ (7.80)$$

$$E_a - I_{a1}(Z_1 + Z_2) = \frac{1}{3}[(a - a^2)I_bZ_f] \quad \text{......... (7.89)}$$

Positive sequence current of phase 'a' is $I_{a1} = \frac{1}{3}\ (I_a + a\ I_b + a^2\ I_c)$

By considering $I_b = -I_c\ and\ I_a = 0$

$$I_{a1} = \frac{1}{3}\ (0 + a\ I_b - a^2\ I_b)$$

$$= \frac{1}{3}\ (a - a^2)I_b \quad \text{......... (7.90)}$$

Substitute in eq. (7.53)

$$E_a - I_{a1}(Z_1 + Z_2) = \frac{1}{3}[(a - a^2)I_bZ_f]$$

$$E_a - \frac{1}{3}\ (a - a^2)I_b(Z_1 + Z_2) = \frac{1}{3}[(a - a^2)I_bZ_f]$$

$$E_a = \frac{1}{3}\ (a - a^2)(Z_1 + Z_2 + Z_f)I_b$$

$$I_b = \frac{3E_a}{(a - a^2)(Z_1 + Z_2 + Z_f)}$$

$$I_b = \frac{-j\sqrt{3}E_a}{Z_1 + Z_2 + Z_f} \quad \text{......... (7.91)}$$

From Eq. (7.78), $I_f = I_b = -I_c$

Short circuit between terminals 'b' and 'c' through fault impedance Z_f, the fault current will be

$$I_f^{LL} = \frac{-j\sqrt{3}E_a}{Z_1 + Z_2 + Z_f} \quad \text{......... (7.92)}$$

7.12.3. Sequence networks of Double line/Line to Line (LL) fault:

It is apparent from equations (7.80), (7.81), (7.83), (7.85) and (7.92) that positive and negative sequence networks are connected in parallel in case of dead short circuit between two phases, whereas, these networks are connected in parallel through fault impedance Z_f in case of fault through impedance. Connection of sequence networks for two cases are shown in Fig.7.21 where Fig.7.21(a) and Fig.7.21(b) show connection of sequence networks without and with fault impedance, respectively.

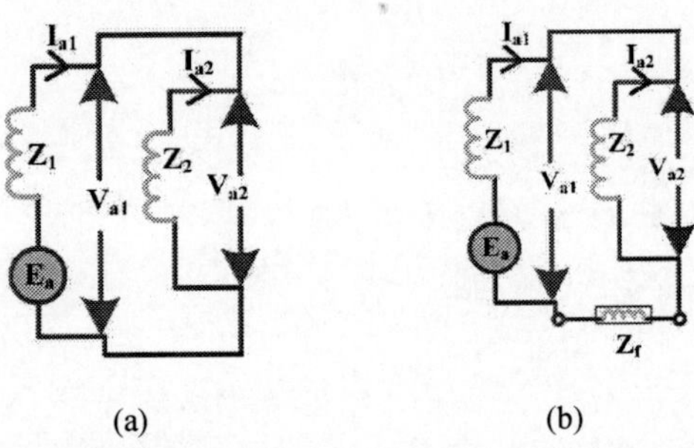

Fig.7.21. Sequence network of Double line/Line to Line (LL) fault

7.13 Double line to ground (LLG) fault:

Let a, b, c be the terminals of an unloaded alternator and fault occurs between terminals ‘b’ and ‘c’ and to ground. Two cases are considered

- Direct short circuit between terminals ‘b’ and ‘c’ to ground.
- Short circuit between terminals ‘b’ and ‘c’ to ground through fault impedance Z_f.

7.13.1. Direct short circuit between terminals ‘b’ and ‘c’ to ground:

Fig.7.22. shows double line to ground (LLG) fault when the neutral is solidly grounded and Fig.7.23. shows LLG fault when the neutral is grounded through impedance. Under this fault condition the currents and voltages can be considered as

$$I_f = I_b + I_c \qquad \text{... ... ... (7.93)}$$

$$I_a = 0,\ V_b = 0, V_c = 0$$

Fig.7.22. **LLG fault:** solidly grounded

Fig.7.23. **LLG fault:** grounded through impedance

The symmetrical voltage components of phase ‘a’ will be $\begin{bmatrix} V_{a0} \\ V_{a1} \\ V_{a2} \end{bmatrix} = \frac{1}{3}\begin{bmatrix} 1 & 1 & 1 \\ 1 & a & a^2 \\ 1 & a^2 & a \end{bmatrix}\begin{bmatrix} V_a \\ 0 \\ 0 \end{bmatrix}$

$$V_{a0} = V_{a1} = V_{a2} = \frac{V_a}{3} \qquad \text{... ... ... (7.94)}$$

The currents in Phase a, b and c are $\begin{bmatrix} I_a \\ I_b \\ I_c \end{bmatrix} = \begin{bmatrix} 1 & 1 & 1 \\ 1 & a^2 & a \\ 1 & a & a^2 \end{bmatrix}\begin{bmatrix} I_{a0} \\ I_{a1} \\ I_{a2} \end{bmatrix}$

$$I_a = I_{a0} + I_{a1} + I_{a2}$$

$$I_b = I_{a0} + a^2\, I_{a1} + a\, I_{a2}$$
$$I_c = I_{a0} + a\, I_{a1} + a^2\, I_{a2}$$

From eq. (7.93) $I_f = I_b + I_c$

$$I_f = I_{a0} + a^2\, I_{a1} + a\, I_{a2} + I_{a0} + a\, I_{a1} + a^2\, I_{a2}$$
$$= 2I_{a0} + (a^2 + a)\, I_{a1} + (a^2 + a)\, I_{a2}$$
$$= 2I_{a0} - I_{a1} - I_{a2} \qquad \text{.........(7.95)}$$

From eq. (7.93) $I_a = 0$

$$I_{a0} + I_{a1} + I_{a2} = 0$$
$$I_{a0} = -I_{a1} - I_{a2} \qquad \text{.........(7.96)}$$

From eq. (7.95)&(7.96)

$$I_f = 2I_{a0} + I_{a0} = 3I_{a0} \qquad \text{.........(7.97)}$$

From eq. (7.94) $V_{a0} = V_{a1} = V_{a2} = \frac{V_a}{3}$

Considering $V_{a0} = V_{a2}$

$$-I_{a0} * Z_0 = -I_{a2} * Z_2$$
$$I_{a0} = \frac{Z_2}{Z_0} * I_{a2} \qquad \text{.........(7.98)}$$

Considering $V_{a1} = V_{a2}$

$$E_a - I_{a1} * Z_1 = -I_{a2} * Z_2$$
$$I_{a2} = -\frac{(E_a - I_{a1} * Z_1)}{Z_2} \qquad \text{.........(7.99)}$$

From eq. (7.98)&(7.99)

$$I_{a0} = \frac{Z_2}{Z_0} * I_{a2}$$
$$= \frac{Z_2}{Z_0}\left[-\frac{(E_a - I_{a1} * Z_1)}{Z_2}\right]$$
$$= -\frac{(E_a - I_{a1} * Z_1)}{Z_0} \qquad \text{.........(7.100)}$$

From eq. (7.96) $I_{a1} = -(I_{a0} + I_{a2})$

$$I_{a1} = \frac{(E_a - I_{a1} * Z_1)}{Z_0} + \frac{(E_a - I_{a1} * Z_1)}{Z_2}$$
$$I_{a1} = E_a\left(\frac{1}{Z_0} + \frac{1}{Z_2}\right) - I_{a1} * Z_1\ \left(\frac{1}{Z_0} + \frac{1}{Z_2}\right)$$
$$E_a\left(\frac{Z_0 + Z_2}{Z_0 Z_2}\right) = I_{a1}\left[1 + Z_1\ \left(\frac{Z_0 + Z_2}{Z_0 Z_2}\right)\right]$$
$$E_a\left(\frac{Z_0 + Z_2}{Z_0 Z_2}\right) = I_{a1}\left(\frac{Z_0 Z_2 + Z_0 Z_1 + Z_1 Z_2}{Z_0 Z_2}\right)$$

$$I_{a1} = \frac{E_a(Z_0 + Z_2)}{Z_0Z_1 + Z_0Z_2 + Z_1Z_2} \quad \text{... ... ... (7.101)}$$

$$I_{a1} = \frac{E_a}{Z_1 + \left(\frac{Z_0Z_2}{Z_0 + Z_2}\right)}$$

$$I_{a1} = \frac{E_a}{Z_1 + (Z_2 \parallel Z_0)} \quad \text{... ... ... (7.102)}$$

Which implies that negative sequence impedance is in parallel with the s zero sequence impedance, this is in series with the positive sequence impedance; therefore I_{a2} *and* I_{ao} will be

$$I_{a2} = \frac{-Z_0E_a}{Z_0Z_1 + Z_0Z_2 + Z_1Z_2} \quad \text{... ... ... (7.103)}$$

$$I_{ao} = \frac{-Z_2E_a}{Z_0Z_1 + Z_0Z_2 + Z_1Z_2} \quad \text{... ... ... (7.104)}$$

From eq. (7.97) $I_f = 3I_{ao}$

Direct short circuit between terminals 'b' and 'c' to ground, the fault current will be

$$I_f^{LLG} = \frac{-3Z_2E_a}{Z_0Z_1 + Z_0Z_2 + Z_1Z_2} \quad \text{... ... ... (7.105)}$$

7.13.2 Short circuit between terminals 'b' and 'c' to ground through fault impedance Z_f

Fig.7.24. shows double line to ground (LLG) fault through fault impedance Z_f when the neutral is solidly grounded and Fig.7.25. shows LLG fault when the neutral is grounded through impedance. Under this fault condition the currents and voltages can be considered as

$$I_f = I_b + I_c, \quad I_a = 0$$

$$V_b = V_c = I_fZ_f = (I_b + I_c)Z_f = 3I_{a0}\,Z_f \quad \text{... ... ... (7.106)}$$

The symmetrical voltage components of phase 'a' will be $\begin{bmatrix} V_{a0} \\ V_{a1} \\ V_{a2} \end{bmatrix} = \frac{1}{3}\begin{bmatrix} 1 & 1 & 1 \\ 1 & a & a^2 \\ 1 & a^2 & a \end{bmatrix}\begin{bmatrix} V_a \\ V_b \\ V_b \end{bmatrix}$

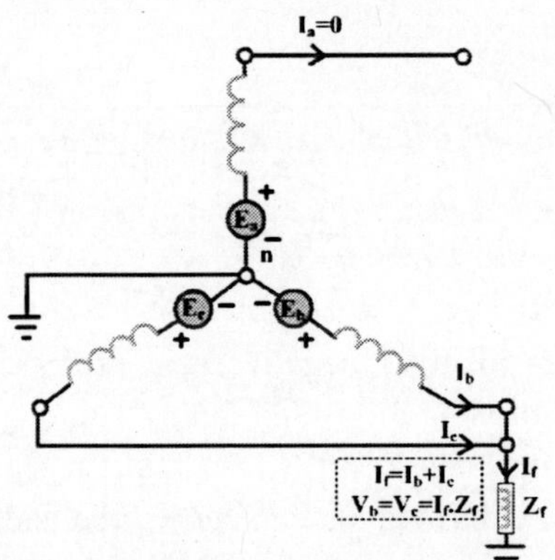

Fig.7.24. LLG fault through Z_f: solidly grounded

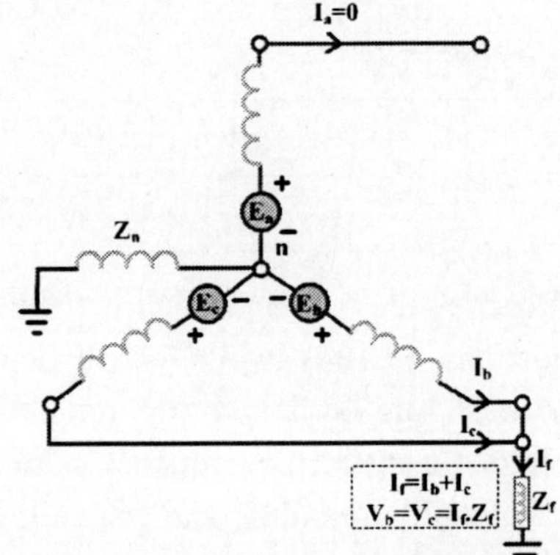

Fig.7.25. LLG fault through Z_f : grounded through Zn

$$V_{a0} = \frac{1}{3}(V_a + 2V_b)$$

$$V_{a1} = V_{a2} = \frac{1}{3}[V_a + (a + a^2)V_b] = \frac{1}{3}(V_a - V_b) \quad \dots\dots\dots(7.107)$$

$$V_{a0} - V_{a1} = \frac{1}{3}(V_a + 2V_b) - \frac{1}{3}(V_a - V_b) = V_b = 3I_{a0}\, Z_f \quad \text{(From eq. (7.105))}$$

$$V_{a0} - V_{a1} = 3I_{a0}\, Z_f$$

$$V_{a0} = V_{a1} + 3I_{a0}\, Z_f \quad \dots\dots\dots(7.108)$$

From eq. (7.107) and (7.108)

$$V_{a1} = V_{a2}$$

$$V_{a0} = V_{a1} + 3I_{a0}\, Z_f \text{ (or) } V_{a2} + 3I_{a0}\, Z_f$$

Which implies that negative sequence impedance is in parallel with the summation of zero sequence impedance and threefold fault impedance, this is in series with the positive sequence impedance; therefore I_{a1}, I_{a2} *and* I_{a0} will be

$$I_{a1} = \frac{E_a}{Z_1 + [Z_2 \parallel (Z_0 + 3Z_f)]} \quad \dots\dots\dots(7.109)$$

$$= \frac{E_a}{Z_1 + \left[\frac{Z_2 * (Z_0 + 3Z_f)}{Z_2 + Z_0 + 3Z_f}\right]}$$

$$= \frac{E_a\,(Z_2 + Z_0 + 3Z_f)}{Z_1(Z_2 + Z_0 + 3Z_f) + Z_2 * (Z_0 + 3Z_f)}$$

$$I_{a1} = \frac{E_a\,(Z_2 + Z_0 + 3Z_f)}{Z_1Z_2 + Z_1(Z_0 + 3Z_f) + Z_2(Z_0 + 3Z_f)} \quad \dots\dots\dots(7.110)$$

$$Similarly,\ I_{a2} = \frac{-E_a\,(Z_0 + 3Z_f)}{Z_1Z_2 + Z_1(Z_0 + 3Z_f) + Z_2(Z_0 + 3Z_f)} \quad \dots\dots\dots(7.111)$$

$$I_{a0} = \frac{-E_a\, Z_2}{Z_1Z_2 + Z_1(Z_0 + 3Z_f) + Z_2(Z_0 + 3Z_f)} \quad \dots\dots\dots(7.112)$$

Fault current $I_f^{LLG} = 3I_{a0}$

Short circuit between terminals 'b' and 'c' to ground through fault impedance Z_f, *the fault current will be*

$$I_f^{LLG} = \frac{-3E_a\, Z_2}{Z_1Z_2 + Z_1(Z_0 + 3Z_f) + Z_2(Z_0 + 3Z_f)} \quad \dots\dots\dots(7.113)$$

7.13.3 Sequence networks of Double line to ground (LLG) fault:

The equations (7.105) and (7.113) demonstrate that the positive, negative, and zero-sequence impedances should be linked in parallel when there is a double line to ground fault. The diagram in Figure 7.26 illustrates the sequence networks of a double line to ground (LLG) fault. It shows the scenarios of solid grounding and grounding through impedance during a direct short circuit and a short circuit through fault impedance. Fig. 7.26 (a) represents the sequence network of a direct short circuit between terminals 'b' and 'c' to ground. Fig. 7.26 (b) represents the sequence network of a short circuit between terminals 'b' and 'c' to ground, with the neutral grounded through an impedance Z_n. Fig. 7.26 (c) represents the sequence network of a short circuit between terminals 'b' and 'c' to ground, with a fault impedance Z_f when the neutral is solidly grounded. Fig. 7.26 (d) represents the sequence network of a short circuit between terminals 'b' and 'c' to ground, with a fault impedance Z_f when the neutral is grounded through an impedance Z_n.

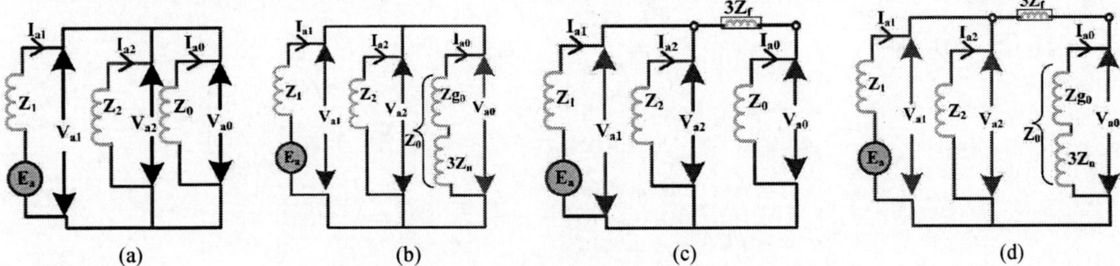

Fig.7.26. Sequence network of Double line to ground (LLG) fault

7.14 Symmetrical (or) 3-Phase (LLL) fault:

Fig.7.27. shows 3-Phase (LLL) fault when the neutral is solidly grounded and Fig.7.28. shows LLL fault when the neutral is grounded through impedance. Sequence networks of symmetrical (or) 3-phase (LLL) fault is given in Fig. 7.29. Let a, b, c be the terminals of an unloaded alternator and fault occurs across all three terminals, then

$$V_a = V_b = V_c \qquad \text{......... (7.114)}$$

$$I_a + I_b + I_c = 0 \qquad \text{......... (7.115)}$$

$$I_b = a^2 I_a, \; I_c = a I_a \qquad \text{......... (7.116)}$$

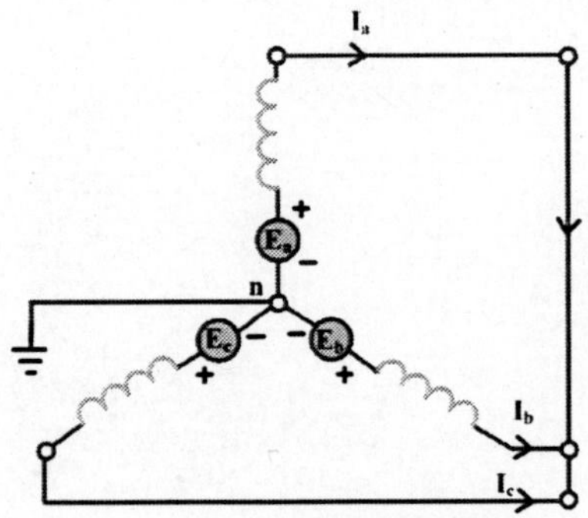

Fig.7.27. **LLL fault:** solidly grounded

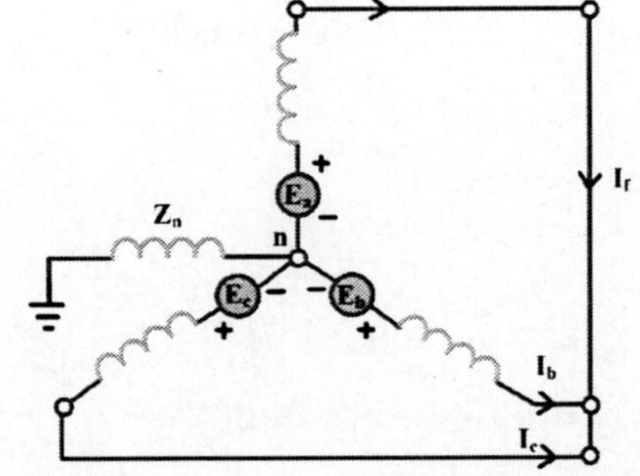

Fig.7.28. **LLL fault:** grounded through impedance

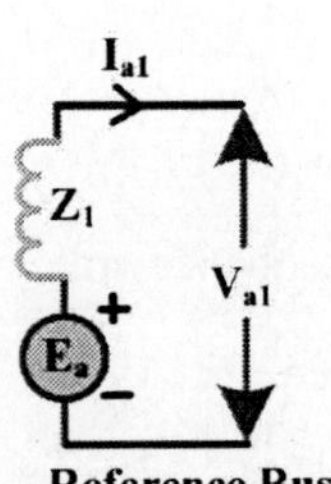

Fig.7.29. Sequence network

Zero sequence current of phase 'a' is $\quad I_{a0} = \frac{1}{3}\,(I_a + I_b + I_c) = 0 \qquad$ .... (7.117)

Positive sequence current of phase 'a' is $\quad I_{a1} = \frac{1}{3}\,(I_a + a\,I_b + a^2\,I_c)$

From eq. (7.116) $\quad I_{a1} = \frac{1}{3}\,(I_a + a\,a^2 I_a + a^2 a\,I_a)$

$$= \frac{1}{3}(I_a + a^3 I_a + a^3 I_a)$$

$$= \frac{1}{3}(I_a + I_a + I_a) \qquad (\because a^3 = 1)$$

$$I_{a1} = I_a \qquad \text{... ... ... (7.118)}$$

Negative sequence current of phase 'a' is $I_{a2} = \frac{1}{3}(I_a + a^2 I_b + a I_c)$

From eq. (7.116) $I_{a2} = \frac{1}{3}(I_a + a^2 a^2 I_a + a a I_a)$

$$= \frac{1}{3}(I_a + a^4 I_a + a^2 I_a)$$

$$= \frac{1}{3}(I_a + a I_a + a^2 I_a)$$

$$= \frac{1}{3}(1 + a + a^2) I_a \qquad (\because 1 + a + a^2 = 0)$$

$$= 0 \qquad \text{... ... ... (7.119)}$$

From eq. (7.117), (7.118) &(7.119), $I_{ao} = 0,\ I_{a1} = I_a$, and $I_{a2} = 0$

For a 3-phase fault, the absence of negative and zero sequence currents indicates that only the positive sequence current, which is similar to the phase current of phase 'a', is present.

Positive sequence voltage of phase 'a' is $V_{a1} = \frac{1}{3}(V_a + a V_b + a^2 V_c)$

From eq. (7.114) $V_{a1} = \frac{1}{3}(V_a + a V_a + a^2 V_a)$

$$= \frac{1}{3}(1 + a + a^2) V_a$$

$$= 0 \qquad \text{... ... ... (7.120)}$$

Negative sequence voltage of phase 'a' is $V_{a2} = \frac{1}{3}(V_a + a^2 V_b + a V_c)$

From eq. (7.114) $V_{a2} = \frac{1}{3}(V_a + a^2 V_a + a V_a)$

$$= \frac{1}{3}(1 + a^2 + a) V_a$$

$$= 0 \qquad \text{... ... ... (7.121)}$$

Zero sequence voltage of phase 'a' is $V_{ao} = \frac{1}{3}(V_a + V_b + V_c)$

From eq. (7.114) $V_{ao} = \frac{1}{3}(0) = 0$ ... ... (7.122)

From eq. (7.120) $V_{a1} = E_a - I_{a1} * Z_1 = 0$

$$I_{a1} = \frac{E_a}{Z_1} \qquad \text{... ... ... (7.123)}$$

From eq. (7.123) $I_{a1} = I_a = I_f = \frac{E_a}{Z_1}$

Short circuit between all three terminals, the fault current will be $I_f^{3\phi} = \frac{E_a}{Z_1}$ ... ... (7.124)

The equation (7.124) shows that only the positive-sequence impedance will be present during symmetrical (or) 3-phase (LLL) fault. Sequence network of three-phase fault is given in Fig.7.29.

Therefore, we can conclude that,

- The positive sequence component is found in all types of faults, including both symmetrical and unsymmetrical faults.
- The presence of negative sequence component is observed in unsymmetrical faults.
- The presence of the zero-sequence component is dependent upon the grounding of the neutral in the system and the occurrence of a fault that includes the ground.
- In a single line to ground fault, the positive, negative, and zero-sequence networks are linked together in a series configuration.
- In line-to-line fault, the positive and negative sequence networks are connected in parallel. If fault is through impedance Z_f, the two networks are connected in parallel through impedance Z_f.
- In a double line to ground fault, the positive, negative, and zero-sequence networks are interconnected in a parallel configuration.
- In a three-phase (LLL) fault, only impedance corresponding to the positive sequence is present.

Example 7.9. The alternator has a solidly grounded neutral and operates at a voltage of 11kV and a power of 50MVA. It has a sub-transient reactance of j0.25 per unit. The reactance for the negative sequence is j0.4 per unit, while the reactance for the zero sequence is j0.2 per unit. A single line to ground fault develops at the terminals of this generator that is not connected to any load. The impedance of the fault is j0.1 per unit. Calculate the fault current and terminal voltages during the fault state. Express all currents and voltages during the fault condition as a percentage of their respective rated values.

Ans. Let a, b, c be the terminals of the alternator and fault occurs between terminal 'a' and ground.

$$Z_1 = j0.25p.u.,\ Z_2 = j0.4p.u.,\ Z_0 = j0.2p.u.,\ Z_f = j0.1 \text{ and } E_a = 1\angle 0^0\ p.u.$$

In a <u>L-G Fault</u>: $I_{a0} = I_{a1} = I_{a2}$ And $I_f = 3I_{a1}$

$$I_{a0}^{pu} = I_{a1}^{pu} = I_{a2}^{pu} = \frac{E_a}{Z_0 + Z_1 + Z_2 + 3Z_f}$$

$$= \frac{1\angle 0^0}{j0.2 + j0.25 + j0.4 + 3 * j0.1}$$

$$= -j0.8695\ p.u.$$

Fault Current $\quad I_f^{LG} = 3I_{a1} = 3 * (-j0.8695) = -j2.6086p.u.$

Symmetrical components of voltages at terminal "a" are V_{a0}, V_{a1} and V_{a2}

$$V_{a0} = -I_{a0} * Z_0 = -(-j0.8695) * j0.2 = -0.1739\ p.u.$$

$$V_{a1} = E_a - I_{a1} * Z_1 = 1 - (-j0.8695) * j0.25 = 0.7826\ p.u.$$

$$V_{a2} = -I_{a2} * Z_2 = -(-j0.8695) * j0.4 = -0.3478\ p.u.$$

Voltage in phase 'a' is $V_a = V_{a0} + V_{a1} + V_{a2}$

$$= -0.1739 + 0.7826 + (-0.3478)$$

$$= 0.2609\ p.u.$$

Voltage in phase 'b' is $V_b = V_{a0} + a^2\, V_{a1} + a\, V_{a2}$

$$= -0.1739 + 1\angle 240^0 * 0.7826 + 1\angle 120^0 * (-0.3478)$$
$$= -0.3913 - j0.9789$$
$$= 1.054\angle -111.78^0\ p.u.$$

Voltage in phase 'c' is $V_c = V_{ao} + a\,V_{a1} + a^2\,V_{a2}$

$$= -0.1739 + 1\angle 120^0 * 0.7826 + 1\angle 240^0 * (-0.3478)$$
$$= 1.054\angle 111.78^0\ p.u.$$

Per unit and Percentage Line-Line Voltages,

$$V_{ab} = V_a - V_b = 0.2609 - 1.054\angle -111.78^0 = 1.176\angle 56.33^0\ p.u. = 117.6\angle 56.33^0\ \%$$
$$V_{bc} = V_b - V_c = 1.054\angle -111.78^0 - 1.054\angle 111.78^0 = 1.957\angle -90^0\ p.u. = 195.7\angle -90^0\ \%$$
$$V_{ca} = V_c - V_a = 1.054\angle 111.78^0 - 0.2609 = 1.176\angle 123.666^0\ p.u. = 117.6\angle 123.666^0\ \%$$

Example 7.10. Three 10MVA, 11kV, 3-phase star connected alternators are operating in parallel. Each has $X_d'' = 15\%, X_2 = 10\%\ and\ X_0 = 5\%$. If an earth fault occurs on busbar, determine the fault current

(i). When all the three alternators are solidly grounded

(ii). When only one of the alternators is solidly grounded and others are isolated

(iii). When only one of the alternators is grounded through an inductor of 0.2Ω and others are isolated

Ans : $X_d'' = Z_1 = 15\% = j0.15 p.u.$, $X_2 = Z_2 = 10\% = j0.1\ p.u$, and $X_0 = Z_0 = 5\% = j0.05\ p.u.$

Consider base values as 11kV, 10MVA

Base Current $I_{base} = \frac{10*10^6}{\sqrt{3}*11*10^3} = 524.86\ A$

(i). <u>When all the three alternators are solidly grounded:</u> Since all three alternators are operating in parallel, the resultant reactance will be one-third

$$Z_1 = \frac{j0.15}{3} = j0.05 p.u.$$

$$Z_2 = \frac{j0.1}{3} = j0.033\ p.u.$$

$$Z_0 = \frac{j0.05}{3} = j0.0166\ p.u.$$

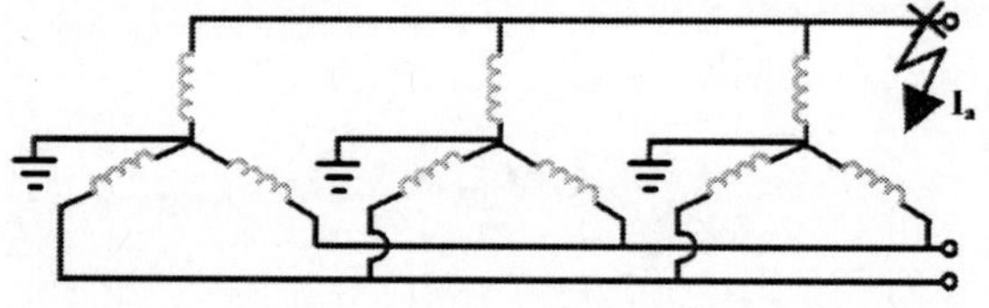

Per unit fault current $I_f^{LG} = 3 * I_{a1}^{pu} = \frac{3E_a}{Z_0 + Z_1 + Z_2} = \frac{3*1\angle 0^0}{j0.0166 + j0.05 + j0.033} = -j30.01\ p.u.$

Actual fault current $I_{f\,actual}^{LG} = I_{f\,p.u.}^{LG} * I_{base} = -j30.01 * 524.86 = 15.752\angle -90^0\ kA$

(ii). <u>When only one of the alternators is solidly grounded and others are isolated</u>

$$Z_1 = \frac{j0.15}{3} = j0.05 p.u.$$

$$Z_2 = \frac{j0.1}{3} = j0.033\ p.u.$$

$$Z_0 = j0.05\ p.u.$$

Per unit fault current $I_f^{LG} = 3 * I_{a1}^{pu} = \frac{3E_a}{Z_0 + Z_1 + Z_2} = \frac{3*1\angle 0^0}{j0.05 + j0.05 + j0.033} = -j22.55\ p.u.$

Actual fault current $I_{f\,actual}^{LG} = I_{f\,p.u.}^{LG} * I_{base} = -j22.55 * 524.86 = 11.835\angle -90^0\ kA$

(iii). When only one of the alternators is grounded through an inductor of 0.2Ω and others are isolated

$Z_1 = \frac{j0.15}{3} = j0.05 p.u.$

$Z_2 = \frac{j0.1}{3} = j0.033\ p.u.$

$Z_0 = Z_{go} + 3 * Z_n = j0.05 + 3 * \left(\frac{j0.2}{11^2} * 10\right) = j0.09958\ p.u.$

Per unit fault current $I_f^{LG} = 3 * I_{a1}^{pu} = \frac{3E_a}{Z_1 + Z_2 + (Z_{go} + 3\,Z_n)} = \frac{3*1\angle 0^0}{j0.05 + j0.033 + j0.09958} = -j16.43\ p.u.$

Actual fault current $I_{f\,actual}^{LG} = I_{f\,p.u.}^{LG} * I_{base} = -j16.43 * 524.86 = 8.623\angle -90^0\ kA$

Example 7.11. Three 10MVA, 11kV, 3-phase star connected alternators are operating in parallel. Each has $X_1 = j0.3\ p.u., X_2 = j0.2\ p.u.\ and\ X_0 = j0.1\ p.u.$ If an earth fault occurs on busbar through a fault impedance $X_f = j0.05\ p.u.$, determine the fault current

(i). When all the three alternators are solidly grounded

(ii). When only one of the alternators is solidly grounded and others are isolated

(iii). When only one of the alternators is grounded through an inductor of 0.2Ω and others are isolated

Ans : $X_1 = j0.3\ p.u., X_2 = j0.2\ p.u., X_0 = j0.1\ p.u.$ and $X_f = j0.05\ p.u.$

Consider base values as 11kV, 10MVA

Base Current $I_{base} = \frac{10*10^6}{\sqrt{3}*11*10^3} = 524.86\ A$

(i). <u>When all the three alternators are solidly grounded:</u>

Since all three alternators are operating in parallel, the resultant reactance will be one-third

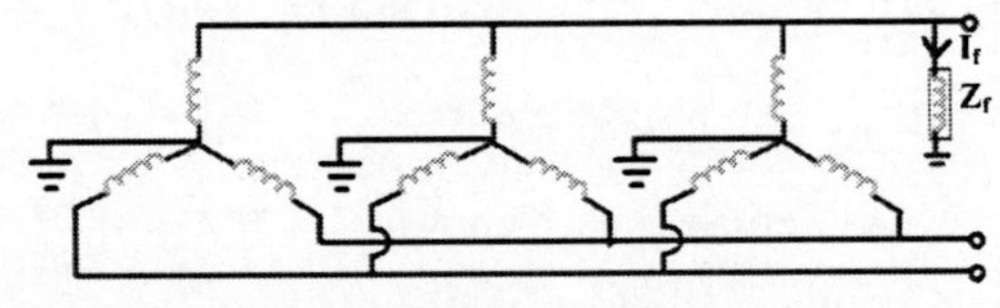

$Z_1 = \frac{j0.3}{3} = j0.1 p.u.$

$Z_2 = \frac{j0.2}{3} = j0.067\ p.u.$

$Z_0 = \frac{j0.1}{3} = j0.033\ p.u.$

Per unit fault current $I_f^{LG} = \frac{3E_a}{Z_0 + Z_1 + Z_2 + 3\,Z_f} = \frac{3*1\angle 0^0}{j0.033 + j0.1 + j0.067 + 3*j0.05} = -j8.57\ p.u.$

Actual fault current $I_{f\,actual}^{LG} = I_{f\,p.u.}^{LG} * I_{base} = -8.57 * 524.86 = 4.498\angle -90^0\ kA$

(ii). <u>When only one of the alternators is solidly grounded and others are isolated</u>

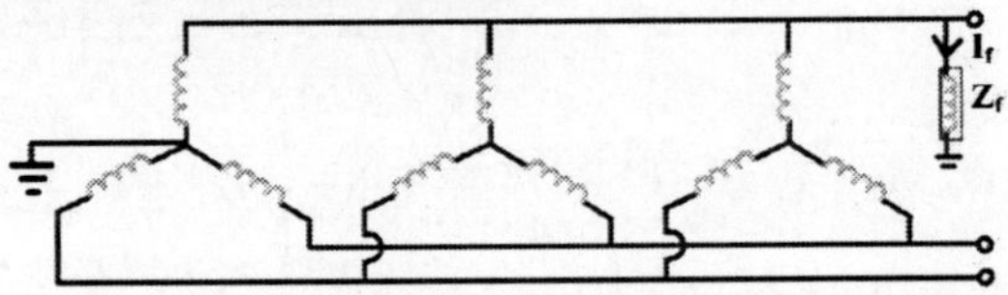

$Z_1 = \frac{j0.3}{3} = j0.1 p.u.$

$Z_2 = \frac{j0.2}{3} = j0.067\ p.u.$

$Z_0 = j0.1\ p.u.$

Per unit fault current $I_f^{LG} = \frac{3E_a}{Z_0 + Z_1 + Z_2 + 3\,Z_f} = \frac{3*1\angle 0^0}{j0.1 + j0.1 + j0.067 + 3*j0.05} = -j7.194\ p.u.$

Actual fault current $I_{f\,actual}^{LG} = I_{f\,p.u.}^{LG} * I_{base} = -j7.194 * 524.86 = 3.776\angle -90^0\ kA$

(iii). When only one of the alternators is grounded through an inductor of 0.2Ω and others are isolated

$$Z_1 = \frac{j0.3}{3} = j0.1 p.u.$$

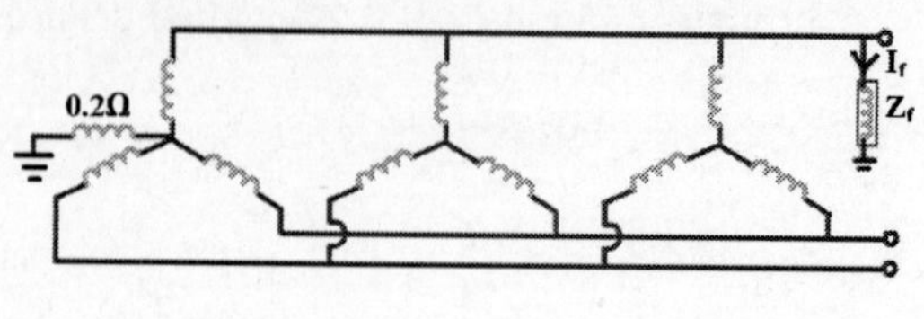

$$Z_2 = \frac{j0.2}{3} = j0.067\ p.u.$$

$$Z_0 = Z_{g0} + 3 * Z_n = j0.1 + 3 * \left(\frac{j0.2}{11^2} * 10\right)$$

$$= j0.1495\ p.u.$$

Per unit fault current $I_f^{LG} = \frac{3E_a}{Z_1 + Z_2 + (Z_{g0} + 3\ Z_n) + 3\ Z_f}$

$$= \frac{3 * 1\angle 0^0}{j0.1\ + j0.067 + j0.1495\ + 3 * j0.05}$$

$$= -j6.43\ p.u.$$

Actual fault current $I_{f\ actual}^{LG} = I_{f\ p.u.}^{LG} * I_{base} = -j6.43 * 524.86 = 3.375\angle -90^0\ kA$

Example 7.12. The per unit values of positive, negative, and zero sequence reactance of a network during a fault are j0.24, j0.16, and j0.12, respectively. Calculate the fault current in the case of a double line to ground fault (LLG).

Ans: Let double line to ground fault (LLG) occurs between the terminals b and c to ground.

$$Z_1 = j0.24,\ Z_2 = j0.16,\ Z_0 = j0.12,\ Z_f = 0 \text{ and } E_a = 1\angle 0^0$$

In a <u>LL-G Fault</u>: $I_f^{LLG} = I_b + I_c$ And $I_a = 0$

'+'ve sequence current $I_{a1}^{pu} = \frac{E_a}{Z_1 + \left(\frac{Z_0 Z_2}{Z_0 + Z_2}\right)} = \frac{1\angle 0^0}{j0.24 + \left(\frac{j0.12 * j0.16}{j0.12 + j0.16}\right)} = -j3.24\ p.u.$

'+'ve sequence voltage $V_{a1} = E_a - I_{a1} * Z_1 = 1 - (-j3.24) * j0.24 = 0.224\ p.u.$

Fault impedance $Z_f = 0$, so in LLG fault $V_{a0} = V_{a1} = V_{a2} = 0.2224\ p.u.$

'-'ve sequence current $I_{a2}^{pu} = \frac{-V_{a2}}{Z_2} = \frac{-0.2224}{j0.16} = j1.39\ p.u.$

Zero sequence current $I_{a0}^{pu} = \frac{-V_{a0}}{Z_0} = \frac{-0.2224}{j0.12} = j1.853\ p.u.$

Current in phase 'a' is $I_a = I_{a0} + I_{a1} + I_{a2}$

$$= j1.853 + (-j3.24) + j1.39$$

$$= 0$$

Current in phase 'b' is $I_b = I_{a0} + a^2\ I_{a1} + a\ I_{a2}$

$$= j1.853\ + 1\angle 240^0 * (-j3.24) + 1\angle 120^0 * j1.39$$

$$= 4.878\angle 145.28^0\ p.u.$$

Current in phase 'c' is $I_c = I_{a0} + a\ I_{a1} + a^2\ I_{a2}$

$$= j1.853\ + 1\angle 120^0 * (-j3.24) + 1\angle 240^0 * j1.39$$

$$= 4.878\angle 34.72^0\ p.u.$$

Per unit fault current $I_f^{LLG} = I_b + I_c = 4.878\angle 145.28^0 + 4.878\angle 34.72^0 = 5.5\angle 90^0\ p.u.$

Example 7.13. The alternator has a power rating of 50MVA and operates at a voltage of 13.2kV. Its neutral is solidly grounded. The sub-transient reactance of the alternator is j0.24 per unit. The reactance for the negative sequence is j0.16 per unit, while the reactance for the zero sequence is j0.12 per unit. Calculate the magnitude of the current flowing during a double line to ground (LLG) fault at the alternator terminals, as well as the voltages between the lines during the fault.

Ans : Let double line to ground fault (LLG) occurs between the terminals b and c to ground.

Consider base values as 50MVA, 13.2kV.

Base Current $I_{base} = \frac{50*10^6}{\sqrt{3}*13.2*10^3} = 2.186\ kA$

In this example $Z_0, Z_1, and\ Z_2$ are considered same as of example 7.12.

So, as determined in example 7.12, $I_{f\ p.u.}^{LLG} = 5.5\angle 90^0\ p.u.$

Actual fault current $I_{f\ actual}^{LLG} = I_{f\ p.u.}^{LLG} * I_{base} = 5.5\angle 90^0 * 2186 = 12.023\ \angle 90^0\ kA$

In LLG $V_{a0} = V_{a1} = V_{a2}$

Voltage in phase 'a' is $V_a = V_{a0} + V_{a1} + V_{a2} = 3 * V_{a1} = 3 * 0.224 = 0.6672\ p.u$

Voltage in phase 'b' and 'c' will be zero $V_b = V_c = 0$

Per unit and actual Line-Line voltages,

$$V_{ab}^{p.u.} = V_a - V_b = 0.6672 - 0 = 0.6672\ p.u.$$

$$V_{bc}^{p.u.} = V_b - V_c = 0 - 0 = 0$$

$$V_{ca}^{p.u.} = V_c - V_a = 0 - 0.6672 = -0.6672\ p.u.$$

$$V_{ab}^{actual} = V_{ab}^{p.u.} * kV_{base} = 0.6672 * (13.2 * 10^3) = 8.807\angle 0^0\ kV$$

$$V_{bc}^{actual} = V_{bc}^{p.u.} * kV_{base} = 0 * (13.2 * 10^3) = 0$$

$$V_{ca}^{actual} = V_{ca}^{p.u.} * kV_{base} = -0.6672 * (13.2 * 10^3) = 8.807\angle 180^0\ kV$$

Example 7.14. A 11kV, 10MVA alternator has $Z_1 = Z_2 = 25\%\ and\ Z_0 = 15\%$. The alternator's neutral is grounded through a 0.75Ω reactance. Calculate currents in all three phases of an alternator when a double-line (LL) fault occurs at its terminals while it is unloaded. The fault impedance has a reactance of 10%. Calculate voltages at alternator's terminals in the event of a fault.

Ans: Let double line fault (LL) occurs between the terminals b and c.

$$I_f = I_b = -I_c\ ;\ I_a = 0$$

$$V_b - V_c = I_b Z_f$$

$Z_1 = Z_2 = j0.25,\ Z_{g0} = j0.15,\ Z_f = j0.1$ and $E_a = 1\angle 0$

$$Z_0 = Z_{g0} + 3 * Z_n = j0.15 + 3 * \left(\frac{j0.75}{11^2} * 10\right) = j0.3359\ p.u.$$

$$I_{a0} = 0;\ I_{a1} = -I_{a2}$$

$$I_{a1} = \frac{E_a}{Z_1 + Z_2 + Z_f} = \frac{1\angle 0^0}{j0.25 + j0.25 + j0.1} = -j1.666\ p.u.$$

$$I_{a2} = -I_{a1} = j1.666\ p.u.$$

Per unit Fault current $I^{LL}_{p.u.} = \frac{-j\sqrt{3}E_a}{Z_1+Z_2+Z_f} = \frac{-j\sqrt{3}*1\angle 0^0}{j0.25+j0.25+j0.1} = -2.886\ p.u. = I_b = -I_c$

Current in phase 'a' is $I_a = I_{a0} + I_{a1} + I_{a2}$

$= 0 + (-j1.666) + j1.666 = 0\ p.u.$

Current in phase 'b' is $I_b = I_{a0} + a^2\, I_{a1} + a\, I_{a2}$

$= 0 + 1\angle 240^0 * (-j1.666) + 1\angle 120^0 * j1.666$

$= -2.886\ p.u.$

Current in phase 'c' is $I_c = I_{a0} + a\, I_{a1} + a^2\, I_{a2}$

$= 0 + 1\angle 120^0 * (-j1.666) + 1\angle 240^0 * j1.666$

$= 2.886\ p.u.$

Symmetrical components of voltages at terminal "a" are $V_{a0}, V_{a1}\ and\ V_{a2}$

$V_{a0} = -I_{a0} * Z_0 = -0 * j0.3359 = 0$

$V_{a1} = E_a - I_{a1} * Z_1 = 1 - (-j1.666\,) * j0.25 = 0.5835\ p.u.$

$V_{a2} = -I_{a2} * Z_2 = -j1.666\ * j0.25 = 0.4165\ p.u.$

To verify $V_{a1} - V_{a2} = I_{a1}Z_f$

$V_{a1} - V_{a2} = 0.5835 - 0.4165 = 0.1666\ p.u.$

$I_{a1}Z_f = -j1.666 * j0.1 = 0.1666\ p.u.$

Voltage in phase 'a' is $V_a = V_{a0} + V_{a1} + V_{a2}$

$= 0 + 0.5835 + 0.4165$

$= 1\ p.u.$

Voltage in phase 'b' is $V_b = V_{a0} + a^2\, V_{a1} + a\, V_{a2}$

$= 0 + 1\angle 240^0 * 0.5835 + 1\angle 120^0 * 0.4165$

$= -0.5 - j0.1446\ p.u.$

$= 0.5204\angle -163.86^0\ p.u.$

Voltage in phase 'c' is $V_c = V_{a0} + a\, V_{a1} + a^2\, V_{a2}$

$= 0 + 1\angle 120^0 * 0.5835 + 1\angle 240^0 * 0.4165$

$= -0.5 + j0.1446\ p.u.$

$= 0.5204\angle 163.86^0\ p.u.$

Per unit line-line voltages,

$V_{ab} = V_a - V_b = 1 - 0.5204\angle -163.86^0 = 1.506\angle 5.50^0\ p.u.$

$V_{bc} = V_b - V_c = 0.5204\angle -163.86^0 - 0.5204\angle 163.86^0 = 0.289\angle -90^0\ p.u.$

$V_{ca} = V_c - V_a = 0.5204\angle 163.86^0 - 1 = 1.506\angle 174^0\ p.u.$

Consider Base voltage = 11kV, Base MVA=10 MVA

$$I_{base} = \frac{kVA_{base}}{\sqrt{3} * kV_{base}} = \frac{10 * 10^6}{\sqrt{3} * 11 * 10^3} = 524.86\ A$$

Actual Fault current $I^{LL}_{actual} = I^{LL}_{p.u.} * I_{base} = -2.886 * 524.86 = 1.514\angle 180^0\ kA$

Actual Current phase 'a' is $I_{a\ actual} = I_{a\ p.u.} * I_{base} = 0 * 524.86 = 0\ A$

Actual Current phase 'b' is $I_{b\ actual} = I_{b\ p.u.} * I_{base} = -2.886 * 524.86 = 1.514\angle 180^0\ kA$

Actual Current phase 'c' is $I_{c\ actual} = I_{c\ p.u.} * I_{base} = 2.886 * 524.86 = 1.514\angle 0^0\ kA$

Actual Voltage phase 'a' is $V_{a\ actual} = V_{a\ p.u.} * V_{base} = 1 * 11000 = 11\angle 0^0\ kV$

Actual Voltage phase 'b' is $V_{b\ actual} = V_{b\ p.u.} * V_{base}$

$$= 0.5204\angle -163.86^0 * 11000 = 5.7224\angle -163.86^0\ kV$$

Actual Voltage phase 'c' is $V_{c\ actual} = V_{c\ p.u.} * V_{base}$

$$= 0.5204\angle 163.86^0 * 11000 = 5.7224\angle 163.86^0\ kV$$

Example 7.15. A 3-phase 10MVA, 11kV, 50Hz generator with solidly earthed neutral has $X''_d = 8\%$, $X'_d = 16\%$, and $X_d = 80\%$ respectively. Negative sequence reactance $X_2 = 15\%$ and zero sequence reactance $X_0 = 5\%$. When a 3-phase fault develops, the generator was operating on an open circuit. If the fault is a 3-phase short with no impedance, calculate the sustained, transient, and sub-transient short circuit currents under the faulty situation. (Consider base values as 11kV, 10MVA)

Ans. $I_{base} = \frac{kVA_{base}}{\sqrt{3} * kV_{base}} = \frac{10*10^6}{\sqrt{3}*11*10^3} = 524.8\ A$

Sub-transient short circuit current $I''_{g\ p.u.} = \frac{E_g}{X''_d} = \frac{1\angle 0^0}{j0.08} = -j12.5\ p.u.$

$$I''_{g\ actual} = I''_{g\ p.u.} * I_{base} = -j12.5 * 524.8 = 6.560\angle -90^0\ kA$$

Transient short circuit current $I'_{g\ p.u.} = \frac{E_g}{X'_d} = \frac{1\angle 0^0}{j0.16} = -j6.25\ p.u.$

$$I'_{g\ actual} = I'_{g\ p.u.} * I_{base} = -j6.25 * 524.8 = 3.28\angle -90^0\ kA$$

Sustained short circuit current $I_{g\ p.u.} = \frac{E_g}{X_d} = \frac{1\angle 0^0}{j0.8} = -j1.25\ p.u.$

$$I_{g\ actual} = I_{g\ p.u.} * I_{base} = -j1.25 * 524.8 = 0.625\angle -90^0\ kA$$

As there is a complete dead short circuit $V_a = V_b = V_c = 0$

Example 7.16. A 11kV, 50MVA alternator has $Z_1 = j0.25, Z_2 = j0.35\ and\ Z_0 = j0.15$. The neutral of the alternator is solidly grounded. The alternator is operating at no-load rated voltage. Calculate the fault currents for

(i) Single line to ground fault (LG)
(ii) Line to line fault (LL)
(iii) Double line to ground fault (LLG)
(iv) Symmetrical 3-phase fault (LLL).

Consider base values as 50MVA and 11kV.

Ans: Base Current $I_{base} = \frac{50*10^6}{\sqrt{3}*11*10^3} = 2.624\ kA$

(i) <u>Single line to ground fault (LG):</u>

Let a, b, c be the terminals of the alternator and fault occurs between terminal 'a' and ground.

Per unit Fault current $I_{f\,p.u.}^{LG} = \frac{3E_a}{Z_0+Z_1+Z_2} = \frac{3*1\angle 0^0}{j0.15+j0.25+j0.35} = -j3.999p.u.$

Actual fault current $I_{f\,actual}^{LG} = I_{f\,p.u.}^{LG} * I_{base} = -j3.999 * 2624 = 10.5\angle -90^0\ kA$

(ii) Line to line fault (LL): Let double line (LL) occurs between the terminals b and c.

Per unit Fault current $I_{f\,p.u.}^{LL} = \frac{-j\sqrt{3}E_a}{Z_1+Z_2} = \frac{-j\sqrt{3}*1\angle 0^0}{j0.25+j0.35} = -2.886\ p.u.$

Actual fault current $I_{f\,actual}^{LL} = I_{f\,p.u.}^{LL} * I_{base} = -2.886 * 2624 = 7.572\angle 180^0\ kA$

(iii) Double line to ground fault (LLG):

Let double line to ground fault (LLG) occurs between the terminals b and c to ground.

Per unit Fault current $I_f^{LLG} = \frac{-3Z_2E_a}{Z_0Z_1+Z_0Z_2+Z_1Z_2} = \frac{-3*j0.35*1\angle 0^0}{j0.15*j0.25+j0.15*j0.35+j0.25*j0.35} = j5.915p.u.$

Actual fault current $I_{f\,actual}^{LLG} = I_{f\,p.u.}^{LLG} * I_{base} = 5.915\angle 90^0 * 2624 = 15.52\angle 90^0\ kA$

(iv) Symmetrical 3-phase fault (LLL): Per unit Fault current $I_{f\,p.u.}^{3\phi} = \frac{E_a}{Z_1} = \frac{1\angle 0^0}{j0.25} = -j4\ p.u.$

Actual fault current $I_{f\,actual}^{3\phi} = I_{f\,p.u.}^{LLL} * I_{base} = -j4 * 2624 = 10.496\angle -90^0\ kA$

Example 7.17. A generator of negligible resistance having 1p.u. Voltage behind transient reactance is subjected to different type of faults

S.No.	Type of fault	Resulting fault current in p.u.
1	3-Phase	5
2	LG	4
3	LL	3

Ignoring resistances calculate the values of three reactance X_0, X_1 and X_2.

Ans: In this example, we need to calculate the values of three reactances X_0, X_1 and X_2 by ignoring resistances. So, in LL fault it is considered as $I_{f\,p.u.}^{LL} = |I_b| = |-I_c|$

Resulting fault currents $I_f^{3\phi} = 5\ p.u., I_f^{LG} = 4\ p.u., and\ I_f^{LL} = 3\ p.u.$

From 3ϕ fault: $I_{f\,p.u.}^{3\phi} = \frac{E_a}{X_1} = \frac{1\angle 0^0}{X_1} = 5\ p.u.$

$$X_1 = \frac{1}{5} = 0.2\ p.u.$$

From LL fault: $I_{f\,p.u.}^{LL} = |I_b| = |-I_c| = \frac{\sqrt{3}E_a}{X_1+X_2} = \frac{\sqrt{3}*1\angle 0^0}{X_1+X_2} = 3\ p.u.$

$$X_1 + X_2 = \frac{\sqrt{3}}{3} = 0.5773\ p.u.$$

$$X_2 = 0.5773 - X_1 = 0.5773 - 0.2 = 0.3773\ p.u.$$

From LG fault: $I_{f\,p.u.}^{LG} = \frac{3E_a}{X_0+X_1+X_2} = 4\ p.u.$

$$X_0 + X_1 + X_2 = \frac{3*1\angle 0^0}{4} = 0.75$$

$$X_0 = 0.75 - X_1 - X_2 = 0.75 - 0.2 - 0.3773 = 0.1727\ p.u.$$

Example 7.18. A 50MVA, 11kV alternator was subjected to the following faults $I_f^{3\phi} = 1600A,\ I_f^{LL} = 1800A,$ and $I_f^{LG} = 2400A$. The neutral of the alternator is solidly grounded. Ignoring resistances calculate the values of three reactance X_0, X_1 and X_2.

Ans: Let a, b, c be the terminals of the alternator and fault impedance be zero.

In this example also, we need to calculate the values of three reactances X_0, X_1 and X_2 by ignoring resistances. So, in LL fault it is considered as $I_{f\,p.u.}^{LL} = |I_b| = |-I_c|$

Resulting fault currents $I_f^{3\phi} = 1600A,\ I_f^{LL} = 1800A,$ and $I_f^{LG} = 2400A$.

From 3ϕ fault: $I_{f\,actual}^{3\phi} = \frac{E_a}{X_1} = \frac{\left(\frac{11000}{\sqrt{3}}\right)}{X_1} = 1600\ A$

$$X_1 = \frac{6350}{1600} = 3.96\ \Omega$$

From LL fault: $I_{f\,p.u.}^{LL} = |I_b| = |-I_c| = \frac{\sqrt{3}E_a}{X_1 + X_2} = \frac{\sqrt{3}*\left(\frac{11000}{\sqrt{3}}\right)}{X_1 + X_2} = 1800A$

$$X_1 + X_2 = 6.111\ \Omega$$

$$X_2 = 6.111 - X_1 = 6.111 - 3.96 = 2.15\ \Omega$$

From LG fault: $I_{p.u.}^{LG} = \frac{3E_a}{X_0 + X_1 + X_2} = 2400A$

$$X_0 + X_1 + X_2 = \frac{3*\left(\frac{11000}{\sqrt{3}}\right)}{2400} = 7.938\ \Omega$$

$$X_0 = 7.938 - X_1 - X_2 = 7.938 - 3.96 - 2.15 = 1.828\ \Omega$$

Example 7.19. A star-connected alternator provides power to a star-connected motor load via a transmission line. The load's star point is grounded, whereas the generator neutral is ungrounded. The load positive, negative and zero sequence reactances are same with a value equal to 0.8 p.u. The positive, negative, and zero sequence reactances of line are j0.2 p.u., j0.2 p.u., and j0.6 p.u. per phase. The generator's positive, negative, and zero sequence reactances are j0.6 p.u., j0.7 p.u., and j0.05 p.u. respectively. A single phase-to-ground fault occurs halfway down the line. Prior to the fault, the network was balanced, and the voltage at the fault site was 1 p.u. Determine the current through the fault path.

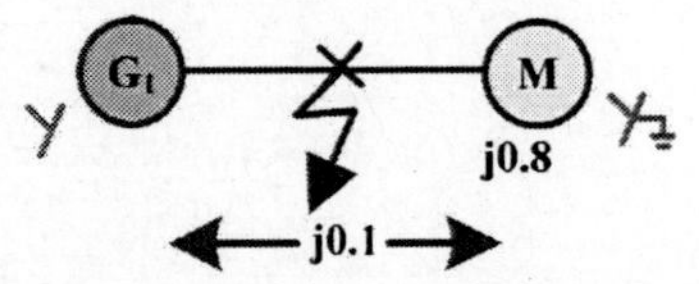

Ans: $X_g^1 = j0.6\ p.u., X_g^2 = j0.7\ p.u., and\ X_g^0 = j0.05\ p.u.$

$X_{line}^1 = X_{line}^2 = j0.2\ p.u., X_{line}^0 = j0.6\ p.u.$

$X_m^1 = X_m^2 = X_m^0 = j0.8\ p.u.$

Positive sequence network:

$Z_1 = [(j0.6 + j0.1) \parallel (j0.1 + j0.8)]$

$= [(j0.7) \parallel (j0.9)]$

$= j0.3937\ p.u.$

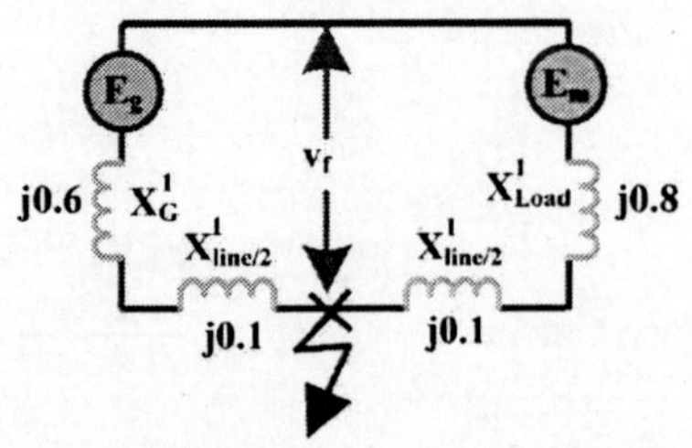

<u>Negative sequence network:</u>

$$Z_2 = [(j0.7 + j0.1) \parallel (j0.1 + j0.8)]$$
$$= [(j0.8) \parallel (j0.9)]$$
$$= j0.4235\ p.u.$$

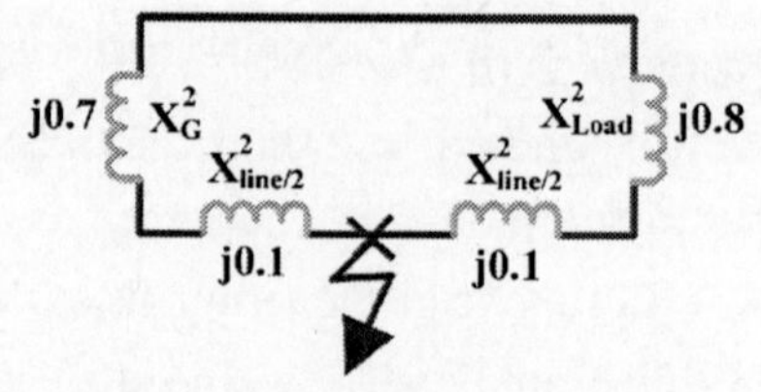

<u>Zero sequence network:</u>

$$Z_0 = j0.8 + j0.3$$
$$= j1.1\ p.u.$$

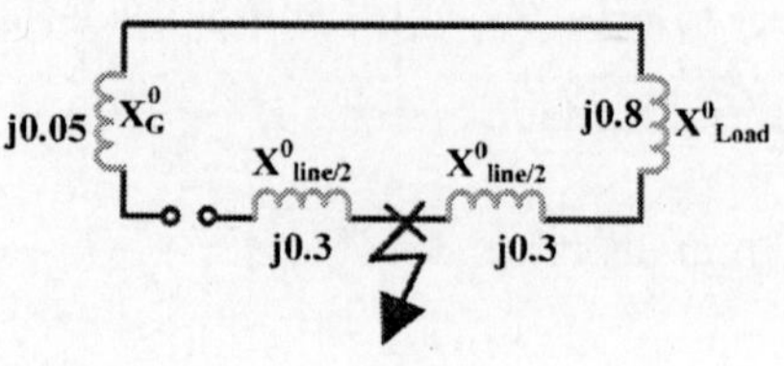

LG Fault current $I_f^{LG} = \dfrac{3E_a}{Z_0 + Z_1 + Z_2}$

$$= \frac{3 * 1\angle 0^0}{j1.1 + j0.3937 + j0.4235}$$
$$= j1.5647\ p.u.$$

Example 7.20. Figure depicts a single line diagram of a power system, with the sequence reactances of the generator, motor, and transformers T_1 and T_2 listed in per unit. The neutrals of the generator and transformers are solidly grounded. The motor neutral is grounded by a reactance X_n=j0.03 p.u. Create positive, negative, and zero sequence networks with reactance values in per unit on a 100 MVA, 11 kV base in the generator's zone. Pre-fault voltage is 1 p.u. Determine the fault current for a three-phase to ground fault on bus 4. The system data are presented below.

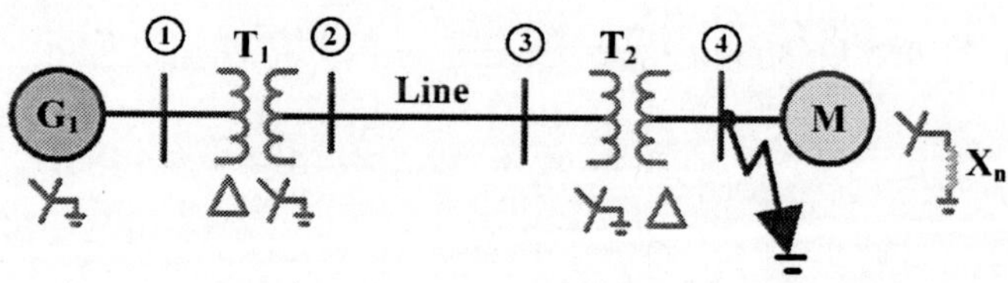

G_1 : 100 MVA, 11 kV; X_1 = j0.15 p. u, X_2 = j0.18 p. u. and X_0 = j0.05 p. u.

T/f_1: 150 MVA, $\dfrac{11\ kV}{110\ kV}$; $X_1 = X_2 = X_0$ = j0.12 p. u.

T/f_2: 150 MVA, $\dfrac{110\ kV}{11\ kV}$; $X_1 = X_2 = X_0$ = j0.12 p. u.

Line : $X_1 = X_2$ = j0.3 p. u., X_0 = j0.9 p. u on 110 kV, 100 MVA base.

M : 100 MVA, 11 kV; X_1 = j0.2 p. u, X_2 = j0.25 p. u., X_{M0} = j0.1 p. u. and X_n = j0.03 p. u.

Ans: Base values are 100MVA, 11kV base in the zone of the generator.

Positive, negative and zero sequence p.u. Reactance of generator motor, transformers and transmission line are as follows

$$X_g^1 = j0.15\ p.u., X_g^2 = j0.18\ p.u., and\ X_g^0 = j0.05\ p.u.$$

$$X_{T/f}^1 = X_{T/f}^2 = X_{T/f}^0 = j0.12 * \left(\frac{11}{11}\right)^2 * \frac{100}{150} = j0.08\ p.u.$$

$$X_m^1 = j0.2\ p.u., X_m^2 = j0.25\ p.u., and\ X_m^0 = X_{mo} + 3X_n = j0.1 + 3 * j0.03 = j0.19\ p.u.$$

Base Current $\quad I_{base} = \dfrac{100*10^6}{\sqrt{3}*11*10^3} = 5.248\ kA$

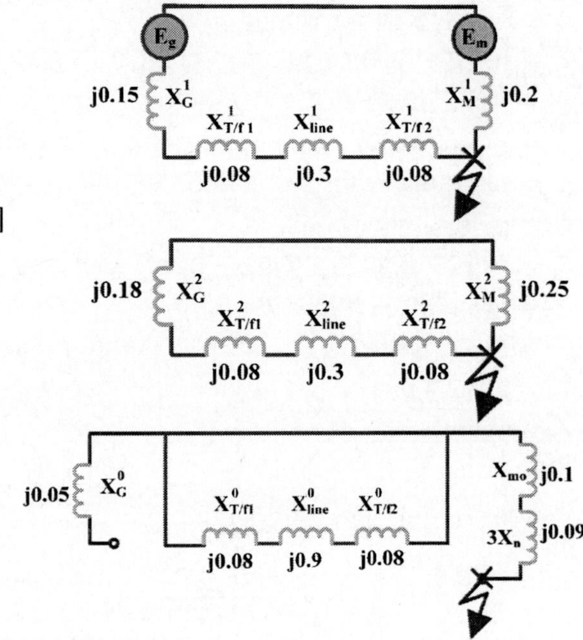

Positive sequence network:

$Z_1 = [(j0.15 + j0.08 + j0.3 + j0.08) \parallel (j0.2)]$

$= (j0.61) \parallel (j0.2)$

$= j0.1506\ p.u.$

Negative sequence network:

$Z_1 = [(j0.18 + j0.08 + j0.3 + j0.08) \parallel (j0.25)]$

$= (j0.64) \parallel (j0.25)$

$= j0.1797\ p.u.$

Zero sequence network:

$Z_0 = (j0.1 + 3 * j0.03)$

$= (j0.1 + j0.09)$

$= j0.19\ p.u.$

A three-phase to ground fault at bus 4,

Per unit Fault current $I^{3\phi}_{f\ p.u.} = \frac{V_f}{Z_1}$

$$= \frac{1\angle 0^0}{j0.1506} = -j6.64\ p.u.$$

Actual fault current $I^{3\phi}_{f\ actual} = I^{3\phi}_{f\ p.u.} * I_{base} = -j6.64 * 5248 = 34.847 \angle -90^0\ kA$

Example 7.21. A single line diagram of a power network is shown in figure. The table shows positive, negative and zero sequence reactances of elements on a common base.

Element	Positive sequence reactance (p.u.)	Negative sequence reactance (p.u.)	Zero sequence reactance (p.u.)
G	j0.2	j0.25	j0.05
T1	j0.3	j0.35	j0.06
T2	j0.4	j0.45	j0.07
M1	j0.5	j0.55	j0.08
M2	j0.6	j0.65	j0.09
Line	j0.2	j0.2	j0.6

Generator grounding reactance is j0.05p.u.

(i) Draw sequence networks

(ii) Find fault currents for a line to line (LL) fault on terminals b&c at bus 2. Assume 1 p.u. Pre-fault voltage throughout.

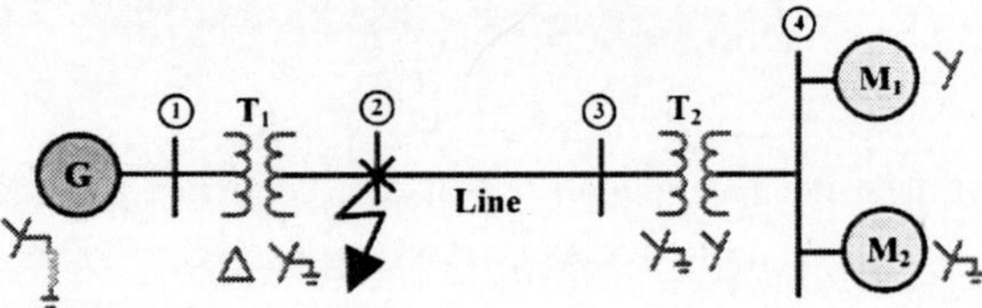

Ans: <u>Positive sequence network:</u>

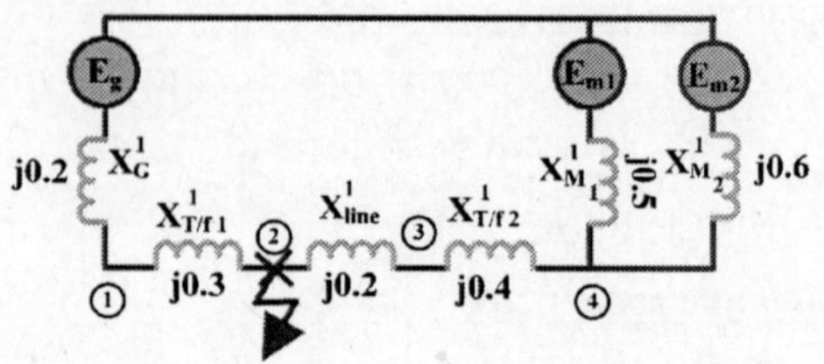

$$Z_1 = \{(j0.2 + j0.3) \parallel [(j0.2 + j0.4) + (j0.5 \parallel j0.6)]\}$$

$$= j0.5 \parallel (j0.6 + j0.2727)$$

$$= j0.5 \parallel j0.8727$$

$$= j0.3178\, p.u.$$

<u>Negative sequence network:</u>

$$Z_2 = \{(j0.25 + j0.35) \parallel [(j0.2 + j0.45) + (j0.55 \parallel j0.65)]\}$$

$$= j0.6 \parallel (j0.65 + j0.2979)$$

$$= j0.6 \parallel j0.9479$$

$$= j0.3674\, p.u.$$

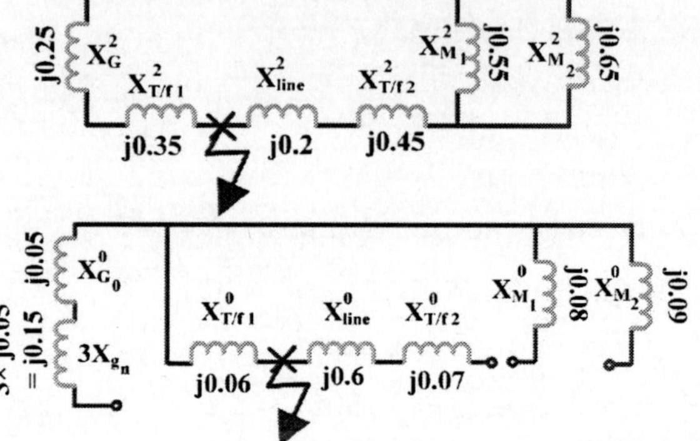

<u>Zero sequence network:</u>

$$Z_0 = j0.06 p.u.$$

LL Fault current $I_f^{LL} = \frac{-j\sqrt{3}E_a}{Z_1 + Z_2}$

$$= \frac{-j\sqrt{3} * 1\angle 0^0}{j0.3178 + j0.3674}$$

$$= -2.5278\, p.u.$$

Example 7.22. A single line diagram of power system is shown in figure

G_1 : 100 MVA, 11 kV; $X_1 = X_2 = j0.3$ p. u. and $X_0 = j0.06$ p. u.

G_2 : 100 MVA, 11 kV; $X_1 = X_2 = j0.35$ p. u., $X_{g0} = j0.075$ p. u. and $X_n = j0.02$ p. u.

T/f$_1$: 150 MVA, $\frac{11\text{ kV}}{110\text{ kV}}$; $X_1 = X_2 = X_0 = j0.4$ p. u.

T/f$_2$: 150 MVA, $\frac{110\text{ kV}}{11\text{ kV}}$; $X_1 = X_2 = X_0 = j0.1$ p. u., $X_n = j0.04$ p. u.

Line 1 : $X_1 = X_2 = j0.2$ p. u., $X_0 = j0.6$ p. u. on 110 kV, 100 MVA base.

Line 2: $X_1 = X_2 = j0.25$ p. u., $X_0 = j0.75$ p. u. on 110 kV, 100 MVA base.

Draw the sequence networks and then calculate the fault currents for

(i) Single line to ground fault (LG)
(ii) Line to line fault (LL)
(iii) Double line to ground fault (LLG)
(iv) Symmetrical 3-phase fault (LLL).

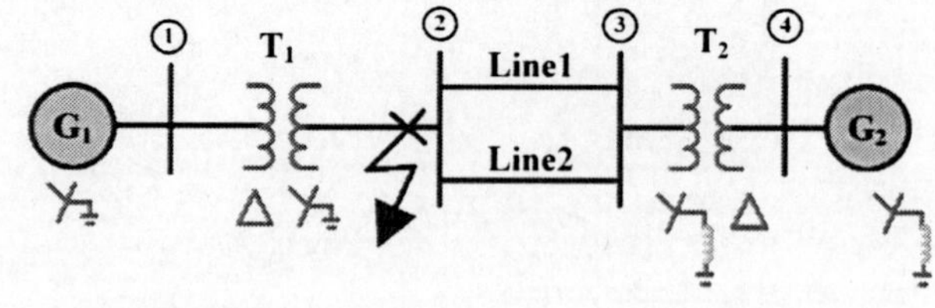

Ans: Consider 100MVA, 11kV as base values. Positive, negative and zero sequence p.u. reactance of generator motor, transformers and transmission line are as follows

$$X^1_{g1} = X^2_{g1} = j0.3 * \left(\frac{11}{11}\right)^2 * \frac{100}{100} = j0.3\, p.u., and\ X^0_{g1} = j0.06\, p.u.$$

$$X^1_{T/f1} = X^2_{T/f1} = X^0_{T/f1} = j0.4 * \left(\frac{11}{11}\right)^2 * \frac{100}{150} = j0.2666\, p.u.$$

$$X^1_{T/f2} = X^2_{T/f2} = X_{\frac{t}{f}0} = j0.1 * \left(\frac{11}{11}\right)^2 * \frac{100}{150} = j0.0666\, p.u.,$$

$$X^0_{T/f2} = X_{\frac{t}{f}0} + 3X_n = j0.1 * \frac{100}{150} + 3 * j0.04 * \frac{100}{150} = j0.0666 + j0.08 = j0.1466\, p.u.$$

$$X^1_{g2} = X^2_{g2} = j0.35 * \left(\frac{11}{11}\right)^2 * \frac{100}{100} = j0.35\, p.u.,$$

$$X^0_{g2} = X_{go} + 3X_n = j0.075 + 3 * j0.02 = j0.135\, p.u.$$

Positive sequence network:

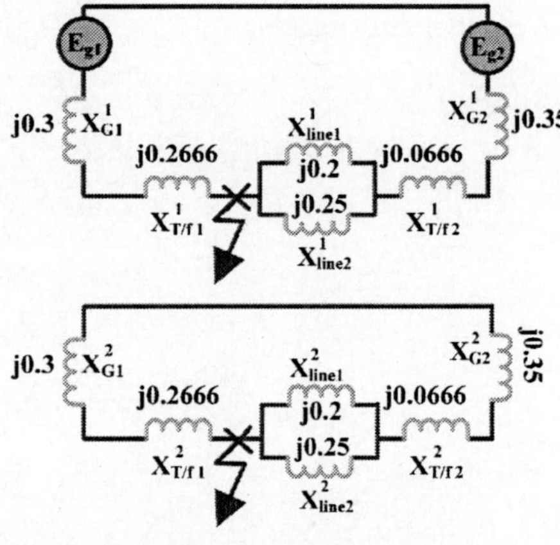

$Z_1 = (j0.3 + j0.2666) \parallel$

$\quad [(j0.2 \parallel j0.25) + j0.0666 + j0.35)]$

$\quad = (j0.5666) \parallel (j0.111 + j0.4166)$

$\quad = (j0.5666) \parallel j0.5276$

$\quad = j0.2732\, p.u.$

Negative sequence network: In the given example, the positive and negative sequence values of all the components in the network are same. $Z_2 = Z_1$

The positive and negative sequence networks will be almost identical, only change is energy source will not present in negative sequence network.

Zero sequence network:

$Z_0 = j0.2666 \parallel$

$\quad [(j0.6 \parallel j0.75) + (j0.0666 + j0.08)]$

$\quad = j0.2666 \parallel (j0.333 + j0.1466)$

$\quad = j0.2666 \parallel j0.4796$

$\quad = j0.1713\, p.u.$

Base Current $I_{base} = \frac{100*10^6}{\sqrt{3}*11*10^3} = 5.248\, kA$

(i) Single line to ground fault (LG):

Let a, b, c be the terminals of the alternator and fault occurs between terminal 'a' and ground.

Per unit Fault current $I^{LG}_{f\,p.u.} = \frac{3E_a}{Z_0 + Z_1 + Z_2} = \frac{3*1\angle 0^0}{j0.1713 + j0.2732 + j0.2732} = -j4.18001\, p.u.$

Actual fault current $I^{LG}_{f\,actual} = I^{LG}_{f\,p.u.} * I_{base} = -j4.180 * 5248 = 21.936 \angle -90^0\, kA$

(ii) Line to line fault (LL): Let double line (LL) occurs between the terminals b and c.

Per unit Fault current $I^{LL}_{f\,p.u.} = \frac{-j\sqrt{3}E_a}{Z_1 + Z_2} = \frac{-j\sqrt{3}*1\angle 0^0}{j0.2732 + j0.2732} = -3.169\, p.u.$

Actual fault current $I^{LL}_{f\,actual} = I^{LL}_{f\,p.u.} * I_{base} = -3.169 * 5248 = 16.635 \angle 180^0\, kA$

(iii) Double line to ground fault (LLG): Let double line to ground fault (LLG) occurs between the terminals b and c to ground.

In LLG fault $\quad I_f^{LLG} = \dfrac{-3Z_2E_a}{Z_0Z_1+Z_0Z_2+Z_1Z_2}$

Per unit Fault current $\quad I_f^{LLG} = \dfrac{-3*j0.2732*1\angle 0^0}{j0.1713*j0.2732+j0.1713*j0.2732+j0.2732*j0.2732} = j4.8717\ p.u.$

Actual fault current $\quad I_{f\ actual}^{LLG} = I_{f\ p.u.}^{LLG} * I_{base} = j4.8717 * 5248 = 25.566\angle 90^0\ kA$

(iv) Symmetrical 3-phase fault (LLL):

Per unit Fault current $\quad I_{f\ p.u.}^{3\phi} = \dfrac{E_a}{Z_1} = \dfrac{1\angle 0^0}{j0.2732} = -j3.66\ p.u.$

Actual fault current $\quad I_{f\ actual}^{3\phi} = I_{f\ p.u.}^{LLL} * I_{base} = -j3.66 * 5248 = 19.209\angle -90^0\ kA$

Example 7.23. A power system is shown in figure with per unit reactances on a common base

(i) Draw sequence networks

(ii) Find fault currents for a single line to ground (LG) fault at bus 2. Assume 1 p.u. Pre-fault voltage throughout.

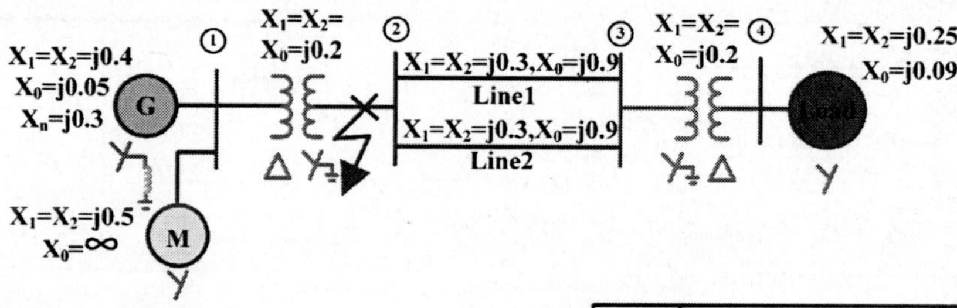

Ans: Positive sequence network:

$Z_1 = [(j0.4 \parallel j0.5) + j0.2] \parallel$

$\qquad [(j0.3 \parallel j0.3) + (j0.2 + j0.25)]$

$= (j0.22 + j0.2) \parallel (j0.15 + j0.45)$

$= j0.42 \parallel j0.6$

$= j0.247\ p.u.$

Negative sequence network:

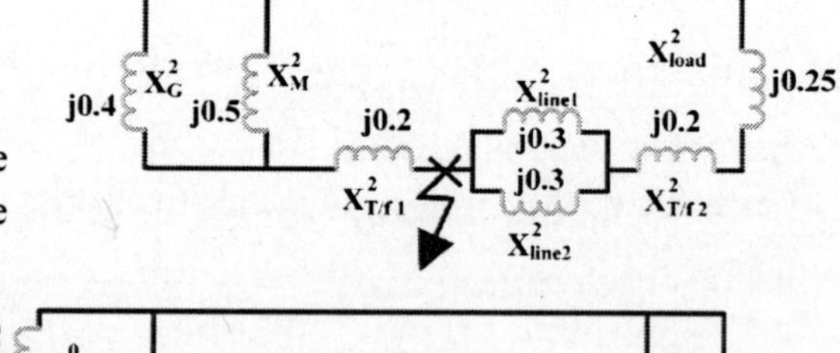

In the given example, the positive and negative sequence values of all the components in the network are same. $Z_2 = Z_1$

Zero sequence network:

$Z_0 = j0.2 \parallel [(j0.9 \parallel j0.9) + (j0.2)]$

$= j0.2 \parallel (j0.45 + j0.2)$

$= j0.2 \parallel j0.65$

$= j0.153\ p.u.$

LG Fault current $\quad I_f^{LG} = \dfrac{3V_f}{Z_0+Z_1+Z_2} = \dfrac{3*1\angle 0^0}{j0.153+j0.247+j0.247} = -j4.636\ p.u.$

Example 7.24. A power system is shown in figure with per unit reactances on a common base

(i) Draw sequence networks

(ii) Find fault currents for a single line to ground (LG) fault at bus 2. Assume 1 p.u. Pre-fault voltage throughout.

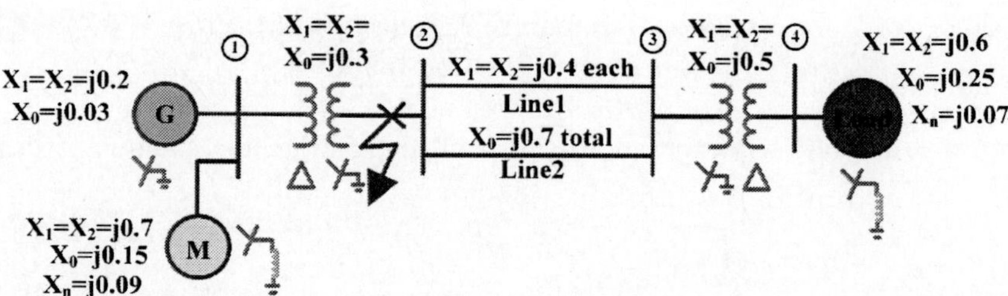

Ans:

<u>Positive sequence network:</u>

$$Z_1 = [(j0.2 \parallel j0.7) + j0.3] \parallel [(j0.4 \parallel j0.4) + (j0.5 + j0.6)]$$

$$= (j0.1555 + j0.3) \parallel (j0.2 + j1.1)$$

$$= j0.4555 \parallel j1.3$$

$$= j0.3373\ p.u.$$

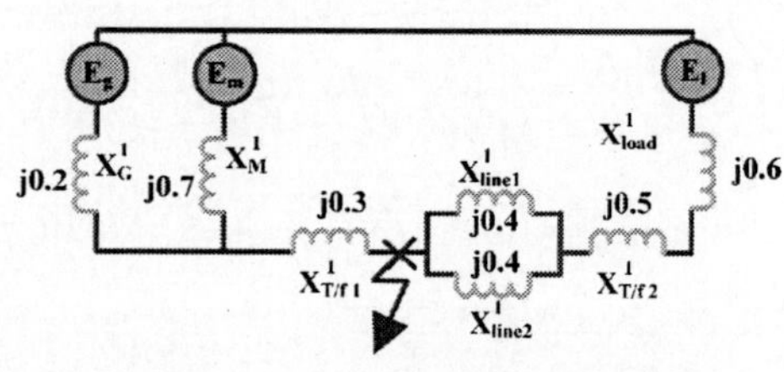

<u>Negative sequence network:</u>

In the given example, the positive and negative sequence values of all the components in the network are same. $Z_2 = Z_1$

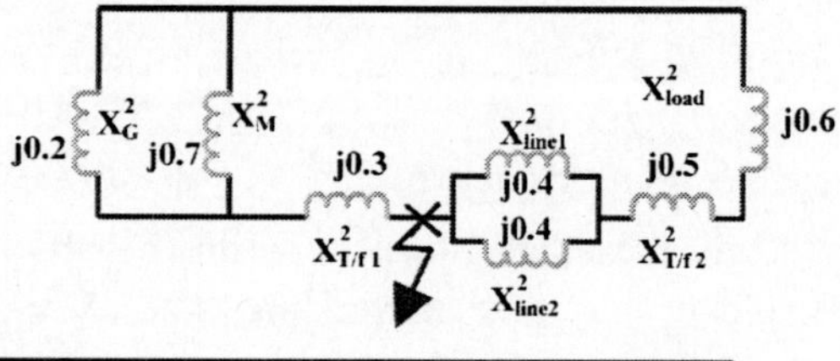

<u>Zero sequence network:</u>

$$Z_0 = j0.3 \parallel (j0.7 + j0.5)$$

$$= j0.3 \parallel j1.2$$

$$= j0.24\ p.u.$$

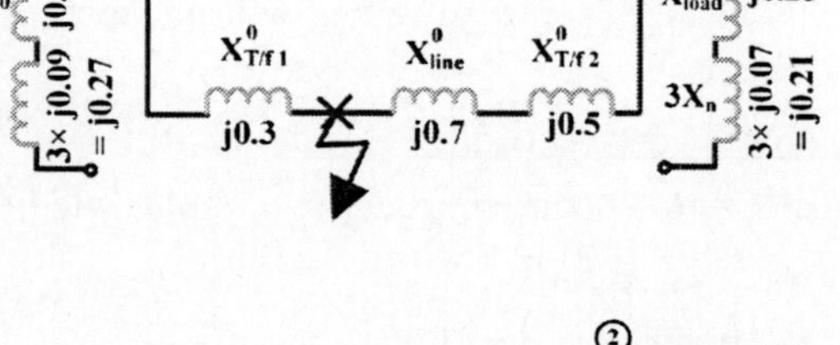

LG Fault current $I_f^{LG} = \dfrac{3V_f}{Z_0 + Z_1 + Z_2}$

$$= \frac{3 * 1 \angle 0^0}{j0.24 + j0.3373 + j0.3373}$$

$$= -j3.2801\ p.u.$$

Example 7.25. A power system is shown in figure, draw sequence networks and determine the per unit fault current at bus 2 for

(i) Single line to ground fault (LG)

(ii) Line to line fault (LL)

(iii) Symmetrical 3-phase fault (LLL).

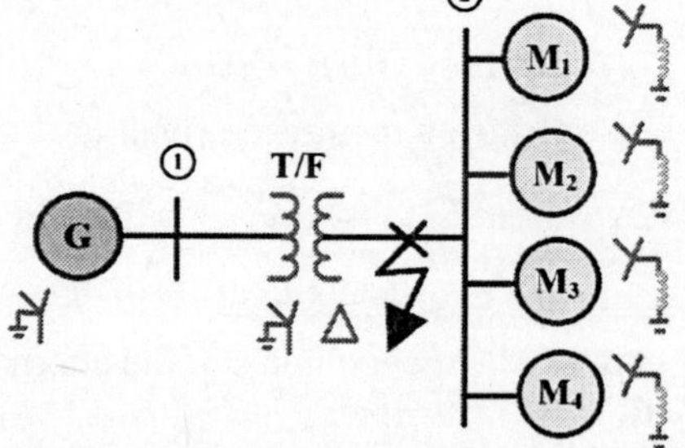

All four motors are star connected and the neutral is earthed through a reactance $X_n = 3\%$ on motor ratings.

Ratings are as follows:

G : 10MVA, 6.6kV; $X_1 = X_2 = 15\%$ and $X_0 = 5\%$

T/f: 10MVA, $\frac{6.6\text{kV}}{430\text{ V}}$; $X_1 = X_2 = X_0 = 10\%$

M : 1500HP each, $\eta = 89.52\%$ at full load UPF, $X_1 = X_2 = 20\%, X_{M0} = 6\%$, and $X_N = 3\%$

Ans: Total kVA of motors $= \frac{total\ HP*0.746}{\eta} = \frac{(1500*4)*0.746}{0.8952} = 5000kVA = 5$ MVA

Positive, negative and zero sequence p.u. reactance of generator, motor, and transformer are as follows

$$X_{G1} = X_{G2} = j0.15\ p.u., and\ X_{G0} = j0.05\ p.u.$$

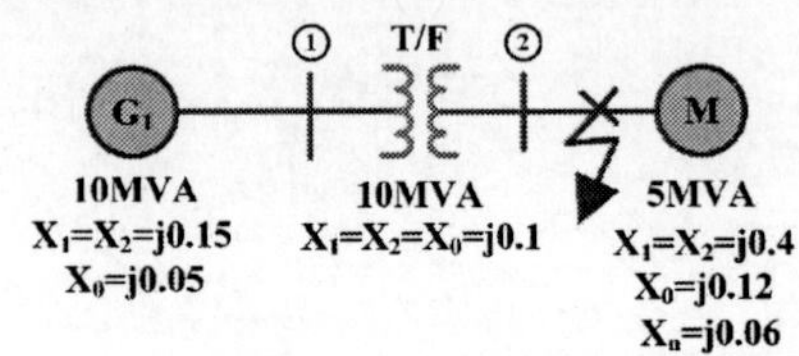

$$X_{T_{f}1} = X_{T_{f}2} = X_{T_{f}0} = j0.1 * \left(\frac{6.6*10^3}{6.6*10^3}\right)^2 * \frac{10}{10} = j0.1\ p.u.$$

$$X_{M1} = X_{M2} = j0.2 * \left(\frac{430}{430}\right)^2 * \frac{10}{5} = j0.4\ p.u.,$$

$$X_m^0 = j0.06 * \left(\frac{430}{430}\right)^2 * \frac{10}{5} = j0.12\ p.u.,$$

$$X_m^n = j0.03 * \left(\frac{430}{430}\right)^2 * \frac{10}{5} = j0.06\ p.u.,$$

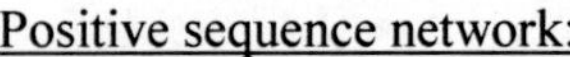

Positive sequence network:

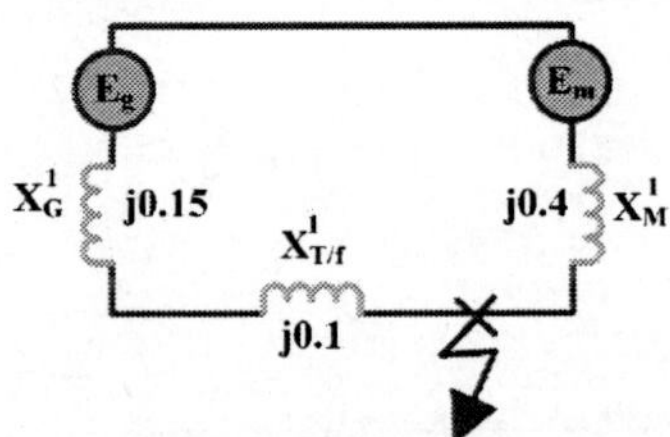

$$Z_1 = (j0.15 + j0.1) \parallel j0.4$$
$$= j0.25 \parallel j0.4$$
$$= j0.1538\ p.u.$$

Negative sequence network:

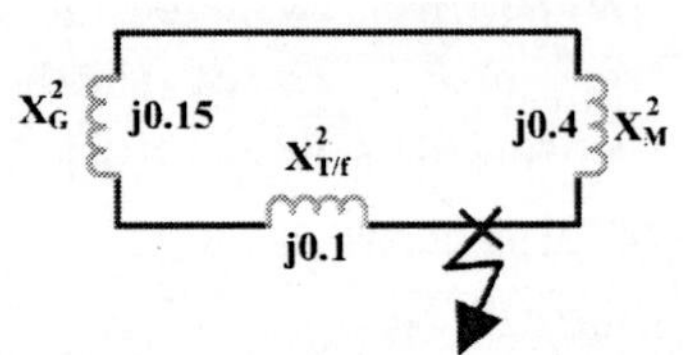

In the given example, the positive and negative sequence values of all the components in the network are same. $Z_2 = Z_1$

The positive and negative sequence networks will be almost identical, only change is energy source will not present in negative sequence network.

Zero sequence network:

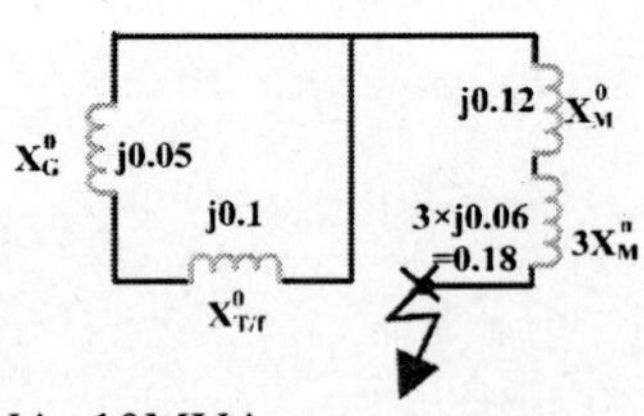

Total zero sequence impedance of motor will be

$$Z_{M0} = X_m^0 + 3X_m^n$$
$$= j0.12 + 3 * j0.06$$
$$= j0.12 + j0.18 = j0.3\ p.u.$$

Base voltage on motor side (fault is at motor terminals) = 430V, Base MVA=10MVA

Base current $I_{base} = \frac{10*10^6}{\sqrt{3}*430} = 13.426\ kA$

Single line to ground fault (LG):

Let a, b, c be the terminals of the alternator and fault occurs between terminal 'a' and ground.

Per unit Fault current $I_{f\ pu}^{LG} = \frac{3*V_f}{Z_0 + Z_1 + Z_2} = \frac{3*1\angle 0^0}{j0.3 + j0.1538 + j0.1538} = -j4.937\ p.u.$

Actual fault current $I^{LG}_{f\,actual} = I^{LG}_{f\,p.u.} * I_{base} = -j4.937 * 13426 = 66.284 \angle -90^0\ kA$

<u>Line to line fault (LL):</u> Let double line (LL) occurs between the terminals b and c.

Per unit Fault current $I^{LL}_{f} = \frac{-j\sqrt{3}*V_f}{Z_1 + Z_2} = \frac{-j\sqrt{3}*1\angle 0^0}{j0.1538 + j0.1538} = -5.63\ p.u.$

Actual fault current $I^{LL}_{f\,actual} = I^{LL}_{f\,p.u.} * I_{base} = -5.63 * 13426 = 75.588 \angle 180^0\ kA$

<u>Symmetrical 3-phase fault (LLL):</u>

Per unit Fault current $I^{3\phi}_{f\,p.u.} = \frac{V_f}{Z_1} = \frac{1\angle 0^0}{j0.1538} = -j6.501\ p.u.$

Actual fault current $I^{3\phi}_{f\,actual} = I^{3\phi}_{f\,p.u.} * I_{base} = -j6.501 * 13426 = 87.282 \angle -90^0\ kA$

Example 7.26. For the given system, draw positive, zero and negative sequence networks and determine the fault current at bus 3 for

(i) Single line to ground fault (LG)
(ii) Line to line fault (LL)
(iii) Double line to ground (LLG)
(iv) Symmetrical 3-phase fault (LLL).

G_1 and G_2: 100MVA, 11kV; $X_1 = X_2 = j0.15$ p.u, $X_0 = j0.05$p.u.

T/f$_1$: 100MVA, $\frac{11kV}{110kV}$; $X_1 = X_2 = X_0 = j0.25$ p.u.; T/f$_2$: 100MVA, $\frac{110kV}{11kV}$; $X_1 = X_2 = X_0 = j0.25$ p.u.

L_1, L_2 and L_3 : $X_1 = X_2 = j0.3$ p.u., $X_0 = j0.9$p.u. on 110kV, 100MVA base.

Ans: <u>Positive sequence network:</u>

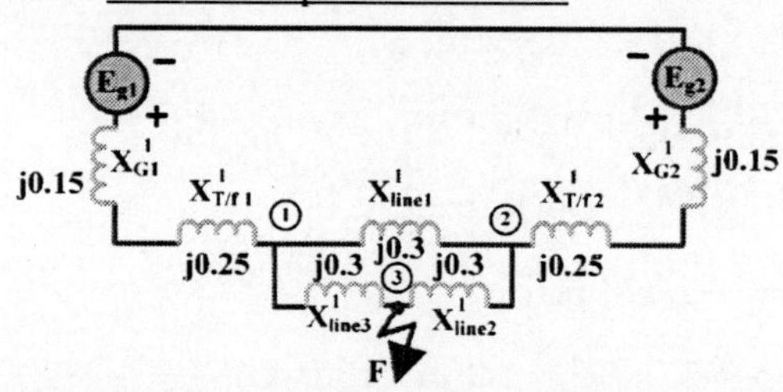

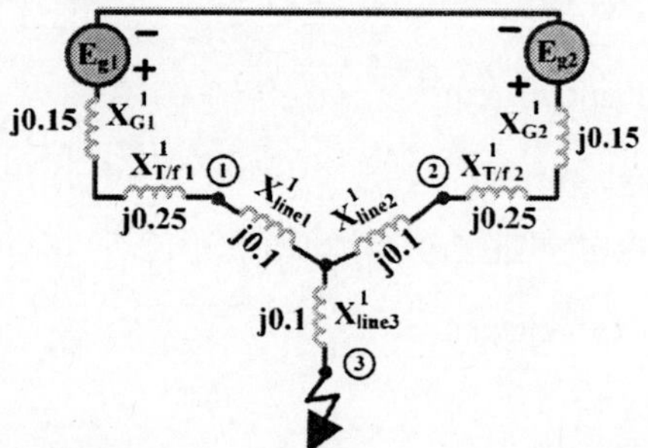

$$Z_1 = [(j0.15 + j0.25 + j0.1) \parallel (j0.1 + j0.25 + j0.15)] + j0.1$$
$$= [(j0.5) \parallel (j0.5)] + j0.1$$
$$= j0.25 + j0.1 = j0.35\ p.u.$$

<u>Negative sequence network:</u>

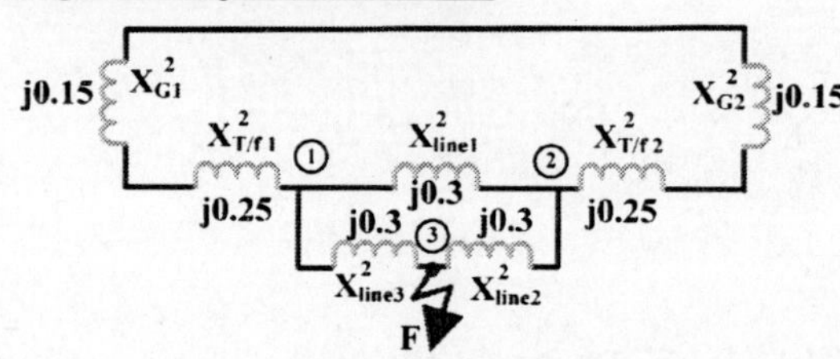

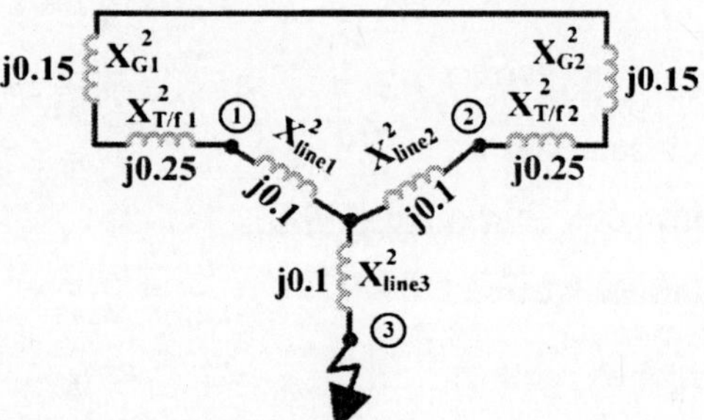

In the given example, the positive and negative sequence values of all the components in the network are same. $Z_2 = Z_1$

Zero sequence network:

$$Z_0 = [(j0.25 + j0.3) \parallel (j0.25 + j0.3)] + j0.3$$
$$= [(j0.55) \parallel (j0.55)] + j0.3$$
$$= j0.275 + j0.3$$
$$= j0.575\ p.u.$$

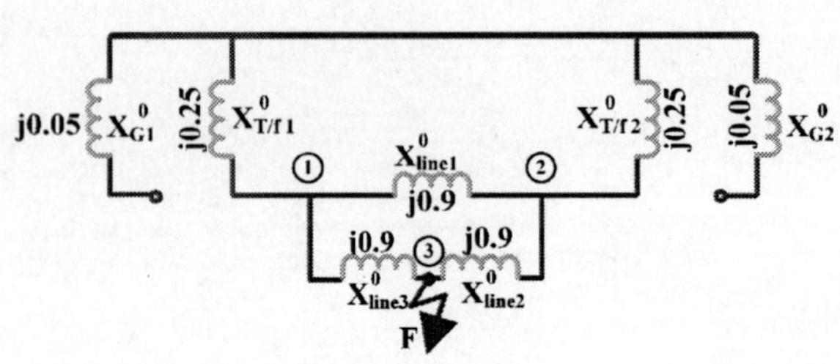

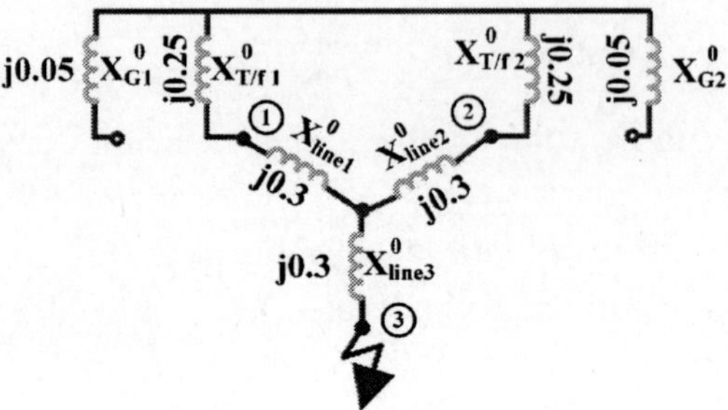

Base voltage = 110 kV (fault is on 110 kV side), Base MVA=100MVA

Base Current $\quad I_{base} = \frac{100*10^6}{\sqrt{3}*110*10^3} = 524.86\ A$

Single line to ground fault (LG):

Let a, b, c be the terminals of the alternator and fault occurs between terminal 'a' and ground.

Per unit Fault current $\quad I_{f\ pu}^{LG} = \frac{3*V_f}{Z_0 + Z_1 + Z_2} = \frac{3*1\angle 0^0}{j0.575+j0.35+j0.35} = -j2.3529\ p.u.$

Actual fault current $\quad I_{f\ actual}^{LG} = I_{f\ p.u.}^{LG} * I_{base} = -j2.3529 * 524.86 = 1.234\angle -90^0\ kA$

Line to line fault (LL):Let double line (LL) occurs between the terminals b and c.

Per unit Fault current $\quad I_f^{LL} = \frac{-j\sqrt{3}*V_f}{Z_1 + Z_2} = \frac{-j\sqrt{3}*1\angle 0^0}{j0.35+j0.35} = -2.474\ p.u.$

Actual fault current $\quad I_{f\ actual}^{LL} = I_{f\ p.u.}^{LL} * I_{base} = -2.474 * 524.86 = 1.298\angle 180^0\ kA$

Double line to ground fault (LLG):

Let double line to ground fault (LLG) occurs between the terminals b and c to ground.

In LLG fault $\quad I_f^{LLG} = \frac{-3Z_2E_a}{Z_0Z_1+Z_0Z_2+Z_1Z_2}$

Per unit Fault current $\quad I_f^{LLG} = \frac{-3*j0.35*1\angle 0^0}{j0.575*j0.35+j0.575*j0.35+j0.35*j0.35} = j2\ p.u.$

Actual fault current $\quad I_{f\ actual}^{LLG} = I_{f\ p.u.}^{LLG} * I_{base} = j2 * 524.86 = 1.049\angle 90^0\ kA$

Symmetrical 3-phase fault (LLL):

Per unit Fault current $\quad I_{f\ p.u.}^{3\phi} = \frac{V_f}{Z_1} = \frac{1\angle 0^0}{j0.35} = -j2.8571\ p.u.$

Actual fault current $\quad I_{f\ actual}^{3\phi} = I_{f\ p.u.}^{3\phi} * I_{base} = -j2.8571 * 524.86 = 1.499\angle -90^0\ kA$

7.15. Unit Summary:

- The symmetrical components approach was initially introduced by C.L. Fortesque in 1918.
- CL Fortesque's theorem states that any unbalanced three-phase system can be decomposed into three balanced systems consisting of phasors representing positive, negative, and zero-sequence components.
- The positive sequence components have equal magnitudes and are spaced 120^0 apart with phase sequence same as that of normal balanced system. The positive sequence components are denoted by the subscript '1'.
- The negative sequence components have equal magnitudes and are spaced 120^0 apart with phase sequence opposite to normal balanced case. The negative sequence components are indicated by the notation '2'.
- The zero sequence components have identical magnitudes and zero phase displacement. The zero sequence components are denoted by the subscript '0'.
- To address an unbalanced system and establish the relationship between phase voltages and/or phase currents, we utilize the sequence operator 'a'.
- The three phasors 1, a^2 *and* a exhibits a balanced and symmetrical. This is due to the fact that the phasors possess identical lengths and are displaced by equal angles of 120^0 degrees relative to each other.
- The induced electromotive force (emf) in series with the positive sequence impedance Z_1 can be used to depict the positive sequence network of a synchronous machine.
- The negative-sequence network for a synchronous machine can be described solely by the negative sequence impedance Z_2, as the synchronous machine does not produce any negative sequence emf.
- Due to the absence of zero sequence electromotive force (emf) generation in synchronous machines, the zero-sequence network associated with such machines can be mathematically represented by the zero-sequence impedance Z_0 and a neutral impedance $3Z_n$.
- The impedances of a transposed transmission line are equal for both positive and negative sequences.
- The impedance of a transposed transmission line in the zero-sequence is approximately 2 to 4 times higher than the impedance in the positive-sequence.
- The positive, negative, and zero-sequence networks in a single line to ground fault are interconnected in series.
- The line-to-line fault involves the connection of positive and negative-sequence networks in parallel configurations.
- The double line to ground fault involves the parallel connection of positive, negative, and zero-sequence networks.
- Only positive-sequence impedance will be present in a three-phase (LLL) malfunction.
- All forms of faults, including both symmetrical and unsymmetrical faults, have a positive sequence component.
- Unsymmetrical faults exhibit the presence of a negative sequence component.
- The zero-sequence component is observed exclusively when the neutral part of the system is connected to the ground and the fault is associated with the ground.

Short and Long Answer Questions

1. Define symmetrical components. Discuss the significance of symmetric components.
2. Discuss the significance of operator 'a' in symmetrical components.
3. Elucidate the process of resolving an unbalanced system consisting of 3-phase voltages into symmetrical components.
4. Determine the relationship between unbalanced voltages and symmetrical components.
5. Describe how to resolve an unbalanced 3-phase current system into symmetrical components.
6. Derive the relation between unbalanced currents and symmetrical components.
7. With neat diagrams, derive the sequence impedances of a synchronous generator.
8. Derive the symmetrical component voltages of a synchronous generator in terms of sequence impedances and currents.
9. Demonstrate that a transposed transmission line's zero-sequence impedance is 2 to 4 times higher than its positive-sequence impedance.
10. Show that the positive and negative sequence impedances of a transposed transmission line are equal.
11. Explain and draw zero-sequence networks for various transformer connection combinations.
12. Calculate total power in 3-phase systems using voltage and current sequences.
13. Determine the fault current from a single line to ground under the following conditions.
    (i) Direct short circuit when neutral is solidly grounded.
    (ii) Direct short circuit when neutral is grounded through an impedance Z_n.
    (iii) Short circuit with fault impedance Z_f when neutral is solidly grounded.
    (iv) Short circuit with fault impedance Z_f when neutral is grounded through an impedance Z_n.
    (v) When neutral is isolated.
14. Determine the fault current for a double line fault under the following conditions.
    (i) Direct short circuit between terminals 'b' and 'c'.
    (ii) Short circuit between terminals 'b' and 'c' through fault impedance Z_f.
15. Calculate the fault current from a double line to ground fault under the following conditions:
    (i) Direct short circuit between terminals 'b' and 'c' to ground.
    (ii) Short circuit between terminals 'b' and 'c' to ground through fault impedance Z_f.
16. Calculate the fault current in a three-phase fault.
17. Draw the sequence networks of unsymmetrical faults when the synchronous machine is solidly grounded and grounded through impedance during direct short circuit and short circuit through fault impedance.
18. Show that the three symmetrical component networks are connected in:
    (i) Series for single line to ground fault.
    (ii) Parallel for double line to ground fault.

Exercises

1. In a 3-phase unbalanced system, the phase currents are as follows: $I_a = 25\angle 10^0 A, I_b = 50\angle 60^0 A, I_c = 75\angle -110^0 A$. Determine the symmetrical variables of electric currents.
2. A delta-connected resistive load is coupled to a balanced 3-phase 11kV supply. $R_{ab} = 150\Omega, R_{bc} = 300\Omega$, and $R_{ca} = 450\Omega$. Find the symmetrical components of the line and delta currents.
3. In a 3-phase unbalanced system, the phase voltages are as follows:$V_a = 50\angle 0^0\ V, V_b = 150\angle 40^0\ V, V_c = 250\angle -160^0\ V$. Determine the symmetrical components of voltages.
4. The symmetrical voltage components of phase 'a' in an unbalanced system are as follows: $V_{a0} = 40\angle 100^0\ V, V_{a1} = 80\angle 0^0\ V, V_{a2} = 120\angle -50^0\ V$. Calculate the phase voltages of a, b and c.
5. A solidly grounded neutral in a 33kV, 25MVA alternator has a sub-transient reactance of j0.33p.u. The reactance values for the negative and zero sequences are given as j0.22p.u. and j0.11p.u., respectively. There is an occurrence of a single line to ground fault at the terminals of the unloaded generator. The measured fault impedance is j0.099 per unit (p.u.). Calculate the fault current and terminal voltages present during a fault state. The currents and voltages under fault conditions are expressed as percentages of their rated values.
6. A 10MVA, 11kV alternator with a solidly grounded neutral has a sub-transient reactance of j0.34p.u. The reactance values for the negative and zero sequences are j0.24p.u. and j0.14p.u., respectively. Calculate the fault current and line-line voltages occurring at the terminals of the alternator during a double line to ground (LLG) fault.
7. A 22kV, 30MVA alternator has $Z_1 = Z_2 = 28\%$ and $Z_0 = 14\%$. The alternator's neutral is grounded with a reactance of 0.75Ω. Calculate the initial symmetrical currents in all three phases of the alternator when a double-line (LL) fault occurs at its terminals in an unloaded situation. The fault impedance is 16% reactive. Calculate the voltages at the alternator's terminals during the fault.
8. A 11kV, 40MVA alternator has $Z_1 = j0.46\ p.u., Z_2 = j0.35\ p.u.$ and $Z_0 = j0.12\ p.u.$ The alternator's neutral is solidly grounded. The alternator is working at its no-load rated voltage. Calculate symmetrical line currents for
   (i) Single line to ground fault (LG).
   (ii) Double line to ground fault (LLG).
   (iii) Line to line fault (LL).
   (iv) Symmetrical 3-phase fault (LLL).

   Consider base values as 50MVA and 11kV.
9. A star-connected alternator powers a star-connected inductive load via a transmission line. The load's star point is grounded, whereas the generator neutral is not grounded. The load reactance is 0.85p.u. each phase. The line's positive, negative, and zero sequence reactance's are j0.25p.u., j0.25p.u., and j0.75p.u. per phase. The generators positive, negative, and zero sequence reactances are j0.65p.u., j0.75p.u., and j0.55p.u. respectively. A single phase to ground fault occurs in phase 'A' halfway down the line. Prior to the failure, the network is balanced, and the voltage at the fault point is 1 p.u. Calculate the current along the fault path.

10. A 100MVA, 33kV alternator was subjected to the following faults $I_f^{3\phi} = 2.34\ kA,\ I_f^{LL} = 3.45\ kA,$ and $I_f^{LG} = 4.56\ kA$. The neutral of the alternator is solidly grounded. Ignoring resistances calculate the values of three reactance X_0, X_1 and X_2.

11. For the given system, draw positive, zero and negative sequence networks. If a fault occurs at bus 3, calculate the fault current for unsymmetrical faults (LG, LL, LLG).

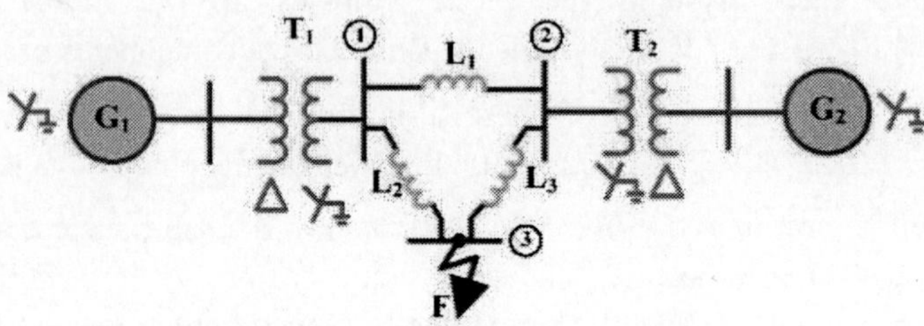

G_1 and G_2: 150 MVA, 33 kV; $X_1 = X_2 = j0.35$ p.u, $X_0 = j0.09$ p.u.

T/f_1: 100 MVA, $\frac{33\text{ kV}}{132\text{ kV}}$; $X_1 = X_2 = X_0 = j0.25$ p.u.

T/f_2: 100 MVA, $\frac{132\text{ kV}}{33\text{ kV}}$; $X_1 = X_2 = X_0 = j0.25$ p.u.

L_1, L_2 and L_3 : $X_1 = X_2 = j0.25$ p.u., $X_0 = j0.75$ p.u on 132 kV, 150 MVA base.

12. Figure depicts a single line diagram of a power system, with the sequence reactances of the generator, motor, and transformers T_1 and T_2 shown per unit. The neutrals of the generator and transformers are solidly grounded. Draw positive, negative, and zero sequence networks with reactance values per unit on a 120 MVA, 22 kV base in the generator's zone. The pre-fault voltage is 1 p.u. Calculate the fault current for a three-phase to ground fault on bus 4.

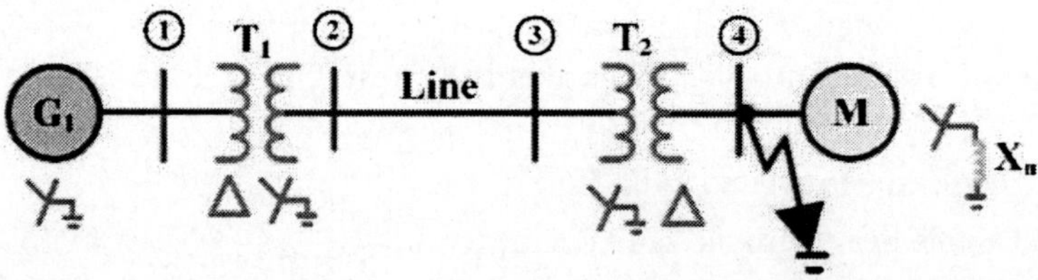

G_1 : 100 MVA, 22 kV; $X_1 = j0.3$ p.u, $X_2 = j0.4$ p.u. and $X_0 = j0.05$p.u.

T/f_1: 150 MVA, $\frac{22\text{ kV}}{220\text{ kV}}$; $X_1 = X_2 = X_0 = j0.2$ p.u.,

T/f_2: 150 MVA, $\frac{220\text{ kV}}{22\text{ kV}}$; $X_1 = X_2 = X_0 = j0.2$ p.u.

M : 240 MVA, 22 kV; $X_1 = j0.4$ p.u, $X_2 = j0.3$ p.u., $X_{mo} = j0.2$ p.u. and $X_n = j0.04$ p.u.

Line : $X_1 = X_2 = j0.2$ p.u., $X_0 = j0.6$ p.u on 220 kV, 100 MVA base.

13. A generator with negligible resistance, 1p.u. voltage, and transient reactance experiences many fault types.

S.No.	Type of fault	Resulting fault current in p.u.
1	3-Phase	5.55 p.u.
2	LG	4.44 p.u.
3	LL	3.33 p.u.

Ignoring resistances calculate the values of three reactance X_0, X_1 and X_2.

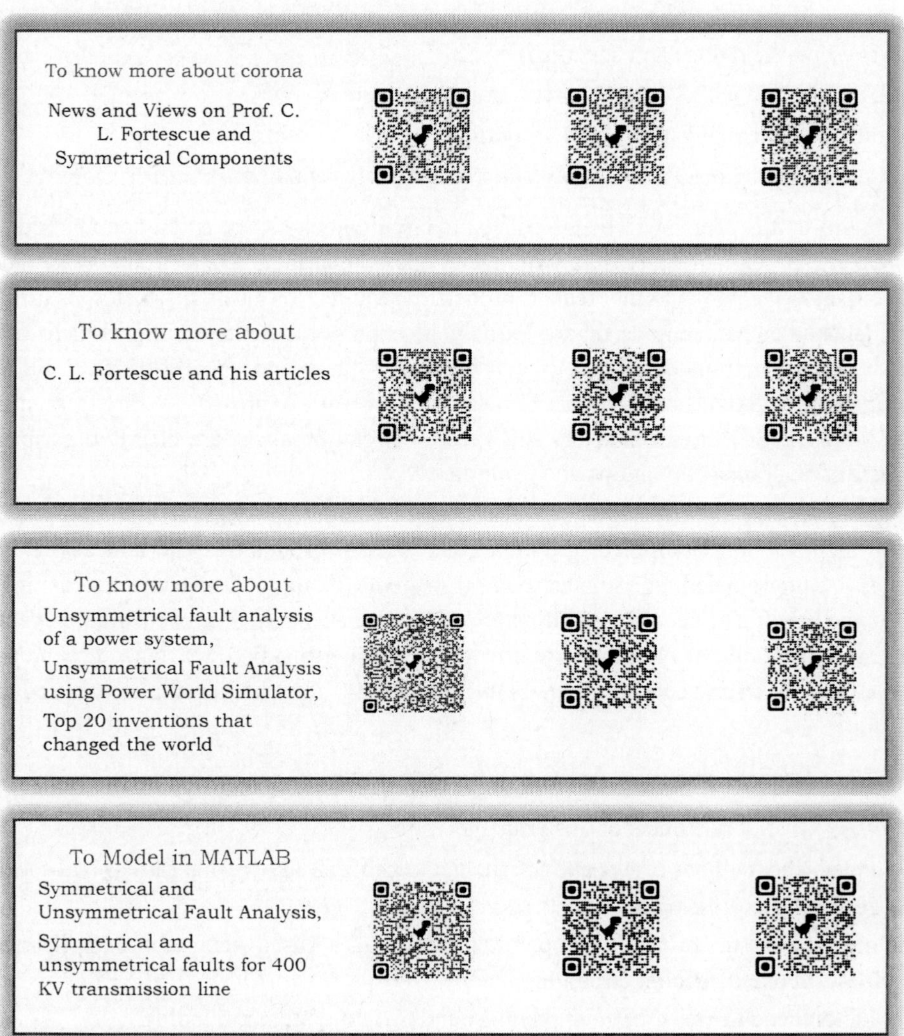

08 CIRCUIT BREAKERS & NEUTRAL GROUNDING

Unit specifics: In this unit, the following topics have been discussed for basic understating of Circuit Breakers and Neutral Grounding:

- Essential features of switchgear, expression for restriking voltage and RRRV.
- Methods of arc extinction, resistance switching and current chopping.
- Circuit breaker ratings and specifications.
- Oil, air blast, SF6 and vacuum circuit breakers.
- Equipment grounding Vs system grounding.
- Solid, resistance, reactance and arc suppression coil grounding.

Rationale: In this unit, students will be introduced to circuit breakers, arc extinction methods and types of circuit breakers. Essential features of a switchgear, operating principle of a circuit breaker, commonly used insulating materials for a circuit breaker, properties required for insulating materials used in circuit breaker, arc phenomenon, methods of arc extinction, expression for restriking voltage and RRRV, resistance switching, current chopping, circuit breaker ratings and specifications, bulk oil circuit breakers, minimum oil circuit breakers, axial blast circuit breakers, cross blast circuit breakers, radial blast circuit breakers, SF6 circuit breakers and vacuum circuit breakers, are clearly described with the help of necessary diagrams, derivations and examples.

This unit also provides an overview of grounding, including its aspects and classification. The classification includes both equipment grounding and system grounding. The unit also examines the disadvantages of ungrounded or isolated neutral systems, notably drawbacks on a single phase. Additionally, the benefits of neutral grounding are discussed. The unit also investigates several types of groundings, including solid or effective grounding, resistance grounding, reactance grounding, and arc suppression coil or Peterson's coil grounding. Diagrams, derivations, and examples are used to improve comprehension.

Pre-Requisites: Basic knowledge of power system components.

Unit Outcomes: List of outcomes of this unit is as follows

U8-O1: To understand methods of Arc extinction, and to know circuit breaker ratings and specifications.

U8-O2: To derive the expression for restriking voltage and RRRV.

U8-O3: To comprehend the principle of operation, advantages, disadvantages and applications of Oil, air blast, SF6 and vacuum circuit breakers.

U8-O4: To comprehend the need of equipment grounding and system grounding.

U8-O5: To recognize the drawbacks of an isolated neutral system and advantages of neutral grounding.

U8-O6: To understand different types of neutral grounding methods including solid, resistance, reactance and arc suppression coil grounding.

Unit-8 outcomes	Expected mapping with course outcomes (1: Weak, 2: medium, and 3: strong correlation)					
	CO-1	CO-2	CO-3	CO-4	CO-5	CO-6
U8-O1	3	2	-	-	3	3
U8-O2	2	-	-	-	2	3
U8-O3	3	2	1	1	3	-
U8-O4	2	-	1	1	1	-
U8-O5	2	2	-	2	1	3
U8-O6	3	2	-	3	3	3

8.1 Introduction to circuit breakers:

A switchgear is a combination of few components/devices used to connect and/or disconnect a part of electrical power system as per the requirement. Switchgear serves two fundamental functions.

(i) Switching during healthy/normal operating conditions for maintenance purpose.

(ii) Switching during faulty/ abnormal operating conditions to protect the power system.

The amount of current to be interrupted during abnormal operating conditions may be much higher compared to that during normal operating condition. This necessitates the need of protective relaying scheme. The equipment used for switching, protecting and controlling the electrical circuits is known as switchgear. Switchgear includes fuses, isolators, relays, switches, circuit breakers, control panels, lightening arrestors, current transformers etc.

8.2 Essential features of a switchgear:

Essential features of a switchgear are

- Discrimination
- Quick Operation
- Reliability
- Manual Control
- Instruments

In the event of a fault arising within the power system, it is imperative that the switchgear possesses the capability to promptly separate the healthy section from the faulty section, so ensuring the preservation of electrical equipment integrity. The incorporation of switchgear into the power system will result in an improvement in reliability. If there is a failure in the electric control system, the required operation can be executed using manual control. It is imperative to include provisions for instruments that may be necessary.

8.3 Circuit Breakers:

Circuit breakers are mechanical switching devices that can open or close its contacts in response to normal or abnormal situations. It can

(i) Automatically break a circuit in the event of a fault.

(ii) Either automatically or manually make a circuit in the event of a fault.

(iii) Either automatically or manually break or make a circuit under normal condition.

8.3.1 Operating principle of a circuit breaker:

A circuit breaker is a device that interrupts and switches electrical current. It primarily consists of fixed and movable contacts. The fixed and movable contacts are usually made of copper materials, with adequate dimensions and cross-sectional area to sustain the continuous flow of the rated current. Typically, these connections remain closed under normal operational circumstances to facilitate the flow of electric currents, and they are automatically opened in the event of malfunctions. To initiate the opening of a CB, a slight amount of force must be exerted on a trigger. In the event of a fault in any component of the power system, the trip coils of the breaker become activated and the mobile contacts are separated from the stationary contact.

In the event of fault conditions, an electric current will be generated between the contacts. The creation of an arc will result in the generation of heat within the CB, potentially leading to damage to both the system and the breaker. In order to prevent any harm to the system, it is imperative to promptly extinguish the arc to mitigate the heat it generates. The primary purpose of the CB is twofold:

(i) It quenches the arc formed between the contacts.

(ii) It provides appropriate insulation between connections.

8.3.2 Commonly used insulating materials for a circuit breaker:

Numerous insulating materials are employed for the purpose of arc extinction, with the selection of the material contingent upon the rating and classification of the CB The widely employed insulating materials for CB are

(i) Oil that generates hydrogen gas for the purpose of arc extinction.

(ii) Air at atmospheric pressure.

(iii) Compressed Air.

(iv) Sulphur hexafluoride (SF6).

(v) Ultra-high vacuum.

8.3.3 Properties required for insulating materials used in circuit breaker:

The insulating materials which are used for a circuit breaker should have following properties

- Arc extinguishing ability.
- High thermal stability.
- Non-flammability.
- High chemical stability.
- High dielectric strength.
- High thermal conductivity.
- Commercial availability at moderate cost.

8.4 Arc Phenomenon:

The CB contacts are separated by the protective mechanism during a fault, allowing a substantial fault current to pass through them. Due to the growing current density caused by the fault, the contact area rapidly decreases as the contacts begin to separate, resulting in an increase in temperature. The heat produced by the medium between the contacts is sufficient to ionize the air. The ionized air acts as a conductor, leading to the formation of an arc between the contacts. A potential difference and ionized particles are the components that keep the arc between the contacts. A sufficient potential difference to maintain the arc exists when the distance between the contacts is small. One potential approach for mitigating the arc is to move the contacts to a distance where the power source becomes inadequate to maintain the arc. Nevertheless, this particular approach is not fitting for high voltage systems that require a significant distance of many meters.

An alternative method for extinguishing the arc involves the deionization of the arc path. This objective can be accomplished through the process of arc cooling or by eliminating the ionized particles present in the space between the contacts.

8.5 Methods of Arc extinction (or) Arc Interruption: Arc interruption can be done using following two approaches

(i). High resistance technique.

(ii). Low resistance or current zero technique.

8.5.1. High resistance technique:

The high resistance approach involves a rise in resistance between the contacts over time, resulting in a reduction in current to a level where the heat generated by the contacts is inadequate to sustain the arc. The increasing of resistance or reduction of current should not be abnormal. The mechanical strength of the CB should be adequate to withstand sudden release of large quantities. This method is restricted to DC circuit breakers only. The resistance of the arc can be increased by

- *Increasing the length of the arc*: The arc's resistance will rise as its length increases. It can be accomplished by increasing the space between the contacts.
- *Reducing the cross-section of the arc:* The arc's resistance will grow when its cross-section is reduced.
- *Cooling the arc:* Cooling helps in de-ionisation of the medium b/w the contacts. This increases the arc resistance.
- *Splitting the arc:* The arc's resistance can be enhanced by breaking it into a series of smaller arcs.

8.5.2. Low resistance or current zero interruption technique:

This technique is exclusively relevant to ac circuit breakers. A 50Hz ac supply results in the current wave traversing a zero point one hundred times per second. This characteristic of an ac circuit is employed to disrupt the flow of current; subsequent to the occurrence of a zero, further increase is not permitted. Current zero interruption or low resistance is further subdivided into two categories.

(i). Slepains Recovery rate theory.

(ii). Cassies Energy balance theory.

8.5.2.1 Slepains Recovery rate theory:

In order to terminate the arc, it is necessary to extract the ionized particles from the gap promptly once the current reaches a state of natural zero. The removal of ionized particles can be achieved by two methods: sweeping away or neutralizing the molecules. The continuity of the arc is disrupted when the ionized particles are extracted from the gap at a rate that exceeds the rate of ionization as shown in Fig. 8.1(a). The extinguishment of the arc in Slepain's Recovery rate theory occurs when the rate of growth in dielectric strength surpasses that of the restriking voltage is shown in Fig. 8.1(b). When the rate of increase in the restriking voltage surpasses that of the dielectric strength, the process of ionization continues and the gap undergoes breakdown, leading to the occurrence of an arc for an additional half cycle as shown in Fig. 8.1(a).

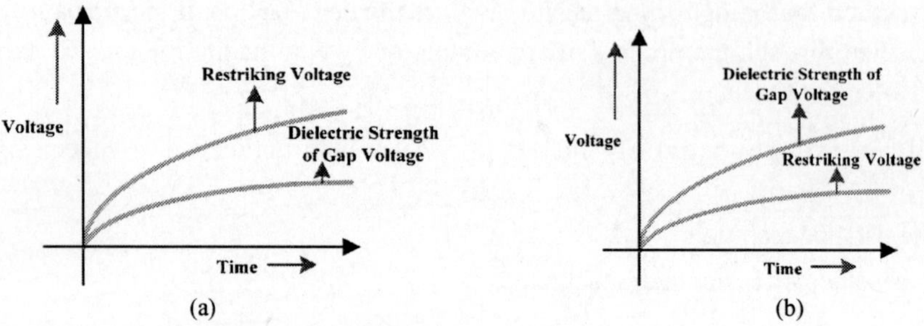

Fig. 8.1 Dielectric strength of CB Vs Restriking voltage (a) Arc does not extinguish and (b) Arc extinguishes

It is possible to deionize the medium by

Lengthening the gap: The length of the gap between contacts is directly proportional to the dielectric strength of the medium. Hence, the quick opening of connections might lead to an increase in the dielectric strength of the medium.

Cooling: The dielectric strength of the medium b/w the contacts can be increased by cooling the arc.

Blast effect/High Pressure: By applying the high pressure in the zone of the arc, the ionised particles will be moved away and the de-ionisation will happen at faster rate and consequently the dielectric strength of the medium between the contacts will be increased.

8.5.2.2 Cassies Energy balance theory:

The extinguishment of an arc in Energy Balance theory occurs when the rate of heat dissipation between the contacts exceeds the rate of heat generation. If the rate of heat/energy dissipation between the contacts is lower than the rate of heat generation, then the arc will rupture. The heat generation fluctuates over time due to changes in the distance between the contacts. At time t=0, the contacts are on the verge of opening, resulting in a restriking voltage of zero and hence the absence of heat generation. When the contacts are completely open, the arc length significantly rises, resulting in an increase in resistance and a decrease in arc current. Consequently, the generation of heat is prevented. Within the range of both boundaries, heat is produced, which gradually increases from zero to maximum and eventually to zero as the distance between the contacts increases, it is clearly depicted in Fig. 8.2.

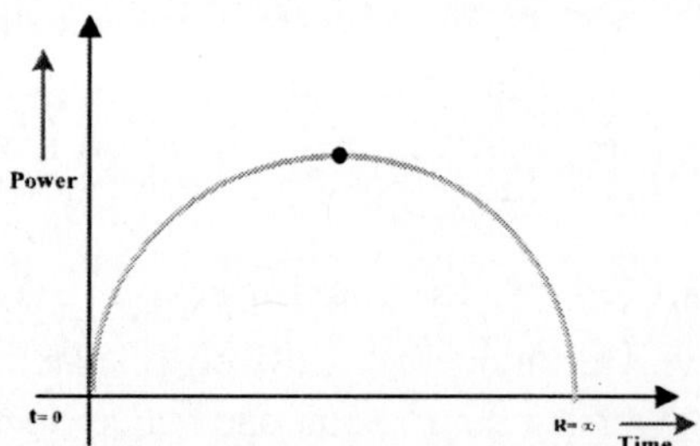

Fig. 8.2 Time Vs Power in Cassie's theory

8.6 Definitions of Arc voltage, Restriking voltage and Recovery voltage:

Arc voltage: The arc voltage refers to the voltage that is observed between the contacts of the circuit breaker (CB) during the arcing phase. The arc voltage has a minimal magnitude, around 3 to 5% of the rated voltage.

Restriking voltage: Restriking voltage refers to the temporary voltage that is observed between the contacts when the current is at or close to zero during the arcing period.

Recovery voltage: The recovery voltage refers to the voltage that is observed across the contacts of the breaker subsequent to the extinguishment of the arc.

8.7 Expression for Restriking Voltage and RRRV:

The power system is comprised of a significant quantity of inductance and a quantity of capacitance. When a failure arises, the system can store a significant amount of energy. The dissipation of stored energy within a circuit breaker (CB) is primarily caused by the interruption of fault current. The remaining energy is dissipated during oscillatory surges in the system. The presence of rhythmic surges is undesirable, thus necessitating the design of the CB to effectively disperse the stored energy. The figure below illustrates a short circuit occurring on a feeder located beyond the CB Single line diagram and equivalent circuit of an electrical power system during fault is shown in Fig. 8.3 (a) & (b) respectively.

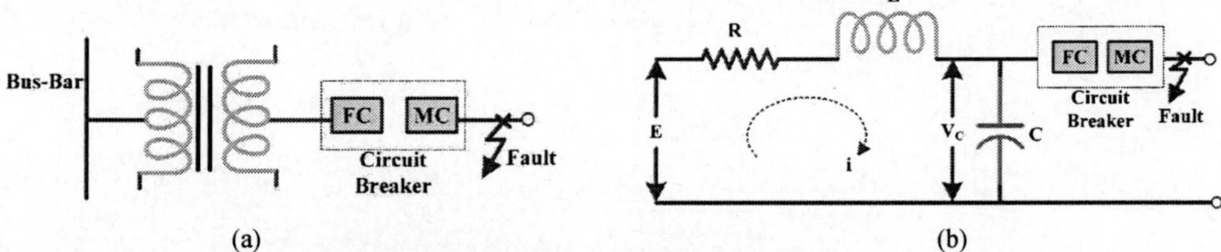

Fig. 8.3 Electrical power system during fault (a) Single line diagram (b) Equivalent circuit

Let L represent the inductance per phase of the system until the fault point,
R represent the resistance per phase of the system until the fault point, and
C represent the capacitance to earth of the CB porcelain bushing.

When the circuit breaker (CB) is closed, a short circuit current passes through the resistors R and L, as well as the contacts of the CB This results in the short circuit of the capacitance C due to the fault. Upon opening the CB contacts and extinguishing the arc, the current 'i' is redirected through the capacitance 'C', leading to a temporary state. The series oscillatory circuit is comprised of inductance and capacitance. The restriking voltage, which represents the voltage across the capacitance, increases and oscillates as depicted in the Fig. 8.4.

Fig. 8.4 Arc voltage, Restriking voltage and Recovery voltage

The natural frequency of oscillation is given as $f_n = \frac{1}{2\pi\sqrt{LC}}$

The angular frequency of oscillation is $\omega_n = \frac{1}{\sqrt{LC}}$ ……. (8.1) $(Since\ \omega_n = 2\pi f_n)$

The mathematical representation of a transient condition is as follows

$$E_{Peak} = L\frac{di}{dt} + \frac{1}{C}\int i\, dt \qquad \text{............(8.2)}$$

Where E is the system voltage, L is Inductance, C is capacitance, i is short circuit current.

We know, $i = \frac{dq}{dt}$ ....(8.3)

but the charge accumulated on capacitor plates is $q = Cv_c$. where, v_c is the voltage across the capacitor.

Now substitute q value in Eq. 8.3, then $i = \frac{d(Cv_c)}{dt} = C\,\frac{dv_c}{dt}$ ....(8.4)

Similarly, $\frac{di}{dt} = \frac{d}{dt}\left(C\,\frac{dv_c}{dt}\right) = C\,\frac{d^2v_c}{dt^2}$ ....(8.5)

$$\int i\,dt = \int C\,\frac{dv_c}{dt}.dt = \int C\,dv_c = C.v_c \qquad \text{............(8.6)}$$

Substitute Eq. (8.5) and Eq. (8.6) in Eq. (8.2), then $E_{Peak} = L\left(C\,\frac{d^2v_c}{dt^2}\right) + \frac{1}{C}(C.v_c)$

$$E_{Peak} = LC\,\frac{d^2v_c}{dt^2} + v_c \qquad \text{............(8.7)}$$

Taking Laplace transform on both sides of Eq. (8.7) by considering initial conditions as zero.

$$L[E_{Peak}] = L\left\{LC\,\frac{d^2v_c}{dt^2}\right\} + L\{v_c\}$$

$$\frac{E_{Peak}}{s} = LC\,s^2\,v_c(s) + v_c(s)$$

$$\frac{E_{Peak}}{s} = [LC\,s^2 + 1]\,v_c(s)$$

$$v_c(s) = \frac{E_{Peak}}{s[LC\,s^2+1]}$$

$$v_c(s) = \frac{E_{Peak}}{s.LC\left[s^2+\frac{1}{LC}\right]}$$

From eq. 8.1 $\omega_n = \frac{1}{\sqrt{LC}} \quad \Rightarrow \quad v_c(s) = \frac{\omega_n^2.E_{Peak}}{s.[s^2+\omega_n^2]}$ ... . (8.8)

The above expression can be written in partial fractions as

$$v_c(s) = \frac{A}{s} + \frac{Bs+C}{[s^2+\omega_n^2]} \qquad \text{... .... (8.9)}$$

$$A = \left.\frac{\omega_n^2.E}{[s^2+\omega_n^2]}\right|_{s=0} = \frac{\omega_n^2\,E}{[\omega_n^2]} = E_{Peak}$$

For calculating B and C, $\frac{\omega_n^2.E_{Peak}}{s.[s^2+\omega_n^2]} = \frac{A(s^2+\omega_n^2)+(Bs+C)s}{s.[s^2+\omega_n^2]}$

$$\omega_n^2.E_{Peak} = A(s^2+\omega_n^2) + (Bs+C)s$$

$$\omega_n^2.E_{Peak} = As^2 + A\omega_n^2 + Bs^2 + Cs$$

$$\omega_n^2.E_{Peak} = (A+B)s^2 + Cs + A\omega_n^2$$

Now compare coefficients of s^2 *and s terms on both sides then A+B=0 and C=0*

⇨ *B=-A=-*E_{Peak} *and C=0*

Substitute A, B and C values in Eq. (8.9), then $v_c(s) = \frac{E_{Peak}}{s} + \frac{-E_{Peak}s}{[s^2+\omega_n^2]}$

$$v_c(s) = E_{Peak}\left(\frac{1}{s} - \frac{s}{s^2+\omega_n^2}\right) \qquad \text{.........(8.10)}$$

Taking inverse Laplace transform on both sides of Eq. (8.10)

$$L^{-1}\{v_c(s)\} = E_{Peak}.L^{-1}\left(\frac{1}{s}\right) - E_{Peak}.L^{-1}\left(\frac{s}{s^2+\omega_n^2}\right)$$

$$v_c(t) = E_{Peak} - E_{Peak}.cos\omega_n t$$
$$v_c(t) = E_{Peak}(1 - cos\omega_n t)$$
$$v_c(t) = E_{Peak}\left(1 - cos\frac{t}{\sqrt{LC}}\right)$$

$\therefore$ *Restriking voltage* $\quad v_c(t) = E_{Peak}\left(1 - cos\frac{t}{\sqrt{LC}}\right)$ ....(8.11)

The maximum value of Restriking voltage is observed at $\frac{t}{\sqrt{LC}} = \pi$

The maximum value of Restriking voltage will be

$$\text{Restriking voltage}_{Max} = E_{Peak}(1 - cos\pi) = E_{Peak}(1 - (-1)) = 2E_{Peak}$$

The maximum value of Restriking voltage $= 2E_{Peak}$ ....(8.12)

The Rate of Rise of Restriking Voltage (RRRV) is

$$RRRV = \frac{dv_c(t)}{dt}$$
$$= \frac{d}{dt}(E_{Peak}(1 - cos\omega_n t))$$
$$= E_{Peak} * \frac{d}{dt}((1 - cos\omega_n t))$$
$$= \omega_n * E_{Peak} * sin\omega_n t$$

The Rate of Rise of Restriking Voltage is $\quad \therefore RRRV = \omega_n * E_{Peak} * sin\omega_n t$ ....(8.13)

The maximum value of RRRV occurs at $\omega_n t = \frac{\pi}{2}$ and its value will be

$$\therefore RRRV_{Max} = \omega_n * E_{Peak}$$ ....(8.14)

RRRV determines whether the arc will restrike. If RRRV exceeds the rate at which dielectric strength rises between the contacts, the arc will restrike. However, if RRRV is smaller than the rate at which dielectric strength rises between the contacts, the arc will not restrike.

8.8 Resistance Switching:

Resistance switching is a technique that involves connecting a sphere gap in series with a resistor across the contacts of a circuit breaker which is shown in Fig. 8.5. This arrangement aims to mitigate the impact of transient oscillations, restriking voltage, and RRRV. The resistance exhibits parallel propagation with the arc.

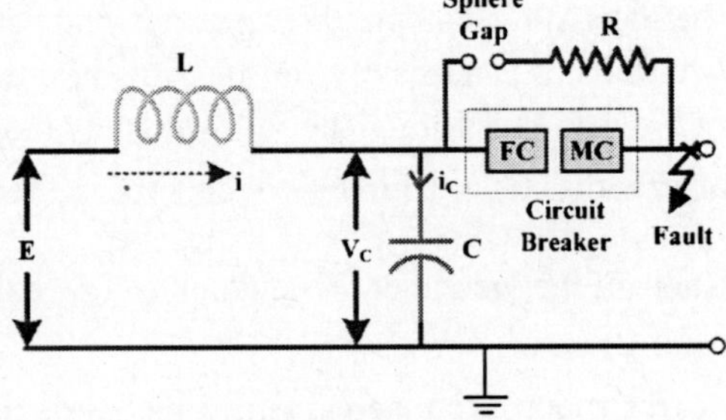

Fig. 8.5 Resistance Switching

During normal operation, a circuit breaker's fixed and movable contacts are closed/connected. As a result, the majority of the current will pass through a circuit breaker's contacts. A small amount of current will pass through the shunt branch, but it is insufficient to conduct the sphere-gap. As a result, the gap stays open, with no discharge across the resistor.

During abnormal conditions, the movable contact moves away from the fixed contact, resulting in an arc between the contacts. A portion of the arc current will pass through the parallel resistance, resulting in a drop in arc current while increasing deionization of the arc path and arc resistance. In order to achieve a higher flow rate of fault current through the shunt path, it is necessary to choose a parallel resistance value that is lower than the resistance of the arc path. The restriking voltage and RRRV will be decreased as a result of the drop in the arc current. The choice of parallel resistance is crucial for minimizing transitory oscillations.

The voltage equation in the closed loop is $E = L\frac{di}{dt} + \frac{1}{C}\int i_c\, dt = L\frac{di}{dt} + V_c(t)$ ....(8.15)

The current equation in the closed loop is $i = i_c + i_R$ ....(8.16)

From (8.15) and (8.16) $$E = L\frac{d}{dt}(i_c + i_R) + V_c(t)$$

$$E = L\frac{di_c}{dt} + L\frac{di_R}{dt} + V_c(t) \qquad \text{..........(8.17)}$$

But we know that, $$i_c = \frac{dq}{dt} = \frac{d}{dt}(CV_c(t)) = C\frac{dV_c(t)}{dt}$$

and $$\frac{di_c}{dt} = \frac{d}{dt}\left(C\frac{dV_c(t)}{dt}\right) = C\frac{d^2V_c(t)}{dt^2} \qquad \text{..........(8.18)}$$

$$\frac{di_R}{dt} = \frac{d}{dt}\left(\frac{V_c(t)}{R}\right) = \frac{1}{R} * \frac{dV_c(t)}{dt} \qquad \text{..........(8.19)}$$

Substitute Eq. (8.18) and Eq. (8.19) in Eq. (8.17), then $E = LC\frac{d^2V_c(t)}{dt^2} + \frac{L}{R}\frac{dV_c(t)}{dt} + V_c(t)$ ....(8.20)

By applying Laplace transform on both sides of above Eq. (8.20), we can get

$$L\{E\} = L\left\{LC\frac{d^2v_c(t)}{dt^2}\right\} + L\left\{\frac{L}{R}\frac{dV_c(t)}{dt}\right\} + L\{V_c(t)\}$$

$$\frac{E}{s} = LCs^2\, v_c(s) + \frac{L}{R} * s\, v_c(s) + v_c(s)$$

$$\frac{E}{s} = v_c(s) * LC\left[s^2 + \frac{1}{RC} * s + \frac{1}{LC}\right]$$

$$v_c(s) = \frac{E}{sLC\left[s^2 + \frac{1}{RC}s + \frac{1}{LC}\right]} \qquad \text{..........(8.21)}$$

All of the equation's roots should be real to ensure that there is no transient oscillation. According to the characteristic equation, there exists a real root at s=0. In order to ensure the realness of the remaining two roots, it is necessary for the roots of the characteristic equation in the denominator to also be real. In order to satisfy this, the following equation must be met. In eq. (8.21), denominator has two terms: one root is at $s = 0$ and two roots for the second term $s^2 + \frac{1}{RC} * s + \frac{1}{LC}$

Roots of the second order equation can be calculated using the formula $\frac{-b \pm \sqrt{b^2 - 4ac}}{2a}$

So, the roots of the second term $s^2 + \frac{1}{RC} * s + \frac{1}{LC}$ will be $\frac{\frac{-1}{RC} \pm \sqrt{\frac{1}{R^2C^2} - \frac{4}{LC}}}{2}$ or $\frac{-1}{2RC} \pm \sqrt{\frac{1}{4R^2C^2} - \frac{1}{LC}}$

For no transient oscillations all the roots should be real. In the above equation discriminant should be greater than or equal to zero. $\sqrt{\frac{1}{4R^2C^2} - \frac{1}{LC}} \geq 0$

$$\frac{1}{4R^2C^2} - \frac{1}{LC} \geq 0$$

$$\Rightarrow \quad \frac{1}{4R^2C^2} \geq \frac{1}{LC}$$

$$\Rightarrow \quad \frac{LC}{4C^2} \geq R^2$$

$$\Rightarrow \quad \frac{1}{4} * \frac{L}{C} \geq R^2$$

$$\Rightarrow \quad R^2 \leq \frac{1}{4} * \frac{L}{C}$$

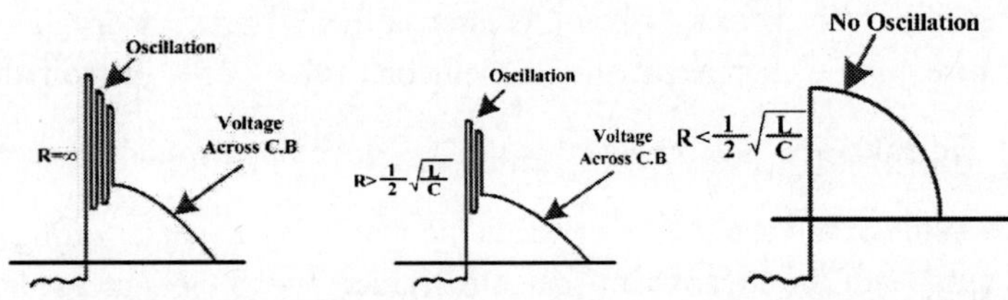

Fig. 8.6 Representation of transient oscillations for different values of resistances

For no transient oscillations $R \leq \frac{1}{2}\sqrt{\frac{L}{C}}$ ....(8.22)

Absence of transient disturbances is guaranteed when the resistance value connected across the circuit breaker's contacts is equal to or less than $\frac{1}{2}\sqrt{\frac{L}{C}}$.

If R is greater than $\frac{1}{2}\sqrt{\frac{L}{C}}$ then there will be oscillations.

If resistance value $R = \frac{1}{2}\sqrt{\frac{L}{C}}$ then this value of resistance is called as critical resistance. At critical resistance, the oscillations will be completely zero.

The frequency of damped oscillation is $f_d = \frac{1}{2\pi}\sqrt{\frac{1}{LC} - \frac{1}{4R^2C^2}}$ ....(8.23)

8.9 Current Chopping:

It is the occurrence of current interruption prior to the zero point of the natural current. A significant drawback of current chopping is that it generates a high voltage transient across the contacts of a circuit breaker. When low inductive current is interrupted and the CB's arc quenching force is more than sufficient to interrupt a low magnitude of current, the current is halted before it reaches its natural zero instant. In this case, the energy held in the magnetic field manifests itself as a high voltage across the stray capacitance, resulting in arc restriking which is shown in Fig. 8.7.

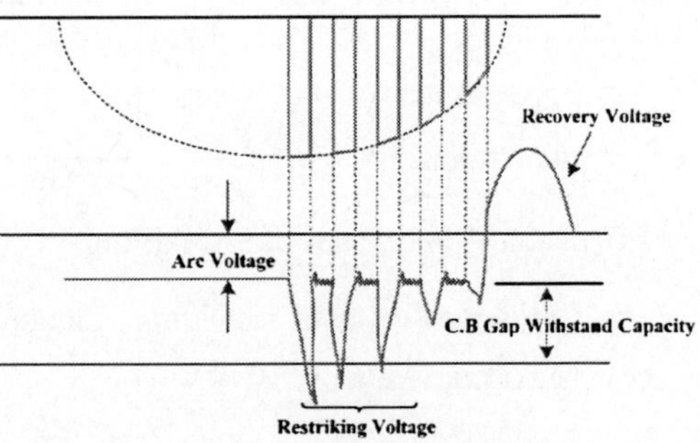

Fig. 8.7 Current Chopping

The energy stored in the magnetic field is $\frac{1}{2}Li^2$ and the electro static energy is $\frac{1}{2}Cv^2$

As these two energies are equal, they can be related as $\frac{1}{2}Li^2 = \frac{1}{2}Cv^2$

$$v = i\sqrt{\frac{L}{C}} \qquad \text{.........(8.24)}$$

For example, consider a 110kV CB interrupting a transformer having magnetising current of 10A. Let the current to be chopped at an instantaneous value of 5A. If the inductance and capacitance values are 50H and $0.003\mu F$, the prospective voltage developed would be $v = i\sqrt{\frac{L}{C}} = 5\sqrt{\frac{50}{0.003*10^{-6}}} = 645kV.$

If the value of the variable 'v' exceeds the maximum capacity of the space between the contacts, the arc will reappear. The stream is once again divided due to the increased quenching force. This phenomenon persists until the value of 'v' falls below the breaking point of the gap.

8.10 Arc recovery Voltage:

Let phase to neutral value of the system voltage is V

The Peak value of line-neutral voltage is $V_{Max} = \sqrt{2}$ V ... ... (8.25)

Recovery voltage is $V_{Max} \sin \varphi$ ... ... (8.26)

The active recovery voltage is $V_r = k_1 k_2 k_3 V_{Max} \sin \varphi$ ... ... (8.27)

Where, k_1 is the demagnetising factor due to which the recovery voltage will be less than system voltage

k_2 is a condition (or) phase factor and is unity in this case since the fault involves ground

k_3 is unity for recovery voltage between phase and neutral and its value is $\sqrt{3}$ for recovery voltage is between the lines

Example 8.1. The reactance and capacitance of a 66kV, 50Hz system are 6Ω and 0.035μF, respectively up to the location of fault. Determine the frequency of transient oscillation, the maximum value of restriking voltage, and the maximum value of RRRV.

Ans. System voltage $E_L = 66$ kV

Phase Voltage $E_{Ph} = \frac{E_L}{\sqrt{3}} = \frac{66\ kV}{\sqrt{3}} = 38.105\ kV$

Peak Voltage $E_{Peak} = \sqrt{2} \times E_{Ph} = \sqrt{2} \times 38.105\ kV = 53.888\ kV$

Reactance $X_L = 6\ \Omega$

Capacitance $C = 0.035\ \mu F$

Inductance $L = \frac{X_L}{2\pi f} = \frac{6}{2\times\pi\times50} = 0.019109\ H$

(a) Natural frequency of oscillation, $f_n = \frac{1}{2\pi\sqrt{LC}} = \frac{1}{2\times\pi\times\sqrt{0.019109\times0.035\times10^{-6}}} = 6.1571\ kHz$

(b) Maximum value of restriking voltage $= 2 \times E_{Peak} = 2 \times 53.888\ kV = 107.776$ kV

(c) Maximum value of RRRV $= \omega_n \times E_{Peak}$

$= 2\pi f_n \times E_{Peak}$

$= 2 \times 3.14 \times 6.1571 \times 10^3 \times 53.888 \times 10^3$

$= 2.084\ kV/\mu s$

Example 8.2. The inductance and capacitance of a 3-ph, 50Hz, 33kV system are 20mH and 0.06μF per phase. Perform the following calculations:

(i). Transient oscillatory frequency.

(ii). Maximum voltage across C. B's contacts when it reaches zero.

(iii). Average voltage rise rate up to the first oscillation peak.

Ans. System voltage $E_L = 33$ kV

Inductance $L = 20$ mH

Capacitance $C = 0.06$ μF

Phase Voltage $E_{Ph} = \frac{E_L}{\sqrt{3}} = \frac{33\ kV}{\sqrt{3}} = 19.052\ kV$

Peak Voltage $E_{Peak} = \sqrt{2} \times E_{Ph} = \sqrt{2} \times 19.052\ kV = 26.944\ kV$

(a) Natural frequency of oscillation, $f_n = \frac{1}{2\pi\sqrt{LC}} = \frac{1}{2\times\pi\times\sqrt{20\times 10^{-3}\times 0.06\times10^{-6}}} = 4.594\ kHz$

First peak (or) peak restriking voltage occurs at time 't'

$$t = \frac{1}{2f_n} = \frac{2\pi\sqrt{LC}}{2} = \pi\sqrt{LC}$$

$$t = \pi \times \sqrt{20 \times 10^{-3} \times 0.06 \times 10^{-6}}$$

$$t = 1.0882 \times 10^{-4}\ sec$$

(b) Maximum voltage across contacts of C.B at the instant it passes through zero i.e.

Maximum value of Restriking voltage $= 2 \times E_{Peak} = 2 \times 26.944\ kV = 53.888$ kV

(c) Average rate of rise of voltage up to first peak of oscillation $= \frac{\text{Max restriking voltage}}{Time\ to\ reach\ first\ peak}$

$$= \frac{2 \times E_{Peak}}{\pi\sqrt{LC}} = \frac{53.888 \times 10^3}{1.0882\times10^{-4}} = 0.494\ kV/\mu s$$

Example 8.3. Determine the natural frequency of the circuit and the average rate of rise of the restriking voltage when the time to reach the initial peak is 25μsec and the peak voltage is 90kV.

Ans. Time to attain the first peak restriking voltage, $t = \pi\sqrt{LC} = 25$ μ sec

(a) Natural frequency of the circuit is, $f_n = \frac{1}{2\pi\sqrt{LC}} = \frac{1}{2\times25\times10^{-6}} = 20$ kHz

(b) Average Rate of Rise of Voltage $= \frac{Max\ restriking\ voltage}{Time\ upto\ first\ peak} = \frac{90\times10^3}{25\times10^{-6}} = 3.6$ kV/μsec

Example 8.4. The inductance of a 50Hz 3-φ alternator is 2mH per phase, and the capacitance to earth between the alternator and the C.B is 0.00356μF each phase. When the rms current value reaches 10,000A during a fault, the breaker opens. Calculate the frequency of oscillations, active recovery voltage, time for maximum RRRV, and maximum RRRV.

Ans. f = 50 Hz, L = 2 mH = 0.002 H, C = 0.00356 μF, Short circuit current = 10,000 A

(a) Frequency of oscillation, $f_n = \frac{1}{2\pi\sqrt{LC}} = \frac{1}{2\times\pi\times\sqrt{0.002\times0.00356\times10^{-6}}} = 59.676\ kHz$

Reactance $X_L = 2\pi fL = 2\pi \times 50 \times 0.002 = 0.628\ \Omega$

Recovery voltage = short circuit current × reactance = 10000 × 0.628 = 6280 $V\ (rms)$

(b) Active recovery voltage, $V_{Max} = \sqrt{2} \times 6280 = 8881.2611\ V\ (Line - Neutral)$

(c) Time to attain maximum RRRV, $t = \frac{\pi\sqrt{LC}}{2} = \frac{\pi\sqrt{0.002\times0.00356\times10^{-6}}}{2} = 4.189$ μs

(d) Maximum value of RRRV,

$$RRRV_{Max} = \frac{V_{Max}}{\sqrt{LC}} = \frac{8881.2611}{\sqrt{0.002\times0.00356\times10^{-6}}} = \frac{8881.2611}{2.668\times10^{-6}} = 3.328\ kV/\mu s$$

Example 8.5 Based on the following statistics for a 50Hz generator: emf to neutral 10kV (rms), reactance of generator and linked system 6 Ω, distributed capacitance to neutral 0.015μF, and negligible resistance: Determine the highest restriking voltage, the frequency of the transient oscillation, and the average rate and rise in voltage up to the first peak of the oscillation.

Ans. f = 50 Hz, V_L = 10 kV, X_L = 6 Ω and C = 0.015 μF

(a) Active recovery voltage, $V_{Max} = \sqrt{2} \times 10\ kV = 14.142\ kV$

Maximum Restriking Voltage $2V_{Max} = 2 \times 14.142\ kV = 28.284\ kV$

(a) Inductance $L = \frac{X_L}{2\pi f} = \frac{6}{2\times\pi\times50} = 0.0191\ H$

Frequency of oscillation, $f_n = \frac{1}{2\pi\sqrt{LC}} = \frac{1}{2\times\pi\times\sqrt{0.0191\times0.015\times10^{-6}}} = 9.405\ kHz$

(c) The restriking Voltage is, $V_{restriking} = V\left(1 - \cos\frac{t}{\sqrt{LC}}\right)$

Max restriking voltage occurs when $\cos\frac{t}{\sqrt{LC}} = -1$

$$\cos\frac{t}{\sqrt{LC}} = \cos\pi \quad or \quad \frac{t}{\sqrt{LC}} = \pi$$

Time to reach the peak restriking voltage $t = \pi\sqrt{LC} = \frac{1}{2f_n} = \frac{1}{2\times9.405\times1000} = 53.163\ \mu s$

Average Rate of Re-striking of Voltage $= \frac{Max\ restriking\ voltage}{Time\ upto\ first\ peak} = \frac{28.284\ \text{kV}}{53.163\mu\text{s}} = 0.5320\ \text{kV}/\mu\text{s}$

Example 8.6. A 50Hz 3-φ generator with 2mH inductance per phase is connected to the bus-bars via an oil CB The capacitance to earth in the circuit between the generator and the breaker is 0.034μF per phase. When the rms value of the current exceeds 8000A, the breaker opens due to a short on the bus bars. Draw a curve representing the re-striking voltage across the breaker and calculate the maximum rate of voltage rise.

Ans. L = 2 mH, C = 0.034 μF, Short circuit current = 8000 A

Reactance $X_L = 2\pi f L = 2\pi \times 50 \times 0.002 = 0.628\ \Omega$

Active Recovery voltage = short circuit current × reactance × $\sqrt{2}$

$= 8000 \times 0.628 \times \sqrt{2}$

$= 7.105\ kV$

The restriking Voltage is, $V_{restriking} = V\left(1 - \cos\frac{t}{\sqrt{LC}}\right)$

$$= 7.105\left(1 - \cos\frac{t}{\sqrt{2\times10^{-3}\times0.034\times10^{-6}}}\right)$$

$$= 7.105\left(1 - \cos\frac{t}{8.24621\times10^{-6}}\right)$$

If 't' is in μsec, $V_{restriking} = 7.105\left(1 - \cos\frac{t}{8.24621}\right)$

$$V_{restriking} = 7.105(1 - \cos 0.121t)\ kV$$

$$V_{restriking} = 7.105(1 - \cos\varphi)\ kV$$

t	Cos φ	1- Cos φ	$V_{restriking}$ =7.105(1- Cos φ)
0	Cos 0°	0	0
1.1	Cos 30°	0.134	0.952
2.2	Cos 60°	0.5	3.5525
3.3	Cos 90°	1	7.105
4.4	Cos 120°	1.5	10.657
5.5	Cos 150°	1.866	13.257
6.6	Cos 180°	2	14.21

Example 8.7. In a 132kV system, the reactance and capacitance up to the circuit breaker are 9Ω and 0.054μF. The CB connections are linked with a resistance of 500Ω. Determine the following.

(i). Natural frequency of oscillation

(ii). Damped frequency of oscillation

(iii). Resistance value required for no transient oscillation.

(iv). The resistance value resulting in damped oscillation is one-fourth of the natural frequency.

Ans. System voltage $E_L = 132kV$

Phase Voltage $E_{Ph} = \frac{E_L}{\sqrt{3}} = \frac{132\ kV}{\sqrt{3}} = 76.210\ kV$

Reactance $X_L = 9\ \Omega$

Resistance $R = 700\ \Omega$

Capacitance $C = 0.054\ \mu F$

Inductance $L = \frac{X_L}{2\pi f} = \frac{9}{2\times\pi\times 50} = 0.0286\ H$

(a) Natural frequency of oscillation, $f_n = \frac{1}{2\pi\sqrt{LC}}$

$$= \frac{1}{2\times\pi\times\sqrt{0.0286\times 0.054\times 10^{-6}}}$$

$$= 4.0519\ kHz$$

(b) Damped frequency of oscillation, $f_n = \frac{1}{2\pi}\sqrt{\frac{1}{LC} - \frac{1}{4R^2C^2}}$

$$f_d = \frac{1}{2\pi}\sqrt{\frac{1}{0.0286\times 0.054\times 10^{-6}} - \frac{1}{4\times(700)^2\times(0.054\times 10^{-6})^2}}$$

$$f_d = \frac{1}{2\pi}\sqrt{647500647.5 - 174967106.2}$$

$$f_d = 3.461\ kHz$$

(c) Critical resistance $= \frac{1}{2}\sqrt{\frac{L}{C}} = \frac{1}{2}\sqrt{\frac{0.0286}{0.054\times 10^{-6}}} = 363.878\ \Omega$

(d) Damped frequency of oscillation $= \frac{1}{4}\times$ (Natural frequency of oscillation)

$= \frac{1}{4}\times$ (4.0519 kHz)

$=$ 1.012975 kHz

$$1012.975 = \frac{1}{2\pi}\sqrt{\frac{1}{0.0286\times 0.054\times 10^{-6}} - \frac{1}{4R^2(0.054\times 10^{-6})^2}}$$

Squaring on both sides

$$(1012.975)^2 = \left(\frac{1}{2\pi}\right)^2 \left(\frac{1}{0.0286 \times 0.054 \times 10^{-6}} - \frac{1}{4R^2(0.054 \times 10^{-6})^2}\right)$$

$$40468465.97 = 647500647.5 - \frac{8.57338 \times 10^{13}}{R^2}$$

$$607032181.5 = \frac{8.57338 \times 10^{13}}{R^2}$$

$$R^2 = \frac{8.57338 \times 10^{13}}{607032181.5} = 141234.3573$$

$$R = 375.81159\ \Omega$$

Example 8.8. At 110kV, a circuit breaker interrupts the magnetizing current of a 150MVA transformer. The Transformer's magnetizing current is 10% of the full-load current. Determine the greatest voltage that will arise across the breaker gap when the magnetizing current is interrupted at 40% of its peak value. Stray capacitance: 2250μF. The inductance equals 25H.

Ans. 150MVA, 110kV T/F, C = 2250μF and L = 25 H

The 3-phase power P = $\sqrt{3}VI$

Full load current of the Transformer = $I = \frac{P}{\sqrt{3} \times V} = \frac{150 \times 10^6}{\sqrt{3} \times 110 \times 10^3} = 787.295\ A$

Magnetizing current = 10% of full load current = 10% of 787.295 A = 78.7295 A

Current chopping occurs at 40% of its peak value = $\frac{40}{100} \times 78.7295 \times \sqrt{2}$ = 44.536 A

The voltage across the C.B contacts $V = i\sqrt{\frac{L}{C}} = 44.536\sqrt{\frac{25}{2250 \times 10^{-6}}} = 4.693\ kV$

Example 8.9. A 220kV C.B has a bushing-to-ground capacitance of 0.025μF and a transformer inductance of 10H. Calculate the voltage that appears between the C.B poles if it interrupts a 15A magnetizing current flowing through the transformer.

Ans. C = 0.025Mf, L = 10 H and i = 15 A

The voltage across the C.B contacts $V = i\sqrt{\frac{L}{C}} = 15\sqrt{\frac{10}{0.025 \times 10^{-6}}} = 300\ kV$

Example 8.10. For a 220KV system, the phase-to-ground capacitance is 0.056μF and the inductance is 7H. The circuit breaker interrupts a 9A (peak) magnetizing current. Determine

(i). the voltage between the C.B contacts after current interruption.

(ii). the resistance required to inhibit restriking voltage across contacts.

Ans. C = 0.056 μF, L = 7 H and i = 9 A (peak)

(b) The voltage across the C.B contacts after the current Interruption

$$V = i\sqrt{\frac{L}{C}} = 9\sqrt{\frac{7}{0.056 \times 10^{-6}}} = 100.623\ kV$$

(b) The value of resistance to be used across the contacts to suppress restriking voltage

$$R = \frac{1}{2}\sqrt{\frac{L}{C}} = \frac{1}{2}\sqrt{\frac{7}{0.056 \times 10^{-6}}} = 5.590\ k\Omega$$

Example 8.11. In a short circuit test on a 3-pole, 220KV, the fault's C.B power factor was 0.62, and the recovery voltage was 0.88 times the full-line value. The breaking current was symmetrically distributed. The restriking voltage oscillation frequency was 20kHZ. Determine the average rate of rise of the restriking voltage. The neutral is grounded, and the fault lies with the earth.

Ans. Line voltage $V_L = 220\ kV$

Phase voltage (or) Line to neutral $V_{LN} = \frac{220}{\sqrt{3}} = 127.017\ kV$

Power factor $\cos\varphi = 0.62$ then $\varphi = \cos^{-1}(0.62)$, so $\varphi = 51.6838^o$ and $\sin\varphi = 0.7846$

Peak value of line-neutral voltage $V_{Max} = \sqrt{2} \times 127.017\ kV = 179.629\ kV$

Recovery voltage is 0.88 times of Full line value.

The active recovery voltage is given as $V_r = k_1 k_2 k_3 V_{Max} \sin\varphi$

Where, k_1 is the multiplying factor due to system voltage and is equal to 0.88

k_2 is a condition (or) phase factor and is unity in this case since the fault involves ground

k_3 is unity for recovery voltage between phase and neutral.

So, recovery voltage (From line to neutral) $V_r = 0.88 \times 1 \times 1 \times 179.629\ kV \times 0.7846$

$V_r = 124.024\ kV\ (Instantaneous)$

Time to reach the peak restriking voltage, $t = \frac{2\pi\sqrt{LC}}{2} = \pi\sqrt{LC} = \frac{1}{2fn} = \frac{1}{2\times20000} = 25\ \mu s$

$$\text{Average RRRV} = \frac{2 \times \text{Recovery voltage}}{t} = \frac{2 \times 124.024 \times 10^3}{25 \times 10^{-6}} = 9.921\ kV/\mu s$$

Example 8.12. The following observations were observed during a short circuit test on a 3-pole, 220KV CB: The fault power factor is 0.74, the recovery voltage is 0.94 times the full-line value, and the breaking current is symmetric. The restriking voltage oscillates at a frequency of 18000 cycles per second. Assume the neutral is grounded, and the fault is not related to ground. Calculate the average rate of rise in restriking voltage.

Ans. Line voltage $V_L = 220\ kV$

Phase voltage (or) Line to neutral voltage $V_{LN} = \frac{220}{\sqrt{3}} = 127.017\ kV$

Peak value of line-neutral voltage $V_{Max} = \sqrt{2} \times 127.017\ kV = 179.629\ kV$

The active recovery voltage is given as $V_r = k_1 k_2 k_3 V_{Max} \sin\varphi$

Where k_1 is the multiplying factor due to system voltage and is equal to 0.94.

k_2 is a condition (or) phase factor and is equal to 1.5 as the fault does not involve ground

k_3 is unity for recovery voltage between phase and neutral.

Power factor $\cos\varphi = 0.74$ then $\varphi = \cos^{-1}(0.74)$ so $\varphi = 42.2685^o$ and $\sin\varphi = 0.6726$

Active recovery voltage $V_r = 0.94 \times 1.5 \times 1 \times 179.629\ kV \times 0.6726$

$V_r = 170.354\ kV\ (Instantaneous)$

Time to reach the peak restriking voltage, $t = \frac{2\pi\sqrt{LC}}{2} = \pi\sqrt{LC} = \frac{1}{2fn} = \frac{1}{2\times18000} = 27.777\ \mu s$

$$\text{Average RRRV} = \frac{2\times\text{Recovery voltage}}{t} = \frac{2\times170.354\times10^3}{27.777\times10^{-6}} = 2.265\ kV/\mu s$$

8.11 Circuit Breaker Ratings and Specifications:

During fault conditions, CB must perform following duties

(i). In the event of a fault, a circuit breaker must be able to break the circuit and isolate the faulty area from the healthy section. This is referred to as breaking capacity of CB

(ii). A CB must be capable of closing on a fault. This is referred to as *making capacity of a CB*

(iii). It must be able to carry fault current for a brief period of time until another circuit breaker clears the fault. This refers to the short-term current rating of a CB

8.11.1 Breaking Capacity of a C.B:

The breaking capacity of a CB is further devided into two types

(i) Symmetrical breaking capacity.

(ii) Asymmetrical breaking capacity.

The symmetrical breaking capacity is defined as the rms value of the ac component of the fault current that the CB is capable of breaking under specified recovery voltage conditions.

The asymmetrical breaking capacity is the rms value of the entire current, which includes both ac and dc components of the fault current, that the CB is capable of breaking under given conditions of recovery voltage.

The S.C. current contains a dc component that gradually decays. Because of the dc component, the S.C. current starts out asymmetrical. When direct current is totally eliminated, the S.C. current becomes symmetrical which is shown in Fig. 8.8.

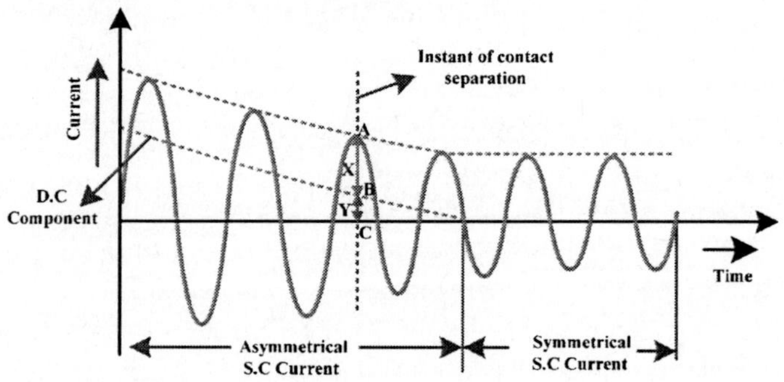

Fig. 8.8 Symmetrical and Asymmetrical Breaking Capacity of a C.B

$$\text{Symmetrical breaking current} = \frac{AB}{\sqrt{2}} = \frac{x}{\sqrt{2}} = \frac{I_{ac}}{\sqrt{2}} \quad \text{.... (8.28)}$$

$$\text{Asymmetrical breaking current} = \sqrt{\left(\frac{AB}{\sqrt{2}}\right)^2 + BC^2} = \sqrt{\left(\frac{x}{\sqrt{2}}\right)^2 + y^2} = \sqrt{\left(\frac{I_{ac}}{\sqrt{2}}\right)^2 + {I_{dc}}^2} \quad \text{.... (8.29)}$$

$$\text{Breaking capacity} = \sqrt{3} * \text{Rated voltage} * \text{Breaking current} \quad \text{.... (8.30)}$$

8.11.2 Making Capacity of a C.B:

The term "rated making current" refers to the maximum value of the current that is closed at the point of a short circuit.

$$\text{Making capacity} = \sqrt{2} * 1.8 * \text{symmetrical breaking current}$$

$$= 2.55 * \text{symmetrical breaking current} \quad \text{.... (8.31)}$$

Multiplying by $\sqrt{2}$ yields the peak value, whereas a gain of 1.8 accounts for the dc component.

8.11.3. Short-time current rating of a C.B: The capacity of a circuit breaker to sustain fault current for a brief duration during the clearing of a fault by another CB is essential. According to British standard,

The short-time is 3 seconds if $\frac{\text{symmetrical breaking current}}{\text{rated current}} \leq 40$ and

the short-time is 1 second if $\frac{\text{symmetrical breaking current}}{\text{rated current}} > 40.$

For more details about how to select a CB, *please go through the previous chapter section 6.6.2.*

Example 8.13. When a 33KV, 300MVA CB closes on a fault, calculate the

(i). symmetrical breaking current and

(ii). the asymmetrical breaking current, assuming a 60% dc component.

(iii). Peak generating current.

(iv). The short-term current rating.

Ans. Operating Voltage = 33 kV and Rupturing capacity = 300 MVA

(a) Symmetrical breaking current $= \frac{\textit{Rupturing capacity}}{\sqrt{3} \times \textit{operating voltage}} = \frac{300 \times 10^6}{\sqrt{3} \times 33 \times 10^3} = 5.2486\ kA$

AC component of short circuit current (X) $= \sqrt{2} \times \textit{Symmetrical breaking current}$

$= \sqrt{2} \times 5.2486 \times 1000$

$= 7.4226\ kA$

DC component of short circuit current (Y) $= 60\%\ of\ X$

$= 0.6 \times 7.4226\ kA$

$= 4.453\ kA$

(b) Asymmetrical breaking current $= \sqrt{\left(\frac{X}{\sqrt{2}}\right)^2 + Y^2}$

$= \sqrt{\left(\frac{7.4226}{\sqrt{2}}\right)^2 + 4.453^2}$

$= 6.882\ kA$

(c) Peak making current $= 2.55 \times \textit{Symmetrical breaking current}$

$= 2.55 \times 5.2486$ kA

$= 13.383\ kA$

(d) Short time current rating $= 6.882\ kA$

Example 8.14. A generator with a 3 cycle C.B coupled to a transformer is rated 10MVA, 11kV, with reactance $X_d'' = 8\%$, $X_d' = 18\ \%$, $X_d = 100\%$. It is set to no load and rated voltage. A 3-Φ short circuit occurs between the breaker and transformer. Determine

(i). The breaker's sustained S.C current

(ii). The first symmetrical rms current of the CB

(iii). Maximum attainable D.C. component of the S.C. current through the CB

(iv). The momentary Current rating of the CB

(v). The current that the circuit breaker is intended to interrupt.

(vi). The kVA that interrupts.

Ans. Rated capacity = 10 MVA, Line Voltage = 11 kV

Sub-transient Reactance $X_d'' = 8\%$

Transient Reactance $X_d' = 18\ \%$

Steady state Reactance $X_d = 100\%$

(a) Sustained SC kVA = $\frac{Rated\ capacity}{X_d} = \frac{10\times10^6}{\left(\frac{100}{100}\right)} = 10MVA$

Sustained SC current = $\frac{Sustained\ SC\ kVA}{\sqrt{3}\times Rated\ voltage} = \frac{10\times10^6}{\sqrt{3}\times11\times10^3} = 524.863\ A$

(b) Sub transient SC kVA = $\frac{Rated\ capacity}{X_d''} = \frac{10\times10^6}{\left(\frac{8}{100}\right)} = 125MVA$

Sub transient or initial symmetric current = $\frac{Rated\ capacity}{\sqrt{3}\times Rated\ voltage\times X_d''} = \frac{10\times10^6}{\sqrt{3}\times11\times10^3\times0.08} = 6.56\ kA$

(c) Maximum permissible DC component $= \sqrt{2}\times Initial\ symmetric\ current$

$= \sqrt{2}\times 6560.798 = 9.278\ KA$

(d) Momentary current rating $= 1.6\times Initial\ symmetric\ current$

$= 1.6\times 6560.798$

$= 10.497\ kA$

(e) For 3 cycles, multiplication factor is 1.2 times of Sub-transient symmetrical current

Current to be interrupted $= 1.2\times Sub\ transient\ symmetric\ current$

$= 1.2\times 6590.798$

$= 7.872\ kA$

(f) Interrupting kVA $= \sqrt{3}\times Current\ to\ be\ interrupted\times Rated\ voltage$

$= \sqrt{3}\times 7872.9576\times 11\times10^3$

$= 149.999\times10^3\ kVA$

8.12 Classification of Circuit Breakers:

The most comprehensive method of categorization is based on the medium employed for arc extinction. Circuit breakers are primarily categorized into the following categories.

(i). Oil circuit breakers

- Bulk Oil or Plain break or Double break circuit breakers
- Minimum Oil circuit breakers

(i). Air blast circuit breakers

- Axial blast circuit breakers
- Cross blast circuit breakers
- Radial blast circuit breakers

(ii). SF6 circuit breakers and

(iii). Vacuum circuit breakers

8.12.1 Oil circuit breakers (OCB):

To extinguish an arc, oil circuit breakers use dielectric oil. Fixed and movable contacts will remain closed under normal working conditions. When a fault occurs, the contacts are separated; the movable contact is moved away from the fixed contact, and an arc is formed between them. Heat will be created across the contacts. At high pressure, the heat generated by the arc causes the surrounding oil to evaporate and separate into a significant amount of gaseous hydrogen. The hydrogen gas has a volume approximately one thousand times greater than that of the degraded oil. Consequently, the oil is displaced from the arc, causing hydrogen gas bubbles to grow around the arc region and the surrounding areas of the contacts.

Advantages of Oil CB:

- ✓ The decomposition of oil involves the absorption of arc energy.
- ✓ The formation of hydrogen gas exhibits exceptional cooling characteristics.
- ✓ Hydrogen gas exhibits a notable capacity for heat absorption and a rapid rate of diffusion.
- ✓ The oil has a significant dielectric strength and facilitates contact disturbance upon the cessation of the arc.

Disadvantages of Oil CB:

- ✗ Oil exhibits a high degree of flammability, hence posing potential fire concerns.
- ✗ As the oil decomposes in the arc, it becomes contaminated with carbon particles, resulting in a decrease in its dielectric strength.
- ✗ It is imperative to engage in regular maintenance and replenishment of oil.
- ✗ There is a possibility of producing an explosive combination with air.

Oil CBs are further sub-divided into two types bulk oil CB and low/minimum oil CB

8.12.1.1 Bulk Oil or plain break or Double break circuit breakers:

Description of Bulk Oil CB: Bulk oil CB comprises a robust metallic container filled with dielectric oil, as well as stationary and mobile contacts. The tank does not contain the dielectric oil in its entirety. Instead, an air cushion is placed on the surface of the oil to accommodate any displaced oil that may occur around the arc. The circuit breaker is commonly referred to as a double break circuit breaker due to its configuration, which includes two fixed contacts and one movable contact which is shown in Fig. 8.9.

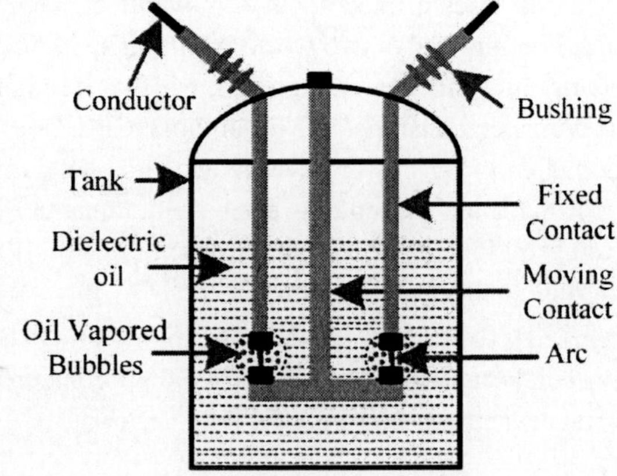

Fig. 8.9 Bulk-oil circuit breaker

The volume of oil filled is exceedingly large. For instance, a 110kV, 3500MVA breaker may require 8000 to 12000 Liters of oil, whereas a breaker with the same rating output for 220kV may require 50,000 Liters of oil. While the amount of oil needed for arc extinction is quite tiny, approximately one-tenth of the overall oil supply. However, following the extinction of the arc, the entire oil, including significant amounts, undergoes carbonization, sludging, and other related processes.

Operation of Bulk Oil CB: Under typical operating conditions, CB contacts are closed. During faulty conditions, the movable contact will move away from the fixed contact, causing an arc to form between the two contacts. Under these circumstances, a significant quantity of thermal energy will be produced between the contacts. Once the temperature surpasses 5000°C, the dielectric oil will undergo rapid vaporization into hydrogen gas, causing the molecules in the surrounding medium to be displaced. The strong heat conductivity of hydrogen gas results in the cooling of the arc, leading to the deionization of the medium between the contacts.

Disadvantages of Bulk Oil CB:

- ✗ High-speed interruption is prohibited by the breakers.
- ✗ The breakers exhibit extended and irregular arcing durations.
- ✗ An unequal distribution of voltage across the breaker results in an uneven distribution of the total interrupting duty.
- ✗ There is no specific control of the arc, save for the expansion in the distance b/w the contacts.

Application of Bulk Oil CB: It is suitable for voltages ranging from 2.5kV to 220kV and breaking capacities ranging from 25 MVA to 5000MVA.

8.12.1.2 Minimum Oil circuit breakers:

Description of Minimum Oil CB: The minimum oil CB comprises of two chambers, a fixed and movable contact, an arc extinction mechanism, an operating rod, a drain valve, a gas vent, and a breather which is shown in Fig. 8.10. The upper chamber is the circuit-breaking chamber, while the lower chamber serves as the sustaining chamber. These two rooms are divided by a partition, which keeps oil from one chamber from mixing with the other. The circuit breaking chamber is a porcelain chamber that is affixed to the upper portion of the central chamber. The container is filled with a limited quantity of oil. The device possesses a fixed contact, movable contact, and arc extinction mechanism. The arc extinction device is comprised of two type of vents, namely axial and radial vents. In order to interrupt lesser arcing currents, an axial vent is employed, whereas a radial vent is utilized to interrupt large arc currents. The circuit breaking chamber of a minimum OCB is not fully filled with dielectric oil. Instead, an air cushion is added on top of the oil level to accommodate any displaced oil that may occur around the arc. The supporting chamber contains a substantial quantity of oil that is physically isolated from the oil in the circuit breaking chamber. The purpose of this oil is to substitute the oil in the circuit breaking chamber according to the specified criteria.

Operation of Minimum Oil CB: The minimum OCB operates in the same way that the bulk OCB does. However, because of the addition of a supporting chamber alongside the circuit breaking chamber, it has several advantages over bulk OCBs.

Advantages of Minimum Oil CB:

- ✓ The circuit breaking chamber requires less amount of oil.
- ✓ There is relatively little oil to be refilled.
- ✓ The time required to refill the oil in the circuit-breaker chamber will be minimal.
- ✓ Fire risk is reduced.
- ✓ It needs less space.
- ✓ Maintenance issues will be minimized.

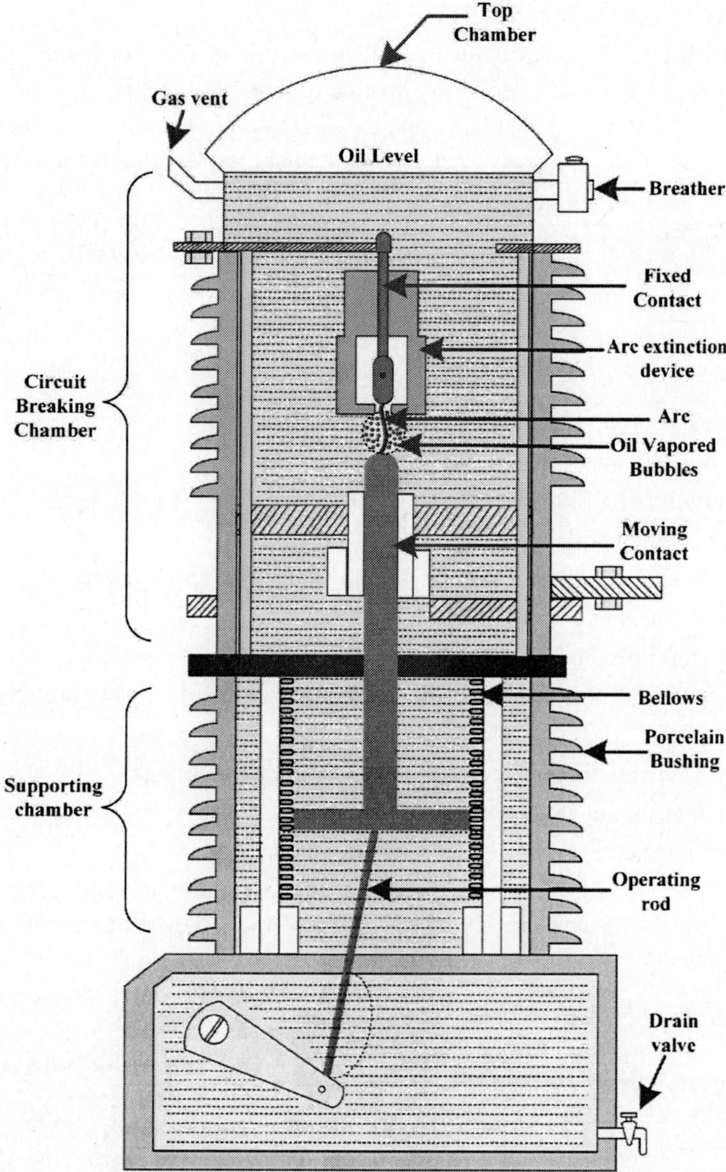

Fig. 8.10 Minimum Oil circuit breaker

Disadvantages of Minimum Oil CB:

- The utilization of a reduced amount of oil in the circuit breaking chamber results in a significant increase in carbonization.
- The quick deterioration of the oil's dielectric strength is attributed to a significant degree of carbonization.

Application of Minimum Oil CB: It is suitable for voltages ranging from 33kV to 220kV and breaking capacities ranging from 1500 MVA to 7500 MVA.

8.12.2. Air blast circuit breakers (ABCB):

The limitations of oil circuit breakers, such as oil deterioration, fire risk owing to inflammable oil, frequent oil replacement, and so on, led to the invention of circuit breakers that use compressed air or gas as the interrupting medium. Depending upon the direction of air-blast in relation to the arc, ABCB are classified into Axial blast CB, Cross blast CB and Radial blast CB which are shown in Fig. 8.11.

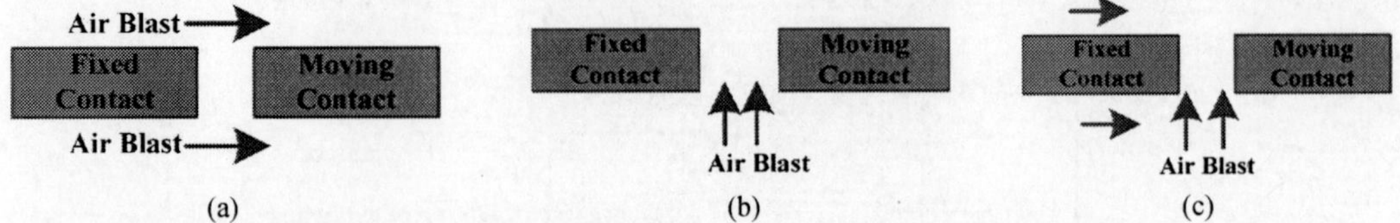

Fig. 8.11. (a) Axial blast circuit breakers, (b) Cross blast circuit breakers and (c) Radial blast circuit breakers.

Advantages of Air blast circuit breakers: Air blast CB has following advantages over oil CB

- ✓ Fire hazard is eliminated
- ✓ Ionised particles in the arcing medium will be completely removed by blast method.
- ✓ Suitable for frequent operation
- ✓ Rapid development in dielectric strength allows for high-speed operation.
- ✓ Negligible maintenance
- ✓ Facility of high speed reclosure
- ✓ Reduced size of the arc extinction gap due to rapid growth of dielectric strength.

Disadvantages of Air blast circuit breakers:

- × Sensitive to the variations in the rate of rise of restriking voltage
- × Relatively inferior arc extinction properties
- × Suitable maintenance is required for the compressor plant.

Application of Air blast CB: Air blast circuit breakers were previously utilized for interior services with voltages up to 15kV and a breaking capability of 2500 MVA, but they are currently employed outdoors for voltages up to 220kV.

8.12.2.1 Axial blast air circuit breakers:

In axial blast air circuit breakers, the air flows longitudinally along the arc. Axial blast CB can be single or double blast which is clearly shown in Fig. 8.12.

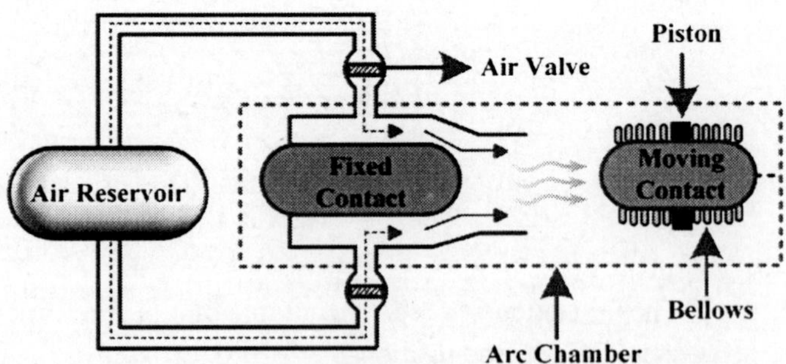

Fig. 8.12 Axial blast air circuit breakers

Description of Axial blast air circuit breakers: Axial blast air circuit breakers consist of an air reservoir, an air valve, fixed and movable contacts, an arcing chamber, a piston, and a series isolator. The air reservoir is connected to the arc chamber by an air valve. The fixed contact is connected to the arching chamber, while the moving contact is connected to the series isolator via the piston.

Operation of Axial blast air circuit breakers: In normal operational circumstances, the contacts of the circuit breaker (CB) will be in a closed state, and the air valve will also be closed. The tripping impulse triggers the opening of the air valve that connects the reservoir to the arc chamber in the event of a malfunction. The introduction of high-pressure air into the arcing chamber results in the displacement of the movable contact. An arc will be struck between the moving contact and the fixed contact. Simultaneously, a high-pressure air blast is sent along the arc, effectively removing the ionized particles. The arc is extinguished, resulting in the interruption of the current flow. The presence of a tiny contact gap following an interruption may result in insufficient clearance for the required system voltage. The isolator switch, attached to the movable contact, will be opened promptly following the fault interruption in order to allow for adequate insulating clearance.

8.12.2.2 Cross blast air circuit breakers:

In cross blast air circuit breakers, the air blast flow is oriented perpendicular to the arc. There exists a significant distinction between the principle employed in cross-blast air CB and axial blast air CB. This approach achieves both the removal of ionized particles and the provision of a sufficient chute for arc extinction.

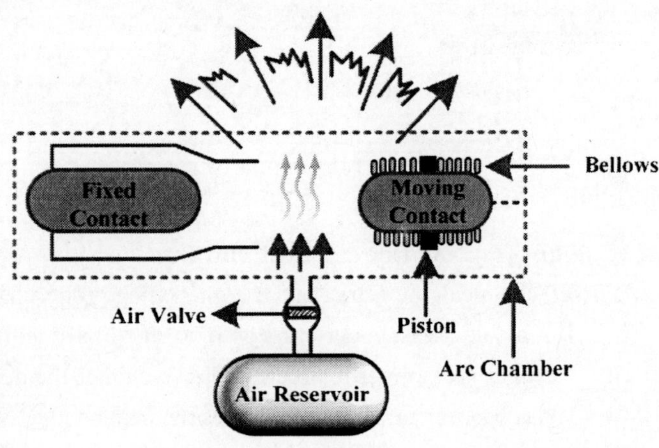

Fig. 8.13 Cross blast air circuit breakers

Description of cross blast air circuit breakers: Cross blast air circuit breakers consist of several essential components, including an air reservoir, air valve, fixed contact, movable contact, arcing chamber, and piston. The air reservoir is linked to the arc chamber via an air valve in order to supply air to the arc which is shown in Fig. 8.13.

Operation of cross blast air circuit breakers: During regular operational circumstances, the contacts of the circuit breaker (CB) will be in a closed state, allowing the breaker to conduct normal current. Additionally, the air valve will be closed. The tripping impulse triggers the opening of the air valve that connects the reservoir to the arc chamber in the event of a malfunction. The introduction of high-pressure air into the arcing chamber results in the displacement of the movable contact. An arc will be struck between the moving contact and the fixed contact. Simultaneously, a high-pressure air blast is directed perpendicular to the arc, so compelling the arc to descend down a chute including arc splitters and baffles. The splitters serve to expand the arc, while the baffles contribute to the cooling process. The flame is extinguished and the passage of electric current is halted. This approach exhibits consistent blast pressure across all currents, hence facilitating effective gap interruption. Consequently, the need for an isolator switch may be avoided.

8.12.2.3 Radial blast air circuit breakers:

Radial blast air circuit breakers allow high-pressure air to flow in all directions which is shown in Fig. 8.14. Breakers with a twin blast arrangement are also referred to as radial blast. It is capable of both axial and cross-blasting. Thus, air moves along and over the arc path. Radial blast air circuit breakers have superior arc quenching capabilities compared to axial and cross blast circuit breakers.

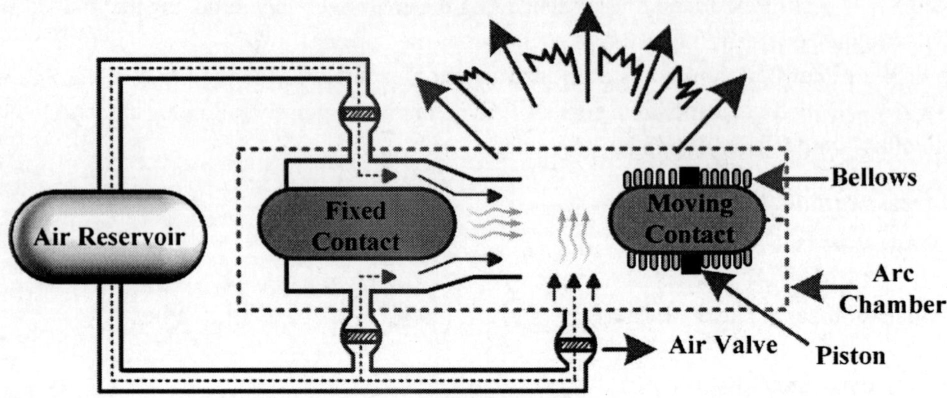

Fig. 8.14 Radial blast air circuit breakers

8.12.3 Sulphur Hexafluoride (SF6) Circuit Breaker:

The physical, chemical, dielectric, and arc-quenching properties of SF6 gas are:

(i). It is a colourless, odourless, nontoxic, and non-flammable gas.

(ii). This gas is extremely stable and has five times the density of air.

(iii). It has approximately 2 to 2.5 times the heat conductivity of air.

(iv). It has improved cooling properties.

(v). SF6 is an electrically negative gas. SF6 gas has a low arc time constant ($\leq 1\ \mu$ sec) and a rapid increase in dielectric strength.

(vi). It is chemically acceptable up to 5000 degrees Celsius.

(vii). SF6 is around 100 times more effective than an air-quenching arc.

Description of SF6 Circuit Breaker: It consists of moving contacts, fixed contacts, an SF6 gas reservoir, a gas outlet, an arc chamber, and insulated rods which is shown in Fig. 8.15. When the circuit breaker's contact is opened, the valve mechanism allows high-pressure SF6 gas from the reservoir to flow into the arc interruption chamber. The movable contact consists of a cylindrical structure with rectangular apertures on its sides, facilitating the release of SF6 gas as it traverses the arc. The cost of SF6 is high; however, it can be mitigated with the implementation of an appropriate auxiliary system following each breaker operation. A single SF6 interrupter is suitable for systems up to 220 kV, a pair of SF6 interrupters is suitable for systems up to 400 kV, and a four-SF6 interrupter is suitable for systems up to 765 kV.

Operation of SF6 Circuit Breaker: During normal operation, both the fixed and movable contacts are closed, and the circuit breaker is carrying the regular circuit current. During this time, the contacts are surrounded by SF6 gas at a pressure of approximately 2.8 kg/cm^2. When a fault occurs, the moving contact separates from the fixed contact, resulting in an arc between the two. The gas valve is then opened, allowing 14 kg/cm^2 of SF6 gas to enter the arc chamber under high pressure. The high-pressure flow of SF6 rapidly absorbs free electrons in the arc path, forming stationary negative ions that are inefficient as charge carriers. The medium between contacts then rapidly accumulates high dielectric strength, causing the arc to extinguish.

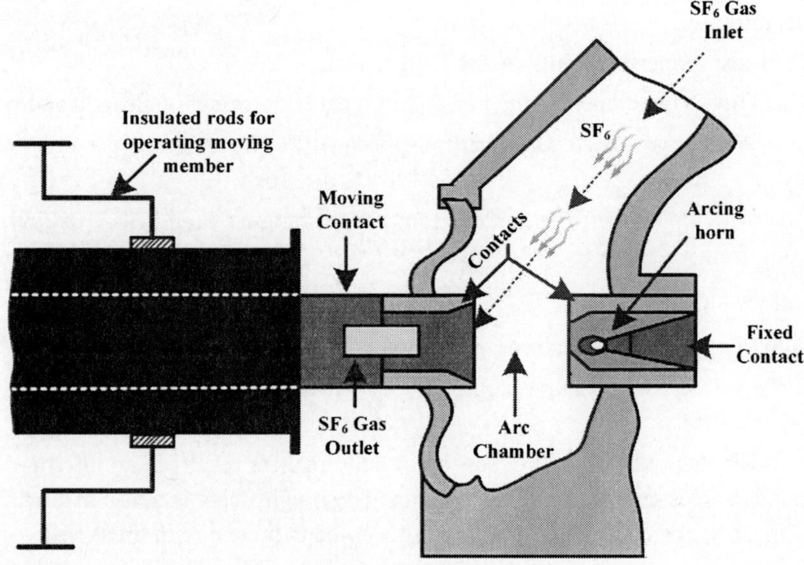

Fig. 8.15 SF6 Circuit Breaker

Advantages of SF6 circuit breaker:

SF6 circuit breakers provide the following advantages over oil and air blast circuit breakers.

- ✓ Excellent insulation, arc extinguishing capabilities, and physical and chemical qualities.
- ✓ The high dielectric strength of SF6 greatly reduces electrical clearances.
- ✓ It is not flammable and has chemical stability. The breakdown products are non-explosive, which means there is no chance of fire or explosion.
- ✓ Its performance is unaffected by atmospheric conditions.
- ✓ It works silently. It makes no noise when in action, unlike an air blast circuit breaker.
- ✓ Arcing time is quite brief.
- ✓ The dielectric strength of SF6 is not reduced since no carbon particles are produced during arching.
- ✓ There is very little upkeep required. The breaker may need maintenance every four to ten years.
- ✓ The sealed (closed) design prevents moisture, dust, and sand contamination.
- ✓ The same gas is recirculated, which reduces the need for SF6 gas.
- ✓ Because of the superheat transferability of SF6 gas, the current carrying capability of SF6 circuit breakers is approximately 1.5 times more than that of air blast circuit breakers for conductors of equal size.

Disadvantages of SF6 circuit breaker:

- SF6 gas is toxic and should not be released.
- The high cost of SF6 makes SF6 breakers expensive.
- SF6 gas must be reconditioned following each use of the breaker addition equipment necessary for this purpose.
- Sealing issue. Imperfect joints cause SF6 gas leaks.
- During periodic maintenance, the inside part must be thoroughly cleaned and dried.
- Special facilities are necessary for transporting SF6, transferring gas, and maintaining gas quality.

Application of SF6 circuit breaker: It is suitable for voltages ranging from 110kV to 220kV and breaking capacities ranging from 10 MVA to 20 MVA.

8.12.4. Vacuum Circuit Breakers:

The high vacuum has two notable features.

(i). It has the greatest insulating strength and significantly superior arc quenching capabilities of any other media.

(ii). In a vacuum, the opening of contacts on circuit breakers results in an interruption at the initial current zero. This interruption is accompanied by a significant increase in the dielectric strength between the contacts, surpassing the rates observed in other circuit breakers by several thousand times.

Vacuum medium: A vacuum is defined as any medium with a pressure below atmospheric, namely 760 mm of Hg. Low pressure is quantified in torr, with one torr equivalent to 1 mm of mercury (Hg). Pressures below approximately 10-5 torr are classified as high vacuum conditions.

Principle: An arc is created when circuit breakers' contacts are separated due to the ionization of metal vapours on them. The utilization of vacuum as an arc extinguishing medium is observed within the range of 10^{-7} to 10^{-5} torr. Nevertheless, the suppression of the arc occurs promptly due to the rapid condensation of metallic vapours, electrons, and ions generated during the arc on the contacts of the circuit breaker. This phenomenon leads to a swift restoration of the dielectric strength. For voltage levels exceeding 36 kV, it is necessary to connect two interrupters in series. Consequently, the vacuum circuit breaker becomes economically inefficient when the rated voltage surpasses 36kV.

Construction of Vacuum Circuit Breakers: The system comprises a fixed contact, a movable contact, an arc shield, bellows, and an insulating tank which is shown in Fig. 8.16. Metal bellows composed of stainless steel are employed for the purpose of facilitating the movement of the contact. The bellows' design holds significant importance as the longevity of the vacuum is contingent upon the component's capacity to execute repetitive actions. The movable contact exhibits a limited displacement ranging from 5 to 10 mm, dependent upon the voltage at which it is in operation.

Operation of Vacuum Circuit Breakers: Under normal operational circumstances. Fixed and movable contacts stay closed, and the breaker maintains normal circuit current. When a fault occurs, the moving contact separates from the fixed contact, resulting in an arc between the two contacts. The arc is rapidly extinguished due to the rapid diffusion of metallic vapours, electrons, and ions generated during the arc.

These species are then captured by the surface of both movable and fixed contacts, as well as arc shields. The arc extinction in a vacuum circuit breaker is facilitated by the rapid recovery of dielectric strength in a vacuum, resulting in a short contact separation.

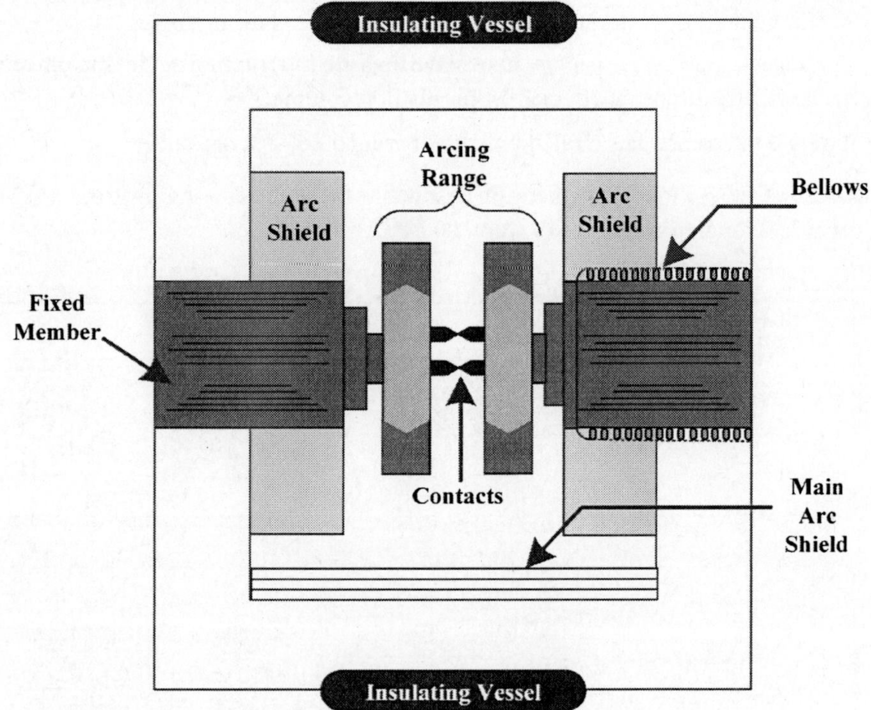

Fig. 8.16. Vacuum Circuit Breaker

Advantages of Vacuum Circuit Breakers: Vacuum circuit breakers has several advantages.

- ✓ Self-contained vacuum circuit breakers don't need oil or gas to function. They do not require an auxiliary air system, oil handling system, or annual refilling.
- ✓ There are no gas emissions, indicating that it is pollution-free.
- ✓ There are no explosives, and the procedure is silent.
- ✓ Rapid recovery of very high dielectric strength after circuit interruption, resulting in just half-cycle or less arching following correct contact separation.
- ✓ Current interruption happens at the first current zero following contact separation with no restriking.
- ✓ With a short contact spacing, very high-power frequency and impulse withstand voltage are achieved.
- ✓ It can be used repeatedly (often).
- ✓ It lasts longer.
- ✓ A vacuum does not cause any gas breakdown.
- ✓ They don't require much maintenance.

Disadvantages of Vacuum Circuit Breakers:

- Loss of vacuum due to transit damage or failure renders the entire interrupter ineffective and cannot be used on the job site.
- Vacuum interrupters are manufactured using advanced technologies.
- The vacuum interrupter is more expensive than the interrupting devices in other types of circuit breakers, and its cost is affected by production volume.
- A single interrupter has a rated voltage of around $36/\sqrt{3}$, or 20 kV.

Application of Vacuum Circuit Breakers: It is suitable for voltages ranging from 11kV to 33kV and breaking capacities ranging from 60 MVA to 100 MVA.

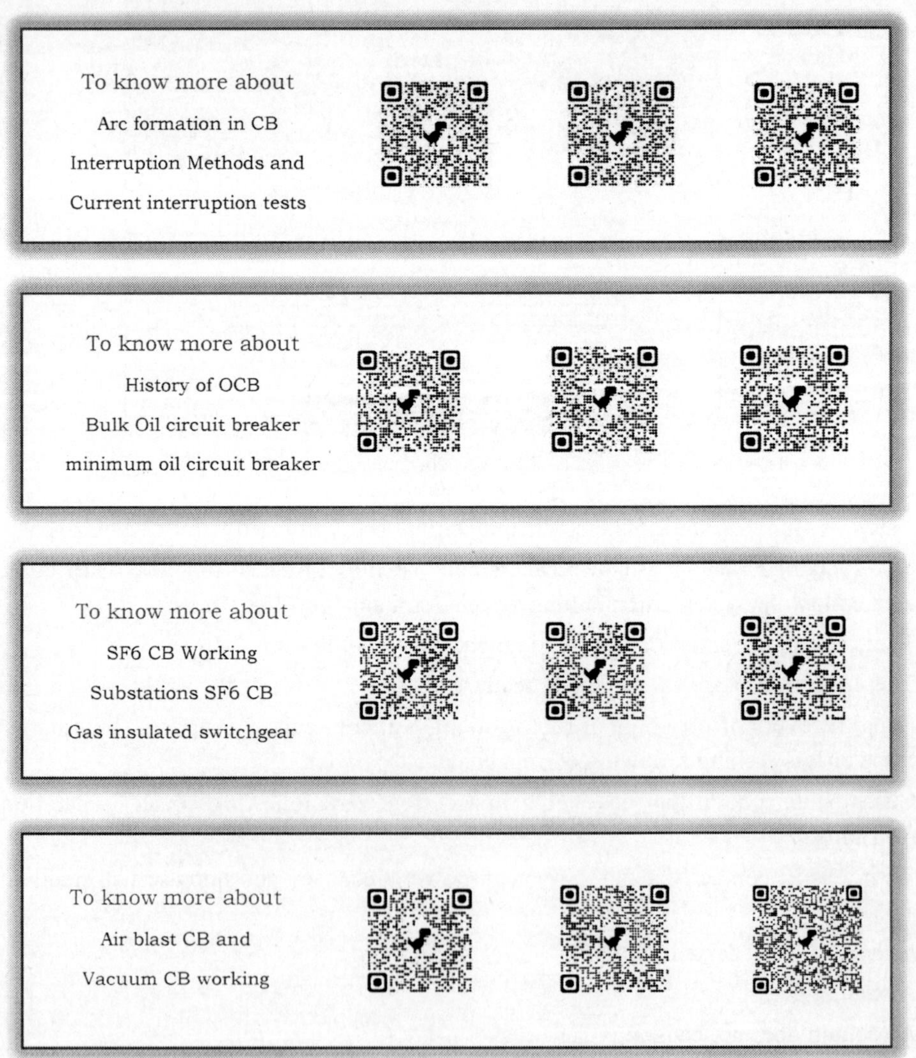

8.13 Introduction to Neutral Grounding:

Earthing or Grounding in a power system is the process of connecting an electrical part of the system (e.g. neutral point in a star-connected 3-Ø system, current carrying conductive parts, etc.) or non-current carrying metallic parts of the system to the earth. The precise earthing prerequisites and methodologies may differ depending on local regulations, equipment type, and electrical behaviour of the system.

In a power system, earthing or grounding offers several features:

- Safety: The use of earthing or grounding protects people and equipment from electric shock by allowing for the safe dissipation of fault currents into the ground.
- Equipment Protection: It safeguards electrical equipment against damage caused by power surges, transient voltages, and lightning strikes by stabilising voltage levels.
- Noise Reduction: It provides a low-impedance path to ground for undesirable electrical signals, which helps to reduce the radio frequency interference (RFI) and electromagnetic interference (EMI).
- Fault Detection: It ensures the identification and location of faults within the system. For example, if a fault occurs in the system, then the earth fault relay can sense the flow of fault current through ground connection. This will enable the identification of faulty section in the system.
- Stability: It improves the stability of system by limiting the operating voltages to rated values and reducing transient voltages during faulty conditions.

8.14 Classification of Earthing or Grounding:

The Earthing or Grounding can be classified based on to whom the protection scheme is employed. According to that it can be categorized into:

1. Equipment Grounding
2. System Grounding

8.14.1 Equipment Grounding:

Equipment grounding is the procedure by which non-current-carrying metallic components (i.e., the metal enclosure) of electrical equipment are connected to earth (i.e., the soil). At this condition if any insulation failure of equipment occurs also, the metal enclosure should remain at earth potential. We frequently come into contact with a wide variety of electrical apparatus, including industrial motors, handheld tools, and household appliances etc. Hence it is necessary to provide protection for the equipment against short circuits, overloading and lightning. To understand the equipment grounding and its benefits, following cases are considered.

8.14.1.1 Ungrounded Metal Enclosure:

Figure 8.17 depicts an ungrounded metal enclosure system with a 1-Ø, 230-V AC source that is properly grounded and powers a motor. Here, the motor is surrounded by a metal enclosure that is not connected to the ground, and there is a virtual resistance (R_a) between the motor terminals and the enclosure. It provides strong resistance when the equipment is in good working condition and lower resistance when the equipment is faulty. During normal operating conditions, if a person touches the metal enclosure, nothing happens. i.e. he did not experience any shock since the virtual resistance (R_a) provides a high resistance, resulting in a small amount of current passing through the person's body. The human body can safely allow a maximum current of 5-10 mA.

During abnormal operating conditions, such as an insulation failure of the motor winding, a high value of fault current (I_L) flows in the motor terminals and circulates through the metal enclosure via virtual resistance (R_a), providing low resistance for the fault current. When a person touches the metal enclosure, a high value of fault current circulates in the system by following a complete circuit from the metal enclosure to ground via the human body, which provides a lower value of body resistance (R_b), and returns to the metal enclosure, as shown in Fig. 8.17. Due to the flow of high value of fault current (I_L) (> 5-10 mA) through human body, the person may experience severe electrical shock, which may result in death. Thus, an ungrounded metal enclosure is dangerous.

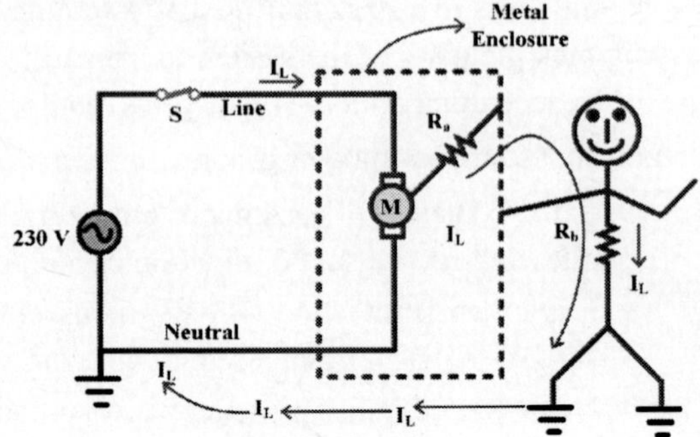

Fig. 8.17 Representation of Ungrounded Metal Enclosure

8.14.1.2 Metal Enclosure connected to Neutral wire:

To address the disadvantages of an ungrounded metal enclosure system, an appropriate component of the metal enclosure is linked to the system's neutral wire, as shown in Fig. 8.18 (a). Because no current flows through the human body during normal operation, there will be no shocks. During abnormal condition, the fault current (I_L) completes the path from motor terminals to metal enclosure to neutral wire. Here the neutral wire is at ground potential hence the metal enclosure is also at ground potential. Because of this, if someone touches the metal enclosure, they will not receive an electrical shock because the majority of the fault current is routed to the ground via the system's neutral.

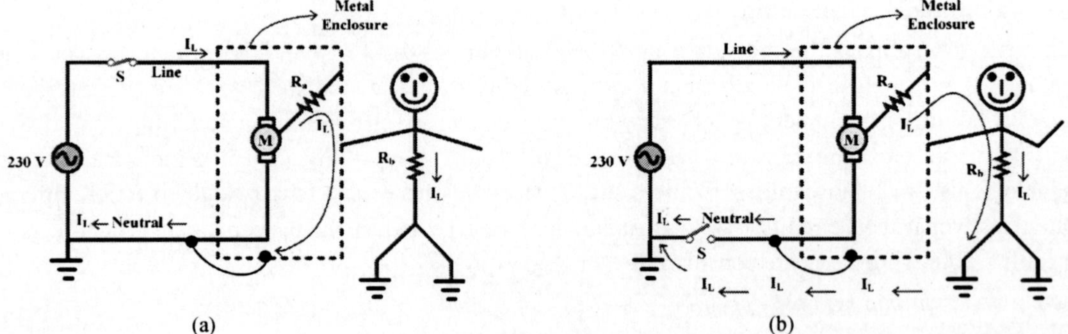

Fig. 8.18 Representation of Metal Enclosure connected to Neutral Wire (a) During normal and abnormal condition when Neutral wire is closed (b) During abnormal condition when neutral wire open circuited.

If the system's neutral wire is accidently or by fault opened, the metal enclosure is no longer at the ground potential and its value rises to that of the live conductor. At this point, if somebody touches the metal enclosure, they will receive an electrical shock, as shown in Fig. 8.18(b). As a result, connecting a Metal Enclosure to an unprotected Neutral wire is dangerous.

8.14.1.3 Ground wire connected to Metal Enclosure:

To overcome the drawbacks of an ungrounded metal enclosure system and a metal enclosure connected to an open neutral wire, a ground wire is connected between the metal enclosure and the ground, as illustrated in Fig. 8.19. Because of this connection, even if a fault occurs on the motor terminals, the fault current is routed to the ground via a metal enclosure and ground wire. Even if somebody touches the metal enclosure, they will not receive a shock. This will provide adequate safety for the equipment.

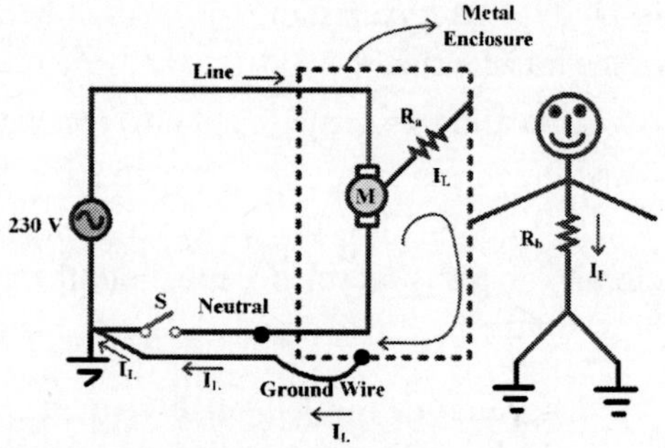

Fig. 8.19 Representation of Ground wire connected to Metal Enclosure

8.14.2 System Grounding:

System grounding is the process of connecting current-carrying electric parts (such as the system's star-connected neutral point or any single wire in the transformer secondary) to earth. Implementing proper system grounding can provide several benefits, including increased security, dependability, and protection of the power system network. Before we get into the many sorts of system grounding, let's talk about why we need them.

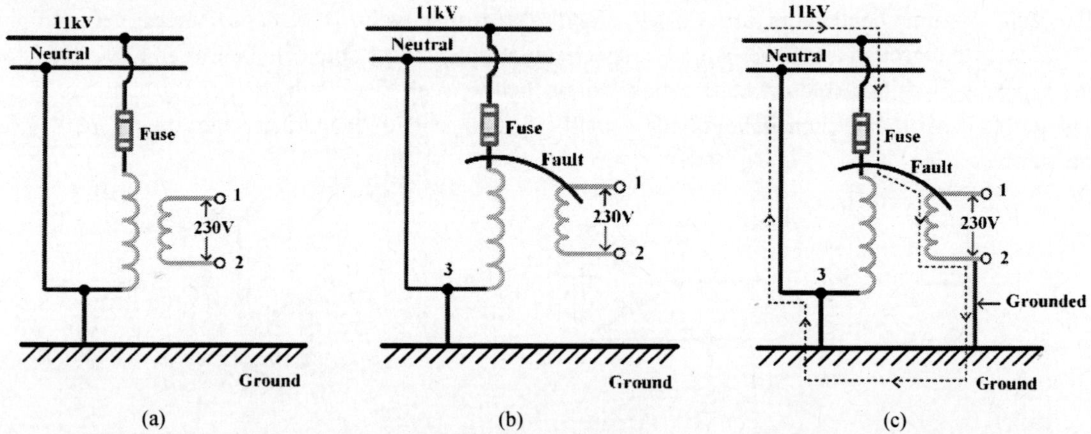

Fig. 8.20 (a) Schematic representation of 11kV/230V Distribution Transformer, (b) while dead short circuit occurred between both primary and secondary without ground connection (c) while dead short circuit occurred between both primary and secondary with ground connection

For this, consider a 11kV/230V distribution transformer with its primary linked between the 11kV line and neutral. The secondary of the transformer is left open, and no ground connection is supplied, as shown in Fig. 8.20(a). If a branch of a tree or pole falls on the 11kV main transformer and the 230V secondary line, a dead short circuit occurs between the primary and secondary windings. Under this scenario, a very high voltage is applied between the secondary of transformer wires and ground. This is due to the presence of a capacitance between the secondary of transformer conductors and ground, as illustrated in Fig. 8.20(b). Because the transformer secondary insulation is only built for 230V, the high voltage between the transformer secondary conductors and ground causes the insulation to be ruptured instantly. This circumstance may result in a major flash over, posing a fire threat on the secondary side of the transformer.

As a result, in this scenario, an ungrounded transformer secondary is dangerous and may cause serious fires. To address the aforementioned issue, the transformer secondary is grounded to earth, as illustrated in Fig. 8.20(c). Even if a dead short circuit exists between the primary and secondary circuits, short circuit current is recycled throughout the system via ground. As a result, the short circuit current will blow off the fuse in that path, isolating the distribution transformer from the 11kV power line.

8.15 Ungrounded (or) Isolated Neutral System:

A power system network is referred to as an ungrounded or isolated neutral system if its neutral is not connected to the ground. This system is also known as the free neutral system or the isolated system. Before the 1950s, many electrical systems did not possess neutral grounding. The primary argument in favour of such a system is the ability to maintain supply continuity in the event that one phase becomes grounded until the faulty line can be disconnected and repaired. This is especially true in an overhead transmission system, where a ground fault on one line is unlikely to propagate to two or more other lines. However, such systems encounter frequent arcing grounds.

Figure 8.21(a) depicts a schematic arrangement of a 3-Ø, 50 Hz AC system with an isolated neutral system. There is a capacitance between line-to-line conductors and line-to-ground. The capacitance between line-line conductors are C_{RY}, C_{YB} and C_{BR} respectively. The capacitances between line conductors and ground are C_R, C_Y and C_B respectively. Capacitances connected between line conductors are typically in delta fashion, whereas those connected between line conductors and ground are in star fashion. In general, line capacitances have a small effect on grounding characteristics, therefore they can be ignored.

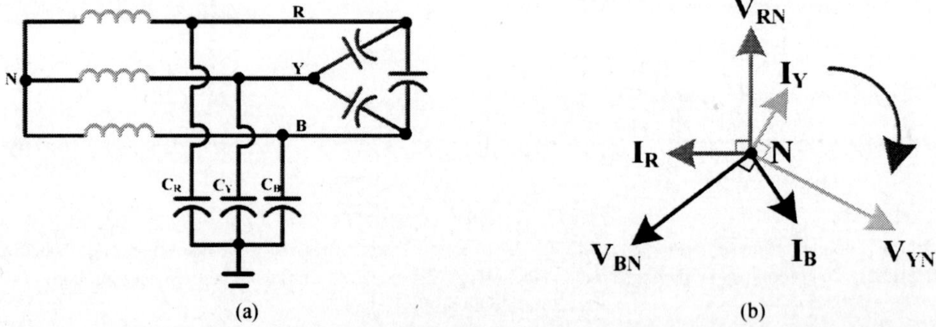

Fig. 8.21 Ungrounded Neutral system: (a) Schematic representation, (b) Phasor diagram when system is at healthy condition.

Consider the 3-Ø transmission line, which is perfectly transposed. i.e. the inductance and capacitance values are the same across the line. This indicates that the capacitance between each line and ground will be the same ($C_R = C_Y = C_B$). Because of the capacitance between each line and ground, a phase voltage appears across each line to ground, represented by V_{RN}, V_{YN} and V_{BN}, causing a capacitive charging current (phase current) to flow in each line to ground, represented by I_{CR}, I_{CY} and I_{CB}, respectively. The phase voltages V_{RN}, V_{YN} and V_{BN} will have equal magnitude and are displaced by 120^0 from each other.

The capacitive charging currents for each line to ground capacitor lead their respective phase voltages by 90^0 and are equal.

$$V_{RN} = V_{YN} = V_{BN} = V_{Ph} \qquad \text{.........(8.32)}$$

$$I_{CR} = \frac{V_{RN}}{X_{CR}} = \frac{V_{Ph}}{X_C}$$

$$I_{CY} = \frac{V_{YN}}{X_{CY}} = \frac{V_{Ph}}{X_C}$$

$$I_{CB} = \frac{V_{BN}}{X_{CB}} = \frac{V_{Ph}}{X_C}$$

$$I_{CR} = I_{CY} = I_{CB} = \frac{V_{Ph}}{X_C} \qquad \text{.........(8.33)}$$

Where, V_{Ph} is the phase voltage and X_C is the capacitive reactance between each line to ground and have same values.

Under normal balanced operating condition, the capacitive charging currents I_{CR}, I_{CY} and I_{CB} are equal in magnitude and displaced by 90^0 to their respective phase voltages V_{RN}, V_{YN} and V_{BN} as shown in Fig. 8.21(b). The algebraic sum of the capacitive charging currents I_{CR}, I_{CY} and I_{CB} is zero and hence no current is passing through the ground.

$$\overrightarrow{I_{CR}} + \overrightarrow{I_{CY}} + \overrightarrow{I_{CB}} = 0 \qquad \text{.........(8.34)}$$

As a result, the neutral potential is the same as the ground potential. As a result, the isolated neutral system causes no problems under balanced working conditions. However, this method will cause some issues in faulty conditions because the operational quantities voltage or current are significantly affected by the fault level.

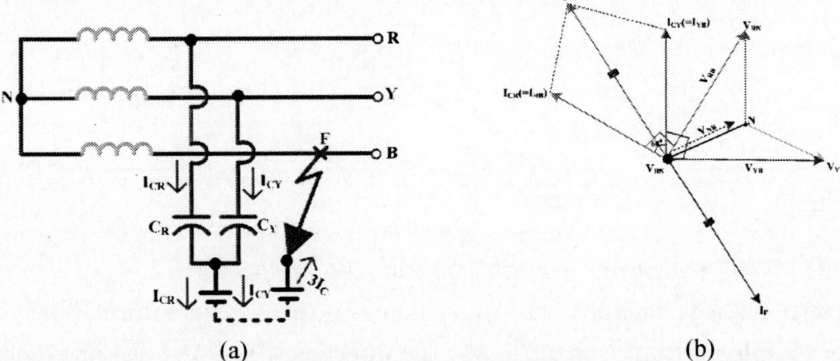

Fig. 8.22 (a) Representation of Ungrounded Neutral system when L-G fault on B-Phase (b) Phasor diagram for fault on B-phase

Let us now consider a single -line to ground fault occurs in any one of the phases. Suppose, line to ground fault occurs in phase B at fault point F as shown in Fig. 8.22(a). Under this condition, the capacitance C_B which is in between phase B to ground will be short circuited results in the voltage of the faulty phase B to be zero ($V_{BN} = V_{Ph} = 0$). i.e. the faulty phase B is held at ground potential whereas, other two healthy phases R and Y voltages are increased from phase values to line values. As the phase B is held at ground potential, the capacitance C_R connection may come in between R-phase and B-phase and because of this the current passing through capacitor C_R is I_{RB}. Similarly, the current passing through capacitor C_Y is I_{YB}. Because of these two currents, voltage will be appeared between R-phase and B-phase i.e. (V_{RB})

and Y-phase and B-phase i.e. (V_{YB}). As the circuit behaviour is pure capacitive, the capacitive charging currents I_{RB} and I_{YB} lead their respective voltages V_{RB} and V_{YB} by 90^0 as shown in Fig. 8.22(b). Therefore, these capacitive charging currents I_{RB} and I_{YB} become unbalanced and circulate from each phase capacitance (C_R & C_Y) to into the fault point and faulty phase through ground.

Thus, the fault current will have two components i.e. I_{RB} and I_{YB}. The phasor sum of these two-charging current will give the fault current.

The capacitive charging currents I_{RB} and I_{YB} can be expressed as

Current in Healthy phases $\overrightarrow{I_{RB}} = \frac{\overrightarrow{V_{RB}}}{X_C} = \frac{\sqrt{3}V_{Ph}}{X_C} = \sqrt{3}I_{CR}$ ……..(8.35)

$$\overrightarrow{I_{YB}} = \frac{\overrightarrow{V_{YB}}}{X_C} = \frac{\sqrt{3}V_{Ph}}{X_C} = \sqrt{3}I_{CY} = \sqrt{3}I_{CR} \quad \text{……..(8.36)}$$

The fault current (I_F) can be expressed at fault point F, as $\overrightarrow{I_F} = \overrightarrow{I_{RB}} + \overrightarrow{I_{YB}}$ ……..(8.37)

From phasor diagram, I_{RB} and I_{YB} are displaced by 60^0.

The phasor sum of vectors $\overrightarrow{I_{RB}}$ *and* $\overrightarrow{I_{YB}}$ will gives the fault current $\overrightarrow{I_F}$

According to the parallelogram theorem,

$$\overrightarrow{I_F}^2 = \overrightarrow{I_{RB}}^2 + \overrightarrow{I_{YB}}^2 + 2\,\overrightarrow{I_{RB}}\,\overrightarrow{I_{YB}}\,cos60^0$$

$$\overrightarrow{I_F}^2 = \overrightarrow{I_{RB}}^2 + \overrightarrow{I_{YB}}^2 + 2\,\overrightarrow{I_{RB}}\,\overrightarrow{I_{YB}} \cdot \frac{1}{2}$$

$$\overrightarrow{I_F}^2 = \overrightarrow{I_{RB}}^2 + \overrightarrow{I_{RB}}^2 + \overrightarrow{I_{RB}}\,\overrightarrow{I_{RB}}$$

$$\overrightarrow{I_F}^2 = 3\overrightarrow{I_{RB}}^2$$

$$\overrightarrow{I_F} = \sqrt{3}\,\overrightarrow{I_{RB}}$$

$$\overrightarrow{I_F} = \sqrt{3}\,(\sqrt{3}I_{CR})$$

Current in faulty phase $\overrightarrow{I_F} = 3\,I_{CR} = 3\frac{V_{Ph}}{X_C}$ ……..(8.38)

The phasor diagram for fault on B-phase can be drawn using following steps:

Step 1: Draw the three phase voltage phasors V_{RN}, V_{YN}, *and* V_{BN} (phase values) displaced by 120^0 apart.

Step 2: Draw the V_{BN} phasor (faulty phase) in reverse direction to the V_{BN} phasor to represent that respective phase voltage is zero.

Step 3: Draw the healthy phase line voltages V_{RB} and V_{YB} (line-line values) using phase voltages V_{RN}, V_{YN}, *and* V_{BN} respectively under fault condition.

Step 4: The phase current and line currents are same in the healthy phases *R* and *Y* due to the phase *B* at ground potential. i.e. $I_{CR} = I_{RB}$ and $I_{CY} = I_{YB}$.

Step 5: Draw the currents (I_{CR} and I_{CY}) with respect to the line voltages (V_{RB} and V_{YB}) by 90^0 leading. i.e. I_{CR} leads V_{RB} by 90^0 and I_{CY} leads V_{YB} by 90^0 respectively.

Step 6: Draw the total capacitive charging current ($I_C = I_{CR} + I_{CY}$) in the system with the help of I_{CR} and I_{CY}.

The following conclusions are drawn from the Ungrounded (or) Isolated Neutral System:

Due to line to ground fault,

1. The voltage of the faulty phase will be zero ($V_{fault} = 0$).
2. The voltages of the healthy phases will be $\sqrt{3}$ times of the phase voltages or line voltages. ($V_{healthy} = \sqrt{3}\, V_{Ph} = V_L$).
3. The capacitive charging currents of healthy phases are increased to $\sqrt{3}$ times of capacitive charging currents which are passing during normal operating condition ($I_{healthy} = \sqrt{3}\, I_C$).
4. The charging current or fault current passing through faulty phase will be 3 time of captative charging current.
5. As the voltage of the healthy phases increases more than their rated values, the breakdown of insulation of respective phases may happen.
6. It offers negligible interference with the neighbouring communication lines as the fault current is small.
7. The arcing grounds are formed due to the fault current (> 4A) is flowing through the ground. The faulty current (> 4A) is sufficient to ionize the air and maintains the arc even though the fault is cleared.
8. The switching transients comes in to the system during fault isolation which will increase the faulty phase voltage 4 to 6 times of normal operating voltage.

Advantages of Neutral Grounding:

- ✓ Voltages of healthy phases are limited to normal values.
- ✓ Surge voltages due to arcing grounds are eliminated.
- ✓ The life span of insulation increases and hence need less maintenance.
- ✓ Continuity of supply is increased.
- ✓ As the fault current is more, earth fault relays can easily detect the faults.
- ✓ The over voltages due to lightning can be easily discharged to ground.
- ✓ The induced charges can easily discharge to ground and hence does not produce any disturbance.
- ✓ Provides safety to the personnel and equipment.

8.16 Neutral Grounding:

Neutral Grounding in a power system is the process of connecting neutral point of 3-Ø star connected system to the ground (earth). The neutral point of the system can either directly connected to the ground or connected to the ground by means of some electrical elements such as resistor, reactor etc.

8.16.1 Types of Neutral Grounding:

Depending on how the neutral point of the system is connected to the ground, the neutral grounding can be classified as:

1. Solid (or) Effective grounding.
2. Resistance grounding.
3. Reactance grounding.
4. Arc suppression coil (or) Peterson's coil grounding.

8.17 Solid (or) Effective grounding:

Solid Grounding (or) Effective grounding in a power system involves connecting the neutral point of a 3-Ø star connected system directly to the ground (earth) via a rod or conductor with a negligible resistance and reactance. Previously, this type of grounding was known as solid grounding, but the IEEE renamed it to effective grounding.

According to the IEEE standards, a system is called as Effective grounding system if the ratio of zero sequence reactance to positive sequence reactance is less than or equal to 3 ($\frac{X_0}{X_1} \leq 3$) and the value of zero sequence resistance to positive sequence resistance is less than 1 ($\frac{R_0}{R_1} < 1$).

A schematic arrangement of 3-Ø, 50 Hz AC system with Solid (or) Effective grounding is shown in Fig. 8.23. In this, the neutral point of the system is directly connected to the ground by means of a metallic conductor with a negligible resistance. Each phase will have line to ground capacitors as C_R, C_Y and C_B respectively. This type of grounding system is less expensive as compared to other type of grounding systems because in the event of line-ground fault, the maximum phase voltage of the healthy system will not exceed 80% of line-line voltage. But in other type of grounding system, the voltage of the healthy phase will increase to 100% of line-line value. Under normal operating conditions, the algebraic sum of the currents in the 3-Ø system is zero and hence the neutral of the system is at ground potential only.

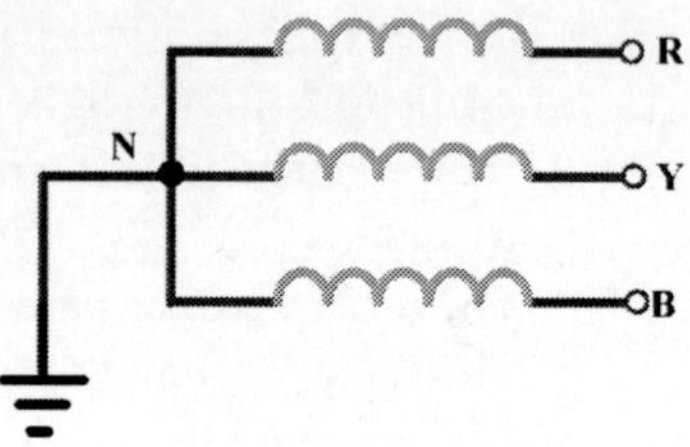

Fig. 8.23 Solid Grounding

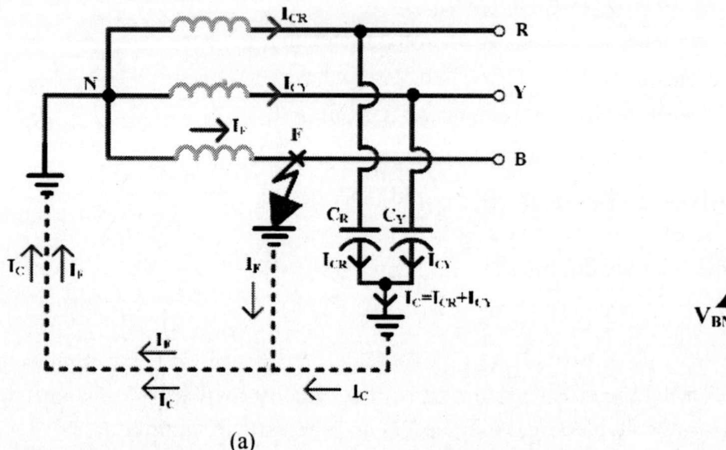

(a)

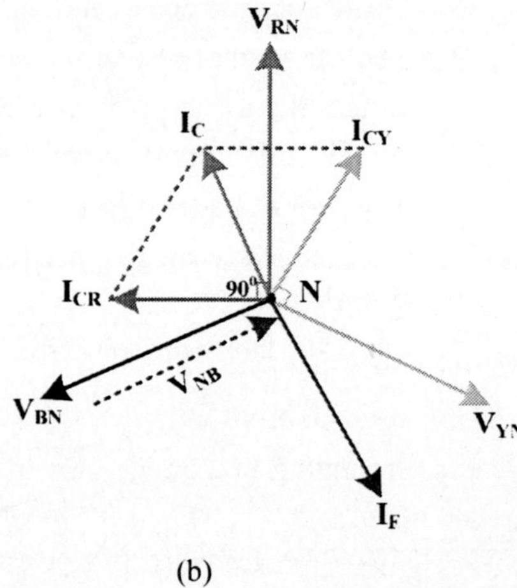

(b)

Fig. 8.24 (a) Schematic representation of Solid Grounding System when L-G fault on B-Phase (b) Phasor diagram for fault on B-phase

Let us now consider a single line to ground fault occurs in any one of the phases. Suppose, a line to ground fault occurs on phase B at fault point F as shown in Fig. 8.24 (a). Which means, the phase B and neutral of the system is at ground potential only. As a result of this fault, the faulty phase to ground voltage will be zero i.e. $V_{BN} = V_{Ph} = 0$ and the remaining two phases will be operated at normal phase voltage values only. Hence the neutral point will not shift in this case. The currents passing through the phases R and Y are I_{CR} and I_{CY} respectively. These currents are pure capacitive nature and the total capacitive charging current in the circuit is $I_C = I_{CR} + I_{CY}$ which will circulate in the system by means of ground and neutral of the system. In addition to the capacitive charging current, the fault on phase B also supplies fault current (I_F) which will circulate in the respective phase through ground and system neutral. The circuit path which will offer the impedance for fault current (I_F) is purely inductive because majority circuits including machines, transformers, and transmission lines are having high inductance with negligible resistance. Hence, the fault current (I_F) lags the respective faulty phase voltage by 90^0. This lagging fault current (I_F) will neutralize the total capacitive charging current and thereby arcing ground phenomena is minimized. The phasor diagram for solid grounding is shown in Fig. 8.24 (b).

The phasor diagram for solid grounding can be drawn using following steps :

Step 1: Draw the three phase voltage phasors (V_{RN}, V_{YN}, *and* V_{BN}) displaced by 120^0 apart.

Step 2: Draw the V_{BN} phasor (faulty phase) in reverse direction to the V_{BN} phasor to represent that respective phase voltage is zero.

Step 3: Draw the phase currents (I_{CR} and I_{CY}) with respect to the phase voltages (V_{RN} and V_{YN}) by 90^0 leading. i.e. I_{CR} leads V_{RN} by 90^0 and I_{CY} leads V_{YN} by 90^0 respectively.

Step 4: Draw the total capacitive charging current ($I_C = I_{CR} + I_{CY}$) in the system with the help of I_{CR} and I_{CY}.

Step 5: Draw the fault current (I_F) lagging the respective faulty phase voltage (V_{BN}) by 90^0.

Step 6: The phasor I_F will be in phase opposition to the total capacitive charging current (I_C) and fully neutralise it.

Advantages of solid grounding system:

- ✓ When a line to ground fault occurs in any of the systems phases, the voltage to earth of the faulty phase drops to zero. However, the healthy phases continue to operate at their regular phase voltages. Thus, equipment should only be insulated for phase voltages.
- ✓ The flow of heavy fault current (I_F) neutralizes the effect of total capacitive charging current (I_C) at fault and so the risk of arcing ground phenomena and overvoltage minimized.
- ✓ The flow of heavy fault current permits the use of discriminative protective gear.
- ✓ The relaying for ground faults is very simple and satisfactory.

Disadvantages of solid grounding system:

- ✗ The fault current level is very high and that may lead to burning of the circuit breaker contacts.
- ✗ Increased earth fault current results in greater influence on neighbouring communication circuits.
- ✗ The ground fault current is very high. Sometimes it will be more than 3-Ø short circuit current.
- ✗ As the ground fault current is large, danger to the personnel in the vicinity of the fault is high.

Applications of solid grounding system:

1. This type of grounding can be used for the systems which will have high circuit impedance.
2. It can be implemented for the systems which are operating at voltage less than 3.3 kV and more than 33 kV.

8.18 Resistance Grounding:

To overcome the drawback of high earth fault current in the solid grounding, resistance grounding can be employed. Resistance Grounding in a power system is the process of connecting neutral point of 3-Ø star connected system to the ground (earth) by means of a resistor with a considerable value of resistance. The value of resistor is selected in such a way that it should be neither low nor high. If it is low, then that type of grounding is very similar to solid grounding. If it is high, then that type of grounding is very similar to isolated neutral system. Usually, two types of resistors are used for this type of grounding. For voltage rating more than 6.6 kV metallic resistors are used for and less than 6.6 kV liquid resistors are used. Hence, the selection of resistor will be critical. A schematic arrangement of 3-Ø, 50 Hz AC system with Resistance Grounding is shown in Fig. 8.25 and 8.26. Each phase will have line to ground capacitors C_R, C_Y and C_B respectively. Under normal operating conditions, the algebraic sum of the currents in the 3-Ø system is zero and hence the neutral of the system is at ground potential only.

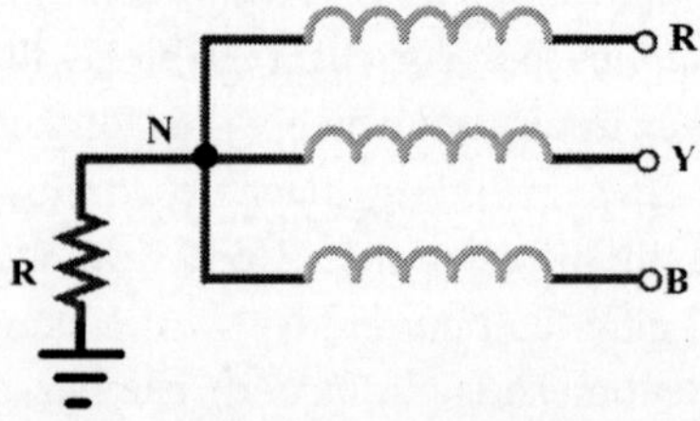

Fig. 8.25 Resistance Grounding System

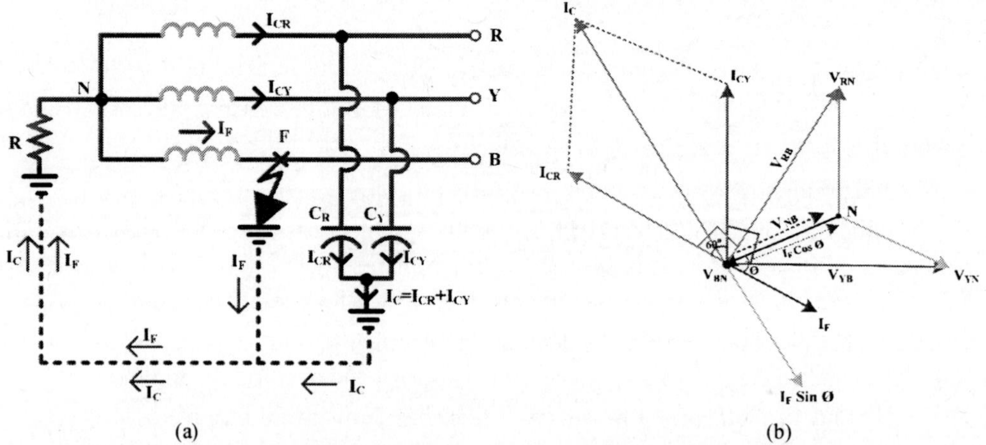

Fig. 8.26 (a) Resistance Grounding System when L-G fault on B-Phase (b) Phasor diagram for fault on B-phase

Suppose, a line to ground fault occurs on phase *B* at fault point F as shown in Fig. 8.26(a). This makes phase *B* coil and grounding resistance R in parallel resulting in voltage V_{NB} (negative of V_{BN}) appearing across this parallel combination. This shifts the neutral with its voltage equal to V_{NB} instead of zero potential. The voltage of healthy phase *R* is the phasor sum of voltage across the *R* coil and the voltage between neutral and ground which becomes equal to line voltage V_{RB} $[(V_{RN}+V_{NB}) = (V_{RN}-V_{BN}) = V_{RB}]$. Similarly, the voltage of healthy phase *Y* becomes equal to line voltage V_{YB}. This is illustrated with phasor diagram shown in Fig. 8.26(b). The currents passing through the healthy phases *R* and *Y* are I_{CR} and I_{CY} respectively. These currents are pure capacitive nature and the total capacitive charging current in the circuit is $I_C = I_{CR} + I_{CY}$ which will circulate in the system by means of ground and neutral of the system.

In addition to the capacitive charging current, the fault on phase B also supplies fault current (I_F) which will circulate in the respective phase through ground and system neutral. The circuit path which will offer impedance to this fault current (I_F) is not pure inductive because a resistor is also included in that path. Hence, the fault current (I_F) lags the respective phase voltage by some angle Ø and its value depends on the value of resistance. The fault current (I_F) can be resolved into two components as $I_F cos\ \emptyset$ which is in phase with the faulty phase voltage and $I_F sin\ \emptyset$ which is lagging the faulty phase voltage by 90^0. The lagging fault current component $I_F sin\ \emptyset$ is in phase opposition to the capacitive charging current. The value of R is so adjusted that, the component $I_F sin\ \emptyset$ will neutralize the total capacitive charging current and thereby arcing ground phenomena is reduced. The phasor diagram for Resistance grounding is shown in Fig. 8.26(b)

The phasor diagram for Resistance grounding can be drawn using following steps :

Step 1: Draw the three phase voltage phasors V_{RN}, V_{YN}, *and* V_{BN} (phase values) displaced by 120^0 apart.

Step 2: Draw the V_{NB} phasor (faulty phase) in reverse direction to the V_{BN} phasor.

Step 3: Draw the healthy phase line voltages V_{RB} and V_{YB} (line-line values) using phase voltages V_{RN}, V_{YN}, *and* V_{BN} respectively under fault condition.

Step 4: Draw the currents (I_{CR} and I_{CY}) with respect to the line voltages (V_{RB} and V_{YB}) by 90^0 leading. i.e. I_{CR} leads V_{RB} by 90^0 and I_{CY} leads V_{YB} by 90^0 respectively.

Step 5: Draw the total capacitive charging current ($I_C = I_{CR} + I_{CY}$) in the system with the help of I_{CR} and I_{CY}.

Step 6: Draw the fault current (I_F) lagging the respective faulty phase voltage (V_{NB}) by some angle Ø.

Step 7: Draw the phasors $I_F cos\ \emptyset$ and $I_F sin\ \emptyset$ by resolving the fault current (I_F) on to: one is in phase with the faulty phase voltage and other is lagging the faulty phase voltage by 90^0.

Step 8: The phasor $I_F sin\ \emptyset$ will be in phase opposition to the total capacitive charging current (I_C) and neutralise charging current.

Advantages of Resistance grounding system:

- ✓ It permits the use of discriminative protective gear.
- ✓ It minimizes the hazard of arcing grounds.
- ✓ The fault current (I_F) is less than the effective grounding and hence less interference on neighbouring communication circuits.
- ✓ The power dissipated in the grounding resistor will improve the system stability because power dissipation reduces the accelerating power.
- ✓ Transient current faults are converted in to controlled current faults.

Disadvantages of Resistance grounding system:

- ✗ There is enormous energy loss in the neutral grounding resistor for dissipation of fault energy.
- ✗ The system neutral is shifted during fault and hence voltage of the healthy phases will become greater than the 0.8 times of line value. This will need more insulation requirement and large rating of surge arrester which incur more cost.
- ✗ It is costlier than solid grounding system.

Applications of Resistance grounding system: It can be employed for the systems with operating voltage rating in between 3.3-33 kV and power rating greater than 500 kVA.

8.19 Reactance Grounding:

Reactance Grounding in a power system is the process of connecting neutral point of 3-Ø star connected system to the ground (earth) by means of a reactor with a negligible value of resistance. i.e. the impedance between neutral point and ground is predominantly reactance. As per the IEEE standard, the value of $\frac{X_0}{X_1}$ is greater than 3 for the reactance grounding. The reactive grounding lies between solid grounding and resonant grounding. The value of reactance in this grounding is selected in such a way that, the earth fault current will keep in safe limit. A schematic arrangement of 3-Ø, 50 Hz AC system with Reactance Grounding is shown in Fig. 8.27. In this type of grounding, the neutral point of the system is connected to the ground by means of a reactor which will have considerable value of reactance and negligible value of resistance. Each phase will have line to ground capacitors C_R, C_Y and C_B respectively. Under normal operating conditions, the algebraic sum of the currents in the 3-Ø system is zero and hence the neutral of the system is at ground potential only.

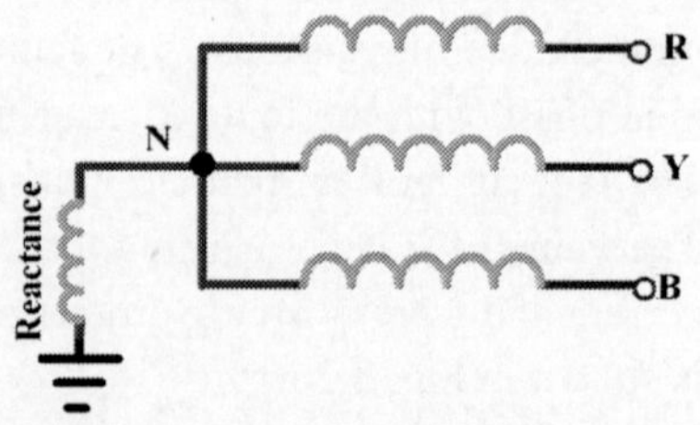

Fig. 8.27 Reactance Grounding

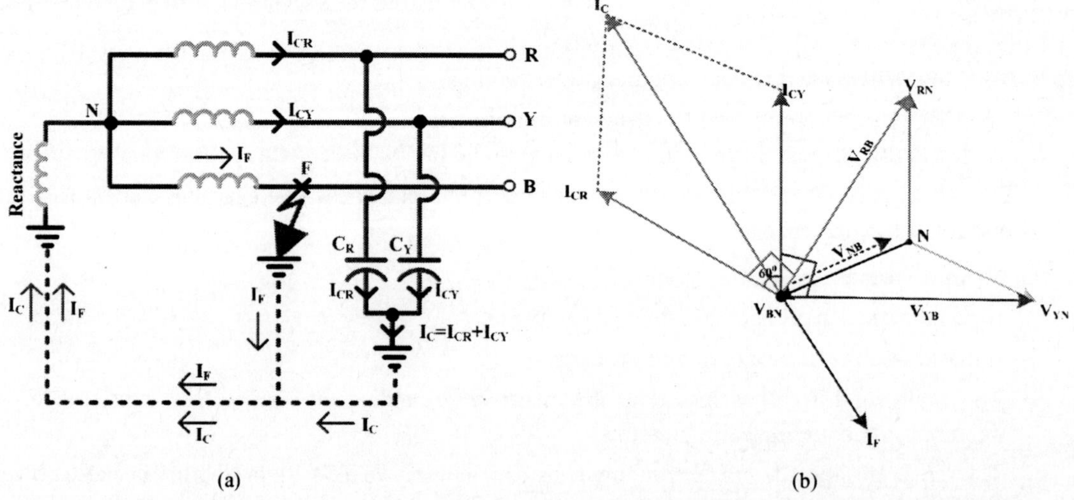

Fig. 8.28 (a) Reactance Grounding System when L-G fault on B-Phase (b) Phasor diagram for fault on B-phase

Suppose, a line to ground fault occurs on phase *B* at fault point *F* as shown in Fig. 8.28(a). This makes phase *B* coil and grounding reactance in parallel resulting in voltage V_{NB} (negative of V_{BN}) appearing across this parallel combination. This shifts the neutral with its voltage equal to V_{NB} instead of zero potential. The voltage of healthy phase *R* is the phasor sum of voltage across the *R* coil and the voltage between neutral and ground which becomes equal to line voltage V_{RB} $[(V_{RN}+V_{NB}) = (V_{RN}-V_{BN}) = V_{RB}]$. Similarly, the voltage of healthy phase *Y* becomes equal to line voltage V_{YB}. This is illustrated with phasor diagram shown in Fig. 8.28(b).

The currents passing through the healthy phases R and Y are I_{CR} and I_{CY} respectively. These currents are of pure capacitive nature and the total capacitive charging current in the circuit is $I_C = I_{CR} + I_{CY}$ which will circulate in the system by means of ground and neutral of the system. In addition to the capacitive charging current, the fault on phase B also supplies fault current (I_F) which will circulate in the respective phase through ground and system neutral. The circuit path which will offer impedance to this fault current (I_F) is pure inductive because a reactor is also included in that path along with the line inductance. Hence, the fault current (I_F) lags the respective faulty phase voltage by 90^0. This lagging fault current (I_F) will neutralize the total capacitive charging current and thereby arcing ground phenomena is minimized. The fault current circulating in this grounding is more than resistance grounding but less than the solid grounding. The phasor diagram for reactance grounding is shown in Fig. 8.28(b).

The phasor diagram for fault on B-phase can be drawn using following steps:

Step 1: Draw the three phase voltage phasors V_{RN}, V_{YN}, *and* V_{BN} (phase values) displaced by 120^0 apart.

Step 2: Draw the V_{NB} phasor (faulty phase) in reverse direction to the V_{BN} phasor.

Step 3: Draw the healthy phase line voltages V_{RB} and V_{YB} (line-line values) using phase voltages V_{RN}, V_{YN}, *and* V_{BN} respectively under fault condition.

Step 4: Draw the currents (I_{CR} and I_{CY}) with respect to the line voltages (V_{RB} and V_{YB}) by 90^0 leading. i.e. I_{CR} leads V_{RB} by 90^0 and I_{CY} leads V_{YB} by 90^0 respectively.

Step 5: Draw the total capacitive charging current ($I_C = I_{CR} + I_{CY}$) in the system with the help of I_{CR} and I_{CY}.

Step 6: Draw the fault current (I_F) lagging the respective faulty phase voltage (V_{NB}) by 90^0.

Advantages of Reactance grounding system:

- ✓ It provides the satisfactory relaying.
- ✓ The earth fault current is less due to the presence of earthing reactor.
- ✓ Reduction in fault current causes the reduced interference with neighbouring communication lines.
- ✓ It has minimum losses during fault condition because no power dissipative element is used in the neutral to ground path.
- ✓ As the fault current is less, the arcing grounds are eliminated.
- ✓ Transient ground faults are converted in to controlled ground faults.

Disadvantages of Reactance grounding system:

- ✗ The voltage of the healthy phases will become greater than the 0.8 times of line value. This requires more cost for the insulation and surge arrester.
- ✗ For identical fault conditions, the protective device in this system requires a greater fault current than resistance grounding.
- ✗ It provides additional reactance to the system.
- ✗ This grounding will be subjected to severe transient voltages under fault conditions.

Applications of Reactance grounding system: **It can be employed for the systems which will have more capacitive currents. Ex: Transmission lines, cables, and synchronous condensers.**

8.20 Arc Suppression coil or Peterson coil or Resonant Grounding:

Arc Suppression coil or Resonant or Peterson coil Grounding is a special case of reactance grounding. In this type of grounding, a reactor with an iron core that is connected between the neutral point of 3-Ø star connected system and ground can be adjusted to match with the capacitance of the system in the event of a line-to-ground (L-G) fault.

It has been observed that, the capacitive currents are the cause of arcing grounds. In this case, the reactance value of reactor is adjusted in such a way that it will be equal to the capacitive reactance of the system i.e. $X_L = X_C$. This equalization of reactances leads to currents with equal magnitude but opposite phase in those paths (i.e. $I_L = I_C$) In this way, the capacitive charging currents are completely neutralized by the inductive current i.e. fault current (I_F).

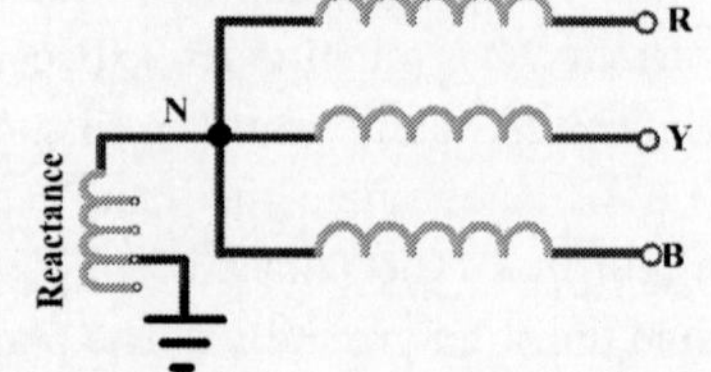

Fig. 8. 29 Arc Suppression coil

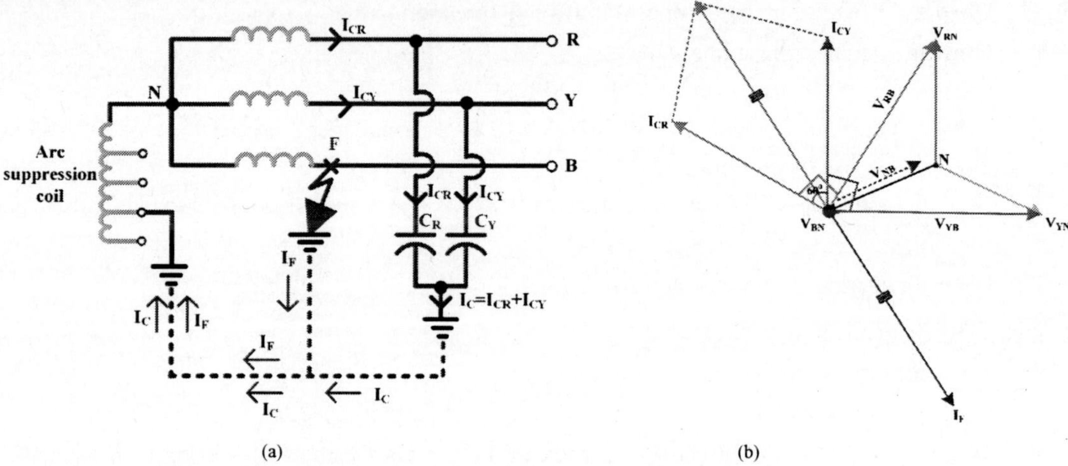

(a) (b)

Fig. 8.30 (a) Peterson coil Grounding System when L-G fault on B-Phase (b) Phasor diagram for fault on B-phase

A schematic arrangement of 3-Ø, 50 Hz AC system with Peterson coil Grounding is shown in Fig. 8.29 In this type of grounding, the neutral point of the system is connected to the ground by means of a reactor which will have the tappings to vary the reactance of the coil. Suppose, a line to ground fault occurs on phase *B* at fault point *F* as shown in Fig. 8.30(a). This makes phase *B* coil and grounding reactance in parallel resulting in voltage V_{NB} (negative of V_{BN}) appearing across this parallel combination. This shifts the neutral with its voltage equal to V_{NB} instead of zero potential. The voltage of healthy phase *R* is the phasor sum of voltage across the *R* coil and the voltage between neutral and ground which becomes equal to line voltage V_{RB} $[(V_{RN}+V_{NB}) = (V_{RN}-V_{BN}) = V_{RB}]$. Similarly, the voltage of healthy phase *Y* becomes equal to line voltage V_{YB}. This is illustrated with phasor diagram shown in Fig. 8.30(b).

The currents passing through the healthy phases *R* and *Y* are I_{CR} and I_{CY} respectively. These currents are pure capacitive in nature and the total capacitive charging current in the circuit is $I_C = I_{CR} + I_{CY}$ which will circulate in the system by means of ground and neutral of the system. In addition to the capacitive charging current, the fault on phase *B* also supplies fault current (I_F) which will circulate in the respective phase through ground and system neutral. The circuit path which will offer impedance to this fault current (I_F) is pure inductive because a reactor is also included in that path along with the line inductance. Hence, the fault current (I_F) lags the respective faulty phase voltage by 90^0. At this context, the tappings provided on the Peterson coil is adjusted in such a way that the fault current become equal to capacitive charging current. Hence the arcing grounds are completely neutralized. In the case of resonant grounding, the system exhibits characteristics similar to an ungrounded neutral system. However, Fig. 8.29(b) drawn for Reactance grounding shows capacitive charging not being fully neutralized, whereas, Fig. 8.30(b) illustrates complete neutralization of capacitive charging current by Peterson coil grounding. The phasor diagram for this grounding is similar to that of the reactance grounding and it is shown in Fig. 8.30(b).

Fault current (I_F) flowing through the Peterson coil is $I_F = \frac{V_{Ph}}{X_L}$ ....(8.39)

Where X_L is reactance of coil and V_{Ph} is the phase voltage.

The total charging current I_C is the phasor sum of I_{CR} and I_{CY}. During the Line to ground fault, healthy phase voltages are increased almost equal to the line values.

Hence, $$I_{CR} = I_{CY} = \frac{\sqrt{3}V_{Ph}}{X_C} \qquad ....(8.40)$$

From phasor diagram,

$$\overrightarrow{I_C}^2 = \overrightarrow{I_{CR}}^2 + \overrightarrow{I_{CY}}^2 + 2\,\overrightarrow{I_{CR}}\,\overrightarrow{I_{CY}}\,cos60^0$$

$$\overrightarrow{I_C}^2 = \overrightarrow{I_{CR}}^2 + \overrightarrow{I_{CY}}^2 + 2\,\overrightarrow{I_{CR}}\,\overrightarrow{I_{CY}}\cdot\frac{1}{2}$$

$$\overrightarrow{I_C}^2 = \overrightarrow{I_{CR}}^2 + \overrightarrow{I_{CY}}^2 + \overrightarrow{I_{CR}}\,\overrightarrow{I_{CR}} \qquad (\because \overrightarrow{I_{CR}} = \overrightarrow{I_{CY}})$$

$$\overrightarrow{I_C}^2 = 3\overrightarrow{I_{CR}}^2$$

$$\overrightarrow{I_C} = \sqrt{3}\,\overrightarrow{I_{CR}}$$

$$\overrightarrow{I_C} = \sqrt{3}\left(\frac{\sqrt{3}V_{Ph}}{X_C}\right) \qquad (\because I_{CR} = I_{CY} = \frac{\sqrt{3}V_{Ph}}{X_C})$$

$$\overrightarrow{I_C} = 3\frac{V_{Ph}}{X_C} \qquad ....(8.41)$$

At resonant condition, $I_F = I_C$

From (8.39) and (8.41) $\frac{V_{Ph}}{X_L} = 3\frac{V_{Ph}}{X_C}$

$$\Rightarrow \frac{1}{X_L} = \frac{3}{X_C}$$

$$\Rightarrow X_L = \frac{X_C}{3}$$

$$\Rightarrow 2\pi fL = \frac{1}{3}\cdot\frac{1}{2\pi fC} \qquad (\because X_L = 2\pi fL \; and \; X_C = \frac{1}{2\pi fC})$$

$$\Rightarrow L = \frac{1}{3}\cdot\frac{1}{(2\pi f)^2C}$$

$$\Rightarrow L = \frac{1}{3\omega^2C} \qquad (\because \omega = 2\pi f) \qquad ....(8.42)$$

∴ Reactance of the Peterson coil is, $X_L = \frac{X_C}{3}$

∴ Inductance of the Peterson coil is, $L = \frac{1}{3\omega^2C}$

Advantages of Arc Suppression coil grounding system:

- ✓ The grounding reactance can be varied.
- ✓ Arcing grounds can be completely eliminated.

Disadvantages of Arc Suppression coil grounding system:

- × The network's capacitance fluctuates periodically due to different operational conditions. Consequently, the inductance L of the Peterson coil necessitates the readjustment.
- × The lines need to be transposed.

Example 8.15. A 220kV, 3-Ø, 50 Hz transmission line having 200km length consists of three conductors of effective diameter of 30mm arranged in a vertical plane with 5m spacing and regularly transposed. Calculate the inductance and kVA rating of the Peterson's coil.

Ans. Given Data: Conductor diameter, D = 30mm

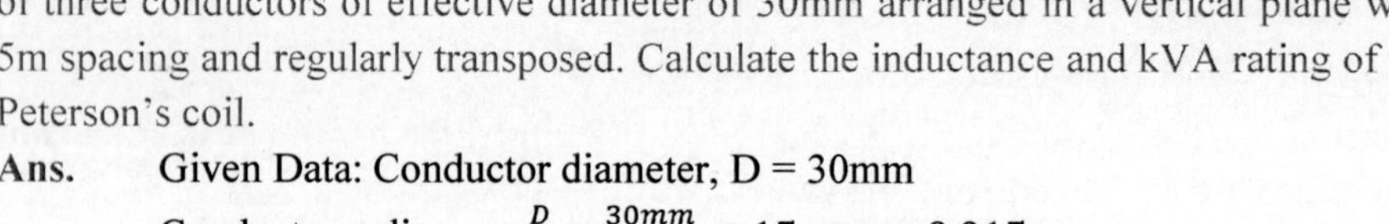

Conductor radius, $r = \frac{D}{2} = \frac{30mm}{2} = 15mm = 0.015m$

Conductor spacings, $D_{AB} = D_{BC}$ = 5m and D_{CA} = 10m

Frequency, f = 50 Hz

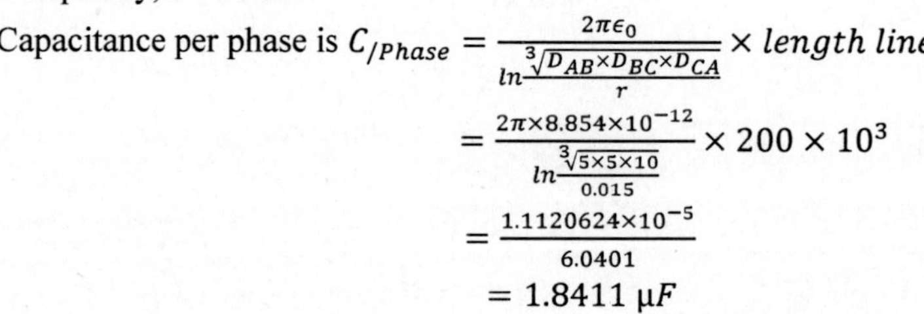

Capacitance per phase is $C_{/Phase} = \frac{2\pi\epsilon_0}{ln\frac{\sqrt[3]{D_{AB}\times D_{BC}\times D_{CA}}}{r}} \times length\ line$

$$= \frac{2\pi\times8.854\times10^{-12}}{ln\frac{\sqrt[3]{5\times5\times10}}{0.015}} \times 200 \times 10^3$$

$$= \frac{1.1120624\times10^{-5}}{6.0401}$$

$$= 1.8411\ \mu F$$

Inductance of Peterson Coil is $L = \frac{1}{3\omega^2 C} = \frac{1}{3\times(2\pi\times50)^2\times1.8411\times10^{-6}} = 1.8362\ H$

Fault Current is $I_L = \frac{V_{Ph}}{\omega L} = \frac{\frac{220\times10^3}{\sqrt{3}}}{2\pi\times50\times1.8362} = 220.298\ A$

Rating of suppressor coil is $= \frac{V_{Ph} I_L}{1000}\ kVA = \frac{220\times10^3}{\sqrt{3}} \times 220.298 \times \frac{1}{1000} = 27.9816\ MVA$

Example 8.16. A 50 Hz overhead transmission line has line to earth capacitance of 2 μF. it is decided to use an earth fault neutralizer. Determine the reactance to neutralize the capacitance of (i) 70% length of the line, (ii) 85% length of the line, and (iii) 100% length of the line.

Ans. Given Data: line to earth capacitance = 2 μF

(i) The inductive reactance of the coil for 70% length of the line is

Reactance of the coil, $\omega L = \frac{1}{3\omega C} = \frac{1}{3\times2\pi\times50\times2\times10^{-6}\times0.7} = 758.265\ \Omega$

(ii) The inductive reactance of the coil for 85% length of the line is

Reactance of the coil, $\omega L = \frac{1}{3\omega C} = \frac{1}{3\times2\pi\times50\times2\times10^{-6}\times0.85} = 624.453\ \Omega$

(iii) The inductive reactance of the coil for 100% length of the line is

Reactance of the coil, $\omega L = \frac{1}{3\omega C} = \frac{1}{3\times2\pi\times50\times2\times10^{-6}\times1} = 530.785\ \Omega$

Example 8.17. A 300 km long, 3-Ø, 50Hz, 110 kV transmission line has a capacitance of 0.05 μF per km per ph. Determine the inductance, inductive reactance, fault current, and kVA rating of an arc suppression coil suited for the line in order to eliminate the arcing ground phenomenon.

Ans. Given Data: $V_L = 110$ kV

f = 50 Hz

C = 0.05 μF/km/ph

Length of line = 300 km

Inductance of arc suppression coil is $L = \frac{1}{3\omega^2 C} = \frac{1}{3\times(2\pi\times 50)^2\times 0.05\times 10^{-6}\times 300\times 10^3} = 0.225\ mH$

Inductive reactance of the coil is $X_L = 2\pi fL = 2\times\pi\times 50\times 0.225\times 10^{-3} = 0.0706\ \Omega$

The fault current in the neutral is, $I_L = \frac{V_{Ph}}{\omega L} = \frac{\frac{110\times 10^3}{\sqrt{3}}}{2\pi\times 50\times 0.225\times 10^{-3}} = 899.554\ kA$

kVA rating of suppressor coil is $= \frac{V_{Ph} I_L}{1000} kVA = \frac{110\times 10^3}{\sqrt{3}} \times 899.554\times 10^3 \times \frac{1}{1000} = 57.129\ MVA$

Example 8.18. A 3-phase 300 km line has three conductors positioned at the corners of a triangle with sides of 1.5m, 3m, and 2.6m. Calculate the inductance, inductive reactance, fault current, and kVA rating of an arc suppression coil in a 3-phase 50Hz, 110 kV system with regularly transposed conductors. Each conductor has a diameter of 1.4 cm.

A
1.5m
2.6m
B
3m
C

Ans. Given Data: $V_L = 110$ kV, $r = \frac{d}{2} = \frac{1.4}{2} = 0.7cm = 0.007\ m$

Spacing b/w conductors D_{AB} = 1.5 m, D_{BC} = 3 m, and D_{CA} = 2.6 m

Equivalent Equilateral spacing, $D_{eq} = \sqrt[3]{D_{AB}\times D_{BC}\times D_{CA}}$

$= \sqrt[3]{1.5\times 3\times 2.6} = 2.27\ m$

We know that, the capacitance per phase is

$$C_{/Phase} = \frac{2\pi\epsilon_0}{ln\frac{\sqrt[3]{D_{AB}\times D_{BC}\times D_{CA}}}{r}} \times length\ of\ line$$

$$= \frac{2\pi\times 8.854\times 10^{-12}}{ln\frac{\sqrt[3]{1.5\times 3\times 2.6}}{0.007}} \times 300\times 10^3$$

$$= \frac{1.66809\times 10^{-5}}{5.7816} = 2.885\ \mu F$$

Inductance of Peterson Coil is $L = \frac{1}{3\omega^2 C} = \frac{1}{3\times(2\pi\times 50)^2\times 2.885\times 10^{-6}} = 1.171 H$

Inductive reactance of the coil is $X_L = 2\pi fL = 2 * \pi * 50 * 1.171 = 367.694\ \Omega$

The fault current in the neutral is, $I_L = \frac{V_{Ph}}{\omega L} = \frac{\frac{110\times 10^3}{\sqrt{3}}}{2\times\pi\times 50\times 1.171} = 172.721\ A$

kVA rating of suppressor coil is $= \frac{V_{Ph} I_L}{1000} kVA = \frac{110\times 10^3}{\sqrt{3}} \times 172.721 \times \frac{1}{1000} = 10969\ kVA$

Example 8.19. Calculate the inductance, inductive reactance, fault current, and kVA rating of an arc suppression coil for a 3-phase 50Hz, 220kV system 150km long and conductors are placed horizontally with spacing $D_{AB} = D_{BC}$ = 1.8m and D_{CA} = 3.6 m. The transposed conductors have a diameter of 2.2 cm.

Ans. $V_L = 220$ kV, $r = \frac{d}{2} = \frac{2.2}{2} = 1.1cm = 0.011m$

Length of Line = 150 km

Line has three conductors placed horizontally with spacing $D_{AB} = D_{BC} = 1.8$m and $D_{CA} = 3.6$ m.

A, B, C; 1.8m, 1.8m; 3.6m

Equivalent Equilateral spacing, $D_{eq} = \sqrt[3]{D_{AB} \times D_{BC} \times D_{CA}} = \sqrt[3]{1.8 \times 1.8 \times 3.6} = 2.267\ m$

Capacitance per phase is $C_{/Phase} = \frac{2\pi\epsilon_0}{ln\frac{\sqrt[3]{D_{AB} \times D_{BC} \times D_{CA}}}{r}} \times length\ of\ line$

$$= \frac{2\pi \times 8.854 \times 10^{-12}}{ln\frac{\sqrt[3]{1.8 \times 3.6 \times 1.8}}{0.011}} \times 150 \times 10^3$$

$$= \frac{8.340 \times 10^{-6}}{5.3283} = 1.565\ \mu F$$

Inductance of Peterson Coil is $L = \frac{1}{3\omega^2 C} = \frac{1}{3 \times (2\pi \times 50)^2 \times 1.565 \times 10^{-6}} = 2.16\ H$

Inductive reactance of the coil is $X_L = 2\pi fL = 2 \times \pi \times 50 \times 2.16 = 678.32\ \Omega$

The fault current in the neutral is, $I_L = \frac{V_{Ph}}{\omega L} = \frac{\frac{220 \times 10^3}{\sqrt{3}}}{2 \times \pi \times 50 \times 2.16} = 187.25\ A$

kVA rating of suppressor coil is $= \frac{V_{Ph} I_L}{1000} kVA = \frac{220 \times 10^3}{\sqrt{3}} \times 187.25 \times \frac{1}{1000} = 23783.94\ kVA$

8.21. Unit summary:

- Switchgear has two primary functions: facilitating maintenance by switching during regular operating conditions and safeguarding the electrical system by switching during abnormal operating conditions.
- There are two ways that can be employed for arc interruption: the High resistance technique and the Low resistance or current zero strategy.
- According to Slepain's Recovery rate theory, the termination of the arc takes place when the rate of increase in dielectric strength exceeds that of the restriking voltage.
- In Energy Balance theory, an arc is extinguished when the rate at which heat is lost between the contacts is greater than the rate at which heat is produced.
- Restriking voltage refers to the temporary voltage that is observed between the contacts when the current is at or close to zero during the arcing period.
- The recovery voltage refers to the voltage that is observed across the contacts of the breaker subsequent to the extinguishment of the arc.
- The maximum value of restriking voltage is $2E_{Peak}$. The Rate of Rise of Restriking Voltage is $\omega_n E_{Peak}$.
- Absence of transient disturbances is guaranteed when the resistance value connected across the circuit breaker's contacts is equal to or less than $\frac{1}{2}\sqrt{\frac{L}{C}}$.
- Air blast circuit breakers were previously utilized for interior services with voltages up to 15kV and a breaking capability of 2500 MVA, but they are currently employed outdoors for voltages up to 220kV.

- Bulk Oil CB. is suitable for voltages ranging from 2.5kV to 220kV and breaking capacities ranging from 25 MVA to 5000MVA. Minimum Oil CB is suitable for voltages ranging from 33kV to 220kV and breaking capacities ranging from 1500 MVA to 7500 MVA.
- SF6 circuit breakers are suitable for voltages ranging from 110kV to 220kV and breaking capacities ranging from 10 MVA to 20 MVA.
- Vacuum Circuit Breakers are suitable for voltages ranging from 11kV to 33kV and breaking capacities ranging from 60 MVA to 100 MVA.
- Earthing or Grounding in a power system is the process of connecting an electrical part of the system (i.e. neutral point in a star-connected 3-Ø system, current carrying conductive parts, etc.) or non-current carrying metallic parts of the system to the earth.
- Equipment grounding is the procedure by which non-current-carrying metallic components (i.e., the metal enclosure) of electrical equipment are connected to earth (i.e., the soil).
- System grounding is the procedure by which current-carrying electric parts (i.e., star connected neutral point of the system or any one conductor in the transformer secondary) of the system is connected to earth (i.e., the soil).
- During Line-Ground fault condition of an ungrounded or isolated neutral system, the voltages of the health phases will be $\sqrt{3}$ times of the phase voltages or line voltages. ($V_{healthy} = \sqrt{3}\, V_{Ph} = V_L$).
- During Line-Ground fault condition of an ungrounded or isolated neutral system, the capacitive charging currents of health phases are increased to $\sqrt{3}$ times of capacitive charging currents ($I_{healthy} = \sqrt{3}\, I_C$) and the current passing through the faulty phase will be 3 time of captative charging current ($I_{Faulty} = 3\, I_C$).
- Neutral Grounding in a power system is the process of connecting neutral point of 3-Ø star connected system to the ground (earth).
- Neutral Grounding can be classified in to Solid (or) Effective grounding, Resistance grounding, Reactance grounding, and Arc suppression coil (or) Peterson's coil grounding.
- According to the IEEE standards, a system is called as Effective grounding system if the ratio of zero sequence reactance to positive sequence reactance is less than or equal to 3 ($\frac{X_0}{X_1} \leq 3$) and the value of zero sequence resistance to positive sequence resistance is less than 1 ($\frac{R_0}{R_1} < 1$).
- Solid grounding system is less expensive as compared to other type of grounding systems because in the event of line-ground fault, the maximum phase voltage of the healthy system will not exceed 80% of line-line voltage.
- Solid grounding can be employed for the systems which will have high circuit impedance.
- Resistance Grounding in a power system is the process of connecting neutral point of 3-Ø star connected system to the ground (earth) by means of a resistor with a considerable value of resistance.
- In Resistance Grounding the system neutral is shifted during fault and hence voltage of the healthy phases will become greater than the 0.8 times of line value.
- As per the IEEE standard, the value of $\frac{X_0}{X_1}$ is greater than 3 for the reactance grounding.

Short and Long Answer Questions

1. What is switchgear? Enumerate the fundamental characteristics of a switchgear panel.
2. Enumerate frequently employed insulating materials utilized in circuit breakers.
3. Discuss the procedures employed for Arc extinction (or) Arc Interruption using clear illustrations.
4. What are the fundamental prerequisites for DC circuit breaking? What issues are commonly linked with direct current (DC) circuit breakers?
5. What distinguishes an isolator from a fuse and a circuit breaker?
6. What is arc? What variables drive the ARC phenomenon? What is the purpose of the arc chute?
7. Define Arc voltage, Restriking voltage and Recovery voltage.
8. Explain how the arc is initiated and sustained in a circuit breaker when the circuit breaker contacts separated with neat sketches.
9. Using schematic diagrams, calculate the expressions for the following: restriking voltage, maximum restriking voltage, RRRV, and maximum RRRV.
10. Describe the following arc interruption procedures in an alternating current system using schematic diagrams: (i) Slepain's theory (ii) Cassie's theory
11. What is current chopping? Explain it with neat diagrams.
12. Demonstrate how to reduce restriking transients by connecting a resistance across CB contacts and calculating critical resistance based on system inductance and capacitance, resulting in no transient oscillations.
13. Compare arc rupture between oil circuit breakers and ABCB (air blast circuit breakers). Summarize the relative advantages and disadvantages of each type of switch gear.
14. Compare and contrast the arc rupture in any two of the following circuit breakers
    (a) Bulk Oil Vs Minimum Oil circuit breakers.
    (b) Axial blast Vs Cross blast Vs Radial blast circuit breakers.
    (c) ABCB Vs SF6 circuit breakers.
15. Describe the construction, operation, benefits, drawbacks, and uses of SF6 circuit breakers. Explain the advantages of SF6 CB against air blast CB.
16. Explain the vacuum circuit breaker's construction, principle of operation, advantages, disadvantages and applications.
17. What is the purpose of earthing? Distinguish between System earthing and Equipment earthing.
18. Using a phasor diagram and mathematical calculations, illustrate the disadvantages of isolated neutral.
19. What is Solid Grounding? Why is it termed effective grounding?
20. Using a schematic and phasor diagram, explain how resistance and reactance grounding operate. Which strategy is most effective, and why?
21. Explain the working of reactance and arc suppression coil grounding using a schematic and phasor diagram. Which strategy is more effective, and why?
22. Explain how resistance grounding works using a schematic and phasor diagram. Explain how the resistance value should be chosen to protect the power system against faults.

23. Explain how reactance grounding works using a schematic and phasor diagram. Explain how to choose the reactance value to safeguard the power system from faults.
24. Explain resonant grounding using a phasor diagram and mathematical calculations. Highlight its benefits.
25. Discuss the benefits, drawbacks, and uses of solid, resistance, reactance, and arc suppression coil grounding techniques.

Exercises

1. A 33kV, 50 Hz system has a reactance of 5Ω and capacitance of 0.025μF up to the fault location. Calculate the frequency of transient oscillation, the maximum restriking voltage, and the maximum RRRV.
2. Determine the circuit's natural frequency and the average rate of rise of the restriking voltage when the initial peak is reached in 55μsec and the peak voltage is 110kV.
3. A 50Hz 3-φ alternator has 6mH inductance per phase and 0.0246μF capacitance to ground between the alternator and the CB During a fault, the breaker opens when the peak rms current reaches 12kA. Determine the frequency of oscillations, active recovery voltage, time for maximum RRRV, and maximum RRRV.
4. For a 50 Hz generator with 11kV (rms) emf to neutral, 9Ω reactance of generator and linked system, 0.045μF distributed capacitance to neutral, and negligible resistance. Determine the maximum restriking voltage, the frequency of the transient oscillation, and the average rate of voltage rise up to the oscillation's first peak.
5. A 50Hz 3-φ generator with 5mH inductance per phase is connected to the bus-bars via an oil circuit breaker. The generator-breaker circuit has a capacitance to ground of 0.0125μF per phase. When the rms current surpasses 5kA, the breaker opens due to a short on the bus bars. Draw a curve reflecting the re-striking voltage across the breaker and determine the maximum rate of voltage rise.
6. For a 110kV system, the reactance and capacitance up to the circuit breaker are 10Ω and 0.045μF, respectively. The CB connectors have a resistance of 450 ohm. Identify the following:
   (i). Natural frequency of oscillation
   (ii). Damped frequency of oscillation
   (iii). Resistance value required for no transient oscillation.
   (iv). The resistance value resulting in damped oscillation having one-fourth of the natural frequency.
7. At 132 kV, a circuit breaker stops the magnetizing current of a 120MVA transformer. The transformer's magnetizing current equals 10% of the full-load current. Determine the maximum voltage that will be generated across the breaker gap when the magnetizing current is interrupted at 50% of its highest value. Stray capacitance: 2150 μF. The inductance equals 30H.
8. A 132kV C.B has 0.012μF bushing-to-ground capacitance and a transformer inductance of 14H. Determine the voltage that emerges between the C.B poles if a 16A magnetizing current flows through the transformer.
9. During a short circuit test of a 3-pole, 220 kV CB, the following observations were made: The fault power factor is 0.86, the recovery voltage is 0.96 times full-line, and the breaking current

is symmetric. The restriking voltage oscillates at a rate of 15000 cycles per second. Assume the neutral is grounded and the fault is not related to the ground. Calculate the average rate of growth in the restriking voltage.

10. When a 66 kV, 500 MVA CB closes on a fault, calculate the
    (i). symmetrical breaking current and
    (ii). the asymmetrical breaking current, assuming a 70% dc component.
    (iii). Peak generating current.
    (iv). The short-term rating current.
11. A 50 Hz, overhead transmission line has a capacitance of 0.1 μF per phase. Determine the inductance of an arc suppression coil to neutralize the effect of capacitance of
    (i) Complete length of the line
    (ii) 90 % length of the line
    (iii) 50 % length of the line
12. Calculate the reactance of a coil that is appropriate for a 33 kV, 3-phase transmission system, where the capacitance to ground of each conductor is 5 μF.
13. A three-phase, 50Hz, 110 kV transmission line, spanning 150 km, is composed of three conductors with an effective diameter of 25mm. The conductors are organized in a vertical plane with a spacing of 15m and are regularly transposed. Calculate the capacitance, inductance, inductive reactance, fault current, and kVA rating of the arc suppression coil in the system.
14. A 250 km long, 3-Ø, 50Hz, 220 kV transmission line has a capacitance of 0.03 μF per km per ph. Determine the inductance, inductive reactance, fault current, and kVA rating of an arc suppression coil suited for the line in order to eliminate the arcing round phenomenon.
15. A 3-phase line has three conductors positioned at the corners of a triangle with sides of 4m, 3m, and 2m. Calculate the inductance, inductive reactance, fault current, and kVA rating of an arc suppression coil in a 3-phase 50Hz, 66 kV system with regularly transposed conductors. Each conductor has a diameter of 1.5 cm.
16. Calculate the inductance, inductive reactance, fault current, and kVA rating of an arc suppression coil for a 3-phase 50Hz, 33kV system with conductors placed horizontally with spacing of D_{31} = 6 m; D_{12} = D_{23} = 3m. The transposed conductors have a diameter of 4 cm.
17. A 3-phase, 50 Hz, 132 kV overhead line has conductors spaced 6 meters apart in a vertical plane. The conductor diameter is 5cm. Calculate the inductance, inductive reactance, fault current, and kVA rating of an arc suppression coil for a 100-kilometer-long line.
18. A double-circuit line has six conductors. Every conductor has a 10mm radius. Six conductors are horizontally placed. The conductors are spaced 5 meters apart. The conductors are listed from left to right: a, b, c, a', b', and c'. Calculate an arc suppression coil's inductance, inductive reactance, fault current, and kVA rating for a 3-phase 50Hz, 220kV system.
19. In a double circuit, each conductor has a radius of 12mm. The six conductors are positioned at the corners of a regular hexagon. The centre-to-centre distance between conductors is as follows: AC'=CA'=7m, BB'=9m, and AC=A'C'=8m. To eliminate the arcing round phenomenon in a 3-phase 50Hz, 110kV system, calculate the inductance, inductive reactance, fault current, and kVA rating of an arc suppression coil appropriate for the line.

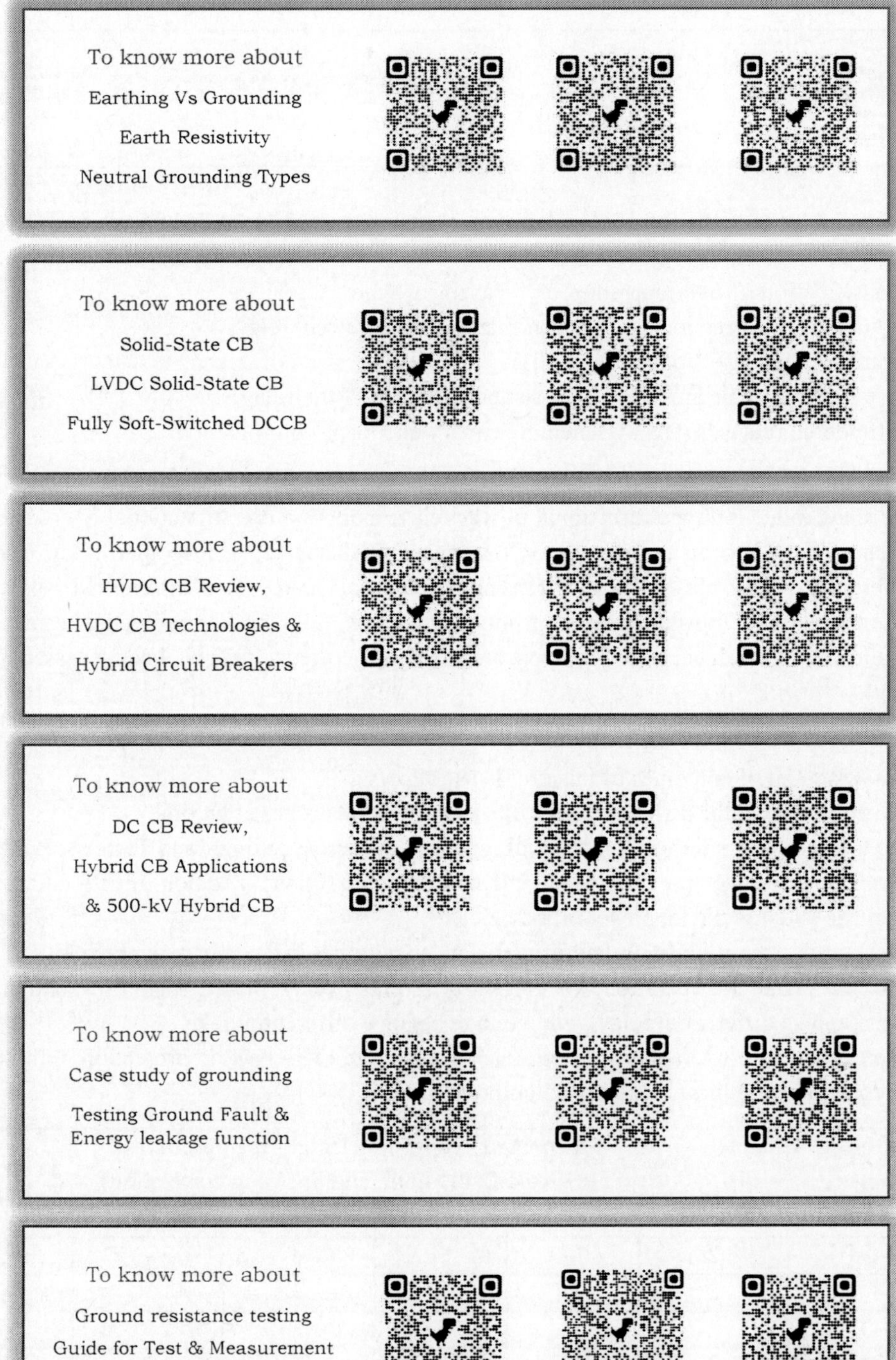
To know more about
Earthing Vs Grounding
Earth Resistivity
Neutral Grounding Types
To know more about
Solid-State CB
LVDC Solid-State CB
Fully Soft-Switched DCCB
To know more about
HVDC CB Review,
HVDC CB Technologies &
Hybrid Circuit Breakers
To know more about
DC CB Review,
Hybrid CB Applications
& 500-kV Hybrid CB
To know more about
Case study of grounding
Testing Ground Fault &
Energy leakage function
To know more about
Ground resistance testing
Guide for Test & Measurement
IEEE Guide for Measuring

09 PROTECTIVE RELAYS

Unit specifics: In this unit, the following topics have been discussed for basic understating of protective relays:

- Classification of over current relays based on time.
- Induction type directional and non-directional overcurrent relays.
- Induction type directional power relay.
- Distance type impedance relays, reactance relay and Mho relay.
- Differential relays, Translay scheme, Primary and back-up protection.

Rationale: In this unit, students will be introduced to protective relays, induction type non-directional overcurrent relay, induction type directional power relay, induction type directional overcurrent relay, definite-distance type impedance relay, time-distance impedance relay, reactance relay, Mho relay, current differential relay, biased beam differential relay (or) percentage differential relay, voltage balance differential relay, translay scheme, primary protection, relay backup protection, remote backup protection, and local breaker backup protection, are clearly described with the help of necessary diagrams and examples.

Pre-Requisites: Basic knowledge of power system components.

Unit Outcomes: List of outcomes of this unit is as follows

U09-O1: To comprehend the different forms of overcurrent relays based on time.

U09-O2: To know different terminologies. Pick-up current, current settings, and Time vs. P.S.M. curve.

U09-O3: To investigate the operation of the induction type directional and non-directional overcurrent relays, and directional power relay.

U09-O4: To examine the operation of definite-distance and time-distance impedance relays.

U09-O5: To investigate the operation of a current differential relay, biased beam differential relay (or percentage differential relay), and voltage balance differential relay.

U09-O6: To evaluate the operation of the primary protection, relay backup protection, remote backup protection, and local breaker backup protection.

Unit-09 outcomes	Expected mapping with course outcomes (1: Weak, 2: medium, and 3: strong correlation)					
	CO-1	**CO-2**	**CO-3**	**CO-4**	**CO-5**	**CO-6**
U09-O1	2	2	-	-	-	-
U09-O2	2	2	-	2	3	3
U09-O3	2	1	1	3	3	-
U09-O4	2	1	1	3	2	-
U09-O5	2	1	2	3	3	-
U09-O6	2	1	-	3	3	-

9.1 Introduction to Protective Relays:

In a power system comprising generators, transformers, transmission, and distribution circuits, it is unavoidable that at some point, a failure will occur within the system. Upon the occurrence of a failure in any component of the system, it is essential to quickly identify and isolate it from the system. There are two main factors contributing to it. Firstly, if the fault is not immediately resolved, it may result in unnecessary disruption of service for the customers. Furthermore, instantly disconnecting faulty equipment minimises the extent of damage and reduces the propagation of fault effects throughout the system.

Fault identification and disconnection of a problematic portion or device can be accomplished by utilising fuses or relays in combination with circuit breakers. A fuse automatically carries out both detection and interruption duties, however it is only suitable for protecting low-voltage circuits. Relays and circuit breakers are used in high voltage circuits, namely those over 3.3 kV, to provide automatic protection. The relays identify the malfunction and provide information to the circuit breaker, which carries out the task of interrupting the circuit.

A protective relay is an apparatus that identifies faults and triggers the activation of the circuit breaker to separate the faulty component from the rest of the system. The relays identify improper conditions in the electrical circuits by continuously measuring the electrical values that vary between normal and faulty states. The electrical parameters that can vary during fault circumstances include voltage, current, frequency, and phase angle. The faults indicate their presence, nature, and location to the protective relays by altering one or more of these parameters. Once the problem is discovered, the relay activates to close the trip circuit of the breaker. As a consequence, the breaker is activated and the faulty circuit is disconnected.

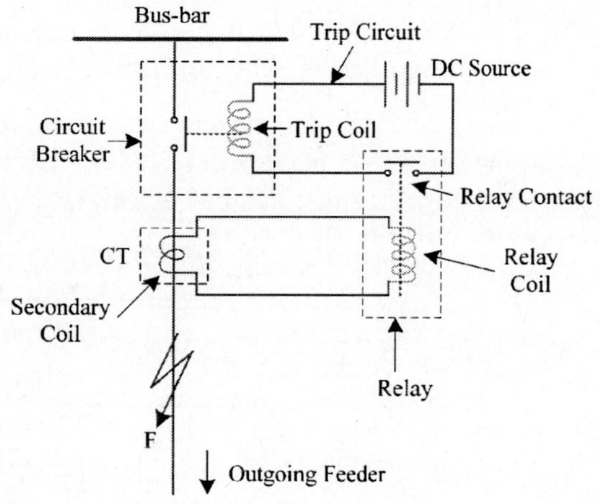

Fig. 9.1 Protective relay circuit

Figure 9.1 depicts a standard relay circuit. This figure illustrates a single phase of a 3-phase system, chosen for the sake of simplicity. The connections of the relay circuit can be categorised into three components. The initial section consists of the main coil of a current transformer that is linked in series with the line that requires protection. The second portion comprises the secondary winding of the current transformer and the coil responsible for activating the relay. The third component is the tripping circuit, which can operate on either alternating current or direct current. The system comprises a supply source, the trip coil of the circuit breaker, and the stationary contacts of the relay.

When a short circuit develops at point F on the transmission line, the current flowing in the line experiences a significant surge. As a consequence, a substantial electric current passes through the relay coil, leading to the activation of the relay as its contacts are closed. As a result, the trip circuit of the breaker is activated, causing the circuit breaker to open and separate the problematic component from the remainder of the system. The relay guarantees the protection of the circuit equipment from harm and the proper functioning of the unaffected part of the system.

9.2 Types of Over current Relays based on time:

The time of operation is a crucial attribute of a relay. The term "time of operation" refers to the duration between the moment when the actuating element is powered and the moment when the relay connections are closed. Occasionally, it is advantageous and essential to regulate the duration of operation for a relay. Relays are utilised in conjunction with mechanical components for this specific function. The over current relays are categorised as follows

(i). Instantaneous over current relay
(ii). Definite over current relay
(iii). Definite time over current relay
(iv). Inverse time over current relay.

9.2.1 Instantaneous over current relay: The Instantaneous Overcurrent Relay is the initial form of relay. This relay type was designed especially to provide protection against extremely high currents within a brief duration, often less than 0.1 seconds. This form of electrical current can arise from malfunctions inside the system, such as a circuitry short-circuit.

9.2.2 Definite over current relay: When the current reaches a predefined level, a definite current relay activates immediately. When the current exceeds the Pick-up value, it operates for a specific amount of time. Its only operating condition is current magnitude (with no time delay). The operating time is constant. Fig. 9.2 (a) illustrates time-current properties of a Definite over current relay.

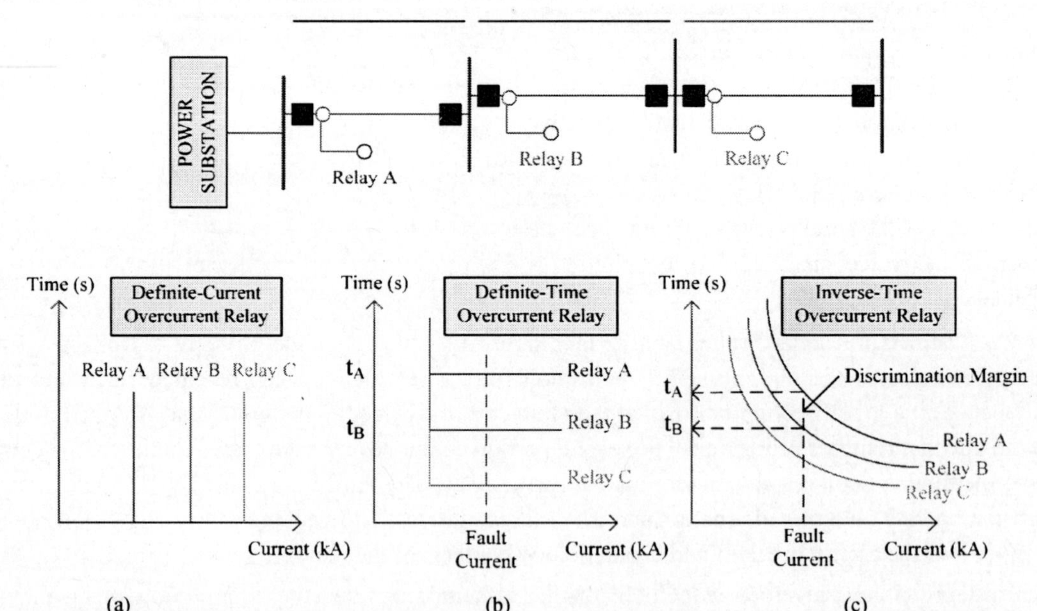

Fig. 9.2 (a) Definite current relay (b) Definite time over current relay and (c) Inverse time over current relay

9.2.3 Definite time over current relay: A definite time over-current relay is one that activates after a certain amount of time when the current exceeds the pickup value. As a result, this relay offers a range of both current and time settings. So, this sort of relay has a set period between the moment of pickup and the closing of relay connections. This time setting is unaffected by the amount of current passing through the relay coil; it remains constant for all current values greater than the pickup value. Fig.9.2(b) illustrates time-current properties of a Definite time over current relay. It is worth noting that almost all inverse-time relays have a specific minimum time characteristic, which ensures that the relay never becomes instantaneous in its activity for extremely lengthy overloads.

9.2.4 Inverse time over current relay: In contrast to instantaneous relays, the inverse-time overcurrent relay has an inversely proportional operation time. This indicates that a larger current reduces the time required for the relay to operate, but a lower current increases it by up to 10 seconds. Figure 9.2(c) depicts the time-current properties of an inverse time over current relay. At current values less than pickup, the relay does not operate. At larger numbers, the relay's time of operation steadily reduces as the current increases.

9.3 Important terms: It is desirable to define and explain several terms that are used in relay-related contexts.

9.3.1 Pick-up current: The pick-up current refers to the minimal amount of current required to initiate the operation of the relay coil. As long as the current flowing through the relay is below the pick-up threshold, the relay remains inactive and the breaker it controls stays closed. Nevertheless, once the current flowing through the relay coil reaches or above the specified pickup threshold, the relay activates and energises the trip coil, resulting in the opening of the circuit breaker.

9.3.2 Current setting: This is referred to as the current setting and is often accomplished by employing tapings on the relay operating coil. The taps are extended to a plug setting bridge, as depicted in Figure 9.3. The plug setting bridge allows for adjusting the number of turns on the relay coil. This alters the rotational force applied to the disc, thereby affecting the duration of the relay's functioning. The values assigned to each tap are expressed as a percentage of the full-load rating of the current transformer (C.T.) that the relay is connected to. These values indicate the point at which the disc starts to revolve and eventually closes the trip circuit. Therefore, the pick-up current can be calculated by multiplying the rated secondary current of the current transformer (C.T.) by the current setting.

Pick-up current = Rated secondary current of C.T. * Current setting .... (9.1)

Consider connecting an overcurrent relay with a current setting of 150% to a supply circuit using a 500/5 current transformer. C.T.'s rated secondary current is 5 amps. The pick-up value will be 50% higher than 5 A, resulting in 7.5 A (5 × 1·5). With the above current setting, the relay will operate for a relay coil current equal to or more than 7.5 A. The current settings for overcurrent relays typically vary from 50% to 200% in increments of 25%, whereas

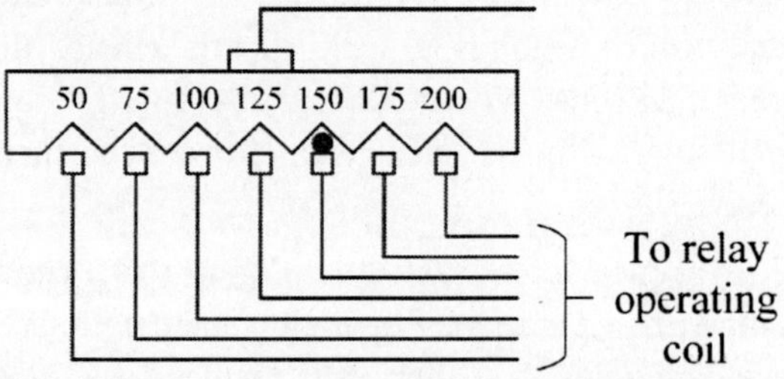

Fig. 9.3 Current Settings of overcurrent relay

the settings for earth leakage relays range from 10% to 70% in increments of 10%. To achieve the desired current setting, one must put a plug into the jaws of a bridge type socket at the needed tap value.

9.3.3 Plug-setting multiplier (P.S.M.):

The plug-setting multiplier is the quotient of the fault current in the relay coil divided by the pick-up current.

$$\text{P.S.M.} = \frac{\text{Fault current in relay coil}}{\text{Pick} - \text{up current}}$$

$$= \frac{\text{Fault current in relay coil}}{\text{Rated secondary current of C.T.} * \text{Current setting}} \quad \text{......(9.2)}$$

For instance, consider a scenario where a relay is linked to a $\frac{400}{5}$ A current transformer and adjusted to a 150% setting. To compute the plug-setting multiplier, use the major fault current value of 1200 A.

Pick-up current = Rated secondary current of C.T. * Current setting = 5 * 1.5 = 7.5 A

$$\text{Fault current in relay coil} = 1200 * \frac{5}{400} = 15\ A$$

$$\text{P.S.M.} = \frac{\text{Fault current in relay coil}}{\text{Pick} - \text{up current}} = \frac{15}{7.5} = 2$$

9.3.4 Time-setting multiplier: A relay typically has a control mechanism that allows for the adjustment of the operation time. This modification is commonly referred to as a time-setting multiplier. The time-setting dial is calibrated with a range from 0 to 1, with increments of 0.05 seconds, as shown in Figure 9.4 These figures are factors that can be utilised to transform the time obtained from the time/P.S.M. curve into the precise operational time. Therefore, if the time setting is 0.6 and the time acquired from the time/PSM curve is 4 seconds.

Fig. 9.4 Time setting multiplier

Actual relay operating time = 4 * 0.6 = 2.4 sec.

In an induction relay, the operating time is regulated by modifying the displacement of the disc from its initial position to its activation point. This is accomplished by modifying the position of a movable barrier that regulates the movement of the disc, hence altering the duration during which the relay will activate its connections for specific fault current levels. This adjustment is facilitated by a time dial, which features a scale that is evenly divided. The current operating time is determined by multiplying the time setting multiplier with the time obtained from the time/PSM curve of the relay.

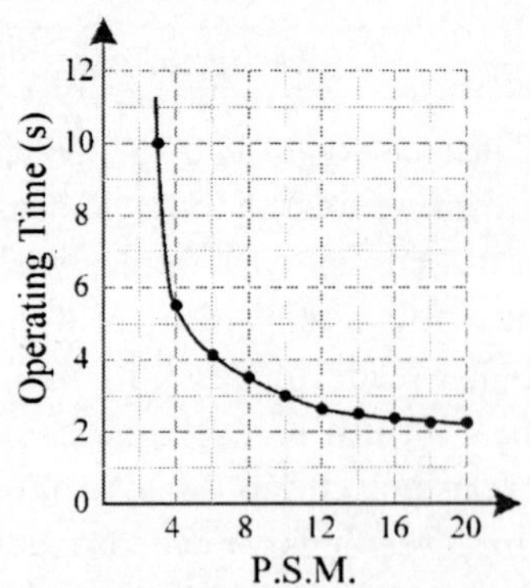

Fig. 9.5 Time Vs P.S.M. Curve

9.3.5 Time Vs P.S.M. Curve: Figure 9.5 illustrates the relationship between the duration of operation and the plug setting multiplier of a standard relay. The horizontal scale is calibrated using the plug-setting multiplier, which indicates the ratio of the relay current to the current setting. The vertical axis is calibrated based on the duration needed for the relay to activate. If the P.S.M is equal to 10, then the duration of the operation (as indicated by the curve) is 3 seconds. The current operational time is calculated by multiplying this time by the time-setting multiplier.

Figure 9.5 clearly demonstrates that at smaller overcurrent values, the time of operation of the relay decreases as the current increases. However, when the current approaches 20 times the full-load value, the operating time of the relay becomes constant. This capability is essential to guarantee the detection of discriminating on extremely high fault currents that run over reliable power lines. Table 9.1 presents the time Vs P.S.M (Plug Setting Multiplier) characteristics of an overcurrent relay with a TMS (Time Multiplier Setting) of 1.

Table 9.1 Time Vs P.S.M characteristics of an overcurrent relay with a TMS of 1.

P.S.M	2	4	6	8	10	12	14	16	18	20
Operating time in seconds	10	5.5	4.2	3.5	3	2.8	2.5	2.4	2.3	2.2

Example 9.1 Calculate the operating period of a 5-ampere, 4-second overcurrent relay with a current setting of 150% and a time setting multiplier of 0.8. The relay is linked to the supply circuit through a 400/5A current transformer, and the circuit carries a fault current of 2500A. Refer to the table 9.1 for Time Vs P.S.M values.

Ans. Rated secondary current of C.T. = 5 A

Pick-up current = Rated secondary current of C.T. * Current setting = 5 * 1.5 = 7.5 A

$$\text{Fault current in relay coil} = 2400 * \frac{5}{400} = 30\,A$$

$$\text{P. S. M.} = \frac{\text{Fault current in relay coil}}{\text{Pick} - \text{up current}} = \frac{30}{7.5} = 4$$

Corresponding to the plug-setting multiplier of 4 (See Table 9.1), the time of operation is 5.5 seconds.

Actual relay operating time $= 5.5 * 0.8 = 4.4$ sec.

Example 9.2 In the figure shown, both the relays R_1 and R_2 are set for 100% plug setting. Assume time grading margin of 0.6sec and TMS for relay 1 is 0.15. Refer to the table 9.1 for Time Vs P.S.M values. Determine the actual operating time of R_1 and TMS for R_2.

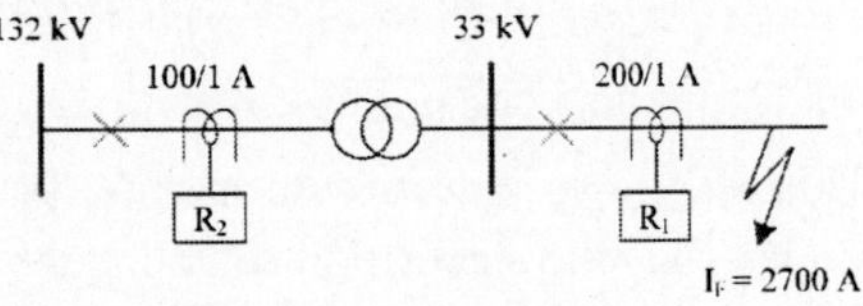

Ans. Rated secondary current of C.T. = 5 A

Plug setting multiplier of relay 1 is

$$\text{P. S. M. of relay 1} = \frac{\text{Fault current in relay coil 1}}{\text{Pick} - \text{up current}}$$

$$= \frac{2700}{1 * 1 * \frac{200}{1}} = 13.5$$

Operating time corresponding to 13.5 PSM is 2.6

Actual operating time of relay 1 is =2.6×TMS=2.6×0.15=0.39sec

$$\frac{V_2}{V_1} = \frac{I_1}{I_2} \quad \Rightarrow \quad \frac{33 * 10^3}{132 * 10^3} = \frac{I_1}{2700}$$

$$\Rightarrow \quad I_1 = 2700 * \frac{33 * 10^3}{132 * 10^3} = 675A$$

Plug setting multiplier of relay 2 is

$$\text{P. S. M. of relay 2} = \frac{\text{Fault current in relay coil 2}}{\text{Pick} - \text{up current}} = \frac{675}{1 * 1 * \frac{100}{1}} = 6.75$$

Operating time corresponding to PSM 6.75 is 3.8 sec.

$$\text{TMS for relay 2 is} = \frac{0.39 + 0.6}{3.8} = 0.2605$$

9.4 Types of Protective Relays:

Currently, most power system relays use electromagnetic attraction or induction. Relays are classified based on their purpose in protecting electrical power circuits, regardless of the underlying concept. An overcurrent relay is one that detects excessive current in a circuit and takes corrective action, regardless of its design. Similarly, an overvoltage relay detects overvoltage in a circuit and conducts corrective actions. Although there are various varieties of special function relays, this chapter will only cover the following major ones:

(i) Induction type non-directional overcurrent relay
(ii) Induction type directional power relay
(iii) Induction type directional overcurrent relay
(iv) Distance or impedance relay
- Definite – distance type impedance relay
- Time-distance impedance relay
- Reactance relay
- Mho (or) Admittance relay

(v) Differential relays
- Current differential relay
- Biased beam differential relay (or) percentage differential relay
- Voltage balance differential relay

(vi) Translay scheme
(vii) Primary and back-up protection

9.5 Induction type non-directional overcurrent relay:

This relay uses induction to initiate remedial steps when the circuit current exceeds a predetermined value. The relays actuation source is the current in the circuit supplied by a current transformer. These relays are only suitable for alternating current circuits and can handle fault current flow in both directions.

Construction: Figure 9.6 depicts the major construction details of a typical nondirectional induction type overcurrent relay. It is made up of a metal (aluminium) disc that can freely revolve between the poles of two electromagnets. The upper electromagnet is made up of two windings: primary and secondary. The primary of a current transformer (C.T.) is linked to the secondary of another C.T. in a power line that requires protection. The primary is also connected on a regular basis. The tapings are connected to a plug-setting bridge, allowing you to change the number of active turns on the relay operating coil. This allows you to achieve your preferred current setting. The secondary winding is energised by electromagnetic induction from the primary winding and connected in series with the lower magnet's winding. The regulating torque is generated by the spiral springs.

When the disc rotates through a predetermined angle, a movable contact bridges two fixed contacts (attached to the trip circuit) on its spindle. This angle can be set to any number between 0° and 360°. This angle can be modified to change the travel of the moving contact, allowing the relay to be set to any time.

Operation: To comprehend the generation of torque in an induction relay, consult the basic configuration depicted in Figure 9.6. The two alternating current fluxes, φ_2 and φ_1, which have a phase difference of angle α, generate electromotive forces (e.m.f.s) in the disc and result in the flow of eddy currents i_2 and i_1, respectively. The currents exhibit a phase shift of 90 degrees with regard to their corresponding fluxes.

Let $\varphi_1 = \varphi_{1\,max} \sin \omega t$

$$\varphi_2 = \varphi_{2\,max} \sin(\omega t + \alpha) \quad \ldots\ldots (9.3)$$

Where φ_1 and φ_2 represent the instantaneous values of fluxes, with φ_2 being ahead of φ_1 by an angle α.

If the rotor currents run through routes with low self-inductance, they will be in phase with the voltages.

$$i_1 = \frac{d\varphi_1}{dt} \propto \frac{d}{dt}(\varphi_{1\,max} \sin \omega t)$$

$$\propto \varphi_{1\,max} \cos \omega t \quad \ldots\ldots (9.4)$$

$$i_2 = \frac{d\varphi_2}{dt} \propto \frac{d}{dt}[\varphi_{2\,max} \sin(\omega t + \alpha)]$$

$$\propto \varphi_{2\,max} \cos(\omega t + \alpha) \quad \ldots\ldots (9.5)$$

$$Force\ F_1 \propto \varphi_1 i_2 \text{ and } F_2 \propto \varphi_2 i_1 \quad \ldots\ldots (9.6)$$

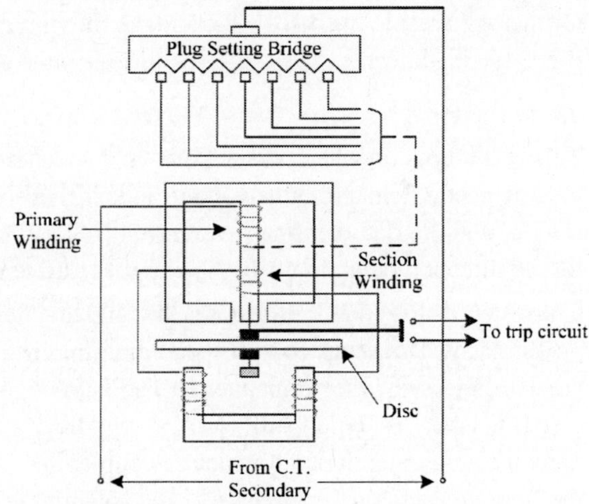

Fig. 9.6 Induction type non-directional overcurrent relay

Figure 9.6 depicts the two opposing forces. The net force F at the given time is

$$F \propto F_2 - F_1$$

$$\propto (\varphi_2 i_1 - \varphi_1 i_2)$$

$$\propto \varphi_{2\,max} \sin(\omega t + \alpha)\varphi_{1\,max} \cos \omega t - \varphi_{1\,max} \sin \omega t \varphi_{2\,max} \cos(\omega t + \alpha)$$

$$\propto \varphi_{2\,max}\varphi_{1\,max}[\sin(\omega t + \alpha) \cos \omega t - \sin \omega t \cos(\omega t + \alpha)]$$

$$\propto \varphi_{1\,max}\ \varphi_{2\,max} \sin\alpha$$

$$F \propto \varphi_1\ \varphi_2 \sin\alpha \quad \ldots\ldots (9.7)$$

where φ_1 *and* φ_2 are the r.m.s. values of the fluxes.

From eq. 10.7, the driving torque 'T' can be derived as

$$T \propto \varphi_1\ \varphi_2 \sin\alpha \quad \ldots\ldots (9.8)$$

Assume the fluxes φ_1 and φ_2 are proportional to the current I in the relay coil.

$$T \propto I^2 sin\alpha \quad \text{...... (9.9)}$$

The following points can be highlighted from Exp. (9.7):

- The higher the phase angle (α) between the fluxes, the larger the net force applied to the disc. The maximum force is produced when the two fluxes are 90 degrees out of phase.
- At each instant, the net force remains constant. This fact is not dependent on the assumptions used to arrive at exp. (9.7).
- The direction of net force and disc motion is determined by which flux is leading.

The induction concept is used to set the driving torque on the aluminium disc. This torque is countered by the spring's restraining torque. Under normal circumstances, the restraining torque exceeds the driving torque generated by the relay coil current. As a result, the aluminium disc is immobile. When the current in the protected circuit reaches the predetermined value, the driving torque exceeds the restraining torque. As a result, the disc spins, and the moving contact bridges the fixed contacts after it has turned a predetermined degree. The trip circuit triggers the circuit breaker, isolating the problematic component.

9.6 Induction Type Directional Power Relay:

This relay turns on when power flows in a specified direction via the circuit. A directional power relay uses magnetic fields from the voltage and current sources of the circuit it safeguards to generate working torque, unlike nondirectional overcurrent relays. This sort of relay functions as a wattmeter, with the torque direction determined by the current relative to the voltage.

Construction: Figure 9.7 illustrates the fundamental components of a standard induction type directional power relay. The setup comprises an aluminium disc that may freely rotate within the magnetic field created by two electromagnets. The upper electromagnet is equipped with a winding, known as the potential coil, which is coupled to the circuit voltage source by a potential transformer.

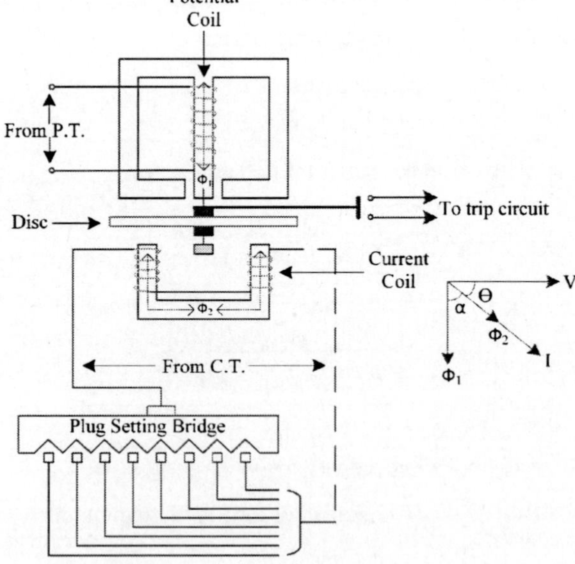

Fig. 9.7 Induction type directional power relay

The lower electromagnet is equipped with an independent winding, known as the current coil, which is connected to the secondary of the current transformer in the line that requires protection. The current coil is equipped with many tapings that are connected to the plug setting bridge. This allows for the selection of any desired current setting. A spiral spring generates the restricting torque. When the disc rotates through a predetermined angle, a moving contact on the spindle bridges two fixed contacts. By altering this angle, the moving disc's path may be altered, allowing the relay to be set to any desired time.

Operation: The current in the potential coil causes a flux φ_1 that is roughly 90° behind the applied voltage. The flux φ_2 generated by the current coil will be nearly in phase with the operational current I. The torque generated by the interaction of fluxes φ_1 and φ_2 with the eddy currents formed in the disc can be expressed as: $T \propto \varphi_1 \; \varphi_2 \; sin\alpha$

Since $\varphi_1 \propto V \, , \; \varphi_2 \propto I \; and \; \alpha = 90 - \Theta$

$$T \propto V I \sin(90 - \Theta)$$

$$T \propto V I \cos \Theta$$

$$T \propto Power \quad \text{...... (9.10)}$$

The direction of driving torque on the disc is determined by the power flow in the circuit connected to the relay. When power flows normally in the circuit, the driving and restraining torques (due to springs) work together to separate the moving and fixed contacts. As a result, the relay continues to be inoperative. However, reversing the current in the circuit changes the direction of the disc's torque. When the reversed driving torque is sufficient, the disc rotates in the opposite direction, and the moving contact closes the trip circuit. This activates the circuit breaker, which disconnects the faulty part.

9.7 Induction type directional overcurrent relay:

Under short-circuit situations, the directional power relay discussed above is inappropriate for use as a directional protective relay. A short-circuit causes the system voltage to drop to a low value and could result in inadequate relay operating torque. The directed overcurrent relay, made to be practically independent of system voltage and power factor, solves this challenge.

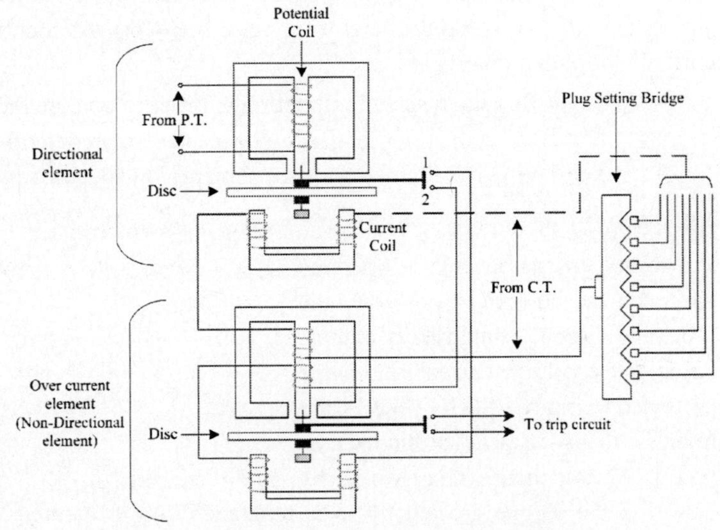

Fig. 9.8. Induction type directional overcurrent relay

Construction: Figure 9.8 illustrates the structural characteristics of a standard induction type directional over current relay. The system comprises of two relay components enclosed in a common casing, namely: (i) a directional element and (ii) a non-directional element.

Directional element: Essentially, it is a power relay that functions when power flows in a particular direction. The potential coil of this component is linked to the system voltage through a potential transformer (P.T.). The circuit current energises the current coil of the element through a C.T. This coil is wrapped around the top magnet of the non-directional component. The directional element's trip contacts (1 and 2) are wired in series with the overcurrent element's secondary circuit. As a result, the later element cannot operate until its secondary circuit is complete. To operate the overcurrent element, the directional element must first operate (i.e., contacts 1 and 2 must close).

Non-directional element: It is an overcurrent element identical to the non-directional overcurrent relay described in Fig. 9.8. The spindle of the disc of this element has a moving contact that closes the fixed contacts once the directional element is operated. It should be noted that the relay also has a plug-setting bridge for current setting, however it is not shown in the figure for clarity and simplicity. The tapings are located on the upper magnet of the overcurrent element and connect to the bridge.

Operation: During normal operation, power flows in the expected direction within the circuit that is safeguarded by the relay. As a result, the directional power relay (upper element) remains inactive, preventing the overcurrent element (lower element) from being energised. Nevertheless, in the event of a short-circuit, there is a tendency for the current or power to flow in the other way. If this occurs, the upper element's disc will rotate in order to connect the fixed contacts 1 and 2. This establishes the electrical pathway for the overcurrent component. The directional element is designed to be highly sensitive, allowing it to generate enough torque from the current winding to successfully activate the element and shut its contacts, even when exposed to the lowest expected voltage during severe fault conditions. The element's disc undergoes rotation, causing the associated moving contact to close the trip circuit. This activates the circuit breaker that separates the defective portion. The two relay elcments are set in such a way that the final tripping of the current, which is regulated by them, does not occur until the following criteria are met:

- current flows in a specific direction in order to operate the directional element.
- the current in the reverse direction exceeds the predetermined value.
- Excessive current remains for the time period specified by the overcurrent element.

9.8 Distance or impedance relay:

Previously discussed relays operated based on the current or power in the protected circuit. Some relays operate based on the voltage-current ratio in the protected circuit. Such relays are referred to as distance or impedance relays. An impedance relay works by opposing the torque created by a current and voltage element. When the V/I ratio falls below a predetermined threshold, the relay activates. Impedance is an electrical measurement of the distance travelled along a transmission line.

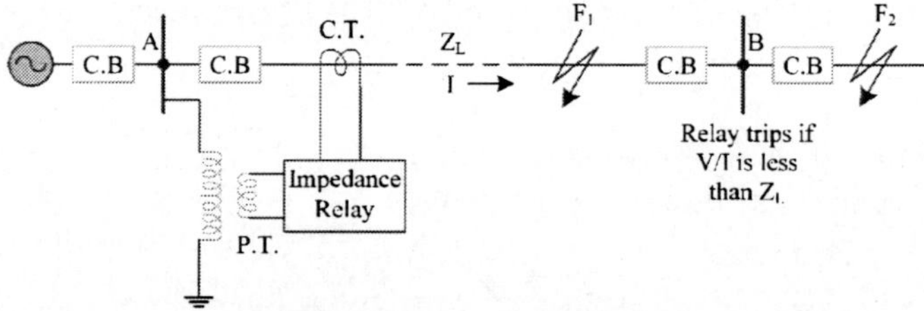

Fig. 9.9 Distance or impedance relay

Figure 9.9 depicts the fundamental operation of an impedance relay. The relay's voltage element is powered by a potential transformer connected to the protected line. The relay's current element is powered by a current transformer connected in series to the line. The protected zone is located on the line section AB. Under typical working circumstances, the shielded zone has an impedance of Z_L. The relay closes contacts when the shielded section's impedance falls below a predetermined value (Z_L in this case).

Assume a fault occurs at point F_1 within the protected zone. The relay activates when the impedance Z (= V/I) between the installation point and the failure location is less than Z_L. If a fault occurs beyond the protected zone (e.g., point F_2), the impedance Z exceeds Z_L, preventing the relay from operating.

A distance or impedance relay is essentially an ohmmeter that activates when the impedance of the protected zone falls below a predetermined threshold. There are two types of distance relays used for power supply protection, namely: definite-distance relay and time-distance relay. A definite-distance relay is designed to instantly detect faults that occur within a predetermined distance from the relay. A time-distance relay operates in proportion to the distance between the fault and the relay location. A fault closer to the relay will activate it faster than a fault farther away. It should be noted here that distance relays are created by changing one of two types of fundamental relays: the balance beam or the induction disc. The major function of Distance Protection/Zone Protection is to protect high-voltage transmission lines. Unit 10 of Section 10.14 contains a detailed explanation accompanied by detailed drawings.

9.8.1 Definite – Distance Type Impedance Relay:

Construction: Figure 9.10 depicts the schematic configuration of an impedance relay of the definite-distance type. The system comprises a pivoting beam F and two electromagnets that are activated by a current and voltage transformer in the circuit being protected.

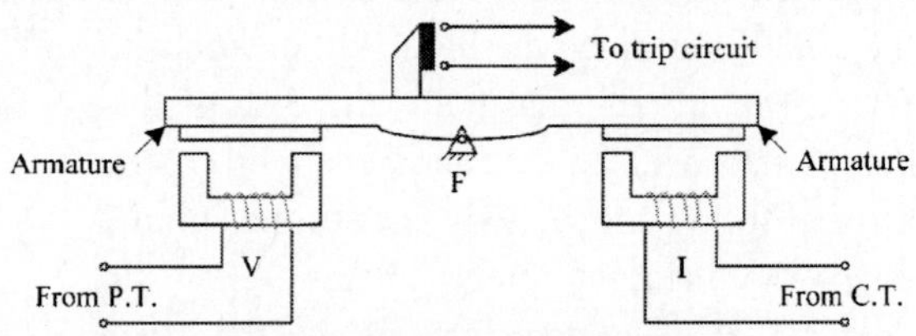

Fig. 9.10 Definite distance type impedance relay

The armatures of the two electromagnets are connected to the beam on opposite sides of the fulcrum through a mechanical coupling. A bridge piece is attached to the beam to connect the trip contacts. The relay is designed in such a way that the torques generated by the two electromagnets are in opposite directions.

Operation: Under normal operating conditions, the pull of the voltage element exceeds that of the current element. Thus, the relay contacts remain open. However, when a fault occurs in the protected zone, the voltage provided to the relay reduces while the current increases. The voltage-to-current ratio (impedance) goes below a predetermined value. As a result, the pull of the current element exceeds that of the voltage element, causing the beam to tilt in the direction of closing the trip connections. The current pull is proportional to I^2, while the voltage pull is proportional to V^2. Consequently, the relay will function when

$$k_1V^2 < k_2I^2$$

$$\frac{V^2}{I^2} < \frac{k_2}{k_1}$$

$$\frac{V}{I} < \sqrt{\frac{k_2}{k_1}}$$

$$Z < \sqrt{\frac{k_2}{k_1}} \qquad \text{...... (9.11)}$$

The constants k_1 and k_2 are proportional to the ampere-turns of the two electromagnets. The relay setting value can be modified by tapping the coils.

9.8.2 Time-Distance Impedance Relay:

A time-distance impedance relay is a type of relay that automatically adapts its operating time based on the distance between the relay and the source of failure. *i.e.*

$$\text{Operating time,} \quad T \propto \frac{V}{I}$$

$$T \propto Z$$

$$T \propto distance \qquad \text{..... (9.12)}$$

Construction: Figure 9.11 depicts the schematic configuration of a standard induction type time distance impedance relay. The device is comprised of a current-driven induction element that bears resemblance to the double winding type induction overcurrent relay. The spindle that holds the disc of this component is attached to another spindle using a spiral spring connection. The second spindle bears the bridging piece of the relay trip contacts. The bridge remains in the open position by the use of an armature that is pressed against the pole face of an electromagnet. The electromagnet is activated by the voltage of the circuit that needs to be safeguarded.

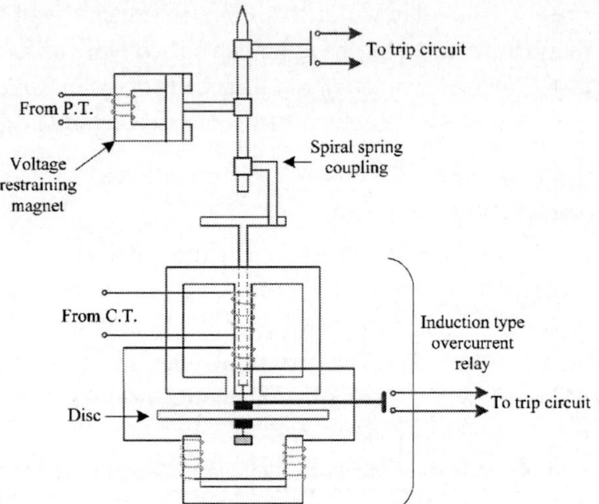

Fig. 9.11 Time-Distance Impedance Relay

Operation: Under normal load circumstances, the pull of the armature exceeds that of the induction element, therefore the trip circuit connections remain open. However, when a short circuit occurs, the disc of the induction current element begins to revolve at a speed determined by the operating current.

The pull of the voltage-excited magnet determines the angle of travel of the disc before the relay operates. The disc will travel further if the pull is stronger. The magnet's pull varies with line voltage. As the line voltage increases, the pull increases and the disc travels further, indicating that operating duration is proportional to voltage. As the disc rotates, the spiral spring coupling is gradually tightened until the spring's strain is strong enough to move the armature away from the pole face of the voltage-excited magnet. As soon as this happens, the spindle that holds the armature and bridging piece travels quickly due to the spring tension, causing the trip contacts to close. This action involves activating the circuit breaker to separate and disconnect the problematic part.

The rotational speed of the disc is proportional to the operating current, assuming the control spring's effect is not considered. The operation period of the relay is directly proportional to the strength of the voltage-excited magnet and therefore to the line voltage V at the relay's connection point. Hence, the operational duration of the relay is directly proportional to the ratio of voltage (V) to current (I), which is equivalent to the impedance (Z) or distance.

9.8.3 Reactance Type Distance Relay:

The reactance relay is a relay that operates at fast speed. This relay comprises two components: an overcurrent element and a current-voltage directional element. The current component generates a positive torque, whereas, the current-voltage component that forms a directional element opposes the current component depending upon the phase angle between the current and voltage. A reactance relay is a type of overcurrent relay that has directional limitation. The directional element is designed to generate the highest amount of negative torque when its current lags behind its voltage by 90°. The induction cup or twin induction loop configurations are most appropriate for activating reactance type distance relays.

Construction: Figure 9.12 illustrates a typical reactance relay that utilises the induction cup structure. The figure below illustrates a four-pole structure that includes operational, polarising, and restraining coils. The operational torque is generated through the interaction of magnetic fluxes resulting from the flow of electric current in coils. Specifically, it is the interaction between the fluxes of coils 2, 3, and 4. On the other hand, the restraining torque is produced by the interaction of magnetic fluxes originating from poles 1, 2, and 4. The operating torque is proportional to the square of the current, and the restraining torque is proportional to $VIcos(\theta - 90^0)$. The desired maximum torque angle is attained using resistance-capacitance circuits, as shown in the Figure 9.12. If the control effect is represented by $-K_3$, the torque equation becomes

$$T = K_1 I^2 - K_2 VI\ cos(\theta - 90^0) - K_3$$

$$T = K_1 I^2 - K_2 VI\ sin\theta - K_3 \qquad ..... (9.13)$$

Where θ is defined as positive when 'I' lag behind 'V'. At the balance point, net torque is zero, and consequently

$$K_1 I^2 - K_2 VI\ sin\theta - K_3 = 0$$

$$K_1 I^2 = K_2 VI\ sin\theta + K_3$$

$$K_1 = \frac{K_2 VI\ sin\theta + K_3}{I^2}$$

$$K_1 = K_2 \frac{V}{I}\ sin\theta + \frac{K_3}{I^2}$$

$$\frac{V}{I}\ sin\theta = \frac{K_1}{K_2} - \frac{K_3}{K_2 I^2}$$

$$Z\ sin\theta = \frac{K_1}{K_2} - \frac{K_3}{K_2 I^2}$$

$$Z\ sin\theta = \frac{K_1}{K_2} - \frac{0}{K_2 I^2} \qquad ..... (9.14)$$

$$Z\ sin\theta = \frac{K_1}{K_2}$$

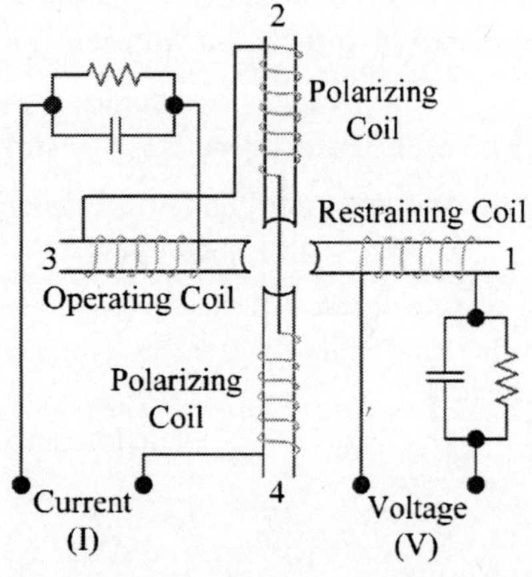

Fig. 9.12. Reactance Type Distance Relay

$$Z\ sin\theta = X \qquad ..... (9.15)$$

In the eq.9.14, the spring control effect is ignored, resulting in $K_3 = 0$

Operation: Figure 9.13 depicts the operational characteristics of a reactance relay. X represents the reactance of the protected line between the relay position and the fault point, whereas R represents the impedance's resistance component. The characteristic indicates that the resistance component of the impedance has no effect on the operation of the relay; the relay reacts only to the reactance component. The area below the working characteristic is known as the positive torque region.

General torque equation is $T = K_1 I^2 - K_2 VI\cos(\theta - \tau) - K_3$ ..... (9.16)

If the value of τ in the general torque equation is changed to any other angle of 90°, a linear characteristic will still be achieved, but it will not be parallel to the R-axis. An angle impedance relay is the term used to describe such a relay.

This particular relay lacks the ability to determine whether a fault has occurred in the portion where the relay is installed or in the adjacent section when it is used on a transmission line. The directional unit employed in conjunction with the reactance relay differs from that utilised with the impedance type relay due to the fact that the restraining reactive volt-ampere in the former case is almost negligible.

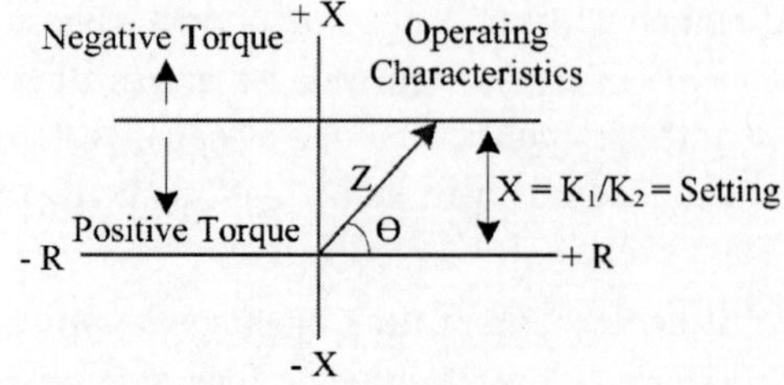

Fig. 9.13. Operating characteristics of Reactance Relay

Hence, the reactance type distance relay necessitates a non-operational directional unit during load situations. The reactance type relay is highly ideal for use as a ground relay for ground faults, as its reach remains unaffected by fault impedance.

9.8.4 MHO (or) Admittance Type Distance Relay:

A mho Relay is a high-speed relay that is also referred to as an admittance relay. The volt-amperes element provides operating torque in this relay, while the voltage element generates the controlling torque. A mho relay is a voltage-controlled directional relay.

Construction: Figure 9.14 depicts an induction cup-based mho relay. The combination of fluxes caused by poles 2, 3, and 4 generates the working torque, while poles 1, 2, and 4 generate the controlling torque.

If the spring's controlling effect is denoted by $-K_3$, the torque equation can be expressed as follows:

$$T = K_1 VI\cos(\theta - \tau) - K_2 V^2 - K_3 \quad ..... (9.17)$$

Where θ and τ are defined as positive when 'I' lag behind 'V'.

At the equilibrium point, the net torque is zero, so the equation becomes

$$K_1 VI\cos(\theta - \tau) - K_2 V^2 - K_3 = 0$$

$$K_1 VI\cos(\theta - \tau) - K_3 = K_2 V^2$$

$$\frac{K_1 VI\cos(\theta - \tau) - K_3}{VI} = \frac{K_2 V^2}{VI}$$

$$\frac{K_1 VI\cos(\theta - \tau) - K_3}{K_2\, VI} = \frac{V}{I}$$

$$\frac{K_1\cos(\theta - \tau)}{K_2} - \frac{K_3}{K_2\, VI} = \frac{V}{I}$$

$$\frac{K_1}{K_2}\cos(\theta - \tau) - \frac{K_3}{K_2\, VI} = Z \quad ..... (9.18)$$

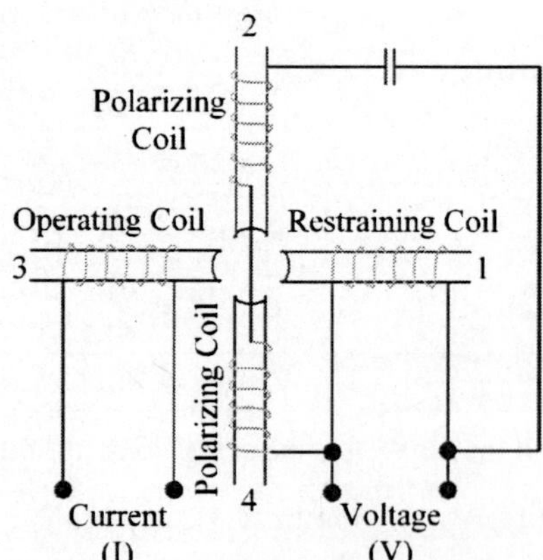

Fig. 9.14. MHO Type Distance Relay

If the spring-controlled effect is ignored, $K_3 = 0$.

$$Z = \frac{K_1}{K_2}\cos(\theta - \tau) - \frac{0}{K_2\, VI}$$

$$Z = \frac{K_1}{K_2}\cos(\theta - \tau) \quad ..... (9.19)$$

Operation: The operational characteristics of the mho relay are depicted in Figure 9.15. The diameter of the circle is essentially unaffected by V and I, except at extremely low levels of voltage and current, where the spring effect comes into play and causes a decrease in diameter. The equation $Z_R = \frac{K_1}{K_2}$ represents the diameter of the circle, where Z_R is the ohmic setting of the relay.

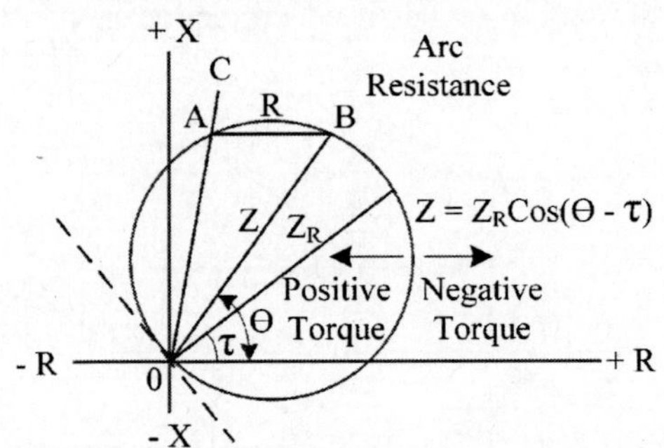

Fig. 9.15. Operating range of MHO distance Relay

Fig. 9.16. Operating characteristics of MHO Relay

The relay operates when it detects an impedance within the circle. The operating characteristic revealed that the circle goes via the origin, making the relay naturally directed. Because of its naturally directed characteristics, the relay requires only one pair of contacts, allowing for quick fault clearance and reducing the VA burden on the current transformer. In Fig. 9.15, line OC represents the impedance angle of the shielded line, which is typically between 60° and 70°. The arc resistance R is represented by the length AB, which is horizontal to OC from the chord Z extremity. By making the τ equal to, or slightly less sluggish than θ, the circle is made to fit around the problematic area, making the relay insensitive to power swings. This makes it particularly useful to the protection of lengthy or heavily loaded lines.

The time constant τ remains constant for a specific relay, and the phasor Y representing admittance will lie on a linear trajectory. The mho relays on the admittance diagram exhibit a characteristic that is represented by a straight line, as depicted in Figure 9.16. The Mho relay is well-suited for strongly loaded transmission lines at Extra High Voltage (EHV) or Ultra High Voltage (UHV) levels. This is because its threshold characteristic in the Z-plane is represented by a circle that passes through the origin, with a diameter of Z_R. As a result, the threshold characteristic is highly condensed, effectively isolating the problematic area and reducing the likelihood of operation during power swing. Additionally, it exhibits directed behaviour.

9.9 Differential Relays:

The majority of the relays that have been discussed so far depended on an abundance of electric current in order to function. These relays have reduced sensitivity as they are unable to accurately differentiate between significant load conditions and small fault conditions. To address this challenge, differential relays are employed. A differential relay works when the phasor difference between two or more similar electrical quantities exceeds a certain threshold.

A current differential relay compares the current entering and leaving a system part. Under normal operating conditions, the two currents are equal; however, when a problem develops, this condition is no longer met. The difference in incoming and outgoing currents is routed through the relay's operational coil. If the differential current equals or exceeds the pickup value, the relay will activate and open the circuit breaker to isolate the problematic part. It should be observed that nearly any kind of relay, when coupled in a specific manner, can function as a differential relay. In simple terms, the differential relay's classification is determined more by the manner in which it is linked within a circuit, rather than its physical structure. There are two primary systems of differential or balanced protection, namely (i) Current balance protection and (ii) Voltage balance protection.

9.9.1 Current Differential Relay:

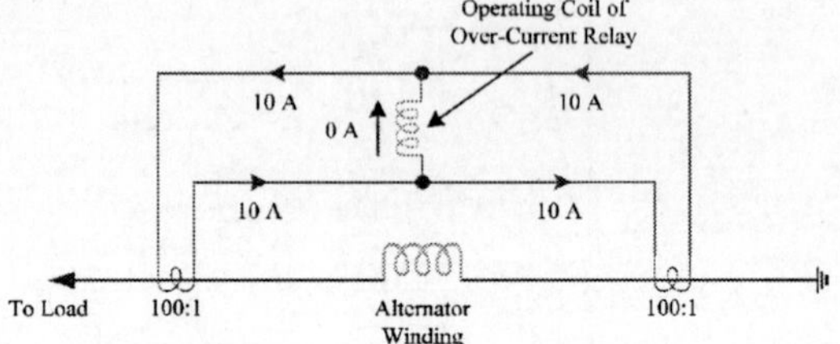

Fig. 9.17 (i). Current Differential Relay w/o fault

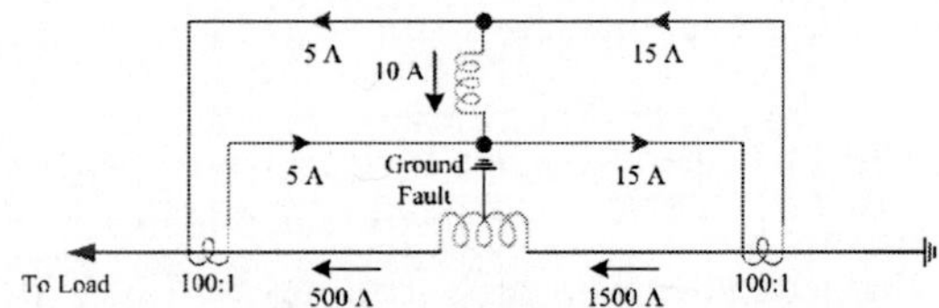

Fig. 9.17 (ii). Current Differential Relay during fault

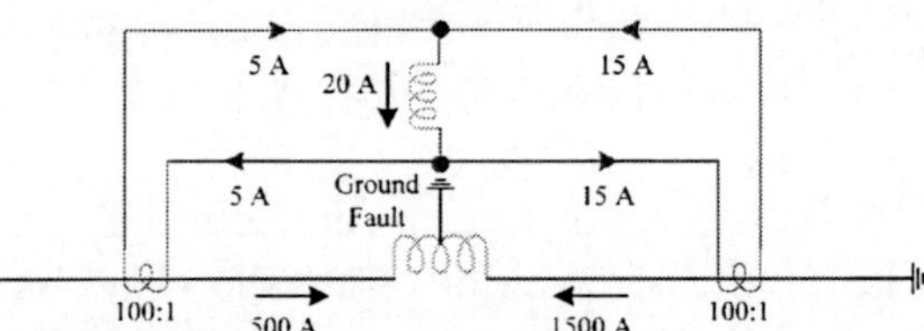

Fig. 9.17 (iii). Current Differential Relay during fault

Figure 9.17 depicts the configuration of an overcurrent relay that is wired to function as a differential relay. Two identical current transformers are installed on each end of the segment that needs protection, specifically the alternator winding in this situation. The secondary windings of current transformers (CTs) are connected in series, ensuring that they carry the induced currents in the same direction. The operational coil of the overcurrent relay is linked to the CT secondary circuit. The differential relay performs a comparison between the current at the two terminals of the alternator winding.

Assuming the alternator is functioning normally, let's consider a scenario where the winding of the alternator carries a current of 1000 A. Subsequently, the currents in the two secondary windings of current transformers (CT's) are identical, as depicted in Figure 9.17(i). The currents will only circulate between the two CTs, and there will be no current passing via the differential relay. Thus, the relay remains non-functional. In the event of a ground fault on the alternator winding, as depicted in Figure 9.17(ii), the two secondary currents will not be of equal magnitude. Consequently, the current will pass through the operational coil of the relay, resulting in the activation of the relay. The current flow through the relay is dependent upon the manner in which the fault is being supplied. In Fig. 9.17(ii), if a current (500 A) flows out one side and a larger current (1500 A) enters the other, the relay will pass the difference of the CT secondary currents (15 - 5 = 10 A). As shown in Fig. 9.17(iii), if current flows to the fault from both sides, the relay will receive the sum of CT secondary currents (10 + 5 = 15 A).

Disadvantages

- The impedance of the pilot cables typically results in a small disparity between the currents at the two ends of the protected section. When the relay is highly sensitive, even a slight differential current passing across it can cause it to activate, even when there is no fault present.

- The presence of capacitance in the pilot cable leads to inaccurate relay functioning when a substantial through-current is present.
- Accurate matching of current transformers is unattainable due to impedance in the pilot circuit.

The aforementioned drawbacks are largely mitigated in biassed beam relay.

9.9.2 Biased or Percentage differential relay:

The biased beam relay is specifically intended to detect and react to the difference in current, relative to the current flowing through the shielded portion, in a fractional manner. Figure 9.18 (i) depicts the Schematic diagram of a biassed beam relay. The device can be described as an overcurrent balanced beam relay with an added restraining coil. The restraining coil generates a counteracting force that opposes the working force.

Under normal load levels, the restraining coil produces a bias force that is larger than the operating force. Thus, the relay remains non-functional. When there is an internal fault, the force required for operation is greater than the force applied to maintain balance. As a result, the trip connections are being closed in order to activate the circuit breaker. The bias force can be modified by altering the number of turns on the restraining coil.

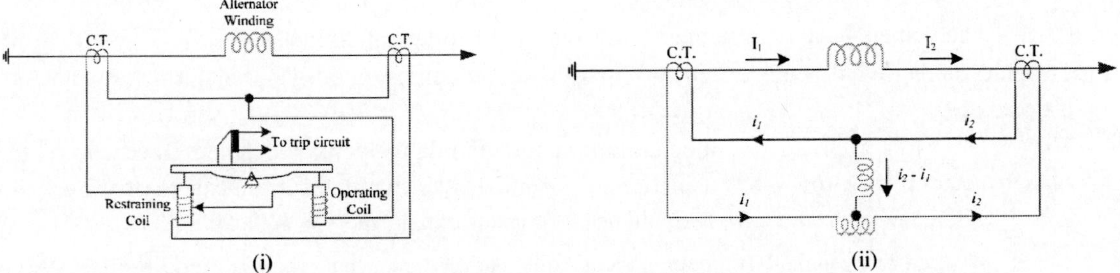

Fig 9.18 Biased or Percentage differential relay (i) Schematic diagram and (ii) Equivalent circuit

Figure 9.18(ii) displays the circuit diagram for a biassed beam relay. The differential current in the operating coil is directly proportional to the difference between i_2 and i_1. Similarly, the equivalent current in the restraining coil is directly proportional to the average of i_1 and i_2, as the operating coil is linked to the midpoint of the restraining coil.

Total ampere-turns on restraining coil of N turns $\propto i_1 \frac{N}{2} + i_2 \frac{N}{2}$

$$\propto \frac{(i_1 + i_2)}{2} N \qquad \text{.....(9.20)}$$

This is equivalent to the scenario when the average current $\frac{(i_1+i_2)}{2}$ passes through the entire restraining coil. The relationship between the current flowing through the restraining coil and the current required in the operating winding to trip the relay is directly proportional. Therefore, when there is a significant amount of weight, a higher amount of electrical current is needed to activate the relay's operational coil compared to when there is a minimal amount of weight. The relay is referred to as a "percentage relay" because the current needed to activate it and induce a trip can be represented as a percentage of the current being carried by the load.

9.9.3 Voltage Balance Differential Relay:

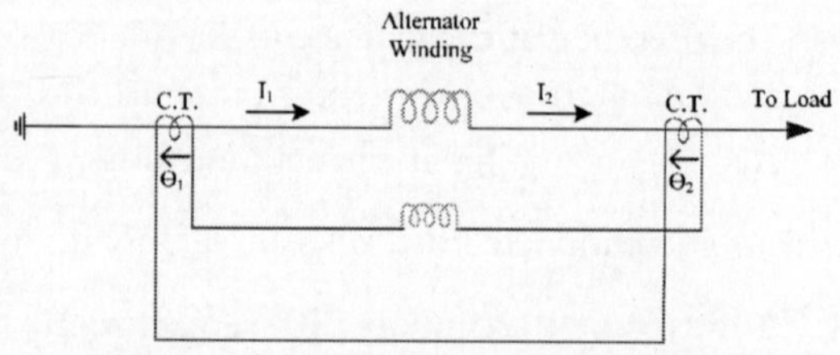

Fig. 9.19 Voltage Balance Differential Relay

Figure 9.19 depicts the configuration of voltage balance protection. In this protection technique, two identical current transformers are linked at each end of the component to be protected, such as an alternator winding, via pilot wires. The secondary windings of current transformers are connected in series with a relay in a manner that ensures their induced electromotive forces (e.m.f.s) are in opposite directions during normal operating conditions.

Under normal circumstances, identical currents ($I_1 = I_2$) pass through both primary windings. Consequently, the secondary voltages of both transformers are equalised, resulting in no current passing through the relay's operational coil. When a fault occurs within the protected region, the currents in the two primary sources will exhibit dissimilarity (i.e. $I_1 \neq I_2$) and their secondary voltages will no longer maintain equilibrium. The voltage disparity will induce an electric current to pass through the operational coil of the relay, so closing the trip circuit. This is the scenario in which a fault is transmitted in a single direction. If the fault is supplied power from both ends, the secondary voltages combine and result in an increase in the out-of-balance current that activates the relay.

Drawbacks: The voltage balance system is hampered by the following limitations:

- To achieve precise balance between current transformer pairs, a multi-gap transformer architecture is necessary.
- The system is appropriate for safeguarding cables of relatively limited lengths because of the capacitance of pilot wires. When using long connections, the charging current can be strong enough to activate the relay, even if a perfect balance of current transformers is achieved.

The aforementioned drawbacks have been successfully addressed in the Translay (modified) balanced voltage system.

9.10 Translay Scheme:

This system is a modified version of the voltage-balance system. While the concept of balanced voltages is still there, this system differs from the previous voltage-balance system in that the balance or opposition occurs between the voltages produced in the secondary coils wound on the relay magnets, rather than between the secondary voltages of the line current transformers. As the current transformers used in the Translay scheme only need to provide power to a relay coil, they can be designed without any air gaps. This allows the system to be utilised for feeders of any voltage. Translay relay is mainly employed for feeder protection, in unit 10 of section 10.23.2 it is explained in detail with neat illustrations. The primary application of Translay relay is to safeguard transmission lines. A comprehensive explanation, accompanied by clear pictures, can be found in unit 10 of section 10.23.2.

9.11 Primary and Secondary Protection Schemes:

In the event of a fault in any component of the electric power system, it is imperative to promptly resolve it to prevent harm or disruption to the remaining parts of the system. A common practice is to categorise the protection strategy into two classes: primary protection and back-up protection.

9.11.1 Primary Protection:

Primary protection scheme is specifically developed to safeguard the individual components of the power system. Referring to Figure 9.20, each line is equipped with an overcurrent relay that serves to protect the line. If a fault arises on any line, it will be promptly resolved by its corresponding relay and circuit breaker. This constitutes the primary or principal protection and functions as the initial barrier of defence.

The service record of the primary relaying is exceptionally high, with a success rate over ninety percent for all operations. Occasionally, errors may remain unresolved by the primary relay system due to issues with the relay itself, the wire system, or the breaker. In such circumstances, the back-up protection fulfils its necessary function. Primary protection may fail for the following reasons.

- Failure of CT or PT operation.
- Failure in relay's operational current or voltage.
- Failure in the circuit breaker's trip mechanism.
- Failure of DC supply to the tripping circuit.
- Failure of main protective relay operation.
- Failure in the wiring of the relay system

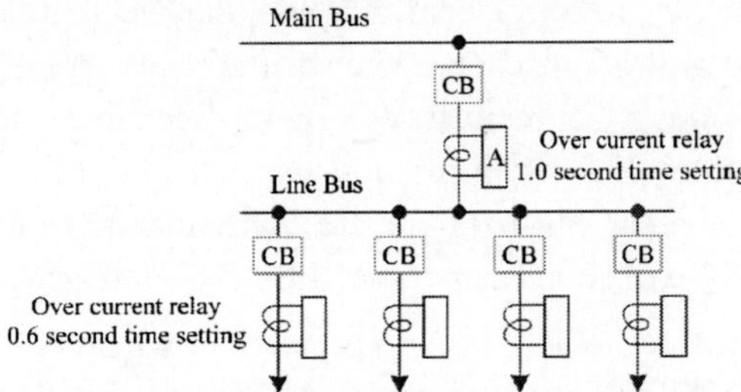

Fig. 9.20 Primary and Secondary Protection Schemes

9.11.2 Back-up protection:

Establishing back-up protection is crucial for maintaining a stable and dependable power system. It is well acknowledged that it is difficult to create a system that is completely secure and efficient due to the tendency for failures in the associated CTs, PTs, circuit breaker, and other components. In the instance of its occurrence, it would result in the complete destruction of our entire switching system.

When the primary protection operation has difficulties, the secondary protection mechanism isolates the faulty part from the system. Additionally, when we disconnect the primary protection for the purpose of testing or maintenance, the secondary or backup protection will function as the primary protection. The design incorporates a suitable time delay to ensure that the primary relaying has adequate time to function, provided it is capable of doing so. Therefore, as depicted in Figure 9.20, relay A offers backup protection for all four lines with a time setting of 1 second. If a line fault persists despite the efforts of its relay and breaker, the relay A on the group breaker will activate after a specific period of time and eliminate the entire group of lines. When the primary relaying performs correctly, a smaller portion is disconnected compared to when the back-up relaying functions. Consequently, it is imperative to prioritise the improved upkeep of primary relays. The types of secondary or backup protection include

(i). Relay Backup Protection

(ii). Remote Backup Protection

(iii). Local Breaker Backup Protection

9.11.2.1 Relay Backup Protection: This is a form of local backup that is achieved by installing an additional protection relay. In this backup protection system, if the primary relay fails, the backup relay will trip the same circuit breaker. Furthermore, the backup relay operates instantaneously without any delay. The relay backup protection is an expensive backup protection system mostly employed in locations where remote backup protection is not feasible. It is crucial to acknowledge that the relay backup protection devices receive power from distinct current and potential transformers.

9.11.2.2 Remote Backup Protection:

Remote backup is defined as the overlapping of primary relays in one protective area onto adjacent areas. In the context of remote backup protection, the backup relays are situated at a neighbouring station. In this backup protection configuration, the backup relays offer comprehensive protection to all components of the protection scheme, such as the relay, circuit breaker, potential transformer, current transformer, and various other elements. Remote backup protection is the predominant form of backup protection employed for transmission lines. It is the most cost-effective and uncomplicated method of backup protection. Furthermore, this particular form of protection is immune to the elements that lead to the failure of the primary protection. This attribute renders it highly coveted as a means of safeguarding.

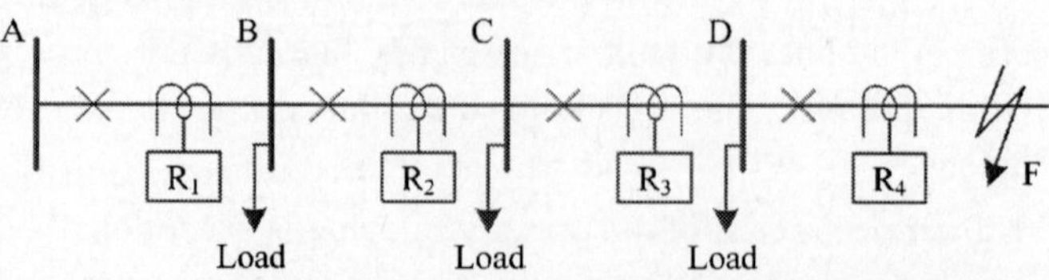

Fig. 9.21 Remote Backup Protection

Consider remote backup protection, which is given by a small-time graded relay, as illustrated in Fig. 9.21. Let F be the fault that occurs on relay R_4. The relay R_4 operates the circuit breaker at D, isolating the defective part. In the event of a failure of circuit breaker D to operate, the relay R_3 at C will be activated to isolate the faulty part. In the event of a failure of circuit breakers D and C to operate, the relay R_2 at B will be activated to isolate the faulty part. In the event of a failure of circuit breakers D, C, and B to operate, the relay R_1 at A will be activated to isolate the faulty part.

Backup protection is used based on economic and technical considerations. For economic reasons, backup protection is typically slower than primary protection.

9.11.2.3 Local Breaker Backup (LBB) Protection:

The breaker backup protection functions as a form of local backup protection, specifically designed for busbar systems that incorporate multiple circuit breakers. In the event of a fault, if a protective relay activates but the circuit breaker fails to trip, the fault is classified as a bus bar fault. LBB is an acronym for Local Breaker Backup. It serves as a backup protection mechanism for the circuit breaker, allowing it to isolate faults in the event that the circuit breaker fails to trip. LBB protection is a crucial component of power system protection.

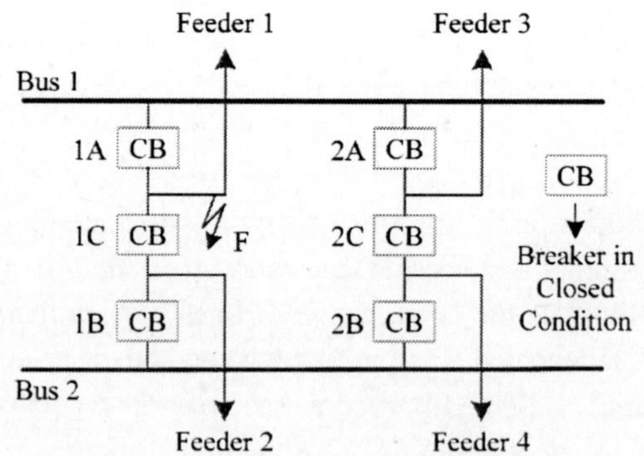

Fig. 9.22 Local Breaker Backup (LBB) Protection

Figure 9.22 depicts the single line schematic of a one and a half breaker system. Assuming that all the Feeders are operational and all the Breakers are in the closed state. Assume a malfunction in Feeder-1 took place. According to the Power System Protection concept, in order to separate the faulty portion, Breaker CB-1A and CB-1C must be deactivated after receiving the directive to deactivate from the Master Trip Relay.

The Master Trip Relay is responsible for issuing the tripping instruction to the Breaker after it receives a trip signal from an individual protection element. In the event of an Earth Fault occurring in Feeder-1, the Earth Fault detecting Relay will not immediately initiate the tripping of the Breakers. Instead, it will send a tripping instruction to the Master Trip Relay, which will then trip the associated Breakers.

Case (i) : Assuming that Breaker CB-1C failed to open after receiving the trip command from the Master Trip Relay, but CB-1A successfully opened. What will occur subsequently?

Since CB-1C did not open, the fault will persist as Bus-2 continues to supply power through CB-1B, which is not ideal for the Power System. In this scenario, it is necessary to identify and separate the fault, which can be accomplished by the implementation of Local Breaker Back-up Protection or Breaker Failure Protection. Therefore, the LBB protection system will isolate the defective portion by activating Breaker CB-1B. Therefore, the Local Breaker Back-up Protection, also known as LBB, triggers the opening of CB-1A and CB-1B in order to fully isolate the problematic part.

There are two crucial concepts to comprehend regarding LBB Protection. One is LBB initiation and other is LBB actuation. The concept of LBB initiation develops as a result of the inherent time delay between the operation of the relevant Fault Sending Relay and Master Trip Relay and the opening of the breaker allow 30ms for the time delay between fault detection and Master Trip Relay activation, and 40ms for Breaker tripping after receiving a trip instruction from the Master Trip Relay. As a result, even after the Master Trip Relay is activated, the LBB Relay will detect an overcurrent for 40ms. So, the LBB Relay is not meant to activate during this 40ms time period. As a result, LBB Relay has an intentionally timed delay. The typical time delay is 200ms.

To resolve this issue, LBB initiation and actuation logic are supplied.

LBB Initiation Logic: The LBB Relay will activate if the Master Trip Relay is activated and an overcurrent condition exists. As a result, the LBB Relay timer will commence.

LBB Actuation Logic: As an intentional time delay (say 200ms) is supplied, after the commencement of LBB logic, if the condition for LBB commencement persists for 200ms, the LBB Relay will activate to trip CB-1A and CB-1B.

Case (ii) : Assume the identical failure occurred in Feeder-1, but *CB-1A failed to open.* In this instance, LBB Protection will trip CB-1C and CB-2A.

9.11.3 Comparison between Primary and Backup Protection:

The table 9.2 outlines the key distinctions between primary and backup protection in a power system.

Table 9.2 Comparison of Primary and Backup Protection

Parameter	Primary Protection	Backup Protection
Main function	The primary protection is designed to provide the initial layer of protection against faults and abnormal circumstances in the power system.	The major purpose of the backup protection is to offer protection against defects in the event that the primary protection fails.

Also referred as	Primary protection is also referred to as "main protection."	Backup protection is also referred to as "secondary protection."
Operations speed	The primary protection mechanism includes high-speed devices that can quickly detect and isolate faults.	The backup protection devices are slower in speed than those employed for primary protection.
Sensitivity	Primary protection is more sensitive in identifying and reacting to abnormal or faulty situations.	Backup protection is designed to have a lower level of sensitivity compared to primary protection. In order to prevent any inadvertent tripping.
Installation location	Primary protection is typically positioned in close proximity to the equipment that requires protection.	Typically, backup protection is built at a remote location apart from the primary protection.
Protection scope	Primary protection offers a more limited range of protection as it is designed to protect a specific portion of the system.	The extent of safeguarding provided by backup protection is broader than that of primary protection. Therefore, it has the ability to safeguard a more extensive portion of the system.
Priority	The detection and clearance of problems are given significant emphasis in primary protection.	Backup protection is of lower priority compared to primary protection. Therefore, it functions as an additional layer of protection.
Cost	The primary protection strategy incurs higher costs due to its high sensitivity, accuracy, and speed.	The cost of implementing a backup protection plan is comparatively lower than that of main protection.

9.12 Summary

- A protective relay is a device that detects faults and activates the circuit breaker to isolate the faulty part from the rest of the system.
- The instantaneous overcurrent relay was specifically developed to defend against extremely high currents in a short period of time, typically less than 0.1 seconds.
- When the current reaches a predetermined threshold, a definite current relay engages immediately.
- The pick-up current refers to the minimal amount of current required to initiate the operation of the relay coil. It can be calculated by multiplying the rated secondary current of the current transformer by the current setting.
- An induction-type non-directional overcurrent relay initiates corrective action when the circuit current exceeds a predetermined limit. These relays are only suitable for alternating current circuits and can handle fault current flow in both directions.
- An induction type directional power relay activates when power flows through the circuit in a specific direction. In contrast to nondirectional overcurrent relays, a directional power relay generates working torque using magnetic fields from the circuit's voltage and current sources.

- The directional power relay cannot be used as a directional protective relay in the event of a short circuit. When a short-circuit develops, the system voltage drops, and the relay may not generate enough torque to operate. This challenge is overcome with the induction type directional overcurrent relay, which is designed to be nearly independent of system voltage and power factor.
- An impedance relay operates by counteracting the torque generated by a current and voltage element. When the V/I ratio goes below a predefined level, the relay activates. Impedance is an electrical measurement of the distance travelled along a transmission line.
- A time-distance impedance relay is a type of relay that automatically adapts its operating time based on the distance between the relay and the source of failure.
- A current differential relay compares the current entering and leaving a system part.
- The biased beam relay is specifically intended to detect and react to the difference in current, relative to the current flowing through the shielded portion, in a fractional manner.
- The voltage balance differential relay employs air core CTs in which voltages generate current. When a fault develops in the protection zone, the current in the CTs becomes unbalanced, causing the voltage in the CTs' secondary to be disturbed. The current starts to flow via the operational coil.
- The Translay scheme is similar to a voltage balancing system, with the main difference being that the balance or opposition is between the voltages produced in the secondary windings wrapped around the relay magnets, rather than between the secondary voltages of the line current transformers.
- The primary protection is designed to provide the initial layer of protection against faults and abnormal circumstances in the power system.
- The major purpose of the backup protection is to offer protection against defects in the event that the primary protection fails.
- In Relay Backup Protection, if the primary relay fails, the backup relay trips the same circuit breaker. Furthermore, the backup relay operates instantly, with no delay.

Short and Long Answer Questions

1. What is a protective relay? Describe its role in an electrical system.
2. Examine the essential criteria for protective relaying.
3. Define and explain the following terms in the context of protective relaying: pick-up current, current settings, plug-setting multiplier and time-setting multiplier.
4. Using a neat diagram, explain the design and operation of a non-directional induction type overcurrent relay.
5. Explain the construction and operation of a directional induction type power relay using a neat schematic diagram.
6. Using a diagram, explain the construction and operation of two types of overcurrent relays: (i) non-directional induction relays and (ii) directional induction relays.
7. Use illustrations to explain the construction and operation of two types of impedance relays: definite distance and time-distance.
8. Explain the structure and operation of the primary and secondary protection schemes using necessary diagrams.

To know more about

How Relays Work
and why we use Relay in
PLC Applications?

To know more about

Distance relay

Primary and Backup
Protection

To know more about

Optimization Techniques
for Directional Overcurrent
Relay Coordination

MATLAB & SIMULINK

To Model & Simulate

- Instantaneous Relay
- IDMT Relay
- Over current Relay

To Model & Simulate

- Differential relay
- PS Protection
- Distance Relay

To Model & Simulate

- Over Voltage Relay
- Under Voltage Relay
- Primary and

10 Protection of Alternators, Transformers, Busbars and Transmission Lines

Unit specifics: In this unit, the following topics have been discussed for basic understating of protection of alternators, transformers, busbars and transmission lines:

- Differential, modified differential protection, biased differential, biased modified differential, stator inter-turn protection, restricted earth fault protection of alternators.
- Differential, modified differential, biased, and biased modified differential protection methods for (a) delta/star, (b) star/delta, (c) star/star, and (d) delta/delta transformers.
- Backup, differential and frame leakage protection of bus-bars.
- Merz-price voltage balance system, translay scheme for protection of transmission lines.
- Distance protection/zone protection for protecting high voltage transmission lines.

Rationale: In this unit, students will learn about alternator faults, differential, modified differential, biased, and biased modified differential alternator protection. Stator inter-turn alternator protection (differential, modified differential, biased differential, and biased modified differential), restricted earth fault protection of alternators; transformer faults, earth fault relay (or) core balance leakage protection of transformers, leakage and overload protection of transformers, problems associated with the application of differential protection to transformers; differential, modified differential, biased, biased modified differential protection of transformers, are clearly described with the diagrams, and examples. In this unit, students will also be introduced to busbar protection, bus-bar configurations, bus zone faults, backup protection for bus-bars, differential protection for busbars, frame leakage/fault bus protection, introduction to protection of lines, time-graded overcurrent protection of feeders, Merz-Price voltage balance system for protecting transmission lines, translay scheme for protection of transmission lines, distance protection/zone protection for protecting high voltage transmission lines, are clearly described with the help of necessary diagrams.

Pre-Requisites: Basic knowledge of power system components.

Unit Outcomes: List of outcomes of this unit is as follows

U10-O1: To understand the types of faults on alternators, transformers, busbars and transmission lines.

U10-O2: To analyse differential, modified differential, biased, biased modified differential protection of alternators; and Stator inter-turn protection methods of alternators

U10-O3: To analyse the differential, modified differential, biased, and biased modified differential protection methods for (a) delta/star, (b) star/delta, (c) star/star, and (d) delta/delta transformers.

U10-O4: To comprehend different faults on bus-bars and transmission lines.

U10-O5: To analyse Backup, differential and frame leakage protection of bus-bars.

U10-O6: To analyse differential pilot-wire protection, Translay scheme, distance protection/zone protection for protection of transmission lines.

Unit-10 outcomes	Expected mapping with course outcomes (1: Weak, 2: medium, and 3: strong correlation)					
	CO-1	CO-2	CO-3	CO-4	CO-5	CO-6
U10-O1	2	1	-	1	1	-
U10-O2	2	2	1	2	3	3
U10-O3	2	2	1	2	2	3
U10-O4	2	-	-	1	1	-
U10-O5	2	2	1	2	3	-
U10-O6	2	2	1	2	3	-

10.1 Introduction:

Electrical Power system consists of alternators, transformers, bus-bars, transmission lines and other associated equipment. It is essential and imperative to safeguard each component from a range of potential malfunctions that may arise in the near or distant future. The protective relays mentioned in the previous chapter can be effectively used to identify any malfunctioning circuit component and activate appropriate correction actions.

The stator winding problems are the most critical issues that alternators might have, necessitating prompt treatment. Transformers experience significant malfunctions primarily as a result of short-circuits within the transformers themselves or in their connections. The primary method employed for safeguarding against these defects is the differential relay scheme, as its differential measurements render it considerably more responsive than alternative protective systems. This unit focuses mainly on the protection strategies for alternators and transformers.

10.2 Alternator Faults:

The generating units, particularly the larger ones, are relatively scarce and have greater individual costs compared to most other equipment. Thus, it is both desirable and essential to implement protection that can address the diverse array of malfunctions that may arise in modern power plants. Several critical failures that can arise on an alternator include:

(a) Prime-mover failure: When the prime-mover malfunctions, the alternator operates as a synchronous motor and consumes a certain amount of current from the power supply system.

(b) Field failure: Alternator field failure is obviously quite unusual. Allowing the alternator to run without a field for a short duration will not result in instant damage. The control room staff may be authorised to manually separate the malfunctioning alternator from the system bus-bars. Therefore, it is a universal practice to not provide automatic protection against this possibility.

(c) Overcurrent: The major cause of this issue is typically the partial breakdown of the insulation in the winding or an excessive load on the power supply system.

(d) Overspeed : The primary factor leading to overspeed is the abrupt reduction or complete elimination of the load on the alternator. Modern alternators typically have mechanical centrifugal devices attached to their driving shafts. These devices are designed to activate the main valve of the prime-mover in the event of a hazardous overspeed.

(e) Overvoltage: The field excitation system of modern alternators is constructed in such a way that it prevents overvoltage problems from occurring during normal running speeds. Overvoltage in an alternator arises when the speed of the prime-mover abruptly increases as a result of a sudden loss of the alternator load.

(f) Unbalanced loading: Unbalanced loading refers to the condition when the phase currents in the alternator are not equal. Unbalanced loading occurs when there are faults in the circuit external to the alternator, either in the form of earth faults or faults between phases. If the imbalanced currents are not addressed, they can cause significant damage to the mechanical fittings of the rotor core or the field winding, potentially resulting in serious burns.

(g) Stator winding faults: Stator winding failures are most commonly caused by stator winding insulation breakdowns. The principal stator winding faults are classified in order of significance, including:

- Phase-to-ground fault (Earth Fault).
- Phase-to-phase fault.
- An inter-turn fault occurs when there is a problem with the turns of the same phase winding.

The defects in the stator winding are highly hazardous and have the potential to cause significant harm on the expensive equipment. Hence, it is essential to implement automatic protection in order to promptly rectify such errors and minimize the magnitude of harm. To protect the alternator against above stated faults, the following protection schemes can be used.

(i) Differential protection (or) Merz-price protection.
(ii) Modified differential protection.
(iii) Biased differential protection.
(iv) Biased Modified differential protection.
(v) Stator inter-turn protection.
(vi) Restricted earth fault protection.

10.3 Differential Protection of alternators:

The prevailing method for protecting against failures in stator windings is based on the circulating-current theory. In this protection technique, the currents at both ends of the protected region are compared. Typically, these currents are the same, but they might become different if a fault occurs in the protected area. The disparity in the electric currents during fault conditions is deliberately directed through the functioning coil of the relay. Subsequently, the relay activates its contacts in order to separate the protected region from the system. The other name for this type of protection is the Merz-Price circulating current scheme.

Construction: Figure 10.1 depicts the schematic configuration of differential protection used for a three-phase alternator. Two identical current transformer pairs, CT_1 and CT_2, are positioned on both sides of each phase of the stator windings. The secondary windings of each set of current transformers are interconnected in a star configuration. The two neutral points and the corresponding terminals of the two-star groups are connected together using a four-core pilot cable. Consequently, there exists a distinct route for the electric currents flowing in each set of current transformers and their corresponding pilot wire P.

The relay coils are connected in a star pattern, with the neutral point connected to the current transformer's common neutral and the outer ends connected separately to each of the other three pilots. To ensure that the load is distributed uniformly throughout each current transformer, the relays are attached to equipotential points along the three pilot wires. The equipotential points are essentially situated in the centre of the pilot wires. Electromagnetic relays are frequently used and are specifically designed for quick response, as it is critical to eliminate flaws as soon as possible.

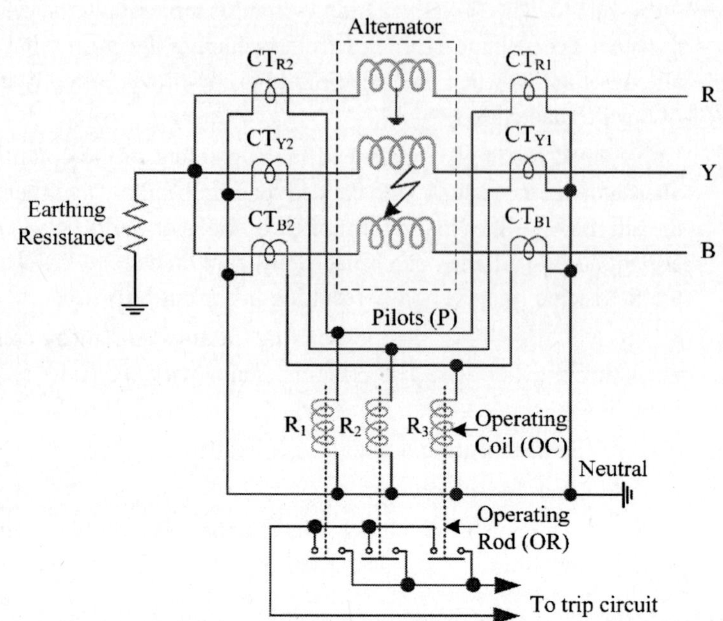

Fig. 10.1 Differential Protection of alternators without balancing resistors

Operation: Under normal operational circumstances, the current at both ends of each winding will be identical, resulting in equal currents in the secondaries of two current transformers (CTs) coupled in either phase. As a result, the pilot wires have an equal and stable flow of current, whereas the operating coils (R_1, R_2, and R_3) of the relays do not have any current passing through them.

Earth Fault: When phase-to-phase fault or an earth fault arises, the existing state becomes invalid, and the differential current passing through the relay circuit causes the relay to activate the circuit breaker. Suppose there is a fault on the **R** phase of the electrical system, resulting from the deterioration of insulation and its connection to the ground, as shown in Figure 10.1. The current flowing through the affected phase winding will pass through the machine's core and frame before grounding. The neutral earthing resistance is used to complete this circuit. When there is an imbalance in the currents flowing through the secondaries of the two current transformers (CTs) in phase R, the resulting disparity in currents will flow through the relay coil (**R_1**) and then return through the neutral pilot. Consequently, the relay serves to initiate the operation of the circuit breaker. Similarly, if there is an earth fault on **Y** phase, relay coil **R_2** will be activated, and if there is an earth fault on **B** phase, coil **R_3** will be activated.

Phase-to-phase fault: Suppose that a short-circuit fault has occurred between the phases Y and B, as depicted in Figure 10.1. The short-circuit current flows through the neutral end connection, the two windings, and the fault, as indicated by the dotted arrows. When there is an issue, the currents in the secondaries of two current transformers (CTs) in each affected phase will become imbalanced. This will cause a differential current to pass through the operating coils of the relays (namely R_2 and R_3) that are connected in these phases. Subsequently, the relay activates its contacts in order to trigger the circuit breaker. Similarly, if there is phase-to-phase fault on **R** and **Y**-Phases, relay coils **R_1R_2** will be activated, and if there is phase-to-phase fault on **B** and **R**-Phases, relay coils **R_3R_1** will be activated. Table 10.1 illustrates the activation of a relay during earth and phase to phase fault in differential Protection of alternators.

Table 10.1 Activation of Relay during earth fault and phase to phase fault in differential Protection of alternators

Differential Protection of alternators					
Type of Fault	**Fault on**	**Relay Coil Triggered**	**Type of Fault**	**Fault between**	**Relay Coils Triggered**
Earth Fault	**R** Phase	**R_1**	Phase-Phase Fault	**R** and **Y** Phases	**R_1 R_2**
	Y Phase	**R_2**		**Y** and **B** Phases	**R_2 R_3**
	B Phase	**R_3**		**B** and **R** Phases	**R_3 R_1**

It should be observed that the relay circuit is designed in such a way that when it is activated, two things happen: (i) the breaker connecting the alternator to the bus-bars is opened, and (ii) the field circuit of the alternator is also opened. An effective approach involves the installation of current transformers CT1 in the neutral connections, usually in the alternator pit, and current transformers CT2 in the switch-gear apparatus. At times, the alternator is situated at a considerable distance from the switchgear. Connecting the relay coils directly to the actual midpoints of the pilots is problematic because the relays are located close to the circuit breaker. To ensure equilibrium, resistances are incorporated into the shorter parts of the pilots under these circumstances. This guarantees that the relay tapping points evenly divide the combined secondary impedance of the two sets of current transformers into two equal halves. Figure 10.2 illustrates this configuration. Typically, these resistances can be adjusted to achieve precise equilibrium.

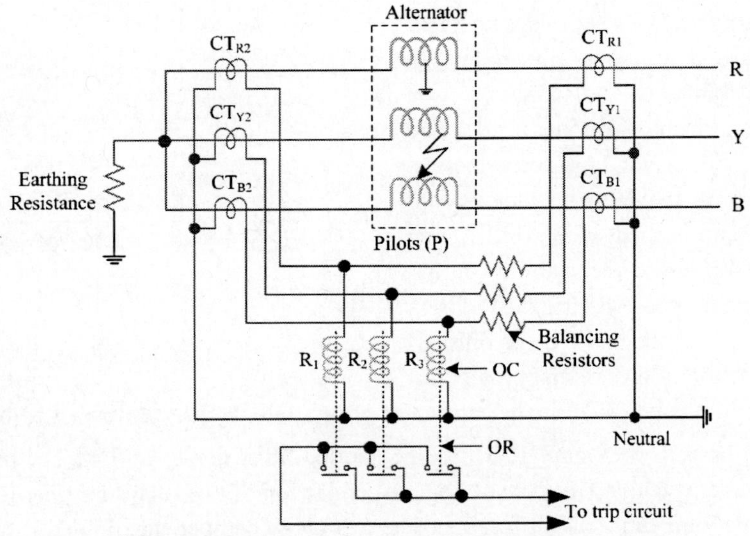

Fig. 10.2 Differential Protection of alternators with balancing resistors

Disadvantages: Both of the depicted circuits for protecting the alternator possess their own respective constraints. Neutral earthing resistance is commonly employed to mitigate the detrimental impact of earth-fault currents. During earth-faults, it is not feasible to fully protect all of the stator windings in a star-connected alternator. If an earth-fault develops in close proximity to the neutral point, there could not be enough voltage across the short-circuited section to generate the required current for the relay to function. The size of the vulnerable area is determined by the earthing resistance and relay setup.

10.4 Modified differential protection of alternators:

When the neutral point of a star-connected alternator is grounded using a high resistance, the protection methods depicted in Figure 10.1 or 10.2 will not offer adequate sensitivity for detecting earth-faults. The high earthing resistance restricts the earth-fault currents to a low level, which requires the use of relays with low current settings in order to adequately safeguard a significant amount of the generator winding. Nevertheless, setting the relay too low is not recommended as it may compromise the stability during severe phase-faults. To address this challenge, a modified version of differential protection is employed, wherein the earth fault settings are decreased without compromising stability.

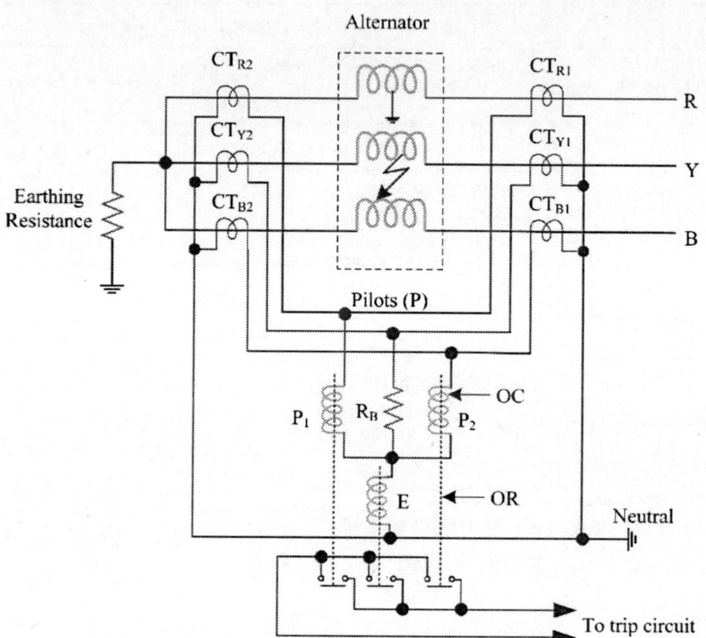

Fig. 10.3 Modified differential Protection of alternators

Construction: Figure 10.3 displays the modified configuration. The modifications exclusively impact the relay connections, involving the connection of two relays for phase-fault protection and a third relay solely for earth-fault protection. The two-phase components, $\mathbf{P_1}$ and $\mathbf{P_2}$, and the balancing resistance, R_B, are connected in a star configuration. The earth relay, **E**, is connected between the star point and the fourth wire of the circulating current pilot-circuit.

Operation: Under normal operational circumstances, the current at both ends of each winding will be identical, resulting in equal currents in the secondaries of two current transformers (CTs) coupled in either phase. As a result, the pilot wires have an equal and stable flow of current, whereas the operating coils (P_1, P_2, and E) of the relays do not have any current passing through them.

Typically, the earth fault relay 'E' is configured to activate when a small amount of fault current flows through it, whereas the phase relays (P_1 and P_2) are configured to activate when a large amount of fault current flows through them.

Earth Fault: Assume that there is a fault on **R** phase of the earth, caused by the insulation breaking down and connecting to the ground, as depicted in Figure 10.3. If there is an imbalance in the currents flowing through the secondary windings of the two current transformers (CTs) in **R** phase, the resulting disparity in currents will flow via the phase relay coil (P_1) and earth relay coil (E), and then return through the neutral pilot. Although the fault current passes through both the phase relay coil (P_1) and the earth relay coil (E), the current setting of the earth relay coil 'E' is adjusted to activate for lower values compared to the phase relay P_1. Therefore, the earth relay **'E'** will be activated.

Suppose there is a fault on the **Y** phase of the earth. If there is an imbalance in the currents passing through the secondary windings of the two current transformers (CTs) in the Y phase, the resulting difference in currents will travel through the braking resistor (R_B) and the earth relay coil (E), and then return through the neutral pilot. In this scenario, the earth relay **'E'** will also be enabled.

Similarly, if there is a fault on **B** phase of the earth, the resulting disparity in currents will flow via the phase relay coil (P_2) and earth relay coil (E), and then return through the neutral pilot. Although the fault current passes through both the phase relay coil (P_2) and the earth relay coil (E), the current setting of the earth relay coil 'E' is adjusted to activate for lower values compared to the phase relay P_2. In this case also the earth relay **'E'** will be activated. In modified differential protection, if an earth fault occurs on any phase of the generator winding, earth relay 'E' will be activated.

Phase-to-phase fault: If a fault occurs between two phases, the fault current will circulate around the two transformer secondary coils P_1, P_2, and R_B without passing through the earth relay 'E'. As a result, only the phase fault relay will operate. Suppose that a short-circuit fault has occurred between the phases **R** and **Y**, the currents in the secondaries of two current transformers (CTs) in each affected phase will become imbalanced. The fault current will circulate through the phase relay coil (P_1) and braking resistor (R_B) without passing through the earth relay 'E'. As a result, only the phase fault relay $\mathbf{P_1}$ will operate.

If short-circuit fault has occurred between the phases **Y** and **B**, the fault current will circulate through the braking resistor (R_B) and phase relay coil (P_2) without passing through the earth relay 'E'. As a result, only the phase fault relay $\mathbf{P_2}$ will operate. In case, if short-circuit fault has occurred between the phases **B** and **R**, the fault current will circulate through both the phase relay coils (P_1 and P_2) without passing through the earth relay 'E'. As a result, either P_1 or P_2 or both P_1 and P_2 will operate. Table 10.2illustrates the activation of a relay during earth and phase to phase fault in modified differential Protection of alternators.

Table 10.2Activation of Relay during earth fault and phase to phase fault in modified differential Protection of alternators

Modified differential protection of alternators					
Type of Fault	**Fault on**	**Relay Coil Triggered**	**Type of Fault**	**Fault between**	**Relay Coils Triggered**
Earth Fault	**R** Phase	**E**	Phase-Phase Fault	**R** and **Y** Phases	$\mathbf{P_1}$
	Y Phase			**Y** and **B** Phases	$\mathbf{P_2}$
	B Phase			**B** and **R** Phases	P_1 and/or $\mathbf{P_2}$

10.5. Biased circulating current protection of alternators:

In order to ensure the functioning of the relay, it is necessary for the current transformers (CTs) at both ends of an alternator winding to have the same level of precision. If there is a significant difference in accuracy, it can lead to a malfunction in the relay. In order to prevent such an operation, biased circulating current protection is employed.

Construction: This protection includes CTs connected in a star pattern on both outgoing sides, as well as a machine winding connection to earth. The restraining coils are powered by the secondary connection of CTs in each phase, while the operating coils are connected to the restraining coils' central tapping and the neutral pilot wire. The constructional details are shown in Figure 10.4.

Operation: Under normal operating conditions, the line CTs' secondary outputs are always equal to the CTs' neutral-end outputs. Thus, the phase pilot wires and relay restraining windings have a balanced circulating current, and no current passes through the operating coils (R_1, R_2, and R_3).

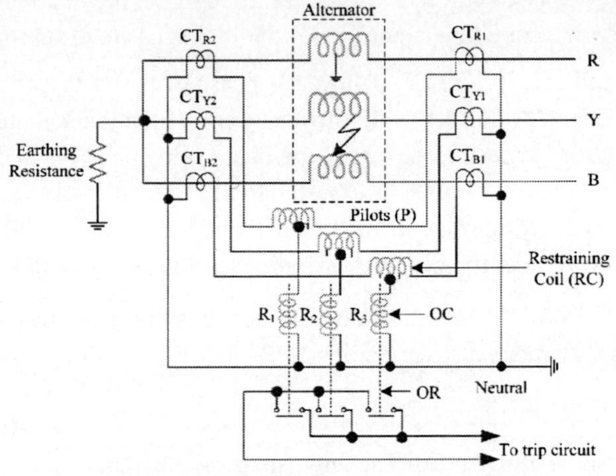

Fig. 10.4 Biased differential Protection of alternators

Earth Fault and Phase-to-phase fault: As previously mentioned, restraining coils are employed to ensure the same level of precision at both ends of the CTs. With the exception of this, biased differential protection will function similarly to differential protection during earth faults and phase-to-phase fault circumstances. Table 10.3 illustrates the activation of a relay during earth and phase to phase fault in biased differential Protection of alternators.

Table 10.3 Activation of Relay during earth fault and phase to phase fault in biased differential Protection of alternators

Biased differential Protection of alternators					
Type of Fault	**Fault on**	**Relay Coil Triggered**	**Type of Fault**	**Fault between**	**Relay Coils Triggered**
Earth Fault	**R** Phase	R_1	Phase-Phase Fault	**R** and **Y** Phases	R_1 R_2
	Y Phase	R_2		**Y** and **B** Phases	R_2 R_3
	B Phase	R_3		**B** and **R** Phases	R_3 R_1

10.6. Biased Modified circulating current protection of alternators:

Construction: Modified biased circulating current protection of alternators is depicted in Fig. 10.5. As previously mentioned, restraining coils are employed to ensure the same level of precision at both ends of the CTs. With the exception of this, modified biased differential protection will function similarly to modified differential protection during earth faults and phase-to-phase fault circumstances.

Table 10.4 Activation of Relay during earth fault and phase to phase fault in biased modified differential Protection of alternators

Biased modified differential protection of alternators					
Type of Fault	**Fault on**	**Relay Coil Triggered**	**Type of Fault**	**Fault between**	**Relay Coils Triggered**
Earth Fault	**R** Phase	**E**	Phase-Phase Fault	**R** and **Y** Phases	$\mathbf{P_1}$
	Y Phase			**Y** and **B** Phases	$\mathbf{P_2}$
	B Phase			**B** and **R** Phases	$\mathbf{P_1}$ and/or $\mathbf{P_2}$

Operation: As mentioned earlier, the earth fault relay 'E' is configured to activate when a small amount of fault current flows through it, whereas the phase relays (P_1 and P_2) are configured to activate when a large amount of fault current flows through them. In biased modified differential protection also, if an earth fault occurs on any phase of the generator winding, earth relay 'E' will be activated.

If phase-to-phase fault occurs between the phases RY, YB and BR then the operating coils $\mathbf{P_1}$, $\mathbf{P_2}$, **&** $\mathbf{P_1}$ and/or $\mathbf{P_2}$ will be activated correspondingly. Table 10.4 illustrates the activation of a relay during earth and phase to phase fault in biased modified differential Protection of alternators.

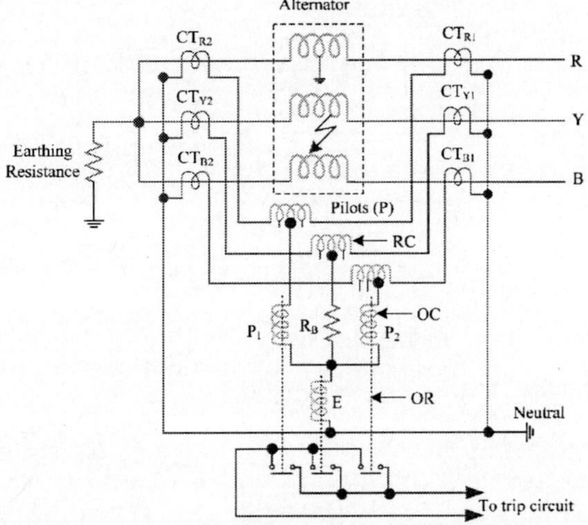

Fig. 10.5 Modified biased differential Protection of alternators

10.7 Stator inter-turn protection of alternators:

Differential protection provides protection against faults that occur between a phase and ground, as well as faults that occur between different phases. It does not provide protection against a problem that occurs between consecutive turns on the same phase winding of the stator. This sort of problem generates a local circuit current between the affected turns, without causing a disparity in the currents entering and exiting the winding at its two ends where current transformers are used. However, it is generally deemed unnecessary to offer protection for inter-turn faults since they eventually progress into earth faults.

Construction: Figure 10.6 depicts a schematic representation of (a) a differential (b) a modified differential stator inter-turn Protection. Figure 10.7 depicts a schematic representation of (a) a biased differential and (b) a Biased Modified differential stator inter-turn Protection. Three additional relay coils (R_X, R_Y, and R_Z) are included to safeguard the system from inter-turn faults, in addition to the current three relay coils. The relay coils (R_X, R_Y, and R_Z) will be utilized to safeguard against inter-turn faults in the R, Y, and B phases, respectively.

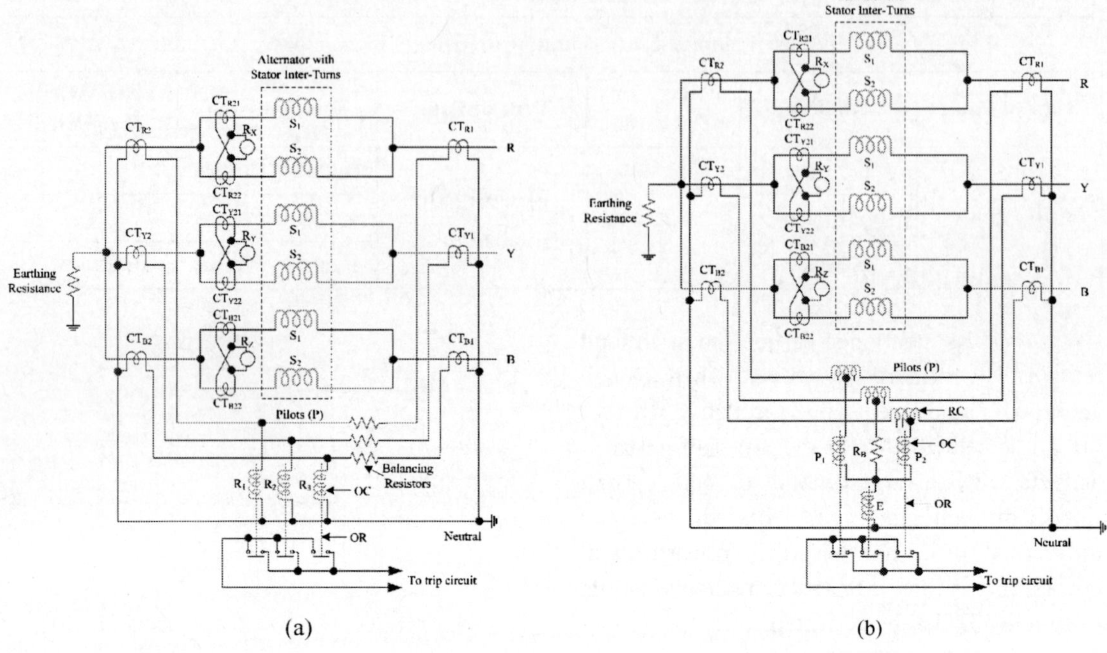

Fig. 10.6 Stator inter-turn Protection (a) Differential and (b) Modified differential

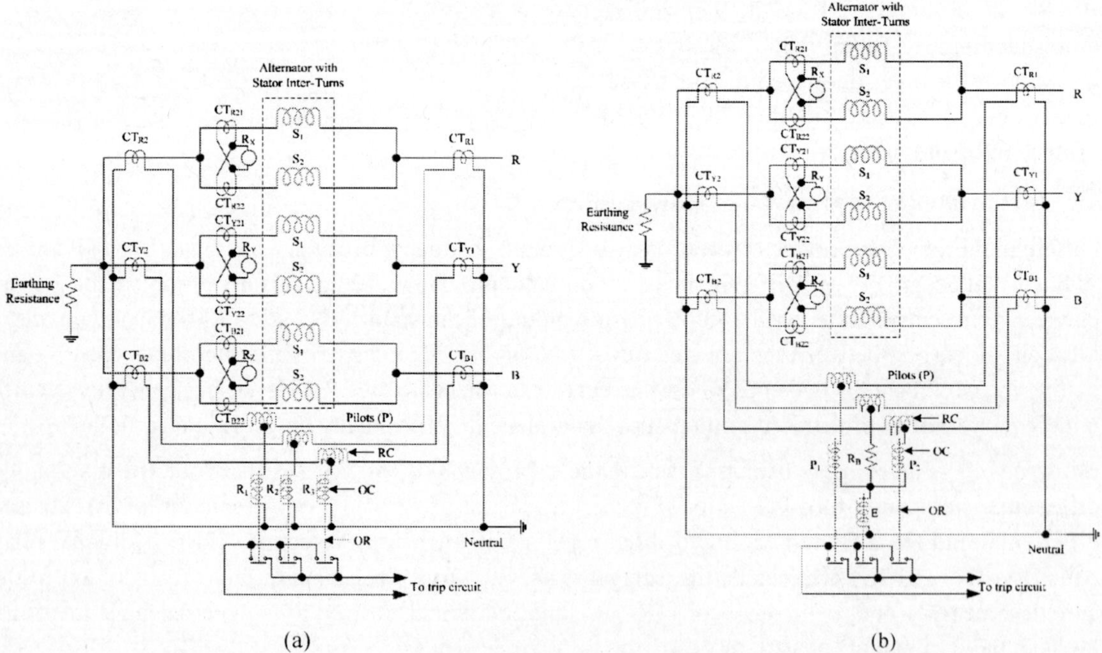

Fig. 10.7 Stator inter-turn Protection (a) Biased Differential and (b) Biased Modified differential

Operation: Under normal operational circumstances, the current at both ends of each winding will be identical, resulting in equal currents in the secondaries of two current transformers (CTs) coupled in either phase. As a result, the pilot wires have an equal and stable flow of current, whereas the operating coils (R_1, R_2, and R_3) and the remaining three relay coils (R_X, R_Y, and R_Z) do not have any current passing through them.

Inter-turn fault: Figure 10.6 illustrates the presence of two stator windings S_1 and S_2 for a single phase, which are designed to prevent inter-turn failures. Two current transformers are interconnected using the circulating-current method. Typically, the currents in the stator windings S_1 and S_2 are identical, resulting in equal currents in the secondaries of both CTs. The secondary current circulating around the loop is uniform at all places, resulting in no current passing through the relay R_X. In the event of a short-circuit occurring between neighbouring turns, namely on S_1, the currents flowing through the stator windings S_1 and S_2 will become unequal. Consequently, the secondaries of CTs will have unequal currents induced, and the disparity between these two currents passes through the relay R_X. Subsequently, the relay activates its contacts in order to eliminate the generator from the system.

Earth Fault and Phase-to-phase fault:

As stated earlier, three additional relay coils (R_X, R_Y, and R_Z) are used to protect the system from inter-turn faults. Except for this, the Stator inter-turn differential protection will operate in a similar manner to differential protection. The Stator inter-turn modified differential protection will also function similar to modified differential protection. Similarly, the Stator inter-turn biased differential protection will operate in a way similar to biased differential protection. Lastly, the Stator inter-turn biased modified differential protection will function similar to biased modified differential protection in cases of earth faults and phase-to-phase fault situations. Table 10.5 illustrates the activation of a relay during earth, phase to phase and inter-turn faults in stator inter-turn differential (or) biased differential protection of alternators. Table 10.6 illustrates the activation of a relay during earth, phase to phase and inter-turn faults in stator inter-turn modified differential (or) biased modified differential protection of alternators.

Consider the following

CT_{R1} = Current flowing through the primary winding of R phase

CT_{R2} = Current flowing through the secondary winding of R phase

CT_{Y1} = Current flowing through the primary winding of Y phase

CT_{Y2} = Current flowing through the secondary winding of Y phase

CT_{B1} = Current flowing through the primary winding of B phase

CT_{B2} = Current flowing through the secondary winding of B phase

CT_{R21} = Current flowing through the secondary winding of first stator inter-turn in R phase

CT_{R22} = Current flowing through the secondary winding of second stator inter-turn in R phase

CT_{Y21} = Current flowing through the secondary winding of first stator inter-turn in Y phase

CT_{Y22} = Current flowing through the secondary winding of second stator inter-turn in Y phase

CT_{B21} = Current flowing through the secondary winding of first stator inter-turn in B phase

CT_{B22} = Current flowing through the secondary winding of second stator inter-turn in B phase

Table 10.5 Activation of a relay during earth, phase to phase and inter-turn faults in stator inter-turn differential protection of alternators.

Stator inter-turn differential (or) Biased differential protection of Alternators			
Type of Fault	**Fault on**	**Condition of current flow through CTs**	**Relay coils triggered**
Inter-turn Fault	R Phase	$CT_{R1} = CT_{R2}$ but $CT_{R21} \neq CT_{R22}$	R_X
Earth fault		$CT_{R1} \neq CT_{R2}$ but $CT_{R21} = CT_{R22}$	R_1
Inter-turn Fault and earth fault		$CT_{R1} \neq CT_{R2}$ and $CT_{R21} \neq CT_{R22}$	R_X and R_1
Inter-turn Fault	Y Phase	$CT_{Y1} = CT_{Y2}$ but $CT_{Y21} \neq CT_{Y22}$	R_Y
Earth fault		$CT_{Y1} \neq CT_{Y2}$ but $CT_{Y21} = CT_{Y22}$	R_2
Inter-turn Fault and earth fault		$CT_{Y1} \neq CT_{Y2}$ and $CT_{Y21} \neq CT_{Y22}$	R_Y and R_2
Inter-turn Fault	B Phase	$CT_{B1} = CT_{B2}$ but $CT_{B21} \neq CT_{B22}$	R_B
Earth fault		$CT_{B1} \neq CT_{B2}$ but $CT_{B21} = CT_{B22}$	R_3
Inter-turn Fault and earth fault		$CT_{B1} \neq CT_{B2}$ and $CT_{B21} \neq CT_{B22}$	R_B and R_3
Phase-to-Phase fault	R and Y phases	$CT_{R1} \neq CT_{R2}$ but $CT_{R21} = CT_{R22}$ $CT_{Y1} \neq CT_{Y2}$ but $CT_{Y21} = CT_{Y22}$	R_1 and R_2
		$CT_{R1} \neq CT_{R2}$ and $CT_{R21} \neq CT_{R22}$ $CT_{Y1} \neq CT_{Y2}$ and $CT_{Y21} \neq CT_{Y22}$	R_1, R_2, R_X and R_Y
	Y and B phases	$CT_{Y1} \neq CT_{Y2}$ but $CT_{Y21} = CT_{Y22}$ $CT_{B1} \neq CT_{B2}$ but $CT_{B21} = CT_{B22}$	R_2 and R_3
		$CT_{Y1} \neq CT_{Y2}$ and $CT_{Y21} \neq CT_{Y22}$ $CT_{B1} \neq CT_{B2}$ and $CT_{B21} \neq CT_{B22}$	R_2, R_3, R_Y and R_Z
	B and R phases	$CT_{B1} \neq CT_{B2}$ but $CT_{B21} = CT_{B22}$ $CT_{R1} \neq CT_{R2}$ but $CT_{R21} = CT_{R22}$	R_3 and R_1
		$CT_{B1} \neq CT_{B2}$ and $CT_{B21} \neq CT_{B22}$ $CT_{R1} \neq CT_{R2}$ and $CT_{R21} \neq CT_{R22}$	R_3, R_1, R_Z and R_X

Table 10.6 Activation of a relay during earth, phase to phase and inter-turn faults in stator inter-turn modified differential protection of alternators.

Stator inter-turn Modified differential (or) Biased Modified differential protection of Alternators			
Type of Fault	**Fault on**	**Condition of current flow through CTs**	**Relay coils triggered**
Inter-turn Fault	R Phase	$CT_{R1} = CT_{R2}$ but $CT_{R21} \neq CT_{R22}$	R_X
Earth fault		$CT_{R1} \neq CT_{R2}$ but $CT_{R21} = CT_{R22}$	E
Inter-turn Fault and earth fault		$CT_{R1} \neq CT_{R2}$ and $CT_{R21} \neq CT_{R22}$	R_X and E
Inter-turn Fault	Y Phase	$CT_{Y1} = CT_{Y2}$ but $CT_{Y21} \neq CT_{Y22}$	R_Y
Earth fault		$CT_{Y1} \neq CT_{Y2}$ but $CT_{Y21} = CT_{Y22}$	E
Inter-turn Fault and earth fault		$CT_{Y1} \neq CT_{Y2}$ and $CT_{Y21} \neq CT_{Y22}$	R_Y and E
Inter-turn Fault	B Phase	$CT_{B1} = CT_{B2}$ but $CT_{B21} \neq CT_{B22}$	R_B
Earth fault		$CT_{B1} \neq CT_{B2}$ but $CT_{B21} = CT_{B22}$	E
Inter-turn Fault and earth fault		$CT_{B1} \neq CT_{B2}$ and $CT_{B21} \neq CT_{B22}$	R_B and E
Phase-to-Phase fault	R and Y phases	$CT_{R1} \neq CT_{R2}$ but $CT_{R21} = CT_{R22}$ $CT_{Y1} \neq CT_{Y2}$ but $CT_{Y21} = CT_{Y22}$	P_1
		$CT_{R1} \neq CT_{R2}$ and $CT_{R21} \neq CT_{R22}$ $CT_{Y1} \neq CT_{Y2}$ and $CT_{Y21} \neq CT_{Y22}$	P_1, R_X and R_Y
	Y and B phases	$CT_{Y1} \neq CT_{Y2}$ but $CT_{Y21} = CT_{Y22}$ $CT_{B1} \neq CT_{B2}$ but $CT_{B21} = CT_{B22}$	P_2
		$CT_{Y1} \neq CT_{Y2}$ and $CT_{Y21} \neq CT_{Y22}$ $CT_{B1} \neq CT_{B2}$ and $CT_{B21} \neq CT_{B22}$	P_2, R_Y and R_Z
	B and R phases	$CT_{B1} \neq CT_{B2}$ but $CT_{B21} = CT_{B22}$ $CT_{R1} \neq CT_{R2}$ but $CT_{R21} = CT_{R22}$	**P_1** and/or **P_2**
		$CT_{B1} \neq CT_{B2}$ and $CT_{B21} \neq CT_{B22}$ $CT_{R1} \neq CT_{R2}$ and $CT_{R21} \neq CT_{R22}$	**P_1** and/or **P_2**, R_Z and R_X

10.8 Restricted earth fault protection of alternators:

In small alternators, it is typical for the neutral ends of the three-phase windings to be connected internally to a single terminal. Therefore, the implementation of the Merz-Price circulating current principle, as previously mentioned, is not possible since there are no provisions for including the necessary current transformers in the neutral connection of each phase winding. Under such circumstances, it is considered acceptable to implement a well-balanced earth-fault protection strategy as the sole means of protecting against earth-faults. This design lacks protection against phase-to-phase failures, unless they develop into earth-faults, which is the scenario for the majority of such failures.

Construction: Figure 10.8 depicts the schematic configuration of a balanced earth-fault prevention system designed for a three-phase alternator. The system comprises three line current transformers, with one installed in each phase. The secondaries of these transformers are connected in parallel with the secondary of a single current transformer in the cable that connects the star point of the alternator to the ground. A relay is linked to the secondary windings of the transformers. The protection against earth faults is confined to the area between the neutral and line current transformers.

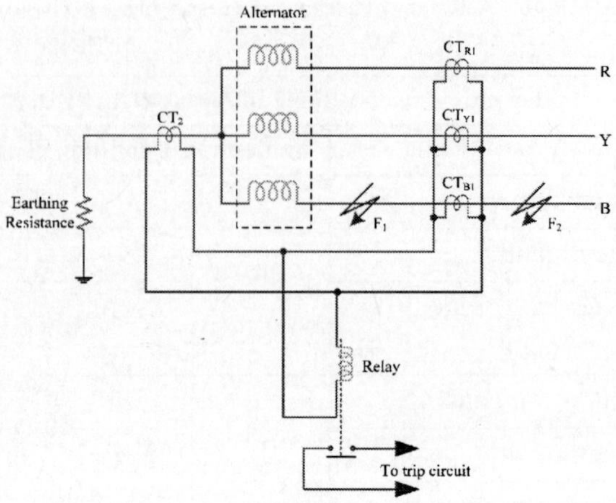

Fig. 10.8 Restricted earth fault protection of alternators

Operation: Under normal operational conditions, the currents flowing through the alternator leads and the currents in the secondaries of the line current transformers neutralise each other, resulting in a net current of zero. Consequently, there is no electrical current flowing through the relay. Moreover, under these conditions, there is no current flowing via the neutral wire, and the secondary of the neutral current transformer does not supply any current to the relay.

If an earth-fault appears at F_2 outside the protected region, the current flowing through the alternator terminals will be exactly the same as the current in the neutral connection. As a result, no current will pass through the relay. When an earth-fault occurs at F_1 or within the protected zone, the currents become unbalanced and the differential current flows through the operating coil of the relay. Afterward, the relay is activated by closing its contacts, which effectively disconnects the alternator from the system.

Example 10.1. The generator is equipped with a 33 kV, 100 MVA capacity and is protected by a differential scheme. The percentage of the generator winding that requires protection against phase to ground faults is 70%. The relay is said to activate when there is a 10% discrepancy in current. Calculate the appropriate resistance to be inserted into the ground connection.

Ans. Given Data: 33kV, 100 MVA,

70 % winding is protected from ground fault. So, % of un-protected winding = 30%

$$Full\ load\ current = \frac{100 \times 10^6}{\sqrt{3} \times 33 \times 10^3} = 1749.546\ \text{A}$$

Minimum fault current which will operate the relay is I = 10 % of Full load current

$$= 0.1 \times 1749.546 = 174.9546\ \text{A}$$

$$\text{Phase voltage } V_{Ph} = \frac{33 \times 10^3}{\sqrt{3}} = 19.0525\ \text{kV}$$

$$\text{\% winding un-protected} = \frac{RI \times 100}{V_{Ph}} \Rightarrow 0.3 = \frac{R \times 174.9546}{19.0525 \times 10^3} \Rightarrow R = 32.67\ \Omega$$

Example 10.2. The neutral point of a 3-phase, 50 megavolt-ampere (MVA), 11 kilovolt (kV) alternator is connected to the ground through a 10-ohm resistance. The relay is configured to activate when there is an imbalance in the current of 2 amperes. The CTs have a ratio of 800/5.

(i). What is the percentage of winding that needs to be protected?

(ii). Determine the grounding resistance necessary to protect 85% of the winding.

Ans. Given data: 3-Ø, 50MVA, 11kV

$$V_{Ph} = \frac{11\times10^3}{\sqrt{3}} = 6350.85 \text{ V}$$

Earth Resistance = 10 Ω

CTs ratio = 800/5

Unbalance current = 2A

The minimum current which will operate the relay during fault conditions: $I = \frac{2*800}{5} = 320\ A$

(i) % winding un-protected = $\frac{RI\times100}{V_{Ph}} = \frac{10\times320\times100}{6350.85}$ =50.39 %

% winding protected = 100% − 50.39% = 49.6063 %

(ii) Given % winding protected = 85 %

% winding un-protected = 15 %

$$0.15 = \frac{R\times320}{6350.85} \Rightarrow R = 2.97\ \Omega$$

Example 10.3. The alternator is connected in a star configuration and has a voltage of 11 kV and a power rating of 25 MVA. It has a reactance of 4 Ω per phase and negligible resistance. Merz-Price protection is employed to safeguard the winding. The neutral grounding resistance is 5 Ω when just 10% of the winding is left unprotected. Determine the setting of the relay.

Ans. Given data: 11kV, 25MVA

$$V_{Ph} = \frac{11\times10^3}{\sqrt{3}} = 6350.85 \text{ V}$$

Neutral grounding resistance = 5 Ω

Reactance = 4Ω

% un–protected winding = 10%

Full load current $= \frac{25\times10^6}{\sqrt{3}\times11\times10^3} = 1312.159$ A

Impedance offered by 10 % of unprotected winding to earthed fault current = 5 + 0.1×j4 = (5 + j0.4) Ω

Voltage induced in the un–protected portion of winding is = 0.1 × 6350 = 635 V

The earth fault current caused by voltage induced in the unprotected winding *is*

$$= \frac{\textit{Voltage induced in un–protected winding}}{\textit{Impedance of un–protected winding}}$$

$$= \frac{635}{5+j0.4} = 126.59 \text{ A}$$

Relay Setting $= \frac{\textit{Earth fault current}}{\textit{Full Load current}} \times 100 = \frac{126.59}{1312.159} = 9.647\ \%$

Example 10.4. The alternator is a star-connected type with a power rating of 10 MVA and a voltage rating of 11 kV. It has a synchronous reactance of 3 Ω per phase and a resistance of 1 Ω. The system is safeguarded by a Merz-Price balanced current system, which activates when the current imbalance exceeds 25% of the full load current. Calculate the proportion of the alternator winding that is unprotected if the star point is grounded through a 9 Ω resistance.

Ans. Given P = 10MVA, V = 11kV Z = 1+ j3, R = 9Ω

$$\text{Full load current} = \frac{10\times10^6}{\sqrt{3}\times11\times10^3} = 524.86\text{A}$$

Out of balance current which can cause relay operation is fault current = 25% of Full load current

$$= 0.25 \times 524.86$$

$$= 131.2159 \text{ A}$$

Let % of un-protected winding = x

Equivalent impedance of un – protected winding = (1+j3) × 0.01 x=0.01 x + j0.03 x

Impedance offered by un-protected winding to earth fault = (Earthing resistance + Equivalent impedance of unprotected winding)

$$= 9 + (1+j3) \times 0.01\, x$$

$$=(9 + 0.01\, x + j0.03\, x)\ \Omega$$

$$\text{Voltage induced in unprotected winding} = \frac{V_{Ph}\times x}{100} = \frac{11\times10^3\times x}{\sqrt{3}\times100} = 63.50\, x$$

The earth fault current will be mainly governed by the earth fault resistance, so neglecting the generator reactance for un–protected winding fault current $= \dfrac{63.50\,x}{9+0.01x}$

$$131.2159 = \frac{63.50\,x}{9+0.01x}$$

$$1180.89 + 1.312\,x = 63.50\,x$$

$$x = \frac{1180.89}{63.5-1.312}$$

$$= 18.9898\%$$

% protected winding = 100% – 18.9898% = 81.01 %.

Example 10.5. The alternator is connected in a star configuration and has a voltage of 33 kilovolts, a power rating of 150 megavolt-amperes, a reactance of 1.8 per unit per phase, and a negligible resistance. The system is safeguarded by a Merz-Price balanced current system, which activates when the current imbalance exceeds 15% of the load current. If the neutral point is connected to the ground through a resistance of 4 Ω, calculate the percentage of the winding that is safeguarded from earth faults.

Ans. Given data: 33kV, 150 MVA, R = 4 Ω, X = 1.8 per unit/phase

Out of balance current exceeds 15 % of load current

Let % unprotected winding is x and Reactance per phase is X

$$V_{Ph} = \frac{33\times10^3}{\sqrt{3}} = 19052.55 \text{ V}$$

$$\text{Full load current} = \frac{150\times10^6}{\sqrt{3}\times33\times10^3} = 2624.32\text{A}$$

Voltage induced in un-protected winding = 19052.55 × 0.01 x = 190.525 x

Earth fault current caused by un-protected winding = $I_{(full\ load)} \times 0.15$

Earth Impedance = $\frac{Voltage\ Induced}{Earth\ fault\ current} = \frac{190.525\ x}{393.65} = 0.484\ x\ \Omega$

Reactance per phase X = $\frac{Reactance\ p.u. \times V_{Ph}}{I} = \frac{1.8 \times 33 \times 10^3}{\sqrt{3} \times 2624.32} = 13.067\ \Omega$

Reactance of un-protected winding = $\frac{13.067 \times x}{100} = 0.13067\ x$

$$Z = \sqrt{R^2 + X^2}$$

$$Z^2 = R^2 + X^2$$

$$(0.484\ x)^2 = 4^2 + (0.13067\ x)^2$$

$$0.2342\ x^2 = 16 + 0.01707 x^2$$

$$x^2 = \frac{16}{0.2342 - 0.01707} = 73.688$$

x = 8.58% (un- protected winding)

% protected winding = 100 – 8.58 = 91.42%

Example 10.6. The relay has a minimum pickup current of 0.4 A and a slope of 10%. A high resistance ground fault occurs in close proximity to the grounded neutral end of the generator winding, with the current distribution depicted as indicated. Consider a current transformer (CT) ratio of 300:6. Determine whether the relay will function or not.

Ans. Given data: I_1 = 480 A, I_2 = 440 A, slope = 10%

CT ratio = 300/6

Min. pick up current = 0.4 A

The difference current = $I_1 - I_2$ = 40 A

The current in the operating coil = $40 \times \frac{6}{300} = 0.8$A

The average sum of 2 currents = $\frac{I_1 + I_2}{2} = \frac{480+440}{2} = 460$ A

The average current through restraining coil = $460 \times \frac{6}{300} = 9.2$ A

Operating current = Min pickup current + Restraining current × slope = 0.4 + (9.2 × 0.1) = 1.32 A

Current through operating coil is 0.8 A which is less than 1.32A i.e; the operating current, so relay will not operate.

Example 10.7. The generator, with a voltage of 13.8 kilovolts and a power rating of 125 megavolt-amperes, is connected to the ground through a resistance of 7 ohms. The C. T's have a ratio of 1000/6. The relay is configured to activate when there is a current imbalance of 2 A. What proportion of the generator winding will be safeguarded by the percentage differential protection scheme?

Ans. Given data : 13.8 kV , 125MVA , R = 7Ω

$$V_{Ph} = \frac{13.8 \times 10^3}{\sqrt{3}} = 7967.43\ \text{V}$$

CTs ratio = 1000:6

Full load current (I) = $\frac{125\times10^6}{\sqrt{3}\times13.8\times10^3}$ = 5229.62 A

I, the min fault current which will operate the relay is = $2\times\frac{1000}{6}$ = 333.333 A

% winding un-protected = $\frac{RI\times100}{V_{Ph}} = \frac{7\times333.33\times100}{7967.43}$ = 29.28 %

% of protected winding = 100% − 29.28 % = 70.714 %

Example 10.8. The alternator stator winding is safeguarded by a percentage differential relay with a typical slope of 20% for the relationship between the difference in currents (I_1 - I_2) and the average of the currents (I_1 + I_2)/2. An earth fault with significant resistance has occurred closer to the grounded neutral end of the generator stator winding while the generator is under load. Given a CT ratio of 800/8 A, will the relay cause the generator circuit breaker to trip at this fault condition?

Ans. Given data: CTs ratio = 800:8

Secondary current of CT_1 = $600\times\frac{8}{800} = 6\ A$

Secondary current of CT_2 = $520\times\frac{8}{800} = 5.2\ A$

Out -of-balance current = $I_1 - I_2 = 6 - 5.2 = 0.8\ A.$

Average of the currents = $\frac{I_1+I_2}{2} = \frac{6+5.2}{2} = 5.6\ A$

Corresponding point on $(I_1 - I_2)$ versus $\left(\frac{I_1+I_2}{2}\right)$ characteristic is 5.6 × 0.2 i.e., 1.12 A.

Since the out-of-balance current is 0.8 A, which is less than 1.12 A, so relay will not operate.

10.9 Transformer faults and protection schemes:

Transformers are stationary apparatuses that are completely enclosed and typically immersed in oil. Consequently, the likelihood of defects arising on them is extremely low. Nevertheless, the repercussions of an infrequent malfunction might be exceedingly severe unless the transformer is promptly withdrawn from the system. This requires the provision of sufficient automated protection for transformers to guard against potential problems.

10.9.1 Transformer faults:

Compared to generators, power transformers are only susceptible to a few specific issues, namely

- open circuits,
- overheating, and
- various types of winding short-circuit such as earth-faults, phase-to-phase faults, and inter-turn faults.

An open circuit in one phase of a 3-phase transformer can cause unwanted heating. However, relay protection is typically not implemented for open circuits as this condition is relatively harmless. In the event of such a fault, the transformer can be manually disconnected from the system. The overheating of the transformer typically arises from prolonged overloads or short-circuits, and in rare occasions, from the malfunction of the cooling system. The relay protection system does not offer protection against this particular situation. Instead, thermal accessories are typically utilized to either emit an alarm or regulate the groups of fans.

Winding short-circuits, also known as internal faults, occur when the insulation of the transformer's winding deteriorates as a result of overheating or mechanical damage. In the event of an internal problem, it is essential to promptly disconnect the transformer from the system to prevent an extended electrical discharge that could potentially ignite an oil fire. Hence, the implementation of relay protection is crucial when it comes to internal failures.

10.9.2 Protection Schemes for Transformer faults:

The Merz-Price circulating-current system is undoubtedly the most effective method for protecting generators. While transformer protection is generally accurate, there are situations where a circulating current system does not provide any specific benefits compared to other systems or becomes impractical due to the challenging conditions caused by the diverse range of voltages, currents, and grounding conditions typically found in power transformers. In such situations, other protective systems are employed that, in numerous instances, are equally as efficient as the circulating-current system.

Transformer protection schemes include:

(i) Earth fault relay (core balance leakage protection)
(ii) Overcurrent relay (leakage and overload protection).
(iii) Differential Protection
(iv) Biased Differential Protection
(v) Buchholz relay.

The functions of the above protective schemes are as follows:

- Earth-fault relays provide protection against earth faults exclusively.
- Overcurrent relays provide protection primarily against phase-to-phase faults and overload.
- Both differential system and biased differential protection protect against earth and phase faults.
- Buchholz relays protect against slow-developing faults, such as winding insulation failure, core heating, and oil level drops caused by leaking joints.

10.10 Earth fault relay (or) core balance leakage protection of transformers:

An earth fault is frequently caused by a partial breakdown of winding insulation to the earth. The leakage current is significantly less than the short-circuit current. Long-lasting earth-faults can cause significant damage before short-circuiting and being eliminated from the system. Earth-fault relays are a cost-effective way to detect and disconnect earth faults or leaks at an early stage. An earth-fault relay is a low-setting overcurrent relay that activates in response to an earth fault or leak.

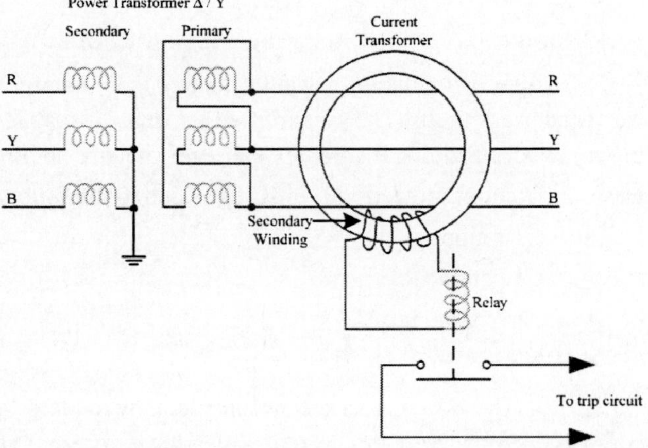

Fig. 10.9 Core balance leakage protection of transformers

Construction: Figure 10.9 illustrates the core-balance leakage protection approach for transformers to prevent earth faults. A power transformer's three primary winding leads pass through the core of a current transformer, which only has one secondary winding. This secondary circuit connects the relay's operational coil.

Operation: Under normal conditions (i.e. no earth fault), the vector sum of three phase currents is zero, resulting in no flux in the core of the current transformer, regardless of load imbalance. As a result, no current passes through the relay, making it inoperable. In the event of an earth fault, the vector sum of three phase currents is no longer zero. The resulting current creates flux in the CT core, causing e.m.f. in the secondary winding. Energizing the relay activates the circuit breaker and disconnects the damaged transformer from the system.

10.11 Overcurrent relay (or) leakage and overload protection of transformers:

The above-mentioned core-balance protection has the disadvantage of not providing overload protection. If there is a defect or leakage between phases, the core-balance relay will not function. Transformers are often protected against both leakage and overload. The earth relay has a low current setting and only functions when there is an earth or leakage problem. Overload relays feature high current settings and prevent problems between phases.

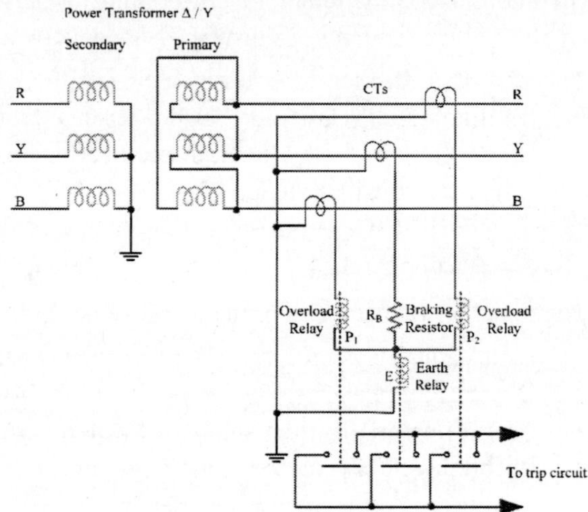

Fig. 10.10 leakage and overload protection of transformers

As depicted, this protective system consists of two overload relays and one leakage or earth relay. The two overload relays guard against phase-to-phase faults, while the earth relay protects against earth faults. Overload relay trip contacts and earth fault relays are linked in parallel. When an overload or earth relay is activated, the circuit breaker will trip.

10.12 Problem associated with the application of differential protection to Transformers:

The Merz-Price circulating-current principle is frequently employed to protect power transformers from earth and phase faults. The system used for transformers is essentially similar to that used for generators, but it includes additional complex elements that are not present in the generator application.

Basic differential protection is unsuitable due to the following limitations:

- CT ratios
- CT connections
- Tap changing
- Magnetising inrush currents.

CT ratios: The currents in the primary and secondary windings of a power transformer need to be compared. Since these two currents often have different values, the utilization of identical transformers (with the same turn ratio) will result in a differential current and activate the relay, even when there is no load present. The disparity in the amount of currents in the primary and secondary windings of a power transformer is balanced by utilizing different turn ratios of current transformers (CTs). The turn ratio of the CTs on the low voltage (LV) side is T times greater than the turn ratio of the CTs on the high voltage (HV) side, where T is the turn ratio of the power transformer. When this requirement is met, the secondary windings of the two current transformers (CTs) will carry the same currents during normal load conditions. As a result, there will be no difference in current flowing through the relay, causing it to remain non-functional.

CT connections: A phase difference typically exists between the primary and secondary currents of a 3-phase power transformer. Despite the employment of CTs with the correct turn-ratio, a relay may nevertheless experience the flow of a differential current during normal conditions, which can trigger its activity. The adjustment for phase difference is achieved through the proper interconnections of current transformers (CTs). The current transformers (CTs) on one side of the power transformer are interconnected in a manner that causes the combined currents sent through the pilot wires to be phase-shifted in the same direction and by the same angle as the phase shift between the primary and secondary currents of the power transformer. Table 10.7 shows the recommended connections for CTs in order to correct for the phase disparity between the primary and secondary currents of a power transformer.

Table 10.7 Primary and Secondary CT connections for different Power Transformer configurations

Current Transformer	Power Transformer		Current Transformer
Primary	Primary	Secondary	Secondary
Delta	Star with Neutral earthed	Star with Neutral earthed	Delta
Delta	Star with Neutral earthed	Delta	Star
Star	Delta	Star with Neutral earthed	Delta
Star	Delta	Delta	Star

It is observed from this table that it is necessary to connect the CTs in a delta configuration on the star side of a star/delta power transformer, and in a star configuration on the delta side.

Tap changing: The majority of transformers are equipped with mechanisms for adjusting the taps, which further complicates this issue. Under normal operating conditions, tap changing will induce differential current across the relay. To address the aforementioned challenge, one can resolve it by modifying the turn-ratio of current transformers (CTs) located on the power transformer side that is equipped with taps.

Magnetising inrush currents: During typical operating conditions, the magnetising current is negligible. Nevertheless, when a transformer is powered on after being taken out of operation, the magnetizing or in-rush current can be exceptionally high for a short period of time. As the magnetizing current flows into the transformer without a comparable current leaving, it is perceived as a fault current by the differential relay, potentially triggering its activation. To address the aforementioned challenge, differential relays are configured to activate when there is a significant level of imbalance. This strategy reduces the sensitivity of the relays. Practically, we exploit the fact that the early surge currents have a noticeable second-harmonic component. Therefore, it is feasible to create a system that utilizes second-harmonic bias characteristics. This system is specifically calibrated to the second-harmonic frequency and is able to exercise control during the activation process in order to prevent malfunction.

10.13 Differential Protection of Transformers:

The differential protection employed for power transformers is founded on the Merz-price circulating current principle. This protection offers protection against both phase-to-phase and phase-to-earth faults and is often utilized for transformers with a rating over 2MVA.

10.13.1 Differential Protection for transformers of delta/star:

Construction: Figure 10.11 illustrates the differential protection mechanism used to safeguard a 3-phase delta/star power transformer against phase-to-ground and phase-to-phase faults. To do this, the current transformers (CTs) on the delta side of the transformer must be linked in a star configuration, while the CTs on the star side should be connected in a delta configuration. This compensates for the phase discrepancy between the primary and secondary of the power transformer. The CTs on both sides are interconnected with pilot wires, with each pair of CTs being controlled by a single relay.

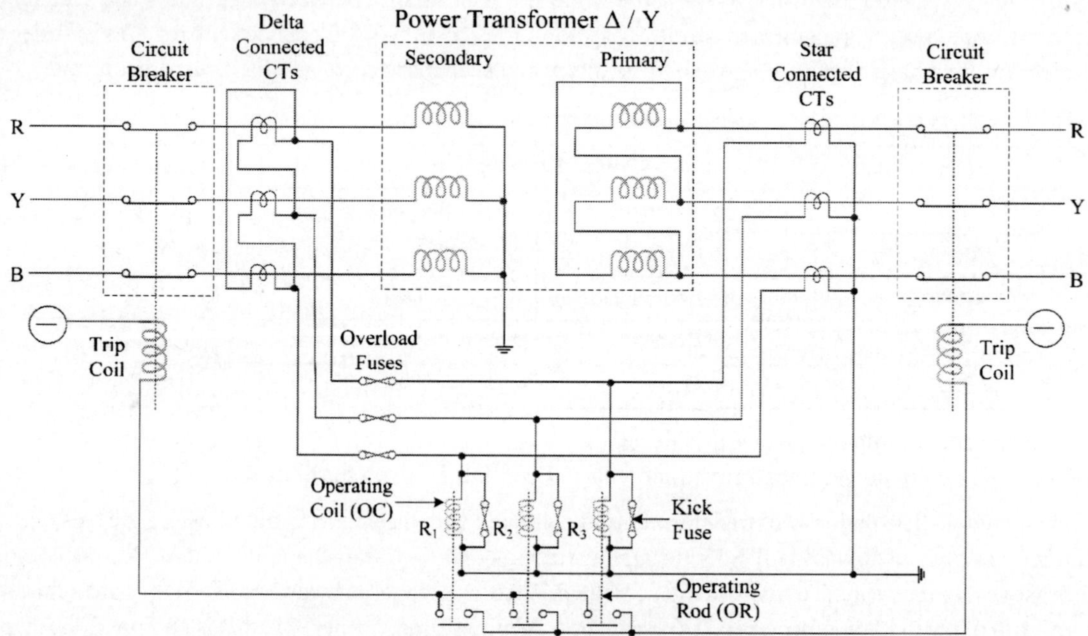

Fig. 10.11 Differential Protection of delta/star transformers

During the activation of a transformer, there is a sudden surge of magnetizing current that flows into the transformer. This surge can be as much as 10 times the normal operating current of the transformer and it decreases gradually over time. Due to the significant flow of current in the primary windings, there is a discrepancy in the output of the current transformers (CTs), leading to incorrect operation of the differential protection system for the transformer. To prevent the activation of differential protection caused by the sudden surge of magnetizing current, kick fuses are installed across the relay coils. These fuses are of the time-limited variety and do not function in short durations. During fault conditions, fuses function by blowing off, allowing the entire current to pass through the relay coils.

Operation: Under normal operational conditions, the secondary windings of current transformers (CTs) carry identical currents. Consequently, the current that enters and exits the pilot wires at both ends is equal, resulting in no current passing through the relays. During an earth or phase fault, the currents in the secondary circuits of current transformers (CTs) become unequal, resulting in the flow of differential current via the relay circuit. The relay circuit will activate the circuit breaker on both ends of the transformer. The circulating current protection mechanism also offers protection against inter-turn failures. A short circuit occurring between turns on the same phase of a power transformer can disrupt the turn ratio and lead to an imbalance between CT pairs. This imbalance triggers the operation of the protection mechanism.

10.13.2 Differential Protection for transformers of star/delta:

Figure 10.12 depicts the differential protection technique employed to safeguard a 3-phase star/delta power transformer against faults occurring between phases and between phases and ground. The primary winding of this star/delta power transformer is connected in a star configuration, while the secondary winding is linked in a delta configuration. The current transformer (CT) on the primary side of the star connection is wired in a delta configuration, while the CT on the secondary side of the delta connection is wired in a star configuration.

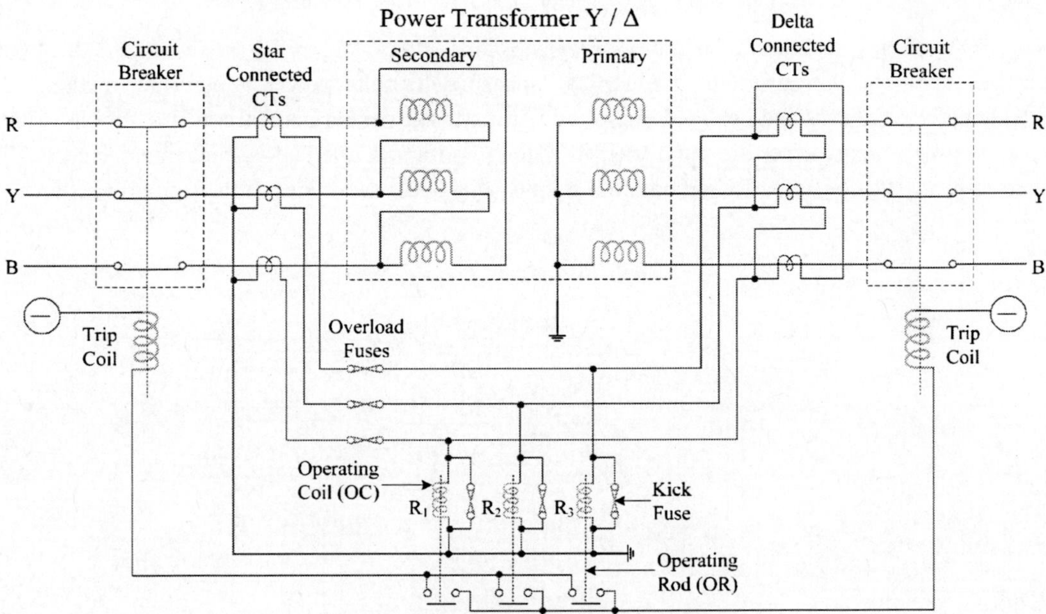

Fig. 10.12 Differential Protection of star/delta transformers

10.13.3 Differential Protection for transformers of star/star: Figure 10.13 depicts the differential protection technique employed to safeguard a 3-phase star/star power transformer against faults occurring between phases and between phases and ground. The primary and secondary windings of this star/star power transformer are connected in a star configuration. The current transformer (CT) is connected in a delta configuration on both the primary and secondary sides of the star connection.

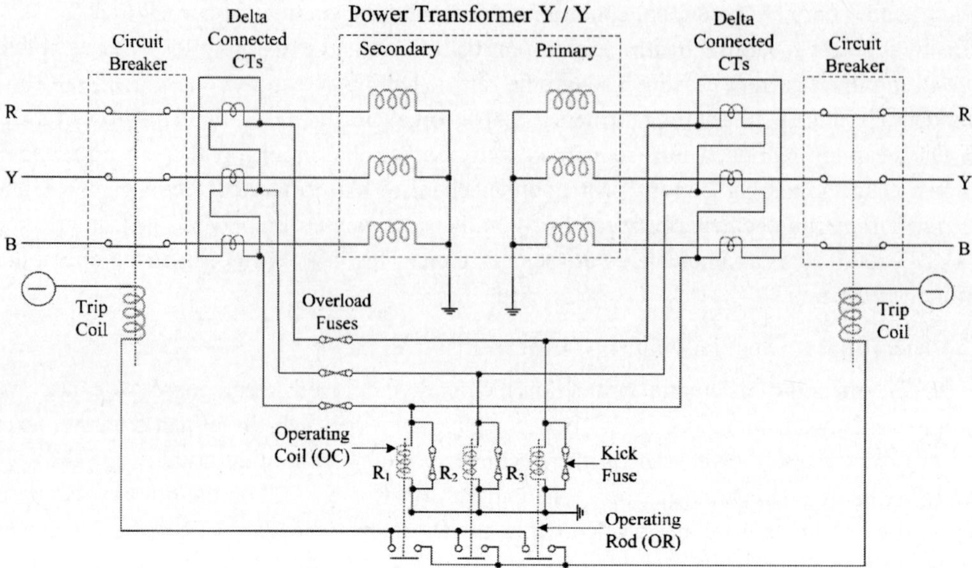

Fig. 10.13 Differential Protection of star/star transformers

10.13.4 Differential Protection for transformers of delta/delta: Figure 10.14 depicts the differential protection technique employed to safeguard a 3-phase delta/delta power transformer against faults occurring between phases and between phases and ground. The primary and secondary windings of this delta/delta power transformer are connected in a delta configuration. The current transformer (CT) is connected in a delta configuration on both the primary and secondary sides of the star connection.

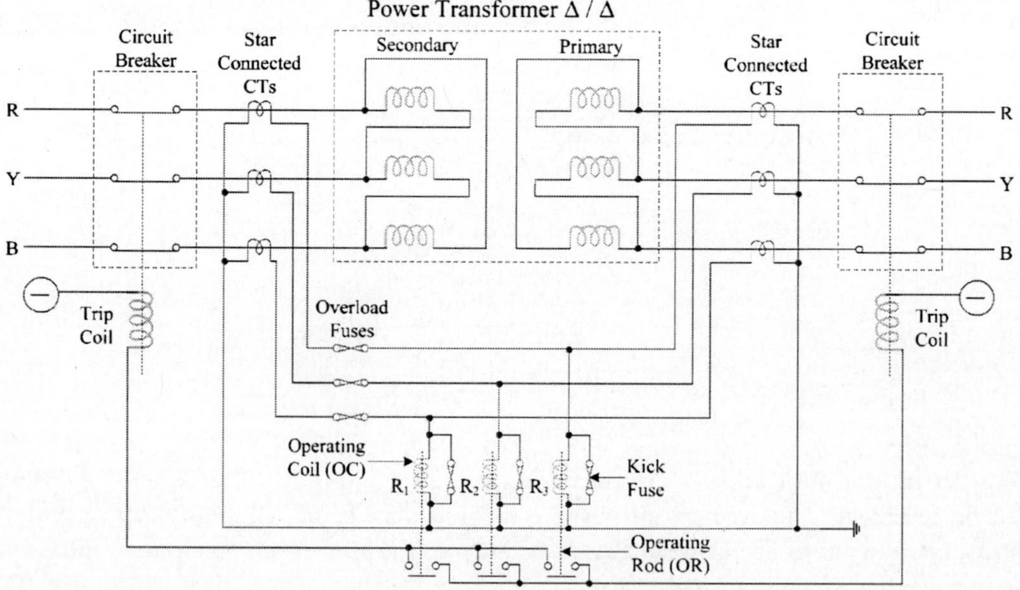

Fig. 10.14 Differential protection of delta/delta transformers

10.14 Modified differential protection of transformers:

As mentioned in the previous section (10.4), modified differential protection employs two relay elements for phase fault protection and a third for earth fault protection relay. Figure 10.15 illustrates the modified differential protection mechanism used to safeguard a 3-phase delta/star, star/delta, star/star and delta/delta power transformer against phase-to-ground and phase-to-phase faults respectively.

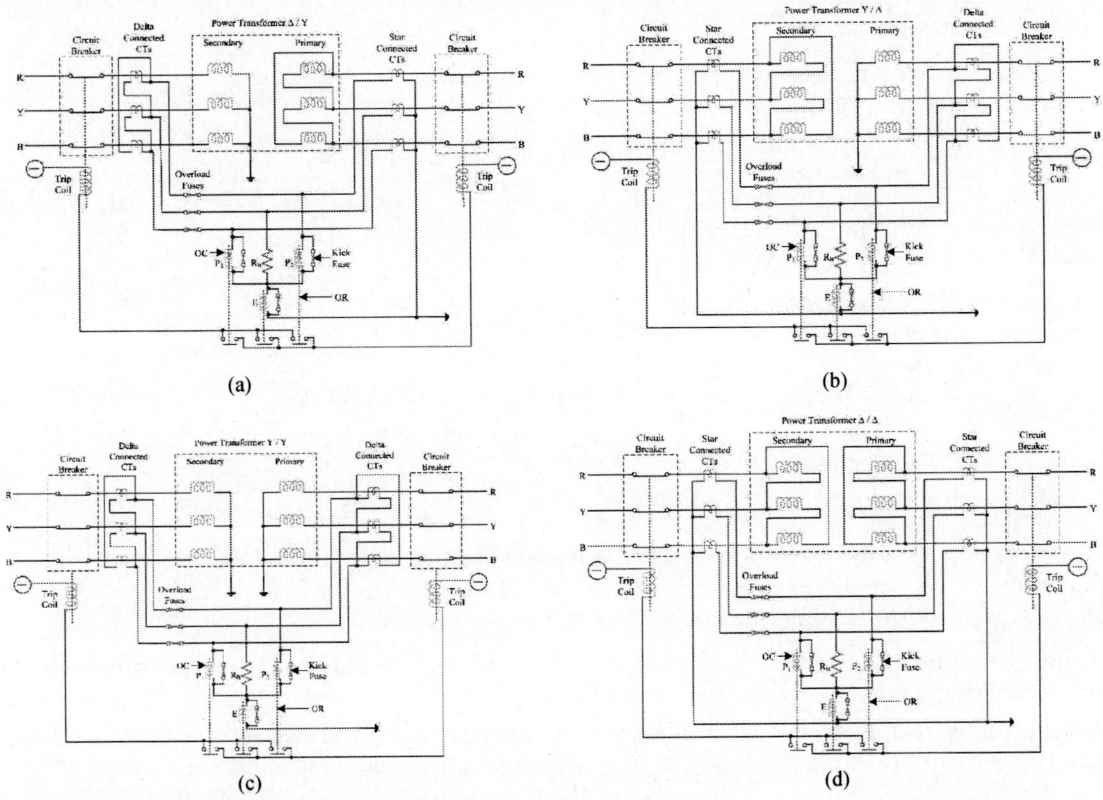

Fig. 10.15 Modified differential protection of (a) delta/star (b) star/delta (c) star/star and (d) delta/delta transformers

10.15 Biased differential protection of transformers:

Biased differential protection in transformers is used to prevent unwanted operation on heavy external faults caused by CT errors and CT ratio changes (due to tap altering). Figure 10.16 illustrates the biased differential protection mechanism used to safeguard a 3-phase delta/star, star/delta, star/star and delta/delta power transformer against phase-to-ground and phase-to-phase faults respectively.

Earth Fault and Phase-to-phase fault:

As previously mentioned, restraining coils are employed to ensure the same level of precision at both ends of the CTs. With the exception of this, biased differential protection of transformers will function similar to differential protection of transformers during earth faults and phase-to-phase fault circumstances.

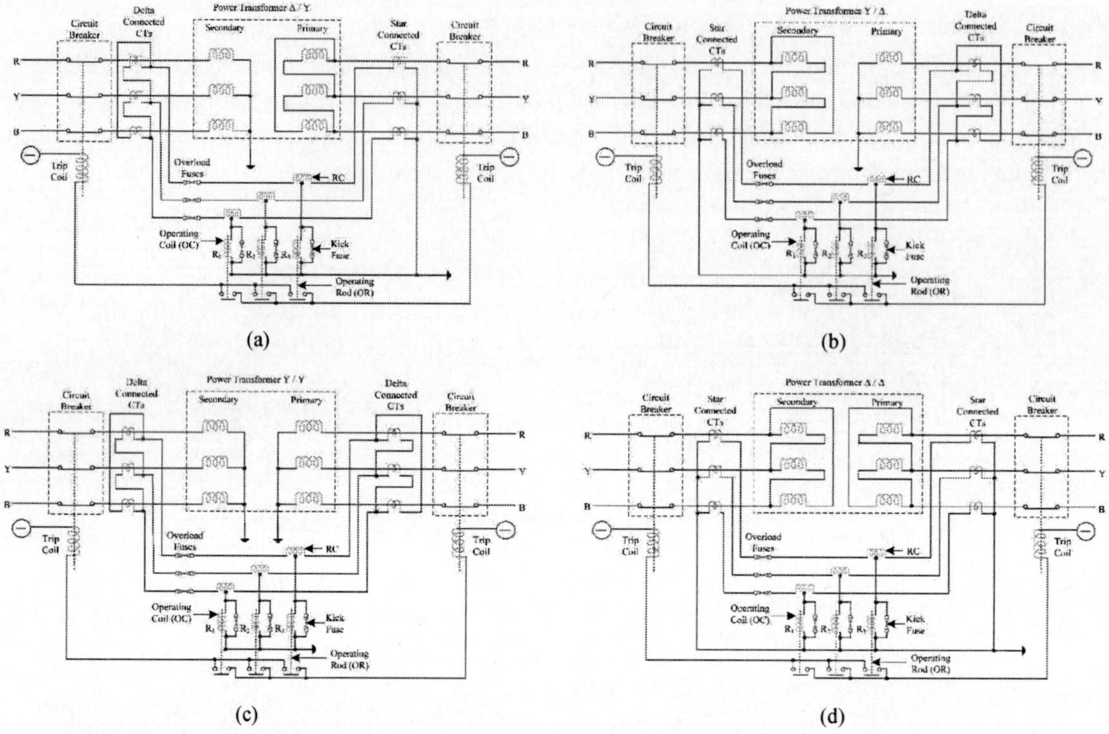

Fig. 10.16 Biased differential protection of (a) delta/star (b) star/delta (c) star/star and (d) delta/delta transformers

10.16 Biased modified differential protection of transformers:

As mentioned in the previous section (10.6), biased modified differential protection employs two relay elements for phase fault protection and a third for earth fault protection. Figure 10.17 illustrates the modified differential protection mechanism used to safeguard a 3-phase delta/star, star/delta, star/star and delta/delta power transformer against phase-to-ground and phase-to-phase faults respectively.

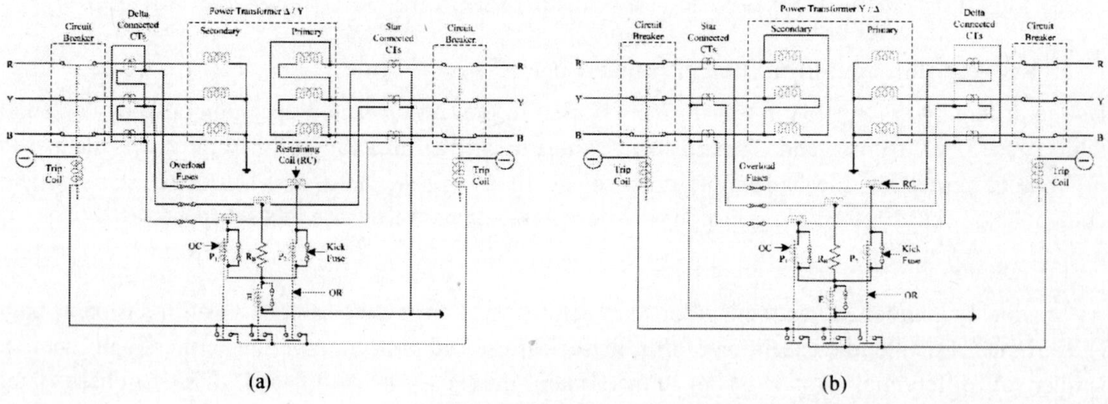

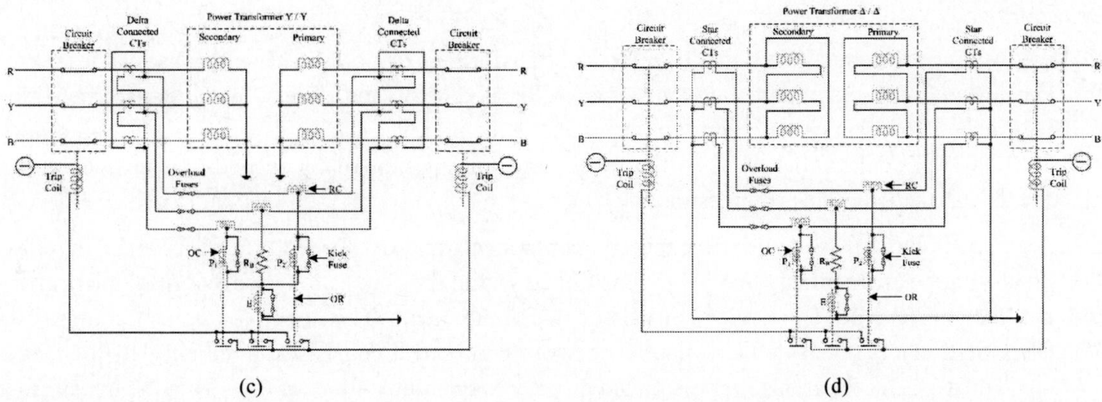

Fig. 10.17 Biased modified differential protection of (a) delta/star (b) star/delta (c) star/star and (d) delta/delta transformers

10.17 Buchholz Relay:

The Buchholz relay is a gas-actuated protective device that is typically used in oil-immersed transformers to safeguard against various types of defects. The Buchholz relay, named after its inventor, is utilized to provide an alert in the event of gradual faults occurring in the transformer, as well as to disconnect the transformer from the power source in the case of severe internal problems. Typically, it is put in the pipe that connects the conservator to the main tank, as depicted in Figure 10.18. Using Buchholz relays is a standard procedure for all oil-immersed transformers with ratings above 750 kVA.

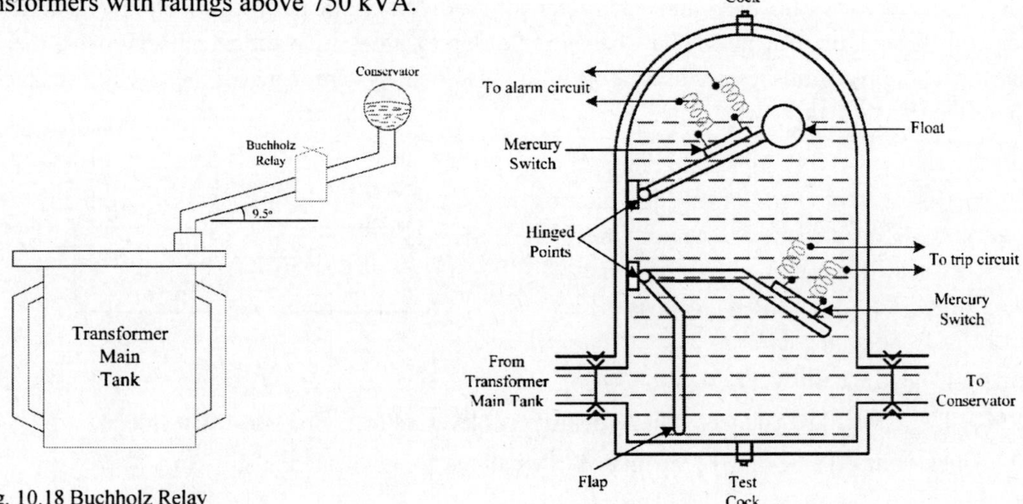

Fig. 10.18 Buchholz Relay

Benefits:

- ✓ It identifies incipient faults at a significantly earlier level compared to alternative protection methods.
- ✓ This is the simplest form of protection for a transformer.

Drawbacks:

- It is exclusively compatible with oil-immersed transformers that are fitted with conservator tanks.
- The device is capable of detecting faults that are located below the oil level in the transformer.
- This relay is ignorant of problems that occur above the oil level.
- The relay has a delayed response, with a minimum working time of 0.1 seconds and an average time of 0.2 seconds. This is undesirable.

Construction: Figure 10.18 depicts the specific components and design features of a Buchholz relay. The structure appears as a domed container positioned within the connection pipe that links the primary tank and the conservator. The device consists of two components. The upper component comprises a mercury switch that is connected to a float. The lower component consists of a mercury switch that is positioned on a hinged flap. This flap is situated directly in the route of the oil flow from the transformer to the conservator. The upper element is responsible for closing the alarm circuit during initial problems, while the bottom element is designed to trip the circuit breaker in the event of serious internal defects.

Operation: The Buchholz relay operates in the following manner: If there are incipient faults in the transformer, the heat generated by the fault leads to the decomposition of some transformer oil in the main tank. The decomposition products consist of over 70% hydrogen gas. The hydrogen gas, due to its low density, attempts to enter the conservator but becomes stuck in the upper section of the relay chamber. When a specific quantity of gas accumulates, it applies enough force on the float to make it tilt and activate the contacts of the mercury switch connected to it. This establishes the alarm circuit in order to activate the alarm sound. In the event of a severe malfunction in the transformer, a substantial quantity of gas is generated within the main tank. The oil from the main tank flows rapidly towards the conservator through the Buchholz relay, causing the flap to tilt and close the contacts of the mercury switch. This action establishes a closed loop in the electrical system, causing the circuit breaker that regulates the transformer to deactivate.

Example 10.9. If a 3-phase, 66kV/11kV transformer is connected in a Y-Δ configuration and the current transformers (CTs) on the low voltage (L.V) side have a ratio of 400:6, what will be the ratio of the CT on the high voltage (H.V) side if Merz-Price protection is to be used?

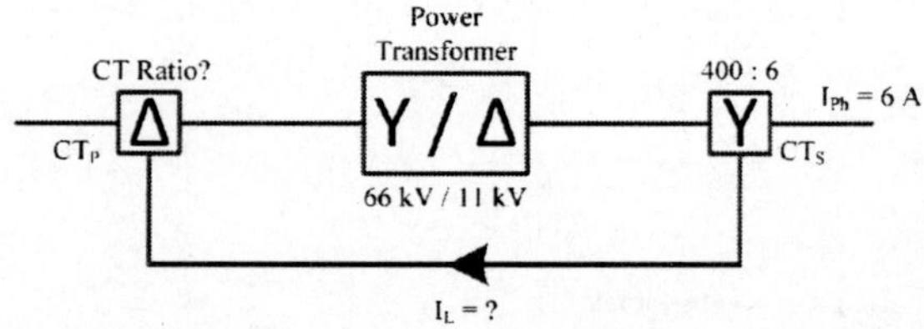

Ans. Given Data: 3Φ, 66kV/11kV (Y-Δ) T/f

For a Y-Δ T/F, CT will be connected in Δ (delta) on 66kV side in Y (star) on 11kV side.

On H.V (high voltage) side of T/F, the main T/F windings are connected in star. So,

$$\text{Phase Voltage} = \frac{\text{Line Voltage}}{\sqrt{3}} = \frac{66000}{\sqrt{3}} = 38.105 \text{ kV}$$

On L.V (low voltage) side of T/F, the main T/F windings are connected in delta. So,

Line Voltage (V_L) =

Phase Voltage (V_{Ph}) = 11 kV

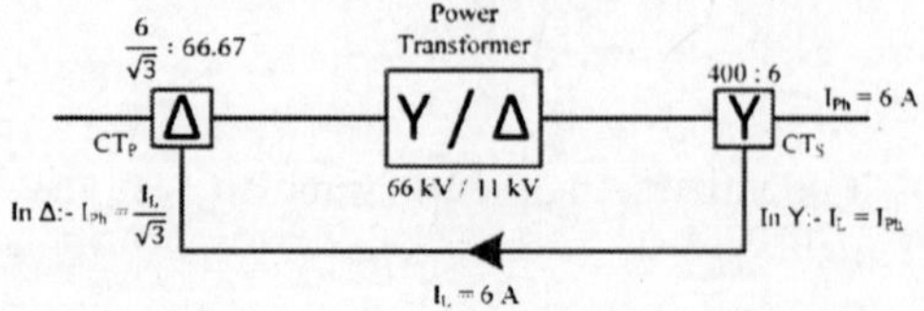

If 'I' is the line current on 66kV side, then:

Primary Apparent power = Secondary Apparent power

$$\sqrt{3} \times (66 \times 10^3) \times I = \sqrt{3} \times (11 \times 10^3) \times 400$$

$$I = \frac{\sqrt{3}\times(11\times10^3)\times400}{\sqrt{3}\times(66\times10^3)}$$

$$I = \frac{200}{3} = 66.67\text{ A} \text{ ----------(1)}$$

Phase current of Y connected CT on primary side of 11kV is $I_{ph\ 1'ry}^{11kV} = 400A$

Phase current of Y connected CT on secondary side of 11kV is $I_{ph\ 2'ry}^{11kV} = 6A$

Line current of Y connected CT on secondary side of 11kV is $I_{L\ 2'ry}^{11kV} = I_{ph\ 2'ry}^{11kV} = 6A$

This current (6 A) flows through the pilot wires. Obviously, this will be the current which flows through the secondary CT on 66 kV side.

Line current of Δ connected CT on secondary side of 66kV is $I_{L\ 2'ry}^{66kV} = 6A$

Phase current of Δ connected CT on secondary side of 66kV is $I_{ph\ 1'ry}^{66kV} = \frac{I_{L\ 2'ry}^{66\ kV}}{\sqrt{3}} = \frac{6}{\sqrt{3}}$ A --(2)

From eq. (1) and (2): Turns ratio of CT on 66kV side (H.V side) is $66.67{:}\frac{6}{\sqrt{3}}$

Example 10.10. The Δ-Y 11kV/6.6kV T/f is safeguarded by a differential protection mechanism. The CT ratio on the 6.6kV side is 400:4, thus calculate the CT ratio on the H.V side (11kV).

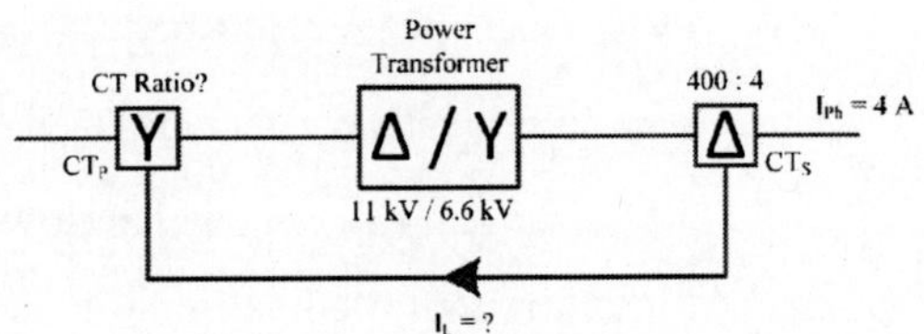

Ans. Given Data:3Φ, 11kV/6.6 kV (Δ-Y) power T/f

CT ratio on 6.6 kV side is 400:4

Let 'I' be the line current on 11 kV side, then:

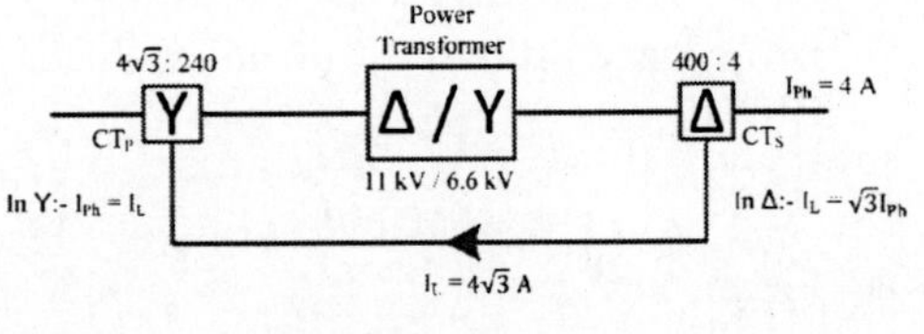

Primary Apparent power = Secondary Apparent power

$$\sqrt{3} \times (11 \times 10^3) \times I = \sqrt{3} \times (6.6 \times 10^3) \times 400$$

$$I = \frac{\sqrt{3}\times(6.6\times10^3)\times400}{\sqrt{3}\times(11\times10^3)}$$

$$I = 240\text{ A} \text{ ----------------(1)}$$

Phase current of Δ connected CT on primary side of 6.6kV is $I_{ph\ 1'ry}^{6.6kV} = 400A$

Phase current of Δ connected CT on secondary side of 6.6kV is $I_{ph\ 2'ry}^{6.6kV} = 4A$

Line current of Δ connected CT on secondary side of 6.6kV is $I_{L\ 2'ry}^{6.6kV} = I_{ph\ 2'ry}^{6.6kV} \times \sqrt{3} = 4\sqrt{3}$ A

This current ($4\sqrt{3}$) flows through the secondary of CT on 11kV side.

Line current of Y connected CT on secondary side of 11kV is $I_{L\ 2'ry}^{11kV} = 4\sqrt{3}$ A

Phase current of Y connected CT on secondary side of 11kV is $I_{ph\ 1'ry}^{11kV} = I_{L\ 2'ry}^{11kV} = 4\sqrt{3}$ A ---(2)

From equations (1) and (2): Turns ratio of CT on 11 kV side is 240: $4\sqrt{3}$

Example 10.11. The differential protection system protects a Y-Y 66kV/11 kV T/F. The CT ratio for the 11kV star connected side is 500:5. Calculate the CT ratio for the H.V side.

Ans. Given Data: Y-Y, 66kV/11kV (Y-Y), Power T/F

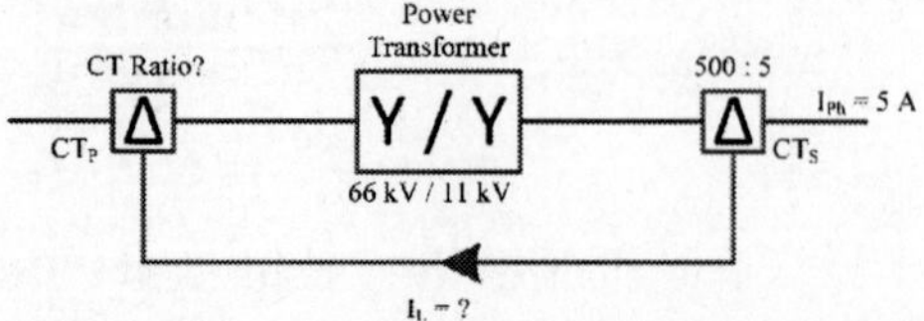

On H.V (high voltage) side of T/F (66 kV), the main T/F windings are connected in star. So,

$$\text{Phase Voltage} = \frac{\text{Line Voltage}}{\sqrt{3}} = \frac{66000}{\sqrt{3}} = 38.105 \text{ kV}$$

On L.V (low voltage) side of T/F (11kV), the main T/F windings are connected in star. So,

$$\text{Phase Voltage} = \frac{\text{Line Voltage}}{\sqrt{3}} = \frac{11000}{\sqrt{3}} = 6.35 \text{ kV}$$

If 'I' is the line current on 66 kV side, then: Primary Apparent power = Secondary Apparent power

$$\sqrt{3} \times (66 \times 10^3) \times I = \sqrt{3} \times (11 \times 10^3) \times 500$$

$$I = \frac{\sqrt{3}\times(11\times10^3)\times500}{\sqrt{3}\times(66\times10^3)} = \frac{250}{3} = 83.33 \text{ A}$$

Phase current of Δ connected CT on primary side of 11kV is $I_{ph\ 1ry}^{11kV} = 500\text{A}$

Phase current of Δ connected CT on secondary side of 11kV is $I_{ph\ 2ry}^{11kV} = 5\text{A}$

Line current of Δ connected CT on secondary side of 11kV is $I_{L\ 2ry}^{11kV} = \sqrt{3} * I_{ph\ 2ry}^{11kV} = 5\sqrt{3}\ A$

This will be the current ($5\sqrt{3}\ A$) which flows through the secondary CT on 66kV side.

Line current of Δ connected CT on secondary side of 66kV is $I_{L\ 2ry}^{66kV} = 5\sqrt{3}$ A.

Phase current of Δ connected CT on secondary side of 66kV is $I_{ph\ 1ry}^{66kV} = \frac{I_{L\ 2ry}^{66kV}}{\sqrt{3}} = \frac{5\sqrt{3}}{\sqrt{3}} = 5\ A$

Therefore, turns ratio of CT on 66kV side is 83.33: 5

Example 10.12. A 3Φ, 33kV/11kV, Δ -Δ connected T/F is protected by Merz-Price protection scheme. The CT on the L.V side has a CT ratio of 330:5. Show that the CT on the H.V side will have a ratio of $110:\frac{5}{\sqrt{3}}$.

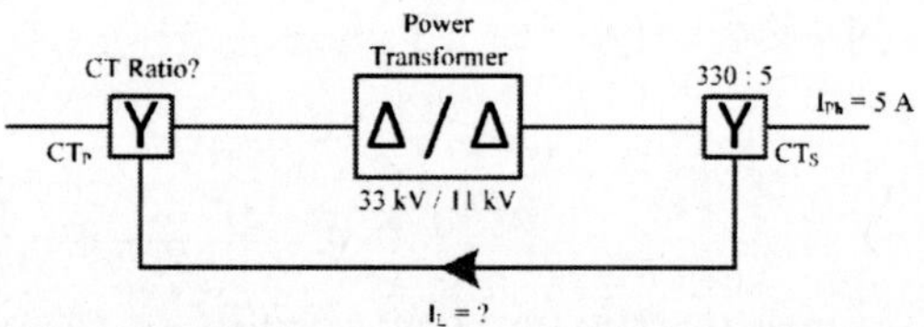

Ans. Given 3Φ, 33kV/11kV (Δ -Δ) power T/f

Let 'I' be the line current on 33 kV side, during normal operating conditions Primary Apparent power and Secondary Apparent power are equal

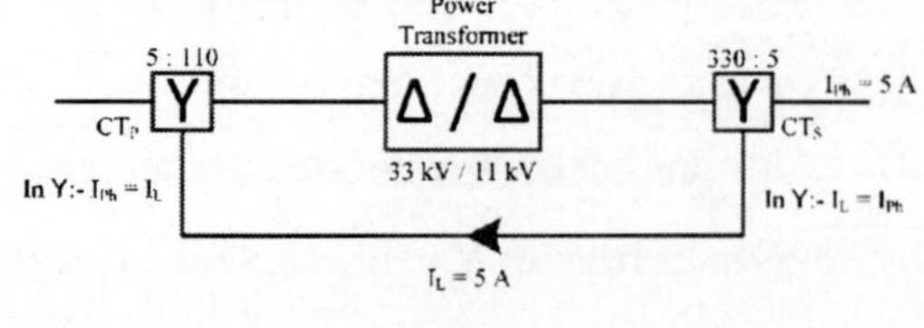

Primary Apparent power = Secondary Apparent power

$$\sqrt{3} \times (33 \times 10^3) \times I = \sqrt{3} \times (11 \times 10^3) \times 330$$

$$I = \frac{\sqrt{3}\times(11\times10^3)\times330}{\sqrt{3}\times(33\times10^3)}$$

$$I = 110\ \text{A} \text{ ------------(1)}$$

Phase current of Y connected CT on primary side of 11kV is $I_{ph\ 1ry}^{11kV} = 330\ \text{A}$

Phase current of Y connected CT on secondary side of 11kV is $I_{ph\ 2'ry}^{11kV} = 5A$

Line current of Y connected CT on secondary side of 11kV is $I_{L\ 2'ry}^{11kV} = I_{ph\ 2'ry}^{11kV} = 5A$

Line current of Y connected CT on secondary side of 33kV is $I_{L\ 2'ry}^{33kV} = 5A$

Phase current of Y connected CT on secondary side of 33kV is $I_{ph\ 2'ry}^{33kV} = I_{L\ 2'ry}^{33kV} = 5A$ --------------(2)

From equations (1) and (2): Turns ratio of CT on 33kV side is $110: 5$

Example 10.13. A 3Φ T/F with a Line-Voltage ratio of 400V/11kV is connected in Y-Δ. Protective T/Fs on the 400V side have a CT ratio of 200:4. What must be the protective T/F ratio on the 11kV side?

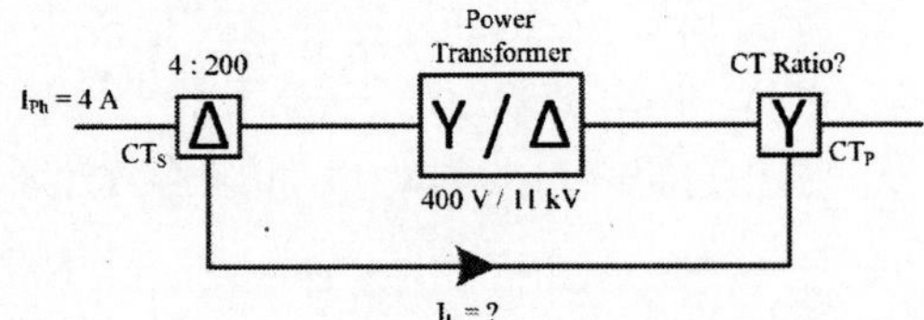

Ans. Given Data: 3Φ, 400V/11kV, Y-Δ T/F

CT ratio on 400 V side is 200:4

If 'I' is the line current on 11 kV side, then Primary power = Secondary power

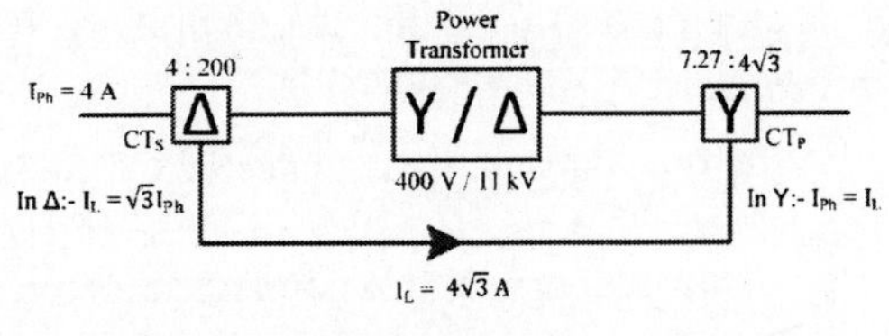

$$\sqrt{3} \times 400 \times 200 = \sqrt{3} \times (11 \times 10^3) \times I$$

$$I = \frac{\sqrt{3}\times(400)\times 200}{\sqrt{3}\times(11\times 10^3)}$$

$$I = 7.27 \text{ A} \text{ --------------(2)}$$

Phase current of Δ connected CT on primary side of 400V is $I_{ph\ 1'ry}^{400V} = 200A$

Phase current of Δ connected CT on secondary side of 400V is $I_{ph\ 2'ry}^{400V} = 4A$

Line current of Δ connected CT on secondary side of 400V is $I_{L\ 2'ry}^{400V} = \sqrt{3} * I_{ph\ 2'ry}^{400V} = 4\sqrt{3}\ A$

This current ($4\sqrt{3}$ A) will flow through the secondary CT on 11kV side.

Line current of Y connected CT on secondary side of 11kV is $I_{L\ 2'ry}^{11kV} = 4\sqrt{3}\ A$

Phase current of Y connected CT on secondary side of 11kV is $I_{ph\ 2'ry}^{11kV} = I_{L\ 2'ry}^{11kV} = 4\sqrt{3}\ A$

Therefore, turns ratio of CT on 11kV side is $7.27: 4\sqrt{3}$

Example 10.14. Percentage differential relays protect a 100 MVA Y-Δ three-phase transformer with 33kV/11kV voltage. If the LV-side current transformers (CTs) have a transformer ratio of 450/6. What will be the CT ratio on the HV side? Also, determine CT ratios on the HV side for Δ-Y, Y-Y, and Δ-Δ combinations.

Ans. Case (i) 3Φ, 33kV/11kV (Y-Δ) power T/f

For a Y-Δ T/F, CT will be connected in Δ on 33kV side and CT will be connected in Y on 11kV side.

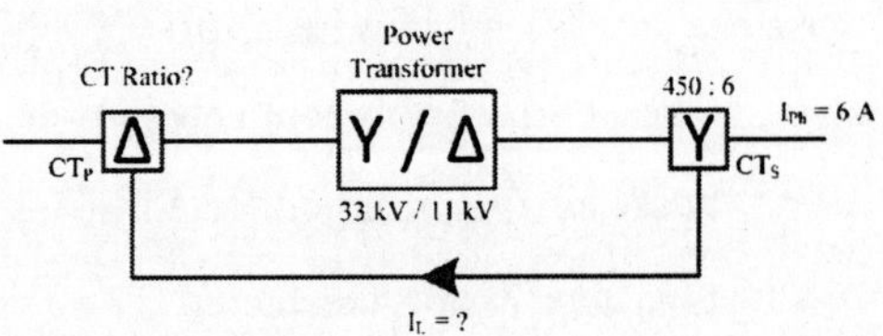

On H.V (high voltage) side of T/F, the main T/F windings are connected in star. So,

$$\text{Phase Voltage} = \frac{\text{Line Voltage}}{\sqrt{3}} = \frac{33000}{\sqrt{3}} = 19.052\text{kV}$$

On L.V (low voltage) side of T/F, the main T/F windings are connected in delta. So,

Line Voltage (V_L) = Phase Voltage (V_{Ph}) = 11kV

If 'I' is the line current on 66kV side, then:

Primary Apparent power = Secondary Apparent power

$$\sqrt{3} \times (33 \times 10^3) \times I = \sqrt{3} \times (11 \times 10^3) \times 450$$

$$I = \frac{\sqrt{3}\times(11\times10^3)\times450}{\sqrt{3}\times(33\times10^3)}$$

$$I = \frac{450}{3} = 150\ A$$

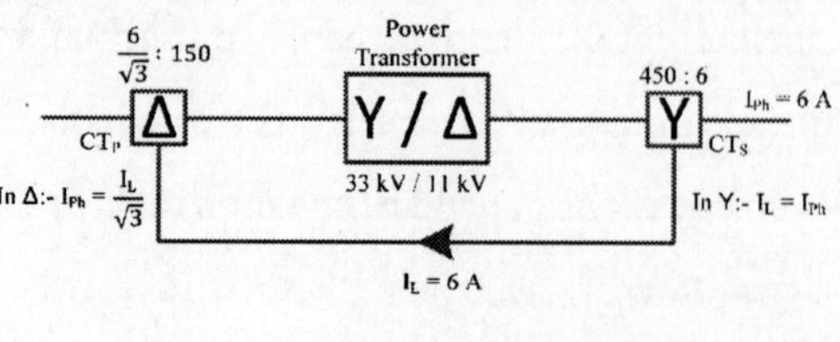

Phase current of Y connected CT on primary side of 11kV is $I_{ph\ 1'ry}^{11\ kV} = 450A$

Phase current of Y connected CT on secondary side of 11kV is $I_{ph\ 2'ry}^{11\ kV} = 6A$

Line current of Y connected CT on secondary side of 11kV is $I_{L\ 2'ry}^{11\ kV} = I_{ph\ 2'ry}^{11\ kV} = 6A$

This current (6A) flows through the pilot wires. Obviously, this will be the current which flows through the secondary CT on 33kV side.

Line current of Δ connected CT on secondary side of 33kV is $I_{L\ 2'ry}^{33kV} = 6A$

Phase current of Δ connected CT on secondary side of 33kV is $I_{ph\ 1'ry}^{33kV} = \frac{I_{L\ 2'ry}^{33\ kV}}{\sqrt{3}} = \frac{6}{\sqrt{3}}\ A$

Therefore, turns ratio of CT on 33kV side is $150:\frac{6}{\sqrt{3}}$

Case (ii) 3Φ, 33kV/11kV (Δ- Y) power T/f

Phase current of Δ connected CT on primary side of 11kV is $I_{ph\ 1'ry}^{11kV} = 450A$

Phase current of Δ connected CT on secondary side of 11kV is $I_{ph\ 2'ry}^{11kV} = 6A$

Line current of Δ connected CT on secondary side of 11kV is $I_{L\ 2'ry}^{11kV} = I_{ph\ 2'ry}^{11kV} \times \sqrt{3} = 6\sqrt{3}\ A$

This current ($6\sqrt{3}A$) flows through the secondary of CT on 33kV side.

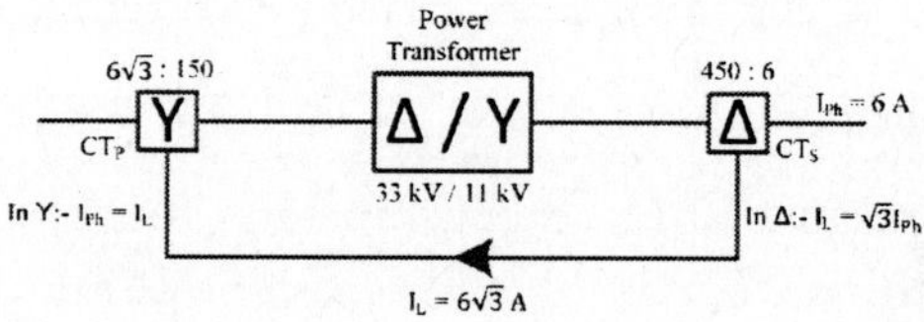

Line current of Y connected CT on secondary side of 33kV is $I_{L\ 2'ry}^{33kV} = 6\sqrt{3}\ A$

Phase current of Y connected CT on secondary side of 33kV is $I_{ph\ 1'ry}^{33kV} = I_{L\ 2'ry}^{33kV} = 6\sqrt{3}\ A$ ---(2)

Therefore, turns ratio of CT on 33kV side is $150: 6\sqrt{3}$

Case (iii) 3Φ, 33kV/11kV (Y-Y) power T/f

Phase current of Δ connected CT on primary side of 11kV is $I_{ph\ 1'ry}^{11kV} = 450\ A$

Phase current of Δ connected CT on secondary side of 11kV is $I_{ph\ 2'ry}^{11kV} = 6A$

Line current of Δ connected CT on secondary side of 11kV is $I_{L\ 2'ry}^{11kV} = \sqrt{3} * I_{ph\ 2'ry}^{11kV} = 6\sqrt{3}\ A$

Line current of Δ connected CT on secondary side of 33kV is $I_{L\,2'ry}^{33kV} = 6\sqrt{3}$ A.

Phase current of Δ connected CT on secondary side of 33kV is $I_{ph\,1'ry}^{33kV} = \frac{I_{L\,2'ry}^{33kV}}{\sqrt{3}} = \frac{6\sqrt{3}}{\sqrt{3}} = 6\,A$

Therefore, turns ratio of CT on 33 kV side (H.V side) is 150: 6

Case (iv) Given Data: 3Φ, 33/11kV (Δ - Δ) power T/f

Phase current of Y connected CT on primary side of 11kV is $I_{ph\,1'ry}^{11kV} = 450$ A

Phase current of Y connected CT on secondary side of 11kV is $I_{ph\,2'ry}^{11kV} = 6A$

Line current of Y connected CT on secondary side of 11kV is $I_{L\,2'ry}^{11kV} = I_{ph\,2'ry}^{11kV} = 6A$

Line current of Y connected CT on secondary side of 33kV is $I_{L\,2'ry}^{33kV} = 6A$

Phase current of Y connected CT on secondary side of 33kV is $I_{ph\,2'ry}^{33kV} = I_{L\,2'ry}^{33kV} = 6A$

Turns ratio of CT on 33 kV side is 150: 6

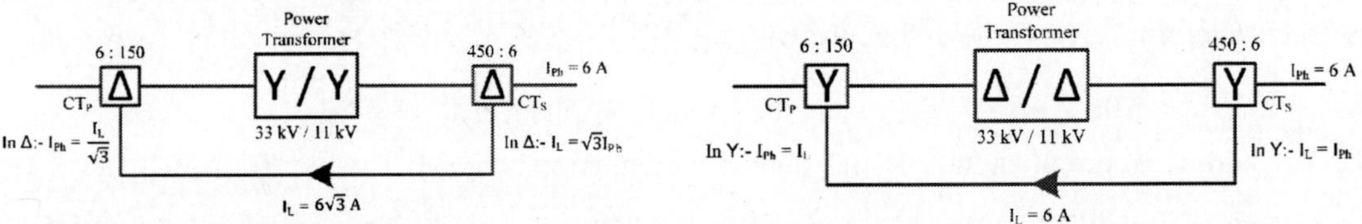

From the above cases we can conclude that

CT ratio of Primary side	Power Transformer	CT ratio of Secondary side
Δ ($\frac{6}{\sqrt{3}}$: 150)	Y- Δ	Y (450:6)
Y ($6\sqrt{3}$: 150)	Δ - Y	Δ (450:6)
Δ (6 : 150)	Y-Y	Δ (450:6)
Y (6 : 150)	Δ - Δ	Y (450:6)

Example 10.15. A three-phase, 300 megavolt-ampere, 22kV/800V transformer is coupled in a delta-star configuration. The protective transformer on the 800V side has a turns ratio of 1000/10. What will be the current transformer (CT) ratio on the high voltage side? Additionally, determine the magnitude of the circulating current in the event of a defect with a current of 1500A on the low voltage side. (i) Earth fault occurring inside the specified protection zone. (ii) Earth fault occurring outside the specified protection zone. (iii) Phase-to-Phase fault occurring within the specified protection zone. (iv) Phase-to-Phase fault occurring outside the specified protection zone.

Ans. **Given Data: 3-Φ, 300 MVA, 22/0.8 kV**

On low-voltage side of transformer the main transformer windings are connected in star,

So, Phase Voltage = $\frac{\text{Line Voltage}}{\sqrt{3}} = \frac{800}{\sqrt{3}} = 0.4618$ kV

On high-voltage side of transformer, the main transformer windings are connected in delta

So, Phase Voltage (V_{Ph}) = Line Voltage (V_L) = 22 kV

The turn – ratio of main transformer $= \frac{22*10^3}{0.4618} = 47.639$

CT on low-voltage i.e., on 0.8 kV side are connected in delta and the turn – ratio $= \frac{1000}{10} = 100$

High voltage side CTs turn ratio $= \frac{Turn - ratio\ of\ CTs\ on\ low - voltage\ side}{Turn - ratio\ of\ main\ transformer} = \frac{100}{47.639} = 2.1\ i.e.,\ \ 2.1{:}1$

Case (i): The earth fault within the protection zone :

Since the primary is isolated therefore, no zero-sequence current will flow in the primary.

The line current on HV side (windings connected in delta) corresponding to 1500A

$$= 1500 \times \frac{0.8}{22} = 54.54\ A.$$

The current ratio of CTs on HV side = 2.1: 1

So, corresponding current through the relay will be $\frac{54.54}{2.1} = 25.97\ A$, and the relay will be energized.

Case (ii): When there is earth fault outside the protection zone :

The distribution of current on the HV side remain the same as in case (i).

The secondary current in CTs on LV side will be $= 1500 \times \frac{10}{1000} = 15\ A$

The current in pilot wires $= 15 \times \sqrt{3} = 25.98\ A$, CTs on LV side being connected in delta.

Thus, the relay will carry no current and thus the circuit breaker for the through fault will not isolate the transformer.

Case (iii): When there is Phase-to- Phase fault with in the protection zone:

It causes flow of current in the two phases of the LV windings, correspondingly there will be flow of current in all the three phases of HV winding which will spill into the relay operating coils. Thus, the relay operating coils will be energized, circuit breakers will get tripped off and the transformer will get isolated.

Case (iv): When there is Phase-to- Phase fault outside the protection zone:

No current will flow through the operating coil of any relay and so the transformer will not be disconnected from the system.

Example 10.16. A 150MVA transformer, linked in a delta-star configuration, with voltage ratings of 22 kV on the delta side and 440 kV on the star side, is to be safeguarded using a percentage differential technique. The CTs used have a ratio of 5000:5 and 300:1, respectively. Draw a comprehensive diagram of the complete scheme. The relays have a maximum rating of 1 ampere.

Ans. Given Data: 150 MVA, Δ/Y -connected, 22kV/440 kV

Rated current for star (440 kV)side $= \frac{MVA\ Rating}{\sqrt{3} \times Rated\ Volatge} = \frac{150 \times 10^6}{\sqrt{3} \times 440000} = 196.823\ A.$

$$\text{Rated current for delta (22 kV)side} = \text{Transformation ratio} \times \text{Y} - \text{side current}$$

$$= \frac{440}{22} \times 196.823 = 3936.46\ A.$$

$$\text{current in LV side CTs} = \frac{3936.46 \times 5}{5000} = 3.9364 \text{ A.}$$

$$\text{current in HV side CTs} = \frac{196.823 \times 1}{300} = 0.6560 \text{ A.}$$

Relays are rated up to 1A hence all the currents exceeding 1A will be reduced further by using auxiliary CTs.

For normal operation of transformer current output of CTs should be balanced (in phase as well as in magnitude) so that relay remains in-operative.

Take 1A current for the rated current in transformer. Auxiliary CTs used on LV side should be of ratio 3.936/1.

On HV side we compensate phase difference because of Y- Δ connection of transformer through using Auxiliary CTs in reverse manner i.e., Δ/Y.

If CTs used are of ratio $x/1$ then we will get $\sqrt{3}$ A output for x amperes in primary winding because of Δ-connected secondary winding of CTs. Hence to get 1A output for 0.6560 A input to CTs, ratio required is $\sqrt{3} \times 0.656 = 1.1363A$

Example 10.17. A 100MVA, Δ/Y-connected, 220kV/110kV three-phase power transformer is safeguarded with percentage differential relays. If the current transformers (CTs) on the delta and star sides of the power transformer are 400/5 A and 1600/5 A, respectively, determine (a) the output current at full load, (b) the relay current at full load, and (c) the minimum relay current setting that allows for 30% overload.

Ans. Given Data: 100 MVA, Δ/Y -connected, 220kV/110 kV

$$\text{Rated current for primary } (\Delta - \text{connected side}) = \frac{MVA\ Rating}{\sqrt{3} \times Rated\ Volatge} = \frac{100 \times 10^6}{\sqrt{3} \times 220000} = 262.431\ A$$

$$\text{Rated current for secondary } (\text{Y} - \text{connected side}) = \frac{MVA\ Rating}{\sqrt{3} \times Rated\ Volatge} = \frac{100 \times 10^6}{\sqrt{3} \times 110000} = 524.86\ A$$

$$\text{Current in HV side CTs} = 262.431 \times \frac{5}{400} = 3.280 \text{ A}$$

$$\text{Current in LV side CTs} = 524.86 \times \frac{5}{1600} = 1.640 \text{ A}$$

(b) Relay current at full load $= 3.280 - 1.640 \times \sqrt{3} = 3.280 - 2.840 = 0.439\ A$

Assuming relay rating as 1A, Percentage of setting = 50%

(c) With 30% overload

Line current in secondaries of CTs connected on HV side is $3.280 \times 1.3 = 4.264$ A and

line current in secondaries of CTs connected in delta on LV side $= \sqrt{3} \times 1.640 \times 1.3 = 3.692$ A

$$\therefore Relay\ current = 4.264 - 3.692 = 0.571 \text{ A}$$

$$\therefore Relay\ setting = \frac{0.571}{1} \times 100 = 57.1\%$$

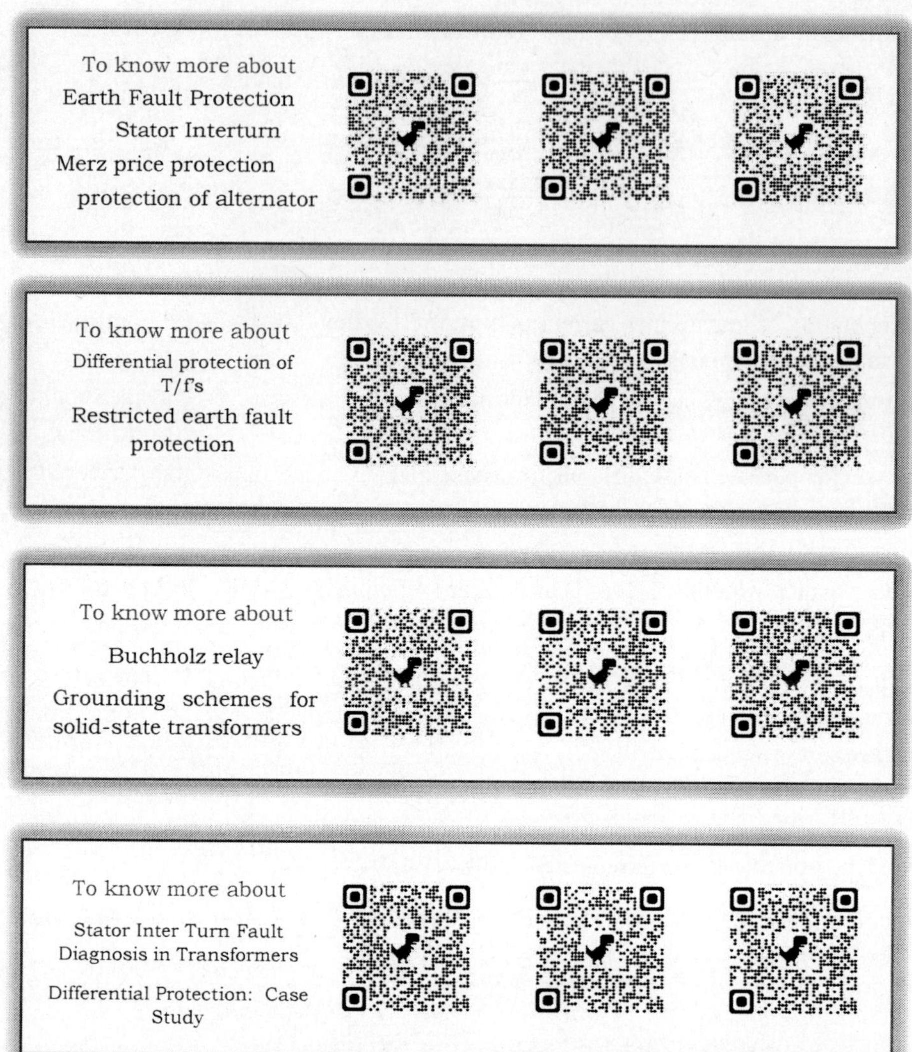

10.18 Introduction to Busbar Protection:

Busbars and transmission lines are critical components of an electrical power system and necessitate the prompt attention of protection engineers to ensure protection against potential faults that may develop on them. The techniques utilized to safeguard generators and transformers can also be applied, with minor adjustments, to protect busbars and lines. The improvements are required to address the security issues that arise from the increased length of lines and the large number of circuits linked to a busbar. While differential protection can be employed, it becomes prohibitively costly for longer lines due to the increased length of pilot wires needed. Fortunately, there are cost-effective alternatives that offer reasonable effectiveness in safeguarding the busbars and lines. This unit will specifically address the techniques used to safeguard busbars and lines.

Busbars in generating stations and sub-stations serve as a crucial connection between the incoming and outgoing circuits. In the event of a fault on a busbar, significant harm and interruption of power will happen unless there is a fast-acting automatic protection system in place to separate the damaged busbar. The busbar zone encompasses the busbars, isolating switches, circuit breakers, and their accompanying connections, all of which are included for protective purposes. Busbars have been constructed to a high security level, resulting in a very low occurrence of bus faults. Nevertheless, the potential for harm and disruption caused by an infrequent bus malfunction is significant enough to warrant increased focus on its safeguard. Enhanced transmission techniques have been developed, minimising the likelihood of improper functioning.

Bus-bar protection requires careful consideration due to the following factors.

- The fault level at the bus-bars is exceedingly high.
- A fault occurring on the bus-bar would lead to a significant disruption in the supply of electricity.
- A problem in the bus zone has a negative impact on the stability of the system.

If a fault occurs on any part of the bus-bar, it is necessary to quickly trip all the circuit equipment linked to that part in order to completely isolate it. This should be done within a very short period, for example, 60 milliseconds, to prevent damage to the installation caused by overheating of the conductors. Line faults occur more frequently than internal bus faults. A bus fault is significantly more serious in terms of worker/operator safety, system stability, and damage. The absence of sufficient bus protection may lead to a significant system shutdown. The bus-zone protection must be quick, stable, and highly dependable, as it is of paramount importance.

The desired qualities of bus protection are listed below:

- Fast operation (less than three cycles).
- Differentiating between faults in the protected section and those elsewhere.
- Provides stability for external failures.
- Eliminate needless operations.
- No operation due to CT saturation or power fluctuations.
- Each circuit breakers trip circuit is controlled separately.
- Use Main and Check protection to ensure isolation only when necessary.
- No automatic reclosure and no single pole tripping of circuit breakers during a bus fault.
- Use interlock overcurrent protection to prevent the generator unit from tripping during bus-zone protection operation.

Some experts believe that local bus protection should be removed and bus faults addressed by backup relays at neighbouring stations, as providing local bus protection would undoubtedly raise the chance of unintended tripping. When local bus protection is given, effort is taken to ensure that two independent protective circuits are met before tripping occurs.

10.19 Bus-Bar configurations:

A bus-bar refers to a primary bar or conductor that carries an electric current and can be connected to multiple devices. The term bus is derived from the word omnibus, which means "collector of things."

Thus, an electrical bus-bar collects electrical energy at a single spot. Bus-bars are just a practical way of connecting switches and other equipment in various configurations. The standard connection configuration in most substations allows you to work on practically any piece of equipment without disrupting incoming or outgoing feeders.

Bus-bars commonly utilised in substations typically consist of bare rectangular cross-section bars. However, they may also adopt alternative shapes like as round tubes, round solid bars, or square tubes. The bus-bars are typically made of aluminium, although copper is occasionally used. Aluminium is chosen as an electric conductor because it has several advantages compared to copper. These include better conductivity per unit weight, lower cost for the same current carrying capability, good resistance to corrosion, and ease of being shaped into desired forms. In order to ensure dependable and secure electrical connections, aluminium buses are coated with silver. Multiple configurations of bus-bars exist. The selection of a certain configuration is contingent upon various elements, including the voltage of the system, the substation's position within the system, the dependability of the power supply, the adaptability of the system, and the associated expenses.

Other technical factors that determine the choice of a specific arrangement include:

- Simplicity is a fundamental characteristic of a reliable system.
- Maintenance can be performed without any interruption to the supply or risk to the operational staff.
- Providing an extension to accommodate the increase in load.
- The installation should prioritise cost-effectiveness while considering the requirements and uninterrupted provision of resources.
- Existence of substitute options in case any of the equipment has a disruption.
- Significance of load and local conditions. Desire for freedom from complete termination and its duration.

In the case of a small to medium-sized station that allows for occasional shutdowns, it is preferable to use a straightforward and uncomplicated single bus-bar system. For large plants, it is crucial to have a sophisticated bus-bar structure to ensure that a backup power source is always accessible in the event of a breakdown. This is necessary to prevent any disruption in the provision of electricity to a wide area. In light of these conditions, the substantial initial expense is warranted. The economy is of utmost importance. Typically, high-voltage equipment like circuit breakers and isolators tend to be expensive.

Therefore, it is advisable not to give equipment that is not needed. It is not advisable to offer extra equipment to handle unexpected situations in remote areas, as it would not only increase the cost of installation but also create vulnerabilities. Busbars can be configured in many ways, they are

(i). Single bus-bar single breaker configuration: with and with-out bus sectionalization.

(ii). Double bus-bar single breaker configuration: with and with-out bus sectionalization.

(iii). Double bus-bar double breaker configuration: with and with-out bus sectionalization.

(iv). Double bus-bar double-breaker ring main configuration.

10.19.1 Single bus-bar single breaker configuration (with-out bus sectionalization):

This is the most basic layout, with a single pair of bus-bars running the entire length of the switchboard, to which all generators, transformers and feeders are connected, as indicated by the single line diagram in Fig. 10.19(a). It is connected to four feeders that include 400V low-tension feeder, 11kV high voltage feeder, 33kV high-tension feeder and 110kV/220kV extra high-tension feeder. A circuit breaker is included in each generator and feeder. Generators, feeders, and circuit breakers can be isolated from bus bars for repair using the isolators. The primary benefits of such a bus-bar configuration are low initial cost, low maintenance, and simplicity of operation.

The most obvious disadvantage of this configuration is that if the bus-bars fail, the entire supply is disrupted, and all healthy feeders are cut-off. Furthermore, any repair on any of the feeder sections or a portion of the bus-bar will cause the entire supply to be disrupted. As a result, such an approach gives the least flexibility and protection against total shutdown. Switchboards, small and medium-sized substations, small power plants, and direct current stations all use this type of bus-bar configuration.

10.19.2 Single bus-bar single breaker configuration with bus sectionalization:

A circuit breaker and isolating switches can be used to sectionalize the bus-bar so that a defect in one component does not result in a complete shutdown. In large generating stations with multiple units, it is usual practice to sectionalize the bus, as illustrated in Fig. 10.19(b).

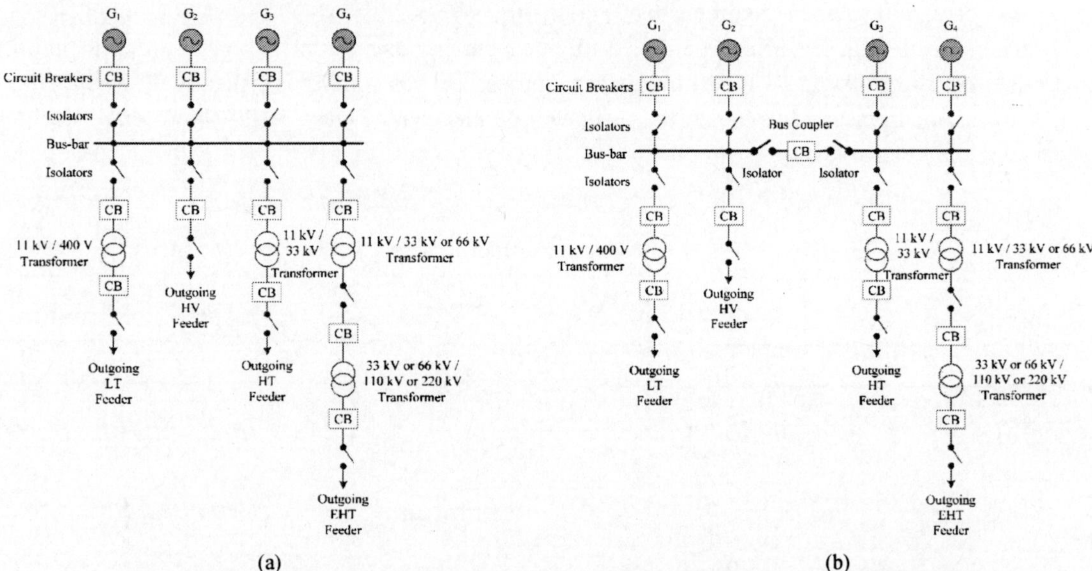

Fig. 10.19 Single bus-bar single breaker configuration (a) with-out and (b) with bus sectionalization

Typically, a substation has 2 to 3 sections of bus-bars. However, the actual number of sections is determined by the short-circuit current that needs to be managed. Only one additional circuit breaker is needed in a sectionalized bus-bar setup, and its cost is relatively low compared to the overall cost of the bus-bar system. This layout offers three primary advantages compared to a basic single bus-bar arrangement. Firstly, if a failure occurs on any part of the bus-bar, that specific area can be isolated without disrupting the power supply to other sections. Furthermore, it is possible to totally halt the operation of one segment for the purpose of maintenance and repairs, without any impact on the supply of the other section(s). Furthermore, the inclusion of a current limiting reactor between the sections allows for a reduction in the fault level (measured in MVA), enabling the use of circuit breakers with lesser capacity.

Sometimes, air-break isolators were employed as bus-sectionalizers instead of circuit breakers due to cost considerations. However, it is important to note that any isolation performed by these isolators must be done when the system is not under load, otherwise it may result in the generation of sparks. It is advisable to use a circuit breaker as a sectionalizing switch in order to properly disconnect the bus-bar during load transfer. However, a double isolation is required when using the circuit breaker as a sectionalizing switch in order to perform maintenance on the circuit breaker while the bus-bars are still energised.

10.19.3 Double bus-bar (main and auxiliary) single breaker configuration:

This configuration is commonly used when the loads and the need for a continuous supply of electricity warrant the extra expenses. This configuration offers enhanced versatility, uninterrupted provision of resources, and the ability to conduct regular maintenance without a complete halt in operations. This design is ideal for a power network that is highly interconnected and requires a high degree of flexibility. Figure 10.20 depicts the main and transfer bus configuration in a generating plant. Such a system consists of two bus-bars, known as the main bus-bar, and the transfer bus-bar which serves as an auxiliary bus bar. Each generator and feeder can be linked to either bus-bar using a bus coupler, which includes a circuit breaker and isolating switches. In this configuration, a bus coupler is typically utilised to allow for the transition from one bus-bar to the next while under load. Figure 10.20 shows only outgoing feeders without voltage levels for simplicity.

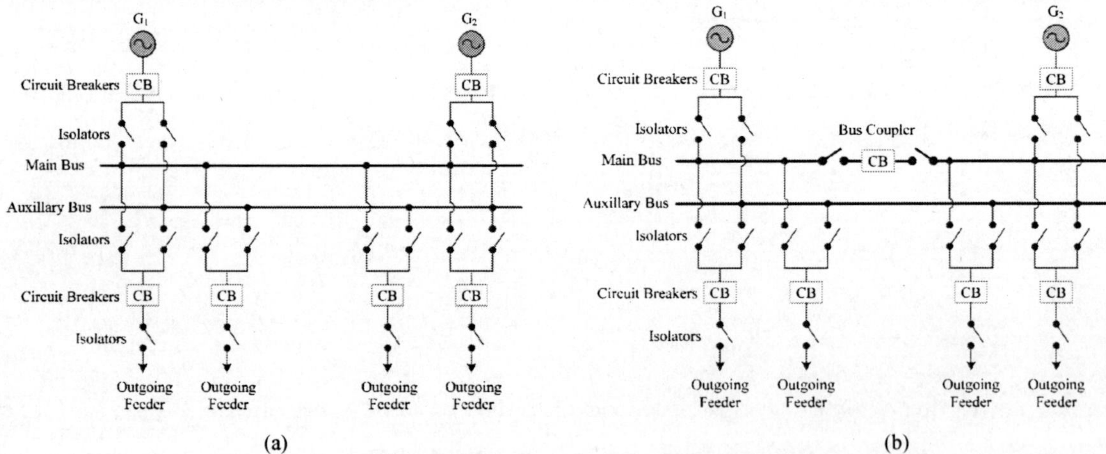

Fig. 10.20 Double bus-bar and single breaker configuration (a) with-out and (b) with bus sectionalization

The following measures can be done while moving the load to the reserve bus.

(i). Close the bus coupler (circuit breaker) so that the two buses have the same potential.

(ii). Close the isolators on the backup bus.

(iii). Open the isolators of main bus.

The load is now transferred to the reserve or auxiliary bus, while the main bus is disconnected. The advantages and disadvantages of the setup are listed below.

Advantages:

- ✓ It ensures that the supply remains uninterrupted in the event of a bus fault. If a fault occurs on one of the buses, the full load can be shifted to the other.
- ✓ Repairs and maintenance on the main bus can be completed without disrupting the supply by transferring the entire load to the auxiliary bus.
- ✓ Each load can be fed from either bus.
- ✓ If necessary, the infeed and load circuits can be separated for operational reasons.
- ✓ Feeder circuit breakers can be tested and maintained on a spare bus while the primary bus remains unaffected.
- ✓ The substation's maintenance cost is reduced.

Disadvantages:

- × Additional costs.
- × To maintain or expand the bus, all circuits are transferred to the auxiliary bus using remote backup relays and breakers to remove faults. If one of the bus circuits fails, the entire station will shut down.

10.19.4 Double bus-bar double-breaker configuration with bus sectionalization:

This configuration employs supplementary bus-bars alongside the primary bus-bar, which are interconnected using a bus coupler, as depicted in Figure 10.21. In this configuration, it is possible to isolate any portion of the bus-bar for maintenance, while also allowing for synchronisation between any sections via the auxiliary bus-bar. For the sake of simplicity, Figure 10.21 only depicts outgoing feeders with no voltage levels.

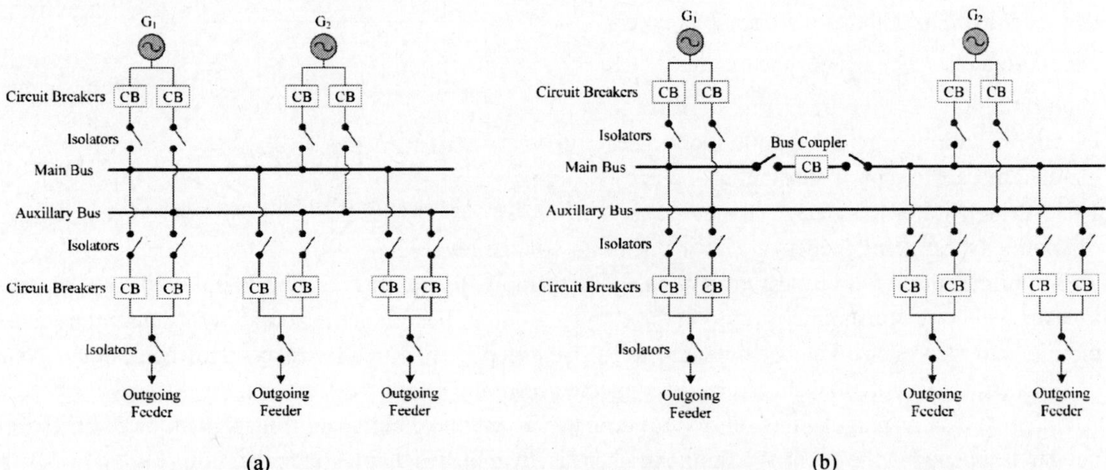

Fig. 10.21 Double bus-bar and double breaker configuration (a) with-out and (b) with bus sectionalization

10.19.5 Double bus-bar double-breaker ring main configuration:

This is an expansion of the sectionalized bus-bar configuration in which the bus-bar ends are returned into themselves to form a ring, as depicted in Fig. 10.22. Figure 10.22 depicts only outgoing feeders without voltage levels for simplicity. This configuration gives more flexibility because each feeder is supplied by two pathways, ensuring that the supply is not interrupted if one part fails. The effect of a single defect is limited to that portion. The remaining components are still operational.

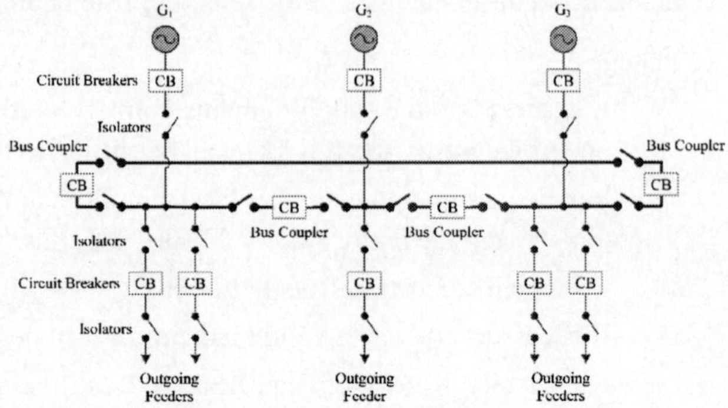

Fig. 10.22. Double bus-bar double-breaker ring main configuration

Circuit breakers can be serviced without affecting the power supply. The cost is also low because the number of breakers utilised is roughly equal to that of a single bus-bar system.

The system's shortcomings are

- difficulty in adding any new circuit to the ring,
- the possibility of overloading the circuits when any portion of the breaker is opened, and
- the requirement of supplying potential to relays separately for each circuit.

10.19.6 One-and-Half breaker configuration:

This arrangement is an advancement over the double bus-bar double breaker configuration and it results in a reduction in the quantity of circuit breakers required. This configuration requires three circuit breakers to accommodate two circuits. The number of circuit breakers per circuit is $1\frac{1}{2}$, hence the name. This configuration is favoured in significant, expansive stations where the amount of power handled by each circuit is substantial. Figure 10.23 demonstrates this design; for the sake of simplicity, outgoing feeders are shown with no voltage levels.

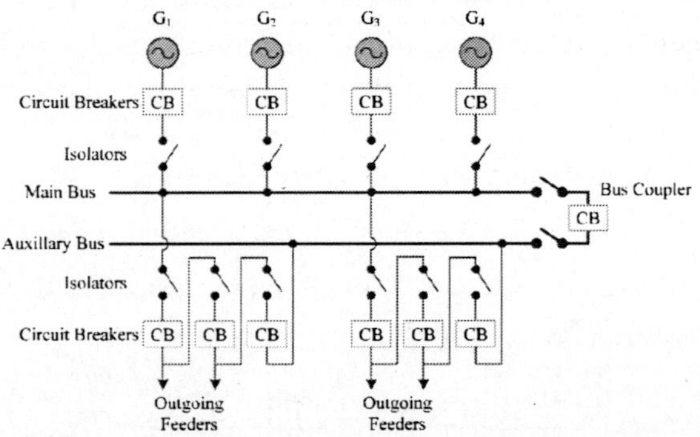

Fig. 10.23. One-and-a-Half Breaker configuration

This configuration offers robust protection against supply disruptions, as a problem in either a bus or a breaker will not cause an interruption in the supply. Another benefit is the potential for integrating more circuits into the system. The bus potential can serve as a power source for relays. However, in the event of a bus fault, this potential should be disconnected from the relay.

An inherent disadvantage of this setup is the complexity involved in transmitting signals, as two breakers need to be opened in the event of a fault. Another disadvantage is that while maintaining circuit breakers, if load shedding is not employed, two breakers must be opened. In this scenario, the other circuit in the lineup will be functioning with only one breaker from one bus. During a fault in the bus, the supply to the other circuit is likewise disrupted. The expense of upkeep is elevated. This configuration has been employed in significant 400 kV and 750 kV substations.

10.20. BUS ZONE FAULTS:

According to statistics, the majority of faults are single-phase in nature and, in some cases, transient. Causes of bus zone defects are:

- Deteriorated support insulator leads to earth fault.
- Excessive over-voltages can result in flash-over across the support insulator.
- Faulty operations carried out by the attending professionals.
- Highly contaminated insulator causes a flashover.
- Foreign objects are accidentally dropping between bus bars.
- Failure of circuit breakers to stop fault current or to clear during thorough fault situations.

To clear a bus fault, it is necessary to open all the circuits that are connected to the faulty bus or bus section, excluding circuits that do not have any back-feed. The predominant methods employed for safeguarding bus-zones are as follows:

(i). Backup protection,

(ii). Differential protection and

(iii). Frame leakage protection.

10.20.1 Backup protection for bus-bars:

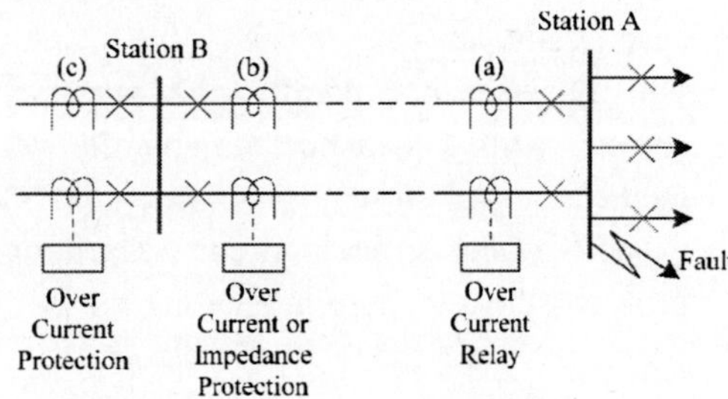

Fig. 10.24. Backup protection for bus-bars

Protecting buses can be achieved by using backup protections on the associated providing parts. These protections should be able to detect and respond to any faults that occur on the buses. If there is no specific protection for the bus, but there is distance protection for the feeders connected to the bus, it is feasible to include the bus-bars within the coverage range of zone 2 of the distance relays. Based on Figure 10.24, bus A is protected by distance protection B in the second phase. Therefore, if there is a malfunction on bus A, the distance protection B will be activated. In this system, the working duration of the second step can be approximately 0.4 seconds. However, due to the slow protection mechanism, there is a possibility of unintended disconnection of all incoming parallel circuits. Distance protection is commonly used to protect power transmission lines, making it cost-effective to also use it for bus protection. The aforementioned approach may be enough for small switchgear installations, but for larger and more significant installations, a distinct bus-zone protection is implemented.

With regards to Figure 10.24

- The overcurrent prevention system at station A serves as the main safeguard for bus-zone A.
- The remote overcurrent protection or impedance protection at station B acts as a secondary safeguard for bus-zone A. If protection 'a' fails, protection 'b' serves as a contingency plan.
- The local overcurrent protection at station B offers main protection to bus B by safeguarding the incoming lines or feeders.

The disadvantages of such protection include: (i) a delay in taking action, (ii) the disconnection of additional circuits if there are several incoming lines, and (iii) the inability to achieve precise classification.

Bus backup protection refers to the situation where, if the breaker fails to respond to a failure on the outgoing line, it is considered a fault in the bus. Subsequently, it should proceed to deactivate all circuit breakers associated with that particular bus. Adequate time delay for backup protection can be achieved by using a timer.

10.20.2 Differential protection for busbars:

The primary approach for busbar protection is the implementation of a differential scheme, where the sum of currents flowing into and out of the bus is calculated. Under normal load conditions, the total of these currents is zero. During the occurrence of a fault, the fault current disrupts the equilibrium and generates a differential current that triggers the operation of a relay.

Figure 10.25 depicts the schematic representation of a current differential scheme used for a station busbar. A generator feeds power to the busbar, which in turn distributes the load to two lines. The secondary windings of the current transformers in the generator lead, specifically in line 1 and line 2, are all interconnected in a parallel configuration. The protective relay is connected in parallel with this connection. All current transformers (CTs) in the scheme must have the same ratio, independent of the capacity of the different circuits. During normal operating circumstances or when there is an external problem, the total current entering the bus is equal to the total current leaving it, resulting in no current flowing through the relay. If a defect arises within the protected area, the currents that enter the bus will no longer be equivalent to the currents that exit it. The disparity between these currents will pass through the relay, resulting in the activation of the generator, circuit breaker, and each of the line circuit breakers.

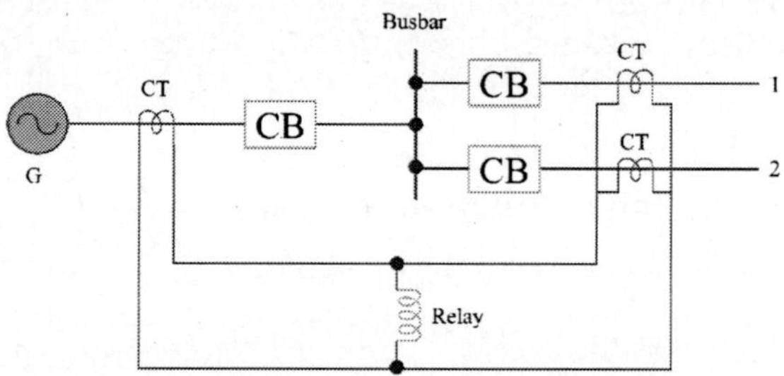

Fig. 10.25. Differential protection for busbars

10.20.3 Frame Leakage/Fault Bus protection:

It is feasible to design a station in such a way that the majority of the faults that occur are primarily earth-faults. To accomplish this, one can implement an earthed metal barrier, also referred to as a fault bus, that encloses each conductor throughout its full length within the bus construction. In this configuration, any potential malfunction must necessarily require a linkage between a conductor and a grounded metallic component. By manipulating the direction of the current caused by faults in the earth, it is feasible to identify and determine the position of these

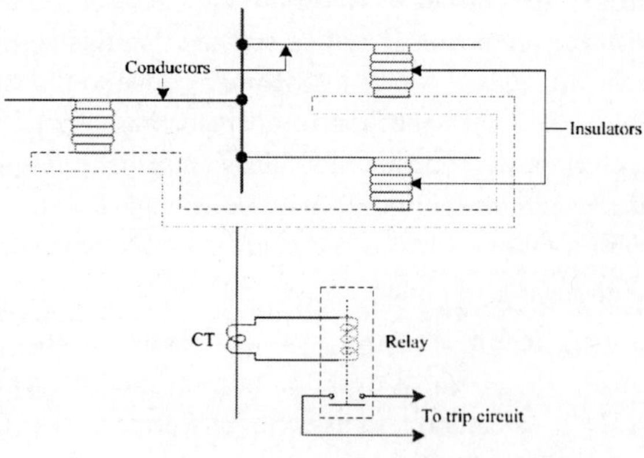

Fig. 10.26 Fault Bus protection

faults. This form of protection is referred to as fault bus protection. Figure 10.26 depicts the schematic configuration of fault bus protection. The metallic support structure or fault bus is grounded using a current transformer. A relay is linked to the secondary winding of this current transformer (CT). During normal operating conditions, there is no electrical current passing from the fault bus to the ground, causing the relay to stay inactive. If there is a fault in the connection between a conductor and an earthed supporting structure, current will flow to the ground through the fault bus, which will cause the relay to activate. The activation of the relay will cause all breakers that connect equipment to the bus to be deactivated.

10.21. Introduction to Protection of Lines:

The longer length and exposure to atmospheric conditions significantly increase the likelihood of faults arising on the lines. Many protective methods have been developed, but they are not applicable to the relatively simple cases of alternators and transformers. The requirements for line protection are:

- In case of a short-circuit, open the circuit breaker closest to the fault and keep all other circuit breakers closed.
- If the nearest breaker fails to open, adjacent circuit breakers can give backup protection.
- Minimise relay operation time to maintain system stability and prevent unwanted circuit tripping.

The protection of transmission lines is a distinct challenge compared to protecting station equipment including generators, transformers, and busbars. Although differential protection is an optimal technique for protecting lines, its implementation is considerably more costly. A significant distance, often spanning several kilometres, separates the two ends of a line. In order to compare the two currents, an expensive pilot-wire circuit is necessary. While this investment may have some justification, it is generally more common to utilise less expensive alternatives. The conventional techniques for safeguarding against line damage include time-graded overcurrent protection, differential protection, and distance protection. Figure 10.27 displays the symbols that represent the different types of relays used in line protection.

- Time-graded overcurrent protection
    - Radial feeder
        - definite time over current relays
        - definite current relays
        - inverse time over current relays
    - Parallel feeders
    - Ring main system
- Differential protection
- Distance protection

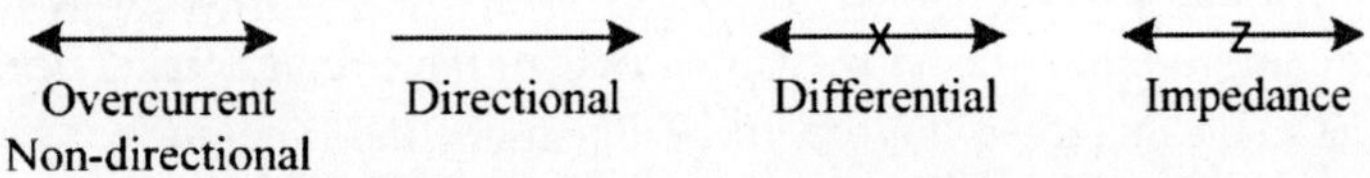

Fig. 10.27 The symbols of different relays.

10.22 Time-Graded Overcurrent Protection of feeders:

The overcurrent protection technique incorporates time discrimination. Relays are timed to isolate the smallest section of the system in case of a malfunction. Time-graded protection for a radial feeder can be accomplished by employing definite time over current relays, definite current relays, and inverse time over current relays.

10.22.1. Definite time over current relay:

Figure 10.28 illustrates the use of definite time relays for overcurrent protection in a radial feeder. The operation time of each relay is predetermined and does not depend on the current it operates with. Relay D has an operating time of 1 millisecond, whereas the time delay for the other relays is increased by 1 millisecond each time. If a fault arises in section DE, it will be promptly resolved within 1 millisecond by the relay and circuit breaker at D, while all other relays have longer operating times. By following this approach, only part DE of the system will be completely separated from the rest. If the relay at D fails to trip, the relay at C will activate with a further time delay of 1millisecond, namely 2 milliseconds after the fault occurs.

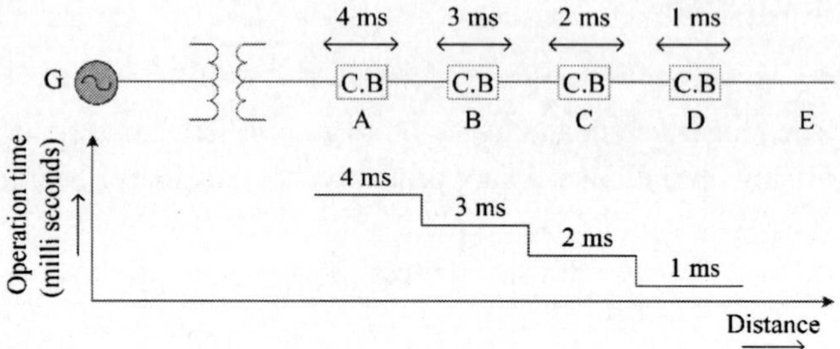

Fig. 10.28 Definite time over current relay

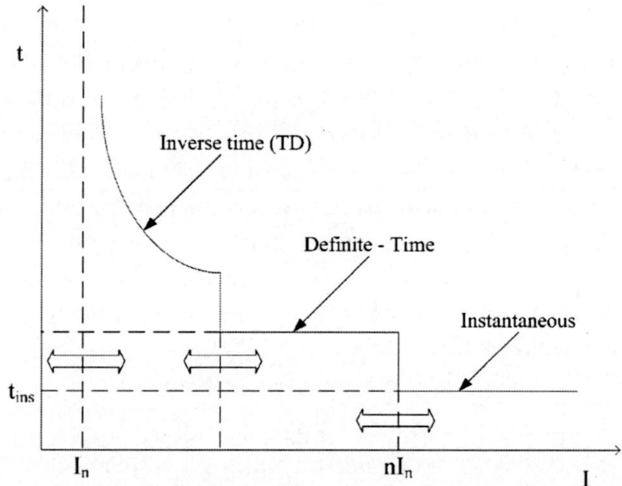

Fig. 10.29 time-current characteristics of inverse time, definite time and instantaneous over current relay

An inherent drawback of this system is that when multiple feeders are connected in a series, the time it takes for the system to trip in the event of a fault occurring near the power supply end is significantly increased, reaching a duration of 4 milliseconds in this particular scenario. Nevertheless, it is typically essential to restrict the maximum duration for tripping to a minimal 4 milliseconds. Using inverse-time relays can effectively mitigate this drawback.

10.22.2. Inverse time over current relay:

In this setup, the time it takes for the relays to operate is inversely related to the current at which they are triggered. In this configuration, the relay running time of the circuit breaker decreases as the distance from the producing station increases. Figure 10.29 illustrates the time-current characteristics of three types of relays: inverse time, definite time and instantaneous over current relay.

10.22.3. Parallel feeders:

If there is a specific need for uninterrupted supply, it is possible to install two parallel feeders. In the event of a fault occurring on one feeder, it is possible to disconnect it from the system and ensure that the supply of electricity is maintained from the other feeder. Non-directional overcurrent relays alone are insufficient for protecting the parallel feeds. Employing directional relays is essential, as well as properly coordinating the time settings of the relays to ensure selective tripping.

Figure 10.30 illustrates a system in which two feeders are interconnected in parallel between the generating station and the sub-station. In order to ensure the security of this system,

- It is necessary for each feeder to be equipped with a non-directional overcurrent relay located at the generator end. These relays should possess an inverse-time characteristic.

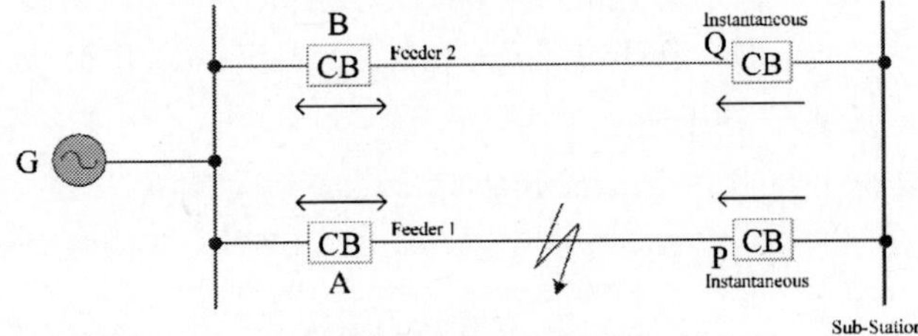

Fig. 10.30 Parallel feeders

- Each feeder is equipped with a reverse power or directional relay at the sub-station end. The relays should be of the instantaneous kind and should only activate when electricity is flowing in the reverse direction, specifically in the direction indicated by the arrow at P and Q.

Assume that a fault develops on feeder 1, as depicted in Figure 10.30. The desired outcome is for only the circuit breakers at A and P to open in order to remove the fault, but feeder 2 should remain unaffected to ensure the uninterrupted supply. Indeed, the aforementioned arrangement successfully fulfils this task. The fault is transmitted by two pathways: (a) directly from feeder 1 through relay A (b) from feeder 2 through B, Q, sub-station, and P.

Consequently, the power flow in relay Q will be in the normal direction, whereas it is reversed in relay P. This results in the activation of the circuit breaker at point P. Furthermore, relay A will function while relay B remains non-functional. The reason for this is that these relays exhibit inverse-time characteristics, and the current passing through relay A exceeds that passing through relay B. Only the faulty feeder is isolated using this approach.

10.22.4 Ring main system:

In a ring main system, multiple power stations or sub-stations are joined through alternate routes, creating a closed ring which is shown in Fig. 10.31. If any piece of the ring is damaged, that section can be detached for repairs. Power will then be supplied from both ends of the ring to ensure uninterrupted supply. Figure 10.32 depicts the schematic diagram of a standard ring main system comprising a single generator G that supplies power to four sub-stations, namely S1, S2, S3, and S4. This configuration allows for bidirectional power flow during fault scenarios. Hence, it is essential to grade in both clockwise and anticlockwise directions around the ring, and to employ directional relays as well.

To ensure that just the problematic piece of the ring is isolated under fault situations, the relays should be of the following types and their time settings should be as specified:

- The two lines departing from the generating station should be fitted with non-directional overcurrent relays (specifically, relays located at points A and J in this instance).
- Reverse power or directional relays should be installed at each sub-station on both the incoming and outgoing lines. In this scenario, relays should be put at sub-stations B, C, D, E, F, G, H, and I.
- The relays should have suitable relative time settings.

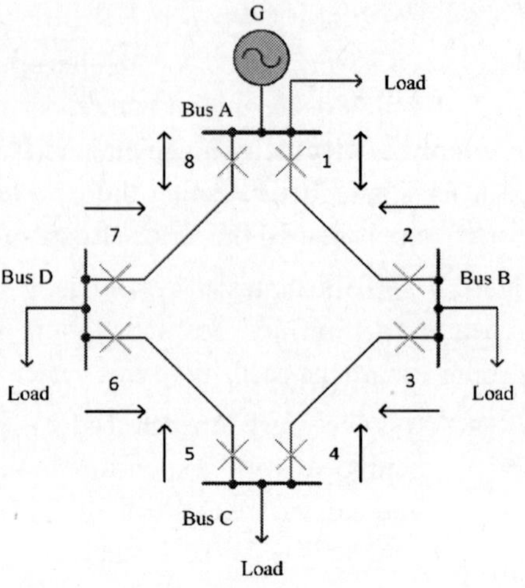

Fig. 10.31 Ring main system

For instance, when moving around the loop G S1 S2 S3 S4 G, the relays at positions A, C, E, G, and I are set with decreasing time limitations. Specifically, relay A is set to 5 milliseconds, relay C is set to 4 milliseconds, and relay E is set to 3 milliseconds. The value of G is equal to 2 milliseconds, while the value of I is equal to 1 millisecond. Similarly, while moving through the loop in the opposite direction (G S4 S3 S2 S1 G), the outgoing relays (J, H, F, D, and B) are similarly configured with a decreasing time limit. For example, J is set to 5 milliseconds, H is set to 4 milliseconds, F is set to 3 milliseconds, D is set to 2 milliseconds, and B is set to 1 millisecond.

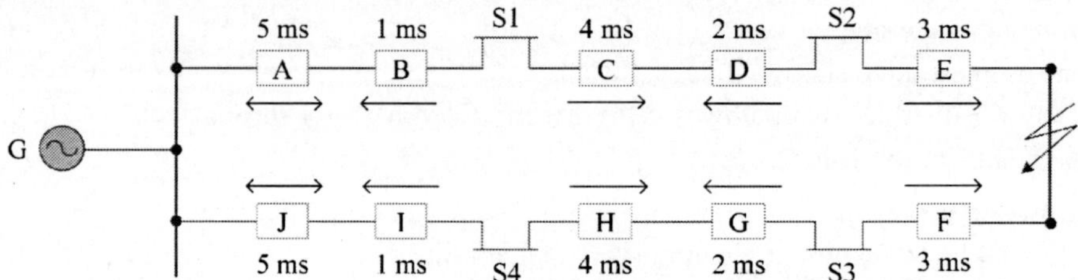

Fig. 10.32 Ring main system

Assume that a short circuit happens at the indicated location in Figure 10.32. To achieve selectivity, it is necessary for only the circuit breakers at points E and F to open in order to clear the fault. The rest of the ring should remain intact to ensure the continuity of power supply. Indeed, the aforementioned arrangement successfully fulfils this task. The power will be supplied to the fault by two ways, namely through the path from G passing through S1 and S2, and through the path from G passing through S4 and S3. It is evident that the relays at A, B, C, and D, as well as J, I, H, and G, will not activate unless relays E & F fail to operate. Thus, due to their lower time-setting, only the relays at E and F will activate before any other relay.

10.23 Differential Pilot-Wire Protection for protecting transmission lines:

The differential pilot-wire protection operates on the idea that, in normal circumstances, the current entering one end of a line is equivalent to the current leaving the other end. Once a fault arises between the two ends, this state ceases to exist and the disparity between the incoming and outgoing currents is directed to pass through a relay, which triggers the circuit breaker to separate the defective line. Multiple differential protection techniques are currently employed for the lines. However, only the Merz-Price voltage balance system and Translay methods will be covered.

10.23.1 Merz-Price voltage balance system for protecting transmission lines:

Figure 10.33 depicts the schematic diagram of the Merz-Price voltage balance system used to protect a three-phase line. Each phase of the line is equipped with identical current transformers at both ends. In each line, the two CTs are connected in series via a relay. This configuration ensures that, in normal circumstances, the secondary voltages of the CTs are equal and opposite, thereby cancelling each other out. In normal circumstances, the amount of electric current entering the line at one end is similar to the amount of current leaving it at the other end. As a result, the secondaries of the CTs at both ends of the line experience equal and opposite voltages. Consequently, there is an absence of electrical current passing through the relays. Assume that a fault occurs at point F on the line, as seen in Figure 10.33. As a result, CT_1 will experience a higher current flow compared to CT_2. As a result, the secondary voltages of the system become imbalanced, leading to the passage of circulating current through the pilot wires and relays. Both circuit breakers at the beginning and end of the line will automatically shut off, thereby isolating the damaged line. Fig. 10.34 shows the connections of Merz-Price voltage balance scheme for all the three phases of the line.

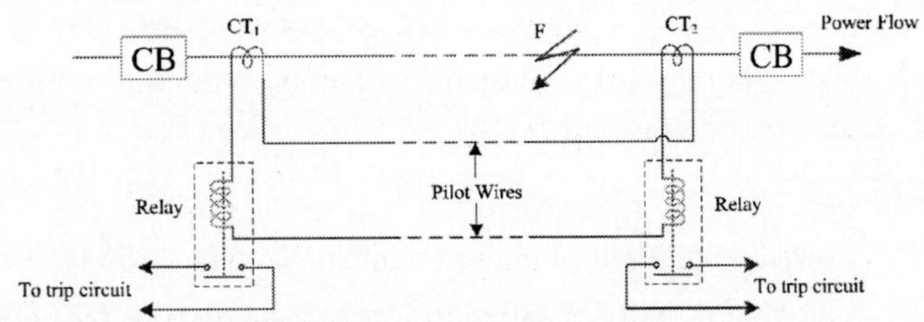

Fig. 10.33 Single line diagram of Merz-price voltage balance system

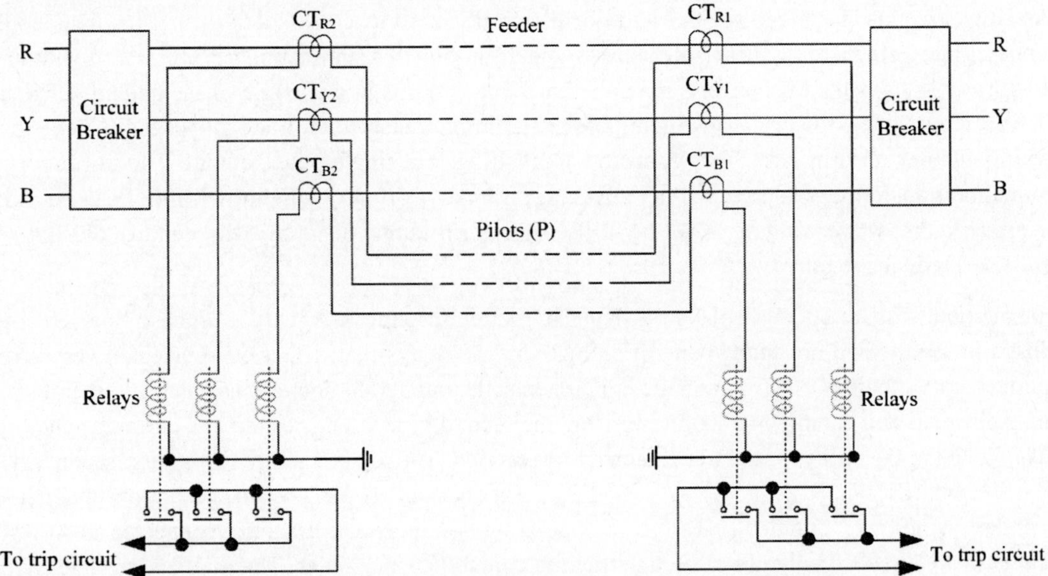

Fig. 10.34. Three-phase Merz-price voltage balance system

Merits:

- ✓ This approach is applicable for both ring mains and parallel feeders.
- ✓ This approach offers immediate safeguarding against ground defects. This reduces the likelihood of these problems affecting other phases.
- ✓ This method offers immediate relaying, which minimises the harm caused to overhead conductors due to arcing problems.

Demerits:

- × Precise alignment of current transformers is crucial.
- × The system will not function if there is a discontinuity in the pilot-wire circuit.
- × This technique is quite expensive due to the longer length of pilot wires required.
- × If there are long lines, the charging current caused by the effects of pilot-wire capacitance may be enough to trigger the functioning of the relay, even under normal situations.
- × The utilisation of this technology is limited to line voltages up to 33kV due to challenges in aligning the current transformers.

10.23.2 Translay scheme for protection of transmission lines:

The Translay scheme is similar to a voltage balancing system, with the key distinction being that the balance or opposition occurs between the voltages induced in the secondary windings wrapped around the relay magnets, rather than between the secondary voltages of the line current transformers. This allows for the utilisation of conventional current transformers and addresses a significant drawback of the original voltage balance system, which was limited to functioning at voltages below 33 kV. To extend this to a 3-phase system, one can apply a single relay at each end of every phase of the 3-phase line. However, it is feasible to achieve additional simplification by consolidating currents obtained from all phases into a single relay at each end, employing the notion of a summation transformer. A summation transformer is a device that converts the polyphase line currents into a single-phase quantity. The three CT lines are linked to the primary winding of the summing transformer. Each line of the CT energises a varying number of turns (from line to neutral), resulting in a single-phase output. The utilisation of a summation transformer has two distinct advantages. Firstly, primary windings 1 and 2 can be utilised for phase faults, while winding 3 can be utilised for earth faults. Secondly, the number of pilot wires required is reduced to only two.

Construction: Figure 10.35 displays the Translay system used to safeguard a 3-phase line. The relays utilised in the method are fundamentally induction type relays that detect overcurrent. Every relay is equipped with a pair of electromagnetic components. The upper component has a coil (either 1 or 1 a) that is powered as a combined transformer from the secondary windings of the line current transformers (CTs) linked to the phases of the line that needs protection. The top component also has a secondary coil (2 or 2 a) that is connected in series with the operational coil (3 or 3 a) on the bottom magnet. The secondary windings 2 and 2a, as well as the operating windings 3 and 3a, are connected in series in a manner that causes the voltages created in them to counteract each other. The diagram has intentionally excluded relay discs and tripping circuits in order to enhance clarity.

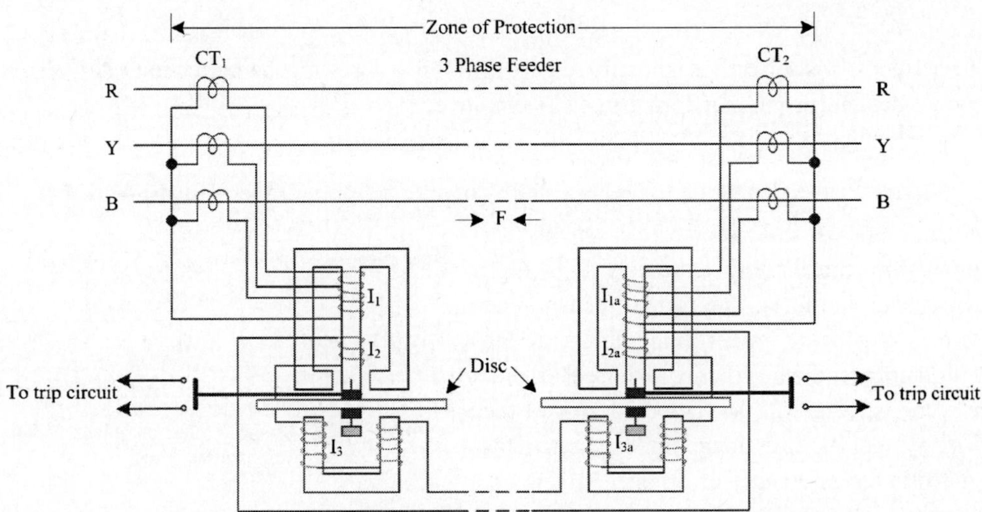

Fig. 10.35. Three-phase Translay scheme for protection of transmission lines

Operation: When the feeder is functioning properly, the currents at both ends of it are equal, resulting in equal secondary currents in both sets of current transformers (CTs). As a result, the currents in the primary winding 1 and 1a of the relay will be the same, causing identical voltages to be induced in the secondary windings 2 and 2a. Due to the opposing connection of these windings, there is no current flowing through them or through the working windings 3 and 3a. If a fault occurs on the protected line, the current at one end of the line must carry a higher magnitude of current compared to the other end. The outcome is that the voltages generated in the secondary windings 2 and 2a will exhibit dissimilarity, leading to the passage of electric current through the working coils 3, 3a, and the pilot circuit. In these circumstances, both the upper and lower components of each relay are powered, resulting in a forward force being exerted on each relay disc. The relays will activate the circuit breakers at both ends of the line, causing them to open.

Assume that there is a fault F between phases R and Y, and it is supplied power from both sides, as depicted in Figure 10.35. Only section 1 of primary windings 1 and 1a will be energised, resulting in the induction of voltages in the secondary windings 2 and 2a. Since these voltages are now cumulative, current will flow through operational coils 3, 3a, and the pilot circuit. This will result in the activation of the relay contacts, leading to the closure and subsequent opening of the circuit breakers located at both ends. A fault occurring between phases Y and B causes section 2 of primary windings 1 and 1a to become energised, whereas a fault between phases R and B energises sections 1 and 2.

Now consider a scenario where a fault occurs on phase R of the electrical system. This will activate sections 1, 2, and 3 of the principal windings 1 and 1a. If a fault originates from both ends, the induced voltages in the secondary windings 2 and 2a combine and generate a current that passes through the working coils 3, 3a. Consequently, the relays function to activate the circuit breakers located at both ends of the line, causing them to open. If there is a failure on phase Y, sections 2 and 3 of primary winding 1 and 1a will become electrified, resulting in the activation of the relays. A failure occurring on phase B will only activate section 3 of the primary windings 1 and 1a of the relay.

Merits:

- ✓ The system is cost-effective as it only requires two pilot wires for the protection of a 3-phase line.
- ✓ Normal-design current transformers can be employed.
- ✓ Relay operation is unaffected by pilot wire capacitance currents.

10.24. Distance Protection/Zone Protection for Protecting high voltage transmission lines:

Neither time-graded nor pilot-wire systems are appropriate for safeguarding extremely long high voltage transmission lines. The former results in an excessively long delay in resolving faults at the producing station end when there are more than four or five sections, whereas, the increased cost of the pilot-wire system requires longer pilot wires. This has resulted in the emergence of distance protection, where the operation of the relay is contingent upon the distance (or impedance) between the installation point of the relay and the location of the fault. This technique offers discriminating protection without the use of pilot wires.

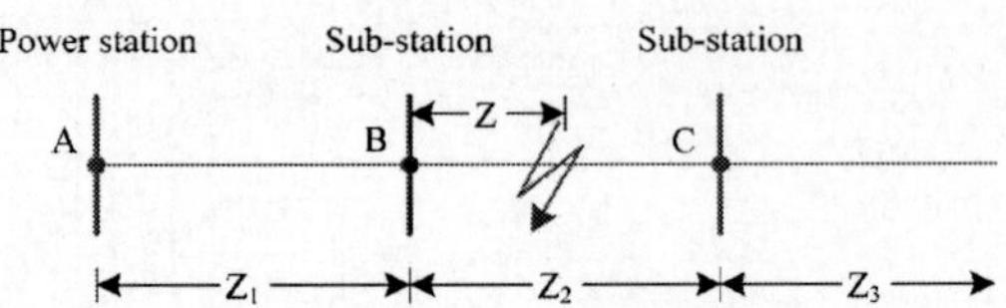

Fig. 10.36. Distance Protection of high voltage transmission lines

Figure 10.36 depicts a basic system comprised of series lines, where power can only travel in one direction, specifically from the left to the right. The relays at points A, B, and C are configured to activate when the impedance is below the thresholds of Z_1, Z_2, and Z_3, respectively. If a fault arises between sub-stations B and C, the fault impedance at power station A and sub-station B will be Z_1+Z and Z respectively. It is evident that first the relay at B will function for the shown portion. However, in the event of a malfunction of relay B, the relay at A will get activated. By using this approach, immediate safeguarding can be achieved for all operational circumstances.

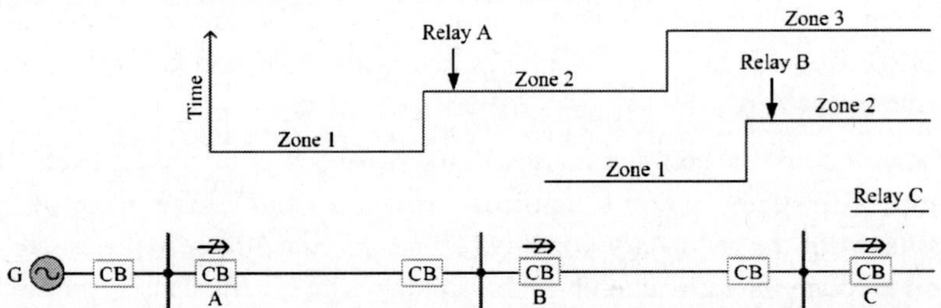

Fig. 10.37. Three-zone protection of high voltage transmission lines

In practice, imperfections in relay components and instrument transformers prevent instantaneous protection for the entire line. The relay at "A" may not reliably discriminate between faults at 99% and 101% of the distance A-B. Figure 10.37 illustrates the use of 'three-zone' distance protection to address this issue. In this protection approach, three distance elements are used at each terminal. The zone 1 element, which covers the first 90% of the line, detects problems immediately. The zone-2 element detects faults in the remaining 10% of the line as well as the following section. However, a time delay prevents the line from being tripped if the fault is in the next section. If the next section's breaker fails to clear a fault, the zone 3 element acts as backup protection.

10.25 Summary

- The alternator can be protected against earth, phase-to-phase, and inter-turn faults using the following protection schemes: differential, modified differential, biased, and biased modified differential, stator inter-turn alternator protection, and restricted earth fault protection.
- Transformer protection techniques include: earth fault relay (core balance leakage protection), overcurrent relay (leakage and overload protection), differential protection, modified differential, biased differential, biased modified differential protection, and Buchholz relay.
- The implementation of a differential protection scheme for transformer protection is not suitable due to the following limitations: the CT ratios, CT connections, tap changes, and magnetizing inrush currents.
- It is necessary to connect the current transformers (CTs) in a delta configuration on the star side of a star/delta power transformer, and in a star configuration on the delta side.
- Biased differential protection in transformers is used to prevent unwanted operation on heavy external faults caused by CT errors and CT ratio changes (due to tap altering).
- The Buchholz relay is a gas-actuated protective device that is typically used in oil-immersed transformers to safeguard against various types of defects.
- For a small to medium-sized station that allows for intermittent shutdowns, a simple single-bus-bar arrangement is preferable. Large plants must have a complex bus-bar system to ensure that a backup power supply is always available in the case of a breakdown.
- Bus-zones are typically protected using backup, differential, or frame leakage approaches.
- Bus backup protection refers to the case in which the breaker fails to respond to a failure on the outgoing line, indicating a malfunction in the bus. Then, it should deactivate all circuit breakers linked with that specific bus.
- In the Distance/Zone protection technique, each terminal has three distance elements. The zone-1 element, which covers the first 90% of the line, finds issues promptly. The zone-2 element detects faults in the remaining 10% of the line, as well as the portion below. A time delay, on the other hand, prevents the line from being tripped if the fault occurs in the following section. If the following section's breaker fails to clear a fault, the zone 3 element serves as backup protection.

Short and Long Answer Questions

1. List the faults that occur on alternators. Describe different alternator protection strategies against stator winding problems.
2. What are the typical protections commonly advised for alternators? How does Differential Protection of alternators differ from Modified differential protection of alternators in terms of effective protection for alternators?
3. With neat diagrams explain the construction and operation of modified differential protection of alternations against earth and phase-to-phase faults.
4. Describe the construction and operation of biased modified differential alternator protection against earth and phase-to-phase faults using neat diagrams. How does biased differential protection of alternators differ from biased modified differential protection of alternators in terms of effective protection for alternators?

5. With neat diagrams explain the construction and operation of biased differential protection of alternator against earth, phase-to-phase and inter-turn faults. How does biased differential protection of alternators differ from differential protection of alternators in terms of effective protection for alternators?
6. List the faults that occur on transformers. Describe different transformer protection strategies for stator winding problems.
7. What factors cause difficulty in applying Merz-price protection to a power transformer? Explain.
8. What are the typical protections commonly advised for power transformers? How does 'core-balance leakage protection' differ from 'combined leakage and overload protection' in terms of effectiveness for power transformers?
9. Describe the differential protection mechanism applied to a power transformer. What are the scheme's shortcomings and how are they overcome?
10. Explain the principle of Merz-piece protection scheme used for power transformers with neat circuit diagram. What are the limitations of this scheme? How they are overcome?
11. Explain with a neat circuit diagram the differential protection scheme used to protect Y-Δ, Δ-Y, Y-Y, Δ-Δ, transformers.
12. Draw a circuit schematic of the biased modified differential protection scheme for Y-Δ, Δ-Y, Y-Y, and Δ-Δ transformers. What distinguishes biased modified differential protection of transformers from modified differential protection of transformers in terms of alternator protection? Which is effective, and why?
13. How do you protect an alternator from turn-to-turn faults on the same phase winding? Explain how it was built and how it works.
14. Describe the construction, principle of operation and application of Gas actuated relay.
15. What are the various bus-bar configurations employed in substations? Support your response with appropriate visual representations.
16. Explain time-graded overcurrent protection for radial feeders, parallel feeders, and ring main systems, using illustrations.
17. What is the purpose of bus bar protection? How is the bus-bar protection scheme stabilised?
18. What is backup protection of bus-bars? Explain it with a neat schematic diagram.
19. Explain differential bus-bar protection. What are the drawbacks of this protective strategy, and how may these be overcome?
20. Describe how voltage differential systems are used to safeguard bus bars. How does it overcome the problem of CT saturation?
21. Explain the translay protection method for feeders using a neat schematic diagram.
22. Explain the distance protection mechanism for feeders with a neat schematic illustration.

Exercises

1. The generator is equipped with a differential protection scheme, operating at 11 kV and 50 MVA. The percentage of the generator winding that requires protection against phase to ground faults is 85%. The relay is activated when there is a 20% discrepancy in current. Calculate the appropriate resistance to be inserted in the ground connection.

2. A 3-ph, 11 kV, 20 MVA star connected alternator is protected by the Merz-Price protection. If the ratio of the current transformer is 1200/5, the minimum operating current of the relay is 0.75 Amps and the neutral point earthing resistance is 6 Ω, calculate the percentage of each phase of the stator winding which is unprotected against earth faults when the machine is operating at normal voltage. Show quantitatively, the effect of varying the neutral earthing resistance (consider min. two values).

3. The neutral point of a three-phase, 100 megavolt-ampere (MVA), 22 kilovolt (kV) alternator is connected to the ground through a 20-ohm resistance. The relay is configured to activate when there is an imbalance in the current of 5 amperes. The CT's have a ratio of 700:5.

   (i) What is the percentage of winding that needs to be safeguarded?

   (ii) Determine the necessary grounding resistance to safeguard 80% of the winding.

4. The alternator is connected in a star configuration and has a voltage of 33 kV and a power rating of 75 MVA. It has a reactance of 10Ω per phase and a negligible resistance. Merz-Price protection is employed to safeguard the winding. The neutral grounding resistance is 6 Ω when just 15% of the winding is left unprotected. Determine the setting of the relay.

5. The alternator is rated at 25MVA and has a voltage of 11kV. It is connected in a star configuration. The synchronous reactance of each phase is 6Ω, while the resistance is 2Ω. The system is safeguarded by a Merz-Price balanced current system, which activates when the current imbalance exceeds 15% of the total load current. Calculate the proportion of the alternator winding that is unprotected when the star point is connected to the earth through an 8Ω resistance.

6. The alternator is a 66kV, 120MVA, star-connected machine with a reactance of 1.5 per unit/phase and a minimal resistance. The system is safeguarded by a Merz-Price balanced current system that activates when the out of balance current is above 10% of the load current. If the neutral point is connected to the ground through a 5Ω resistance, calculate the percentage of the winding that is safeguarded against an earth fault.

7. If Merz-Price protection is to be employed for a 3-phase, 33/11 kV transformer linked in Y-Δ with current transformers (CTs) on the low voltage (L.V) side with a ratio of 450:5, what will be the ratio of the CT on the high voltage (H.V) side?

8. A three-phase delta-star, 33/11 kV transformer is safeguarded using a differential protection mechanism. Given that the CT ratio on the 11 kV side is 350:5, calculate the CT ratio on the H.V side (33 kV).

9. A 50 MVA, 132/66 kV, Δ-Y three-phase transformer is safeguarded by percentage differential relays. Determine the output current at full load, relay current at full load, and minimum relay current setting for 25% overload if the current transformers (CTs) on the Δ and Y sides of the power transformer are 300/5 A and 1200/5 A, respectively.

10. A 100 MVA, 110 kV/33kV, Y-Δ, three-phase transformer is protected by percentage differential relays. If the current transformer (CT) located on HV side have a transformer ratio of 70/5, what will be CT ratio on LV side? Also find ratios of CT on HV side for (i)Δ-Y, (ii)Δ-Δ and (iii)Y-Y configurations.

MATLAB/SIMULINK

To Model and Simulate

Alternator protection,
Over voltage protection and
Power System Protection

To Model and Simulate
Differential Protection of a three phase Transformer
Power System Protection Simulations

To Model & Simulate

Busbar differential relay

Tx. Line differential Relay

PSCAD & Proteus

To Model and Simulate
Differential relay in PSCAD
3-Ph T/f line current in proteus
Remote Monitoring oil level, temperature and earth

Hardware/Experimental

To Design & Implement
Phase Fault Detection System Using Arduino
GSM based T/f Fault Detection
IoT based T/f health Monitoring

11 HVDC TRANSMISSION

Unit specifics: In this unit, the following topics have been discussed for basic understating of basic concepts:

- Evolution of HVDC transmission system.
- Comparison of HVAC and HVDC systems, Applications of the HVDC system.
- List of significant HVDC projects worldwide, including line, multi-terminal, LCC, and VSC.
- Line Commutating Converter (LCC) and voltage source converter (VSC) based HVDC links.
- Types of HVDC links, Real power flow control in the HVDC link.
- Milestones of HVDC development in India.

Rationale: In this unit, students will be introduced to the evolution of the HVDC system, its advantages and disadvantages, comparison of HVDC and HVAC systems, limitations and application of HVDC systems, List of significant HVDC projects worldwide, including line, multi-terminal, LCC, and VSC., role of Short Circuit Ratio in HVDC link, LCC and VSC-based HVDC system, real power flow control in HVDC link, types of HVDC links and Milestones of HVDC development in India, are clearly described with the help of necessary diagrams, derivations, and examples.

Pre-Requisites: Basic knowledge of basic electrical engineering.

Unit Outcomes: The list of outcomes of this unit is as follows:

U11-O1: To understand the evolution of HVDC transmission.

U11-O2: To understand the real power flow control in the HVDC link.

U11-O3: Analyse the LCC and VSC-based HVDC systems.

U11-O4: To understand the types of HVDC links and the application of HVDC links.

U11-O5: To analyse the concept of Short Circuit Ratio in HVDC.

U11-O6: To acquire knowledge about the various HVDC projects that have been established around the world.

Unit-11 outcomes	Expected mapping with course outcomes (1: Weak, 2: medium, and 3: strong correlation)					
	CO-1	**CO-2**	**CO-3**	**CO-4**	**CO-5**	**CO-6**
U11-O1	3	2	1	-	3	-
U11-O2	2	2	-	-	3	-
U11-O3	2	2		-	3	-
U11-O4	2	2	-	-	3	2
U11-O5	2	2	-	-	3	-
U11-O6	3	-	-	-	3	-

11.1 Introduction:

High Voltage Direct Current (HVDC) transmission is a crucial technology that plays a vital role in the efficient and reliable transfer of electrical power over long distances. Unlike traditional AC transmission, which dominates most power distribution systems, HVDC operates by converting AC power into DC power for transmission and then converting it back to AC at the receiving end. This process of converting power allows HVDC systems to overcome various limitations associated with AC transmission, making it particularly suitable for long-distance transmission, interconnecting asynchronous grids, integrating renewable energy sources, and transmitting power across undersea or underground cables. For power transmission above 500 km, HVDC is useful because it reduces transmission losses. It required a minimum of three conductors in AC transmission and two conductors in HVDC transmission. The AC transmission is frequency-dependent. The line reactance increases with the frequency. The line resistance also increases due to the skin effect. Fig. 11.1 shows the HVDC link, which is sending power from the generating stations to the AC loads. The generation units are solar PV, thermal stations, hydro power plant and wind turbine, the generated power is converted to DC and transmitted through DC transmission lines. Further DC power is converted to AC using DC to AC converter station.

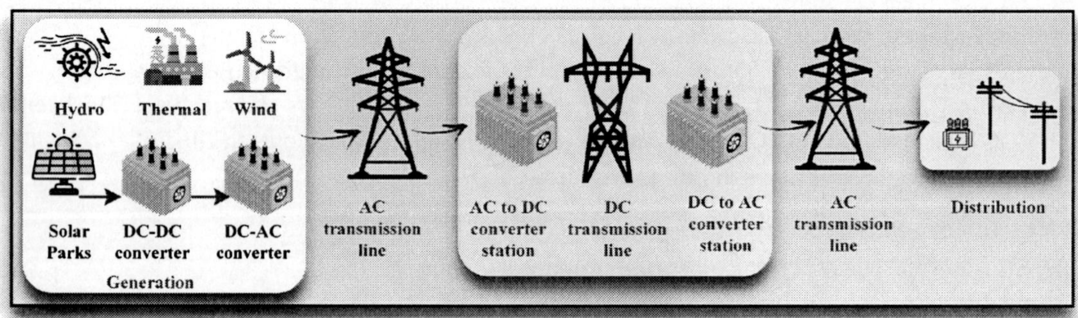

Fig. 11.1 HVDC application in power system.

The initial commercial implementation of HVDC occurred in 1954 with the connection between Gotland Island and the Swedish mainland. Over the next 55 years, significant advancements in HVDC technology have emerged alongside increased economic opportunities associated with HVDC systems. A notable milestone in India was the commissioning of the Rihand-Dadri HVDC link in 1991. This link connected the thermal power plant in Rihand, Uttar Pradesh (Northern Grid), with Dadri (situated in the Western Part of the Northern Grid). Spanning approximately 816 km, this HVDC line was constructed by ABB and is currently under the ownership of PGCIL.

The development of HVDC transmission dates back several decades, with significant advancements in technology, design, and application. HVDC systems are characterized by their ability to control power flow independently, lower transmission losses, reduce right-of-way requirements, and improve stability in interconnected power grids. These features have made HVDC transmission a preferred choice for addressing modern energy challenges, including integrating renewable energy, enhancing grid resilience, and supporting international and intercontinental power exchange.

11.2 Evolution of HVDC Transmission:

The history of HVDC transmission spans several decades, marked by significant technological advancements and milestones that have transformed the way electrical power is transmitted over long distances. Here's a brief overview of the key events in the history of HVDC transmission:

1950s - Early Developments: The concept of HVDC transmission began to take shape in the early 1950s with experimental projects and theoretical studies. In 1954, the first commercial HVDC link was established between the island of Gotland and mainland Sweden, spanning about 96 kilometres.

1960s - Mercury Arc Rectifiers: During the 1960s, HVDC technology saw advancements with the introduction of mercury arc rectifiers, which improved the efficiency of the conversion process. HVDC systems using mercury arc valves were deployed in various projects, demonstrating the feasibility of long-distance DC transmission.

1970s - Thyristor-Based Converters: The 1970s marked a significant shift with the development and deployment of thyristor-based converters in HVDC systems. Thyristor valves offered higher reliability, improved control, and efficiency compared to mercury arc rectifiers, leading to the widespread adoption of HVDC technology.

1980s - Advancements in Control Systems: In the 1980s, HVDC systems benefited from advancements in control systems and semiconductor technology. Digital control systems and advanced thyristors allowed for more precise and efficient operation of HVDC transmission networks.

1990s - Expansion and Integration: The 1990s witnessed a surge in the construction of HVDC transmission projects worldwide. HVDC links were used for interconnecting power grids, integrating renewable energy sources, and improving grid stability.

2000s - Voltage Source Converters (VSCs): The 2000s saw the emergence of VSC technology in HVDC systems. VSC-based HVDC offered enhanced controllability, improved performance in weak AC systems, and compatibility with renewable energy integration.

2010s - Growth and Modernization: In recent years, HVDC transmission has experienced continued growth and modernization. Projects involving ultra-high-voltage HVDC transmission, offshore wind farm connections, and international interconnections have become more prevalent.

11.3 Advantages and Disadvantages of HVAC Transmission:

HVAC transmission, while widely used and effective in many scenarios, has several disadvantages compared to HVDC transmission. Here are some of the key drawbacks of HVAC transmission:

- ✓ *Reactive Power Consumption:* HVAC transmission systems require reactive power to maintain voltage levels across the network. This reactive power does not contribute to useful work but is necessary for the operation of inductive loads and the stability of the system.
- ✓ *Line Losses and Efficiency:* Due to the inherent properties of AC transmission, such as skin effect and proximity effect, HVAC systems experience higher line losses compared to HVDC systems. These losses occur primarily due to resistance in the conductors and increase with the transmission distance.
- ✓ *Grid Stability Concerns:* HVAC grids face challenges related to grid stability, especially during high-demand periods or in the presence of fluctuating loads. Maintaining stable voltage and frequency across a large AC network requires careful management and control.

✓ *Limited Power Transfer Capability:* HVAC systems have limitations on the amount of power that can be transferred over long distances without significant losses. This limitation is influenced by factors such as line impedance, voltage drop, and reactive power flow.

✓ *Complex Grid Interconnections:* Connecting large sections of AC grids or integrating asynchronous AC networks can be challenging due to synchronization issues, grid imbalances, and the risk of cascading failures. This complexity adds to the overall operational challenges of HVAC transmission.

✓ *Charging Current and Capacitance Effects:* Capacitive effects in HVAC transmission lines lead to charging currents, especially in long-distance overhead lines and cables. Managing these charging currents and mitigating the associated losses require additional equipment and infrastructure.

Disadvantages of HVDC:

× Complexity: HVDC systems are generally more complex than HVAC systems, requiring advanced control algorithms and communication systems for operation and monitoring.

× Converter Station Limitations: The converter stations in HVDC systems are critical components and can be vulnerable to failures, leading to potential system downtime and maintenance challenges.

× Limited Voltage Control: Unlike HVAC systems, HVDC systems have limited capability for voltage control, especially under varying load conditions and system disturbances.

× Harmonics and Filtering: HVDC systems can introduce harmonics into the power grid, requiring additional filtering and compensation equipment to maintain grid stability and power quality.

× Ground Electrode Requirements: Some HVDC configurations, especially those using ground or sea return paths, may require extensive grounding infrastructure, which can add to the overall complexity and cost.

× Limited Distance for Monopolar Operation: In monopolar operation, where one conductor is used for current return, there are distance limitations due to voltage drop and power losses.

× Environmental Impact: The construction and operation of HVDC transmission lines and converter stations can have environmental impacts, such as land use, electromagnetic interference, and visual aesthetics.

11.4 Comparison of AC and DC Transmission:

As the load demand increases the load on transmission lines also increases. So, expansion of the lines is done by the energy demand. There are three different aspects to compare the AC and DC transmission.

- Technical Aspect
- Economical
- Reliability

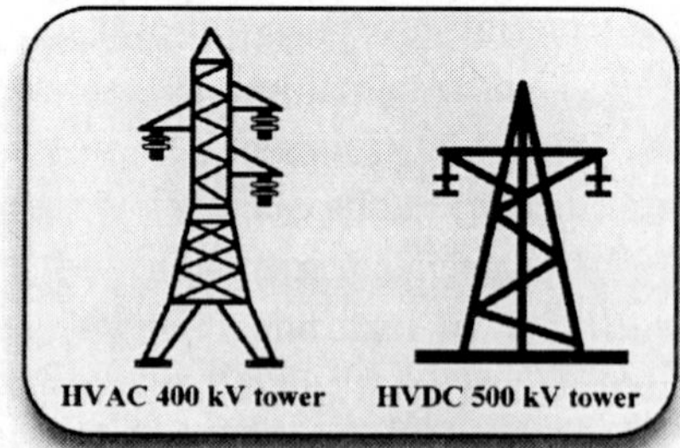

Fig. 11.2 Comparison of HVAC and HVDC tower structure.

11.4.1 Technical Aspects:

11.4.1.1 Power Capacity and Transmission Distance:

AC Transmission: The power capacity of AC transmission lines is influenced by reactive power requirements and stability limits. It is difficult to transmit reactive power over long distances. Apart from this as the transmission distance increases, the efficiency of AC lines decreases due to higher losses.

DC Transmission: DC lines can transmit power over longer distances more efficiently with lower losses compared to AC lines. DC systems do not require transmission of reactive power, making them suitable for long-distance and high-capacity transmission. Fig 11.2 shows that the height of the HVAC tower is greater for the same power transfer compared to the HVDC tower. HVAC contains bundle conductors to reduce the line inductance and skin effect. But HVDC is free from such losses, it has only ohmic loss.

11.4.1.2 Voltage Control and Conversion:

AC Transmission: AC systems can easily change voltage levels using transformers, providing flexibility in distribution and integration with existing networks.

DC Transmission: HVDC systems require complex and expensive converter stations to convert AC to DC and vice versa. These stations are necessary at both ends of the transmission line.

11.4.1.3 Harmonics and Power Quality:

AC Transmission: AC systems are prone to issues related to harmonics, voltage sags, and phase imbalances, which can affect power quality.

DC Transmission: HVDC converters generate harmonics, but advanced filtering techniques are used to mitigate their impact on power quality.

11.4.2 Economical Aspect:

The cost of the AC transmission is increased when the distance is more than 500 km. It includes the operational and initial investment costs. The investment cost includes the capital investment in building transmission lines, substations, and other equipment required for power transmission. The costs depend on factors such as the distance of transmission, voltage levels, terrain, and environmental considerations. The operational cost contains the cost of losses. Fig. 11.3 shows the comparison graph of cost versus distance of transmission line for AC and DC supply.

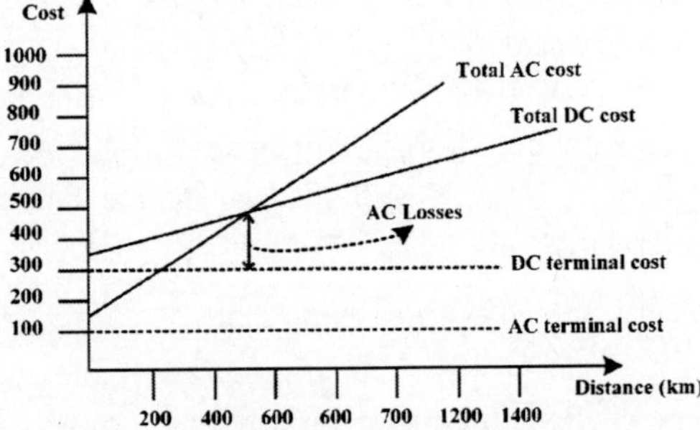

Fig. 11.3 Comparison graph of cost versus distance of transmission line for AC and DC.

11.4.2.1 Initial Investment Costs:

AC Transmission: The initial costs for AC transmission lines are generally lower compared to DC lines. However, as transmission distance increases, the costs of AC lines rise significantly due to the need for multiple substations and reactive power compensation equipment.

DC Transmission: HVDC lines have higher initial costs due to expensive converter stations. Despite this, HVDC lines become more economical for long-distance transmission due to lower line and land acquisition costs. Fig. 11.4 shows the power transfer capability versus distance for HVAC and HVDC. It shows that after a certain distance, the power transfer capability by the AC line reduces due to the reactive loss, skin effect, etc, that increases the cost of transmission. So, for the long distance, HVDC transmission is economical compared to HVAC.

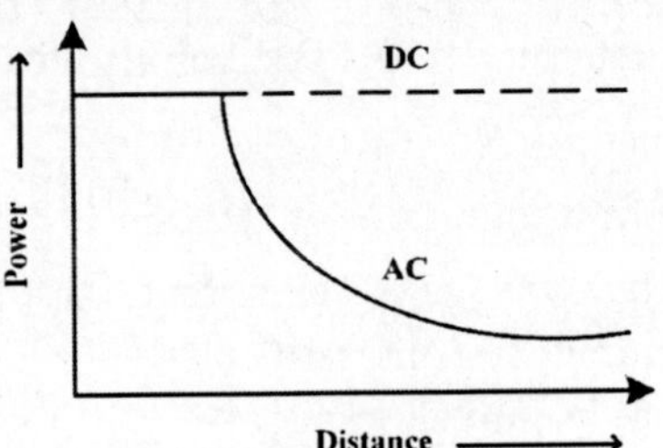

Fig. 11.4 Power transfer capability Vs distance for HVAC and HVDC.

11.4.2.2 Operational and Maintenance Costs:

AC Transmission: Operational and maintenance costs for AC systems can be higher due to the need for frequent maintenance of substations and reactive power equipment.

DC Transmission: HVDC systems typically have lower operational and maintenance costs. The lower losses and reduced need for reactive power compensation contribute to overall cost savings.

11.4.2.3 Break-Even Distance:

AC Transmission: AC transmission is cost-effective for shorter distances. However, the efficiency decreases as distance increases.

DC Transmission: HVDC becomes more cost-effective beyond a certain break-even distance, typically around 500 km for overhead lines and 50 km for submarine cables. This makes HVDC ideal for long-distance and undersea power transmission.

11.4.3 Reliability:

11.4.3.1 Stability and Fault Management:

AC Transmission: AC systems are susceptible to stability issues, such as voltage collapse and frequency instability, especially over long distances. Faults in AC systems can propagate quickly, affecting the entire network.

DC Transmission: HVDC systems offer better stability and control over power flow. They do not contribute to short-circuit currents in the AC network, making them more reliable in maintaining system stability. HVDC systems can isolate faults more effectively, minimizing the impact on the overall network.

11.4.3.2 Integration with Renewable Energy:

AC Transmission: Integrating renewable energy sources, such as wind and solar, into AC systems can be challenging due to variability in power generation and the need for grid stability.

DC Transmission: HVDC systems are better suited for integrating renewable energy sources, as they can efficiently transmit power over long distances from remote generation sites to load centres. The controllability of HVDC allows for better management of variable power inputs from renewable sources.

11.4.3.3 Environmental Impact:

AC Transmission: AC lines require wider right-of-way, which can have a significant environmental impact, including land use and visual intrusion.

DC Transmission: HVDC lines typically require a narrower right-of-way, reducing the environmental footprint. HVDC systems also have a lower visual impact and can utilize existing infrastructure more effectively.

11.5 Limitations of HVDC:

High voltage direct current transmission systems offer several advantages over traditional Alternating current systems, particularly for long-distance and high-capacity power transfer. However, HVDC also has its limitations:

- High Initial Costs: The initial investment for HVDC systems is significantly higher than for AC systems, primarily due to the expensive converter stations needed to convert AC to DC and vice versa. This includes costs for transformers, inverters, rectifiers, and other associated equipment.

- Complexity of Converter Stations: HVDC systems require complex converter stations at both ends of the transmission line to manage the conversion between AC and DC. These stations are technologically advanced and require skilled personnel for maintenance and operation.
- Inability to Use Transformers for Voltage Transformation: Unlike AC systems, which can easily change voltage levels using transformers, HVDC systems cannot directly use transformers. This lack of flexibility can complicate integration with existing AC networks and infrastructure.
- Generation of Harmonics: HVDC systems can generate harmonics due to the switching operations in the converters. These harmonics can interfere with other electronic equipment and require additional filtering equipment to mitigate.
- Reactive Power Requirements: HVDC converters need reactive power to operate, which must be supplied locally. This can necessitate additional equipment such as capacitors or synchronous condensers to provide the required reactive power support.
- Complexity of Controls: The control systems for HVDC are more complex than those for AC systems. Advanced digital control systems and sophisticated algorithms are required to manage the operation and ensure the stability of the HVDC transmission.
- Limited Multi-Terminal Configurations: While HVDC technology allows for multi-terminal configurations, such setups are more complex and less common compared to point-to-point HVDC systems. Managing and controlling power flow in multi-terminal HVDC systems can be challenging.
- Environmental and Aesthetic Concerns: HVDC lines and converter stations can have a significant visual impact on the landscape. The large and bulky converter stations, in particular, may face opposition from local communities due to their appearance and footprint.
- DC Circuit Fault Management: Handling and isolating faults in HVDC systems can be more challenging compared to AC systems. HVDC systems require sophisticated protection schemes to detect and clear faults quickly, which adds to the complexity and cost of the system.

11.6 Types of the HVDC Links:

11.6.1 DC Circuit: The DC circuit is a fundamental element in HVDC systems, playing a crucial role in power transmission and control. The simplified equivalent circuit depicted in Fig. 11.5 represents the basic structure of a DC circuit in an HVDC pole. It shows the voltage sources V_1 and V_2, which typically correspond to the DC voltages at the sending and receiving ends, respectively. The overall DC voltage is controlled by the voltage difference between V_1 and V_2, determining the magnitude of the current flowing through the circuit.

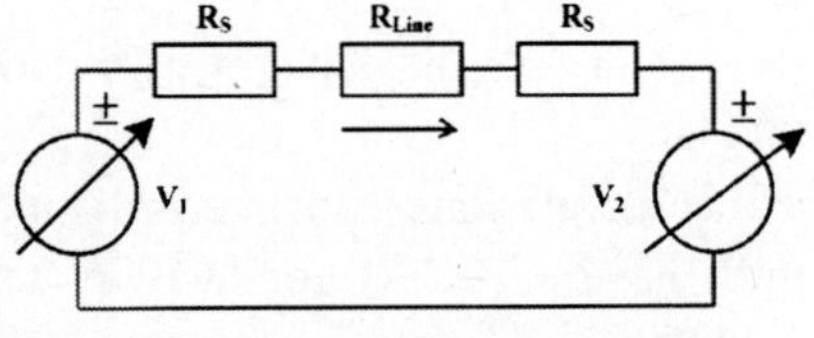

Fig. 11.5 Equivalent DC circuit for HVDC link.

One of the key characteristics of the DC circuit in HVDC systems is the fixed direction of current flow. Unlike AC systems where the direction of current alternates periodically, in DC circuits, the current flows consistently in one direction. This unidirectional current flow simplifies the control of power flow, with the voltage polarity serving as the means to control the direction of power transfer.

11.6.2 Back-to-Back Converters:

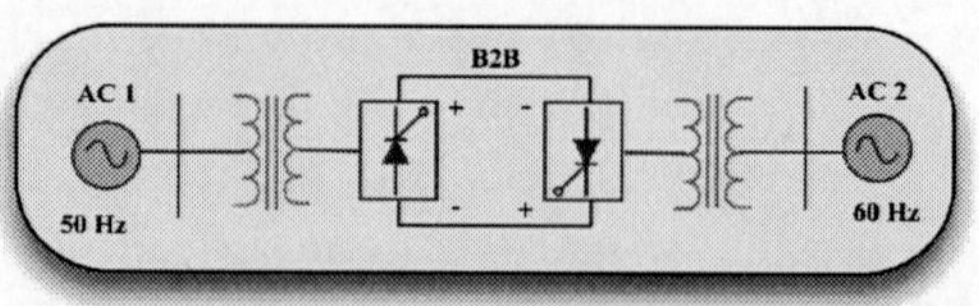

Fig. 11.6 Block diagram of back-to-back HVDC link.

These converters serve the critical function of linking two AC systems that may have differing parameters, such as frequency, voltage levels, or phase angles, which would otherwise prevent direct connection. The setup typically involves two converter stations, each equipped with AC/DC and DC/AC converters, semiconductor devices like thyristors or IGBTs for efficient power conversion, transformers, filters, and sophisticated control and protection systems. Fig. 11.6 shows the equivalent DC circuit for the back-to-back HVDC link.

The operation of back-to-back converters follows a systematic process: the first station converts incoming AC power to DC, which is then transmitted through a DC link to the second station. Here, the DC power is converted back to AC with the required parameters for integration into the receiving AC system. This process allows for seamless power exchange between the interconnected systems while maintaining stability and quality. Applications of back-to-back converters are diverse, ranging from grid interconnection and renewable energy integration to power quality improvement and grid stability enhancement.

11.6.3 Monopolar:

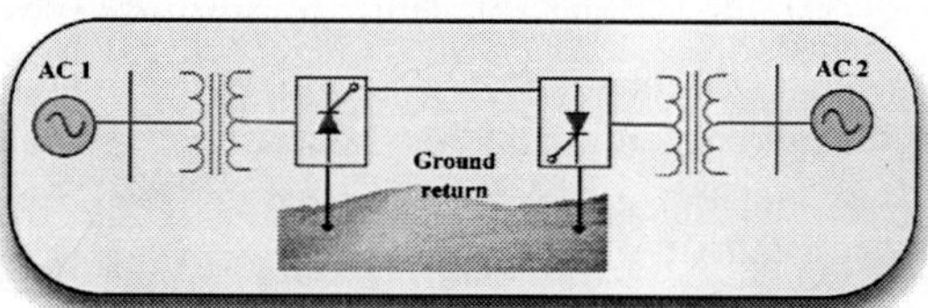

Fig. 11.7 Block diagram of monopolar HVDC link.

It refers to a configuration in HVDC power transmission where only one conductor is utilized for the transmission of electrical power. In such a system, the return path for the current is provided through the ground or a conducting medium such as seawater in the case of submarine cables. Monopolar systems are often used for long-distance transmission, especially in scenarios where a dedicated return conductor is not feasible or economical. Fig. 11.7 shows the block diagram of the monopolar HVDC link.

One of the key advantages of monopolar HVDC transmission is its cost-effectiveness compared to bipolar systems, as it requires fewer conductors and infrastructure. However, there are also limitations and considerations associated with monopolar transmission. For instance, the ground or sea return path must have sufficient conductivity to handle the current without excessive losses or voltage drops. Additionally, issues such as grounding and electromagnetic interference need to be carefully managed in monopolar systems.

11.6.4 Bipolar: Bipolar HVDC link is a type of high-voltage direct current transmission system consisting of two conductors: one with positive polarity and the other with negative polarity. This configuration is in contrast to monopolar HVDC systems, where one conductor carries the current and the return path is through the ground or

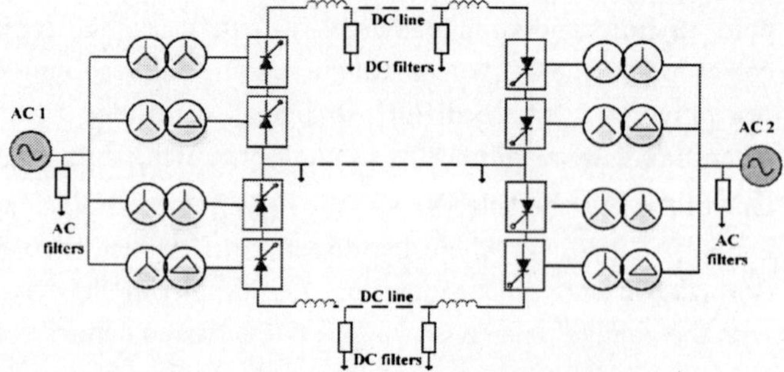

Fig. 11.8 Block diagram of bipolar HVDC link.

another medium. In a bipolar HVDC link, each conductor carries half of the total power and the current flows in opposite directions in the two conductors. This allows for the transmission of higher power levels compared to monopolar systems, as the total current is divided between the positive and negative conductors. Fig. 11.8 shows the block diagram of the bipolar HVDC link.

11.6.5 Homopolar: A homopolar HVDC link uses multiple conductors, all carrying a current of the same polarity (either all positive or all negative) to the ground. The return path for the current is provided through the ground or sea, similar to monopolar systems. Fig. 11.9 shows the block diagram of the homopolar HVDC link.

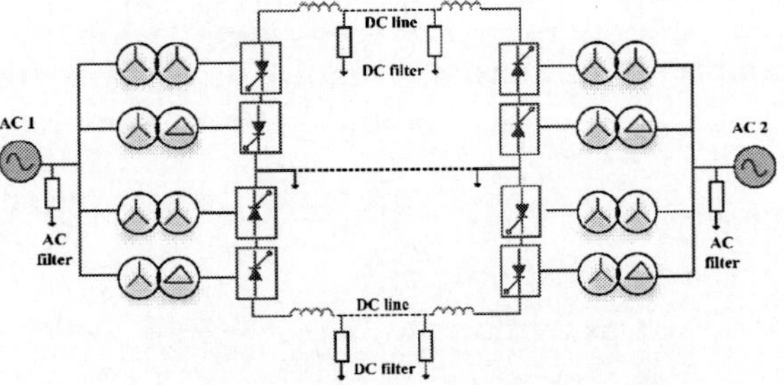

Fig. 11.9 Block diagram of homopolar HVDC link.

11.6.6 Multi-terminal:

The exploration of multi-terminal HVDC systems dates back to the first HVDC installation in the 1950s, although their practical implementation has been limited. Nonetheless, there are compelling reasons to consider adding a third terminal to an existing HVDC setup, especially when dealing with long DC lines or proximity to major load/generation centres as shown in Fig.11.10. POWERGRID is installing the first multi-terminal project, +/-800 kV, 6000 MW HVDC multi-terminal system of approx. length of 1728 km from the North Eastern Region to Agra. One Rectifier station in Biswanath Chariali (North Eastern Region), a second one in Alipurduar (Eastern Region), and an Inverter station at Agra (Northern Region). The converter stations at Biswanath Chariali and Alipurduar each handle a power of 3000 MW and the converter station at Agra handles 6000 MW of power. In a parallel multi-terminal HVDC system, several key properties are notable: All terminals maintain the same DC voltage level, with tight control assumed over this voltage. DC voltage polarity remains constant, except in certain small systems where coordinated polarity changes across terminals are possible. One terminal takes on the role of DC voltage control, while others regulate local DC. Terminals controlling current should also incorporate DC voltage support via local droop feedback. As all terminals must handle the full DC voltage, terminals with lower power capacity can be more expensive.

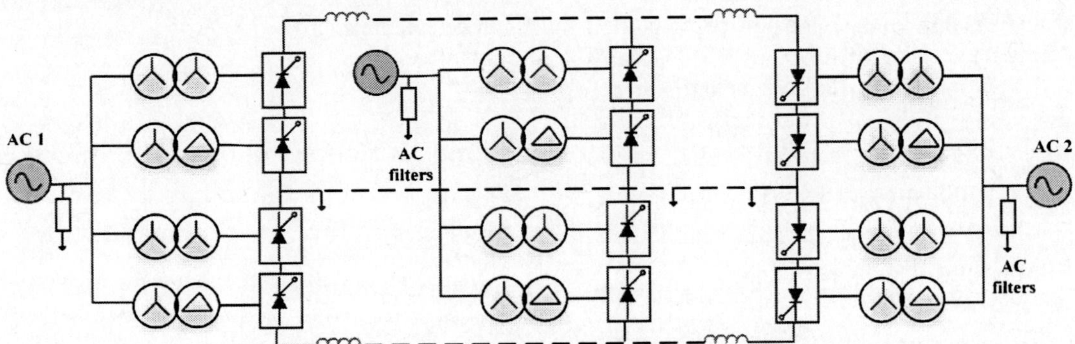

Fig. 11.10 Block diagram of multi-terminal HVDC link.

A fault in DC voltage or converters results in power interruption across all terminals. Commutation failure at any terminal leads to DC voltage collapse and subsequent power loss across all terminals. Instantaneous power reversal is not feasible at any terminal due to the need for voltage polarity reversal. Slow offline power reversal can be achieved using special mechanical switches, or bidirectional thyristor valves can be employed at higher costs for faster current polarity reversal.

Example 11.1. In a monopolar HVDC link energized by a 3-phase, 50 Hz, 400 kV source, the DC current is 1.5 kA, and the rectifier (six-pulse bridge converter) end DC voltage is 500 kV. For a delay angle of 10°,

a) Find the commutation resistance.

b) Find the commutation angle μ.

c) If AC voltage is reduced to 220 kV, find the commutation angle μ. Assume the DC current is constant.

Ans. a) DC voltage output of a six-pulse rectifier bridge is: $V_o = \frac{3\sqrt{2}}{\pi} V_{ml} \cos\alpha - I_{dc} R_c$

$$500 = \frac{3\sqrt{2}}{\pi} 400 \times \cos 10 - 1.5R_c$$

$$= 531.98 - 1.5R_c$$

$$531.98 - 500 = 1.5R_c$$

$$31.98 = 1.5R_c \Rightarrow R_c = 21.32\ \Omega$$

b) The output voltage with the source inductance $V_o = \frac{3\sqrt{2}}{2\pi} V_{ml}[\cos\alpha + \cos(\alpha + \mu)]$

$$500 = \frac{3\sqrt{2}}{2\pi} \times 400[\cos 10 + \cos(10 + \mu)]$$

$$= 270.09[0.984 + \cos(10 + \mu)]$$

$$0.867 = \cos(10 + \mu)$$

$$10 + \mu = \cos^{-1} 0.867 = 29.88$$

$$\mu = 19.88°$$

c) If the AC voltage is reduced then the DC link voltage will also change

$$V_o = \frac{3\sqrt{2}}{\pi} V_{ml}[\cos\alpha] = \frac{3\sqrt{2}}{\pi} \times 220 \times \cos 10 = 292.59 \text{ kV}$$

For this DC link voltage, the new commutation angle can be calculated as

$$V_o = \frac{3\sqrt{2}}{2\pi} V_{ml}[\cos\alpha + \cos(\alpha + \mu')]$$

$$292.59 = \frac{3\sqrt{2}}{2\pi} \times 220[\cos 10 + \cos(10 + \mu')] = 148.55[0.984 + \cos(10 + \mu)]$$

$$0.985 = \cos(10 + \mu)$$

$$10 + \mu' = \cos^{-1} 0.985 = 9.93 \text{ then } \mu' = -0.063°$$

Example 11.2. The output voltage of single-phase full-bridge VSI is controlled using PWM with one pulse per half cycle. Determine the required approximate pulse width so that fundamental RMS component is 80% of the DC input voltage.

Ans. The output voltage $V_{01} = \frac{4V_{dc}}{\pi} \sin(\frac{\delta}{2}) \sin\omega t$

Where δ is the pulse width,

$$V_{o1peak} = \frac{4V_{dc}}{\pi}\sin(\frac{\delta}{2})$$

$$V_{o1rms} = \frac{V_{o1peak}}{\sqrt{2}} = \frac{4V_{dc}}{\sqrt{2}\,\pi}\sin(\frac{\delta}{2})$$

The output fundamental voltage $V_{o1rms} = 0.80\times V_{dc}$

$$\frac{4V_{dc}}{\sqrt{2}\,\pi}\sin(\frac{\delta}{2}) = 0.80\times V_{dc}$$

$$\sin(\frac{\delta}{2}) = \frac{0.80\times\sqrt{2}\,\pi}{4} = 0.888$$

$$\frac{\delta}{2} = \sin^{-1}(0.888) = 62.96° \text{ then } \delta = 125.38°$$

Example 11.3. A monopolar HVDC system transmits 2000 MW at ±800 kV over a 1000 km transmission line with a line resistance of 0.05 Ω/km. Calculate: (a) The DC current in the system. (b) The power loss in transmission. (c) The efficiency of power transmission.

Ans : Total Power Transmitted: $P_{output} = 2000\ MW$, Voltage = ±800 kV and Distance = 1000 km

Line Resistance per km: R=0.05 Ω/km

Total Resistance = R * Distance = 0.05×1000 = 50 Ω

$$\text{Current} = I = \frac{P}{V} = \frac{2000*10^6}{800*10^3} = 2500\ A$$

$$\text{Power Loss in Transmission } P_{loss} = I^2 * R = 2500^2 * 50 = 312.5\ MW$$

$$\text{Efficiency of power transmission} = \frac{P_{output}}{P_{input}} * 100 = \frac{P_{output}}{P_{output} + P_{loss}} * 100$$

$$= \frac{2000 * 10^6}{2000 * 10^6 + 312.5 * 10^6} * 100 = 86.5\ \%$$

11.7 Applications of HVDC:

HVDC technology finds numerous applications across different sectors due to its unique advantages over conventional AC transmission. Here are some key applications of HVDC:

Long-Distance Power Transmission: One of the primary applications of HVDC is long-distance bulk power transmission. HVDC is mainly advantageous for transmitting electricity over vast distances, such as across continents or undersea cables. It incurs lower transmission losses compared to AC transmission over long distances. Fig. 11.11 shows the block diagram of the HVDC link for long-distance power transmission. It contains the two grids AC 1 and AC 2 which are interconnected by an HVDC transmission line.

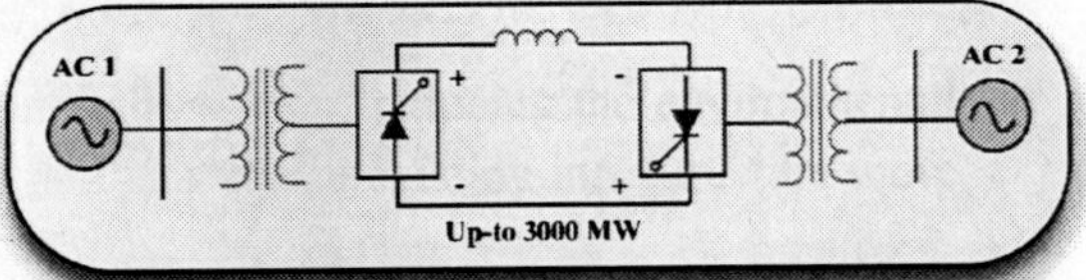

Fig. 11.11 HVDC link for long-distance power transmission.

Submarine and Underground Cables: HVDC is well-suited for submarine and underground cable applications. Submarine cables, especially for interconnecting islands or crossing large bodies of water, benefit from HVDC due to reduced losses, lower voltage drop, and enhanced control over power flow. The block diagram of the HVDC link with a submarine cable to carry the DC power is shown in Fig. 11.12. The HVDC link with an underground cable is shown in Fig. 11.13.

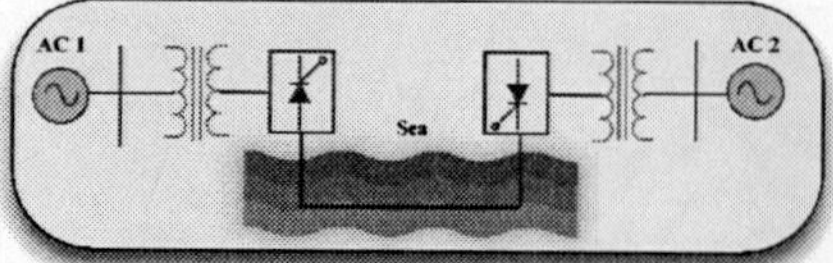

Fig. 11.12 Block diagram of HVDC link with submarine cable.

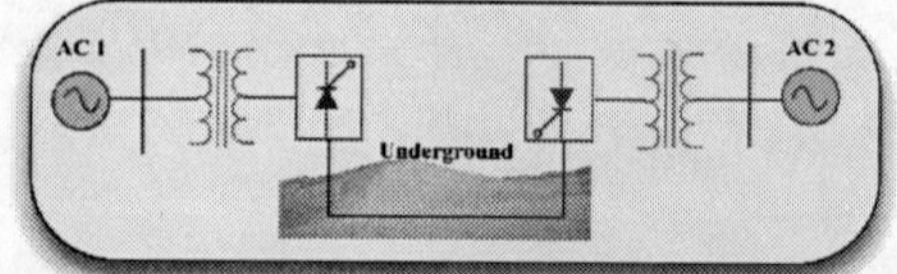

Fig. 11.13 Block diagram of HVDC link using underground cable.

Interconnecting Asynchronous Grids (Back-to-back): HVDC facilitates the interconnection of asynchronous AC grids operating at different frequencies. This is essential for international grid interconnections where neighbouring countries may have distinct AC systems. HVDC links enable power exchange between these grids without synchronization issues. Fig. 11.14 shows the block diagram of the back-to-back HVDC link. It connects the two grids operating at different frequencies to transfer the power from one grid to another grid to enhance the stability.

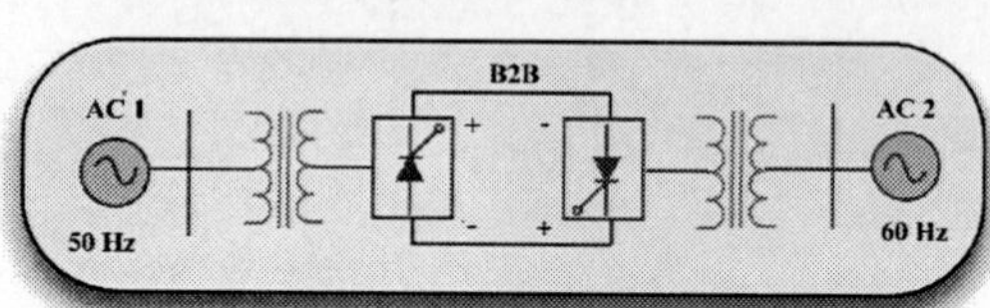

Fig. 11.14 Block diagram of back-to-back HVDC link.

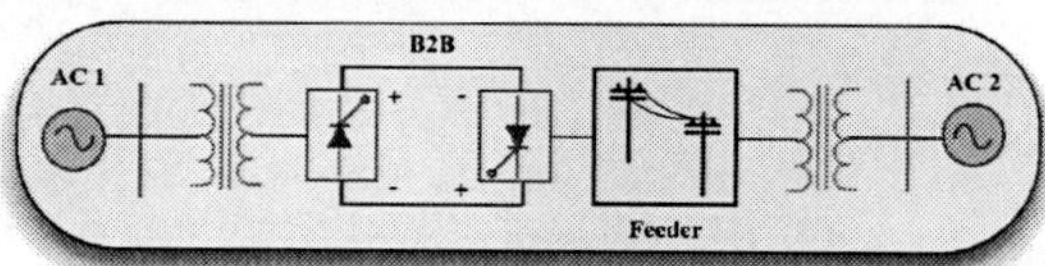

Fig. 11.15 Block diagram of back-to-back HVDC with feeder.

Stabilizing AC Grids: HVDC systems can contribute to grid stability by providing dynamic power control and improving system response during contingencies. They help in stabilizing AC grids by managing power flow, enhancing voltage stability, and mitigating grid disturbances. Fig. 11.15 shows the block diagram of back-to-back HVDC with a feeder.

Renewable Energy Integration: HVDC plays a crucial role in integrating renewable energy sources like wind farms and solar power plants into the grid. HVDC lines can connect remote renewable energy generation sites to population centres efficiently, overcoming grid constraints and minimizing power losses. Fig. 11.16 shows the VSCs are connected to the wind turbines to feed power into the utility grid. VSCs are connected in parallel using the DC bus.

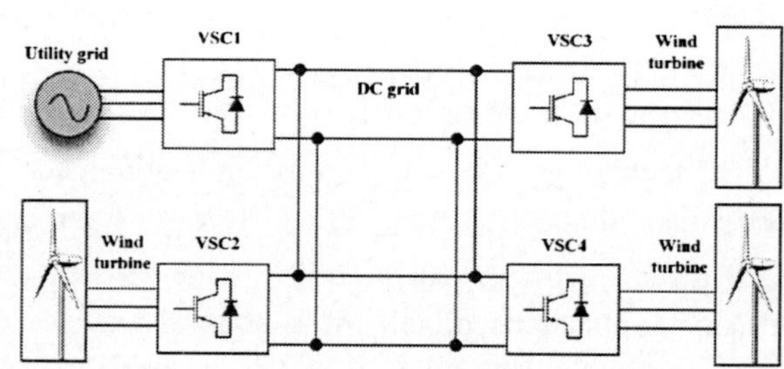

Fig. 11.16 Block diagram of VSC-based HVDC system.

Island Grids and Remote Areas: HVDC technology is suitable for powering island grids and remote areas with limited access to conventional power sources. It allows for efficient transmission of electricity from mainland grids or large power plants to isolated regions, improving energy accessibility and reliability.

Industrial Applications: HVDC systems are utilized in various industrial applications, including aluminium smelting, steel production, and chemical processing.

Table 11.1 Worldwide multi-terminal HVDC links.

Project Name	Location	Capacity (MW)	Voltage (kV)	Year	Description
North Sea Wind Power Hub	North Sea	12,000	±525	Planned for 2030s	A proposed project aims to create an offshore energy hub that will connect multiple offshore wind farms and distribute electricity to several European countries.
Tres Amigas SuperStation	USA	5,000	±345	Planned	A proposed project aiming to connect the Eastern, Western, and Texas grids of the United States, facilitating renewable energy distribution and grid stability.
Pacific DC Intertie Expansion	USA	4,800	±500	Initial: 1970, Expansion : Ongoing	Originally a point-to-point HVDC system, it is being expanded to accommodate more renewable energy sources and additional terminals.
Zhangbei HVDC Grid	China	9,000	±500	2021	One of the world's largest multi-terminal HVDC projects, integrating renewable energy sources from Zhangbei, Kangbao, and other areas into the Beijing-Tianjin-Hebei grid.
NEMO Link	UK/ Belgium	1,000	±400	2019	Connects the UK and Belgium, and is the first HVDC link between the two countries. It's a point-to-point link with the potential for multi-terminal expansion.
Bipole III	Canada	2,000	±500	2018	Enhances the reliability and capacity of the Manitoba Hydro network, and although currently point-to-point, it is designed with the potential for multi-terminal upgrades.
INELFE (France-Spain Interconnector)	France/ Spain	2,000	±320	2015	An underground HVDC link to enhance electricity exchange between France and Spain, with potential future expansion to a multi-terminal configuration.
Quebec - New England Phase II	Canada/ USA	2,000	±450	1990	A multi-terminal HVDC system connecting hydroelectric power from Quebec to the New England grid in the USA.
Nelson River Bipole I and II	Canada	3,200	±450	1972, 1985	One of the earliest multi-terminal HVDC systems, transmitting hydroelectric power from Northern Manitoba to Southern Manitoba.

Table 11.2 Details of major worldwide HVDC links.

Project Name	Country	Length (km)	Capacity (MW)	Commissioned	Description
North Sea Link	UK-Norway	720 (submarine)	1,400	2021	The world's longest subsea interconnector, enabling power exchange between the UK and Norway.
NordLink	Germany-Norway	623 (submarine and underground)	1,400	2020	Connects the German and Norwegian grids, facilitating renewable energy integration and enhancing grid stability.
Changji-Guquan	China	3293	12000	2019	Transmits power from Zhongwei to Guiyang, showcasing China's advanced HVDC infrastructure.
Bipole III	Canada	1,388	2,000	2018	Runs from northern Manitoba to southern Manitoba, providing redundancy and reliability.
North East Agra	India	1,728	6,000	2017	Enhances the power supply to the northern region of India, reducing transmission losses.
Inga-Kolwezi	Democratic Republic of Congo	1,700	1000	1982 (upgraded in 2017)	Transmits power from the Inga Dam to the Katanga region, supporting the mining industry.
Pacific DC Intertie	USA	965	3800	1970 (upgraded in 2016)	Connects the Pacific Northwest to Los Angeles, using overhead lines.
Rio Madeira	Brazil	2,375	7,100	2013	Connects hydropower plants on the Madeira River to the southeastern region of Brazil, integrating renewable energy.
UHVDC Jinping-Sunan	China	2,090	7,200	2012	Transmits power from Jinping to Sunan, highlighting China's leadership in UHVDC technology.
Xiangjiaba-Shanghai	China	2,071	6,400	2010	Transmits hydroelectric power from Xiangjiaba Dam to Shanghai.
Chandrapur-Padghe	India	752	1,500	1999	Connects Chandrapur in Maharashtra to Padghe near Mumbai, enhancing grid stability and capacity.
Inga-Shaba	Democratic Republic of Congo	1,700	560	1982	Transmits power from the Inga Dam to the Shaba mining region.

Table 11.1 lists the worldwide multi-terminal HVDC links. Table 11.2 contains the details of major worldwide HVDC links. It shows the project name, country, length of the transmission line, commissioning year, and the description of the project. Fig. 11.17 shows the maps of the HVDC links for long-distance transmission and back-to-back converter stations in India. Table 11.3 lists the worldwide back-to-back available HVDC links. The zhongwei-Guiyang project in China is transmitting the highest 12000 MW of power with a distance of 3293 km. The world's first multi-terminal UHVDC transmission link is developed by Hitachi Energy chosen by Power Grid Corporation of India Limited from Agra to the northeast (Biswanath Charali) in India. The ±800 kV North-East Agra UHVDC link, with a converter capacity of 6,000 MW (including a 2,000 MW redundancy), transmits clean hydroelectric power from India's northeast region to the city of Agra over a distance of 1728 km.

Table 11.3 Worldwide back-to-back HVDC links.

Project Name	Location	Capacity (MW)	Voltage (kV)	Year of Completion	Description
Backbone Power Project	Brazil	2,250	±400	2018	Strengthens the Brazilian grid by linking different regional grids, enhancing reliability and capacity.
Murraylink	Australia	220	±150	2002	Connects the electricity grids of South Australia and Victoria, enabling power transfer and grid support.
Eagle Pass	USA	36	±15	2000	A smaller-scale project connecting the power grids of Texas and Mexico, enhancing cross-border electricity exchange.
Chandrapur Back-to-Back	India	1,000	±205	1999	Connects the western and southern regional grids of India, improving grid stability and reliability.
Vyborg Converter Station	Russia	1,420	±85	1981,1994	Facilitates power exchange between the Russian and Finnish power grids, enhancing energy security for both countries.
New England - New York Interconnection	USA	500	±150	1988	Improves power transfer capability between the New England and New York power grids, supporting grid stability.
Sakuma Frequency Converter	Japan	300		1965	One of the first back-to-back HVDC projects, connecting the 60 Hz grid of western Japan to the 50 Hz grid of eastern Japan.

11.8 Short Circuit Ratio (SCR):

The term "SCR" in the context of HVDC systems typically refers to Short-Circuit Ratio. The Short-Circuit Ratio is a critical parameter that determines the stability and performance of HVDC systems, particularly when integrating them with AC networks.

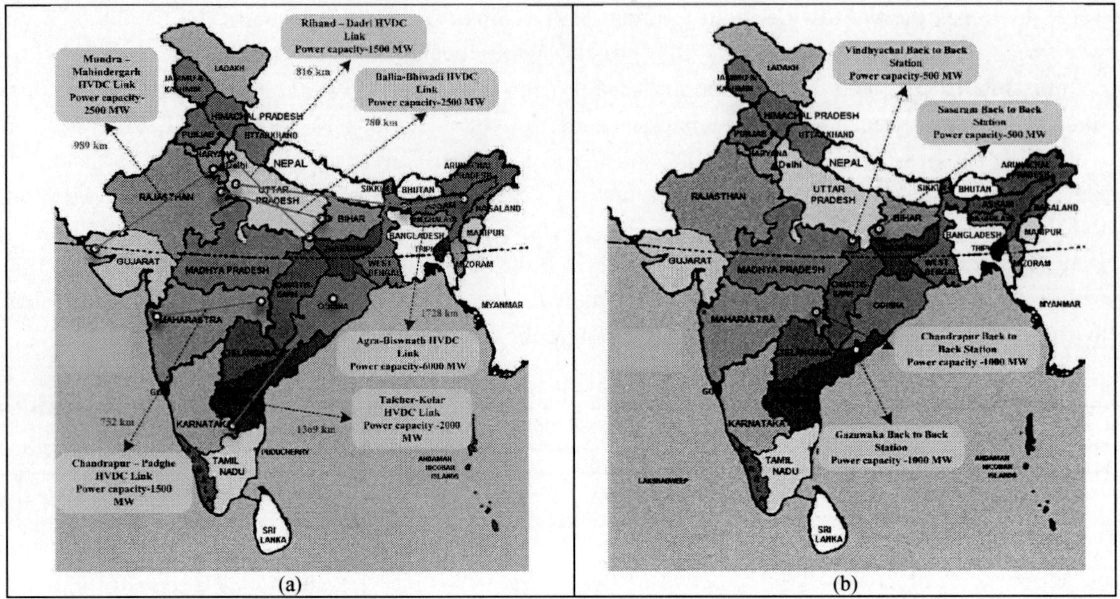

Fig. 11.17 (a) HVDC links available in India (b) Back-to-back HVDC stations in India.

Definition and Importance of SCR : SCR is defined as the ratio of the short-circuit capacity of the AC system at the point of common coupling (PCC) to the rated power of the HVDC link. Short Circuit Power (SCP) or Short Circuit Capacity (SCC) is used in this case to represent the strength of the AC system connected to an HVDC transmission line. Mathematically, it is expressed as:

$$SCR_{ac} = \frac{V_{pcc}^2}{Z_{ac}} \qquad \text{.........(11.1)}$$

Where V_{pcc} is the AC voltage at the PCC, Z_{ac} is the AC system impedance.

For the AC system connected to the HVDC transmission with rated DC power of P_{dc} in megawatts (MW), the SCC definition will be

$$SCC = \frac{SCR_{ac}}{P_{dc}} = \frac{\text{Short Circuit Ratio on Converter Side}}{\text{DC Power Rating}} \qquad \text{.........(11.2)}$$

$$SCC = \frac{V_{pcc}^2}{Z_{ac}P_{dc}} = \frac{1}{Z_{pu}} \qquad \text{.........(11.3)}$$

As a result, the SCC is inversely proportional to the per unit impedance. The lower the SCC value, the weaker the system, and the greater the interactions between DC and AC systems. Very weak system: The AC/DC system with SCC values less than 2.

Role of SCR in HVDC Systems

a. System Stability:

- **High SCR (≥ 3)**: When the SCR is high, the AC system is considered strong. This implies that the AC network can support the HVDC link without significant voltage fluctuations or stability issues. A high SCR indicates that the AC system has sufficient capacity to absorb the reactive power and control disturbances caused by the HVDC link.

- **Low SCR (< 3)**: A low SCR indicates a weak AC system. In such cases, the HVDC link can cause voltage instability, harmonics, and control challenges. Special control strategies and reactive power compensation are often required to maintain system stability.

b. Reactive Power Management:

- HVDC converters, especially LCC, consume reactive power. The ability of the AC system to supply this reactive power without significant voltage drops is crucial for stable operation.
- In weak systems (low SCR), additional reactive power compensation devices, such as Static Var Compensators (SVCs) or STATCOMs, may be necessary to support the reactive power requirements of the HVDC converters.

c. Control Strategies:

- In systems with high SCR, standard control methods for HVDC converters are usually sufficient to maintain stable operation.
- For systems with low SCR, advanced control techniques, such as supplementary damping controllers or adaptive control strategies, are employed to ensure stable operation. These methods help in mitigating the adverse effects of a weak AC network.

d. Converter Design:

- The design of HVDC converters and their control systems takes into account the SCR of the interconnected AC system. For weak systems, VSC are often preferred over LCCs due to their better performance in low SCR conditions. VSCs can provide independent control of active and reactive power, which enhances system stability.

e. System Planning and Expansion:

- During the planning phase of HVDC projects, the SCR of the target AC system is a critical factor. Ensuring a sufficient SCR is crucial for the successful integration and operation of HVDC links.
- When expanding existing HVDC links or adding new ones, the impact on the SCR must be evaluated to ensure that the AC system can support the additional power transfer without compromising stability.

11.9 Line Commutating Converter (LCC) based HVDC:

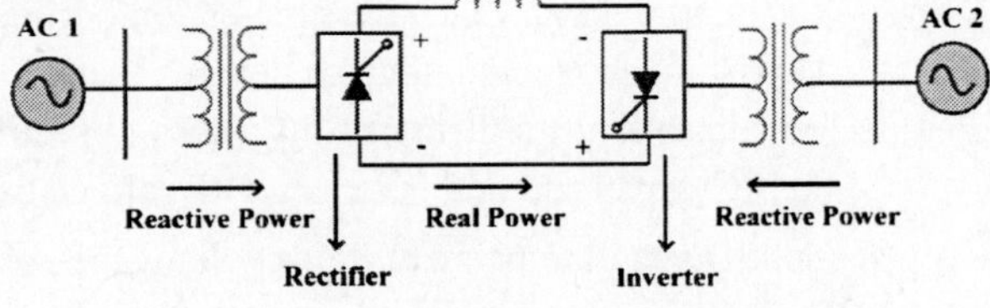

Fig. 11.18 Block diagram of line commutated converter.

LCC, also known as a current source converter (CSC), utilizes thyristor-based technology for its conversion process. Thyristors are silicon semiconductor devices with four layers of N and P-type material acting as bi-stable switches. They are triggered with a gate pulse and remain in the on condition until the next current zero crossing. However, LCC requires a very high synchronous voltage source for commutation, which limits its use for black start operations. With LCC's current rating reaching up to 6.250kA and a blocking voltage of 10kV, it boasts the highest voltage and power rating among all HVDC converter technologies.

Fig. 11.18 shows the LCC-based HVDC system. It contains the thyristors for the rectifier and inverter. Table 11.4 shows the major LCC projects worldwide. LCC controls its operation by regulating the firing angle $\bar{\alpha}$ on both the rectifier and inverter sides. It operates with a unidirectional line-commutated flow of DC current injected into a receiving AC network, hence the term CSC because the output current is maintained at a constant level. Power reversal from one station to another is achieved by inverting the DC voltage polarity in both stations while keeping the current direction constant. The long-distance and back-to-back HVDC projects are discussed in Unit 1.

Table 11.4 Details of major worldwide LCC-based projects.

Project Name	**Location**	**Characteristic (MW)**	**Voltage (kV)**	**Year**	**Distance (km)**
Zhundong – Sichuan	China	10000	±1100	2015	2600
Xiluodu – Zhejiang	China	8000	±800	2014	1688
Southern Hami – Zhengzhou	China	8000	±800	2014	2200
Biswanath – Agra	India	6000	±800	2014	1728
Rio – Madeira	Brazil	2x3150	±600	2013	2375
Xiluodu – Guangdong	China	6400	±500	2013	1251
Nuozhadu – Guangdong	China	5000	±800	2013	1451
Jinpin – Sunan	China	7200	±800	2012	2093
Mundra – Haryana	India	2500	±500	2012	960
Rio – Madeira	Brazil	800	100	2012	Back-to-Back
UK - Netherlands	UK, Netherlands	1000	±400	2011	260

This technology is known for its reliability, minimal maintenance requirements, and suitability for transmitting bulk power using high-voltage transmission lines. Among them, the Zhundong-Sichuan scheme stands out with the highest voltage, power, and longest distance project in China. Overall, LCC technology remains popular among HVDC schemes due to its unique features and proven performance in power transmission. Here's an overview of a Line Commutated Converter:

Operation Principle: In an LCC, the conversion between AC and DC is achieved using thyristor-based circuits. Thyristors are semiconductor devices that can control the flow of electrical current. The LCC operates by controlling the firing of thyristors to rectify AC voltage into DC voltage during one half-cycle of the AC waveform and inverting DC voltage back to AC during the other half-cycle.

Rectification: During the positive half-cycle of the AC waveform, the thyristors in the rectifier bridge are triggered to conduct, allowing current to flow from the AC system to the DC system. This rectified DC voltage is then smoothed using filters to reduce harmonics and ripple.

Inversion: During the negative half-cycle of the AC waveform, the thyristors in the inverter bridge are triggered to conduct, which reverses the polarity of the voltage and allows current to flow from the DC system to the AC system. This process enables bidirectional power flow in the HVDC system.

Commutation: The term "line-commutated" refers to the method of commutation used in LCCs. Commutation is the process of switching from one thyristor to another to maintain the flow of current. In an LCC, commutation is achieved naturally when the AC voltage reverses polarity during each half-cycle of the waveform. This natural commutation is known as line-commutation.

Characteristics: LCC-based HVDC systems have certain characteristics, such as being suitable for long-distance transmission, high power transmission capacity, and robust performance under steady-state conditions. However, they have limited controllability compared to newer technologies like VSCs.

11.10 Voltage Source Converter (VSC) based HVDC:

VSC utilizes insulated gate bipolar transistor technology. This technology allows the current to be switched on and off independently of the AC voltage, enabling it to generate its AC voltages during black-start scenarios. VSC converters operate at high frequencies using pulse width modulation, which enables simultaneous adjustment of the amplitude and phase angle of the converter while maintaining voltage constancy. This technology offers a high degree of flexibility with inherent capabilities to control both active and reactive power, particularly advantageous in urban power network areas. A basic VSC-HVDC system consists of two converter stations utilizing VSC topologies. The simplest VSC topology is the conventional two-level three-phase bridge.

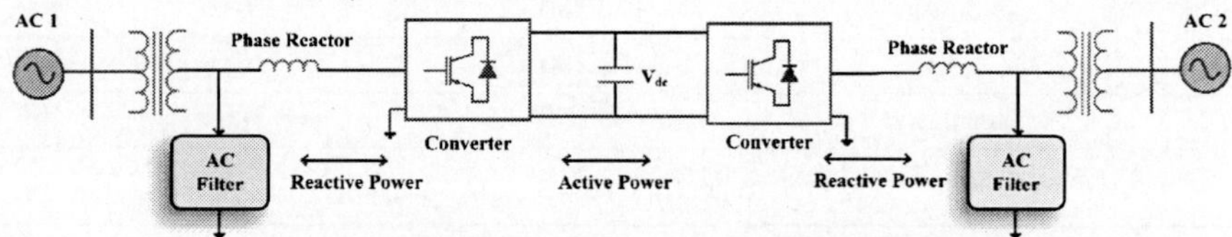

Fig. 11.19 Block diagram of voltage source converter-based HVDC link.

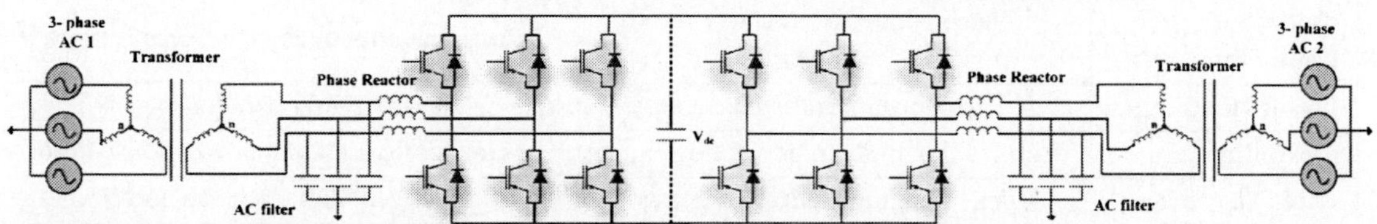

Fig. 11.20 Circuit diagram of voltage source converter-based HVDC system.

In practical applications, each semiconductor switch in the VSC Bridge is typically composed of multiple series-connected IGBTs to achieve a higher blocking voltage capability, thus increasing the DC bus voltage level of the HVDC system. Each IGBT is paired with an antiparallel diode to ensure the four-quadrant operation of the converter. Additionally, the DC bus capacitor in the system provides the necessary energy storage to control the power flow and also helps in filtering the DC harmonics. Other VSC topologies can also be employed to build a VSC based HVDC system, depending on the specific requirements of the application. Fig. 11.19 shows the VSC-based HVDC system-based HVDC link. It consists of a rectifier and inverter system with filters. The rectifier is a six-pulse converter. Its output frequency is six times of supply frequency. When the firing angle is less than or equal to 60° then its output voltage is $\frac{3V_{mL}}{\pi}\cos\alpha$. Fig. 11.20 shows the circuit diagram of a voltage source converter-based HVDC system. The output voltage of the six-pulse converter is $V_o = \frac{3V_{mL}}{\pi}\cos\alpha$.

VSC-HVDC was developed in the 1990s, and ABB commissioned the first project in 1997. However, its adoption has been limited due to capacity constraints, including low device ratings, increased power losses, and heightened dielectric stress on equipment insulation, compared to its counterpart LCC schemes. Current applications of VSC-HVDC are typically around 1800MW at 500kV. An illustration is the Nordlink project, a 1400MW, ±525kV HVDC link connecting the Statnett grid in Norway with TenneT in Germany over a distance of 623km. Ongoing research aims to address these limitations, particularly focusing on fault ride-through capability. Table 11.5 shows the details of VSC-based projects worldwide. Table 11.6 shows the comparison between LCC and VSC.

Table 11.5 Detail of VSC-based projects worldwide.

Project Name	Location	Voltage (kV)	Year	MW	km
Skagerrak 4	Norway	±500	2014	700	244
Inelfe	France	±320	2013	1000	65
EWIC	UK	±200	2012	500	261
Caprivi link	Namibia	±350	2010	300	951
Transbay	USA	±200	2010	400	85
Borwin 1	Germany	±150	2009	400	200

Table 11.6 Comparison between LCC and VSC.

Aspect	LCC	VSC
Technology	Thyristor-based	Insulated Gate Bipolar Transistor (IGBT)
Control Capability	Limited control over, active/reactive power	Precise control over, active/reactive power
Power Handling	High power handling capability	Moderate power handling capability
Flexibility	Limited flexibility in power flow control	Greater flexibility in power flow control
Fault Tolerance	Limited fault tolerance, system-wide	Improved fault tolerance, localized
Complexity	More complex control and operation	Generally simpler control and operation
Cost	Lower initial cost but higher operating	Higher initial cost but lower operating and maintenance costs
Application	Bulk power transmission over long distances	Frequencies regulation, renewable integration

11.10.1 12-Pulse VSC Converter:

A 12-pulse rectifier is an advanced type of rectifier used to convert AC to DC with the primary advantage of significantly reducing harmonic distortion. This rectifier configuration employs two sets of six-pulse rectifiers, each connected to secondary windings of a transformer that are phase-shifted relative to each other. Specifically, one set of windings produces a phase shift of 0 degrees, while the other set produces a phase shift of 30 degrees. This phase shift effectively cancels out many of harmonics that would otherwise be present in a simple rectifier setup shown in Fig. 11.21.

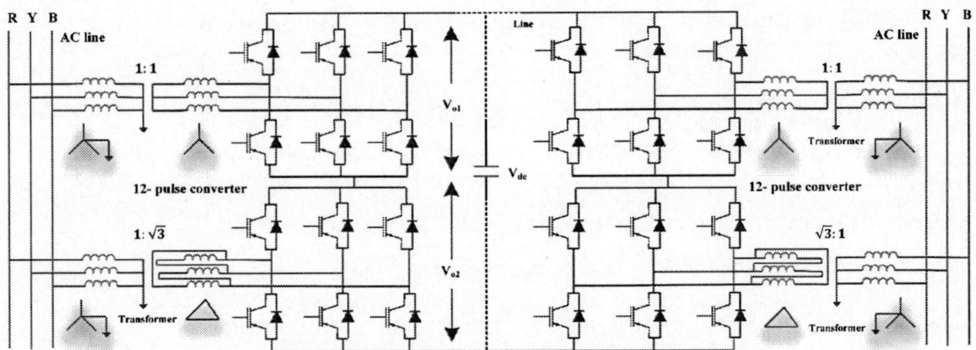

Fig. 11.21 Circuit diagram of 12-pulse converter-based HVDC system.

The 12-pulse rectifier operates by combining the outputs of these two six-pulse rectifiers. Each rectifier converts its AC input into a DC output, and then these outputs are merged to form a single, more stable DC output. This combined output not only enhances the quality of the DC power by reducing ripple and harmonics but also increases the overall DC voltage.

The output voltage of upper six-pulse converter $V_{o1} = \frac{3V_{mL}}{\pi}\cos\alpha_1$ ……… (11.4)

The output voltage of lower six-pulse converter $V_{o2} = \frac{3V_{mL}}{\pi}\cos\alpha_2$ ……… (11.5)

The output voltage of 12-pulse converter $V_{dc} = V_{o1} + V_{o2} = \frac{3V_{mL}}{\pi}(\cos\alpha_1 + \cos\alpha_2)$ ……… (11.6)

11.11 Real power flow control in a DC Link:

In the HVDC system power flows from the generating station to the load. The generating voltage is AC. This voltage is converted to DC using the rectifier and then transmitted through the transmission lines. Further, this DC voltage is converted to AC voltage as shown in Fig. 11.22. The equivalent circuit diagram of the HVDC system is shown in Fig. 11.23.

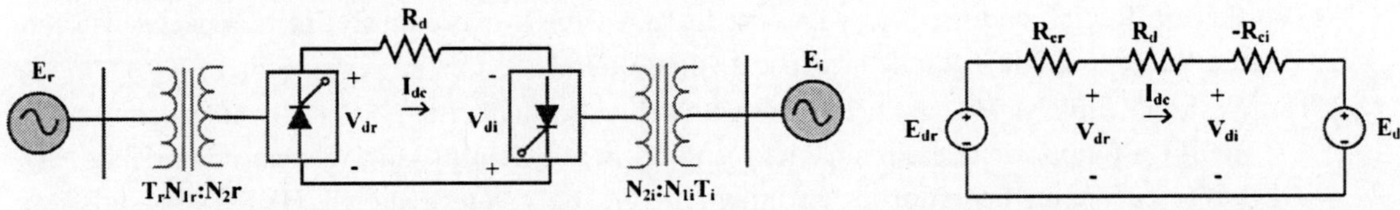

Fig. 11.22 Overview of HVDC system having rectifier, inverter, and transformers.

Fig. 11.23 Equivalent circuit diagram of HVDC power flow.

When the rectifier is turned ON, its resistance is R_{cr}. The resistance offered by the inverter is R_{ci}. R_d is the resistance of the transmission line. E_r is the R.M.S value of AC generation supply voltage at the rectifier side. E_i is the R.M.S value of the voltage at the inverter side appearing across AC load terminals. T_r and T_i are the tapping settings of transformers used at the rectifier and inverter side respectively. N_{1i}, N_{2i}, N_{1r}, and N_{2r} are the turns ratio of the transformers used at the inverter and rectifier sides. E_{dr} and E_{di} are voltages on DC side of rectifier and inverter, respectively, whereas, V_{dr} and V_{di} represent DC side voltage of rectifier and inverter, respectively, considering drop in converter resistance R_{cr} and R_{ci}, respectively.

The voltage at the output of the transformer at the rectifier side $E_v^{rec} = \frac{N_{2r}}{T_r N_{1r}} E_r$ ……… (11.7)

The voltage at the input of the transformer at the inverter side $E_v^{inv} = \frac{N_{2i}}{T_i N_{1i}} E_i$ ……… (11.8)

The voltage at the output of the rectifier after resistive drop $V_{dr} = E_{dr} - R_{cr} I_{dc}$ ……… (11.9)

The voltage at the input of the inverter $V_{di} = E_{di} - R_{ci} I_{dc}$ ……… (11.10)

The voltage at the DC side of rectifier $E_{dr} = \frac{3\sqrt{2}}{\pi} \frac{N_{2r} E_r}{T_r N_{1r}} \cos\alpha_r = A_r \frac{E_r}{T_r} \cos\alpha_r$ ……… (11.11)

The voltage at the DC side of inverter $E_{di} = \frac{3\sqrt{2}}{\pi} \frac{N_{2i} E_i}{T_i N_{1i}} \cos\alpha_i = A_i \frac{E_i}{T_i} \cos\alpha_i$ ……… (11.12)

Assume that $A_r = \frac{3\sqrt{2}}{\pi} \frac{N_{2r}}{N_{1r}}$; $A_i = \frac{3\sqrt{2}}{\pi} \frac{N_{2i}}{N_{1i}}$

Transmission line current $I_{dc} = \frac{E_{dr} - E_{di}}{R_{cr} + R_d - R_{ci}}$ ……… (11.13)

Putting the value of E_{dr} and E_{di} from equations (11.11) and (11.12), in equations (11.13), we get

$$I_{dc} = \frac{A_r \frac{E_r}{T_r} \cos\alpha_r - A_i \frac{E_i}{T_i} \cos\alpha_i}{R_{cr} + R_d - R_{ci}} \quad \text{……… (11.14)}$$

I_{dc} is controlled by the four variables T_r, T_i, α_i, and α_r.

11.12 Milestones of HVDC development in India:

1948: The Electricity (Supply) Act of 1948 provided for the establishment of the Central Electricity Authority (CEA) and the State Electricity Boards.

1989: Development of HVDC back-to-back System.

1990: Introduction of HVDC bipolar line.

1992: The Eastern Region and the North-Eastern Region were synchronously interconnected via the Birpara-Salakati 220 KV double-circuit transmission line.

2003: The Western Region was synchronously interconnected with the ER-NER system through the 400 kV Rourkela-Raipur double-circuit line, resulting in the Central India system comprising ER-NER-WR becoming operational. Additionally, the bulk inter-regional HVDC transmission system (Talcher-Kolar HVDC link, 2000 MW) was established.

2016: The North-Eastern Region (NER) was directly connected to the Northern Region (NR) via the longest 6000 MW HVDC line (±800 kV) running from Bishwanath Chariali in NER to Agra in NR, facilitating the dispersal of power from NER to NR and Western Region.

2017:

- +800 kV Champa-Kurukshetra HVDC Bipole-I, commissioned in September 2017
- +800 kV Alipurduar-Agra HVDC Bipole, commissioned in September 2017.

2020:

- Pole 4 of the Champa-Kurukshetra HVDC Station was commissioned in March 2020.

- The 800 kV HVDC Raigarh (HVDC Station) - Pugalur (HVDC Station) Bipole Link (3531 km) was commissioned in September 2020.
- The Raigarh-Pugalur Station HVDC Terminal (Pole-I) was charged in September 2020.

2021:

- Bipole-I (3000 MW) of the ±800 kV Raigarh-Pugalur HVDC transmission line was commissioned.
- The Raigarh-Pugalur Station HVDC Terminal Pole-II was commissioned in March 2021.
- For the first time in India, Voltage Source Converter (VSC) technology was implemented: Monopole 1 & 2 of ±320 kV VSC-based HVDC terminals and the associated ±320 kV HVDC Pugalur-North Thrissur transmission line were established.

11.13 Unit Summary:

- The HVDC transmission is economical when the power transmission distance is long (more than 500 km).
- The first commercial HVDC implementation was in 1954, connecting Gotland Island and the Swedish mainland.
- In India, a notable HVDC milestone was the commissioning of the Rihand-Dadri link in 1991, spanning 816 km and connecting the Rihand thermal power plant with Dadri.
- HVDC has several advantages compared to HVAC transmission like no reactive power requirement to maintain the voltage of the line, less line losses, asynchronous grid interconnection, grid stability, and higher efficiency.
- HVDC line requires a narrow right of way compared to HVAC.
- HVDC applications include submarine or underground power transfer, long-distance power transmission, renewable integration, and back-to-back HVDC transmission respectively.
- SCR should be greater than 3 for the strong connection of HVDC and HVAC. It supports the HVAC reactive power.
- Line Commutated Converter, a Current Source Converter, employs thyristor-based technology for power conversion. LCC requires a very high synchronous voltage source for commutation, limiting its use in black start operations.
- LCC technology is widely used in long-distance and back-to-back HVDC projects. It has unidirectional current-carrying capability.
- Voltage Source Converter technology utilizes IGBTs, allowing independent switching of current irrespective of AC voltage. It has the bidirectional current carrying capability.
- The 12-pulse converter has less voltage ripple and it improves the power quality by reducing the harmonics.
- In monopolar HVDC configuration the return path for the current is provided through the ground or a conducting medium such as seawater in the case of submarine cables. Monopolar systems are often used for long-distance transmission, especially in scenarios where a dedicated return conductor is not feasible or economical.
- Multiterminal HVDC systems involve multiple HVDC converter stations interconnected through a common DC transmission network.

- The North-East Agra multi-terminal UHVDC (800 kV) link is the world's first UHVDC multi-terminal link, with a capacity of 6,000 MW and a length of 1,728 km.
- The Changji-Guquan UHVDC link in China transmits 12,000 MW of electricity over 3293 kilometers at 1100 kV, setting new world records for voltage level, transmission capacity, and distance.

Short and Long Answer Questions

1. What do you understand by the HVDC system? What is the need for HVDC links if HVAC is available?
2. Brief the evolution of the HVDC transmission.
3. Write the comparison between the HVAC and HVDC transmission system including all aspects.
4. Explain the procedure for controlling the DC power in the HVDC link.
5. Derive the formula for the DC current flowing in the HVDC link.
6. Explain the significance of the AC and DC filters in the HVDC systems.
7. What do you mean by the power quality in the HVDC system?
8. What do you understand about the short circuit ratio in the HVDC link?
9. Explain the difference between LCC and VSC and write their advantages.
10. Explain the workings of the LCC-based HVDC link with a neat and clean diagram.
11. Explain the workings of the VSC-based HVDC link with a neat and clean diagram.
12. With the help of the necessary diagram explain the types of HVDC links.
13. Explain the working of a 12-pulse-based HVDC link with a neat diagram and waveform.

Exercises

1. In a monopolar HVDC link energized with a 3-phase, 50 Hz 400 kV source, the commutation reactance is 15 ohms and the rectifier (six pulse bridge converter) end DC voltage is 500 kV. For a delay angle of 10°:
   a) Find the DC current in the link
   b) Find the commutation angle μ
   c) If AC voltage is reduced to 220 kV, find the commutation angle μ. Assume the DC current is constant
2. A six-pulse inverter is operating at a firing angle of 10°. The commutation reactance is 5 ohm and the DC link voltage is 500 kV. The line is carrying 500 MW of power. The AC supply voltage for the converter is 400 kV. Find the following:
   a) Find the DC current in the link
   b) Find the commutation angle μ
   c) If AC voltage is reduced to 220 kV, find the commutation angle μ. Assume the DC current is constant
3. A back-to-back HVDC link with one bridge at each end is transmitting 200 MW with $V_d = 110\ kV$. If $\alpha = 12°$, $\gamma = 15°$. Find V_d^{rec}, reactive power of rectifier and inverter. Assume that $R_{cr} = R_{ci} =$ 10 Ω.

To know more about

An Overview of HVDC Technology

HVDC Vs HVAC

HVDC Applications

To know more about

Hybrid LCC/VSC HVDC Systems

To know more about

Back-to-Back HVDC Systems

To know more about

MMC-based HVDC systems

List of HVDC Projects

World's first multi-terminal ±800kV 6000 MW HVDC station at Agra

List of HVDC Projects

HVDC-MMC Interconnection

To Design

VSC based HVDC system

Thyristor based HVDC system

HVDC-MMC Interconnection

12 PHOTOVOLTAIC SYSTEMS

Unit specifics: In this unit, the following topics have been discussed for basic understating of photovoltaic systems:

- Principle of PV, Mathematical modelling and manufacturing technologies of a PV cell.
- Classification of PV systems, Partial shading condition and its effects.
- Effect of irradiance and temperature on P-V and I-V characteristics.
- PV array configurations, reconfigurations, architectures, MPPT techniques and
- PV system applications.

Rationale: This unit will introduce students to the concept of photovoltaic (PV) energy conversion, the mathematical modelling of PV cells, and the many manufacturing procedures used to make different types of PV cells. Photovoltaic systems are classified as either stand-alone, grid-connected, or hybrid. The impacts of partial shadowing circumstances on solar systems are also taken into consideration. The effect of irradiance and temperature on voltage-current characteristics, strategies for mitigating the effects of partial shadowing, and configurations of solar arrays are discussed. It also covers PV array reconfiguration, PV system topologies, maximum power point tracking methodologies, and a variety of PV system applications. The description is supported with relevant illustrations, derivations, and examples.

Pre-Requisites: Basic knowledge of power system components.

Unit Outcomes: List of outcomes of this unit is as follows

U12-O1: To derive the expression for currents and voltages of PV cell and Array.
U12-O2: To analyse stand-alone and grid connected PV Systems.
U12-O3: To understand the effect of partial shading.
U12-O4: To analyse various PV configurations.
U12-O5: To know about partial shading mitigation techniques and PV applications.
U12-O6: To analyse the photo-voltaic system numerically.

Unit-12 outcomes	Expected mapping with course outcomes (1: Weak, 2: medium, and 3: strong correlation)					
	CO-1	CO-2	CO-3	CO-4	CO-5	CO-6
U12-O1	2	3	2	-	2	3
U12-O2	2	2	-	-	2	-
U12-O3	2	-	2	2	2	-
U12-O4	2	2	2	-	2	3
U12-O5	2	2	2	-	2	-
U12-O6	2	2	-	-	2	3

12.1 Introduction:

Developing countries are leading the way in the growth of global demand for electrical energy. An expanding economy, increasing population, urbanization and greater access to power are all contributing factors to this growth. As a result of severe environmental, economic, and social consequences of relying on fossil fuels, governments are looking for better alternatives that are more sustainable to meet the energy demand. Rapid technological advancement and continuous government assistance has resulted in Renewable Energy Sources (RESs), examples of which are solar, hydro, wind, geo-thermal, and bio-fuels becoming more feasible and cost-effective solutions to this challenge. Among the various RESs, solar photovoltaic (SPV) power generation has a significant potential to meet the global energy demand due to its numerous advantages in terms of environmental and economic considerations.

The benefits of SPV power generation are as follows:

- It is abundant in nature, freely available and generates clean and green energy.
- It does not emit any green-house gases and is eco-friendly.
- The cost for operational and maintenance of PV panels is negligible as compared to other RESs.
- It requires bare minimum maintenance and hence less requirement in manpower.
- It offers easy installation and noise-less power generation suitable for residential and urban applications.
- It shows versatility. i.e., PV panels can generate electricity anywhere where the sun is available.
- It does not have any moving parts; hence losses are very less.
- It has almost no material depletion.

It is evident that from the last two decades, solar PV power generation has grown at an annual rate of 20 percent to 25 percent. It is due to dropping prices of PV modules, incentives provided by the government, and innovative models for standalone and grid-connected PV systems. Many countries have reached grid priority with solar PV systems, and many plans are proposed to reach 100 percent utilization of green energy sources by 2050. Furthermore, in the next ten years, the cost of renewables is projected to undercut the price of fossil fuels. Nevertheless, PV system is one of the most preferred solutions for meeting future energy demand. The factors for increasing the adoption of PV are: increase in demand for energy, government policies towards renewable energy sources, concerns related to environment, rapid development in PV technologies, etc.

History of Photovoltaics:

The word 'photovoltaic' is derived from the Greek words 'Phos' (light) and 'voltaic' (electrical), which means the generation of electricity. Among English speakers, the term "photovoltaic" has been in widespread usage since 1849. Bequerel discovered PV effects in 1839, but it was not until the late 1950s that commercial PV applications were found. The prices of generating PV power have historically been greater than for comparable forms of traditional power sources. Due to a significant drop in pricing, the development of new semiconductor technologies, and government subsidies depending on the market and system location, PV power generation has gained a lot of importance in these days.

12.2 Principle of obtaining electricity from light using PV cell:

PV cell is the main constituent in the PV energy conversion. The main function of the PV cell is to directly convert the incident solar light energy into electricity by photovoltaic effect. Basically, a solar PV cell is a P-N junction device and is built from two layers of silicon that have been lightly doped with impurity atoms. In the n-layer, there are atoms called donors that have one additional valence electron. In the p-layer, there are atoms called acceptors that have one fewer valence electron. The initial layer is N-type and exhibits a negative charge as a result of an excess of electrons. The other layer is P-type and is positively charged due to a higher number of holes. When these two layers are placed in contact with each other P-N junction is formed. When photons of light energy fall on the surface of the solar PV cell, electron-hole pairs are generated in the depletion layer.

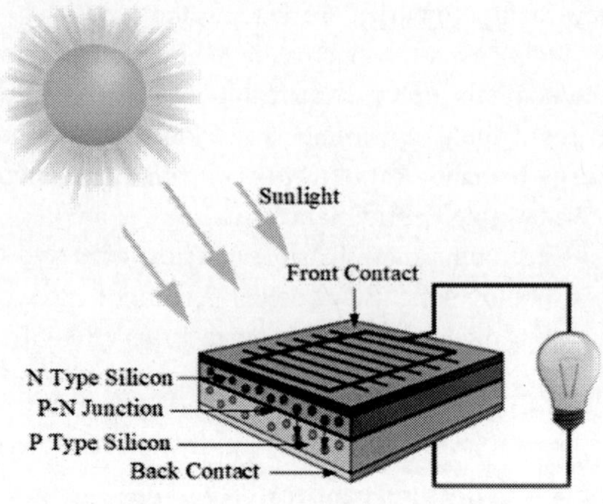

Fig. 12.1 Conversion of solar light energy to electricity

The produced electrons and holes exhibit opposite directional movement as a result of the junction field. The electrons migrate towards the N-side, while the holes migrate towards the P-side. The movement of charge carriers in a specific direction leads to the flow of electric current when an external load is connected to the PV cell, as illustrated in Fig. 12.1. The current produced by the PV cell is directly proportional to the solar insolation it receives. Typically, a standard PV cell generates a voltage of just 0.5 V. Consequently, many PV cells are interconnected in both series and parallel configurations to fulfil the desired power output demands. To achieve the desired voltage level, the PV module can be created by linking multiple PV cells in a series configuration. A PV array is created by connecting multiple PV modules together, either in series or in parallel, in order to obtain the desired power output. Structure of a PV Cell, module, and an array is shown in Fig. 12.2

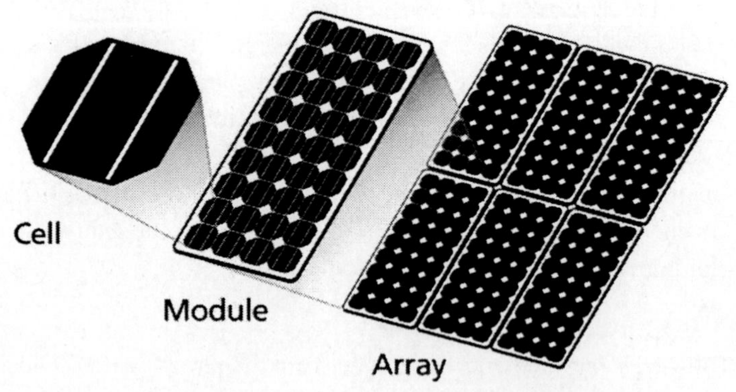

Fig. 12.2 Structure of a PV Cell, module, and an array

PV cell : It is a semiconductor device that transforms solar energy into electrical energy.

PV module: It is created by connecting many PV cells in series, parallel, or a combination of both.

PV array : It is created by connecting many PV modules in series, parallel, or a combination of both.

Fig. 12.3 shows the effect of adding multiple PV cells in parallel. By looking at the I-V characteristics obtained, it is evident that parallel connection increases the current of the system. Similarly, Fig. 12.4 shows the effect of adding multiple PV cells in series. It is seen from the I-V characteristics that the total voltage produced increases keeping the current same. Fig. 12.5 illustrates the formation of a PV module by connecting multiple PV cells in series and parallel combination and further, the formation of a PV array by connecting multiple such modules in series and parallel combinations. In Fig. 12.5(a), 5 PV cells having voltage of 0.5 V and current of 3 A each are connected in parallel and 8 such PV cells are connected in series to produce a total voltage of 4V and total current of 15A in a PV module. This is further extended to forming a PV array producing 8V and 30A using 4 such modules with series combination of two modules in parallel with series combination of remaining two modules.

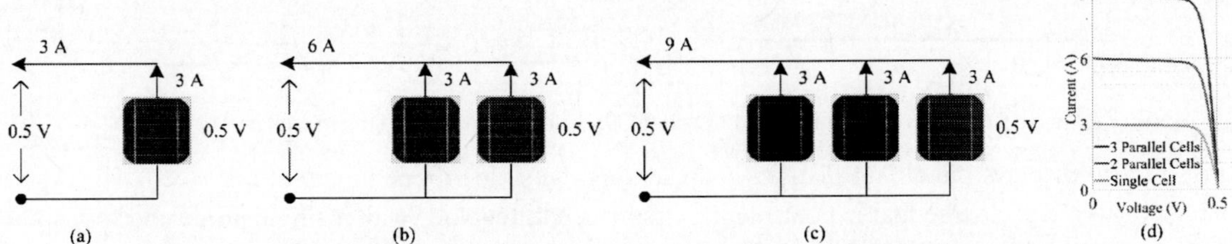

Fig. 12.3. Parallel connection of solar cells: (a) single cell, (b) 2 cells, (c) 3 cells, (d) I-V Characteristics

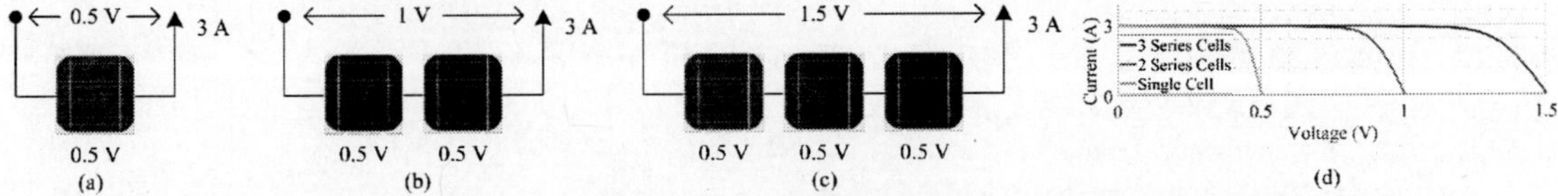

Fig. 12.4. Series connection of solar cells: (a) single cell, (b) 2 cells, (c) 3 cells, (d) I-V Characteristics

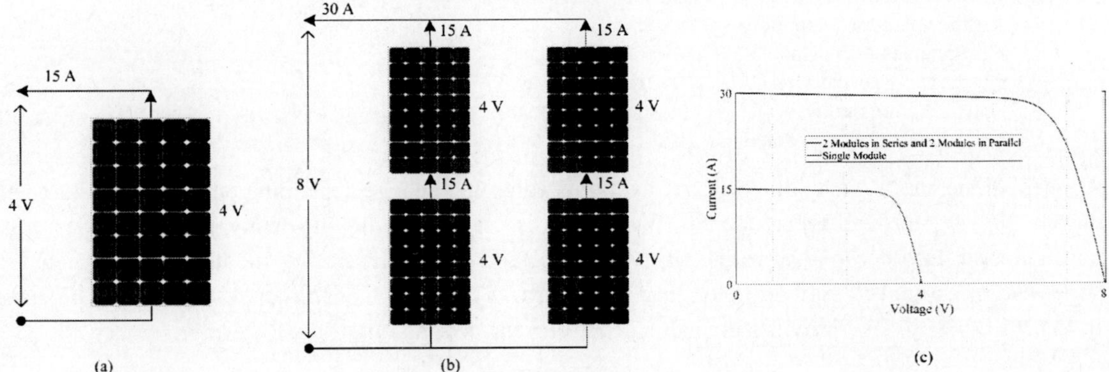

Fig. 12.5. (a) Formation of PV module from multiple PV cells, (b) Formation of PV array from multiple PV modules, (c) I-V Characteristics

12.3 Mathematical Modelling of a PV Cell:

The primary constituent of the photovoltaic system is the PV cell. The primary purpose of the PV cell is to convert incoming solar irradiance into electrical energy. Before examining the behaviour of a PV cell, it is essential to construct a model of the PV cell. Various mathematical models have been proposed in literature, commonly utilising the single diode and two diode models. Due to the increased computational requirements, the single diode model is typically preferred over the two-diode model for solving nonlinear equations. Furthermore, the output characteristics of the two-diode model PV cell closely resemble those of the single diode model PV cell.

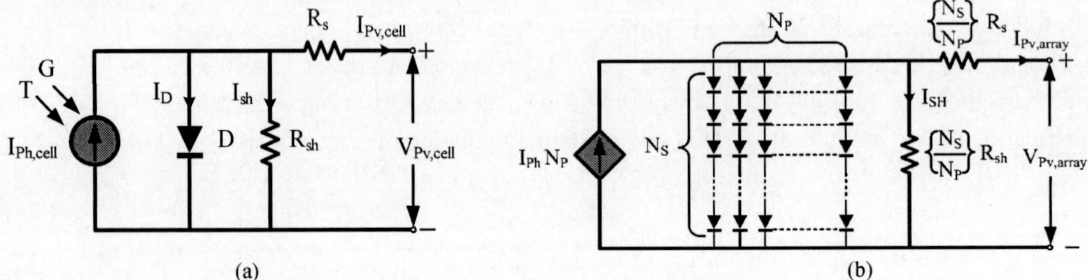

Fig. 12.6 (a) Equivalent circuit of a PV cell, (b) Equivalent circuit of a PV array with $N_S \times N_P$ modules

Fig. 12.6(a) displays the circuit diagram representing the single diode model of a PV cell. The system employs a voltage source that is controlled by the current, together with a single diode and resistances connected in series and shunt. The PV cell terminal current is given by Eq. 12.1.

$$I_{Pv,cell} = I_{Ph,cell} - I_r\left[e^{\left(\frac{V_{Pv,cell}+R_S I_{Pv,cell}}{V_{T,cell}a}\right)} - 1\right] - \frac{V_{Pv,cell} + R_S I_{Pv,cell}}{R_{sh}} \quad \text{...... (12.1)}$$

Here, $V_{T,cell} = \frac{KT}{q}$

Where $I_{PV\,cell}$ is terminal current of PV cell in [A],
$I_{Ph\,cell}$ is photon current due to irradiance,
I_r is the reverse saturation current of diode [A],
$V_{PV\,cell}$ output voltage of PV cell in [V],
$V_{T\,cell}$ is the thermal voltage of PV cell,
T is the operating temperature of PV cell,
K is Boltzmann's constant [$1.3806503 \times 10^{-23}$ J/K],
q is the charge of electron [$1.60217646 \times 10^{-19}$ C],
R_s and R_{sh} are the series and shunt resistances of PV cell [Ω],
a is the ideality factor of diode.

An ideal diode has 'a' value of unity; however, this value can change depending on the temperature and current flowing through the diode. A PV module is created by linking many PV cells in a series configuration. In order to achieve the desired voltage and current levels, PV modules are connected in series and/or parallel configurations, creating a PV array. The equivalent circuit of this array is depicted in Fig. 12.6(b). Eq. 12.2 provides the value of the current flowing out of the PV array.

$$I_{Pv,array} = I_{Ph}N_P - I_r N_P\left[e^{\left(\frac{q\left(V_{Pv,array}+R_S\left(\frac{N_S}{N_P}\right)I_{Pv,array}\right)}{n_S kTa}\right)} - 1\right] - \frac{V_{Pv,array}+R_S\left(\frac{N_S}{N_P}\right)I_{Pv,array}}{R_{sh}\left(\frac{N_S}{N_P}\right)} \quad \text{...... (12.2)}$$

Where N_S is the number of series connected modules,
N_P is the number of parallel connected modules,
$I_{Pv\,array}$ is the PV array output current,
$V_{Pv\,array}$ is the PV array output voltage

12.4 Manufacturing technologies of a PV Cell:

The efficiency of a PV cell is a measure of the amount of incident solar light energy is converted into electricity. Generally, the efficiency of a PV cell depends on so many factors but one of the main factors is type of material used for the design. Due to the various technological advancements in photovoltaics in recent years, the average panel efficiency has improved from 15% to more than 20%. With such a significant gain in efficiency, a standard-sized panel's average power rating has increased from 250 W to 370 W. For many years, silicon was the only raw material used in designing of PV cells.

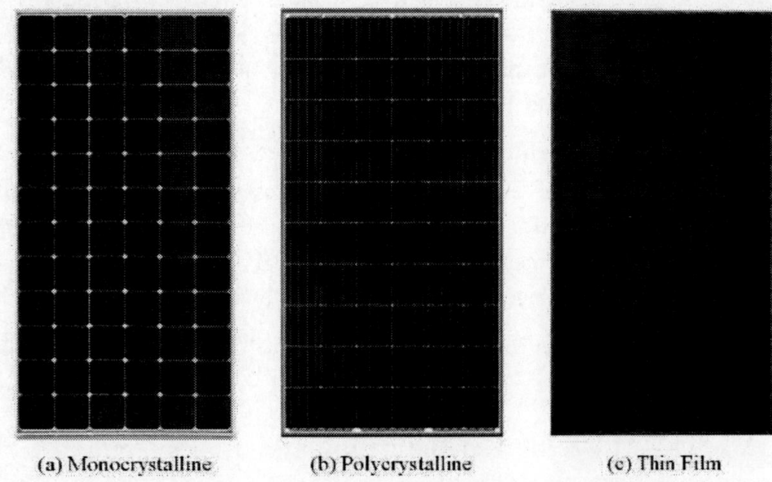

Fig. 12.7 Commercially available types of solar PV panels

Although alternative materials and processes have been created, silicon still accounts for more than 80% of production. The two primary types of silicon solar cells are monocrystalline and polycrystalline. Amorphous silicon, albeit a third form, is less frequently employed owing to its poorer efficiency. Cadmium Telluride (CdTe) and Copper Indium Gallium Selenide (CIGS) are two newly created forms of solar cells. Despite extensive research and development endeavours to develop novel materials, there are yet no commercially accessible substitutes for the aforementioned types of solar cells. Fig. 12.7 displays the many types of solar panels that are now offered in the market.

12.4.1 Mono-Crystalline Silicon:

A mono-crystalline PV cell is the most efficient solar PV technology available to generate power. PV cells of this type have been around for a long time and are still widely utilized in both residential and commercial applications. A mono-crystalline PV cell outperforms all other types of PV cells in terms of efficiency. Pure single-crystal silicon is used in the fabrication of these PV cells. A typical oval shape is found in these silicon crystals. In addition, the solar panel is constructed using a grid-like arrangement of mono cells. This type of PV cell is easily identifiable because of its uniform dark appearance and round edges. The efficiency of these PV cells ranges from 19% to 20%. Due to the usage of pure silicon, these PV cells are more durable at high temperatures. Nevertheless, the high purity of the silicon used in these PV cells makes them a bit pricey and it requires less space for installation.

12.4.2 Poly-Crystalline Silicon:

In recent years, poly-crystalline PV cells have been the most popular option for consumers. Unlike other PV cells, these ones have squares and uncut angles as their specialty. Poly-crystalline solar panels are fabricated by heating, melting, and fusing together shards of raw silicon crystals. The cost of these PV cells is cheaper than mono-crystalline PV cell. The surface of this type of PV cells has a blueish tint, and they also have a speckled/distinctive appearance. High temperatures in the surrounding environment affect the performance of this type of PV cells. Poly-crystalline PV cells are less efficient than mono-crystalline because of various imperfections in raw silicon. The efficiency of Poly-crystalline PV cell is roughly around 16% to 17%.

12.4.3 Amorphous Thin-film Silicon:

The thin film PV cell technology has the lowest market share in India and other nations. This type of solar panel is made with thin film solar cells. The portability of these PV cells makes them an excellent choice for solar projects that require a great deal of flexibility in installation and as well as in where low power generation in necessary. It is fabricated by layering semiconducting materials such as amorphous silicon, cadmium, or copper on top of each other in thin-film form. Compared to mono and poly-crystalline PV cells, these are less expensive or more affordable because of their ease of fabrication. These cells are also less sensitive to temperature.

12.4.4 Other Cells and Materials:

Recently research is going on the fabrication of nano-crystal, polymer-based, dye-sensitized, and concentrated based PV cells. These are the most recent and most promising technologies. However, they are still under development. The main advantage of these technologies includes their satisfactory performance under even low and dim sunlight conditions. Especially, the materials used for dye-sensitized PV cells are abundant, non-toxic, and relatively its fabrication is very easy.

12.5 PV System Components:

Solar PV systems typically comprise of six main components: solar PV array, battery bank, charge controller, inverter, energy meter, and the electricity grid. The inclusion of a charge controller and battery bank is discretionary in grid fed systems. While these two components enhance the storage and utilisation of solar energy, they also contribute to the overall cost of the installation. In standalone systems, use of battery and charge controller becomes mandatory. The schematic of various components used in a PV system is shown in Fig. 12.8.

Solar PV array: A solar PV array is composed of multiple interconnected PV panels. The solar PV array produces direct current (DC) electricity by harnessing sunlight. It is crucial to consider that photovoltaic systems require installation on sturdy mounting structures capable of supporting the weight of the array and enduring weather conditions such as wind, rain, and the resultant corrosion. Solar PV arrays or panels are affixed to rooftops using a mounting system that incorporates a combination of railings, frames, and either tile or tin feet. The majority of mounting systems consist of aluminium structures with stainless steel components and are specifically designed to accommodate various types of solar modules on various roof surfaces. Premium mounting systems are fabricated using superior grades of stainless steel and aluminium, leading to reduced roof weight and decreased corrosion levels over an extended duration. High-quality mounting rails may also include sturdy anchoring points and design options that accelerate the installation process of your solar system. Acquiring a robust and meticulously designed mounting solution is the prudent approach to safeguard the financial commitment made in solar system.

In addition, the balance of system, consisting of cables, switchboards, junction boxes, meters, earthing system, circuit breaker, fuses etc. are required. It is recommended to use high quality materials and high-quality workmanship for long lasting reliability.

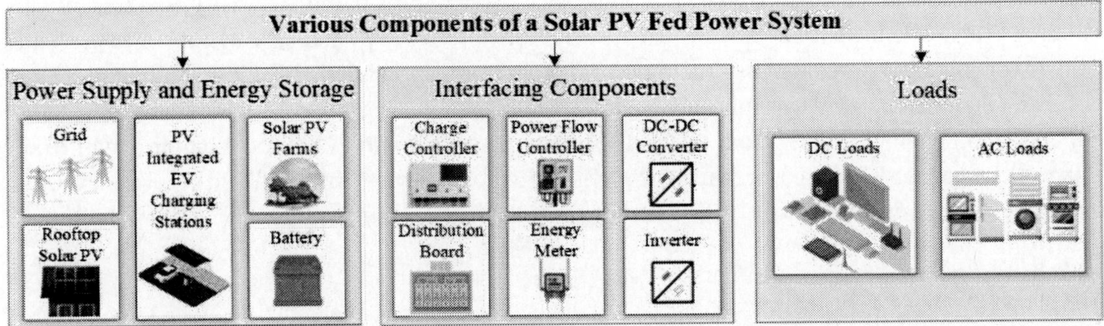

Fig. 12.8 Various components of a solar PV fed power system

Charge controller: Charge controllers regulate and oversee the movement of DC power from the PV panels to either the battery bank, DC loads or the inverter to further feed the AC loads. It prevents overcharging of the batteries. Charge controllers can be classified into two separate categories: Maximum Power Point Tracking (MPPT) type and Pulse Width Modulation (PWM) type. PWM is a traditional and suitable choice for smaller solar systems and battery banks. MPPT charge controllers are more suitable for larger systems. A charge controller is essential only if a battery bank is used.

Power flow controller: A more advanced version of the charge controller. It is used when there are multiple power sources which have to be interfaced with loads and battery bank.

Battery bank: A battery bank enables to optimize utilization of solar energy and serves as a dependable contingency plan during instances of power grid failures. The energy generated by the PV array is stored for later use instead of being immediately consumed. It is discretionary to incorporate a battery bank into solar PV systems.

DC-DC converters: DC-DC converters are used to convert the fluctuating voltage output of a module, which is affected by factors like time of day and weather conditions, into a consistent voltage output. This constant voltage output can be employed for tasks such as battery charging or as an input for an inverter in a grid-connected system.

Inverter: An inverter plays a crucial role in the solar PV power system. It converts the DC output from a PV panel into AC. AC energy can be delivered to home to operate household appliances. Based on the setup of the home, any excess electricity generated in home can be either sent to the power grid via power lines or stored in batteries for later use.

Bi-directional energy meter: Irrespective of the type of solar PV system, every building will have a service connection point equipped with an energy meter to measure the amount of electricity consumed. The surplus solar energy generated by the solar PV system, which is not stored or used on-site, will be injected into the public electrical grid. Bi-directional meters monitor both the electricity consumed from the public grid and the excess solar energy that is sent back to the public grid.

Distribution board: A distribution board is used to regulate and control the flow of power between the grid and the domestic loads.

Electric grid: While solar PV enables off-grid electricity sourcing, the majority of homeowners and organisations choose to utilise a combination of solar PV and grid-based electricity sourcing. When a building is linked to the public grid, any excess electricity produced by solar PV system once the battery bank is completely charged will be sent to the grid. Therefore, during periods of low sunlight or overcast weather, when the solar PV system is unable to generate enough energy for the building's needs, there is the option to obtain electricity from the grid. The exported solar energy from the building will receive financial compensation based on the state's regulations. This compensation can be obtained by net metering, net feed-in, or gross feed-in accounting mechanisms.

12.6 Classification of PV Systems:

The scope of solar PV systems has significantly increased in recent years, and there is a possibility for additional growth. The solar PV system offers a significant advantage due to its modular architecture. A modular system architecture allows for easy scalability in response to fluctuations in power demands. The categorisation of PV systems is determined by their operational and functional requirements, component arrangement, and the manner in which the equipment is connected to power sources and loads. The three main categories are standalone, grid-connected, and hybrid PV systems.

12.6.1 Stand-Alone PV System:

Stand-alone PV systems are often called as off-grid PV systems and these are relying on solar power only. Standalone photovoltaic (PV) systems are beneficial for remote areas facing challenges in accessing the grid system.

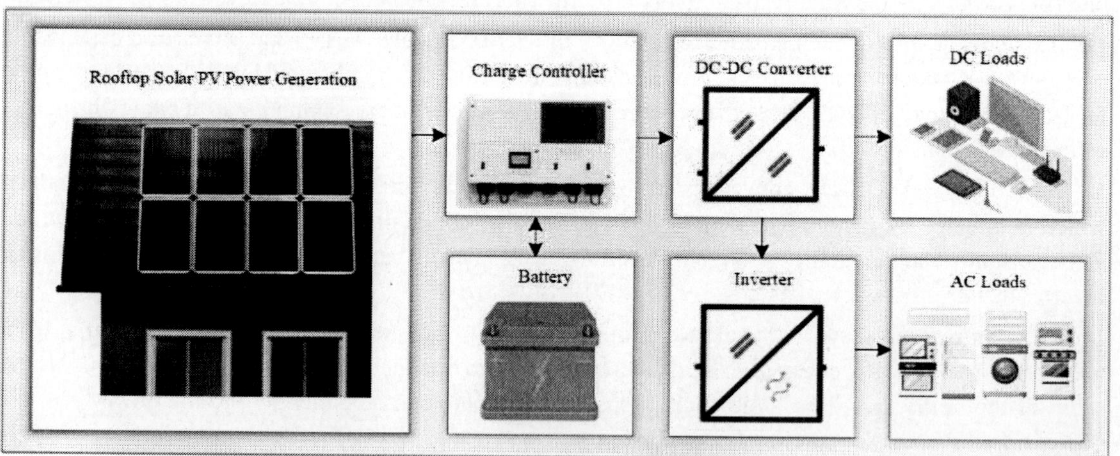

Fig. 12.9 Stand-Alone PV system architecture

In general, there are two ways for forming the stand-alone PV system: one with PV modules directly connected to the loads without energy storage referred to as direct-coupled stand-alone PV system and other with PV modules directly connected to the loads with energy storage referred to as stand-alone PV system with battery storage powering DC and AC loads as shown in Fig. 12.9. This kind of PV systems can be suitable for powering applications such as solar thermal heating systems, water pumps, ventilation fans etc.

These systems can be comprised of photovoltaic PV modules and a load, or they can incorporate batteries for the purpose of energy storage. When utilising batteries, charge regulators are incorporated to deactivate the PV modules once the batteries reach full charge and may deactivate the load to avoid the batteries from being depleted below a specific threshold. Regular maintenance is necessary for this system, including monitoring battery electrolyte levels and addressing terminal corrosion. The batteries must possess sufficient capacity to accumulate the energy generated throughout the day for use at night time and during inclement weather conditions when solar PV power generation is hampered.

12.6.2 Grid-Connected PV System:

Grid-connected solar PV systems are becoming more and more common due to their incorporation into building applications. A grid-connected PV system has solar panels that gather solar radiation and transform it into DC power. The DC is then used by the solar system's inverter to convert the DC energy into alternating current (AC). Household devices can consume AC power in a way similar to their reliance on a grid system. Fig. 12.10 displays the diagrammatic layout of a photovoltaic system connected to the power grid. A grid connected PV system is seen in all power ranges, starting from utility scale solar farms which generate power in the range of Giga Watts, medium scale solar parks and rooftop systems in large buildings which produce power in the range of Mega Watts and to rooftop solar PV on individual houses and small buildings producing power in the range of Kilo Watts. The rooftop solar systems installed in small houses and residential buildings use central converter and inverter topology. The medium scale solar parks in residential and academic complexes, and rooftop solar systems of large buildings generally use string integrated converter and inverter topology while large utility scale solar farms generally use multi string converter and inverter topology. These PV system architectures are explained in detail in the section 12.10.5.

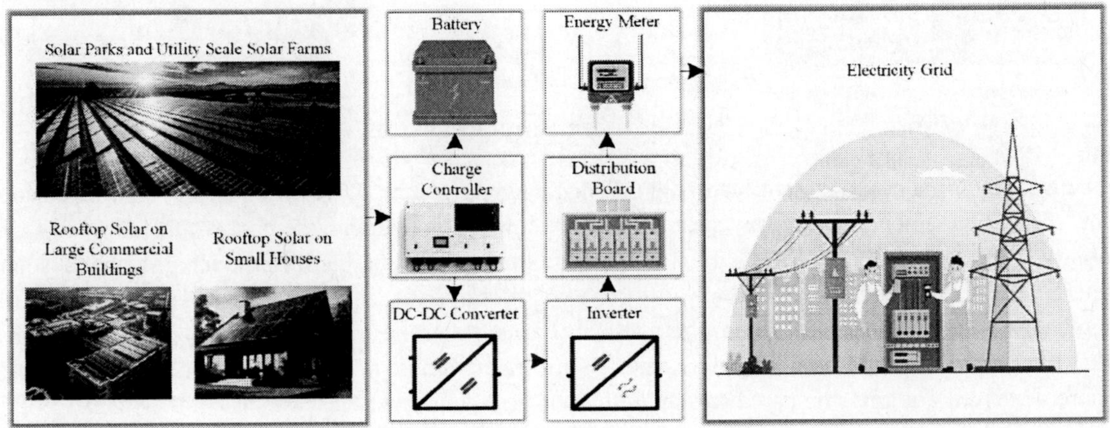

Fig. 12.10 Grid-connected PV system architecture

The inverter is linked to the distribution board in residential houses. The PV system can either transfer the generated power to the energy grid or supply it directly to AC appliances in the residence. These systems are connected to the power grid and hence the inclusion of a battery is not mandatory. However, addition of a battery helps in timing the sourcing and feeding of power from and to the grid depending on the power needs of the grid. They use the grid as a buffer to manage surplus solar electricity and access power from the grid when solar generation is inadequate. It leads to reduced dependence on electricity from the grid, while guaranteeing access to grid electricity during periods when the solar system is not producing enough energy. An inherent benefit of using a grid-connected system is its cost-effectiveness in comparison to other solar PV systems. Furthermore, it offers flexibility in designing the solar PV system layout as the system does not need to provide electricity to every load in the household. A fundamental constraint of a grid-connected system is its inability to prevent power outages.

12.6.3 Hybrid PV System:

A hybrid PV system is a configuration that integrates different power sources in order to improve the accessibility and utilisation of electricity. This system has the capability to harness energy from several sources, including multiple solar PV sources, backup diesel generators, etc. In addition, hybrid PV systems are commonly equipped with a battery as a backup to optimise the system's efficiency. The schematic arrangement of hybrid PV system having multiple solar PV sources and a diesel generator as backup power supply is shown in Fig. 12.11.

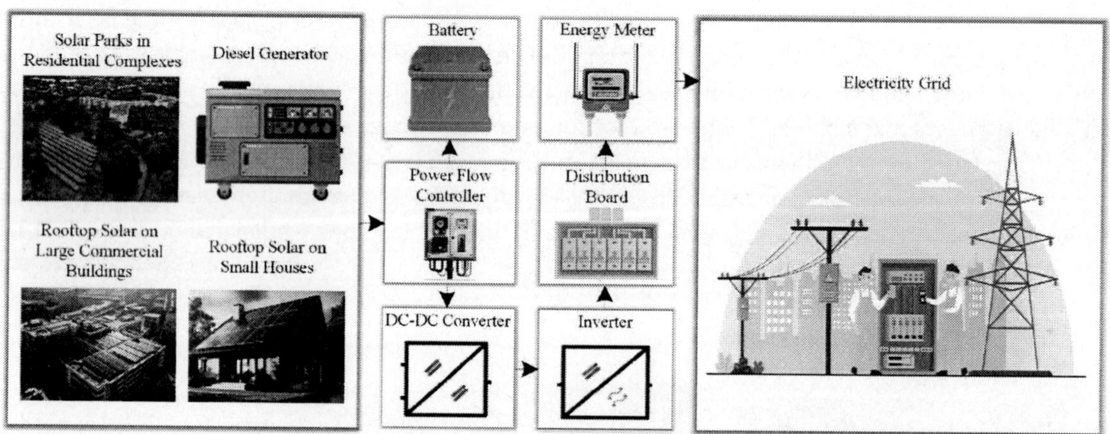

Fig. 12.11 Hybrid PV system architecture

There are multiple benefits associated with utilising a hybrid system. Diversified energy sources ensure that the system is not reliant on any specific energy source. For example, the photovoltaic array has the ability to recharge the battery when the weather conditions are favourable for producing sufficient solar energy. In the same manner, when there are strong winds or overcast conditions, a wind turbine can fulfil the battery's charging needs. Hybrid photovoltaic (PV) systems are most suitable for remote locations that have restricted access to the power grid. Hybrid photovoltaic (PV) systems generally necessitate more advanced controls compared to stand-alone or grid-connected PV systems. In the scenario of a PV/diesel system, the diesel engine is initiated once the battery reaches a specific level of discharge and is halted once the battery attains a satisfactory level of charge. The backup generator has the capability to either replenish batteries alone or provide power to the load. However, there are several challenges that come with a hybrid system. The requirement of a complex power flow controller to regulate power flow between the multiple power sources to the loads and battery can be expensive. Furthermore, the utilisation of various energy sources can lead to higher initial expenses.

12.7 Partial shading condition and its effects:

Partial shading conditions (PSC) happen when certain photovoltaic (PV) cells and/or modules are shaded as a result of various factors. It may arise from factors associated with the site itself, although these can be resolved early on during the installation of the PV system. For instance, selecting appropriate and optimal locations for installing the PV system based on site attributes such as temperature and irradiance, and ensuring there are no nearby buildings or towers, can greatly enhance the power production and overall efficiency of the PV system. Alternatively, it might arise as a result of unavoidable factors such as nearby tall structures, vegetation, airborne particles, deterioration from time, and shifting clouds, snow or other elements.

Fig. 12.12 Partial shading scenarios due to: (a) snow, (b) passing clouds, (c) trees, and (d) neighbouring buildings

The PV cells/modules located in the shaded area are forced to endure the increased current flowing through the unshaded cells/modules. Consequently, they consume energy and operate as a load rather than producing energy. Partial shadowing of solar panels reduces power generation and raises the likelihood of hot spot problems, which can ultimately lead to overheating and failure of shaded photovoltaic (PV) modules. Therefore, the existence of shaded PV cells or panels has a negative impact on the amount of power collected and the efficiency of the partially shaded solar system. The PV system has substantial power losses due to partial shadowing or incorrect global maximum power tracking, which can exceed 70% of the total power generated. Hence, it is crucial to oversee the worldwide peak power to achieve best power efficiency, reduce power losses, and maximise output power. Fig. 12.12 displays many instances of partial shading that can occur in natural environments.

12.8 Effect of irradiance and temperature on P-V and I-V characteristics:

Various parameters, such as temperature and irradiance, can influence the efficiency of solar PV modules. The open circuit voltage (V_{OC}) of a PV module varies in response to variations in cell temperature. With a rise in temperature, whether due to environmental changes or internal heat generated during energy production, the V_{OC} reduces significantly while the short circuit current (I_{SC}) experiences a slight increase. As a result, the power output drops as the temperature increases. When building a solar PV system, it is crucial to take into account the temperature coefficient of the PV module. This entails comparing the expected average temperature of the cell in the system's operational environment with the STC (Standard Test Conditions) data used to calculate the module's output. Similarly, changes in irradiance will directly affect the performance of the module, leading a fall in the amount of sunlight to primarily lead to a decrease in current, which in turn results in a reduced power output. Fig. 12.13 provides the performance parameters (I-V and P-V) of a PV cell under two different conditions: constant irradiance and variable temperature, and constant temperature and changing irradiance.

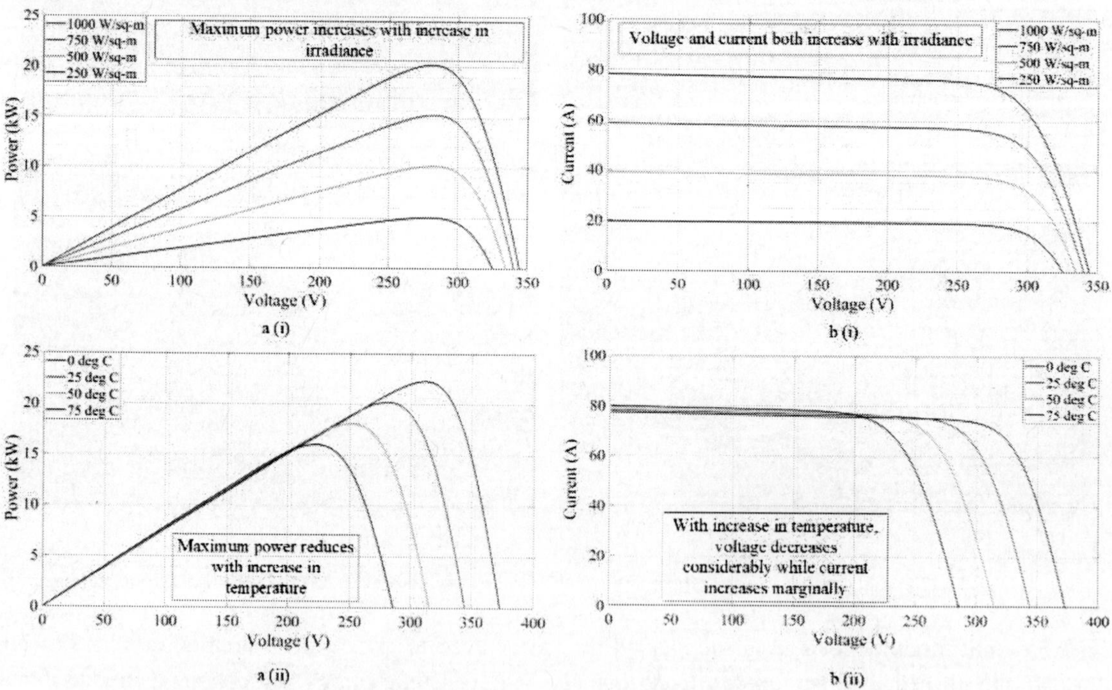

Fig. 12.13 (a) P-V and (b) I-V characteristics of a PV cell at (i) variable irradiance and constant temperature, (ii) variable temperature and constant irradiance

12.9 Mitigation of Partial shading conditions:

Partial shading of PV array results in significant power reduction. In order to mitigate this power reduction, researchers have devised multiple techniques to resolve the issue of partial shading. The strategies for mitigating power system disturbances have been categorised into two main groups: PV system design topologies and MPPT techniques, as illustrated in Fig. 12.14. PV systems incorporate many topologies, including bypass and blocking diodes, diverse PV system architectures, PV array layout, and the capacity to reconfigure the PV array. The strategies for maximum power point tracking (MPPT) encompass classical, intelligent, optimisation, and hybrid methods.

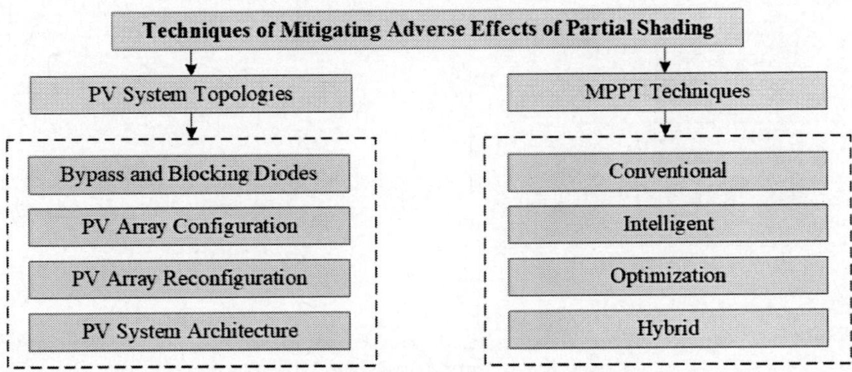

Fig. 12.14. Partial shading mitigation techniques.

12.10 PV System Topologies:

The PV system topologies are classified based on various factors such as the use of bypass and blocking diodes, the arrangement of PV arrays, the type of connections between modules, and the application of algorithms to ensure even shade dispersion. Additionally, PV system architectures and PV array reconfiguration based on protection mechanism during PSCs are also considered.

12.10.1 Bypass and Blocking diodes:

Bypass diodes are linked in parallel with the PV modules and the blocking diode is connected in series with the PV string or array to prevent the reverse passage of current through shaded modules. Under homogeneous irradiation conditions, the bypass diode will not conduct as it will be reverse biased. In the case of non-uniform irradiation or PSCs, the bypass diode will be in a state of conduction mode, as shown in Fig. 12.15.

The bypass diode serves two fundamental roles, namely:

- Ensuring the PV module is safeguarded from the hot spot issue.
- Minimises the voltage drop in the shaded cells of the modules, so restricting it to the reverse voltage of the diode (0.4 – 0.7 V). The diode's reverse voltage enhances the overall output voltage of the module. Furthermore, the power consumption of the shadowed cells will be diminished. The decrease in power consumption results in a decrease in the amount of heat generated in the shadowed area, hence increasing the lifespan of the modules. Consequently, it significantly reduces the adverse consequences of PSC.

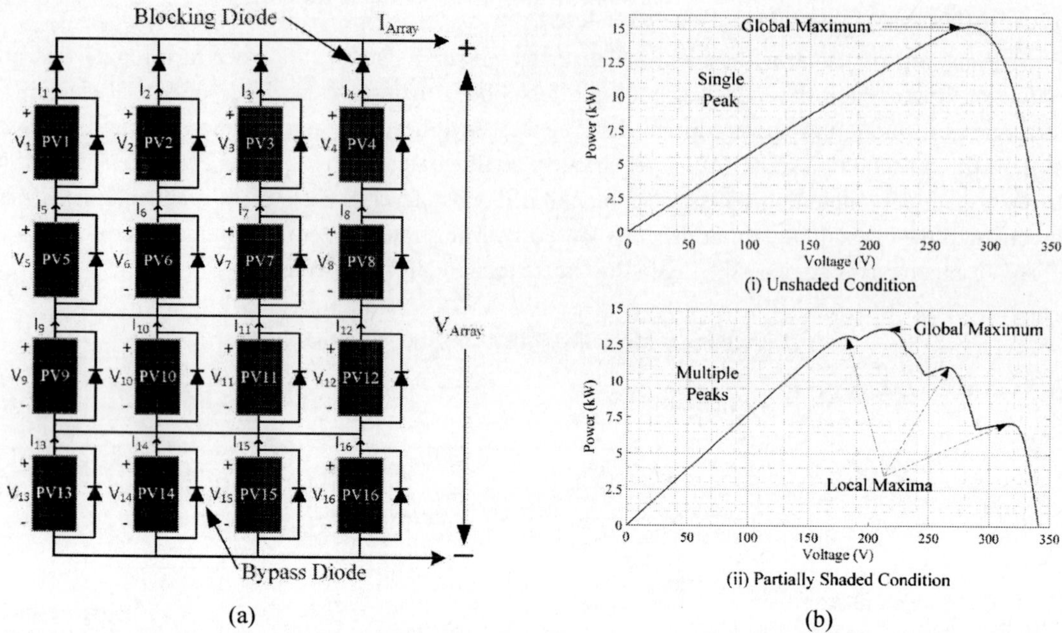

Fig. 12.15 Illustration of (a) PV array with Bypass and Blocking diodes, and (b) P-V curves under unshaded and partially shaded condition

12.10.2 PV Array Configurations:

Changing the arrangements and interconnections of PV modules in an array can mitigate the impact of partial shading. Various PV array configurations, including series, parallel, series-parallel (SP), bridge-linked (BL), honeycomb (HC) and total-cross-tied (TCT), configurations, have been described in recent literature for this. These PV array configurations are illustrated in Fig. 12.16. The configurations are developed for 16 PV modules. These configurations can be developed for larger number of modules as well. The PV array voltage (V_{Array}), current (I_{Array}) and power (P_{Array}) obtained for each configuration are given in Table 12.1.

12.10.2.1 Series PV Array configuration (S): The series PV array design is a fundamental setup where all the PV modules are connected in a series. Due to the series connection, the current flowing through all PV modules is same, indicating that the current in each PV module is equivalent to the total output current of the PV array. Moreover, the output voltage of the PV array is equivalent to the total sum of voltages from each individual module.

12.10.2.2 Parallel PV Array configuration (P): Parallel PV array arrangement is a straightforward method of connecting PV modules, in which all the PV modules are connected in parallel. In this arrangement, the cumulative output voltage of the PV array is equivalent to the voltage of a single module, whereas the cumulative output current of the PV array is the sum of the individual module currents.

12.10.2.3 Series-Parallel Array PV configuration (SP): This configuration is widely used in PV system applications because to its simplicity in design and lack of unnecessary connections and cross-ties between modules. In the SP configuration, PV modules are connected in series to form strings to achieve the desired output voltage of the array, and these strings are then connected in parallel to achieve the desired output current of the array. The shown SP PV array arrangement consists of four strings, each including four series-connected modules.

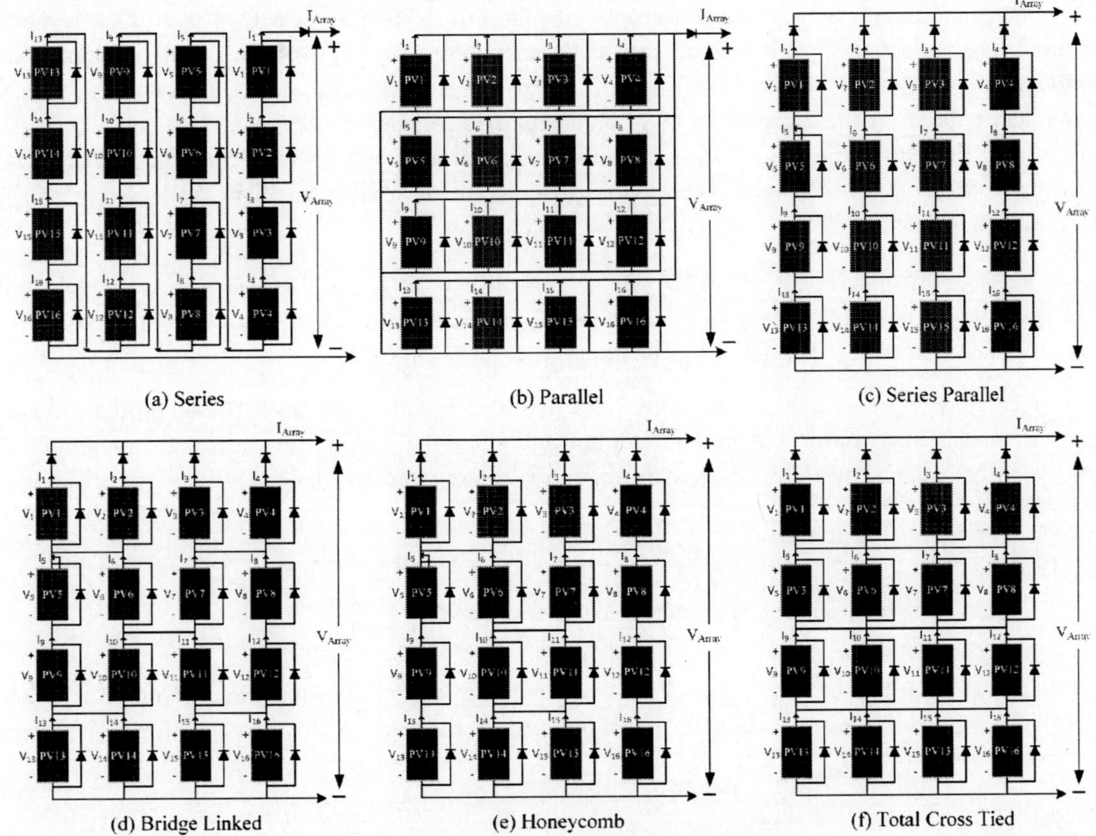

Fig. 12.16 Illustration of PV Array Configurations

12.10.2.4 Bridge-Link PV Array configuration (BL): The given arrangement involves the formation of a bridge rectifier structure by linking every four modules, with all the bridges joined through cross ties. The voltages in a series circuit are combined, while the currents in a parallel circuit are combined to get the output voltage and current of the array. The BL arrangement is characterized by two modules connected in series within each bridge, and this combination is then connected in parallel with another two series-connected modules.

12.10.2.5 Honeycomb PV Array configuration (HC): The HC configuration is taken on the hexagonal structure of honeybee hives. The configuration consists of six interconnected modules arranged in a hexagonal pattern, similar to honeybee hives. All hexagonal constructions are connected to each other using cross-ties. The voltages in the series circuit are summed, while the currents in the parallel circuit are summed to obtain the output current and voltage of the array. The HC configuration is characterized by three modules arranged in series within a hexagonal shape, and this arrangement is then connected in parallel to another set of three series-connected modules.

12.10.2.6 Total-Cross-Tied PV Array configuration (TCT): The TCT configuration is created by connecting all the columns of modules in the PV array together via cross-tied links. The modules inside a row are connected in parallel using the cross-tied connection. The voltage across each row is identical to the voltage across each module in the row, and the output voltage of the array equals the sum of the individual row voltages. The output current of the array is the sum of the currents generated by each individual column in the array. The TCT architecture has numerous advantages, such as extended operational lifespan, minimised multi-peak impact, exceptional fault tolerance, and improved power generation capability in different shading scenarios due to the provision of multiple current bypass pathways for shaded modules.

12.10.3 Performance Indicators of PV Array configurations:

The main parameters that are used to characterise the performance of solar PV array configurations are the Mismatch Power Loss (ΔP_{MMPL}), Fill Factor (FF), and efficiency.

Mismatch Power Loss (ΔP_{MMPL}): It is the ratio of difference between peak power produced under uniform insolation condition and partial shading condition to the peak power generated during uniform insolation. Mismatch power loss is given by eq. (12.3) and is expressed in percentage terms.

$$\Delta P_{MMPL}(\%) = \frac{P_{MP,UIC} - P_{MP,PSC}}{P_{MP,UIC}} \times 100 \qquad \text{……… (12.3)}$$

Where ΔP_{MMPL} is the mismatch power loss

$P_{MP,UIC}$ is the peak power developed under uniform insolation

$P_{MP,PSC}$ is the peak power developed under the given PSC.

Fill Factor (FF): It is the ratio of global peak power to the product of open circuit voltage and short circuit current and it is given by eq. (12.4). It is preferred to have the FF closer to unity for a PV system.

$$FF = \frac{P_{Max}}{V_{OC} \times I_{SC}} = \frac{V_{MP} \times I_{MP}}{V_{OC} \times I_{SC}} \qquad \text{……… (12.4)}$$

Where, P_{Max} is the maximum power

V_{MP} is the Voltage at maximum power

I_{MP} is the Current at maximum power

V_{OC} is the open circuit voltage

I_{SC} is the short circuit current

Efficiency (η): It is the ratio of global peak output power to the input solar power and is computed by eq. (12.5).

$$\eta = \frac{V_O \times I_O}{L \times A} \qquad \text{……… (12.5)}$$

Where, η is the efficiency of the module

V_O is the output voltage of the PV panel

I_O is the output current of the PV panel

L is the solar radiation intensity on the PV panel per square – metre

A is the area of the PV panel in square – metre.

Table 12.1. Output Voltage, Current and Power of Different PV Array Configurations

PV Configurations	Array Voltage (V_{Array})	Array Current (I_{Array})	Array Power (P_{Array})
Series	$\sum_{PV=1}^{16} V_{PV} = 16 \times V_{PV}$	I_{PV}	$16 \times V_{PV} \times I_{PV}$
Parallel	V_{PV}	$\sum_{PV=1}^{16} I_{PV} = 16 \times I_{PV}$	$16 \times V_{PV} \times I_{PV}$
Series Parallel	$V_4 + V_8 + V_{12} + V_{16}$	$\sum_{PV=1}^{4} I_{PV} = 4 \times I_{PV}$	$16 \times V_{PV} \times I_{PV}$
Bridge Linked	$V_4 + V_8 + V_{12} + V_{16}$	$\sum_{PV=1}^{4} I_{PV} = 4 \times I_{PV}$	$16 \times V_{PV} \times I_{PV}$
Honeycomb	$V_4 + V_8 + V_{12} + V_{16}$	$\sum_{PV=1}^{4} I_{PV} = 4 \times I_{PV}$	$16 \times V_{PV} \times I_{PV}$
Total Cross Tied	$V_4 + V_8 + V_{12} + V_{16}$	$\sum_{PV=1}^{4} I_{PV} = 4 \times I_{PV}$	$16 \times V_{PV} \times I_{PV}$

Example 12.1. A single solar cell of dimension (20 cm×20 cm) produces a voltage of 0.8 V and current of 3.5 A. The intensity of light absorbed by the cell is 600 W/m². Determine the efficiency of the cell.

Ans. Given data: Area (A) = 20 cm×20 cm = 400 cm² = 400×10^{-4} m²

Output voltage (V_O) = 0.8 V

Output current (I_O) = 3.5 A

Intensity, L = 600 W/m²

$$\text{Efficiency } (\eta) = \frac{P_O}{P_{in}} \times 100$$

$$= \frac{V_O \times I_O}{L \times A} \times 100 = \frac{0.8 \times 3.5}{600 \times 400 \times 10^{-4}} \times 100 = \frac{2.8}{24} \times 100 = 11.66\%$$

Example 12.2. A solar cell has efficiency of 15% and surface area of 6 cm². Find the electrical energy generated in one second, by 4000 such cells connected to form a solar panel, given that 740 J of solar energy is incident on an area of one-meter square of solar panels, in one second.

Ans. Given data: Efficiency (η) = 15%

Area of single solar cell (A) = 6 cm² = 6×10^{-4} m²

Total surface area of solar cells on which solar energy is incident

$$= \textit{area of each cell} \times \textit{number of cells} = 6 \times 4000\ cm^2 = 2.4\ m^2$$

The solar energy falling on 4000 cells = 2.4 × Solar energy incident on one-metre square area

$$= 2.4 \times 740\,J = 1776\,J$$

$$\text{Efficiency}\ (\eta) = \frac{\text{Electrical Energy generated (or) Output}}{\text{Incident energy (or)Input}} \times 100$$

$$\text{Electrical Energy generated} = \text{Efficiency} \times \text{Incident energy} = \frac{15}{100} \times 1776\,J = 266.4\,J$$

Example 12.3. 10 kW of power is used by a household. Solar energy is incident on the horizontal surface at an average rate of 400 W per square meter. If 25% of this energy is converted to electrical energy, how much area of solar panels is required to supply 10 kW?

Ans. Given data: Power used by the household, P = 10 kW = 10×10^3 W

Per square metre solar energy incident = 400 W

Efficiency of solar energy to electrical energy conversion = 25%

Area of solar panels required to generate the electricity required = A

According to the data given in the question, we get:

$$10 \times 10^3 = \frac{25}{100} \times (A \times 400)$$

$$\therefore A = 100\ m^2$$

Example 12.4. What is the value of Fill Factor if maximum power (P_{max}) is 50W, open circuit voltage (V_{OC}) is 20 V and short circuit current (I_{SC}) is 3.5 A?

Ans. Given data: P_{max} = 50W, V_{OC} = 20 V and I_{SC} =3.5 A

$$\text{Fill Factor} = \frac{P_{Max}}{V_{OC} \times I_{SC}} = \frac{50}{20 \times 3.5} = 0.76$$

Example 12.5. The fill factor of a photovoltaic module is 0.74. a single cell in a module has an open circuit voltage of 0.6V and short circuit current of 8.2 A. The module has 54 cells connected in series. If the voltage at maximum power is 0.8 times of open circuit voltage, find the current at maximum power.

Ans. Given data: FF = 0.74

Number of cells connected in series = 54

Open circuit voltage of single PV cell = 0.6 V

Short circuit current of single PV cell = 8.2 A

Open circuit voltage of PV module is (V_{OC}) $= 54 \times 0.6 = 32.4\ V$

$$V_m = 0.8 \times V_{OC} = 0.8 \times 32.4 = 25.92\ V$$

$$\text{We know that, } FF = \frac{V_m \times I_m}{V_{OC} \times I_{SC}} \Rightarrow 0.74 = \frac{25.92 \times I_m}{32.4 \times 8.2} \Rightarrow I_m = \frac{32.4 \times 8.2 \times 0.74}{25.92} = 7.585\ A$$

Example 12.6. A solar PV panel generates a global maximum power of 500 W at standard test conditions (W=1000 W/m^2 and T=25^{0}C). A shadow of tall building is falling on the PV panel which reduces the power generation capability of PV panel. Due to the shadow of tall building the PV panel output is around 300 W then calculate the mismatch power loss in the system.

Ans. Given data: Maximum power at STC = 500 W

Maximum power under shadowing conditions = 300 W

$$Mismatch\ Power\ loss(\Delta P_{MMPL}) = \frac{P_{MP@STC} - P_{MP@PSC}}{P_{MP@STC}} \times 100$$

$$= \frac{500-300}{500} \times 100 = 40\%$$

Example 12.7. A PV system is to be designed to produce 350 W at 34 V. The voltage at maximum power of the PV cell is 0.7 V and the current density at maximum power point of the cell is 340 A/m². Using the data of PV cell, design the PV module, considering operation at the maximum power point, if each cell is 10 cm² in area.

Ans. Given data: $V_{max} = 0.7$ V.

The current density at maximum power point is = 340 A/m².

Therefore, the PV cell current is, $I_{max} = 340 \text{ A/m}^2 \times 10 \text{ cm}^2 = 3400 \times 10^{-4} \text{ A} = 0.34$ A

Thus, power obtained per PV cell = 0.7 × 0.34 = 0.2 W.

Number of PV cells required in the module $= \frac{350}{0.2} = 1750$

Number of cells in series $= \frac{\text{System voltage}}{\text{Voltage per cell}} = \frac{34}{0.7} = 48.57 \approx 49$ (With 49 cells, Voltage = 34.3 V).

Number of strings of 49 cells each, connected in parallel $= \frac{1750}{49} = 35.71 \approx 36$

(This PV module yields 36 × 49 × 0.2 =352.8 W).

Example 12.8. Design a PV array configuration which can generate maximum power point current of 9A at maximum power of 5 kW at STC. Also, mention what kind of configuration is used and draw the same. Use the following PV module specifications.

Maximum Power, P_{max}	: 315 W
Voltage at maximum power, V_{mp}	: 35 V
Current at maximum power, I_{mp}	: 9 A
Open Circuit Voltage, V_{OC}	: 43 V
Short Circuit Current, I_{SC}	: 9.77 A

Ans. Given data:

PV array has to generate a maximum current of 9 A

PV array has to generate a maximum power of 5 kW

Single PV module specifications:

$P_{max} = 315$ W, $V_{mp} = 35$ V, and $I_{mp} = 9$ A

Single PV module generates a maximum power of 315 W

Number of PV modules required to generate 5kW power is

$= \frac{5\ kW}{315} = 15.87 \approx 16\ modules$

But the PV array generates a maximum power of 5kW at a maximum current of 9 A

i.e. the terminal current from the PV array is = 9A which is same as the module current.

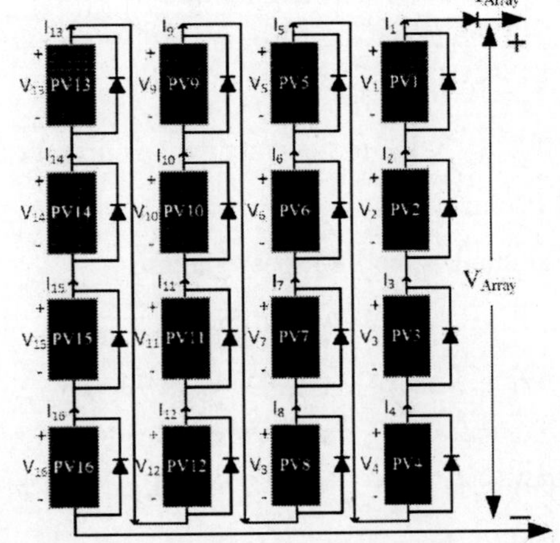

That means to get the terminal current of PV array as 9A, all 16 modules have to be connected in series. i.e. in series configuration.

Example 12.9. Design a PV array configuration which is generating maximum current of 83.36 A at maximum power of 4.1 kW. Also, mention what kind of configuration is used and draw the same. Use the following PV module specifications:

Maximum Power, P_{max}	: 255.29 W
Voltage at maximum power, V_{mp}	: 49 V
Current at maximum power, I_{mp}	: 5.21 A
Open Circuit Voltage, V_{OC}	: 59.8 V
Short Circuit Current, I_{SC}	: 5.55 A

Ans. Given data: PV array has to generate a maximum current of 83.36 A

PV array has to generate a maximum Power of 4.1 kW

Single PV module specifications:

$$P_{max} = 255.29 \text{ W}, V_{mp} = 49 \text{ V, and } I_{mp} = 5.21 \text{ A}$$

Single PV module generates a maximum power of 255.29 W Number of PV modules required to generate 4.1kW power is $= \frac{4.1\ kW}{255.29} = 16.060 \approx 16$

But the PV array generates a maximum power of 4.1 kW at a maximum current of 83.36 A

i.e. the terminal current from the PV array is = 83.36 A, but single PV module can generate maximum current of 5.21 A.

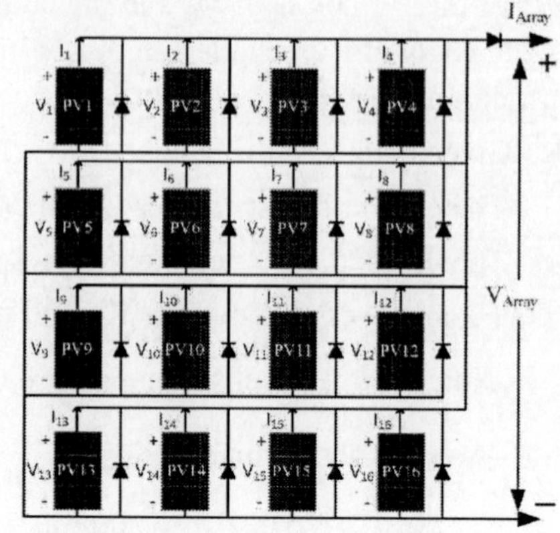

Number of modules to be connected in parallel $= \frac{\text{Terminal current of the PV array}}{\text{Current rating of single PV module}} = \frac{83.36}{5.21} = 16$

That means to get the terminal current of PV array as 83.36 A, all 16 modules have to be connected in parallel manner. i.e. parallel configuration.

Example 12.10. Design a 7.2 kW PV array which can generate a maximum voltage of 158 V and a maximum current of 46 A at STC. Also, mention what kind of configuration is it and draw the same. Use the following PV module specifications:

Maximum Power, P_{max}	: 200 W
Voltage at maximum power, V_{mp}	: 26.3 V
Current at maximum power, I_{mp}	: 7.61 A
Open Circuit Voltage, V_{OC}	: 32.9 V
Short Circuit Current, I_{SC}	: 8.21 A

Ans. Given data:

PV array has to generate a max current of 46 A

PV array has to generate a max voltage of 158V

Single PV module specifications:

$$P_{max} = 200 \text{ W}, V_{mp} = 26.3 \text{ V, and } I_{mp} = 7.61 \text{ A}$$

Single PV module generates a max power = 200W

Number of PV modules required to generate 7.2kW power is $= \frac{7.2\ kW}{200} = 36\ modules$

But the PV array generates a maximum power of 7.2 kW at a maximum voltage of 158 V and a maximum current of 46 A.

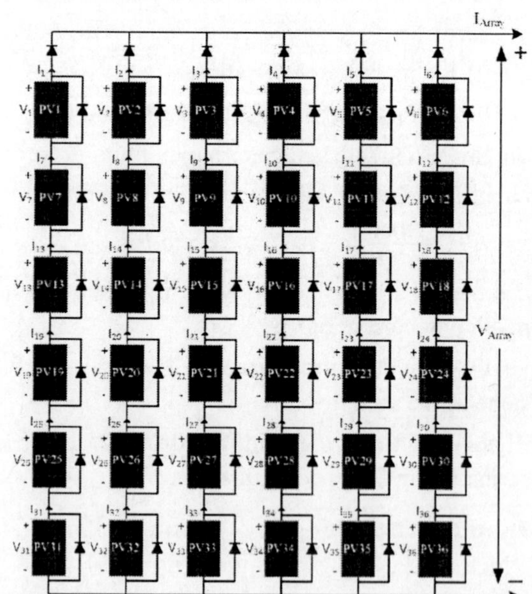

Number of modules connected in series is

$$= \frac{\text{Terminal voltage of the PV array}}{\text{Voltage rating of the single PV module}} = \frac{158}{26.3} \approx 6$$

Number of stings connected in parallel is $= \frac{\text{Terminal current of the PV array}}{\text{Current rating of the single PV module}} = \frac{46}{7.61} \approx 6$

That means to get 7.2 kW at a maximum voltage of 158 V and a maximum current of 46 A, six modules are connected in series to form a string and six such strings are connected in parallel to form the array. i.e. series-parallel configuration.

Example 12.11. A PV array consists of 9 modules connected series-parallel configuration with three modules connected in a series string and three such strings in parallel. The array is subjected to non-uniform irradiation with the modules in the string to the left being irradiated at 500 W/m^2, the modules in the central string being irradiated at 700 W/m^2 and the modules in the string to the right being irradiated at 1000 W/m^2. The array is illustrated below. The specification of the PV module used in the array is given below. Calculate the output power of the PV array given that the ambient temperature is 25 °C.

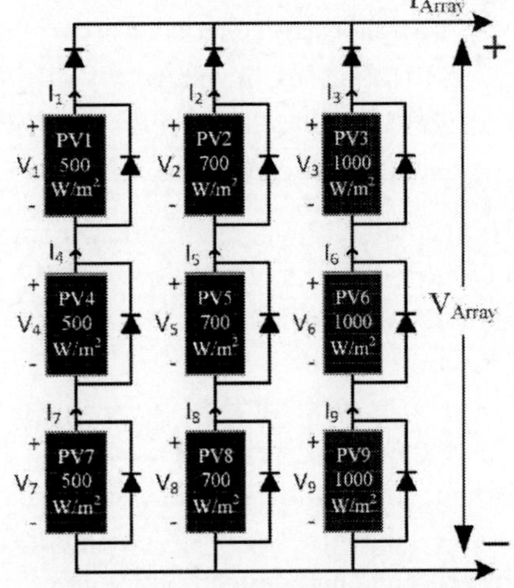

Maximum Power, P_{max}	: 200 W
Voltage at maximum power, V_{mp}	: 26.3 V
Current at maximum power, I_{mp}	: 7.61 A
Open Circuit Voltage, V_{OC}	: 32.9 V
Short Circuit Current, I_{SC}	: 8.21 A
Efficiency, η	: 18.92 %
Length of module surface	: 1660 mm
Breadth of module surface	: 1003 mm

Ans. There are a total of nine modules in the array with three modules each being irradiated at 500W/m^2, 700 W/m^2 and 1000 W/m^2, respectively. Since the irradiation in each individual string is not same, we can calculate the output power of each string separately and add it up to find the total power output from the array.

Area of module in m^2 = $1660 * 1003 = 1.66498$ m^2

Power output of each string = 3 × Input Power of the module × Efficiency of the module

Input power of each module = Irradiation on the module in W/m^2 × Area of module in m^2

Input power of the modules in the left side string = $500 \times 1.66498 = 832.49\ W$

Hence, power output of the left side string = $3 \times 832.49 \times 0.1892 = 472.52\ W$

Input power of the modules in the central string = $700 \times 1.66498 = 1165.48\ W$

Power output of the central string = $3 \times 1165.48 \times 0.1892 = 661.52\ W$

Input power of the modules in right side string = $1000 \times 1.66498 = 1664.98\ W$

Power output of the right-side string = $3 \times 1664.98 \times 0.1892 = 945.04\ W$

Total power output of the array = Sum of the power outputs of each string

$= 472.52 + 661.52 + 945.04$

$= 2079.08\ W$

Example 12.12. A PV module has an I-V characteristic having the X- axis intercept as 36V and the Y-axis intercept as 8.2 A. Calculate the short-circuit current and open circuit voltage for the

a) Series-parallel array topology with 26 modules in series and 8 strings as parallel.
b) Total-cross-tied topology with 8 rows and each row comprises of 26 modules.
c) Total-cross-tied topology with 10 strings and each string comprises of 20 modules.

Ans: (a) Short circuit current = current intercept × Number of strings in parallel = 8.2 * 8 = 65.6 A

Open circuit voltage = Voltage intercept × Number of modules in series = 36 * 26 = 936 V

(b) Short circuit current = current intercept × Number of strings in parallel = 8.2 * 26 = 213.2 A

Open circuit voltage = Voltage intercept × Number of modules in series = 36 * 8 = 288 V

(c) Short circuit current = current intercept × Number of strings in parallel = 20 * 36 = 720A

Open circuit voltage = Voltage intercept × Number of modules in series = 10 * 8.2 = 82 V

Example 12.13. A PV module generates the output voltage and current of (25V, 5A) at 1000W/m^2 and the same PV module generates (22V, 3.5A) at a particular shading condition as shown in Fig.? From the given data, what are the sources of variations in the voltage and current? Also, elaborate on the variations in voltage and current when PV modules are connected in series, and parallel with neat circuit diagrams. Assume the number of modules connected in series and parallel combination.

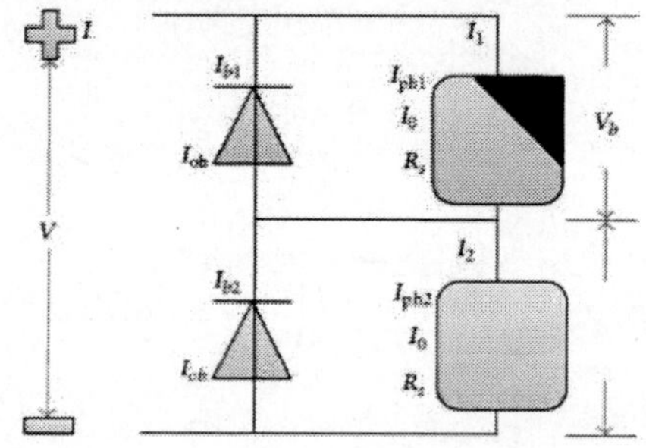

Ans: Due to shading of module-1, the irradiation on the module is less than that under standard conditions. As a result, the voltage and current produced by the module decreases. As the total voltage 'V' is the sum of the voltages produced by modules 1 and 2, the total output voltage 'V' reduces to 22V under shading from 25V under unshaded conditions. Decrease in irradiation has a greater effect on current compared to voltage and as a result, the current of the array drops by 30% from 5A under unshaded conditions to 3.5A under shaded condition.

When we connect modules in series, the net effect is that their voltages get added up while their current remains the same. This is illustrated in Fig. A1 where the I-V curves of a single module, two modules in series and three modules in series are shown. It is observed that the voltage multiplies depending on the number of series connected modules.

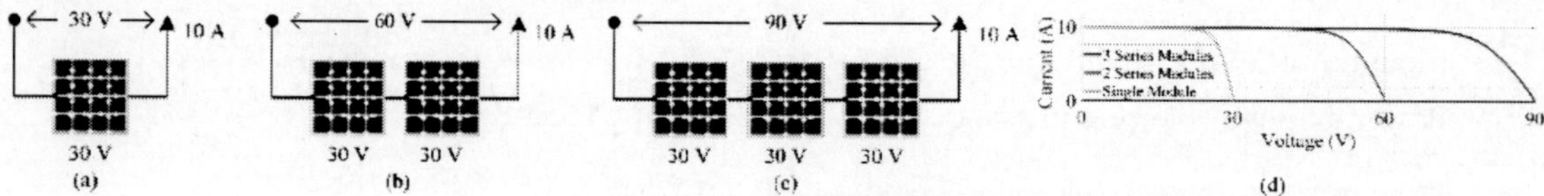

Fig. A1. Series connection of PV modules

Conversely, when modules are connected in parallel, their currents get added up while their voltage remains the same. This is illustrated in Fig. A2 where the I-V curves of a single module, two modules in parallel and three modules in parallel are shown. It is observed that the current multiplies depending on the number of parallel connected modules.

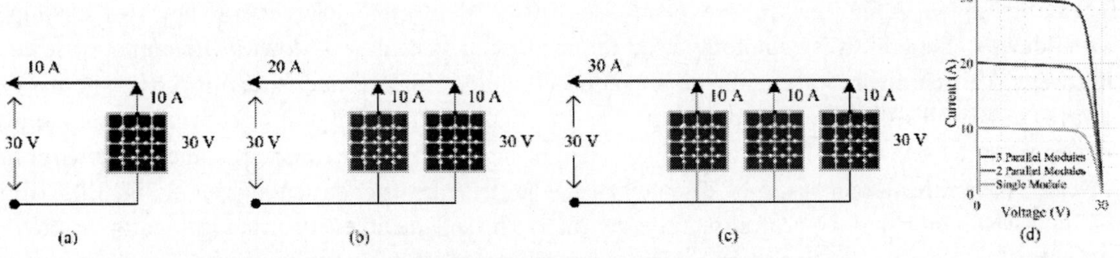

Fig. A2. Parallel connection of PV modules

12.10.4 *PV Array Reconfiguration:*

The reconfiguration method effectively addresses the issue of partial shading. During reconfiguration, the arrangement of the PV modules or the interconnections between them are modified according to the levels of irradiation, with the goal being to disperse the shade uniformly across the array. The reconfiguration techniques can be broadly categorised into two groups based on their implementation: static and dynamic. The static reconfiguration methods involve physically adjusting the position of the panels in the array during installation such that whenever shading occurs, the same would be distributed within the array. This strategy eliminates the need for extra switching devices. The static reconfiguration is done by testing the performance of the PV array under a wide range of PSCs and then using various algorithms to find the appropriate positions of the panels in the array to maximize power, such as Non-Symmetrical, Zig-Zag technique, SuDoKu puzzle-based methods, Futoshiki puzzle technique, Dominant Square (DS), Magic Square (MS), Odd-Even techniques, and Skyscraper puzzle technique, etc.,

Dynamic reconfiguration is a more effective approach to increase power generation from a photovoltaic (PV) system during partial shading conditions (PSC) because positions of panels in an array are changed dynamically based on the shading condition. In this strategy, switches are used within the PV array to alter the interconnections between PV panels, grouping PV panels based on shaded panels to prevent loss due to shading, and adjusting the series and parallel connections of panels to equalise the row current using a switching matrix. During PSC, the PV array is reconfigured dynamically using the control technique by receiving the switching conditions from the switching matrix. The different methods developed under dynamic reconfiguration methods are Electrical Array Reconfiguration (EAR), Irradiation Equalization (IE) reconfiguration, Adaptive Array Reconfiguration (AAR), and Meta-heuristic optimization algorithms.

12.10.5 *PV system Architectures:*

PV system architectures refer to the procedure of creating the layout of the PV system. Multiple PV setups possess the ability to observe and adapt to the global maximum power point (GMPP) of each individual PV module. Thus, PV systems equipped with maximum power point tracking (MPPT) for each module are better suited for circumstances with partial shadowing. Nevertheless, it is crucial to acknowledge that incorporating this functionality will lead to an increased overall expense for the system, therefore this module integrated converter topology is not implemented practically, and its use is limited to research work. The PV system topologies can hence be classified into three categories: central, string and multi-string as depicted in Fig. 12.17.

The design shown in Fig. 12.17(a) uses centralised DC-DC converter and inverter to link a large number of modules to the grid. It is commonly used for small solar systems with low-power output of several kilowatts. This configuration entails the interconnection of a single converter and inverter with the PV array. The centralised topology offers a significant advantage in terms of cost-effectiveness compared to other topologies. However, it's disadvantage is that in case of any failure or fault in the converter or the inverter, the entire system has to be disconnected. The string inverter structure seen in Fig. 12.17(b) is an improved iteration of the centralised inverter. In this arrangement, every string is directly linked to a converter and inverter, leading to an enhanced overall dependability of the systems. The input voltage may be above the threshold necessary for voltage amplification. In this specific scenario, the adverse consequences resulting from partial shade are minimised due to the ability of each string to operate at its individual maximum power point. The use of string topology improves the flexibility of the PV system by enabling the smooth integration of extra strings, hence improving its power capacity. This setup is generally used in medium power range (in Mega Watts) solar PV applications.

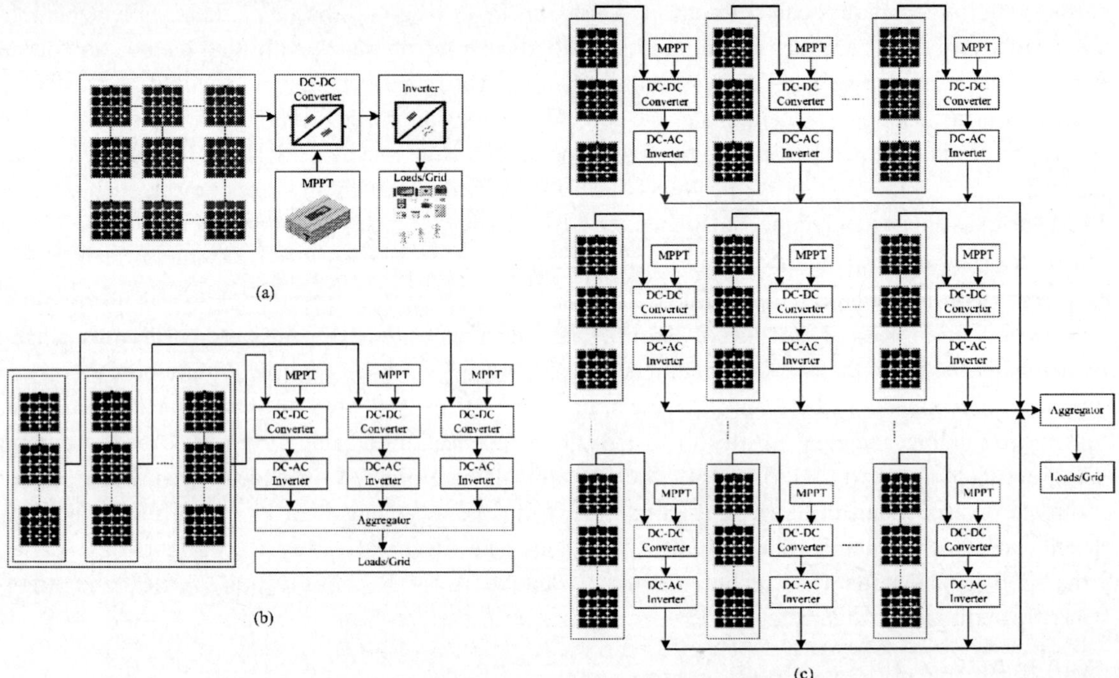

Fig. 12.17 PV system topologies: (a) Central, (b) String and (c) multi-string.

The multi-string topology shown in Fig. 12.17(c) is an improved iteration of the string topology. The system includes many strings, each linked to an individual DC-DC converter and inverter which are then connected in series and parallel combinations. This offers a distinct benefit since it enables independent control of each string and enables repair of any one string and its components without affecting power output from the rest of the system.

12.11 Maximum Power Point Tracking Techniques (MPPT): Variations in external factors, such as temperature and sun irradiation, result in fluctuations in the power output of the PV system. As previously mentioned, the phenomenon of uneven distribution of radiation is referred to as partial shading. Under the uniform condition, the P-V curve of the module has only one point, denoted as Pmax, where the module produces the maximum power. However, in cases of partial shadowing, each PV array is exposed to different levels of radiation, leading to differential power outputs throughout the arrays. In addition, the use of bypass diodes to protect shaded PV modules/arrays from the hot spot condition leads to the creation of many peaks, including one global maximum power point (GMPP) and several local maximum power points (LMPP). Consequently, there is a decrease in power generation, as illustrated in Fig. 12.15. Hence, the aforementioned causes contribute to a disparity between the installed energy capacity and the actual energy generation.

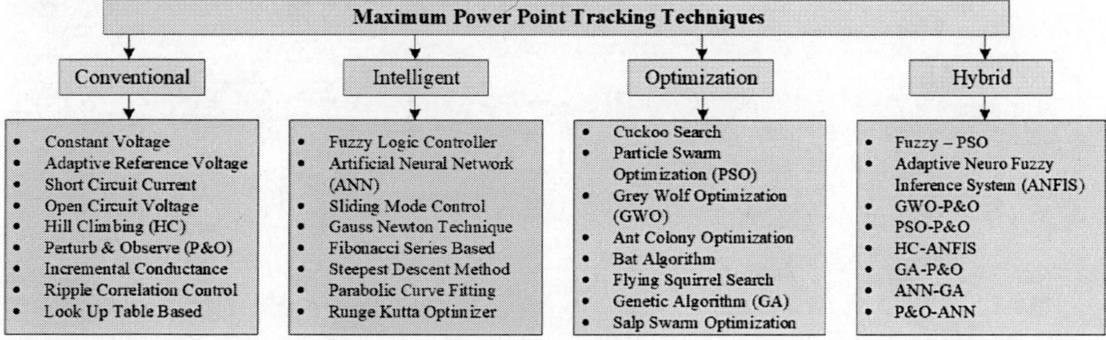

Fig. 12.18 Classification of PV MPPT techniques.

In order to mitigate the effects of PSCs and enhance power generation, researchers have focused their endeavours on the development of MPPT systems. Various MPPT techniques have been recorded in the literature to enhance the performance of PV modules at their highest power output. However, the efficacy of the particular technique depends on its ability to effectively monitor amongst fluctuating weather conditions. The PV MPPT approaches are categorised as Conventional, Intelligent, Optimisation, and Hybrid algorithms depending on their tracking characteristics. The classification, along with corresponding examples, of each sort of MPPT technique is depicted in Fig. 12.18. Traditional methods are easy to apply and can quickly identify the maximum power point in situations with consistent shading. However, they are unable to accurately track the overall highest power output in photovoltaic solar cells. Intelligent and optimization-based MPPT systems have the ability to track the global highest power output under PSCs. However, they experience fluctuations in the operating point. Hybrid MPPT techniques involve the integration of two or more MPPT techniques.

12.12 Solar Energy Applications: Solar energy systems have also found their place in a wide range of applications. A few applications are briefly mentioned here:

Solar Water Pumping: Solar energy is frequently utilised for water pumping applications in rural areas. The solar panel harnesses energy to power the pump, which elevates water from a lower to a higher level. Water pumping is utilised for many purposes such as small-scale remote irrigation, stock watering, household use, isolated communities, and marine sump pumps. The schematic depiction of solar water pumping system is shown in Fig. 12.19(a).

Solar Lighting: Solar photovoltaic lighting systems are suitable for illuminating street lights and rural areas. Compact panels have the capacity to efficiently capture sufficient energy to illuminate a street light and LEDs. An example of solar street lightning is shown in Fig. 12.19(b).

Solar PV Systems in Space: In space applications, solar PV is the fundamental source of electricity. The solar PV arrays positioned in the space station generate an excess amount of power compared to what is needed for the space station so that the excess power can be used during times when there is no sunlight. About 60 percent of the electricity produced by the solar arrays is dedicated to charging the station's batteries during periods of abundant sunlight. PV panels on spacecraft provide power firstly, to operate the sensors, active heating, cooling, and telemetry and secondly, for propulsion. The schematic depiction of usage of solar panels in space is shown in Fig. 12.19(c).

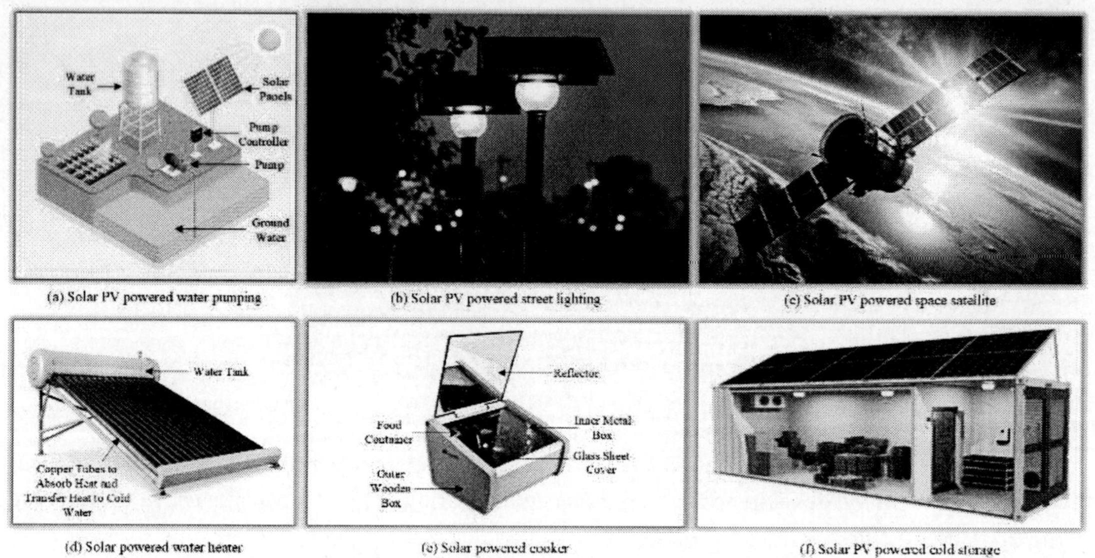

(a) Solar PV powered water pumping (b) Solar PV powered street lighting (c) Solar PV powered space satellite

(d) Solar powered water heater (e) Solar powered cooker (f) Solar PV powered cold storage

Fig. 12.19 Solar Energy Applications

Solar Water Heater: Solar water heaters and air heaters have been widely used for many decades, predating the existence of PV cells. These again use the thermal energy present in sunlight. Solar water heaters significantly minimise energy use. These devices capture and retain the thermal energy emitted by the sun, and store it as hot water in the tank. Schematic representation of a solar water heater is given in Fig. 12.19(d).

Solar Cooker: Solar cookers make use of the thermal energy present in sunlight for cooking food. Commercially available solar cookers are user-friendly and need minimal effort for operation and maintenance. India has established the world's largest solar kitchen in Taleti, located close to Mount Abu, located at an elevation of 1219 m above sea level in Rajasthan. The solar steam cooking system used there is equipped with six modules and a total of 84 parabolic dish concentrators with shell type receivers. The schematic depiction of solar cooker is shown in Fig. 12.19(e).

Solar Cold Storage: Solar energy can be utilised for both cold storage and air conditioning purposes. For these goals, solar photovoltaic panels can be utilised in a vapour compressor system, while thermal collectors can be employed in a vapour absorption system. The schematic depiction of solar cold storage is shown in Fig. 12.19(f).

12.13. Unit Summary:

- Several nations have achieved grid parity with solar photovoltaic (SPV) systems, and other strategies have been suggested to achieve complete reliance on renewable energy sources by 2050.
- SPV power generation offers several advantages: it is abundant in nature, readily accessible, produces clean and sustainable energy, has zero greenhouse gas emissions, and is environmentally benign.
- The term 'photovoltaic' is derived from the Greek terms 'Phos' (meaning light) and 'voltaic' (related to electricity), indicating the production of electricity.
- The primary purpose of the PV cell is to convert solar light energy into electricity through the photovoltaic effect.
- The mono-crystalline PV cell has an efficiency of approximately 19% to 20%, whereas the poly-crystalline PV cell has an efficiency of roughly 16% to 17%.
- Grid-interactive solar PV systems consist of six main components: a solar PV array, a charge controller, a battery bank, an inverter, a utility meter, and the energy grid.
- The three main categories of PV system are standalone, grid-connected, and hybrid PV systems.
- Partial shading condition (PSC) occurs when specific photovoltaic (PV) cells and/or modules are shaded due to various circumstances. Possible causes include the presence of tree shadows, birds, clouds, and tall buildings.
- Variables, such as temperature and irradiance, can impact the efficiency of solar PV modules.
- When the temperature rises, the power output from the PV panel decreases. However, when the amount of irradiance grows, the amount of electrical power generated by the photovoltaic panel increases.
- The methods for reducing the impact of partial shading can be classified into two primary categories: PV system design topologies and MPPT approaches.
- PV system design includes various topologies, such as bypass and blocking diodes, alternative PV system architectures, PV array configuration, and the capability to reconfigure the PV array. The strategies used for maximum power point tracking (MPPT) include traditional, intelligent, optimisation, and hybrid techniques.
- The bypass diode has two primary functions: protecting the PV module from the hot spot issue and reducing the voltage drop in the shaded cells of the modules, limiting it to the reverse voltage of the diode (0.4 - 0.7 V).
- The performance analysis of PV array topologies takes into account the following factors: mismatch power losses, fill factor, and efficiency.
- The PV system topologies can be classified into four categories: central inverter, string inverter, multi-string inverter, and micro-inverters.
- A PV array reconfiguration involves physically relocating the PV modules or adjusting the interconnections between them based on the levels of irradiation.
- The reconfiguration strategies can be classified into two main types based on their implementation: static and dynamic.
- An MPPT technique involves running the photovoltaic (PV) system at the global maximum power point, regardless of the type of partial shading circumstance.
- PV systems can be used for various applications such as power generation, water pumping, and lighting.

Short and Long Answer Type Questions

1. Briefly discuss about the different Energy Conversion Technologies. Also, mention their merits and demerits.
2. Define photo voltaic effect. Briefly explain the principle of operation of PV cell with a neat sketch.
3. Briefly explain the mathematical modelling of a PV cell with necessary equations. Also draw the different mathematical models of a PV cell.
4. How the PV system can be classified? Briefly explain the different PV system classifications with a neat sketch.
5. Discuss the following:
   (i). Distinguish by-pass diode with blocking diode in a PV module.
   (ii). Brief the reason for mismatch in solar cell/module.
   (iii). What is the reason form formation of hot spots in PV modules?
6. What is partial shading condition? How it can affect the PV system performance? Briefly explain different partial shading condition mitigation techniques.
7. What is meant by on-grid PV system? Briefly explain the working mechanism of on-grid PV system with a neat sketch.
8. Why are multiple peaks caused in the P-V characteristics of solar P-V systems under PSCs? How can you operate the PV system at global maximum power point? And explain any two methods with a neat sketch.
9. What is the need of MPPT controller in PV system? Briefly discuss the various MPPT techniques used for PV system applications.

Exercises

1. Irradiance on a PV module (1.5 m x 2.0 m) is 700 W/m^2. If the efficiency of the cells is 14 %, what is the power output of the module?
2. A consumer is in need of a 2-kW solar photovoltaic energy system with the output voltage of 120 V. For the design of PV system, he uses PV modules whose output voltage is 10 V and output current is 1.66 A at maximum power point condition. Design the suitable PV array topology and draw the approximate output characteristics.
3. A solar PV panel generates a global maximum power of 800 W at standard test conditions (W=1000 W/m^2 and T=25^{0}C). A shadow of tall tree is falling on the PV panel which reduces the power generation capability of PV panel. Due to the shadow of tall tree the PV panel output is around 200 W then calculate the mismatch power loss in the system.
4. Design a PV array configuration which is generating the maximum current of 121.76 A at maximum power of 3.2 kW. Also, mention what kind of configuration is it? Draw the same. Use the following PV module specifications.

Maximum Power, P_{max}	: 200 W
Voltage at maximum power, V_{mp}	: 26.3 V
Current at maximum power, I_{mp}	: 7.61 A
Open Circuit Voltage, V_{OC}	: 32.9 V
Short Circuit Current, I_{SC}	: 8.21 A

5. Design a 16.3 kW PV array configuration which can generate a maximum voltage of 392 V and the maximum current of 41.68 A at STC. Also, mention what kind of configuration is it? Draw the same. Use the following PV module specifications.

Maximum Power, P_{max}	: 255.29 W
Voltage at maximum power, V_{mp}	: 49 V
Current at maximum power, I_{mp}	: 5.21 A
Open Circuit Voltage, V_{OC}	: 59.8 V
Short Circuit Current, I_{SC}	: 5.55 A

To know more about

Hybrid and Conventional PV Array Configurations

To know more about

Static Reconfiguration Strategies

To know more about

Hybrid, optimal, intelligent and classical PV MPPT techniques

To know more about

Inverter topologies for PV system architectures and Novel H5, H6, HERIC Transformer-less Inverter topologies

To know more about

How to write a Research Paper, Smart Electric Grid, and Simulation of PV Configurations

MATLAB/SIMULINK

To Model & Simulate

PV Home On-Grid Solar System, and 2-MW PV Farm Connected to a 25-kV Distribution System

13 WIND ENERGY

Unit specifics: In this unit, the following topics have been discussed for basic understating of wind energy:

- Evolution of wind energy utilization, Efficiency limit of wind energy.
- Aerodynamics of wind rotor, Working of wind power plant
- Characteristics curves: Power speed and torque speed curves of wind turbine.
- Conversion of energy from mechanical to electrical.
- Type of wind turbines, Application of wind energy.
- Advantages and disadvantages of horizontal axis wind turbine and vertical axis wind turbine.
- Numericals related to the wind energy.

Rationale: In this unit, students will be introduced to the evolution of the wind energy system, its efficiency limit for power conversion, aerodynamics of wind turbines including control of turbine, cut-in and cut-out speed; Working of wind power plant, characteristics of wind turbine: torque speed and power speed, characteristics of wind power plant, types of generators used in energy conversion, induction generator, synchronous generator, types of wind turbines: HAWT, VAWT, small turbines; electrical torque generation in induction generator and applications of wind energy are clearly described with the help of necessary diagrams, derivations, and examples.

Pre-Requisites: Knowledge of basic electrical engineering.

Unit Outcomes: The list of outcomes of this unit is as follows:

U13-O1: To understand the evolution of wind energy.
U13-O2: To understand the efficiency limit for wind energy conversion.
U13-O3: Analyse the equivalent circuit diagram of induction generator and synchronous generator.
U13-O4: To understand the working and characteristics of wind turbine.
U13-O5: To analyse the concept of power flow in induction generator.
U13-O6: To acquire knowledge about the various wind turbines and their applications.

Unit-13 outcomes	**Expected mapping with course outcomes** (1: Weak, 2: medium, and 3: strong correlation)					
	CO-1	**CO-2**	**CO-3**	**CO-4**	**CO-5**	**CO-6**
U13-O1	3	2	-	-	2	-
U13-O2	2	2	-	-	3	2
U13-O3	2	2	2	-	3	2
U13-O4	3	3	-	-	3	-
U13-O5	2	2	-	-	3	2
U13-O6	3	2	-	-	3	-

13.1 Introduction to Wind Energy:

Wind energy harnesses the power of wind to generate electricity, offering a renewable and sustainable alternative to fossil fuels. Utilizing wind turbines, which convert kinetic energy from the wind into mechanical power and subsequently into electrical energy, wind energy systems have seen significant advancements in technology and efficiency. This chapter delves into the intricacies of wind energy, beginning with an overview of the different types of wind turbines, including horizontal-axis and vertical-axis designs, each with unique characteristics and applications. We will explore the working principles of these turbines, elucidating how aerodynamic forces and rotor dynamics are harnessed to generate power. The wind power generation and sharing in India, state-wise shown in Fig. 13.1. It shows that Gujarat has the maximum generation of wind energy. It shares 25.87% of total generation as of 31-01-2025. Tamil Naidu holds second position with 23.66% power generation.

Further, the chapter will examine the various types of generators used in wind energy systems, such as synchronous and asynchronous generators, highlighting their operational principles and suitability for different scales of wind energy applications. The characteristics of wind energy, including its intermittency, capacity factors, and the impact of site selection, will be discussed to provide a holistic understanding of the challenges and opportunities within the wind energy sector.

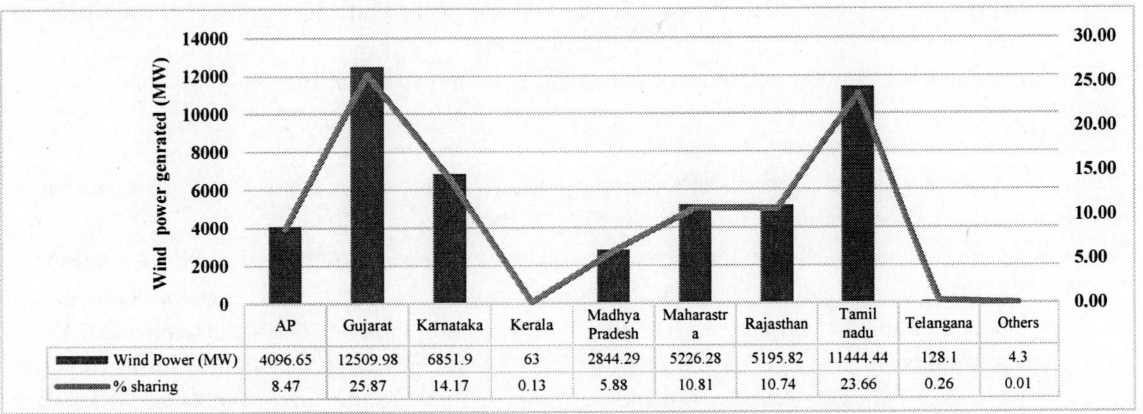

	AP	Gujarat	Karnataka	Kerala	Madhya Pradesh	Maharastr a	Rajasthan	Tamil nadu	Telangana	Others
Wind Power (MW)	4096.65	12509.98	6851.9	63	2844.29	5226.28	5195.82	11444.44	128.1	4.3
% sharing	8.47	25.87	14.17	0.13	5.88	10.81	10.74	23.66	0.26	0.01

Fig. 13.1 Bar graph of state-wise wind power generation in India as on 31-01-2025. (https://mnre.gov.in/en/physical-progress/)

13.2 Evolution of wind energy:

The evolution of wind energy is a fascinating story of technological innovation and increasing efficiency. Here's a brief overview of how wind energy has developed over time:

(i) Early uses and concepts:

- Ancient Times: Wind energy has been harnessed since ancient civilizations. The earliest known use was in ancient Persia (modern-day Iran) around 500-900 AD with the construction of vertical-axis windmills for grinding grain and pumping water.
- Medieval Europe: In the Middle Ages, horizontal-axis windmills were developed in Europe, particularly in the Netherlands and England. These were used for similar purposes as their predecessors—pumping water and milling grain.

(ii) Industrial revolution:

- **19th Century**: During the Industrial Revolution, windmill technology saw some improvements, but its use remained relatively limited compared to other forms of energy like coal and steam. The focus was more on industrial applications rather than electricity generation.

(iii) Early 20th Century:

- **1900s**: The early 20th century saw some pioneering work in wind turbine design. In 1931, the Danish engineer Poul la Cour developed a more efficient wind turbine, which included the use of aerodynamic blades. His work laid the groundwork for future advancements in wind technology.

(iv) Post-World War II:

- **1950s-1970s**: After World War II, there was a growing interest in renewable energy sources due to increasing concerns about fossil fuel depletion and environmental issues. During this period, there was some development in wind turbine technology, but progress was slow and intermittent.

(v) Modern Era:

- **1980s-1990s**: The energy crises of the 1970s spurred renewed interest in renewable energy. The development of more efficient wind turbines began, with significant advances in materials and aerodynamics. The first modern wind farms started to appear, and research institutions and companies began focusing on improving turbine design and efficiency.

(vi) 21st century:

The early 2000s saw rapid growth in wind energy technology, driven by concerns about climate change, government incentives, and advances in technology. Modern wind turbines are much larger and more efficient than their predecessors as illustrated in Fig. 13.2. Advances include:

- **Larger turbines**: Modern turbines can reach heights of over 150 meters and have blades that can span more than 80 meters. The largest wind turbine installed in India is by the Adani group and has a radius of 80 meters. It is located in Mundra, Gujarat. As of September 2024, the world's largest wind turbine is the MySE 18.X-20MW, which is located in Hainan, China. It has a rotor radius of around 130m and can generate up to 20 MW of power. Table 13.1 shows how significant a role is played by the length of the wind blades in the amount of wind power captured.
- **Offshore wind farms**: The development of offshore wind farms has become a major area of growth, with floating turbines allowing for installations in deeper waters where wind speeds are higher and more consistent. Windmills located on land also cause constantly flickering shadows on the ground and noise which can get irritating. Due to these constraints, large windmills are often installed offshore, although this can be considerably more expensive. Amongst the offshore installations, windmills located near the shore and in shallow water areas have their base foundation installed in the bed of the water body. For deep sea installations, it is easier to make a floating installation for the windmills. Different types of windmill installations are illustrated in Fig. 13.3.
- **Integrated systems**: Innovations in control systems, materials science, and energy storage have improved the efficiency and reliability of wind energy systems.

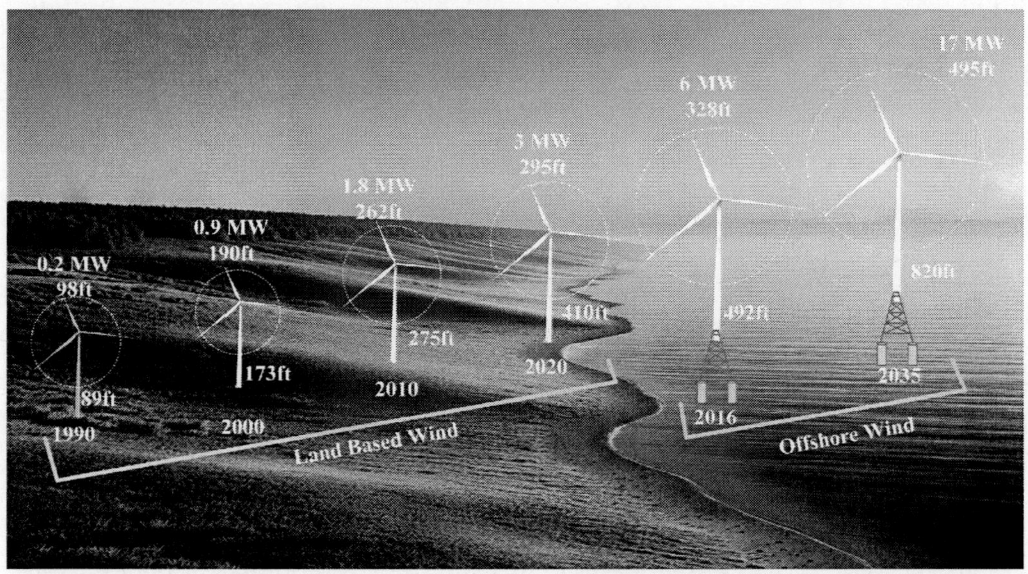

Fig. 13.2 Evolution of windmill sizes over the years on-shore and off-shore [https://www.energy.gov/eere/articles/wind-turbines-bigger-better]

Table 13.1 Variation in wind power captured with change in length of blades of a wind turbine when

$$Air\ density\ (\rho) = 1.225\frac{Kg}{m^3},$$

$$and\ wind\ velocity\ (v) = 10\ m/sec$$

Length of the blades (m) = r	Area covered by blades (sq.m) $A = \pi r^2$	Wind Power captured (kW) $P = 0.5 * \rho.A.v^3$
5	78.53	48.10
10	314.16	192.42
20	1256.63	769.68
60	11309.73	6927.21
100	31415.92	19242.25
130	53092.91	32519.40

Fig. 13.3 Different types of windmill installations: on the ground, on the bed of the sea, floating in the sea

13.3 Efficiency limit for wind energy conversion:

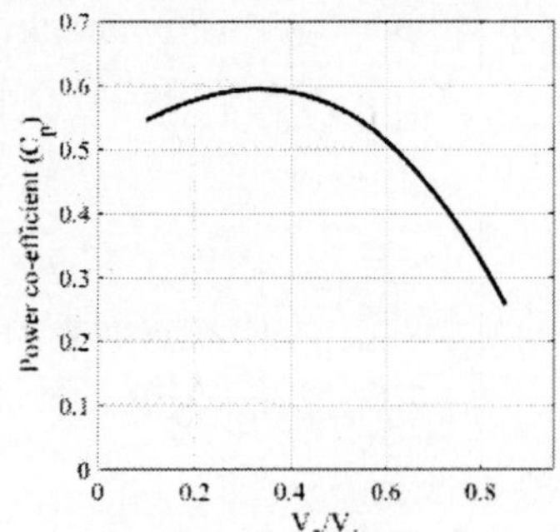

Fig. 13.4 Graph between power co-efficient and speed ratio.

The efficiency limit for wind energy conversion is primarily governed by the **Betz limit**, which states that no wind turbine can capture more than 59.3% (or 16/27) of the kinetic energy in wind. The factor (16/27) or 59.3% is also called **Betz coefficient.** This limit is based on the principle that a wind turbine cannot capture all the wind's energy because if it did, the wind would have to stop completely behind the turbine, preventing more wind from flowing through. Fig. 13.4 shows the graph plotted between power co-efficient and speed ratio of a wind turbine. It shows that the power coefficient is maximum at velocity ratio, 1/3. In practical terms, modern wind turbines achieve efficiencies of 30-50% due to various losses (aerodynamic, mechanical, and electrical). The Betz limit is derived from the principles of conservation of mass and momentum for a stream of air passing through a wind turbine. Consider a wind turbine with an incoming wind speed v_1 and an outgoing wind speed v_2. The area swept by the turbine blades is A m^2.

The mass inflow rate of wind coming to turbine is given by:

$$M_i = \rho A v_1 \quad \text{.........(13.1)}$$

Where, ρ is the air density.

The power available with wind entering the blades of turbine can be given by:

$$P_{wind} = \frac{1}{2}(mass\ inflow\ rate).v_1^2 = \frac{1}{2}.\rho A v_1^3 \quad Watt/m^2 \quad \text{.........(13.2)}$$

The average mass flow rate of wind through turbine blades can be written as:

$$M_{avg} = \rho A v_{avg} \quad \text{.........(13.3)}$$

Where v_{avg} represents average wind speed in the turbine and is equal to:

$$v_{avg} = \frac{v_1 + v_2}{2} \quad \text{.........(13.4)}$$

The power extracted $P_{extracted}$ from the wind is the difference of kinetic energy associated with incoming wind and outgoing wind from turbine and can be given by:

$$P_{extracted} = \frac{1}{2}(average\ mass\ flow\ rate).(v_1^2 - v_2^2) = \frac{1}{2}\rho A v_{avg}(v_1^2 - v_2^2) \quad \text{......(13.5)}$$

From eq. 13.4 & 13.5, $P_{extracted} = \frac{1}{2}\rho A(v_1^2 - v_2^2)\left(\frac{v_1+v_2}{2}\right) \quad Watt/m^2$ ....(13.6)

The power coefficient C_p is defined as the ratio of extracted power to the power available with incoming wind $\left(i.e.\ C_p = \frac{P_{extracted}}{P_{wind}}\right)$.

From equations (13.2) and (13.6), $C_p = \frac{P_{extracted}}{P_{wind}} = \frac{\frac{1}{2}\rho A(v_1^2-v_2^2)\left(\frac{v_1+v_2}{2}\right)}{\frac{1}{2}\rho A v_1^3}$ ....(13.7)

$$C_p = \frac{\frac{1}{2}(v_1^2-v_2^2)(v_1+v_2)}{v_1^3} = \frac{1}{2}\left[1 - \left(\frac{v_2}{v_1}\right)^2\right]\left(1 + \frac{v_2}{v_1}\right) C_p = \frac{1}{2}(1 - x^2)(1 + x) \quad \text{.........(13.8)}$$

Where $x = \frac{v_2}{v_1}$

In order to extract maximum power from the wind, $\frac{dC_p}{dx}$ has to be zero.

From equation (13.8), $\frac{dC_p}{dx} = \frac{d}{dx}\left[\frac{1}{2}(1 - x^2)(1 + x)\right] = 0$

$$= \frac{d}{dx}\left[\frac{1}{2}(1 + x - x^2 - x^3)\right] = \frac{1}{2}(1 - 2x - 3x^2) = \frac{1}{2}(1 - 3x)(1 + x) = 0$$

After solving the above equation we get, $x = -1\ and\ x = \frac{1}{3}$ ....(13.9)

x = -1 represents infeasible solution as it makes outgoing wind speed negative of incoming wind speed resulting in C_p to become zero.

By putting the value of $x = \frac{1}{3}$ *in equation* (13.8)

$$C_p = \frac{1}{2}(1 - x^2)(1 + x) = \frac{1}{2}\left(1 - \frac{1}{9}\right)\left(1 + \frac{1}{3}\right) = \frac{16}{27} = 0.593 \quad \text{.........(13.10)}$$

Equation (13.10) indicates that maximum power that can be extracted from the wind is 0.593 times the available power. This limit to extract power is known as Betz limit which represents maximum theoretical efficiency of a wind turbine.

13.4 Aerodynamics of wind rotors:

The aerodynamics of wind rotors is a critical factor in determining the efficiency and performance of wind turbines. Wind rotors are designed to capture the kinetic energy of the wind and convert it into mechanical energy, which is then transformed into electrical energy. The blades of a wind turbine are shaped such that the upper and lower surfaces have different curvature as shown in Fig. 13.5 (a). This results in the air moving across the upper surface of the blade having to cover a greater distance than the air moving across the lower surface of the blade. This causes a difference in pressure between the top and bottom portions of the blade with the upper portion having a low-pressure area and the lower portion having a high-pressure area. This leads to lift being created which makes the blade move from the high-pressure to the low-pressure area thereby creating rotatory motion of the turbine. Here's a detailed look at the key aerodynamic principles involved in wind rotor design:

(i) Blade Air foil Design: Wind turbine blades are shaped like air foils (similar to airplane wings) to maximize lift and minimize drag. The air foil shape allows the rotor to capture wind energy efficiently by exploiting differences in air pressure. The definitions of lift and drag forces are given below. The goal is to maximize the lift-to-drag ratio, which enhances the efficiency of energy conversion.

- Lift is the force that acts perpendicular to the direction of the wind flow. It is generated by the pressure difference on the upper and lower surfaces of the blade.
- Drag is the force that acts parallel to the wind flow and opposes the motion of the blade.

(ii) Angle of Attack: The angle of attack is the angle between the incoming wind and the chord line of the blade (the straight line connecting the leading and trailing edges of the blade). This is illustrated in Fig. 13.5(b).

- **Optimal Angle:** For most wind turbine blades, the optimal angle of attack is around 5 to 15 degrees, where the lift force is maximized, and drag is minimized.
- **Stall:** If the angle of attack is too high, the airflow can separate from the blade surface, causing a stall, which drastically reduces lift and increases drag, leading to a loss of efficiency.

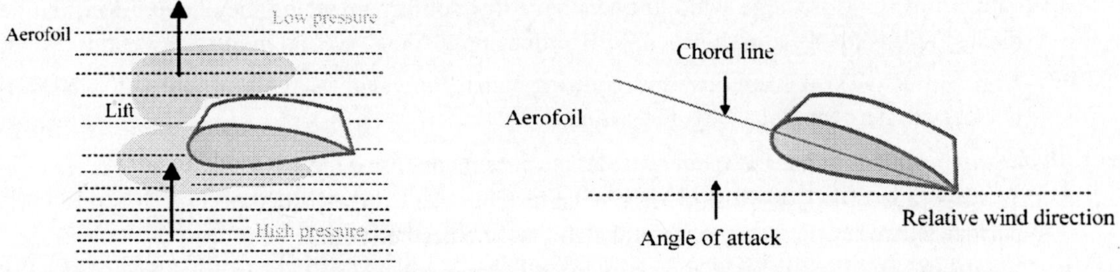

Fig. 13.5 (a) Creation of low- and high-pressure zones around the blade. (b) Illustration of angle of attack

(iii) Blade Twist and Taper: The shape of the blade is not uniform along its length; it is twisted and tapered. It is illustrated in Fig. 13.6 how the shape, thickness and angle of attack of the blade of the turbine varies across its length. It is designed in such a way that optimum wind power is captured by every part of the blade as when we go from the hub towards the tip of the blade, the angular speed of the blade surface increases. This is illustrated in Fig. 13.7 where a 25 m blade length is considered to be rotating at 15 rpm. The section of the blade labelled 'A' is located at a distance of 1.5m from the centre and has a tangential speed of 2.355 m/s while the section labelled 'B' of the same blade located at the tip and 25m from the centre has a much higher tangential speed of 39.26 m/s. As a result of this variation in speed, the optimum angle of attack also varies for both the sections. Table 13.2 illustrates how the tip speed of the wind turbine blade is different from the speed of the inner part of the blade located 2m from the centre for different turbine sizes and the same rotating speed of 15 rpm.

- **Twist:** The blade is twisted so that the angle of attack is optimized along the entire length of the blade. This is necessary because the wind speed relative to the blade increases with distance from the hub due to the rotational speed.
- **Taper:** The blade is also tapered, meaning it is wider at the base (near the hub) and narrower at the tip. This design reduces the mass of the blade and optimizes the distribution of aerodynamic forces, reducing bending moments and improving structural efficiency.

(iv) Tip Speed Ratio (TSR): TSR is the ratio of the blade speed to the wind speed.

$$TSR = \frac{\text{Blade speed}}{\text{Wind speed}} \qquad \ldots\ldots\ldots (13.11)$$

- **Optimal TSR:** Each turbine design has an optimal TSR where the rotor operates most efficiently. If the TSR is too high, drag forces increase, reducing efficiency. If too low, the rotor does not capture sufficient wind energy.
- **Solidity:** Solidity is defined as the ratio of the projected blade area of the wind rotor to the area of the wind intercepted by the rotor. Mathematically, it is expressed as:

$$\sigma = \frac{A_{blade}}{A_{wind}} \qquad \ldots\ldots\ldots (13.12)$$

Where: A_{blade} is the total projected area of the rotor blades and A_{wind} is the area of the wind intercepted by the rotor.

(v) Rotor Solidity: Solidity is the ratio of the total blade area to the swept area of the rotor. It affects how much wind the rotor can capture and the overall performance.

- **Low Solidity:** Most large wind turbines have low solidity, meaning they have fewer, thinner blades. This is ideal for high-speed, low-torque applications, typical in modern wind turbines.
- **High Solidity:** Used in applications requiring high torque, such as in water-pumping windmills, but less efficient for electricity generation.

(vi) Blade Number: The number of blades affects the aerodynamics of the rotor.

- **Three-Bladed Design:** Most modern wind turbines use a three-bladed design. This provides a balance between efficiency, structural stability, and aesthetic considerations.
- **Single and Double Blades:** Fewer blades reduce drag and weight but can cause more vibrations and lower efficiency.
- **Multi-Blade Designs:** More blades can capture more wind but increase drag and structural complexity, making them less efficient for large-scale electricity generation.

Table 13.2 Comparison of tangential speed of different sized blades close to the centre and at the tip of the blades when wind turbine is rotating at 15 rpm

Length of the blade (m)	Tangential velocity of the part 1.5m away from the center of the blades (m/s)	Tangential velocity of the blade tips (m/s)
25	2.355	39.26
50	2.355	78.53
75	2.355	117.80
100	2.355	157.06
125	2.355	196.34
150	2.355	235.5

If wind turbine is rotating at 15 rpm and length of the blade is 25m; then the tangential velocity of the blade can be calculated as = $2\pi r * rev/sec$

$$= 2\pi * 25 * \frac{15}{60} = 39.26\ m/sec$$

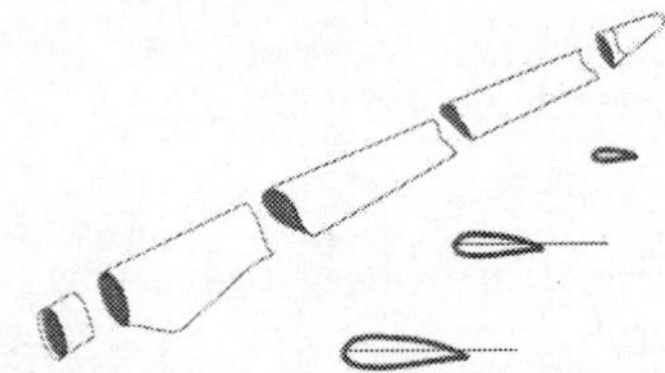

Fig. 13.6 Variation of shape, thickness, twist and taper along the length of a wind turbine blade

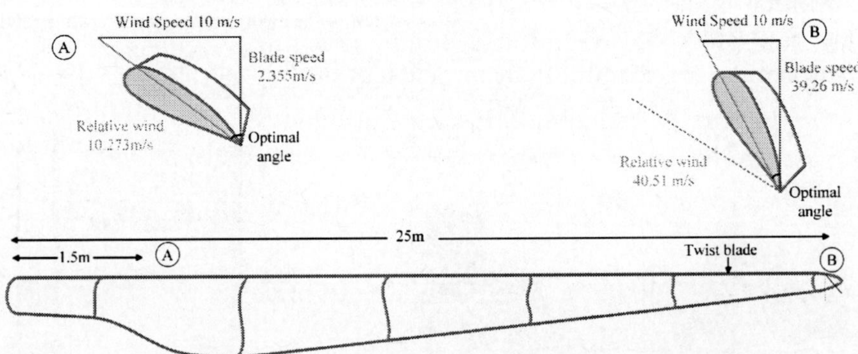

Fig. 13.7 Variation of tangential speed and angle of attack of the blade along the length of the blade

(vii) Yaw Control: Yaw control involves rotating the entire rotor assembly around the vertical axis to keep the blades facing into the wind.

- Active Yaw Control: Most modern turbines have an active yaw control system that continuously adjusts the rotor orientation to maximize energy capture.
- Passive Yaw Control: Some older or smaller turbines use passive systems, relying on wind vane mechanisms, which are less precise.

(viii) Blade Materials: Modern wind turbine blades are typically made of composite materials such as fiberglass-reinforced polyester or carbon fibre. These materials offer a good balance of strength, weight, and flexibility, which is essential for aerodynamic efficiency and durability.

13.5 Working of wind power plant:

The key characteristics and components of a wind power plant are illustrated in Fig. 13.8 and discussed below.

13.5.1 Wind Turbine Components:

Rotor Blades: Capture the kinetic energy of the wind and convert it into rotational energy. The length and design of the blades affect the efficiency and power output. They are made using reinforced glass fibre making them strong and light enabling them to be of large size without breaking. Previously, metal and wooden blades were also being used but they are too heavy and also prone to corrosion in the case of metal blades and rotting in the case of wooden blades. Heavy blades also make it difficult for them to start and stop quickly due to increased inertia.

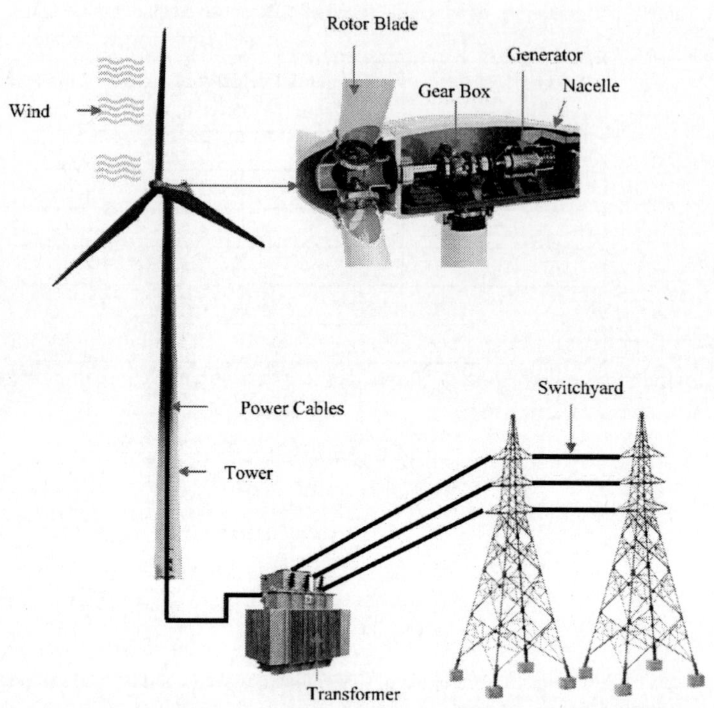

Fig. 13.8 Overview of the main components of a wind power plant

Fig. 13.9 Role of a gearbox in a wind turbine

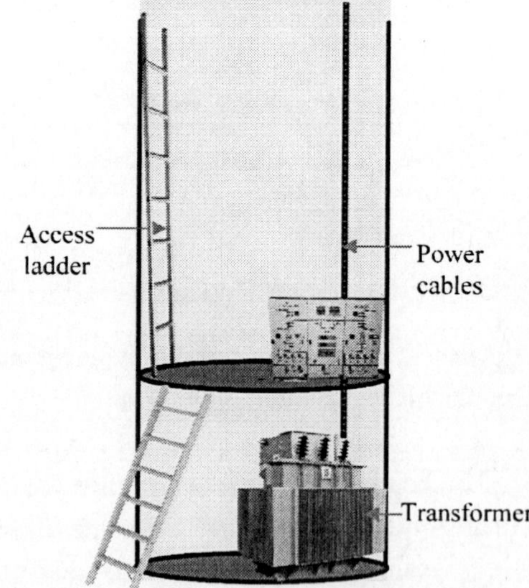

Fig. 13.10 Base part of the tower on which the wind turbine is mounted

Nacelle: The bed plate, generator, gearbox, electrical control panel and hydraulic control panel along with other associated components are covered with a fiberglass housing which forms the nacelle. This protects the components from the weather elements. The nacelle also has a wind vane to determine the direction of the wind and an anemometer to determine the wind speed mounted on it.

Gearbox: It is used to convert the low speed, high torque mechanical power from the turbine on the main shaft to a high speed, low torque input for the electrical generator. For example, a wind turbine rotating at 15 rpm would require a gearbox with 1:100 gear ratio to drive a 4-pole generator at 1500 rpm to generate electricity at 50 Hz. This is illustrated in Fig. 13.9.

Generator: Converts mechanical energy into electrical energy. It can be an induction generator, synchronous generator, or permanent magnet synchronous generator.

Tower: The hollow structure on which the wind turbine and all other components of the windmill are mounted. There is a ladder inside the tower using which the wind turbine, generator and other components can be accessed at the top. At the base of the tower, a transformer and electrical control panel are usually installed which help in controlling power flow to the grid from the windmill as shown in Fig. 13.10. The height of the tower affects the amount of wind energy captured, with taller towers generally being more efficient.

Bed plate: It is the structure at the top of the steel tubular tower on which components such as the wind turbine, generator, gearbox, electrical control panel, hydraulic control panel and other associated components are mounted. A set of yaw motors are used to rotate the bed plate and hence the turbine in the direction of the wind.

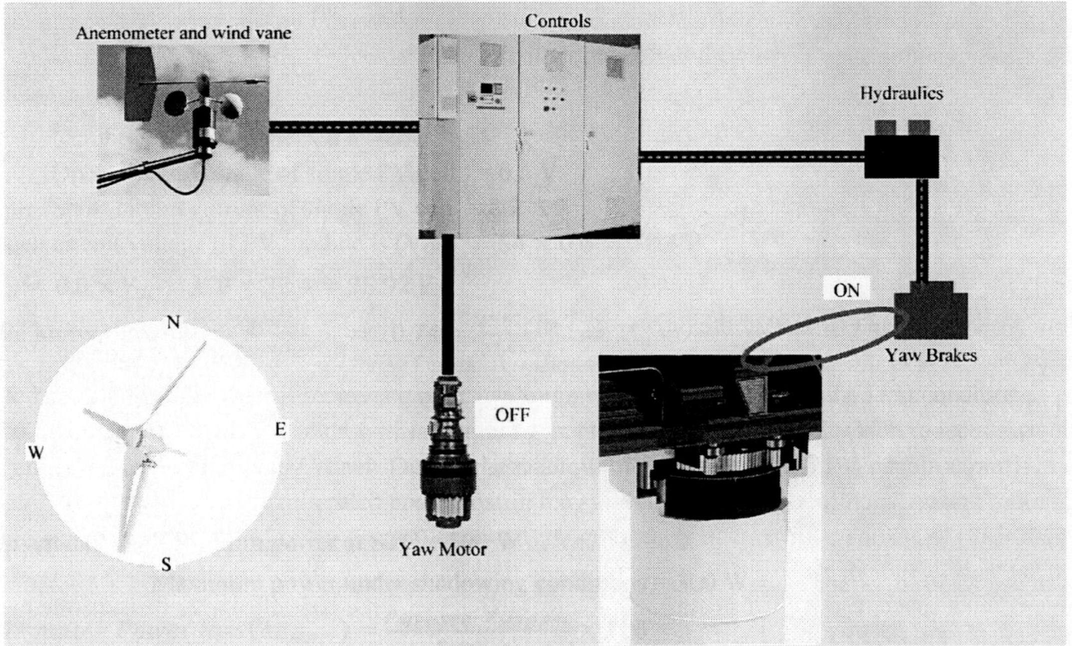

Fig. 13.11 Yaw control mechanism

Fig. 13.11 shows how the direction of the wind turbine is adjusted based on input from the wind vane using hydraulic controls. The yaw motor is used to move the bed plate on which the turbine is mounted, and the yaw brakes are used to stop the bed plate from moving.

Yaw and Pitch Control System: Yaw control system adjusts the direction of the turbine to face the wind while the pitch control system adjusts the angle of the rotor blades to control the turbine's speed and optimize performance. The wind power captured by the wind turbine can be controlled by adjusting the blade pitch angle as it changes the angle of attack of the wind on the blades. The blade pitch angle therefore plays a critical role in deciding both, the power captured and the speed of the wind turbine. The wind turbine needs to rotate at a constant speed to maintain the frequency of electricity generated constant, although the wind speed may vary.

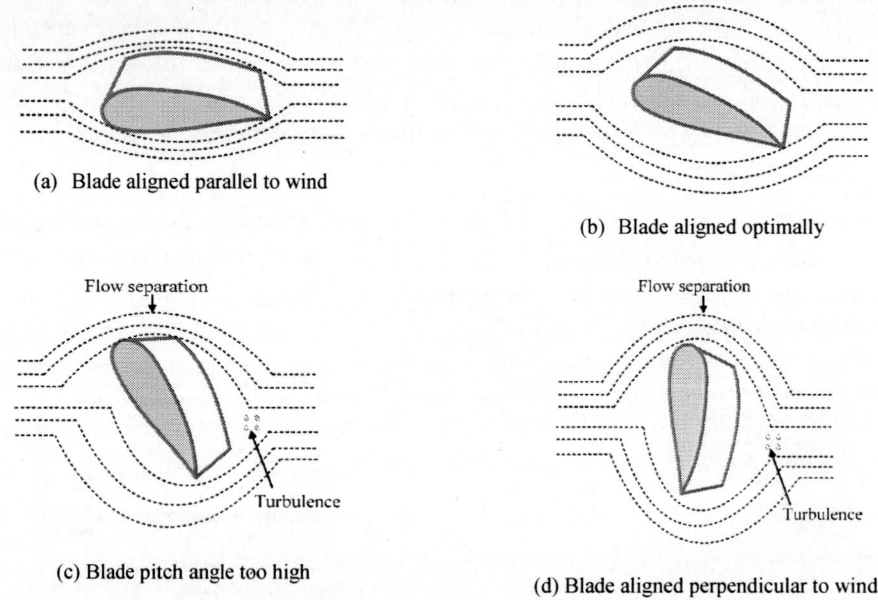

Fig. 13.12 Different blade pitch angles and their effect on generating lift and causing turbulence

If the blade is aligned parallel to the direction of wind flow as shown in Fig. 13.12(a), the lift created is negligible because there is not enough difference of pressure created between the upper and lower areas of the blade to create rotation. Conversely, if the blade is aligned perpendicular to the wind direction as shown in Fig. 13.12(d) or having a blade pitch angle too high as shown in Fig. 13.12(c), there is a lot of turbulence and drag created by the wind as it passes the blade which is again not ideal for capturing optimum power. Therefore, what we desire is an optimum blade pitch angle as shown in Fig. 13.12(b) which leads to sufficient difference in air pressure between the upper and lower portions of the blade and at the same time doesn't cause any turbulence in the wind. This leads to optimum capturing of wind power. Wind speed data is fed from the anemometer to the hydraulic controller which adjusts the blade pitch angle of the wind turbine for optimum power and speed.

- **Control System:** Monitors and manages the operation of the turbine, including start-up, shut-down, and performance optimization.

13.5.2 Operational Characteristics:

- **Wind Speed:** Wind turbines require a certain minimum wind speed to operate efficiently (cut-in wind speed). They also have a cut-out wind speed, beyond which they shut down to prevent damage.
- **Capacity Factor:** The ratio of actual output to the maximum possible output if the turbine operated at full capacity all the time. It varies with wind conditions and turbine efficiency.
- **Efficiency:** Wind turbines are generally 30-50% efficient at converting wind energy into electrical energy. The overall efficiency of a wind farm depends on turbine performance, wind conditions, and layout.

13.5.3 Performance Metrics:

- **Rated Power Capacity:** The maximum electrical power output the turbine can produce under optimal wind conditions. This is typically between 1 MW and 3 MW for onshore turbines and higher for offshore turbines.
- **Cut-in Wind Speed:** The minimum wind speed at which the turbine starts generating electricity, usually around 3-5 m/s (6.7-11.2 mph).
- **Rated Wind Speed:** The wind speed at which the turbine generates its rated power, typically around 12-14 m/s (26.8-31.3 mph).
- **Cut-out Wind Speed:** The wind speed at which the turbine shuts down to avoid damage, usually around 25 m/s (56 mph).

13.5.4 Environmental and Site Characteristics:

- **Wind Resource:** The availability and consistency of wind at the site. Wind power plants are typically located in areas with high average wind speeds.
- **Topography:** The physical features of the land, which can affect wind patterns and turbine placement.
- **Noise:** Wind turbines generate noise from the blades and gearbox, which can be a concern for nearby communities.
- **Wildlife Impact:** Turbine placement and operation can impact local wildlife, particularly birds and bats. Environmental assessments are conducted to mitigate negative effects.

13.5.5 Types of Wind Power Plants:

- **Onshore Wind Farms:** Located on land. They are more common and less expensive but can be limited by land availability and local regulations.
- **Offshore Wind Farms:** Located in bodies of water. They benefit from higher and more consistent wind speeds but involve higher costs and technical challenges related to installation and maintenance.

13.6 Power-Speed and Torque-Speed Characteristics:

The power-speed and torque-speed characteristics of a wind turbine are crucial for understanding its performance and behaviour under different wind conditions. These characteristics help in designing the turbine's control systems and selecting appropriate generators. Here's an overview:

13.6.1 Power-Speed Characteristics: The power output of a wind turbine is a function of the wind speed and is given as:

$$P_{extracted} = P_{wind} * C_p = \frac{1}{2}\rho A {v_1}^3 * C_p \quad \text{......... (13.13)}$$

as explained in section 13.3.

The rotor speed in m/sec is equal to the product of its angular speed, ω in radians/sec and the radius, r of the rotor. TSR (λ) is defined in Section 13.4 as the ratio of blade speed (i.e. the rotor speed) and wind speed

$$\text{i.e. } TSR\ (\lambda) = \frac{\omega.r}{v} \quad \text{......... (13.14)}$$

It is observed from equation (13.14) that TSR is proportional to rotor RPM and inversely proportional to wind speed. The power co-efficient, C_p is a function of TSR (λ) and blade pitch angle, β. A generic model of $C_p(\lambda, \beta)$ obtained based on turbine characteristics is given as:

$$C_p(\lambda, \beta) = C_1(C_2 - C_3\beta - C_4\beta^2 - C_5)e^{-C_6} \quad \text{......... (13.15)}$$

Where, the coefficients are, $C_1 = 0.5, C_2 = \frac{116}{\gamma}, C_3 = 0.4, C_4 = 0, C_5 = 5, C_6 = \frac{21}{\gamma}$

Where, $\frac{1}{\gamma} = \frac{1}{\lambda + 0.08\beta} - \frac{0.035}{1+\beta^3}$

Plots of power output vs. rotor speed of the wind turbine for different wind speeds are shown in Fig. 13.13(a), whereas, plots of power co-efficient, C_p Vs. TSR, λ for different blade pitch angles, β are shown in Fig. 13.13(b). These curves are obtained using equations (13.13) to (13.15).

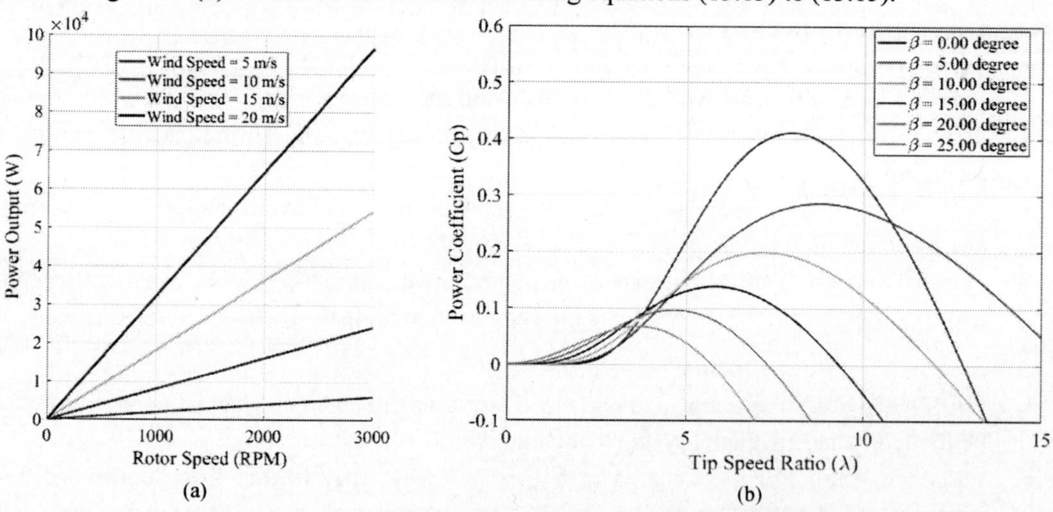

Fig. 13.13 (a) Power output Vs rotor speed of the wind turbine. (b) Power co-efficient Vs tip speed ratio.

13.6.2 Power Vs Wind Speed:

- **Cut-in Speed:** At very low wind speeds (below the cut-in speed, typically 3-4 m/s), the turbine does not generate any power because the wind energy is insufficient to overcome mechanical friction and inertia.

- **Rated Speed:** As the wind speed increases, the power output increases rapidly (since power is proportional to the cube of the wind speed) until it reaches the rated wind speed. At this point, the turbine generates its maximum rated power.
- **Cut-out Speed:** If the wind speed exceeds a certain limit (cut-out speed, typically 25 m/s), the turbine is usually shut down to prevent damage from excessive mechanical loads. Fig. 13.14 shows the curve of turbine output power with varying wind velocity. When the wind speed is less than the cut-in speed, the power output is zero. As the wind speed increase the power output increases linearly and turbine operates at its rated power at rated speed. If wind speed is increased further, the blade angle changes to maintain the turbine power output constant. When the wind speed reaches at cut-out speed then turbine output power becomes zero to protect the turbine from heavy wind speed.

13.6.3 Torque-Speed Characteristics:

The torque generated by the wind turbine is also a function of the wind speed and rotor speed. The torque speed curves are shown in Fig. 13.15.

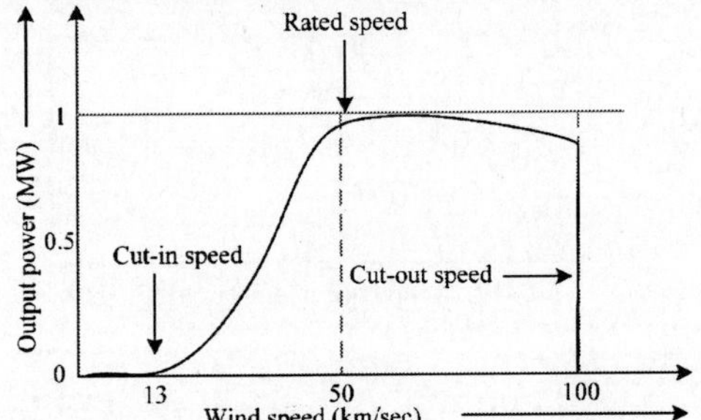

Fig. 13.14 Turbine output power Vs Wind speed.

Fig. 13.15 Torque Vs rotor speed for different wind velocities.

It shows that when the wind velocity increases, torque also increases. Torque T and power P are related by:

$$T = \frac{P}{\omega} \quad \text{......... (13.16)}$$

$$P = \frac{1}{2}\rho . A . C_p . v^3 \quad \text{......... (13.17)}$$

$$T = \frac{1}{2} . \frac{\rho . A . C_p . v^3}{\omega} \quad \text{......... (13.18)}$$

From the equation 13.18, we can say that torque produced by the turbine is proportional to the cube of wind velocity and inversely proportional to the rotor speed.

13.7 Conversion of energy from mechanical to electrical:

The energy conversion from mechanical to electrical energy in wind turbine, takes place with the help of induction and synchronous generators. Both the generators have their advantages and disadvantages. Synchronous generator requires the DC excitation to generate the magnetic field to produce the E.M.F in the stator winding or permanent magnet is required to produce the magnetic field. On the other hand, induction generators required reactive power from the grid or capacitors bank. The complete detail about both the generators are given below.

13.7.1 Induction generator:

An induction generator, also known as an asynchronous generator, is a type of electric generator that converts mechanical energy into electrical energy using the principles of electromagnetic induction. Unlike synchronous generators, induction generators do not require an external source of DC for excitation; instead, they rely on the induction of voltage in the rotor from the stator's magnetic field. This makes them simpler, more robust, and less expensive to manufacture and maintain. Induction generators are widely used in various applications, particularly in wind turbines, due to their ability to operate efficiently over a range of speeds, making them well-suited for harnessing the variable nature of wind energy. There are two main types of induction generators: Squirrel Cage Induction Generators (SCIGs) and Doubly-Fed Induction Generators (DFIGs). SCIGs are known for their simplicity and reliability, but they require external reactive power compensation and are typically used in older or smaller wind turbines. DFIGs, on the other hand, are more complex and allow for variable-speed operation, enabling greater energy capture from varying wind speeds. They also offer the flexibility to control both active and reactive power, which is beneficial for grid stability.

13.7.1.1 Equivalent circuit diagram of induction generator:

The equivalent circuit of an induction generator shown in Fig. 13.16 is similar to that of an induction motor but operates with a negative slip (i.e., the rotor speed is greater than the synchronous speed). Below is a description of the equivalent circuit and the components involved.

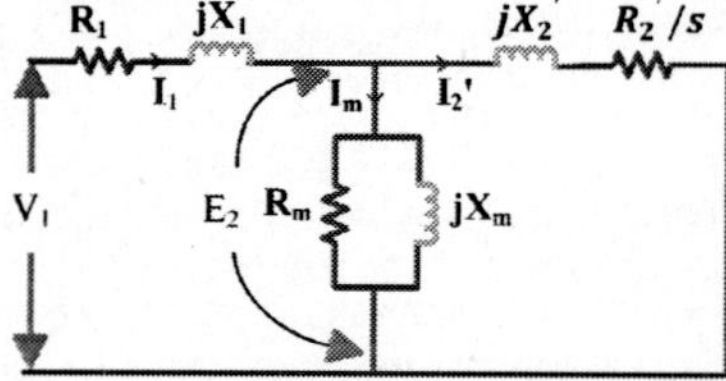

Fig. 13.16 the equivalent circuit diagram of the induction generator.

Applying the K.V.L at stator side circuit

$$V_1 = I_1.(R_1 + jX_1) + E_2 \quad \text{......... (13.19)}$$

$$E_2 = I_m.jX_m \quad \text{......... (13.20)}$$

Where, V_1 = per phase stator voltage, E_2 = induced E.M.F in the stator winding, R_1 = Resistance of stator winding, X_1 = Leakage reactance of the stator winding, I_m = magnetizing current, and I_1 = Stator current.

The induced voltage in the rotor for slip s, $E_2' = s.E_2$ .... (13.21)

The rotor current refereed to stator side, $I_2' = \dfrac{E_2'}{\frac{R_2'}{s} + jX_2'}$ .... (13.22)

The air –gap power is, $P_g = I_2'^2 \frac{R_2'}{s}$ .... (13.23)

Where R_2' and jX_2'represent rotor resistance referred to stator side and rotor leakage reactance referred to stator side, respectively and 's' is the slip.

The electromagnetic torque is developed under generator action

$$T_e = \frac{3.P_g}{\omega_s} = \frac{3.I_2'^2 \frac{R_2'}{s}}{\omega_s} \quad \text{......... (13.24)}$$

$$T_e = \frac{3.\left(\frac{E_2'}{\frac{R_2'}{s} + jX_2'}\right)^2 \frac{R_2'}{s}}{\omega_s} \quad \text{......... (13.25)}$$

Table 13.3 Parameters used for the torque Vs. slip curve of induction machine.

Parameter	Value
Rated Power	15 kW
Supply Voltage and frequency	400 V and 50 Hz
Number of poles	2
Stator resistance in ohms	0.2 Ω
Rotor resistance referred to stator	0.15 Ω
Stator reactance	0.5 Ω
Rotor reactance referred to stator	0.4 Ω
Magnetizing reactance	25 Ω

Table 13.3 shows the parameters used for the torque Vs. slip curve of induction machine. The value of slip is negative for induction generator. When the rotor rotates at higher speed than the stator magnetic field, the active power flows from rotor to stator (generator action). The main drawback of the induction generator is that it takes reactive power from the supply line. It required a capacitor bank to make it, self-excited machine. The torque slip characteristic is plotted by inserting a resistance (R_x) in series with rotor resistance shown in Fig. 13.17. It shows that when the rotor resistance is high, starting torque is also high.

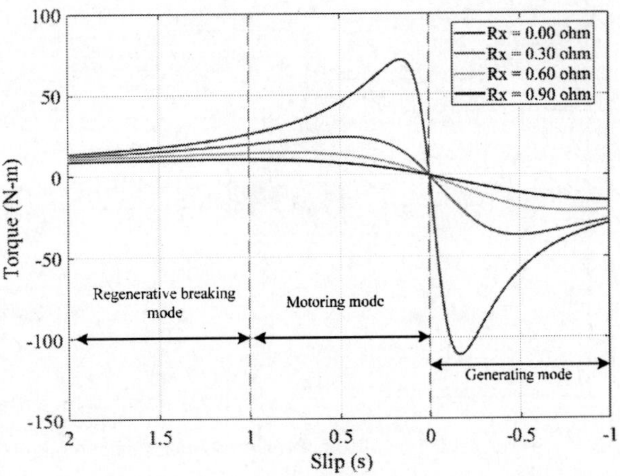

Fig. 13.17 The torque Vs slip curve of induction machine for different rotor resistances.

13.7.1.2 Operation of Grid connected Squirrel Cage Induction Generator:

In a grid-connected system, the induction generator relies on the external power grid to supply reactive power as shown in Fig. 13.18. This is because, unlike synchronous generators, induction generators do not inherently produce reactive power; they consume it. The grid provides the necessary reactive power to excite the induction generator. As the rotor speed of the induction generator exceeds the synchronous speed of the stator's rotating magnetic field, the machine starts to generate electrical power. The power generated is fed into the grid. The induction generator's stator produces AC that is synchronized with the grid frequency and voltage. The voltage and frequency of the power generated are regulated by the grid. The induction generator operates within the constraints of the grid's voltage and frequency. The reactive power is supplied to the generator from external capacitor bank.

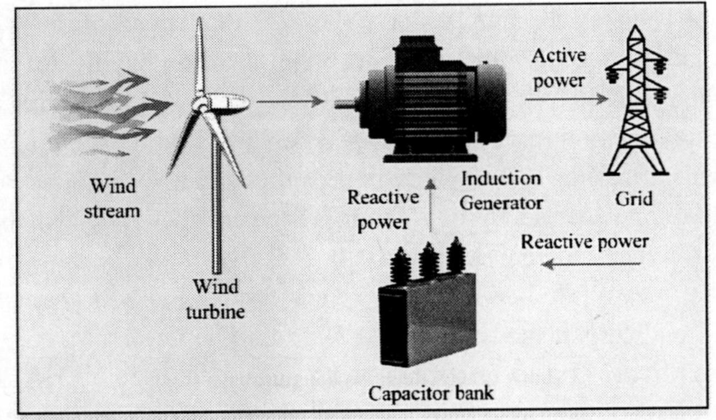

Fig. 13.18 SCIG-based wind turbine, connected to grid with capacitor bank.

13.7.1.3 Operation of Grid connected Double Fed Induction Generator:

Fig. 13.19 shows the working of the control system in grid connected operation. The stator windings are connected to the grid. According to the reference active and reactive power, controller controls the output of rectifier and inverter to control the reactive and active power flow in DFIG. The reference active power is calculated from the wind speed and other parameters.

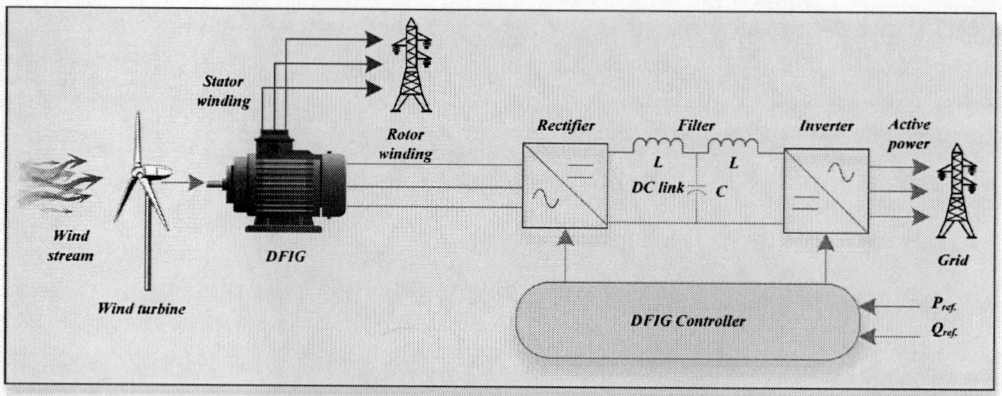

Fig. 13.19 Block diagram represents the control of Double Fed Induction Generator.

In certain cases, the wind speed is such that despite using blade pitch control, it is not possible to maintain the speed of the wind turbine at a value that corresponds to the generator rotating at the speed that generates electricity at 50Hz. The active power flow into grid depends on the phase angle of voltage at the inverter output side. It should lead from the grid phase angle to flow active power from inverter to the grid. The flow of reactive power depends on the magnitude of voltages at inverter side and grid side. Let us take an example where the optimum speed of the wind turbine is 15 rpm. The wind turbine is connected to a 4-pole generator via a gearbox having 1:100 gear ratio that drives the generator input shaft at 1500 rpm to produce electricity at 50 Hz. In the first case, if the speed of the wind turbine drops below 15 rpm to 13 rpm, the speed obtained on the shaft connecting the generator would be 1300 rpm. This translates to electricity being produced at a frequency of 43.33 Hz. Therefore, we would need an additional frequency of rotation of the magnetic field of 6.67 Hz on the rotor coils to make up the difference and generate electricity at 50 Hz. It will be done by the DFIG controller. When speed reduces that means wind speed also reduces further P_{ref} will change to maintain the frequency 50 Hz. In the second case, if the speed of the wind turbine is 17 rpm, the speed of the generator shaft would be 1700 rpm translating to electricity being generated at 56.67 Hz. In this case, we would need a frequency of rotation of magnetic field in the rotor coils of 6.67 Hz in the reverse direction to make the frequency of electricity generated 50 Hz. This control of frequency of the rotation of magnetic field in the rotor coils is done by the DFIG controller such that electricity is always generated at 50 Hz irrespective of the variation of speed of the wind turbine.

Advantages of grid connected induction generator:

- ✓ **Simple Construction:** Fewer components compared to synchronous generators.
- ✓ **Robust and Reliable:** Proven technology with a long history of use.
- ✓ **Self-Protecting:** Under certain conditions, it can handle transient conditions and faults well than synchronous generators.

Challenges of grid connected induction generator:

- ✗ **Reactive Power Demand:** It requires a stable grid to provide the necessary reactive power.
- ✗ **Limited Control:** Less flexibility in controlling power output compared to other generator types.

13.7.1.4 Operation of self-excited or standalone induction generator:

Self-excited induction generators (SEIG) are capable of operating in standalone or isolated systems. They generate their own reactive power through the use of capacitors as shown in Fig. 13.20. To operate as a generator, the SEIG needs a certain level of reactive power, which is provided by external capacitors. These capacitors are connected to the stator windings and help create the necessary magnetic field for power generation. Once the rotor speed exceeds synchronous speed, the induction machine starts generating electrical power. The amount of power generated depends on the rotor speed, the load, and the capacitance. In self-excited mode, the SEIG must be carefully managed to maintain stable voltage and frequency. The system's performance can vary with load changes and rotor speed fluctuations.

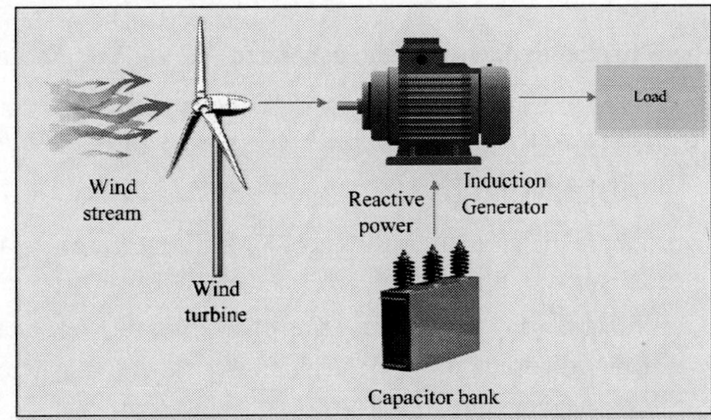

Fig. 13.20 Self-excited induction generator-based wind turbine, connected to local load.

Advantages of self-excited induction generator:

- ✓ **Independent Operation:** This can function independently of a grid, making it suitable for remote or off-grid applications.
- ✓ **Simple and Cost-Effective:** Fewer complex controls and components compared to synchronous generators.

13.7.2 Synchronous Generator used in Wind Turbines:

A synchronous generator operates on the principle of electromagnetic induction, where a rotating magnetic field (produced by the rotor) induces an AC in the stator windings. The frequency of the AC output is directly proportional to the speed of the rotor and the number of poles in the generator. The wind turbine blades capture kinetic energy from the wind and convert it into rotational mechanical energy, which drives the rotor of the synchronous generator.

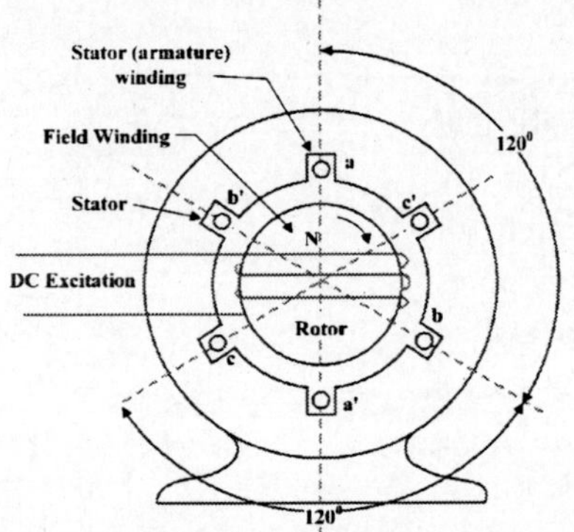

Fig. 13.21 Wound rotor synchronous generator.

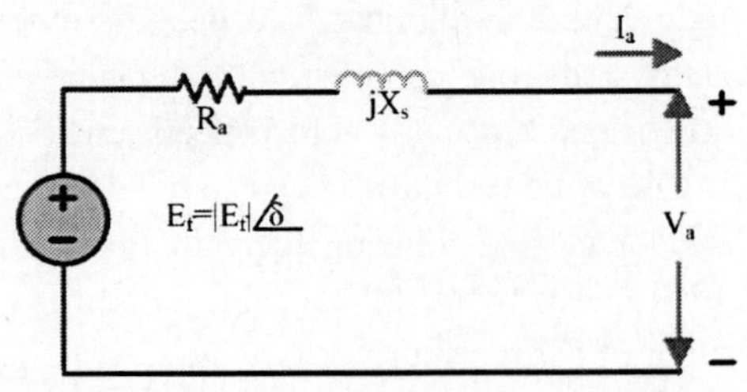

Fig. 13.22 Equivalent circuit diagram of synchronous generator.

The supply voltage frequency is determined by the rotor speed and the number of poles:

$$f = \frac{N.P}{120} \quad \ldots\ldots\ldots(13.26)$$

Where f is the frequency of the generated AC voltage, N is the rotor speed in RPM, P is the number of poles. In grid-connected applications, the generator speed must be controlled to maintain synchronization with the grid frequency. The equation of back E.M.F is given by 13.27.

$$E_f = I_a(R_a + jX_s) + V_a \quad \ldots\ldots\ldots(13.27)$$

Where, R_a is the armature resistance, X_s is the synchronous reactance and V_a is the terminal voltage of the generator.

13.7.2.1 Types of synchronous generators used in wind turbines: There are two types of synchronous generators are used in wind turbine, wound rotor synchronous generator and permanent magnet synchronous generator.

13.7.2.2 Wound rotor synchronous generators (WRSG): WRSGs use electromagnets in the rotor, which are excited by a DC supplied via slip rings and brushes. The rotor's magnetic field interacts with the stator to produce electricity with a frequency that matches the rotor's rotational speed. Fig. 13.23 shows the working of the synchronous generator-based wind turbine. The output of the generator is variable. It is converted from AC to DC with the help of a rectifier. The rectified output is passed through the LC filter to smoothen the DC. This DC is given to the inverter, which provides the constant voltage and frequency.

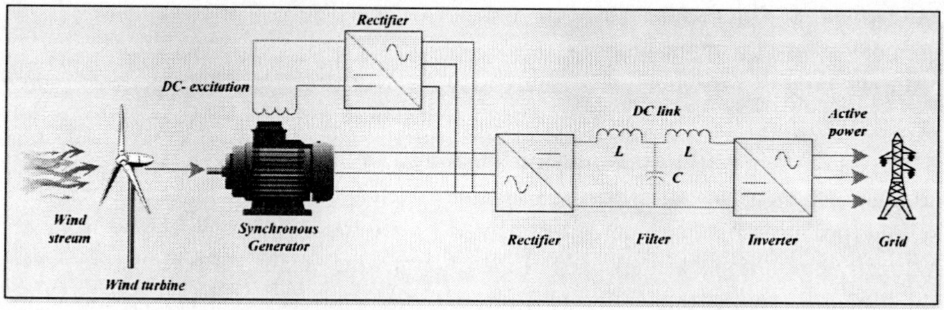

Fig. 13.23 Block diagram of wound rotor type synchronous generator.

13.7.2.3 Permanent Magnet Synchronous Generators (PMSG): PMSGs use permanent magnets embedded in the rotor to create a magnetic field. This eliminates the need for external excitation and slip rings, reducing complexity and maintenance. PMSGs are highly efficient, especially at low speeds, making them suitable for direct-drive wind turbines. They can have a large number of poles (e.g., 60 or more), allowing them to operate effectively at low rotational speeds without the need for a gearbox. Fig. 13.24 shows the layout of permanent rotor type generator. The stator windings are 120 degrees apart to each other in space. The magnetic flux is generated by the permanent magnet rotating in the rotor. Fig. 13.25 shows the block diagram of permanent magnet

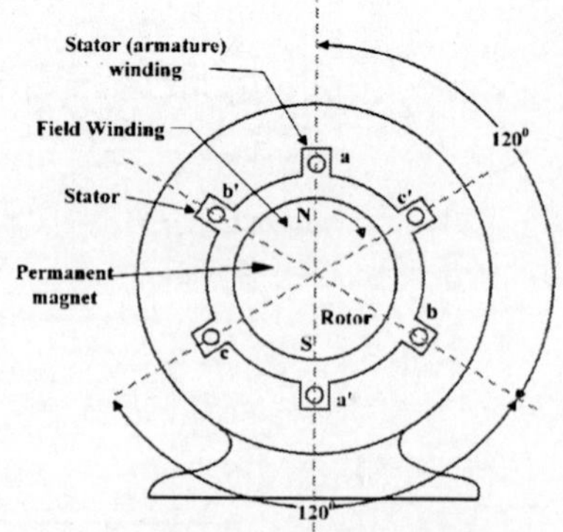

Fig. 13.24 Layout of synchronous generator with permanent magnet type rotor.

synchronous generator. This generator does not require reactive power from grid or capacitor bank. Table 13.4 gives the comparison between PMSG and WRSG.

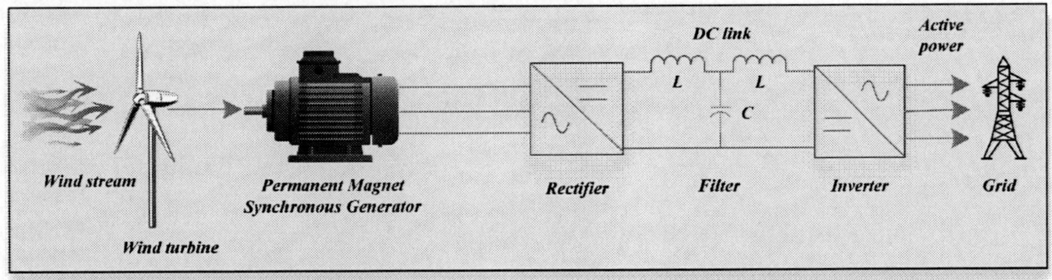

Fig. 13.25 Block diagram of permanent magnet synchronous generator.

Variable Speed Operation: PMSGs are often used in variable-speed wind turbines, where the generator's output is rectified into DC and then inverted back to AC for grid connection. This allows the turbine to adjust its speed according to wind conditions, maximizing energy capture.

Table 13.4 Comparison between PMSG and WRSG.

Aspect	PMSG	WRSG
Excitation System	No external excitation is needed (uses permanent magnets).	Requires external DC excitation via slip rings and brushes.
Maintenance	Low maintenance due to the absence of brushes and slip rings.	Higher maintenance due to wear and tear of brushes and slip rings.
Efficiency	High efficiency, especially at low speeds.	Efficient at synchronous speed, but less efficient at varying wind speeds.
Speed Operation	Suitable for variable-speed operation with power electronics technology.	Typically operates at a fixed speed, limiting efficiency under variable wind conditions.
Complexity	Simpler design without sliprings and external excitation.	More complex due to the need for external excitation and additional components.
Cost	Generally higher initial cost due to permanent magnets and advanced materials.	Lower initial cost but higher long-term costs due to maintenance.
Pole Number	Can have a high number of poles, enabling direct-drive operation without a gearbox.	Fewer poles, often require a gearbox for grid synchronization.
Control of Reactive Power	No control over excitation, limiting reactive power control.	Allows control of reactive power by adjusting the rotor excitation.
Power Quality	Can produce high power quality with appropriate power electronics technology.	High power quality, but typically at the cost of fixed-speed operation.
Application Suitability	Ideal for direct-drive or low-speed wind turbines.	Suitable for medium to high-speed turbines with gearboxes.

13.8 Types of Wind Turbines:

Wind turbines are devices that convert the kinetic energy of wind into electrical energy. They come in various types, primarily categorized based on their orientation, number of blades, and the specific technology used. Below are the main types of wind turbines:

13.8.1 Horizontal Axis Wind Turbines (HAWTs):

Fig. 13.26 Upwind and downwind type horizontal axis wind turbines

HAWTs are the most widely used type of wind turbine, recognizable by their iconic three-blade design. In these turbines, the rotor shaft is positioned horizontally, parallel to the ground, and the blades rotate perpendicular to the wind direction. HAWTs are highly efficient at converting wind energy into electricity, making them the preferred choice for large-scale wind farms. However, they require complex mechanisms such as a yaw system to continuously orient the blades toward the wind. Their tall towers and large blades also necessitate significant space and infrastructure, contributing to higher installation and maintenance costs. They can also be designed for functioning upwind or downwind as shown in Fig. 13.26. Upwind turbines have a greater efficiency range of 30-50% compared to downwind turbines which have an efficiency range of 20-40% typically. Upwind turbines need to be stronger in construction as the wind hits the blades directly unlike in downwind turbines where the wind hits the nacelle first and, in the process loses some of its velocity.

13.8.2 Vertical Axis Wind Turbines (VAWTs):

VAWTs feature a vertical rotor shaft, allowing them to capture wind from any direction without the need for orientation systems like those used in HAWTs. This design offers several advantages, including easier maintenance since the gearbox and generator are typically located near the ground. VAWTs are also better suited for turbulent wind conditions, making them ideal for urban or complex terrain environments. However, they generally have lower efficiency of around 15% compared to HAWTs that have an efficiency of up to 50%, and their design can lead to increased mechanical stress and wear. VAWTs are often used in niche applications where space constraints or wind variability make them more practical than their horizontal-axis counterparts. HAWT is shown in Fig. 13.27(a) and two types of VAWTs are shown in Fig. 13.27(b) and (c). Table 13.5 gives the comparison of VAWT and HAWT.

13.8.2.1 Savonius turbines: Savonius turbines are simple in design, consisting of two or more semi-cylindrical buckets that rotate a vertical shaft due to the drag force. They are known for their constructive simplicity, low cost, low visual impact, and ability to start at low wind speeds regardless of wind direction. However, their efficiency is generally lower than other conventional rotors, making them suitable mainly for small-scale applications or use in developing countries.

13.8.2.2 Darrieus turbines: These use air foil-shaped blades, typically three, to generate lift force and rotate the main shaft. These turbines can have different configurations, such as the "egg beater," H-shape, or helical shape. Darrieus turbines are more efficient at higher rotational speeds but have lower starting torque. Some hybrid designs combine the strengths of both Savonius and Darrieus turbines to improve performance.

(a) Horizontal Axis Wind Turbine

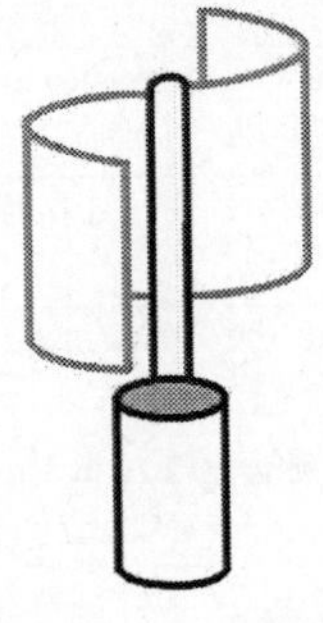

(b) Savonius Turbine

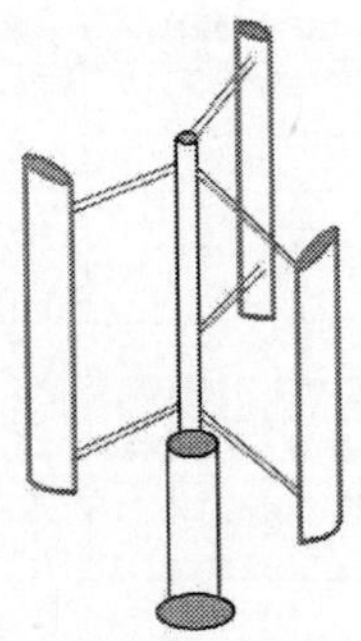

(c) H-Darrieus Turbine

Fig. 13.27 Types of turbines (a) Horizontal Axis Wind Turbine, (b) Savonius Turbine and (c) H-Darrieus Turbine

13.8.3 Ducted or Shrouded Wind Turbines:

Ducted or Shrouded Wind Turbines incorporate a surrounding duct or shroud around the rotor blades, designed to accelerate the wind passing through the turbine. This configuration can significantly increase the efficiency and power output of the turbine by concentrating the wind flow. These turbines are often used in situations where maximizing energy capture in a small area is crucial. However, the complex design of the duct or shroud increases the cost of production and installation, making them less common in large-scale wind farms. Their application is usually limited to specific scenarios where the benefits of increased efficiency outweigh the higher costs.

13.8.4 Direct-Drive Wind Turbines:

Direct-Drive Wind Turbines eliminate the need for a gearbox by directly coupling the rotor to the generator. This design reduces mechanical complexity, resulting in fewer moving parts and lower maintenance requirements. The absence of a gearbox also enhances the overall reliability of the turbine, as there is less wear and tear. However, direct-drive turbines tend to be bulkier and heavier, making transportation and installation more challenging. They are often more expensive to produce due to the need for high-performance magnets and materials. Despite these drawbacks, direct-drive turbines are gaining popularity for their durability and reduced operational costs, particularly in offshore installations where maintenance access is limited.

13.8.5 Small Wind Turbines:

Small Wind Turbines are designed for residential or small-scale commercial use, offering an accessible and cost-effective way to generate renewable energy. These turbines are typically much smaller in size compared to utility-scale turbines and can be installed on rooftops, in backyards, or on small plots of land. Small wind turbines are ideal for providing power in remote locations or for reducing electricity bills in areas with suitable wind conditions. However, their power output is limited, making them less suitable for large energy demands.

Table 13.5 Comparison between VAWT and HAWT.

Feature	VAWT	HAWT
Wind Direction Sensitivity	Omnidirectional; no need for yaw mechanism.	Needs to face the wind; requires a yaw mechanism.
Performance at low wind speeds	Effective at lower wind speeds.	Generally, requires higher wind speeds for optimal performance.
Installation	Easier maintenance with ground-level installation.	Components are higher up, making maintenance more difficult.
Noise levels	Typically quieter, suitable for urban environments.	Can generate more noise, especially in larger installations.
Aesthetic impact	Less visual impact, can be integrated into buildings.	More visible, larger footprint.
Efficiency	Generally, less efficient due to lower tip speed ratios.	Higher efficiency with optimal wind direction.
Starting torque	Requires higher starting torque; may need an external system.	Lower starting torque; typically self-starting.
Structural stress	More stress on components due to cyclic loading.	More balanced stress distribution.
Complexity	Simpler design with fewer moving parts.	More complex design with advanced mechanisms.
Scalability	Typically, smaller scale, suited for residential/commercial.	Better scalability for large-scale power generation.
Wind farm suitability	Can be placed closer together due to less wake effect.	Requires more spacing to avoid wake interference.
Commercial use	Less common in large-scale applications.	Dominates commercial wind energy production.

13.9 Applications of Wind energy:

Wind energy, harnessed through wind turbines, has a wide range of applications. It provides a clean, renewable source of power that has become increasingly important as part of global energy strategies. Here are some key applications of wind energy that are shown in Fig. 13.28 and given below:

(i) Electricity Generation:

- *Onshore Wind Farms:* These are large installations of wind turbines located on land. Wind energy generated from these farms is transmitted to the grid to supply electricity to homes, industries, and businesses.

- *Offshore Wind Farms:* Located in oceans or seas, these wind farms harness stronger and more consistent winds compared to land-based turbines. Offshore wind is becoming a crucial component for countries aiming to decarbonize their energy grids.

(ii) Hybrid Energy Systems: Wind energy can be integrated with other renewable energy sources, such as solar power, to form hybrid systems. The combination of wind and solar works wonderfully as cloudy days are windiest whereas bright sunny days are less windy. Therefore, these two generation sources can complement each other thus increasing the reliability of power supply.

(a) Bulk electricity generation in wind farms

(b) Distributed electricity generation

(c) Water pumping

(d) Bulk electricity generation in offshore wind farms

(e) Hybrid solar wind electricity generation

(f) Wheat grinding

Fig. 13.28 Applications of wind energy

(iii) Distributed Power Generation: Small wind turbines are used in remote locations or individual homes to provide localized power, reducing dependence on the central grid. These distributed wind energy systems can be installed in rural areas where extending the grid infrastructure is less feasible.

(iv) Water pumping and wheat grinding: Wind energy can be used to power mechanical systems, such as pumps for water supply for irrigation and wheat grinding. Wind-powered water pumps are particularly useful in rural or off-grid areas for agricultural applications, reducing the need for diesel or electric pumps. Wheat grinding is the first known application of wind power historically.

(v) Wind Energy for desalination: Wind energy is being increasingly utilized in desalination plants to provide the power needed to convert seawater into fresh water. This is especially useful in arid regions with access to coastal winds but limited freshwater resources.

(vi) Hydrogen production: Wind energy can be used to produce hydrogen through electrolysis, where electricity from wind turbines powers the splitting of water molecules into hydrogen and oxygen. This green hydrogen can be used as an energy carrier for various industrial applications, including transport, heating, and chemical production.

(vii) Industrial Applications: Some industries use wind energy directly to power their operations, particularly those located near windy areas. Wind turbines can supply the energy needs of large-scale industrial processes, reducing carbon emissions and reliance on fossil fuels.

(viii) Wind Energy for Offshore Platforms: Wind turbines are being used to power offshore oil and gas platforms, reducing their reliance on diesel generators. Wind energy provides a cleaner, more sustainable alternative for powering these remote installations.

Example 13.1. A wind turbine has a rotor diameter of 50 meters and is operating at a site where the wind speed is 10 m/s. Calculate the power extracted from the wind at this site. Assume air density is 1.225 kg/m³ and the power coefficient (Cp) of the turbine is 0.4.

Ans. Given: Rotor diameter (d)=50m,

Wind speed $(v) = 10\ m/s$

Air density (ρ) =1.225 kg/m³

Power coefficient of the turbine $(C_p) = 0.4$

Area swept by the wind blades (A) $= A = \pi r^2 = \pi\left(\frac{d}{2}\right)^2 = \pi\left(\frac{50}{2}\right)^2 = 1963.49\ m^2$

Power extracted, $P = \frac{1}{2}\rho . A . C_p . v^3 = \frac{1}{2} * 1.225 * 1963.49 * 0.4 * 10^3 = 481.05\ kW$

Example 13.2. A three-phase induction generator with a 6-pole configuration operates at 1050 rpm, delivering power to a 410 V, 50 Hz system. The generator is rated for a power output of 30 kW and operates at 400 V. The electrical parameters of the generator are:

Stator resistance, R_1=0.40 Ω Rotor reactance, X'_2=1.19 Ω

Stator reactance, X_1=1.40 Ω Core loss resistance, R_f=340 Ω

Rotor resistance, R'_2=0.25 Ω Magnetizing reactance, X_m=40.0 Ω

Find the active and reactive power supplied by the induction generator.

Ans. The synchronous speed is: $N_s = \frac{120f}{P} = \frac{120 * 50}{6} = 1000\ rpm$

The slip is: $s = \frac{N_s - N_r}{N_s} = \frac{1000-1050}{1000} = -0.05$

$Z_0 = \frac{R'_2}{s} + jX'_2 = \frac{0.25}{-0.05} + j1.19 = -5 + j1.19 = 5.14\angle 166.61°\ \Omega$

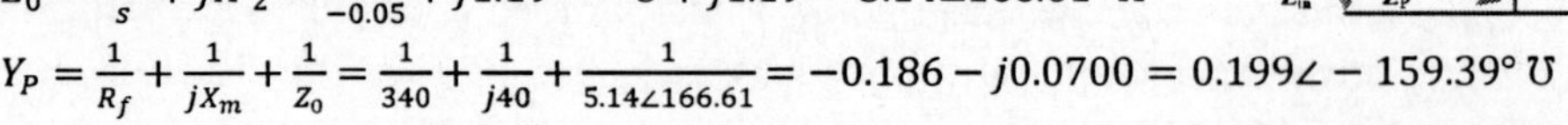

$Y_P = \frac{1}{R_f} + \frac{1}{jX_m} + \frac{1}{Z_0} = \frac{1}{340} + \frac{1}{j40} + \frac{1}{5.14\angle 166.61} = -0.186 - j0.0700 = 0.199\angle -159.39°\ \mho$

$Z_P = \frac{1}{Y_P} = -4.7035 + j1.7688 = 5.023\angle 159.39°\ \Omega$

$Z_{in} = R_1 + jX_1 + Z_P = 0.4 + j1.4 - 4.7035 + j1.7688 = -4.3035 + j3.168 = 5.34\angle 143.62°\ \Omega$

$I_1 = \frac{V_1}{Z_{in}} = \frac{400/\sqrt{3}}{5.34\angle 143.62} = \frac{230.94\angle 0}{5.34\angle 143.62} = 43.247\angle -143.62°\ A$

MVA at the terminals is $S_1 = P + jQ = \sqrt{3}\, V_1 I^*_1$

$= \sqrt{3} * 400 * 43.247\angle 143.62°$

$= -24122.75 + j17771.83\ VA$

Active power supplied by the generator is 24.122 kW and it requires 17.771 kVAR reactive power to generate active power.

Example 13.3. A 3-phase, 400V, 50Hz squirrel-cage induction motor is running as an induction generator. The machine has the following parameters:

Rated power: 15 kW — Stator resistance $R_s = 0.2\ \Omega$

Rated speed: 1500 RPM — Stator reactance $X_s = 1.0\ \Omega$

Magnetizing reactance $X_m = 40\ \Omega$

Rotor resistance referred to stator $R_r = 0.25\ \Omega$

Rotor reactance referred to stator $X_r = 1.2\ \Omega$

The generator operates at a speed of 1600 RPM, and the no-load power factor is 0.8 lagging. A capacitor bank is connected across the generator terminals to supply the required reactive power for self-excitation and to improve the power factor to unity. Calculate the value of the capacitor bank required for the self-excitation at no load.

Ans. $Q_s = P * tan\emptyset = P * \tan[\cos^{-1}(power\ factor)]$

$$Q_s = 15 * \tan(\cos^{-1} 0.8) = 15 * 0.75 = 11.25\ kVAR$$

Per phase, the reactive power requirement is $Q_{ph} = \frac{Q_s}{3} = \frac{11.25 * 10^3}{3} = 3750\ VAR$

Relation between reactive power and capacitance is given by

$$Q_{ph} = V_{ph} I_c = V_{ph} \frac{V_{ph}}{X_c} = V_{ph}^2 * 2\pi * f * C_{ph}$$

$$3750 = \left(\frac{400}{\sqrt{3}}\right)^2 * 2\pi * 50 * C_{ph}$$

$$C_{ph} = \frac{3750}{53333.33 * 2\pi * 50} = 223.81\ \mu F$$

Example 13.4. A wind turbine has a rotor diameter of 50 meters and operates at a wind speed of 12 m/s. The efficiency of the wind turbine is 45%. The turbine is connected to a synchronous generator that operates at 1500 RPM and generates power at 400V. The generator has a rated power factor of 0.85.

Given: Air density, $\rho = 1.225\ kg/m^3$ — Rotor diameter, D = 50 m

Wind speed, v = 12 m/s — Turbine efficiency, $\eta = 45\% = 0.45$

Find: a) The mechanical power extracted from the wind by the turbine.

b) The electrical power output of the synchronous generator.

c) The current drawn by the generator under full load conditions.

Ans. a) The wind power, $P_{wind} = \frac{1}{2}\rho . A . v^3 = \frac{1}{2} * 1.225 * \pi \left(\frac{50}{2}\right)^2 * 12^3 = 2078.163\ kW$

The mechanical power, $P_m = \eta * P_{wind} = 0.45 * 2078.163 = 935.17\ kW$

b) Electrical power generated = mechanical power developed by the turbine, if there is no mechanical loss.

$$P_{electrical} = P_{wind} = 935.17\ kW$$

c) The electrical power output of the generator neglecting copper loss is

$$P_{electrical} = \sqrt{3} V_a I_a \cos\varphi = \sqrt{3} * 400 * I_a * 0.85$$

$$I_a = \frac{935.17 * 10^3}{\sqrt{3} * 400 * 0.85} = 1588\ A$$

13.10 Unit Summary:

- As on 31-01-2025, the installed capacity of wind energy in India is 48.365 GW. Gujarat holds the first position with a 25.87% share. Tamil Nadu holds the second position with 23.66% of the share.
- Wind energy was first harnessed in ancient Persia (500-900 AD) with vertical-axis windmills for grinding grain and pumping water.
- Horizontal-axis windmills were developed in Europe (Netherlands and England) for similar uses as ancient windmills.
- In 1931, Danish engineer Poul la Cour advanced wind turbine design with aerodynamic blades, laying the groundwork for future innovations.
- 21st Century (2000s-present): Rapid growth in wind energy technology due to climate change concerns and government incentives, including:
    - Larger turbines with heights over 150 meters and blade spans over 60 meters.
    - Growth in offshore wind farms, including floating turbines for deeper waters.
    - Innovations in control systems, materials science, and energy storage enhancing efficiency and reliability.
- Horizontal-Axis Turbines: Common design with blades aligned parallel to the wind direction.
- Vertical-Axis Turbines: Blades are perpendicular to the wind direction.
- The maximum theoretical efficiency for wind energy conversion is 59.3% (or 16/27), known as the Betz coefficient. This limit arises because a wind turbine cannot capture all the wind's energy without stopping the wind flow entirely behind it.
- The power coefficient versus speed ratio of a wind turbine, shows that the power coefficient is maximized at a velocity ratio of 1/3.
- Modern wind turbines achieve efficiencies between 30-50% due to various losses (aerodynamic, mechanical, and electrical), but the Betz limit remains the theoretical maximum.
- Wind turbine blades are designed as air foils, similar to aeroplane wings, to maximize lift and minimize drag.
- Lift acts perpendicular to the wind flow, generated by pressure differences on the blade's surfaces. Drag acts parallel to the wind flow and opposes the blade's motion.
- Blades are twisted to optimize the angle of attack along their length, accommodating varying wind speeds at different radial distances from the hub.
- The ratio of the blade tip speed to the wind speed is called TSR.
- Active Yaw Control: Modern turbines use active systems to adjust rotor orientation continuously for optimal wind capture.
- Passive Yaw Control: Older or smaller turbines may use passive systems, like wind vanes, which are less precise.
- SCIGs are simple, reliable, and less expensive but need external reactive power to produce magnetic field.
- DFIGs are more complex, allow for variable-speed operation, and control both active and reactive power, enhancing grid stability.

Short and Long Answer Questions

1. How have advancements in wind turbine technology since the early 20th century contributed to the growth and efficiency of modern wind energy systems?
2. Discuss the impact of site selection on the capacity factor and overall efficiency of wind energy systems. What factors should be considered to optimise wind energy generation?
3. What are the primary differences between HAWTs and VAWTs, and how do they affect their respective applications?
4. Explain the Betz limit and its significance in wind energy conversion. How does the Betz coefficient (59.3%) relate to the theoretical maximum efficiency of a wind turbine, and what are the implications for practical turbine design?
5. Discuss the key aerodynamic principles involved in the design of wind turbine blades. How do blade airfoil design, angle of attack, and blade twist and taper contribute to optimizing the lift-to-drag ratio and overall efficiency of wind turbines?
6. Explain the significance of the TSR in wind turbine performance. How does TSR affect the efficiency of energy capture, and what are the implications of having a TSR that is either too high or too low?
7. Analyse the power-speed and torque-speed characteristics of wind turbines. How do these characteristics influence turbine design and control strategies? Specifically, describe how power output and torque are related to rotor speed and wind speed, and explain the role of control strategies such as pitch control and yaw control in optimizing turbine performance.
8. What are the primary components of a wind turbine, and how does each contribute to the overall function of a wind power plant? Specifically, discuss the roles of the rotor blades, nacelle, gearbox, generator, tower, yaw system, pitch system, and control system.
9. Describe the operational characteristics of wind turbines, including the significance of cut-in wind speed, rated wind speed, and cut-out wind speed. How do these parameters influence the efficiency and performance of a wind turbine?
10. Compare and contrast the use of induction generators and synchronous generators in wind turbines. Highlight the advantages and challenges associated with each type of generator, and explain how they affect the overall energy conversion process.
11. Explain the role of environmental and site characteristics in the design and performance of a wind power plant. How do factors such as wind resources, topography, noise, and wildlife impact the planning and operation of wind farms?
12. How does an induction generator operate in a grid-connected system, and what are the advantages and challenges associated with its use in wind turbines? Specifically, discuss the role of reactive power in the operation of induction generators and how external components, such as capacitor banks, support their functionality.
13. Discuss the advantages and disadvantages of using PMSG compared to WRSG in wind turbine applications.
14. For a synchronous generator connected to the grid, why is it important to control the rotor speed, and how does this control ensure synchronization with the grid frequency?
15. Derive the relation between active power and power angle for the synchronous generator.

Exercises

1. A wind turbine has a rotor diameter of 40 meters and is operating at a site where the average wind speed is 8 m/s. Calculate the power extracted from the wind at this site. Assume air density is 1.225 kg/m³ and the power coefficient (Cp) of the turbine is 0.35.
2. A wind turbine generates 2 MW at a wind speed of 12 m/s. If the wind speed changes to 15 m/s, calculate the new power output, assuming the power output follows a cubic relationship with the wind speed.
3. A 3-phase, 400V, 50Hz squirrel-cage induction motor is running as an induction generator. The machine has the following parameters: Rated power: 25 kW

   The generator operates at a speed of 1500 RPM, and the no-load power factor is 0.85 lagging. A delta capacitor bank is connected across the generator terminals to supply the required reactive power for self-excitation and to improve the power factor to unity. Calculate the value of the capacitor bank required for the self-excitation at no load.
4. A 3-phase, 6-pole wind turbine induction generator is connected to a 400V, 50Hz grid. The machine has the following specifications:

   Rated power = 100 kW   Stator resistance = 0.3 Ω
   Rated speed =1000 RPM   Stator reactance = 1.0 Ω
   Magnetizing reactance = 30 Ω
   Rotor resistance referred to stator side = 0.4 Ω
   Rotor reactance referred to stator side = 1.5 Ω

   The wind turbine operates at 1100 RPM, and the power factor of the induction generator is 0.85 lagging. Calculate:

   a) The slip of the machine.
   b) The power generated by the turbine.
   c) The reactive power consumed by the generator.
5. A wind farm has 10 wind turbines, each with a rated power of 1.5 MW. If the wind turbines operate at 80% capacity factor for a year (8760 hours), calculate the total energy produced by the wind farm in a year.
6. A wind turbine has an installation cost of $3 million and a capacity of 2 MW. If the turbine operates at an average capacity factor of 35% and generates electricity at a rate of $0.07 per kWh, calculate the total revenue generated by the turbine over a year and the payback period.
7. A wind turbine generator operates at 1.2 MW with a power factor of 0.8 lagging. Calculate the reactive power (in kVAR) supplied by the generator and the apparent power (in kVA).
8. A synchronous generator operates with a terminal voltage of 11 kV and generates a power of 6 MW with a power factor of 0.85 lagging. If the synchronous reactance of the generator is 0.5 per unit, calculate the internal voltage of the generator. Neglect armature resistane.
9. Two synchronous generators, G_1 and G_2, are connected in parallel. Generator G_1 has a voltage of 12 kV and a frequency of 50 Hz, while Generator G_2 has a voltage of 11.8 kV and a frequency of 49.8 Hz. If G_1 is running at 1500 RPM, calculate the RPM at which G_2 should operate to synchronize with G_1.

10. A wind turbine has a rated power output of 3 MW and operates at a wind speed of 12 m/s. The power curve of the wind turbine is approximately linear between 6 m/s (where it generates 0.5 MW) and 12 m/s (where it generates 3 MW). If the wind speed drops to 8 m/s, calculate the new power output of the turbine.
11. A synchronous generator with a rated power of 10 MVA operates with a power factor of 0.9 lagging. The generator's efficiency is 92%. Calculate the real power output and the reactive power supplied by the generator. Assume the generator is fully loaded.
12. A wind farm with a total capacity of 20 MW operates at a power factor of 0.8 lagging. The farm uses synchronous generators that need to improve the power factor to 0.95 lagging. Calculate the required reactive power compensation.

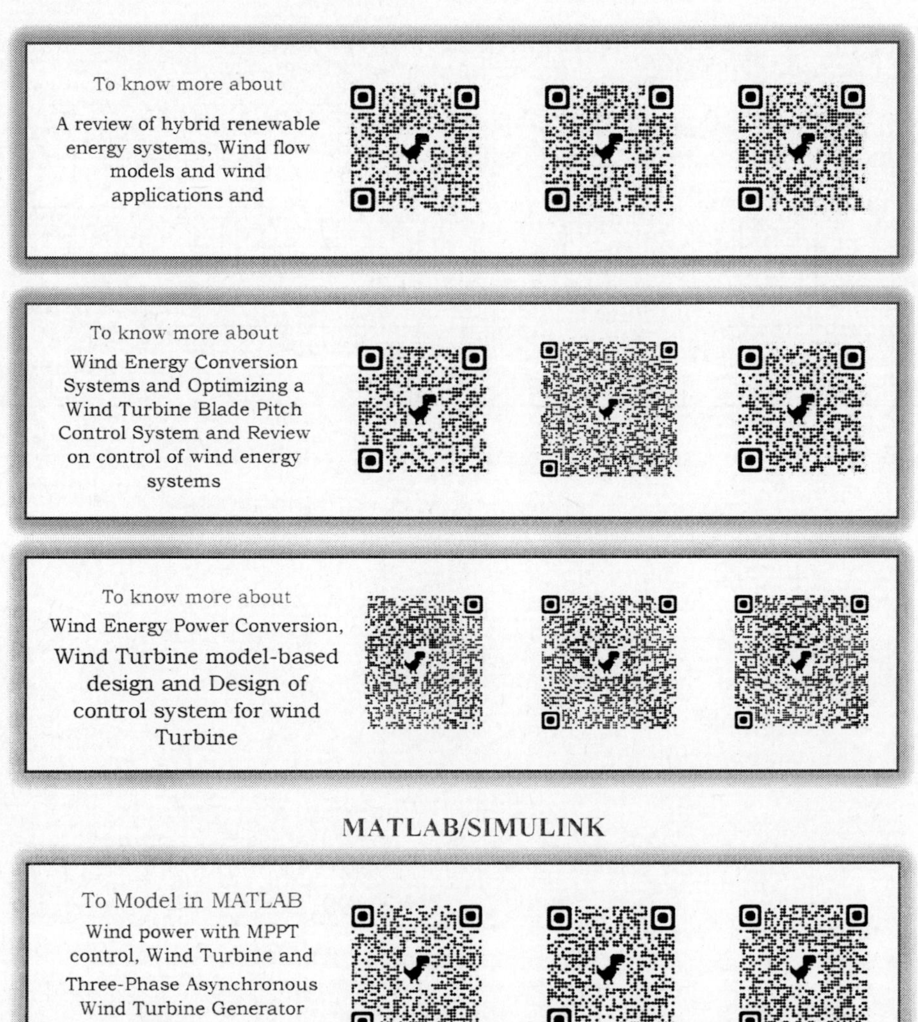

CO AND PO ATTAINMENT TABLE

Course outcomes (COs) for this course can be mapped with the programme outcomes (POs) after the completion of the course and a correlation can be made for the attainment of POs to analyze the gap. After proper analysis of the gap in the attainment of POs necessary measures can be taken to overcome the gaps.

Table for CO and PO attainment

Course Outcomes	Attainment of Programme Outcomes *(1- Weak Correlation; 2- Medium correlation; 3- Strong Correlation)*											
	PO-1	PO-2	PO-3	PO-4	PO-5	PO-6	PO-7	PO-8	PO-9	PO-10	PO-11	PO-12
CO-1												
CO-2												
CO-3												
CO-4												
CO-5												
CO-6												

The data filled in the above table can be used for gap analysis.

Index Terms

A

B

C

I

L

M

R